UNITEXT for Physics

UNITEXT for Physics series, formerly UNITEXT Collana di Fisica e Astronomia, publishes textbooks and monographs in Physics and Astronomy, mainly in English language, characterized of a didactic style and comprehensiveness. The books published in UNITEXT for Physics series are addressed to upper undergraduate and graduate students, but also to scientists and researchers as important resources for their education, knowledge and teaching.

More information about this series at http://www.springer.com/series/13351

Alexandre Obertelli · Hiroyuki Sagawa

Modern Nuclear Physics

From Fundamentals to Frontiers

 Springer

Alexandre Obertelli
Institut für Kernphysik
Technische Universität Darmstadt
Darmstadt, Germany

Hiroyuki Sagawa
Nishina Center
RIKEN
Wako, Saitama, Japan

ISSN 2198-7882   ISSN 2198-7890   (electronic)
UNITEXT for Physics
ISBN 978-981-16-2291-5   ISBN 978-981-16-2289-2   (eBook)
https://doi.org/10.1007/978-981-16-2289-2

This Springer imprint is published by the registered company Springer Nature Singapore Pte Ltd.
The registered company address is: 152 Beach Road, #21-01/04 Gateway East, Singapore 189721, Singapore

Preface

The beauty of nature originates in its diversity and its regular patterns. The association of these two characteristics, opposite at first sight, can be the source of troubling and deep feelings. The enjoyment of a walk in the forest lies in the dense variety of trees as much as in the pleasure to recognize each species from its trunks and leaves. The forest is enjoyed as an environment or can be seen as multiple complex small-scale microcosms. Nuclear physics exhibits these features at its own scales of space and energy. To understand nature, symmetry is an essential concept. In physics or mathematics, a symmetry is a feature of a system that remains unchanged under a certain transformation. In parallel, symmetry breaking is widely encountered and is the source of diversity in physics. Deformation in nuclei and the mass of particles emerge from the breaking of symmetries of the relevant Hamiltonian or Lagrangian in quantum mechanics. In nature, the dynamics of macroscopic and microscopic systems are governed by a few basic equations such as the Newton and Schrödinger equations of motion. In quantum many-body mechanics, the main difficulty is to solve the Schrödinger equation for interacting particles at a chosen resolution scale. Symmetries give a guidance for adequate approximations to solve problems with clear physics ideas, while the nucleus is a multi-scale object.

This book aims at providing the basic knowledge of both theoretical and experimental nuclear physics, especially for young students and non-experts, as well as an up-to-date introduction to current research. Our intention, as authors, is to provide a faithful description of what is nuclear physics today and its exciting challenges. It is written in a self-contained format. Its structure is the following; after an introduction to quantum mechanics illustrated with examples from low-energy nuclear physics in Chap. 1, the basic concepts of nuclear forces and the major different theoretical approaches to describe the nucleus and related phenomena are given in Chaps. 2 and 3, respectively. The observables which allow to study nuclear structure and the related experimental methods are described in Chap. 4. In Chap. 5, the concept of energy shells in nuclei is discussed. The most recent challenges offered by the study of radioactive nuclei are presented in Chap. 6. Chapter 7 focuses on nuclear deformation. The large variety of nuclear reactions and the different approaches to describe them are introduced in Chap. 8. The application of nuclear physics to astrophysics and the standard model are detailed in Chaps. 9 and 10, respectively. A summary at the end of each chapter highlights the messages to be taken away. Problems are given in each chapter, together with solutions.

We, the authors, met within an international collaboration between experimentalists and theorists at the RIKEN Nishina Center, Japan, a world leading experimental institute in nuclear physics. An evening in the suburb of Tokyo, we recognized that despite the several high-quality nuclear physics textbooks and modern reviews, none gathers the main facets of modern nuclear physics while being accessible to young researchers. The motivation for the present textbook was born.

This would not have started without Shinsuke Ota, assistant professor at the Center of Nuclear Study, University of Tokyo, and Tomohiro Uesaka, chief scientist at the RIKEN Nishina Center. We are grateful to them.

We are indebted to the kind feedback given by our esteemed colleagues Profs. Yoshiko Kanada En'yo, Koichi Hagino and Kenichi Matsuyanagi from the University of Kyoto, Profs. Thomas Aumann, Wilfried Nörtershäuser, Robert Roth and Dr. Frank Wienholtz from the Technical University of Darmstadt, Dr. Takumi Doi, Dr. Hajime Sotani and Dr. Tetsuo Hatsuda from RIKEN, Dr. Peter Möller from Hawaii, USA, Prof. Toshio Suzuki from Nihon University, Prof. Atsushi Tamii from Osaka University, Prof. Laura Fabbietti from the Technical University of Munich, Dr. Xavier Roca-Maza and Prof. Gianluca Colò from the University of Milano and Prof. Alfredo Poves from the University of Madrid and Dr. Guillaume Hupin from the Irène Joliot Curie laboratory (IJClab). This textbook would serve its purpose only if it is appreciated by students and young researchers. We had the chance to benefit from the precious comments and suggestions on the different chapters and exercises from excellent young colleagues: Tomoya Naito from the University of Tokyo, Dr. Meytal Duer, Dr. Mario Gòmez, Clara Klink, Moritz Schlaich, Simone Velardita and Sabrina Zacarias from the Technical University of Darmstadt, and Dr. Nancy Paul from the Laboratoire Kastler Brossel (LKB). We thank them deeply.

Many illustrations were performed by Ai Kubota, Manga drawer, during a stay in Germany. We thank her very much for her commitment to this walk across nuclear physics.

Darmstadt, Germany
Wako, Japan
March 2021

Alexandre Obertelli
Hiroyuki Sagawa

Contents

About the Authors

Alexandre Obertelli received his Ph.D. degree from the University of Paris XI, France, in 2005. He is Alexander-von-Humboldt professor at the Technical University of Darmstadt, Germany. He carries experiments on the structure of radioactive nuclei at the Radioactive Isotope Beam Factory of RIKEN, Japan, at CERN, Switzerland, and GSI/FAIR, Germany.

Hiroyuki Sagawa obtained his Ph.D. degree in 1975 from Tohoku University, Japan. After he worked at the Niels Bohr Institute, Danemark, University of Paris-Sud, France, University of Tokyo, Japan, and the National Superconducting Cyclotron Laboratory, USA, he obtained a full-professor position at the University of Aizu, Japan. He is now a senior visiting scientist of RIKEN and Professor Emeritus of the University of Aizu. He has been working mainly on nuclear structure theory and published seven books and over 250 papers in international journals.

Chapter 1
Concepts of Quantum Mechanics from a Nuclear Physics Viewpoint

Abstract We present basic concepts of quantum mechanics from a nuclear-physics point of view. In particular, spin and isospin, the Heisenberg uncertainty relation and the Schrödinger and the Dirac equations are introduced. The analytical solution of the Schödinger equation for bound states in one-dimensional potential and quantum tunnelling are discussed. Quantum entanglement and the concept of Einstein–Podolsky–Rosen (EPR) pair are presented with the experimental evidence of nuclear observations. The importance of symmetries and symmetry breaking is illustrated.

Keywords Schrödinger's equation · Spin and isospin · Quantum entanglement · Square-well potential · Quantum tunnelling · Heisenberg's uncertainty principle · Dirac equation · Symmetry and symmetry breaking

1.1 Genesis of Quantum Physics

The formulation of classical physics started at the beginning of the 17th century by the Italian scientist Galileo Galilei. He discovered not only the isochronal nature of pendulums and the laws of bodies in motion, but also the phases of Venus, the four biggest satellites of Jupiter and sunspots from astronomical observations. Through to the 19th century, many experimental and theoretical discoveries were made in physics. Classical mechanics was established by Galileo Galilei and Isaac Newton, electromagnetism by Michael Faraday and James Maxwell and thermodynamics by Ludwig Boltzmann. At the end of the 19th century, many physicists thought that most physical aspects of our world were solved and that our description of the universe was complete. However, there were still few unsolved issues. The three most important of them were (i) the measurement of the referential independence of the speed of light, (ii) the enigma of the black-body radiation spectrum and (iii) the nature of matter at very small scales. The study of these three problems led to major discoveries in the history of science: special relativity and quantum mechanics. Albert Einstein contributed to these discoveries by publishing four seminal works in 1905 [1–4].

Classical physics can solve the trajectory of particles such as falling bodies and motion of planets around the sun. The propagation of waves, such as light or sound, can also be treated in this context. In classical physics, particles are assumed to have well defined positions and momenta. Conceptually, the position and momentum of each particle are considered as objective quantities and can be measured simultaneously with the desired accuracy. Classical physics has thus a deterministic character. A precise prediction of the evolution of the world is given by mathematical equations without any ambiguity: the equation of motion (the second law of Newton) in the classical mechanics reads

$$\sum F = ma, \tag{1.1}$$

where $\sum F$ is the sum of the forces acting on an object of mass m causing an acceleration a. This Eq. (1.1) determines the motion of an object when the initial conditions for the position and the velocity are provided. In addition, the result of observations is always assumed to be independent of the observer in the classical world. Nature behaves very different at a quantum level. There exists no such deterministic equation in quantum mechanics since there, nature is not described by a deterministic approach but with a statistical one. The equation which determines the behaviour of an object in the quantum world is the so-called Schrödinger equation.

The quantum world is characterized by the Plank constant h or the so-called reduced Planck constant $\hbar$

$$\hbar = \frac{h}{2\pi} = 1.0546 \times 10^{-34} \text{ J} \cdot \text{s} = 6.5821 \times 10^{-16} \text{ eV} \cdot \text{s}. \tag{1.2}$$

It has the unit of an action. For any physical system, the product of a relevant distance and energy can be given in units of

$$\hbar c = 3.162 \times 10^{-26} \text{ J} \cdot \text{m} \tag{1.3}$$

or[1]

$$\hbar c = 1973 \text{ eV} \cdot \text{Å} = 197.3 \text{ MeV} \cdot \text{fm}. \tag{1.4}$$

The notations of units are $1 \text{ Å} = 10^{-10}$ m, $1 \text{ fm} = 10^{-15}$ m and $1 \text{ MeV} = 10^{6}$ eV.

Quantum effects prevail on classical physics for a system or phenomenon with characteristic kinetic energy E (momentum p) and size r when

$$Er \lesssim \hbar c. \tag{1.5}$$

At the human scale, for example, a walking individual (typically with momentum $p = 70 \text{ kg} \cdot 1$ m/s) walking in a room (typically size of $r = 5$ m) exceeds the value of $\hbar c$ by 28 orders of magnitude. On the other hand, in atoms, typical scales are

---

[1] The reduced Planck constant is known with a precision of 12×10^{-9}, and its value from experiment is $\hbar c = 197.326978$ MeV · fm.

nanometers (nm) in size and electron volts (eV) in energy. The former corresponds to the size of the hydrogen atom, while the latter is the typical energy of X rays from the excited states of the hydrogen atom. The ratio $Er/\hbar c$ is 10^{-2}, which is much smaller than unity, as implied by the quantum nature of the atom. Nuclei are much more compact systems characterized few fm in size and nucleons have a typical intrinsic momentum of $100\,\mathrm{MeV/c}$. The ratio $Er/\hbar c$ is about 0.1, which is again in agreement with the condition of Eq. (1.5). Note that quantum phenomena are not restricted to microscopic scales and do not depend on the number of constituents of the system. Famous examples of quantum mechanical phenomena visible at macroscopic scales are the superfluidity of helium or superconductivity. The characteristics of classical and quantum phenomena by momentum p, energy E and size r are summarized in Table 1.1.

The sizes of atoms and molecules were firstly observed in "Brownian motion", which is the random motion of particles suspended in a fluid. This motion is named after the botanist Robert Brown, who observed the phenomenon in 1827, looking through a microscope at vegetal pollen grains immersed in water. 80 years later, Albert Einstein published a paper where he modeled the motion of the pollen as being moved by individual water molecules, one of his first major scientific contributions. This explanation of Brownian motion served as convincing evidence that atoms and molecules exist and was further verified experimentally by Jean Perrin in 1908. Three different Brownian motions of the same mastic grain (0.53 μm diameter) on water surface are demonstrated in the left panel of Fig. 1.1. Direct observation of atoms is possible today by using electron microscopy as shown in the right panel of Fig. 1.1.

The breakdown of the classical description of the world was realized at the beginning of the 20th century. Light played a key role to develop the concept of quantum mechanics. Light behaves not only as emitted and absorbed waves, but appears also to act like particles named "photons". This is the so-called "wave-particle duality". Max Planck introduced the quantization hypothesis (characterized by Planck's constant h) for the energy of light and he obtained a unified equation to explain the black-body radiation. Albert Einstein proposed in 1905 the existence of "light quanta", i.e., individual photons, to explain the photoelectric effect. This quantization of energy gave its name to quantum mechanics. At that time, electrons were still considered as particles only. One of the most striking discoveries by Ernest Rutherford at the

Table 1.1 Characterization of classical and quantum phenomena by momentum p, kinetic energy E and size r

Phenomenon	p	E	r	$Er/\hbar c$
Man walking in a room	$70\,\mathrm{kg \cdot m/s}$	$35\,\mathrm{J}$	$5\,\mathrm{m}$	6×10^{27}
Electron accelerated in a nanotube	$10\,\mathrm{keV/c}$	$10\,\mathrm{eV}$	$20\,\mathrm{\AA}$	0.1
Electron orbiting in an atom	$3\,\mathrm{keV/c}$	$1\,\mathrm{eV}$	$1\,\mathrm{nm}$	5×10^{-2}
Nucleon in a nucleus	$100\,\mathrm{MeV/c}$	$5\,\mathrm{MeV}$	$4\,\mathrm{fm}$	0.1

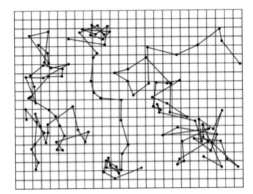 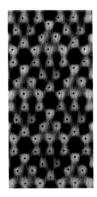

Fig. 1.1 (Left) Three different Brownian motions of the same mastic grain (0.53 μm diameter) on water surface. The position is measured every 30 s and the mesh size is 3.2 μm. Figure from the work of Jean Perrin [5] reported in his book "Les Atomes", 1914. This experiment contributed to validate Einstein's prediction for the Brownian motion based on the assumption of the molecular structure of water [1]. Mastic grains are constantly hit by H_2O molecules and show an erratic motion which follows the statistical law of diffusion. Jean Perrin obtained the Nobel prize in 1926 for his contribution in proving the discontinuity of matter, i.e., the evidence of atoms. (Right) Early image of individual Si atoms resolved by electron microscopy based on quantum tunnelling. Figure reprinted with permission from [6]. ©2021 by the American Physical Society

beginning of 20th century, in 1911, was that a hydrogen atom consists of a positively charged small heavy nucleus (proton), surrounded by a negatively charged light particle (electron). According to classical electromagnetism, the electron accelerated in the circular orbit would emit radiations and collapse immediately into the nucleus. There was an enigma still to be solved.

In 1913, Niels Bohr proposed a model for the hydrogen atom which explained both the stability of the atom and the existence of discrete energy spectra. The model depicts the atom as a small, positively charged nucleus surrounded by electrons that travel in circular orbits around the nucleus by an analogy of the structure to the solar system. Although the Bohr model is a relatively primitive model, it explained and predicted many atomic phenomena surprisingly well, such as the Rydberg formula for the spectral emission lines of atomic hydrogen. While the Rydberg formula had been known experimentally quite a long time before 1913, the Bohr model explained for the first time the structure of the Rydberg formula and gave a justification for its empirical results in terms of fundamental physical constants. In 1925, Werner Heisenberg developed a quantum mechanical formalism in which the variables of position and momentum are presented by matrices and non-commutable algebra. This idea was further developed to the uncertainty principle of two non-commutable observables in 1927 when P. A. M. Dirac introduced the idea that physical observables are represented by operators and physical states are vectors in a Hilbert space.[2] In 1926, Erwin Schrödinger proposed a completely different idea of quantum mechanics: a differential-equation formalism for the quantum mechanical laws. Basic mathematical tools of the two formalisms looked completely different, but later the equivalence was proven mathematically. In 1927, at the Bell laboratory (USA), it was found by Clinton Davisson and Lester Germer that electrons display diffraction patterns as characteristics of waves as previously proposed by Louis de Broglie in 1924.

Niels Bohr proposed a statistical interpretation of quantum measurements to understand the duality of particle and wave. In quantum mechanics, any physical system can be described by a wave function $\Psi(\mathbf{r}, t)$ and the observation of a phenomenon at a given position of space $\mathbf{r}$ and time t is predicted by a statistical probability $|\Psi(\mathbf{r}, t)|^2$ obtained from the wave function. It was a great success to understand with no ambiguity all the observed quantum phenomena of that time. Einstein, however, did not accept Bohr's idea and opposed it strongly by claiming "God does not play dice". He proposed many "Gedanken" (thought) experiments to deny the statistical interpretation. The most famous was the Einstein–Podolsky–Rosen paradox (EPR), i.e., the problem of entanglement of two quantum states (explained in more details in Sect. 1.3). Einstein proposed also a new idea of quantum mechanics called "Local Hidden Variable Theory". The debate between Einstein and Bohr appeared to be a philosophical problem as well as a physics issue since there was no proposal for measurable quantities to distinguish between the two theories. In 1966, John Stewart Bell proposed an experiment to distinguish two theories by the later-named Bell's inequality. Einstein's theory was based on the idea of "action through medium", while Bohr claimed the idea of "action at distance" in quantum measurements. Eventually the measurement of Bell's inequality proved that the Bohr's idea was correct. Important historical milestones of quantum and nuclear physics are given in Table 1.2.

From the particle-wave duality, the motion in the microscopic world can be understood only in a probabilistic interpretation such that the wave function Ψ gives the probability of finding the corresponding particle in space-time. This is the concept called the Copenhagen interpretation of quantum mechanics, developed by Niels Bohr and his group. Albert Einstein, on the other hand, opposed strongly to this probabilistic interpretation and insisted that the laws of nature must be deterministic like Newton's laws. He never accepted the Copenhagen interpretation. Today, quantum mechanics is still considered by some physicists to remain conceptually an incomplete theory because of the probabilistic interpretation. However, all phenomena ever known in the microscopic world are satisfactorily explained by the Copenhagen interpretation of quantum mechanics. In other words, in practice there is no conceptual show-stopper for understanding and predicting quantum mechanical microscopic phenomena, even though not intuitive for us, humans, who evolve

---

[2] A Hilbert space is a generalization of Euclidian space with any number of dimensions. The wave functions of quantum mechanics are square-integrable functions and a linear combination of any functions is also integrable. One can define a scalar product for a wave function $\phi(\mathbf{r})$ in this space as $\langle \phi, \phi \rangle \equiv \int \phi^*(\mathbf{r})\phi(\mathbf{r})d\mathbf{r}$ which is equivalent to the norm in Euclidian space.

Table 1.2 A brief chronological table of quantum and nuclear physics

Year	Name	Achievement
1900	M. Planck	Quantization of light energy
1905	A. Einstein	Light quanta hypothesis Atom hypothesis for the Brownian motion
1908	J. Perrin	Experimental evidence of atoms
1911	E. Rutherford	Discovery of the atomic nucleus
1913	N. Bohr	Model of hydrogen atom
1926	E. Schrödinger	Wave equation of particle
1927	W. Heisenberg	Uncertainty principle
1932	J. Chadwick	Discovery of the neutron
1935	H. Yukawa	Meson theory of strong interaction
1938	R. Meitner and O. Hahn	Discovery of nuclear fission
1949	S. Tomonaga, R. Feynman and J. Schwinger	Renormalization of QED[1]
1964	M. Gell-Mann and G.Zweig	Quark hypothesis in hadrons
1968	G. Veneziano	String theory for quantum gravity
1973	D. Politzer, F. Wilczek and D. Gross	Asymptotic freedom in QCD[2]
1984	J. Schwarz and M. Green	Superstring theory

[1] QED (Quantum Electrodynamics)

[2] Asymptotic freedom is a characteristic feature of QCD (Quantum Chromodynamics, the quantum field theory of the strong interaction mediated by quarks and gluons). Quarks interact weakly at high energies, allowing perturbative calculations of cross sections in deep inelastic processes of particle physics, but they interact strongly at low energies, making possible the binding of baryons with three quarks and mesons with two quarks in finite nuclei and also in nuclear matter

in a world ruled by classical mechanics. This peculiarity was claimed by Richard Feynman as "If you think you understand quantum mechanics, you don't understand quantum mechanics." in his lecture about "The Character of Physical Law".

1.2 Spin and Isospin Quantum Numbers

1.2.1 The Spin

In addition to position and momentum, the wavefunction of quantum systems, e.g. elementary particles or nuclei, are also characterized by degrees of freedom called "quantum numbers" which do not have a correspondence in the classical world. The spin is such a quantum number. It has the unit and form of an angular momentum and it is usually given in units of the reduced Planck constant $\hbar$. It has a different nature than the orbital angular momentum with no good classical analogy. Although it has been historically mentioned as a spinning motion around an intrinsic axis of the particle, this picture was shown to be incorrect and the spin can be distinguished from the orbital angular momentum by several observations:

1. the spin quantum number can take half-integer values,
2. although the direction of spin of a particle can be changed, it cannot be increased nor decreased. It is an intrinsic quantum number of the particle,
3. a charged elementary particle with a non-zero spin has a magnetic dipole moment μ that differs from the value expected from the circulation of particle on its orbit (see Eq. (1.12)). In the classical picture, this would only be possible if the particle had a charge distribution different from the mass distribution, and impossible for a structureless elementary particle. Even non-charged particles (for example neutrons) have a non-zero magnetic dipole moment. This suggests that a neutron has a finite size with an extended charge distribution, although its net charge is zero.

All elementary particles have a spin 0, $\hbar/2$ or $\hbar$. Protons and neutrons, like electrons, have a spin $s = \hbar/2$. In the following, the Plank constant $\hbar$ will not be written explicitly, except when misleading notations. We will then refer to spin values of 0, 1/2, 1, etc.

The spin is a vector of three spatial coordinate components s_x, s_y, and s_z which satisfy the commutation relations,

$$s_i s_j = -s_j s_i = i\varepsilon_{ijk} s_k/2, \tag{1.6}$$

$$\left[s_i, s_j\right] \equiv s_i s_j - s_j s_i = i\varepsilon_{ijk} s_k, \tag{1.7}$$

where ε_{ijk} is the Levi–Civita symbol.[3] The commutation relation is an intrinsic relation of operators in quantum mechanics. If two operators do not commute, it is proved that there is no simultaneous eigenstate of the two operators. The commutation relation has a deep meaning of observations in quantum mechanics, as well as the uncertainty relation (see Sect. 1.7). The *eigenstate* is described by two quantum numbers: the total spin quantum number s and the projection of spin onto one coordinate axis, for instance the projection s_z of the spin on the z axis. The state vector $|s s_z\rangle$ is an eigenvector of the operators s^2 and s_z, following the same rules for angular momentum in quantum mechanics,

---

3

$$\varepsilon_{ijk} = \begin{cases} +1 & \text{if (i,j,k) is (1,2,3), (2,3,1) or (3,1,2)} \\ -1 & \text{if (i,j,k) is (3,2,1), (1,3,2) or (2,1,3)} \\ 0 & \text{if i = j, or j = k, or k = i} \end{cases} \tag{1.8}$$

$$\mathbf{s}^2|ss_z\rangle = s(s+1)|ss_z\rangle, \tag{1.9}$$

$$s_z|ss_z\rangle = s_z|ss_z\rangle. \tag{1.10}$$

The Austrian physicist Wolfgang Pauli introduced the theoretical concept of spin for the electron in December 1924 as a pragmatic *ansatz* to solve the puzzle of the splitting of atomic spectral lines in a magnetic field, discovered by Pieter Zeeman and thereafter named as Zeeman splitting in quantum mechanics. He introduced at the same time the so-called Pauli exclusion principle for fermions (particles with half-integer spin) which states that two or more identical fermions cannot occupy simultaneously the same quantum state with specified particular quantum numbers, such as the angular momentum and its projection. This principle was first formulated for electrons in 1925, and extended to all fermions as the spin-statistics theorem later, in 1940. W. Pauli obtained the Nobel prize in 1946 for the concept of spin and the subsequent exclusion principle among fermions.

According to the spin-statistics theorem, particles can be divided into two families depending on their spin, leading to fundamental differences in their properties. Particles with half-integer spins are called *fermions*, as stated above, while those with integer spins are called *bosons*. While fermions obey the Fermi statistics, bosons obey the rules of the Bose–Einstein statistics and have no exclusion restriction: many bosons can occupy the identical states, and can lead to a boson condensation at the origin of phenomena such as superfluidity or superconductivity.

1.2.2 The Stern–Gerlach Experiment

Due to the simple nature of spin (e.g. electrons with spin $1/2$ can only take two projection values $-1/2$ and $+1/2$ along a given axis), essential features of quantum mechanics were proven and illustrated by spin-based experiments. The so-called Stern–Gerlach experiment was performed in 1922 by Otto Stern and Walther Gerlach in Frankfurt am Main, Germany, which evidenced the existence of the magnetic spin. The initial motivation of the experiment was to prove experimentally some spatial quantization predicted in 1916 by a theory derived by Sommerfeld. The result of the experiment, which will be presented in the coming paragraphs, showed some inconsistency in the model but evidenced a clear spatial separation of silver atoms after passing through a magnetic field gradient along the direction of this gradient. It is only two years later that the effect would be understood as an effect of spin.

The setup of the Stern–Gerlach experiment is shown in Fig. 1.2. Neutral atoms of silver (49 protons and 49 electrons) are sent through an inhomogeneous magnetic field, whose gradient ∇B_z is created by a pair of magnets. The transmitted atoms are detected on a position sensitive plane and the image pattern is recorded. To determine what would be expected from a classical picture point of view, the notion of magnetic moment is now introduced.

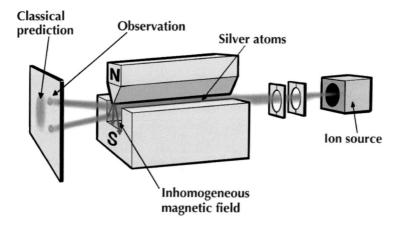

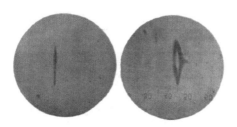

Fig. 1.2 (Top) Schematic view of the experimental setup used in 1922 by Stern and Gerlach to evidence the existence of spin angular momentum of silver nuclei by use of the deflection of neutral silver atoms in an inhomogeneous magnetic field. (Bottom) Deflection pattern observed by Stern and Gerlach with no magnetic field (left) and magnetic field (right). Note that in the original figure from Stern and Gerlach, the magnetic field gradient and the spin separation is horizontal, while it is shown vertical on the top figure. The figure from the original work by Stern and Gerlach is reprinted with permission from [7]. ©2021 by Springer-Verlag

In a magnetic field, a neutral particle is not deflected. On the other hand, if there is a spatial field gradient the particle may experience a non-zero force if it has a magnetic moment. In the case of an silver atom, its magnetic properties are governed by its last electron while all other electrons are paired to zero angular momentum. A magnetic moment comes from the relative motion of two charges spatially separated from each other by r with a relative momentum p so that a relative angular momentum L is given by

$$L = r \times p = m_e r v u, \tag{1.11}$$

where m_e and v are the mass and the velocity of the electron, respectively, and u is a unit vector perpendicular to the motion plane, as illustrated in Fig. 1.3.

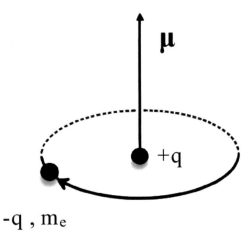

Fig. 1.3 Classical view of an electron with charge $-q$ and mass m_e moving on a circular orbit around a charge $+q$

A classical magnetic moment μ is defined as the product of the electric current I and a flat surface S delimited by the current. In the case of the electron moving around a nucleus it is given by

$$\mu = I S u = \frac{-qv}{2\pi r}\pi r^2 u = \frac{-q}{2m_e}L. \tag{1.12}$$

In the case of an atom with a magnetic moment proportional to the angular momentum, the atom would experience the so-called Larmor precession when passing through the magnetic field following the fundamental equation of dynamics

$$\frac{dL}{dt} = \mu \times B, \tag{1.13}$$

i.e., the magnetic moment will precess around the magnetic field with a frequency of $\omega = \frac{q}{2m_e}B$. This frequency is too high to lead to a deflection of the silver atoms in a preferred direction in the Stern–Gerlach experiment, the measured silver atoms after the non-uniform magnetic field actually show a clear splitting of the atoms in two components with the equal portion in the direction of the field gradient as illustrated in Fig. 1.2. This was a great experimental finding and it could not be understood from classical physics and from what was known at that time, while art can be explained by the existence of the spin of the electron.

The magnetic field between the two magnets with highly odd shapes is non-uniform so that a significant gradient of the magnetic field dB_z/dz is experienced by silver atoms along their path. The interaction energy of the magnetic moment with a magnetic field is given by

$$H_M = -\mu \cdot \mathbf{B}. \tag{1.14}$$

When the atoms go between the magnets in the direction of y-axis, they experience the force,

$$\mathbf{F} = \nabla_z(\boldsymbol{\mu} \cdot \mathbf{B}) = \mu_z \frac{\partial B_z}{\partial z} \hat{e}_z. \qquad (1.15)$$

In quantum mechanics, the magnetic moment (1.12) will take only discrete values in the direction of the z-axis: $L_z = \pm L, \pm(L-1), \pm(L-2), \ldots, 0$. Then the force (1.15) has also discrete values and the atomic beams will be bent in the specific directions affected by the force (1.15). If the spin angular momentum $\mathbf{L} = 1$, L_z takes three values $L_z = \pm 1, 0$ so that the atomic beams will split into three directions. If $\mathbf{L} = 2$, L_z takes five values $L_z = \pm 2, \pm 1, 0$ so that the atomic beams will split into five directions. However, the Stern–Gerlach experiment showed the splitting into two directions, inconsistent with the effect of a magnetic moment due to an angular momentum of integer value projection.

Only few years later, in 1924, the splitting was understood as coming from the z-axis projection s_z of intrinsic spin quantum number $s = 1/2$. The observation demonstrates clearly that quantum laws do not follow the same rules as classical physics: the intrinsic spin $s = 1/2$ of the unpaired electron of an silver atom can only take two values $\pm 1/2$ in the z direction and so the resulting magnetic moment is quantized, while a classical magnetic moment can take any direction in space. The Stern–Gerlach experimental scheme provides an efficient way to manipulate spin, since the direction of the field gradient can be chosen from the design of the experiment. This was extensively used from the 1920s to determine and validate the fundamental laws of the quantum world, such as quantum entanglement described in Sect. 1.3.

1.2.3 The Isospin

The isospin is a quantum number related to the strong interaction and is particularly important in nuclear physics. It was introduced by Werner Heisenberg, a german Physicist, in 1932 when the neutron was discovered. The near equality of the mass of protons m_p and neutrons m_n ($\Delta m = (m_n - m_p)/\overline{m} = 1.4 \times 10^{-3}$; $\overline{m} = (m_n + m_p)/2$) immediately suggests a deep similarity between them.

The neutrons and protons are assigned to the doublets of a Lie algebra SU(2),[4] which is the same fundamental representation of particles with spin 1/2, i.e., similar to a spin 1/2 particle, neutrons and protons are classified as a doublet of isospin $t = 1/2$

---

[4] In mathematics, a Lie group is a group with smooth structure, the so-called differentiable manifold. A manifold is a space that locally resembles Euclidean space. Lie groups and their associated Lie algebras play an important role in modern physics to describe a symmetry of a physical system. Groups whose representations are of particular importance are, for examples, the rotation group SO(3), the special unitary group SU(2) for the spin and also for the isospin, and the special unitary group SU(3) for the quark model.

particle associated with different isospin projections $t_z = 1/2$ and $t_z = -1/2$, respectively. The quantum mechanical description is similar to that of angular momentum and spin. This is the reason why the name "*isospin*" was introduced. In particular, the isospin couples in the same way as the spin. For example, a proton-neutron pair can couple the total isospin $T = 0$ and $T = 1$ with $T_z = (N - Z)/2 = 0$. Under the isospin symmetry, the interactions between neutron-neutron, proton-proton and neutron-proton $T = 1$ pairs are equivalent, while the Coulomb interaction *explicitly* breaks the isospin symmetry.

The spin has the same dimension as the angular momentum and can couple to the total angular momentum $\mathbf{j} = \boldsymbol{\ell} + \mathbf{s}$. The single-particle state in the shell model is commonly denoted by the total angular momentum and its projection $|jj_z\rangle$. It should be noticed that the isospin is a dimensionless quantity unlike the spin and is introduced as a quantum number to provide a classification scheme of many-body systems of protons and neutrons. While the mathematical formalism of the isospin is identical to the one developed for the spin, the isospin is independent and distinct from other quantum numbers. The isospin has three components t_x, t_y, and t_z and satisfies the commutation relation of the SU(2) Lie algebra,

$$[t_i, t_j] = i\varepsilon_{ijk}t_k, \tag{1.16}$$

which is exactly the same as that of spin given in Eq. (1.7).

A nucleus with neutron and proton numbers N and Z has the isospin projection $T_z = \sum_{i=1}^{A} t_z(i) = (N - Z)/2$. The total isospin is also defined as $\mathbf{T} = \sum_{i=1}^{A} \mathbf{t}(i)$. If the nuclear interactions conserve the isospin, all states have good isospin quantum numbers T and T_z;

$$|N, Z : \alpha\rangle = |T_\alpha, T_z\rangle, \tag{1.17}$$

where α is an index of given state. Since T_z is positive in stable heavy nuclei, the expected isospin in the lowest isospin $T = T_z$ for the ground state, and some excited states may have different isospin $T = T_z + 1, T_z + 2, \ldots$. In heavy nuclei, the Coulomb interaction is strong and may violate largely the isospin symmetry. Surprisingly, even in heavy nuclei, the isospin symmetry is preserved well and the so-called *Isobaric analog state* (IAS)[5] was found as a sharp resonance just above the broad Gamow–Teller (GT) resonance[6] in many nuclei by charge exchange reactions such as (p, n) or $(^3\mathrm{He}, t)$ reactions, as illustrated in Fig. 1.4.

The isospin symmetry plays also an important role in particle physics. In the quark model, the up and down quarks have an isospin 1/2 with the different isospin projections $\pm 1/2$. Then all hadrons are classified by the isospin and its projection. The isospin gives a subset of the flavor symmetry of quarks, seen more broadly in the

---

[5] IAS has the same isospin $T' = T$ as the mother (initial) state of the reaction, $T = T_z = (N - Z)/2$, but the isospin projection T'_z is smaller than that of mother by one unit $T'_z = (N - Z - 2) = T - 1$ by the nature of (p, n) reaction.

[6] The Gamow–Teller resonance is a collective state excited by the spin-isospin operator σt_-.

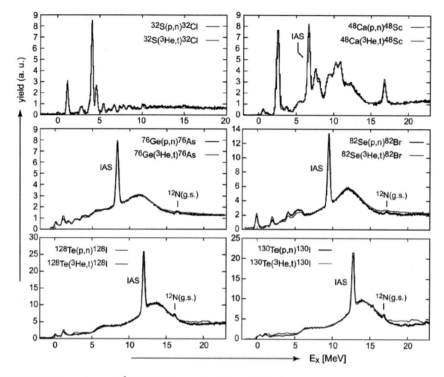

Fig. 1.4 Comparison of $(^3\text{He}, t)$ (red curve) and (p, n) (black curve) spectra at 140 MeV/A and 134 MeV on ^{32}S, ^{48}Ca, ^{76}Ge, ^{82}Se, ^{128}Te and ^{130}Te. The sharp IAS peaks are seen in both the $(^3\text{He}, t)$ and (p, n) spectra below broad Gamow-Teller peaks. This figure is reprinted with permission from [8]. @2021 by Springer Nature

interactions of baryons and mesons. More detailed discussions of quarks are found in Chap. 10.

1.3 Quantum Entanglement

We consider the spin of a two-particle system, $\mathbf{S} = \mathbf{s}_1 + \mathbf{s}_2$, in which each particle has a spin 1/2. The total spin of the system is either $S = 0$ or 1. The wave function of the $S = 0$ state (with therefore $S_z = s_{1z} + s_{2z} = 0$) is written as

$$|\Psi_{12}\rangle = \frac{1}{\sqrt{2}}\{|\uparrow_1\rangle|\downarrow_2\rangle - |\downarrow_1\rangle|\uparrow_2\rangle\}, \tag{1.18}$$

where $|\uparrow\rangle(|\downarrow\rangle)$ is a notation for spin up (spin down) state describing the projection $s_z = 1/2$ ($s_z = -1/2$). This pair is called an *entangled pair* since the projection of the spin of one of the two particles "drives" the projection of the second, since their sum is fixed by definition to be $S_z = 0$ for a $S = 0$ pair. It is also called an EPR

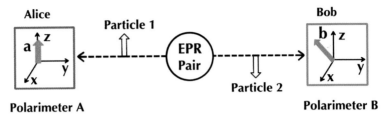

Fig. 1.5 Geometry of correlation measurements of EPR pair by two polarimeters A and B. Alice located at the Moon has a polarimeter A placed to the direction of the North pole (z-axis) assigned by a unit vector **a**, while Bob on the Earth has another polarimeter B inclined by an angle θ from the North pole **a** in x-z plane assigned by another unit vector **b**

pair and it was originally proposed by A. Einstein, B. Podolsky and N. Rosen for photon pairs [9] to demonstrate the incompleteness of statistical interpretation of quantum observations by the Copenhagen group led by Niels Bohr. Later this idea was extended to electron pairs by David Bohm.

We discuss first the quantum measurement of two entangled particles. We consider two detectors (polarimeters) A and B: Alice has a polarimeter A on the Moon, while Bob has another polarimeter B on the Earth. The polarimeters A and B are assigned their special directions by unit vectors **a** and **b** and measure the orientations of the spins of particles 1 and 2, respectively. The spins of the two particles can be measured by the detectors and allow the study of the correlations between two measurements. Suppose that Alice is specialized in measuring $\sigma_z = 2s_z$ of particle 1, while Bob observes σ_z of particle 2.[7] If Alice finds σ_z positive for particle 1, then one can predict the result of Bob's measurement, even before Bob performs any measurement: Bob must find σ_z is negative for particle 2. On the other hand, if Alice does no measurement, Bob has a fifty-fifty chance of getting σ_z positive or negative.

Let us consider the measurement of EPR pairs by two polarimeters placed very very far away. If Alice finds $\boldsymbol{\sigma}_1 \cdot \mathbf{a}$ to be positive; Bob's measurement of particle 2 with the polarimeter B in the same direction $\boldsymbol{\sigma}_2 \cdot \mathbf{a}$ will be 100% negative because of the perfect anti-correlations between particles 1 and 2 in a EPR pair (1.18). Now let's consider the measurements of two polarimeters A and B oriented in different directions. The detector A is placed to the direction of the North Pole (z-axis), while the direction of the detector **b** inclines by an angle θ from **a** in the x-z plane as shown in Fig. 1.5. The quantum measurement of correlation between the two measurements is expressed as

$$C_{QM} = \langle \Psi_{12} | (\mathbf{a} \cdot \boldsymbol{\sigma}_1)(\mathbf{b} \cdot \boldsymbol{\sigma}_2) | \Psi_{12} \rangle \tag{1.19}$$

where σ is a Pauli spin matrix and total spin is expressed as $\mathbf{S} = (\boldsymbol{\sigma}_1 + \boldsymbol{\sigma}_2)/2$ (see Eq. (1.91) for explicit representations of spin matrices). Because of the total spin $S = 0$ of the EPR pair, the two spins have opposite directions, i.e.

---

[7] In quantum computing and quantum information, the operator σ_z is commonly adopted instead of s_z to implement the quantum bit (qbit). Since bits are characterized by integer numbers, the eigenvalue of $\sigma_z = \pm 1$ are more convenient to specify the qbits.

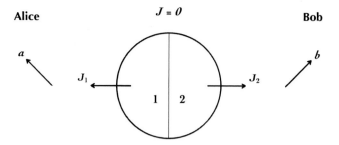

Fig. 1.6 Fission of a nucleus. Two fragments fly with angular momenta J_1 and J_2 in opposite directions. Alice measures fragment 1 with a detector oriented in the direction a, while Bob observes fragment 2 with another detector oriented in the direction b

$$\sigma_{1i}|\Psi_{12}\rangle = -\sigma_{2i}|\Psi_{12}\rangle, \quad (i = x, y, z). \tag{1.20}$$

Using the identity

$$(\mathbf{a} \cdot \boldsymbol{\sigma})(\mathbf{b} \cdot \boldsymbol{\sigma}) = \mathbf{a} \cdot \mathbf{b} + i(\mathbf{a} \times \mathbf{b}) \cdot \boldsymbol{\sigma}, \tag{1.21}$$

we obtain the correlation function

$$\begin{aligned} C_{QM} &= \langle \Psi_{12}|(\mathbf{a} \cdot \boldsymbol{\sigma}_1)(\mathbf{b} \cdot \boldsymbol{\sigma}_2)|\Psi_{12}\rangle \\ &= -\langle \Psi_{12}|(\mathbf{a} \cdot \boldsymbol{\sigma}_1)(\mathbf{b} \cdot \boldsymbol{\sigma}_1)|\Psi_{12}\rangle = -(\mathbf{a} \cdot \mathbf{b}) = -\cos(\theta), \end{aligned} \tag{1.22}$$

where the second term of Eq. (1.21) is dropped since $\langle \Psi_{12}|\sigma_1|\Psi_{12}\rangle = 0$.

Problem

1.1 Prove Eq. (1.21).

We now discuss the classical correlation by taking the example of the fission of an atomic nucleus with angular momentum $J = 0$, as shown in Fig. 1.6. While fissioning, a nucleus breaks into two pieces; one with angular momentum J_1 and the other with J_2. The conservation law of angular momentum gives $J = J_1 + J_2 = 0$, which means equal magnitude ($J_1 = J_2$) and opposite signs ($J_{1i} = -J_{2i}$, ($i = x, y, z$)) of two angular momenta. In the case of $J_1 = J_2 = \frac{1}{2}$, we can identify the two fragments as an EPR pair.

How can we evaluate the correlations of measurements by Alice and Bob of the fission fragments in the classical way? Alice measures the angular momentum of the first fragment with the polarimeter A oriented in the direction of the North Pole $\mathbf{a}$, while Bob observes another fragment with the polarimeter B oriented to the direction $\mathbf{b}$ which inclines by an angle θ from the North Pole. We record the measurements of the variables $(\mathbf{a} \cdot \mathbf{J}_1)$ and $(\mathbf{b} \cdot \mathbf{J}_2)$ by their sign, $+1$ or -1 and call them a and b, respectively. Namely, we have the results,

$$a \equiv \text{sign}(\mathbf{a} \cdot \mathbf{J}_1) = \text{sign}(\mathbf{a} \cdot \boldsymbol{\sigma}_1) = \pm 1, \tag{1.23}$$
$$b \equiv \text{sign}(\mathbf{b} \cdot \mathbf{J}_2) = \text{sign}(\mathbf{b} \cdot \boldsymbol{\sigma}_2) = \pm 1. \tag{1.24}$$

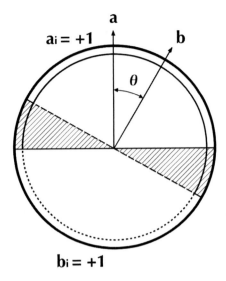

Fig. 1.7 Geometry of correlation of measurements by two polarimeters A and B oriented to the unit vectors **a** and **b**, respectively. The polarimeter A is placed in the direction of the North Pole, while the direction **b** of polarimeter B inclines by an angle θ from the North Pole **a**. This figure shows cross sections divided by equatorials of two spheres whose North Poles are assigned by **a** and **b**, respectively. On the sphere with the pole **a**, the value $a = +1$ if σ_1 or $\mathbf{J}_1$ points to the northern hemisphere (shown by a semicircle of a solid line); otherwise $a = -1$. On the sphere with the pole **b**, $b = +1$ if σ_2 or $\mathbf{J}_2$ points to the southern hemisphere (shown by a semicircle of a broken line); otherwise $b = -1$. In the shaded area $ab = 1$; otherwise $ab = -1$. See the text for more details

We repeat the measurement M times, denoting the i-th event as a_i, b_i. The averages are defined as

$$C_a = \sum_{i=1}^{M} a_i/M, \qquad\qquad C_b = \sum_{i=1}^{M} b_i/M. \qquad (1.25)$$

Because $\mathbf{J}_1$ and $\mathbf{J}_2$ are both pointed to arbitrary directions, their average values approach to 0 as M becomes large enough:

$$C_a \to 0, \qquad C_b \to 0 \qquad (M \to \infty). \qquad (1.26)$$

How about the correlation

$$C_{ab} = \sum_{i=1}^{M} a_i b_i/M \qquad (1.27)$$

between a_i and b_i? If the vector **b** is positioned to be parallel to **a**, we have $b_i = -1$ whenever $a_i = 1$ and $b_i = 1$ whenever $a_i = -1$, since $\mathbf{J}_1$ is anti-parallel to $\mathbf{J}_2$. In this case, we get the complete anti-correlation

$$C_{ab} = -1. \tag{1.28}$$

We now derive the correlation when detectors A and B are placed at angle θ. Let us imagine a sphere where the vector $\mathbf{a}$ is in the direction of the North Pole and the detector B inclines by an angle θ from $\mathbf{a}$ as shown in Fig. 1.7. On the sphere with the North Pole specified by the orientation of polarimeter $\mathbf{a}$, $a = +1$ if $\mathbf{J}_1$ points to the northern hemisphere (shown by a semicircle of a solid line); $a = -1$ if $\mathbf{J}_1$ points to the southern hemisphere. On another sphere with the North Pole specified by the orientation of polarimeter $\mathbf{b}$, $b = +1$ if $\mathbf{J}_2$ points to the southern hemisphere (shown by a semicircle of a broken line); $b = -1$ if $\mathbf{J}_2$ points to the northern hemisphere. Therefore, the sphere is divided by the two equatorial planes into four sectors with alternating signs of ab; the shadowed areas of Fig. 1.7 give $ab = +1$, whereas the other areas correspond to $ab = -1$. Thus, we obtain the correlation coefficient

$$C_{ab} = \frac{(2\theta - (2\pi - 2\theta))}{2\pi} = \frac{2\theta}{\pi} - 1. \tag{1.29}$$

It is clear that this result is different from the quantum mechanical correlation (1.22),

$$C_{QM}(\mathbf{a}, \mathbf{b}) = -\mathbf{a} \cdot \mathbf{b} = -\cos\theta \tag{1.30}$$

for projections of spins of EPR pairs on the directions of detectors $\mathbf{a}$ and $\mathbf{b}$. Results in Fig. 1.8 show that quantum mechanical correlation is always stronger than classical correlation.

Nuclear experiments were performed using the reaction $^1H(d, ^2He)n$ by H. Sakai et al. in 2006 to measure the correlations of ^2He, which has the total spin $S = 0$ and can be considered as an entangled EPR pair,[8] consisting of two protons [11]. The angular correlations were measured in the direction $\theta = (0, 90)$ degree very precisely and display clearly quantum correlations as shown in the shaded area in Fig. 1.8. The classical correlation might be interpreted as a result of the locality principle of classical mechanics, i.e., "action through medium", so that the correlations can not be transmitted faster than by the speed of light. On the other hand, in quantum mechanics, the correlations may be transmitted by the principle "action at distance" and the stronger correlations are implemented as shown in Fig. 1.8. The experimental data validate the quantum mechanical description of the correlation function.

---

[8] A similar experiment by W. Mittig published in 1976 had similar objectives, but with lower precision [10]. This pioneer experiment used the nuclear reaction (p, p') scatterings to prepare the entangled pair of protons.

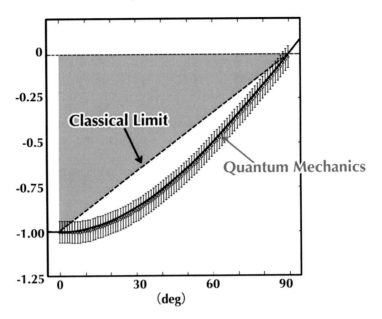

Fig. 1.8 Correlation function C_{ab} on measurement of an EPR pair by two detectors A and B along the directions **a** and **b**, respectively. The dashed line shows classical correlation (1.29), while the line curve shows the quantum mechanical one (1.30). The dots are the observation of spin-spin correlations of di-protons with the total spin $S = 0$ in the ^{1}H(d,^{2}He)n reaction published by Sakai et al. [11]. Courtesy H. Sakai, RIKEN

1.4 The Schrödinger Equation and Its Physical Meaning

Niels Bohr proposed a model for the hydrogen atom with quantized electron orbitals. His model was able to relate the observed spectral lines in the emission spectrum of hydrogen to previously known constants such as the electron mass and the fine structure constant. However, the Bohr model did not have any power to explain why the electrons move around the nucleus in quantized circular orbits, nor why these orbits are stable. It was Erwin Schrödinger, an Austrian physicist, who found an equation (the Schrödinger equation) which allows to predict the motion of electrons in an atom [12].

First, we study how the Schrödinger equation was based on the quantum hypothesis. In free space without any potential, the kinetic energy E relates to the momentum p through the non-relativistic relation

$$E = \frac{p^2}{2m}, \tag{1.31}$$

where m is the mass of the particle. For quantal objects, the Einstein and the de Broglie relations read

$$E = \hbar\omega, \quad p = \hbar k, \tag{1.32}$$

where $\hbar = h/2\pi$ is the reduced Planck's constant, and ω and k are the frequency and the wave number of the object, respectively. Substituting (1.32) into (1.31), we have the relation between the momentum and the wave number,

$$\hbar\omega = \frac{\hbar^2 k^2}{2m}. \tag{1.33}$$

Consider the matter wave Ψ which is moving along the x direction

$$\Psi = e^{i(kx - \omega t)}. \tag{1.34}$$

The relation (1.33) can be recast in the operator equation,

$$i\hbar \frac{\partial \Psi}{\partial t} = -\frac{\hbar^2}{2m} \frac{\partial^2 \Psi}{\partial x^2}. \tag{1.35}$$

Note that the Eq. (1.35) is mathematically complex which is the essential feature of quantum mechanics as it shall be seen in the following sections. The equation in free space can be extended to the case of the motion under a potential V. The Hamiltonian then reads

$$H = \frac{p^2}{2m} + V. \tag{1.36}$$

The Schrödinger equation for the system is then given in it most well-known formulation as

$$i\hbar \frac{\partial}{\partial t} \Psi = H\Psi, \tag{1.37}$$

where Ψ is called the wave function. Expressed in the 3-dimensional space, the equation reads

$$i\hbar \frac{\partial}{\partial t} \Psi(\mathbf{r}, t) = -\frac{\hbar^2}{2m} \left(\frac{\partial^2}{\partial x^2} + \frac{\partial^2}{\partial y^2} + \frac{\partial^2}{\partial z^2} \right) \Psi(\mathbf{r}, t) + V(\mathbf{r}, t)\Psi(\mathbf{r}, t). \tag{1.38}$$

In the static case, the wave function is expressed in a separable form of time and position coordinates,

$$\Psi(\mathbf{r}, t) = \psi(\mathbf{r})e^{-i\omega t}, \tag{1.39}$$

and the Schrödinger equation is expressed as

$$H\psi(x) = E\psi(x) \tag{1.40}$$

where $E = \hbar\omega$ is called *eigenenergy* for the eigenstate $\psi(x)$.

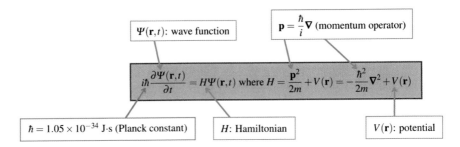

Fig. 1.9 Main components of the Schrödinger equation

The Schrödinger equation is the key equation in quantum mechanics to explain essential features of quantum many-body physics. The main aspects of the Schrödinger equation are summarized in Fig. 1.9. In Schrödinger's picture of quantum mechanics, the wave function gives the most fundamental description of the state of a particle. The wave function $\Psi(\mathbf{r}, t)$ depends on position $\mathbf{r}$ and time t. The Schrödinger equation is a mathematical object which describes the evolution of the wave function, from which properties of the particle may be obtained. The position and the velocity will be provided through the probability $P = |\Psi(\mathbf{r}, t)|^2$ in space and time. In principle, the Schrödinger equation predicts all the quantum phenomena except relativistic ones, which require its more general relativistic formulation: the Dirac equation (see Sect. 1.8). In practice, it is solvable with high accuracy only for the hydrogen atom and the helium atom. Otherwise, for many-body systems, the equation becomes complicated and needs high-performance computing power to solve. For most systems, approximations are necessary.

1.5 Solving the Schrödinger Equation in One Dimension

The Schrödinger equation in one dimension is analytically solvable and gives some important insight into realistic quantum mechanical problems. Let us consider the time independent potential $V(x)$. The wave function $\Psi(x, t)$ can be written in a separable form in space and time as

$$\Psi(x, t) = \psi(x)e^{-i\omega t} \tag{1.41}$$

because $V(x)$ has no time dependence. The wave function $\psi(x)$ obeys the equation

$$-\frac{\hbar^2}{2m}\left(\frac{d^2}{dx^2}\psi(x)\right) + V(x)\psi(x) = E\psi(x). \tag{1.42}$$

In the free space $V(x) = 0$, the solution of (1.42) is easily obtained as

$$\psi(x) = Ae^{ikx} + Be^{-ikx}, \tag{1.43}$$

where $k = \sqrt{2mE/\hbar}$, and A and B are complex factors. Then the time-dependent wave function (1.41) is expressed as

$$\Psi(x,t) = Ae^{i(kx-\omega t)} + Be^{-i(kx+\omega t)}, \tag{1.44}$$

where the first and the second terms of the right-hand side (r.h.s.) correspond to the outgoing and incoming plane waves, respectively. The solution of Eq. (1.43) can be expressed in an alternative way as

$$\psi(x) = C\sin(kx) + D\cos(kx), \tag{1.45}$$

which is useful to solve some boundary potential problems such as the infinite square well potential.

For the infinite square well potential in Fig. 1.10, the wave function (1.45) is constrained with the boundary conditions $\psi(x = 0) = \psi(x = L) = 0$, and the eigenenergy and the eigenfunction of a particle of mass m are obtained as

$$E_n = \frac{\hbar^2\pi^2 n^2}{2mL^2} \qquad n = 1, 2, 3, \ldots, \tag{1.46}$$

and

$$\psi_n(x) = \sqrt{\frac{2}{L}}\sin k_n x \qquad n = 1, 2, 3, \ldots, \tag{1.47}$$

with $k_n = n\pi/L$, respectively (see Problem 1.2).

The system in the finite size potential such as Fig. 1.10 shows characteristic features of quantum mechanics:

(1) The eigenenergies are quantized. The energy in the square well potential is quantized to be

$$E_n = Dn^2 \qquad n = 1, 2, 3, \ldots, \tag{1.48}$$

while that in the harmonic oscillator potential is written as

$$E_n = E(N + 1/2) \qquad N = 0, 1, 2, 3, \ldots, \tag{1.49}$$

where D and E are constants which are determined uniquely for each quantum system.

(2) The lowest energy is finite, but not zero. This is referred as "quantum fluctuation" of the particle in its ground state. Even in the ground state, the particle moves.

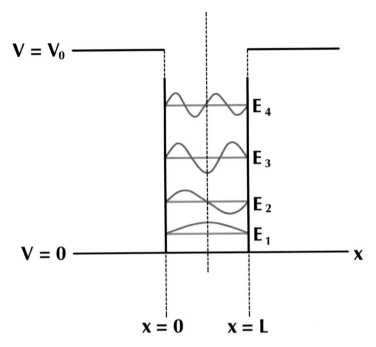

Fig. 1.10 One-dimensional infinite square well potential with $V = 0$ in the region $0 \leq x \leq L$ and $V_0 \to \infty$ in the region $x < 0$ and $L < x$. The boundary conditions impose that the solution wave functions vanish at $x = 0$ and $x = L$. The solutions are indexed by their quantum number n which gives the number of nodes, where $\psi_n(x) = 0$. For $n = 1$, the wave function shows no node, for $n = 2$ the wave function has one node at $x = L/2$, etc.

(3) The wave functions show the nodes and the maxima so that the existence of the particle is forbidden at the node positions since the probability $|\psi(x = \text{node})|^2 = 0$. This phenomenon does not occur in classical mechanics where the particle can explore the entire region of $0 \leq x \leq L$ with $V = 0$, for any energy.

In quantum mechanics, the "particle-in-a-box model" (also known as the infinite potential well or the infinite square well potential) describes a particle movement in a small space surrounded by impenetrable barriers. Assuming one-dimensional infinite square well potential with a very narrow width (on the scale of a nanometer), quantum effects become important and the energy is quantized, i.e., the particle can not move freely and may never be detected at certain positions, known as spatial nodes. In classical systems, on the contrary, a ball (particle) trapped inside a large box on the scale of one meter can move freely in space at a given speed and can be found everywhere in the box. This feature of quantum mechanics will be shown more quantitatively in the followings.

Problem

1.2 Solve the Schrödinger equation (1.42) for a particle of mass m in an infinite square well potential, Fig. 1.10, defined as having zero potential energy inside the region $0 \leq x \leq L$, and infinite potential energy outside that region

$$V(x) = \begin{cases} 0, & 0 \leq x \leq L, \\ \infty, & \text{otherwise.} \end{cases} \tag{1.50}$$

Problem

1.3 Calculate the lowest energy of an electron in the one-dimensional infinite square well potential of width $L = 0.4$ nm $= 0.4 \times 10^{-9}$ m.

Problem

1.4 Prove the wave function (1.47) to satisfy the orthonormal condition

$$\int_0^L \psi_n(x)^* \psi_m(x) dx = \delta_{nm}, \tag{1.51}$$

where δ_{nm} is the Kronecker delta defined by

$$\delta_{nm} = \begin{cases} 1 & n = m, \\ 0 & \text{otherwise.} \end{cases} \tag{1.52}$$

The probability interpretation is unavoidable for any quantum phenomenon. In general, one needs large enough statistics to observe quantum phenomena. In radioactive beam experiments, at least few hundred events are usually required to consider the data set reliable. However, nature is not always that kind. Let us detail an example of how quantum mechanics does "play dice", in opposition to the famous Albert Einstein's claim. At the RIKEN research center, Kosuke Morita and his group have evidenced a new superheavy element (SHE) $Z = 113$ starting from the beginning of 21st century. It was officially named Nihonium (symbol Nh) after the Japanese name for Japan: "Nihon". They use the reaction

$$^{209}\text{Bi} + ^{70}\text{Zn} \rightarrow ^{278}\text{Nh} + \text{n}.$$

The elements were created by sending a huge number of ^{70}Zn nuclei on a ^{209}Bi target. Many different reactions take place in the target while only a very small proportion ends up into the fusion of the two colliding nuclei and leads to the formation of a ^{278}Nh nucleus. This probability is given by the fusion cross section σ_f.[9] It was really

---

[9] The cross section of a nucleus is used to characterize the probability that a nuclear reaction will occur. The cross section can be quantified in terms of "geometrical area" where the reaction takes place. Namely a larger area means a larger probability of interaction to induce nuclear reactions, such as fusion or fission.

an exciting moment when they observed the first event on July 23, 2004, after 80 days of experiment [13]. The second event came to the detector nine months later, on April 2, 2005. Morita thought he could observe one or two events of $Z = 113$ SHE element every year, while in reality it took more than 8 years to observe the third event (Fig. 1.11)! As a result, we can say scientifically that the cross section of ^{209}Bi$+^{70}$Zn reaction creates one $Z = 113$ SHE element every 4 years. One needs 100 trillion (100,000,000,000,000) heavy ion collisions to create one SHE with $Z = 113$ protons from ^{70}Zn and ^{209}Bi! Morita and his group have been patient enough and spent more than 10 years to observe these rare events! Eventually they won against a capricious nature. The probability interpretation is quite right at the microscopic level. However when one faces a macroscopic time interval, it is very much confusing and may require patience and many efforts from observers. In November 2016, four new superheavy elements obtained official names and symbols approved by the International Union of Pure and Applied Chemistry (IUPAC). They are:

- Element 113: nihonium (Nh)
- Element 115: moscovium (Mc)
- Element 117: tennessine (Ts)
- Element 118: oganesson (Og)

Two other SHE with $Z = 114$ and $Z = 116$ were named officially in May 2012 as

- Element 114: flerovium (Fl)
- Element 116: livermorium (Lv)

The periodic table of elements is now completed up to $Z = 118$ and next challenges for SHE search are nuclei with $Z = 119$ and 120.

1.6 Quantum Tunnelling and Probability of Wave Function

The probability $P(\mathbf{r}, t)$ to find a particle at a given position $\mathbf{r}$ and time t is given by its wave function Ψ:

$$P(\mathbf{r}, t) = |\Psi(\mathbf{r}, t)|^2.$$

In principle, this formula tells us that the probability is non-zero everywhere in the universe while it is extremely small when we measure the wave function of bound particles (the size of a nucleus is of the order of 1 fm $= 10^{-15}$ m) in a macroscopic distance (1 m or more). In a microscopic level, it is sometimes non-negligible. In classical mechanics, a particle is not allowed to go through a potential barrier which is higher than the kinetic energy of the particle. However, quantum mechanics allows particles to penetrate the barrier even if its height exceeds the kinetic energy of particles. This is called *tunnelling* in quantum mechanics.

To demonstrate this peculiar phenomenon, we consider the square barrier potential problem in one-dimension. The potential barrier is given by (see Fig. 1.12)

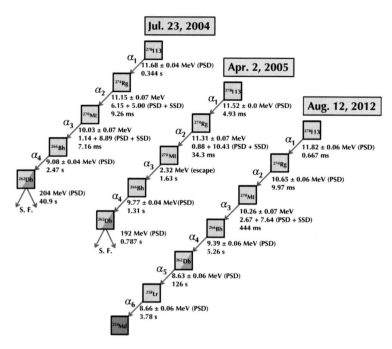

Fig. 1.11 Decay chains of new super-heavy element (SHE) $Z = 113$ observed by K. Morita et al., at RIKEN Radioactive Ion Beam Factory (RIBF), Japan. The fusion reaction $^{209}\text{Bi} + ^{70}\text{Zn} \rightarrow ^{278}\text{Nh} + n$ was used

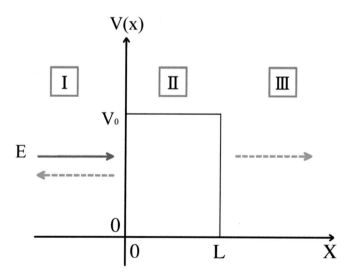

Fig. 1.12 Square barrier potential and quantum tunnelling. E is the kinetic energy of the incident particle, while V_0 and L are the barrier height and width, respectively

$$V(x) = \begin{cases} 0 & x < 0 \quad \text{(region I)}, \\ V_0(> 0) & 0 \le x \le L \quad \text{(region II)}, \\ 0 & x > L \quad \text{(region III)}. \end{cases} \tag{1.53}$$

The time-independent Schrödinger equation for the wave function $\psi(x)$ reads

$$H\psi(x) = \left[-\frac{\hbar^2}{2m}\frac{d^2}{dx^2} + V(x) \right]\psi(x) = E\psi(x), \tag{1.54}$$

in the same way as the infinite square well potential problem. The barrier divides the space in three parts ($x < 0, 0 \le x \le L, x > L$). In any of these parts, the potential is constant, meaning that the particle is quasi-free, and the solution of the Schrödinger equation can be written in general as a superposition of left and right moving waves. In region I and III, the wave function reads

$$\psi_{\mathrm{I}}(x) = e^{ik_0 x} + Re^{-ik_0 x} \quad x < 0, \tag{1.55}$$

$$\psi_{\mathrm{III}}(x) = Te^{ik_0 x} \quad x > L, \tag{1.56}$$

where the wave number k_0 is related to the energy as

$$k_0 = \sqrt{2mE/\hbar^2} \qquad x < 0 \quad \text{or} \quad x > L \tag{1.57}$$

In region I, the initial right-going wave is normalized to unity. The coefficients R and T stand for the reflection and the transmitted waves by the barrier, respectively. In region III, there is no incoming wave and only the outgoing wave may exist. The behavior of waves in region II is different either exponential for $E < V_0$ or sinusoidal for $E > V_0$;

$$\psi_{\mathrm{II}}(x) = Ae^{\kappa x} + Be^{-\kappa x} \quad E < V_0 \tag{1.58}$$

$$\psi_{\mathrm{II}}(x) = Ce^{ik_1 x} + De^{-ik_1 x} \quad E > V_0 \tag{1.59}$$

where the wave numbers are related to the energy via

$$\kappa = \sqrt{2m(V_0 - E)/\hbar^2} \quad E < V_0, \tag{1.60}$$

$$k_1 = \sqrt{2m(E - V_0)/\hbar^2} \quad E > V_0. \tag{1.61}$$

The coefficients R, T, A, B, C, D have to be found from the boundary conditions of the wave functions at $x = 0$ and $x = L$. The wave function and its derivative have to be continuous everywhere so that

$$\psi_I(0) = \psi_{II}(0), \tag{1.62}$$

$$\frac{d}{dx}\psi_I(0) = \frac{d}{dx}\psi_{II}(0), \tag{1.63}$$

$$\psi_{II}(L) = \psi_{III}(L), \tag{1.64}$$

$$\frac{d}{dx}\psi_{II}(L) = \frac{d}{dx}\psi_{III}(L). \tag{1.65}$$

Inserting the wave functions (1.59) for $E > V_0$, the boundary conditions give the following conditions on the coefficients

$$1 + R = C + D,$$
$$ik_0(1 - R) = ik_1(C - D),$$
$$Ce^{iLk_1} + De^{-iLk_1} = Te^{iLk_0},$$
$$ik_1(Ce^{iLk_1} - De^{-iLk_1}) = ik_0Te^{iLk_0}.$$

We then eliminate the coefficients C, D from these equations and solve for T and R. The results are

$$T = \frac{4k_0k_1e^{-iL(k_0-k_1)}}{(k_0 + k_1)^2 - e^{2iLk_1}(k_0 - k_1)^2}, \tag{1.66}$$

$$R = \frac{(k_0^2 - k_1^2)\sin(Lk_1)}{2ik_0k_1\cos(Lk_1) + (k_0^2 + k_1^2)\sin(Lk_1)}. \tag{1.67}$$

In the case $E > V_0$, the transmission probability and the reflection probability are

$$P_T \equiv |T|^2 = \frac{4E(E - V_0)}{4E(E - V_0) + V_0^2\sin^2(k_1L)} \quad E > V_0, \tag{1.68}$$

$$P_R \equiv |R|^2 = \frac{V_0^2\sin^2(k_1L)}{4E(E - V_0) + V_0^2\sin^2(k_1L)} \quad E > V_0 \tag{1.69}$$

$$= 1 - |T|^2. \tag{1.70}$$

For $E < V_0$, we can repeat the same procedure for the wave functions (1.58) and obtain the transmission probability

$$P_T = |T|^2 = \frac{4E(V_0 - E)}{4E(V_0 - E) + V_0^2\sinh^2(\kappa L)} \quad E < V_0, \tag{1.71}$$

for the particle to be transmitted through the barrier, with $\kappa = \sqrt{2m(V_0 - E)/\hbar^2}$. This effect differs from the classical case and is the quantum tunnelling. The transmission is exponentially suppressed with the barrier width, which can be understood from the functional form of the wave function; outside of the barrier it oscillates with wave number k_0, whereas within the barrier it is exponentially damped over a

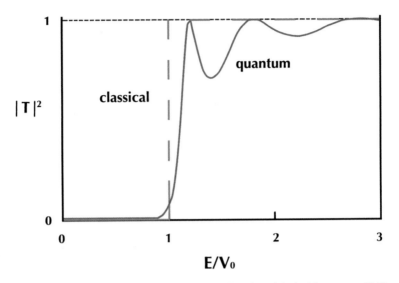

Fig. 1.13 Variation of the transmission probability as a function of the incident energy E. The solid curve shows the quantum mechanical transmission probability, while the dashed one corresponds to the classical one. The potential width is given by a coefficient $2m V_0 L^2/\hbar^2 = 50$

distance $1/\kappa$. Equally surprising is that for energies larger than the barrier height, $E > V_0$, the particle may be reflected from the barrier with a non-zero probability $|R|^2 = 1 - |T|^2$ in Eq. (1.70). The variation of the transmission probability as a function of the incident energy E is shown in Fig. 1.13.

Nuclear physics provides many examples of the tunnelling effect. The α decay is one of the typical example of tunnelling effect (see Fig. 1.11 where the SHE decays consecutively emitting several α particles).

Problem

1.5 Evaluate the nuclear size r_A for a nucleus of the mass A with a constant nuclear density $\rho_0 = 0.15$ fm^{-3};

$$\rho(r) = \begin{cases} \rho_0 & \text{for } r < r_0 \\ 0 & \text{for } r > r_0 \end{cases} \tag{1.72}$$

Problem

1.6 Following the steps of Gamow in 1928, let us consider that the α-decay process results from a α particle, pre-existing in the mother nucleus, which escapes from quantum tunnelling through the Coulomb barrier [14]. Consider the model sketched in Fig. 1.14. Evaluate the α decay life time of ^{212}Po. The energy of emitted α from ^{212}Po is 8.78 MeV experimentally.

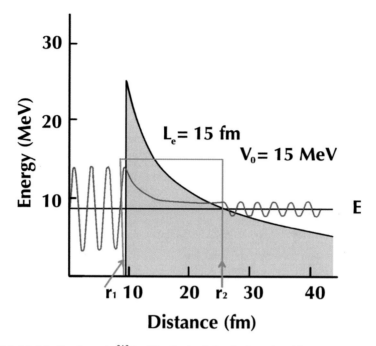

Fig. 1.14 Model of α decay in ^{212}Po. The Coulomb barrier is replaced by a rectangular barrier with the width $L_e = 15$ fm and the height $V_0 = 15$ MeV

1.7 Uncertainty Relation

One of the most important concept of quantum mechanics is the "uncertainty relation", also called Heisenberg's uncertainty principle which was introduced in 1927 by Werner Heisenberg, a German physicist. The idea of the uncertainty principle came originally from problems encountered in measurements with subatomic objects. It states that the more precisely the position of a particle is determined, the less precisely its momentum can be known, and vice versa. The formal inequality relating the standard deviations of position Δx and the momentum Δp was formulated as

$$\Delta x \, \Delta p \geq \hbar/2. \tag{1.73}$$

It has become clear now that the uncertainty principle is inherent to the properties of wave nature of subatomic systems. Thus, the uncertainty principle actually states a fundamental property of quantum systems, and not just a problem of simultaneous measurement of two observables for a single object.

In quantum mechanics, the position **r** and the momentum **p** act as operators upon the wave function $\Psi(\mathbf{r}, t)$. The momentum operator is expressed by the differentiation ∇ as

$$\mathbf{p} = \frac{\hbar}{i}\nabla. \tag{1.74}$$

For any pair of operators A, B, the *commutation relation* is defined as

$$[A, B] \equiv AB - BA. \tag{1.75}$$

For A, B, one can consider observable operators such as the position and momentum of a particle. For x and p_x, when we $[x, p_x]$ operates on a wave function $\psi(x)$, it leads to

$$[x, p_x]\psi(x) = (xp_x - p_x x)\psi(x) = xp_x\psi(x) - p_x(x\psi(x))$$
$$= xp_x\psi(x) - p_x(x)\psi(x) - xp_x\psi(x) = i\hbar\psi(x), \tag{1.76}$$

where we use $p_x \equiv \frac{\hbar}{i}\frac{\partial}{\partial x}$ in Eq. (1.74). Equation (1.76) can be applied also for $[y, p_y]$ and $[z, p_z]$ and we can derive the relations,

$$[x, p_x] = i\hbar, \quad [y, p_y] = i\hbar, \quad [z, p_z] = i\hbar. \tag{1.77}$$

The relation between the operators given by Eq. (1.77) is *non-commutable*. This means that one cannot measure the position of a particle and then its momentum, with the same probability if the measurement were performed the other way around. Other commutation relations with $\mathbf{r}$ and $\mathbf{p}$, such as $[x, y]$, $[p_x, p_y]$, $[y, p_x]$, are all 0, and are called *commutable* and thus the probability to measure, e. g., x position and y position, is independent of the order in which they are measured.

We consider two operators that are independent of each other. Let us show that the operators A and B are commutable if all the eigenstates of A are at the same time the eigenstates of B. Taking a common eigenstate ψ, whose eigenvalue of A is a while that of B is b:

$$A\psi = a\psi, \qquad B\psi = b\psi. \tag{1.78}$$

The commutation relation $[A, B]$ with ψ gives

$$[A, B]\psi = (AB - BA)\psi = (ab - ba)\psi = 0. \tag{1.79}$$

Since the common eigenstates of A and B constitute a complete set,[10] the equality $[A, B] = 0$ holds. Conversely, it can be shown that there exists a complete set of common eigenstates of A and B if $[A, B] = 0$.

Problem

1.7 Prove that there exist common eigenstates of ψ_a if $[A, B] = 0$.

---

[10] If any eigenstate ϕ can be expressed by a linear combination of a set of eigenstates, the set is said "complete". Since the set $\{\psi\} = \{\psi_1, \psi_2, \psi_3 \ldots\}$ includes all the eigenstates of the operator A, any eigenstate ϕ can be expressed by a linear combination of the states in the set $\{\psi\}$ so that $\{\psi\}$ is a complete set.

What about the case where A and B do not commute? Let us consider, in particular, the case with $A = x$ and $B = p_x$. If there were a common eigenstate ψ of A and B, it would satisfy the equation: $[A, B]\psi = 0$. This equation contradicts the relation obtained from the commutation relation: $[A, B]\psi = i\hbar\psi$. Thus, it is concluded that there is no common eigenstate of the position x and the momentum p_x. In other words, no state can have a definite position and a definite momentum simultaneously. This conclusion implies that if x is determined by measurements without any ambiguity, a fundamental uncertainty steaming from quantum mechanics arises in the measurements of p_x. Alternatively, if p_x is determined precisely, x fluctuates seriously and its values can not be determined. This relation is the *uncertainty principle* between position and momentum. Let us define the uncertainty in the observation of an operator A as the standard deviation ΔA,

$$(\Delta A)^2 \equiv \langle (A - \langle A \rangle)^2 \rangle = \langle A^2 \rangle - \langle A \rangle^2. \tag{1.80}$$

The uncertain relation between position and momentum is expressed as

$$\Delta x \cdot \Delta p_x \geq \tfrac{\hbar}{2},$$
$$\Delta y \cdot \Delta p_y \geq \tfrac{\hbar}{2}, \tag{1.81}$$
$$\Delta z \cdot \Delta p_z \geq \tfrac{\hbar}{2}.$$

Problem

1.8 Assume that two Hermitian operators A and B satisfy the commutation relation $[A, B] = i\hbar$. Prove the uncertainty relation $\Delta A \cdot \Delta B \geq \tfrac{\hbar}{2}$

As can been understood from the small value of the reduced Planck constant $\hbar = 1.06 \times 10^{-34}$ J·s, this uncertainty relation is an inherent feature of microscopic quantum objects. We should emphasize again that the uncertainty principle is an intrinsic property of the wave nature of quantum objects, but not a problem of accuracy of experimental measurements. We will come back to this point in relation with a halo nucleus which has a large extension of wave function in the coordinate space in Chap. 5. This means the wave function for a halo nucleus will be very compact in the momentum space according to the uncertainty relation Eq. (1.81). This peculiar feature was confirmed by experimental observation of the momentum distribution of breakup reactions of ^{11}Li.

There is not only a uncertainty relation between momentum and position, but also between energy and time,

$$\Delta E \cdot \Delta t \geq \frac{\hbar}{2}, \tag{1.82}$$

which plays an important role to understand the range of nuclear interaction induced by the exchange of mesons with finite mass, as further detailed in Chap. 2.

1.8   The Dirac Equation

In nuclear physics, the dynamics is in general non-relativistic, i.e., the relative velocity of nucleons is mostly small compared to the speed of light. However, the relativistic effect appears in some observables such as magnetic moments or spin-orbit interactions. The relativistic quantum physics is described by the Dirac equation which was proposed by Paul Dirac in 1928 [15]. We can consider the Dirac equation as the extension of the Schrödinger equation (1.38) to account fully for special relativity in the context of quantum mechanics. In the following Sect. 1.8.1, the derivation of the Dirac equation from the relativistic kinematics is given. Next, the physical meaning of the negative energy solution of the Dirac equation is detailed in Sect. 1.8.3. At last, in the Sect. 1.8.4, the spin of the Dirac particle (fermion) will be derived as an outcome of the spinor wave function of the Dirac equation.

1.8.1   Relativistic Kinematics and Dirac Equation

The Schrödinger equation (1.38) is based upon the non-relativistic energy-momentum relation

$$E = \frac{p^2}{2m}. \tag{1.83}$$

This relation is an approximation of the relativistic relation

$$E^2 = p^2c^2 + m^2c^4, \tag{1.84}$$

where c is the speed of light and m is the mass of the particle. Dirac proposed an equation which allows the atom to be treated in a consistent way with relativity, and to explain the behavior of the relativistically moving electron. To this end, he adopted the operator equivalents of the energy and momentum from the Schrödinger theory,

$$E \rightarrow i\hbar\frac{\partial}{\partial t}, \; \mathbf{p} \rightarrow \frac{\hbar}{i}\nabla, \tag{1.85}$$

and substituted them into Eq. (1.84). From this prescription, he obtained an equation describing the propagation of waves constructed from a relativistically invariant relation,

$$\left(-\frac{1}{c^2}\frac{\partial^2}{\partial t^2} + \nabla^2\right)\phi = \frac{m^2c^2}{\hbar^2}\phi, \tag{1.86}$$

which is called the Klein–Gordon equation for a scalar field ϕ without any spin and isospin dependence. The Schrödinger equation and Klein–Gordon equation involve the second derivatives of time and/or space coordinates. Dirac wanted to invent a type of covariant equation which is invariant under the 4 dimensional space-time Lorenzian transformation of relativistic theory. The covariant form should involve only the first derivatives of space and time coordinates. In order to satisfy the covariant condition, Dirac proposed the so-called Dirac equation:

$$i\hbar\frac{\partial\Psi(\mathbf{r},t)}{\partial t} = \left(c\boldsymbol{\alpha}\cdot\mathbf{p} + \beta mc^2\right)\Psi(\mathbf{r},t) \tag{1.87}$$

where $\Psi(\mathbf{r},t)$ is the wave function for the electron of rest mass m with space-time coordinates $\mathbf{r}$, t, and α_i and β are constant quantities. Taking the square root of operators in both sides of Eq. (1.87) on Ψ, we obtain

$$-\hbar^2\frac{\partial^2}{\partial t^2}\Psi = \left(c^2(\boldsymbol{\alpha}\cdot\mathbf{p})^2 + c(\boldsymbol{\alpha}\cdot\mathbf{p})\beta mc^2 + \beta mc^2 c(\boldsymbol{\alpha}\cdot\mathbf{p}) + \beta^2 m^2 c^4\right)\Psi, \tag{1.88}$$

where we keep the ordering of α_i and β as they appear in the original products. To compare with the Klein–Gordon equation (1.86), the quantities α_i and β should satisfy twelve equations

$$\alpha_i\alpha_j + \alpha_j\alpha_i = 2\delta_{ij},$$
$$\alpha_i\beta + \beta\alpha_i = 0, \tag{1.89}$$
$$\beta^2 = 1,$$

for $\alpha_i \equiv (\alpha_x, \alpha_y, \alpha_z)$ and β. To satisfy all the equations, the α_i, β should be 4×4 matrices and represented as

$$\alpha_x = \begin{pmatrix} 0 & \sigma_x \\ \sigma_x & 0 \end{pmatrix}, \alpha_y = \begin{pmatrix} 0 & \sigma_y \\ \sigma_y & 0 \end{pmatrix}, \alpha_z = \begin{pmatrix} 0 & \sigma_z \\ \sigma_z & 0 \end{pmatrix}, \beta = \begin{pmatrix} I & 0 \\ 0 & -I \end{pmatrix}, \tag{1.90}$$

where $\sigma_x, \sigma_y, \sigma_z$ are 2×2 Pauli spin matrices and I is the 2×2 unit matrix defined as

$$\sigma_x = \begin{pmatrix} 0 & 1 \\ 1 & 0 \end{pmatrix}, \sigma_y = \begin{pmatrix} 0 & -i \\ i & 0 \end{pmatrix}, \sigma_z = \begin{pmatrix} 1 & 0 \\ 0 & -1 \end{pmatrix}, I = \begin{pmatrix} 1 & 0 \\ 0 & 1 \end{pmatrix}. \tag{1.91}$$

It is now clear that the α_i and β are not simple numbers, but 4×4 matrices so that they are not commutable in general. It is natural that the Dirac wave function $\psi(\mathbf{r},t)$ will be a 4 component spinor wave function[11]

---

[11] The non-relativistic wave function of a spin 1/2 particle is expressed by a two-component vector with spin up and spin down components ψ_+ and ψ_-;

$$\Psi = \begin{pmatrix} \psi_+ \\ \psi_- \end{pmatrix}. \tag{1.92}$$

This vector has an analogy to that of ordinary space which has the geometric entry with three components and transforms one into another under rotation. This is also the case for the two component vector, whose components transform one into another by rotation. This vector is called *spinor* or *two-component spinor*. The relativistic wave function of a spin 1/2 particle has four components.

$$\Psi(\mathbf{r}, t) = \begin{pmatrix} \phi_1 \\ \phi_2 \\ \phi_3 \\ \phi_4 \end{pmatrix} \tag{1.93}$$

to satisfy Eq. (1.87). Multiplying by β/c on Eq. (1.87), we have

$$\left(i\hbar\beta \frac{1}{c}\frac{\partial}{\partial t} + +i\hbar\beta\boldsymbol{\alpha}\cdot\boldsymbol{\nabla} - mc \right)\Psi(\mathbf{r}, t) = 0, \tag{1.94}$$

where relations $\mathbf{p} = -i\boldsymbol{\nabla}$ and $\beta^2 = 1$ are used. The Dirac equation (1.94) can be written in a covariant form,[12]

$$(i\hbar\gamma_\mu\partial^\mu - mc)\Psi(\mathbf{r}, t) = 0 \tag{1.95}$$

where $\gamma_\mu(\mu = 0, 1, 2, 3)$ is the 4×4 matrix,

$$\gamma_\mu = (\gamma_0, -\boldsymbol{\gamma}), \text{ with } \gamma_0 = \beta = \begin{pmatrix} I & 0 \\ 0 & -I \end{pmatrix}, \boldsymbol{\gamma} = \beta\boldsymbol{\alpha} = \begin{pmatrix} 0 & \boldsymbol{\sigma} \\ -\boldsymbol{\sigma} & 0 \end{pmatrix}, \tag{1.96}$$

and 0 and I are 2×2 null and unit matrices, respectively. The 4 component derivative is also defined as

$$\partial^\mu \equiv \frac{\partial}{\partial x_\mu} = \left(\frac{1}{c}\frac{\partial}{\partial t}, -\boldsymbol{\nabla} \right). \tag{1.97}$$

Fundamental characters of the Dirac equation will be discussed in more details in the next section.

Problem

1.9 Derive the Klein–Gordon equation (1.86) from the Dirac equation (1.88) by using the orthonormal relations (1.89).

1.8.2 Continuity Equation of Dirac Wave Function

We derive a covariant form of the continuity equation in this subsection. This is an important requirement to express physical quantities in the covariant form in relativistic quantum mechanics. The density is expressed as

$$\rho(\mathbf{r}, t) = \Psi^\dagger(\mathbf{r}, t)\Psi(\mathbf{r}, t), \tag{1.98}$$

---

[12] When the same suffix appears in the superscript and subscript of symbols in the same equation (see Eq. (1.95)), this implies the summation of all components of the symbols; $\gamma_\mu\partial^\mu \equiv \sum_{\mu=\{0,1,2,3\}} \gamma_\mu\partial^\mu$.

which is the same definition in both the non-relativistic and the relativistic formalisms. We express the Hermite conjugate of Dirac equation (1.87) as

$$i\hbar\frac{\partial \Psi^\dagger(\mathbf{r}, t)}{\partial t} = -ic\nabla\Psi^\dagger(\mathbf{r}, t) \cdot \boldsymbol{\alpha} - \Psi^\dagger(\mathbf{r}, t)\beta mc^2. \tag{1.99}$$

Multiplying $\Psi^\dagger(\mathbf{r}, t)$ on Eq. (1.87) from the left and $\Psi(\mathbf{r}, t)$ on Eq. (1.99) from the right, and adding the two equations, we obtain

$$i\frac{\partial}{\partial t}(\Psi^\dagger(\mathbf{r}, t)\Psi(\mathbf{r}, t)) = -i\nabla\left(\Psi^\dagger(\mathbf{r}, t)\boldsymbol{\alpha}\Psi(\mathbf{r}, t)\right). \tag{1.100}$$

Equation (1.100) can be rewritten as

$$\frac{\partial \rho(\mathbf{r}, t)}{\partial t} + \nabla \cdot \mathbf{j}(\mathbf{r}, t) = 0, \tag{1.101}$$

where the vector $\mathbf{j}(\mathbf{r}, t)$ is defined as

$$\mathbf{j}(\mathbf{r}, t) = \Psi^\dagger(\mathbf{r}, t)\boldsymbol{\alpha}\Psi(\mathbf{r}, t). \tag{1.102}$$

Equation (1.101) is nothing but the continuity equation for the Dirac wave function. The continuity equation can be expressed in a covariant form by introducing the Dirac adjoint wave function $\overline{\Psi}(\mathbf{r}, t) = \Psi^\dagger(\mathbf{r}, t)\gamma^0$ as

$$\Psi^\dagger(\mathbf{r}, t)\Psi(\mathbf{r}, t) = \Psi^\dagger(\mathbf{r}, t)\gamma^0\gamma^0\Psi(\mathbf{r}, t) = \overline{\Psi}(\mathbf{r}, t)\gamma^0\Psi(\mathbf{r}, t), \tag{1.103}$$

$$\Psi^\dagger(\mathbf{r}, t)\boldsymbol{\alpha}\Psi(\mathbf{r}, t) = \Psi^\dagger(\mathbf{r}, t)\gamma^0\gamma^0\boldsymbol{\alpha}\Psi(\mathbf{r}, t) = \overline{\Psi}(\mathbf{r}, t)\gamma^\nu\Psi(\mathbf{r}, t), \tag{1.104}$$

where $\gamma^\nu \equiv \gamma^0\alpha^\nu$ ($\nu = 1, 2, 3$). Then the continuity equation (1.101) is expressed in a covariant form as

$$\partial_\mu j^\mu = 0, \tag{1.105}$$

where $j^\mu \equiv \overline{\Psi}(\mathbf{r}, t)\gamma^\mu\Psi(\mathbf{r}, t)$ ($\mu = 0, 1, 2, 3$). In the relativistic formalism, the density ρ and the current vector $\mathbf{j}$ are the time and the space components of 4 dimensional vector, respectively. The adjoint wave function $\overline{\Psi}(\mathbf{r}, t) = \Psi^\dagger(\mathbf{r}, t)\gamma^0$ is commonly adopted to express physical quantities in the covariant form.

1.8.3   Negative Energy and Spin of the Dirac Wave Function

The Dirac equation is an equation with a 4×4 spin matrix and the eigenfunction is given by a four-component spinor as the solution of the spin matrix. Let us study a plane wave of Dirac eigenfunction to illustrate physical implications of the Dirac equation. The plane wave function reads

$$\Psi(\mathbf{r}, t) = \begin{pmatrix} \phi_1 \\ \phi_2 \\ \phi_3 \\ \phi_4 \end{pmatrix} \exp\left\{ \frac{i}{\hbar}(\mathbf{p} \cdot \mathbf{r} - Et) \right\}, \tag{1.106}$$

where $\phi_i (i = 1, 2, 3, 4)$ are independent of $\mathbf{r}$ and t. A four-component spinor Ψ satisfies the Dirac equation

$$\begin{aligned} E\Psi &= (c\boldsymbol{\alpha} \cdot \mathbf{p} + \beta mc^2)\Psi \\ &= \begin{pmatrix} Imc^2 & c\boldsymbol{\sigma} \cdot \mathbf{p} \\ c\boldsymbol{\sigma} \cdot \mathbf{p} & -Imc^2 \end{pmatrix} \Psi. \end{aligned} \tag{1.107}$$

It is convenient to write the eigenstate Ψ as

$$\Psi = \begin{pmatrix} \varphi \\ \chi \end{pmatrix}, \tag{1.108}$$

where

$$\varphi = \begin{pmatrix} \phi_1 \\ \phi_2 \end{pmatrix}, \chi = \begin{pmatrix} \phi_3 \\ \phi_4 \end{pmatrix}. \tag{1.109}$$

The two-component vectors (1.109) will satisfy coupled equations of the Dirac equation (1.107)

$$E\varphi = mc^2\varphi + c\boldsymbol{\sigma} \cdot \mathbf{p}\chi, \tag{1.110}$$
$$E\chi = c\boldsymbol{\sigma} \cdot \mathbf{p}\varphi - mc^2\chi \tag{1.111}$$

or

$$\varphi = \frac{c\boldsymbol{\sigma} \cdot \mathbf{p}}{E - mc^2}\chi, \tag{1.112}$$
$$\chi = \frac{c\boldsymbol{\sigma} \cdot \mathbf{p}}{E + mc^2}\varphi. \tag{1.113}$$

To eliminate χ from Eq. (1.112), we obtain

$$\varphi = \frac{c\boldsymbol{\sigma} \cdot \mathbf{p}}{E - mc^2}\frac{c\boldsymbol{\sigma} \cdot \mathbf{p}}{E + mc^2}\varphi,$$

which can be rewritten

$$(E^2 - (mc^2)^2)\varphi = c^2(\boldsymbol{\sigma} \cdot \mathbf{p})^2\varphi = c^2 p^2 \varphi.$$

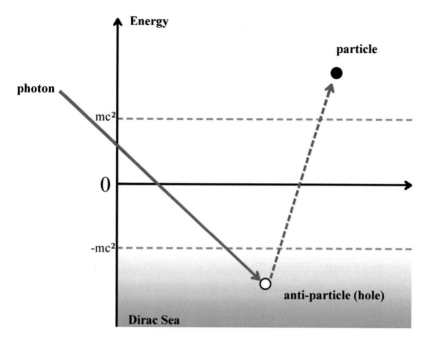

Fig. 1.15 Dirac sea and pair creation of particle and anti-particle by a high-energy photon with the energy $\hbar\omega > 2mc^2$. The quantum states with the energy $E < -mc^2$ are fully occupied in the initial vacuum (the Dirac sea). The energy region between $-mc^2 < E < mc^2$ is forbidden to create a particle because of the energy conservation law

This equation guarantees that the relativistic energy-momentum relation $E^2 = c^2 p^2 + m^2 c^4$ holds in the Dirac equation (1.107).

Problems

1.10 Prove $(\boldsymbol{\sigma} \cdot \mathbf{p})^2 = p^2$.

1.11 Derive the non-relativistic limit of Dirac equation with a scalar central potential $V(r)$.

The possible eigenvalue of Eq. (1.107) may have positive and negative values

$$E = \pm E_p = \pm\sqrt{p^2 c^2 + (mc^2)^2}. \tag{1.114}$$

The negative-energy solutions are problematic since it is natural to think that all eigenenergies (the particle mass) are positive. Furthermore, if there is an interaction between the electron and the electromagnetic field, any electron placed in higher energy will decay to lower energy state, i.e., a positive-energy eigenstate will decay into negative-energy eigenstates by emitting photons. Real electrons obviously do not behave in this way. It was thought at the beginning of relativistic theory that negative-energy solutions are unphysical and it was a serious shortcoming of the Dirac theory.

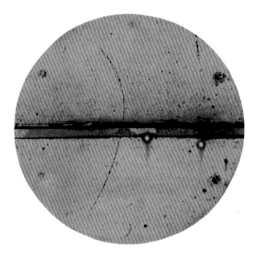

Fig. 1.16 Picture of a positron track, an almost-continuous chain of droplets triggered by the energy loss of the positron on its path in the cloud chamber. The chamber was located inside a magnetic field which causes the curvature of the trajectory. A 6-mm thick lead plate was placed in the chamber, visible on the picture as the horizontal bar. The energy loss of the positron in the plate and the increase of the curvature tell that the observed track was from a charged particle going from bottom to top. The orientation of the magnetic field implied that the particle was of positive electric charge. The historical deduction by Anderson can be found in [16] and is further detailed in [17]. Figure reprinted with permission from [17]. ©2021 from the American Physical Society

To cure this problem, Dirac introduced the hypothesis, known as "hole theory", that the vacuum is a many-body quantum state in which all the negative-energy electron eigenstates are occupied. This description of the vacuum as a "sea" of electrons is called the Dirac sea as illustrated in Fig. 1.15. When all the negative energy states are occupied, any additional electron would be forced to occupy a positive-energy eigenstate due to the Pauli exclusion principle and positive-energy electrons would never decay into negative-energy eigenstates since these states are already occupied by other electrons. Dirac further made a revolutionary hypothesis that if the negative-energy eigenstates are incompletely filled, each unoccupied eigenstate called a "hole state" would behave like a positively charged particle. Dirac initially thought that the hole might be the proton, but Hermann Weyl pointed out that the hole should behave as if it has the same mass as an electron with a positive charge, whereas the proton is over 1800 times heavier. The hole was eventually identified experimentally by Carl Anderson in 1932 as the "positron" with the same mass as the electron but with a positive electric charge [16]. The first published image of such a positron track taken with a cloud chamber by Carl Anderson is shown in Fig. 1.16. It was a surprise and a great success that the negative energy solution raised the prediction of positron creation as a hole in the Dirac sea.

1.8.4 Spin of Dirac Particle

Let us discuss now another physical implication of the four-component spinor eigenstate. We can see in the Dirac Hamiltonian (1.107) that the so-called "helicity" operator $\hat{h} \equiv \boldsymbol{\sigma} \cdot \mathbf{p}/p$ commutes with the Hamiltonian so that the state ψ can be an eigenstate of both the Hamiltonian and the helicity operator, simultaneously. Taking the direction of $\mathbf{p}$ to be the z-axis, the helicity operator will be $\hat{h} = \sigma_z$ and may have two eigenvalues of ± 1. Thus, we obtain four pairs of eigenvalues

$$(+E_p, +1), (+E_p, -1), (-E_p, +1), (-E_p, -1) \tag{1.115}$$

for the Dirac hamiltonian. Let us construct the spinor wave functions of four eigenvalues (1.115). The spin eigenstates of $s_z = \sigma_z/2$ can be written as

$$\varphi_1 = \begin{pmatrix} 1 \\ 0 \end{pmatrix}, \quad \varphi_2 = \begin{pmatrix} 0 \\ 1 \end{pmatrix}, \tag{1.116}$$

with the eigenvalues $\pm 1/2$, respectively. Taking these functions for the upper components of spinor wave functions (1.108), the Eq. (1.113) gives the lower components as

$$\chi_1 = \begin{pmatrix} \frac{cp_z}{E+mc^2} \\ \frac{c(p_x+ip_y)}{E+mc^2} \end{pmatrix}, \quad \chi_2 = \begin{pmatrix} \frac{c(p_x-ip_y)}{E+mc^2} \\ -\frac{cp_z}{E+mc^2} \end{pmatrix}. \tag{1.117}$$

We can show that these wave functions have positive energies (see Problem 1.11 and its solution). For the negative solutions, we take spin wave functions (1.116) as the lower components of Dirac spinors. Then Eq. (1.112) determines the upper components of the wave functions. Finally, we get the four eigenstates,

$$\Phi_1 = \begin{pmatrix} 1 \\ 0 \\ \frac{cp_z}{E_++mc^2} \\ \frac{c(p_x+ip_y)}{E_++mc^2} \end{pmatrix}, \Phi_2 = \begin{pmatrix} 0 \\ 1 \\ \frac{c(p_x-ip_y)}{E_++mc^2} \\ -\frac{cp_z}{E_++mc^2} \end{pmatrix}, \Phi_3 = \begin{pmatrix} \frac{cp_z}{E_--mc^2} \\ \frac{c(p_x+ip_y)}{E_--mc^2} \\ 1 \\ 0 \end{pmatrix}, \Phi_4 = \begin{pmatrix} \frac{c(p_x-ip_y)}{E_--mc^2} \\ -\frac{cp_z}{E_--mc^2} \\ 0 \\ 1 \end{pmatrix}, \tag{1.118}$$

where we discard the normalization factor $\sqrt{(E_p + mc^2)/2E_p}$ and the phase factor of plane wave.

Problem

1.12 Considering $\mathbf{p} = (0, 0, p_z)$, verify that the states (1.118) are the eigenstates of energy and spin.

1.9 Symmetries and Symmetry Breaking in the Nucleus

Symmetry has a precise definition in mathematics and physics. It is a property of invariance to a transformation, such as rotation in space or translation in time. In everyday language, symmetry refers to a sense of harmonious and beautiful proportion and balance. Geometric symmetry is commonly associated with shapes, i.e., reflection symmetry, rotational symmetry and translational symmetry. In physics, while these geometric symmetries play an important role, the term "symmetry" has been generalized to represent all sorts of invariance, i.e., lack of change under any kind of transformation. This concept has become one of the most powerful tools of theoretical physics. Physics laws have not only geometrical symmetries but also internal symmetries such as isospin (proton and neutron, up and down quark symmetry), flavor and color[13] form the fundamental structure of the standard model,[14] or symmetries of space time. In fact, this role inspired the Nobel laureate Philip Warren Anderson to write in his widely known article "More is Different" in 1972 that "it is only slightly overstating the case to say that physics is the study of symmetry" [18].

1.9.1 Explicit and Spontaneous Symmetry Breaking

Spin and isospin generate symmetries in nuclei. Nucleons such as protons and neutrons also exhibit a quantum spin $\hbar/2$, which is an intrinsic degree of freedom inherent to elementary particles. The spin can have different orientations in space, the same as the angular momentum of particles. The spin is generally represented by the Pauli spin matrix (1.91), $s = \hbar\sigma/2$. Another important quantum number of nucleons is the isospin quantum number which was proposed by Werner Heisenberg in 1932 to explain symmetry of strong nucleon–nucleon interaction. The strength of the nucleon-nucleon interaction does in reality depend on the spin and the isospin quantum numbers as well as the relative momentum of the two nucleons and the total angular momentum of the nucleon pair.

The concept of symmetry breaking is as important as symmetries themselves. In nuclear physics, the Hamiltonian or the Lagrangian has a well-defined geometrical symmetry, i.e., a spherical symmetry. However, due to a dynamical effect of single-particle motions, nuclear systems will be deformed in the intrinsic frame. This is an example of spontaneous symmetry breaking in quantum many-body systems analogous to the Jahn–Teller effect in molecules and ions which induces a geometrical

---

[13] The internal degrees of freedom of quarks and gluons are classified by the names of flavor and color, and the symmetry is imposed by a combination of these degrees of freedom.

[14] The standard model is a theory for the electromagnetic, weak, and strong nuclear interactions, as well as classifying all known subatomic particles. The model consists of the quantum chromodynamics for strong interaction and Weinberg–Salam model for electroweak interaction. It was developed throughout the latter half of the 20th century and completed in the mid-1970s upon experimental confirmation of the existence of quarks. Recent observation of the Higgs boson (2013) has given further credit to the Standard Model.

Table 1.3 Symmetry and symmetry breaking in nature

Symmetry	Outcome of symmetry breaking
Space and time symmetry	
Rotational symmetry	Deformation (prolate, oblate)
Translational symmetry	
Reflection symmetry in space	Parity violation in β decay
Charge conjugation parity symmetry (CP-symmetry)	Electric dipole moment (EDM)
Flavor SU(2) quark symmetry	Hyperons
Isospin symmetry	Isospin mixing in nuclei
Chiral symmetry	Masses of baryons

deformation of these objects. As a consequence, the deformed mean field or the deformed shape gives rise to typical phenomena of deformed objects such as rotational bands in nuclear many-body systems. The deformation of nuclei is discussed in Chap. 7. Various symmetries and their associated symmetry breakings are given in Table 1.3.

1.9.2 The Chiral Symmetry of QCD

Another important example of symmetry breaking in nuclear physics is the "chiral symmetry breaking". The Lagrangian which governs the dynamics of baryons has the chiral symmetry. The masses of protons and neutrons come from a spontaneous breaking of the chiral symmetry, while their difference comes from an explicit symmetry breaking due to the difference of quark masses. The Standard Model does not actually refer to protons and neutrons, but to the particles of which they are composed, known as quarks and gluons. The proton consists of three quarks: two quarks of a type called "up" and one quark of a type called "down". Similarly, the neutron consists of two down quarks and an up quark.

In the Standard Model, the only thing that can violate the proton-neutron symmetry is the masses of the quarks. The up and down quark masses are not equal, i.e., the down quark is nearly twice as heavy as the up quark while the masses of two quarks are very small, $m_u \sim 2\,\text{MeV}/c^2$, $m_d \sim 5\,\text{MeV}/c^2$, respectively. In an exact meaning, the chiral symmetry between u and d quarks is violated in the Lagrangian due to the mass difference, although this violation is very small because of negligible mass difference (an order of a few MeV) compared with hadron masses (an order of GeV). This is called "explicit" chiral symmetry breaking. On the other hand, the bulk of proton and neutron masses comes from the chiral symmetry breaking which is called the "spontaneous" symmetry breaking (SSB). Spontaneous symmetry breaking is a mode of symmetry breaking in a physical system, where the underlying Hamiltonians or Lagrangians are invariant under a symmetry transformation, but the system as a whole breaks the symmetry under such transformations. This symmetry breaking is responsible to convert very light bound quarks into 100 times heavier constituents of baryons. The mechanism that gives a mass to particles is discussed in Chap. 10.

Summary

Quantum effects prevail on classical physics for a physical system with characteristic kinetic energy E and size r when

$$\frac{Er}{\hbar c} \lesssim 1.$$

For a human walking in a room, this ratio is $\sim 10^{28}$, while the ratio is ~ 0.1 for a nucleon in a nucleus and ~ 0.01 for an electron in an atom.

A non-relativistic quantum system is described by the Schrödinger equation

$$i\hbar \frac{\partial}{\partial t} \Psi(\mathbf{r}, t) = -\frac{\hbar^2}{2m} \nabla^2 \Psi(\mathbf{r}, t) + V(\mathbf{r})\Psi(\mathbf{r}, t),$$

where V is the interaction potential. Its solution, the wave function $\Psi(\mathbf{r}, t)$, makes a link between the microscopic theory of particle motion and the probability, expressed by $|\Psi|^2$, of finding the particle in space-time through the statistical interpretation.

The eigenenergy and the eigenfunction of a particle of mass m in a one-dimensional infinite square well potential with width L are obtained as

$$E_n = \frac{\hbar^2 \pi^2 n^2}{2mL^2}, \qquad n = 1, 2, 3, \ldots$$

and

$$\psi_n(x) = \sqrt{\frac{2}{L}} \sin k_n x, \qquad n = 1, 2, 3, \ldots$$

with $k_n = n\pi/L$, respectively.

Quantum tunnelling is a unique phenomenon where quantum objects have a non-zero probability to exist in regions forbidden by classical mechanics. α-decay of nuclei, the fission, fusion and halo phenomena become possible in microscopic world with this effect. The transmission probability of quantum tunnelling of the square barrier potential with the height $V_0 > 0$ and width L reads

$$P_T = \frac{4E(V_0 - E)}{4E(V_0 - E) + V_0^2 \sinh^2(\kappa L)} \quad \text{with } E < V_0,$$

for the particle with the energy E to be transmitted through the barrier, with $\kappa = \sqrt{2m(V_0 - E)/\hbar^2}$ in Eq. (1.71).

The Heisenberg's uncertainty principle in the quantum world is deeply related to the measurement and the probability interpretation of microscopic phenomena. Quantum mechanics correlates and limits the precision for the measurement of momentum and position, and the precision for the measurement of energy and duration. The uncertainty relation between position and momentum and between energy and time are given by

$$\Delta x \cdot \Delta p \geq \frac{\hbar}{2}, \qquad \Delta E \cdot \Delta t \geq \frac{\hbar}{2}. \qquad (1.119)$$

A feature of quantum mechanics is *entanglement*: a physical state can be a superposition of several intrinsic states. The measurement is equivalent to the projection of the wave function onto one of the intrinsic states. In the case of a pair of particles with non-zero spin (EPR pair), the entangled state is written by

$$|\Psi_{12}\rangle = \frac{1}{\sqrt{2}}\{|\uparrow_1\rangle|\downarrow_2\rangle - |\downarrow_1\rangle|\uparrow_2\rangle\}, \qquad (1.120)$$

where $|\uparrow\rangle(|\downarrow\rangle)$ denotes spin up (spin down) state. The angular correlation $C(\theta)$ between the two spins of the pair are expressed in quantum mechanics as,

$$C_{\text{QM}} = -\cos\theta,$$

while the correlation in classical mechanics gives

$$C_{\text{Classical}} = \frac{2\theta}{\pi} - 1.$$

The Dirac equation is the relativistic version of the Schrödinger equation and reads

$$i\hbar\frac{\partial\psi(\mathbf{r}, t)}{\partial t} = (c\boldsymbol{\alpha} \cdot \mathbf{p} + \beta mc^2)\,\psi(\mathbf{r}, t),$$

where α and β are 4×4 matrices. The solutions of the Dirac equation for a fermion are called *spinors* and have four components

$$(\phi(E)\chi_{\uparrow}, \phi(E)\chi_{\downarrow}, \bar{\phi}(-E)\chi_{\uparrow}, \bar{\phi}(-E)\chi_{\downarrow}),$$

where ϕ and $\bar{\phi}$ are the wave functions of a particle and antiparticle of energy E, respectively, $\chi_{\uparrow}$ and $\chi_{\downarrow}$ are the two projections of spin.

A symmetry in the microscopic world is defined as the invariance of the Hamiltonian (or Lagrangian) of the considered system under a certain transformation. A symmetry is "explicitly" broken if the Hamiltonian is not invariant under the corresponding transformation. In the case of an invariant Hamiltonian, a symmetry is broken "spontaneously" by dynamical correlations. Chiral symmetry is almost verified for the QCD Lagrangian (the mass of quarks is very small compared to the meson and baryons masses): the explicit symmetry breaking is weak. On the other hand, spontaneous symmetry breaking is important and is responsible for the mass of mesons and baryons.

Solutions of Problems

1.1   We decompose the scalar product as

$$(\mathbf{a} \cdot \boldsymbol{\sigma})(\mathbf{b} \cdot \boldsymbol{\sigma}) = \sum_{i,j=x,y,z} a_i \sigma_i b_j \sigma_j = \sum_{i=j} a_i b_j + \sum_{i \neq j} a_i b_j \sigma_i \sigma_j$$

$$= \sum_i a_i b_i + \sum_{i \neq j \neq k} a_i b_j i \varepsilon_{ijk} \sigma_k, \qquad (1.121)$$

where the identities $\sigma_i^2 = 1$ and $\sigma_i \sigma_j = \varepsilon_{ijk} \sigma_k$ are used (see Eq. (1.6)). The quantity $\varepsilon_{ijk} a_i b_j$ is equivalent to the kth component of vector product of $\mathbf{a}$ and $\mathbf{b}$ and then we obtain

$$(\mathbf{a} \cdot \boldsymbol{\sigma})(\mathbf{b} \cdot \boldsymbol{\sigma}) = \mathbf{a} \cdot \mathbf{b} + i(\mathbf{a} \times \mathbf{b}) \cdot \boldsymbol{\sigma}. \qquad (1.122)$$

1.2   Inside the square well $V(x) = 0$, therefore Eq. (1.42) becomes

$$-\frac{\hbar^2}{2m}\left(\frac{d^2}{dx^2}\right)\psi(x) = E\psi(x) \quad \Longrightarrow \quad \frac{d^2}{dx^2}\psi(x) = -k^2\psi(x) \qquad (1.123)$$

where $k = \sqrt{2mE/\hbar^2}$. The solution, as presented in Eq. (1.45), can be written as

$$\psi(x) = C\sin(kx) + D\cos(kx), \qquad (1.124)$$

with the boundary conditions,

$$\psi(x = 0) = \psi(x = L) = 0. \qquad (1.125)$$

Therefore $D = 0$, and the condition $\sin(kL) = 0$ gives the quantization condition,

$$kL = \pi n \quad \Longrightarrow \quad E = \frac{\hbar^2 \pi^2 n^2}{2mL^2}, \quad n = 1, 2, 3 \ldots \qquad (1.126)$$

To get the factor C, we normalize the wave function such that $|\psi(x)|^2 = 1$:

$$\int_{-\infty}^{\infty} |\psi(x)|^2 dx = \int_{0}^{L} C^2 \sin^2(kx)dx = \int_{0}^{L} C^2 \left(\frac{1 - \cos(2kx)}{2} \right) dx$$

$$= C^2 \left(\frac{x}{2} - \frac{1}{4k}\sin(2kx) \right)\Big|_{0}^{L} = C^2 \left(\frac{x}{2} - \frac{1}{2k}\sin(kx)\cos(kx) \right)\Big|_{0}^{L} = 1$$

$$\Longrightarrow C = \sqrt{\frac{2}{L}}. \tag{1.127}$$

Then the normalized solution is expressed as

$$\psi_n(x) = \sqrt{\frac{2}{L}}\sin\left(\frac{n\pi}{L}x\right) \quad n = 1, 2, 3\ldots \tag{1.128}$$

1.3 The lowest eigenenergy is given by Eq. (1.46) with $n = 1$,

$$E_1 = \frac{\hbar^2 \pi^2}{2m_e L^2} = \frac{(\hbar c)^2 \pi^2}{2m_e c^2 L^2}, \tag{1.129}$$

where m_e is the electron mass. Inserting the electron mass $m_e c^2 = 5 \times 10^5$ eV, $\hbar c = 200$ eV.nm and $L = 0.4$ nm, one gets the lowest energy

$$E_1 = \frac{(200\,\text{eV} \cdot \text{nm})^2 \pi^2}{2 \times 5 \times 10^5\,\text{eV}(0.4\,\text{nm})^2} = 2.5\,\text{eV}. \tag{1.130}$$

1.4 Substituting the wave function (1.47) into Eq. (1.51), we get

$$\int_{0}^{L} \psi_n(x)^* \psi_m(x)dx = \frac{2}{L}\int_{0}^{L} \sin\left(\frac{n\pi}{L}x\right) \sin\left(\frac{m\pi}{L}x\right) dx. \tag{1.131}$$

Using the trigonometric identity $2\sin(\alpha)\sin(\beta) = \cos(\alpha - \beta) - \cos(\alpha + \beta)$ we can write Eq. (1.131) as

$$\frac{2}{L}\int_{0}^{L} \frac{1}{2}\left[\cos\left(\frac{(n-m)\pi}{L}x\right) - \cos\left(\frac{(n+m)\pi}{L}x\right)\right] dx, \tag{1.132}$$

and performing the integral gives

$$\left[\frac{1}{\pi} \cdot \frac{\sin\left(\frac{(n-m)\pi}{L}x\right)}{(n-m)} - \frac{1}{\pi} \cdot \frac{\sin\left(\frac{(n+m)\pi}{L}x\right)}{(n+m)}\right]\Bigg|_{0}^{L} = 0. \tag{1.133}$$

As can be seen for the case $n \neq m$, both terms vanish at $x = 0$, L since n and m are integer numbers. For $n = m$, the first term in Eq. (1.132) simply equals to 1 and therefore we have

$$\frac{2}{L} \int_0^L \frac{1}{2} \left[1 - \cos\left(\frac{(n+m)\pi}{L} x \right) \right] dx = \frac{2}{L} \cdot \frac{1}{2} \cdot x \Big|_0^L = 1, \qquad (1.134)$$

as requested.

1.5   The mass number A can be evaluated as

$$\int_0^{r_a} \rho(r) dr = \rho_0 \frac{4\pi}{3} r_A^3 = A. \qquad (1.135)$$

The radius r_A is obtained as

$$r_A = \left(\frac{3}{4\pi\rho_0} \right)^{1/3} A^{1/3} \approx 1.2 \times A^{1/3} \text{fm}. \qquad (1.136)$$

1.6   We evaluate the α decay life time of ^{212}Po by using a simple rectangular barrier tunnelling model. In a realistic model, the barrier between the core nucleus ^{208}Pb and α particle in ^{212}Po is determined by the Coulomb potential

$$V_C = \frac{e^2 Z_1 Z_2}{r_1}, \qquad (1.137)$$

where Z_1 and Z_2 are proton numbers of ^{208}Pb and α and r_1 is the distance between two fragments. The distance may be evaluated to use a standard formula for radius of nucleus $r_A = 1.2 \times A^{1/3}$ fm. Then the distance is $r_1 = 1.2 \times 208^{1/3} + 1.2 \times 4^{1/3} = 9.0$ fm. The height of Coulomb barrier at $r = 9.0$ fm is given by Eq. (1.137) as

$$V_C = e^2 \times 2 \times 82/9 = (\hbar c)e^2/(\hbar c) \times 2 \times 82/9 = 26 \text{ MeV}, \qquad (1.138)$$

where $e^2/(\hbar c) = 1/137$ is the fine structure constant and $\hbar c = 200$ MeV $\cdot$ fm. The distance at which the Coulomb potential drops to the energy of the observed alpha can be obtained by using Eq. (1.137) again,

$$r_2 = (\hbar c)e^2/(\hbar c) \times 2 \times 82/8.78 = 27 \text{ fm}. \qquad (1.139)$$

So the width of barrier $L = r_2 - r_1 = 18$ fm. If we use the transmission formula of the rectangular potential (1.71) naively with $L = 18$ fm and $V_0 = 26$ MeV, we will get too much hindrance of the α decay as you can suspect from Fig. 1.14. So we adopt effective values of $L_e = 15$ fm and $V_0 = 15$ MeV, we obtain the value $\kappa = \sqrt{2m_\alpha(V_0 - E)/\hbar^2} = 0.11$ fm^{-1} and the transmission probability (1.71),

$$P_T = |T|^2 = 2.05 \times 10^{-14}. \qquad (1.140)$$

In addition to the transmission probability, the alpha decay rate depends upon how many times an alpha particle with this energy inside the nucleus will hit the walls. The velocity of the alpha can be calculated from the non-relativistic formula $E = m_\alpha v_\alpha^2/2$ to be $v_\alpha/c = 0.0069$ and $v_\alpha = 2.1 \times 10^7$ m/s. The frequency of hitting the walls is then

$$f = v_\alpha/2r_1 = 2.1 \times 10^{22} \text{ fm/s}/18 \text{ fm} = 1.17 \times 10^{21} \text{ s}^{-1}. \tag{1.141}$$

The decay probability of a α particle may be given by the product of $f \times P_T$:

$$-\frac{dN/dt}{N} \equiv \lambda = f \times P_T. \tag{1.142}$$

Then, the decay rate and the half life will be evaluated by

$$N = N_0 e^{-\lambda t}, \tag{1.143}$$
$$T_{1/2} = 0.693/\lambda. \tag{1.144}$$

With $\lambda = f \times |T|^2 = 1.17 \times 10^{21} \text{ s}^{-1} \times 2.05 \times 10^{-14} = 2.4 \times 10^7 \text{ s}^{-1}$, the half life is determined to be

$$T_{1/2} = 0.693/\lambda = 0.29 \times 10^{-7} \text{ s} = 0.029 \,\mu\text{s}. \tag{1.145}$$

Compared with the observed half life $T_{1/2} = 0.3\,\mu\text{s}$. Although our estimate is one order of magnitude smaller, it can be qualified as quite close to the experimental value since the decay probability is easily changed by few orders of magnitude by different value of potential shape and the effective barrier length L. We emphasize here that the assumption that the α particle pre-exists in the mother nucleus before the decay is a simplification. The mother nucleus is a many-body system in which all nucleons are highly correlated with each other. It is predicted that α particles cluster at the low-density surface of nuclei.

1.7 Let ψ_a be an eigenstate of A with an eigenvalue a. Since A and B are commuting, we have

$$AB\psi_a = BA\psi_a = aB\psi_a. \tag{1.146}$$

$B\psi_a$ is also the eigenstate of A with the eigenvalue a. If there is only one eigenstate with the eigenvalue a, $B\psi_a$ should be a linearly dependent state to ψ_a,

$$B\psi_a = b\psi_a, \tag{1.147}$$

where b is a constant and identified as the eigenvalue of B for the state ψ_a. Thus, the state ψ_a is the eigenstate of two operators of A and B with the eigenvalues a and b, respectively. In the above derivation, we assume that $a \neq b$. One can prove the existence of common eigenstate even if there are degenerate eigenstates for a given eigenvalue $a = b$.

1.8 Define the following state vector by applying two Hermitian operators $A - \langle A \rangle$ and $B - \langle B \rangle$ to the state vector $|\psi\rangle$:

$$|\tilde{\psi}\rangle = (A - \langle A \rangle)|\psi\rangle + i\lambda(B - \langle B \rangle)|\psi\rangle, \tag{1.148}$$

where λ is a real number. The bra vector $\langle \tilde{\psi}|$ conjugate to the state (1.148) is written to be

$$\langle \tilde{\psi}| = (|\tilde{\psi}\rangle)^{\dagger} = \langle \psi|(A - \langle A \rangle) - i\lambda\langle\psi|(B - \langle B \rangle). \tag{1.149}$$

The scalar product of the bra (1.149) and the ket (1.148) becomes

$$\langle \tilde{\psi}|\tilde{\psi}\rangle = \langle \psi|(A - \langle A \rangle)^2 + i\lambda[A, B] + \lambda^2(B - \langle B \rangle)^2|\psi\rangle. \tag{1.150}$$

Since the scalar product (norm) of the same state vector must be positive or null, (1.150) satisfies the condition

$$\langle \tilde{\psi}|\tilde{\psi}\rangle = (\varDelta A)^2 - \lambda\hbar + \lambda^2(\varDelta B)^2 \geq 0. \tag{1.151}$$

Note that the inequality hold for any real number λ. Considering (1.151) as a quadratic equation with respect to λ, its discriminant must be negative or zero:

$$\hbar^2 - 4(\varDelta A)^2(\varDelta B)^2 \leq 0, \tag{1.152}$$

leading to the uncertainty relation

$$\varDelta A \cdot \varDelta B \geq \frac{\hbar}{2}. \tag{1.153}$$

1.9 Since the momentum operators are commutable $p_i p_j = p_j p_i$, the first term of Eq. (1.88) will be

$$-\hbar^2 c^2 \sum_{i,j} \alpha_i p_i \alpha_j p_j = -\frac{\hbar^2 c^2}{2} \sum_{i,j}(\alpha_i\alpha_j + \alpha_j\alpha_i)p_i p_j = -\hbar^2 c^2 p^2, \tag{1.154}$$

where we use the first relation in Eq. (1.89). The second and third terms can be combined into the form,

$$-i\hbar c m c^2 \sum_i (\alpha_i\beta + \beta\alpha_i)p_i = 0, \tag{1.155}$$

where we use the second relation in Eq. (1.89). The third term will be

$$\beta^2 m^2 c^4 = m^2 c^4, \tag{1.156}$$

where we use the third relation in Eq. (1.89). Thus, we can prove that the Dirac equation (1.88) is identical to Klein–Gordon equation (1.86) under the condition (1.89).

1.10 Using the anti-commutation relation of spin operators $\sigma_i \sigma_j + \sigma_j \sigma_i = 2\delta_{ij}$, we get

$$
\begin{aligned}
(\boldsymbol{\sigma} \cdot \mathbf{p})^2 &= \sum_{i,j} \sigma_i p_i \sigma_j p_j = \sum_i \sigma_i^2 p_i^2 + \sum_{i \neq j} \sigma_i p_i \sigma_j p_j \\
&= \sum_i p_i^2 + \sum_{i > j} p_i p_j \{\sigma_i \sigma_j + \sigma_j \sigma_i\} \\
&= p^2.
\end{aligned}
\tag{1.157}
$$

1.11 The Dirac equation with the central potential reads

$$
\begin{aligned}
E\psi &= (c\boldsymbol{\alpha} \cdot \mathbf{p} + \beta(mc^2 + V(r))\psi \\
&= \begin{pmatrix} I(mc^2 + V) & c\boldsymbol{\sigma} \cdot \mathbf{p} \\ c\boldsymbol{\sigma} \cdot \mathbf{p} & -I(mc^2 + V) \end{pmatrix} \psi.
\end{aligned}
\tag{1.158}
$$

This equation gives rive to coupled equations of two wave functions Ψ and χ as

$$
(mc^2 + V)\Psi + c\boldsymbol{\sigma} \cdot \mathbf{p}\chi = E\Psi \tag{1.159}
$$
$$
c\boldsymbol{\sigma} \cdot \mathbf{p}\Psi - (mc^2 + V)\chi = E\chi. \tag{1.160}
$$

In a non-relativistic limit, the eigenvalue E is slightly different from the rest mass so that one can write

$$
E = mc^2 + \varepsilon. \tag{1.161}
$$

Equation (1.160) is rewritten to be

$$
\chi = \frac{c\boldsymbol{\sigma} \cdot \mathbf{p}}{2mc^2 + \varepsilon + V} \Psi. \tag{1.162}
$$

Substituting Eq. (1.162) into Eq. (1.159), we obtain

$$
(c\boldsymbol{\sigma} \cdot \mathbf{p} \frac{1}{2mc^2 + \varepsilon + V} c\boldsymbol{\sigma} \cdot \mathbf{p} + V)\Psi = \varepsilon \Psi. \tag{1.163}
$$

In Eq. (1.163), the first term of l.h.s. corresponds to the kinetic energy, while the second term is the potential energy. In the case of constant potential $V(r) \equiv V_0$, the first term is expressed by an effective mass $2m^* = 2m + \varepsilon + V$ (in natural units, i.e., $c = 1$) and Eq. (1.163) becomes

$$
\left(\frac{p^2}{2m^*} + V_0 \right) \Psi = \varepsilon \Psi, \tag{1.164}
$$

by use of the Eq. 1.157. The effective mass is often introduced in many-body physics to renormalize the energy and the momentum dependences of kinetic energy into a single quantity m^*. which has exactly the same structure as the Schrödinger equation with an effective mass m^*.

1.12 Since the momentum has only a z component, the wave functions (1.118) become

$$\psi_1 = \begin{pmatrix} 1 \\ 0 \\ \frac{cp_z}{E_+ + mc^2} \\ 0 \end{pmatrix}, \psi_2 = \begin{pmatrix} 0 \\ 1 \\ 0 \\ -\frac{cp_z}{E_+ + mc^2} \end{pmatrix}, \psi_3 = \begin{pmatrix} \frac{cp_z}{E_- - mc^2} \\ 0 \\ 1 \\ 0 \end{pmatrix}, \psi_4 = \begin{pmatrix} 0 \\ -\frac{cp_z}{E_- - mc^2} \\ 0 \\ 1 \end{pmatrix}.$$

$$(1.165)$$

Dirac equation is expressed by a matrix form to be

$$c\alpha_z p_z + \beta mc^2 = \begin{pmatrix} mc^2 & 0 & cp_z & 0 \\ 0 & mc^2 & 0 & -cp_z \\ cp_z & 0 & -mc^2 & 0 \\ 0 & -cp_z & 0 & -mc^2 \end{pmatrix}, \tag{1.166}$$

We will make a multiplication of 4×4 matrix and the four-column vector. Using

$$mc^2 + cp_z \frac{cp_z}{E_+ + mc^2} = \frac{E_+ mc^2 + m^2 c^4 + c^2 p_z^2}{E_+ + mc^2} = \frac{E_+ mc^2 + E_+^2}{E_+ + mc^2} = E_+$$

and

$$cp_z - mc^2 \frac{cp_z}{E_+ + mc^2} = E_+ \frac{cp_z}{E_+ + mc^2},$$

we can have an eigenvalue for ψ_1 as

$$(c\alpha_z p_z + \beta mc^2)\psi_1 = E_+ \psi_1.$$

It is straightforward to obtain the same eigenvalue E_+ for ψ_2. Next we can calculate the eigenvalue for ψ_3. For the multiplication of matrices, we use the relation

$$cp_z \frac{cp_z}{E_- - mc^2} - mc^2 = \frac{c^2 p_z^2 - E_- mc^2 + m^2 c^4}{E_- - mc^2} = \frac{E_- - E_- mc^2}{E_- - mc^2} = E_-,$$

and find

$$(c\alpha_z p_z + \beta mc^2)\psi_3 = E_- \psi_3. \tag{1.167}$$

The eigenvalue for ψ_4 is also evaluated to be E_- in the same way.

Next, we calculate the eigenvalue for the spin operator

$$\mathbf{S} = \frac{1}{2}\Sigma = \frac{1}{2}\begin{pmatrix} \sigma & 0 \\ 0 & \sigma \end{pmatrix}, \tag{1.168}$$

which is a 4×4 matrix. Since the spin is quantized to the z-direction, we consider the eigenvalue of

$$S_z = \frac{1}{2}\Sigma_z = \frac{1}{2}\begin{pmatrix} 1 & 0 & 0 & 0 \\ 0 & -1 & 0 & 0 \\ 0 & 0 & 1 & 0 \\ 0 & 0 & 0 & -1 \end{pmatrix}. \tag{1.169}$$

This is a straightforward matrix multiplication to obtain the eigenvalue as

$$S_z\psi_1 = \frac{1}{2}\psi_1, \ S_z\psi_2 = \frac{1}{2}\psi_2, \ S_z\psi_3 = -\frac{1}{2}\psi_3, \ S_z\psi_4 = -\frac{1}{2}\psi_4. \tag{1.170}$$

In this way we can estimate the eigenvalues (1.115) for the eigenfunctions (1.165).

Books for Further Readings

"Quantum mechanics" by A. Messiah (Dover Publications, 2014) is a classical textbook (first edition in 1959) resulting from a course given at the Center for Nuclear Studies at Saclay, which trained generations of physicists.

"Modern quantum mechanics" by J. J. Sakurai and J. Napolitano (Cambridge University Press, third edition, 2020) is a modern quantum mechanics textbook (first edition in 1967) covering the main quantum mechanics concepts in a clear and pedagogical manner.

"Quantum mechanics: a new introduction" by K. Konishi and G. Paffuti (Oxford University Press, 2009) provides a modern and comprehensive introduction to quantum mechanics, including modern-physics examples and numerical analyses.

References

1. A. Einstein, Annalen der Physik **17**, 132 (1905)
2. A. Einstein, Annalen der Physik **17**, 549 (1905)
3. A. Einstein, Annalen der Physik **17**, 891 (1905)
4. A. Einstein, Annalen der Physik **18**, 639 (1905)

5. J. Perrin, Ann. Chim. Phys. (VIII) **18**, 5 (1909)
6. G. Binning et al., Phys. Rev. Lett. **50**, 120 (1983)
7. O. Stern, W. Gerlach, Zeitschrift für Physik **9**, 349 (1922)
8. D. Frekers, M. Alanssari, Eur. Phys. J. A **54**, 177 (2018)
9. A. Einstein, B. Podolsky, N. Rosen, Phys. Rev. **47**, 777 (1935)
10. M. Lamehi-Rachti, W. Mittig, Phys. Rev. D **14**, 2543 (1976)
11. H. Sakai et al., Phys. Rev. Lett. **97**, 150405 (2006)
12. E. Schrödinger, Phys. Rev. **28**, 1049 (1926)
13. K. Morita et al., J. Phys. Soc. Jpn. **73**, 2593 (2004)
14. G. Gamow, Nature **122**, 805 (1928)
15. P.A.M. Dirac, Proc. R. Soc. Lond. **778**, 610 (1928)
16. C.D. Anderson, Science **76**, 238–239 (1932)
17. C.D. Anderson, Phys. Rev. **43**, 491 (1933)
18. P.W. Anderson, Science **177**, 4047 (1972)

Chapter 2
Nuclear Forces

Abstract In this chapter, we describe the interactions of constituent nucleons inside the nucleus. The nucleon-nucleon (*NN*) interaction was firstly derived from meson exchange potentials between nucleons such as the pion exchange potential. The phase shift of nucleon-nucleon scattering data is discussed in terms of *NN* interactions. QCD based Chiral effective field theory (ChEFT) is introduced to describe low-energy physics phenomena in nuclei. In ChEFT, nucleons and mesons are introduced as degrees of freedom instead of quarks and gluons. ChEFT presents a hierarchy of many-body forces expanded and ordered in powers of momentum in a perturbative way. The origin and role of many-body interactions is introduced.

Keywords One-pion exchange potential (OPEP) · Nucleon-nucleon scattering · Phase shifts · Chiral effective field theory (ChEFT) · Three-body forces

2.1 Fundamental Interactions

The interactions between objects or particles are carried by specific particles, messengers of the interactions. These specific particles are called the *gauge bosons*. There are four fundamental interactions[1] in nature: the gravitational, electromagnetic, weak and strong nuclear interactions.

According to the Big Bang hypothesis, in early universe just after the Big Bang explosion ($t \leq 10^{-43}$ s), the universe is extremely dense and hot (temperature T $\sim 10^{32}$ K), and interactions between particles are governed by a single force. In other words, the four fundamental forces merge into one at these energies (see Fig. 2.1). In the inflationary phase ($t = 10^{-43}$ to 10^{-35} s), the universe expands exponentially, and gravity separates from the other forces. The universe cools to approximately T $= 10^{27}$ K. During the electroweak epoch ($t = 10^{-35}$ to 10^{-12} s), the universe continues to expand, and the strong nuclear force separates from the electromagnetic and weak nuclear forces (or electroweak force). Soon after, during the quark epoch

---

[1] A fundamental interaction, or fundamental force, is not reducible to more basic interactions. Frictional force or force of viscosity, for example, are not fundamental, but can be traced back to the fundamental electromagnetic force at a more microscopic level.

© Springer Nature Singapore Pte Ltd. 2021
A. Obertelli and H. Sagawa, *Modern Nuclear Physics*, UNITEXT for Physics,
https://doi.org/10.1007/978-981-16-2289-2_2

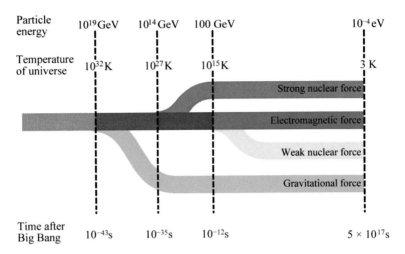

Fig. 2.1 The separation of the four fundamental forces in the early universe. In early universe just after the Big Bang explosion ($t \leq 10^{-43}$ s), the universe is extremely dense and hot (temperature T $\sim 10^{32}$ K), and interactions between particles are governed by a single force. In the inflationary phase ($t = 10^{-43}$ to 10^{-35} s), the universe expands exponentially, and gravity separates from the other forces. During the electroweak epoch ($t = 10^{-35}$ to 10^{-12} s), the universe continues to expand, and the strong nuclear force separates from the electromagnetic and weak nuclear forces (or electroweak force). Soon after, during the quark epoch ($t = 10^{-12}$ to 10^{-6} s), the weak nuclear force separates from the electromagnetic force

($t = 10^{-12}$ to 10^{-6} s), the weak nuclear force separates from the electromagnetic force. The universe is at this point a hot soup of quarks, leptons, photons, and other particles. In the hadron epoch which includes our current era ($t = 10^{-6}$ to 5×10^{17} s), we experience four different kinds of fundamental forces in nature and life as an outcome of the galactic history of 13.7 billion years ($\sim 5 \times 10^{17}$ s) after the Big Bang explosion. These four forces, and only these four forces exist in our current knowledge of physics. We are routinely in contact with direct manifestations of these fundamental interactions as illustrated in Fig. 2.2.

The electroweak interaction is the unified description of two of the four known fundamental interactions of nature: electromagnetism and the weak interaction. Although these two forces appear very different in everyday low energy phenomena, the theory tells us that they are two different aspects of the same force. Above the unification energy, on the order of 100 GeV, they would merge into a single electroweak force. Thus, if the universe is hot enough (approximately T $= 10^{15}$ K, a temperature exceeded until shortly after the Big Bang), then the electromagnetic force and weak force merge into a combined electroweak force.

The weak force is well known for its role in nuclear beta (β) decay. The weak interaction plays a crucial role in the formation of elements at the heart of stars. Its intensity is weaker by orders of magnitude compared to the electromagnetic and strong interactions, which explains its name. It has very short range and is the only force to interact with electrons, with neutrinos, as well as with quarks. The

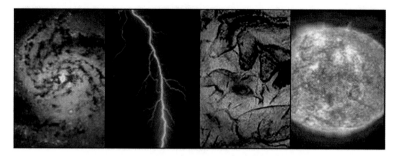

Fig. 2.2 Illustration of physics effects provoked by each of the four fundamental forces. (From left to right) Stars and planets attract each other by the gravitational interaction, a thunder is created by an electric discharge in the air, the dating of ancient objects is made possible by measuring ^{14}C whose radioactivity is due to the weak interaction, the stars shine and resist to gravity thanks to nuclear reactions originating from the strong interaction. Courtesy P. Chomaz, CEA

weak force is communicated via exchange particles, i.e., gauge bosons, like other forces. The exchange particles for the weak force are the charged $W^{\pm}$ bosons and the neutral Z^0 boson. These bosons have very heavy masses, $m_W c^2 = 80.4$ GeV and $m_Z c^2 = 91.2$ GeV. The three bosons and the photon are considered to be members of the same symmetry group, the so called "electroweak symmetry" and the different masses are produced by the spontaneous symmetry breaking (SSB) caused by the Higgs mechanism (see more details of SSB in Chap. 7). Sheldon Glashow, Abdus Salam, and Steven Weinberg were awarded the 1979 Nobel Prize in Physics for their contributions to the unification of the weak and electromagnetic interactions between elementary particles. The existence of the electroweak interaction was experimentally confirmed by the finding of neutral current (the exchange of Z bosons) in the neutrino scattering in 1974 at CERN in Europe first [1] and then at Fermilab, USA, and the discovery of the W and Z bosons in proton-antiproton collisions about ten years later in 1983 at CERN [2, 3].

While the gravitational and electromagnetic forces act on microscopic and macroscopic objects, the weak and strong interactions are short range and concern exclusively atomic and nuclear systems.

The electromagnetic force has a long range and acts among particles with electric charge. Electromagnetism gathers electric and magnetic processes both led by the same interaction. It is repulsive for particles with the same electric charge and attractive for particles with opposite signs of electric charges. At the microscopic scale, it is much stronger than the gravitational interaction. It is the only interaction needed to explain chemical phenomena. The electromagnetic interaction is carried by exchange of photons as the gauge bosons.

Gravity is also a long range interaction that acts on all massive objects of the universe with a strength varying as $\propto 1/r$ with the relative distance r of the two interacting massive objects. Gravitation is always attractive since no object with negative mass is found, and it explains phenomena at the stellar scale such as galaxy formation or the evolution of the solar system. Nevertheless, at microscopic scales,

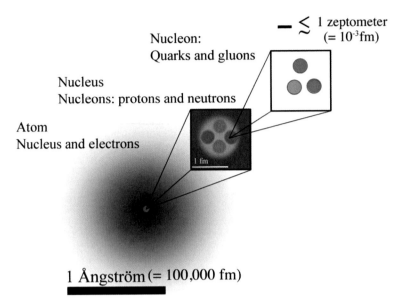

Fig. 2.3 Constituents of the atom. Quarks are today the smallest elementary brick of matter as described by the standard model, although not individually observable

such as the atom, the gravitational force is the weakest of all interactions. The gauge boson for gravitation is called *graviton* which has not been found so far in any observations, while gravitational waves were observed for the first time on September 14, 2015 from a merger of a pair of black holes [4], and also from a merger of binary neutron stars on August 17, 2017 [5].

The fourth interaction, the strong force, is responsible for the binding of the nucleus and its constituents. The strong interaction is the source of heat in stars and of nuclear energy. From 1960s, we know that protons and neutrons, altogether called nucleons, are made of elementary particles "quarks", although they can not be observed directly. The quarks are inside nucleons and interact with each other by exchanging gluons. A gluon can interact also directly with other gluons. The interaction of gluons with themselves is a unique property of the strong force. The field theory of strong interaction with quarks and gluons is known as the quantum chromodynamics (QCD). Constituents of the atom are illustrated according to their sizes in Fig. 2.3.

In the advanced theory of fundamental interactions, the so called "Grand Unified Theory (GUT)", the electromagnetic, weak and strong interactions, are merged as a single force. Some predictions of GUT are the proton decay, the electric dipole moments of elementary particles, and the existence of magnetic monopoles. There is so far no confirmed experimental evidence of the predictions by GUT models.

2.2 Energy Resolution and Effective Theories

In everyday life, we experience nature at our scale, i.e., distances larger than ~ 100 micrometers (μm), with effective concepts: we manipulate objects (an apple, a chair) without resolving their internal structure. The famous article "More is different", 1972, by P. W. Anderson states the *holistic* view of physics: a new world appears at every energy scale and one needs knowledge of the laws of physics at the relevant energy scale to understand the new world, although the fundamental bricks of matter are of great interest by themselves. An analogy between a nucleus and sumo fighters is given in Fig. 2.4.

Momentum and spatial scales are intimately related. Physics is indeed a matter of resolution and degrees of freedom adopted to the processes to study. In quantum physics, following the de Broglie relation, a probe with momentum Q probes the distances $R \geq \hbar/Q$.

The basic idea of any effective theory is to introduce active light particles, while heavy particles are frozen as static sources. The dynamics is described by an effective

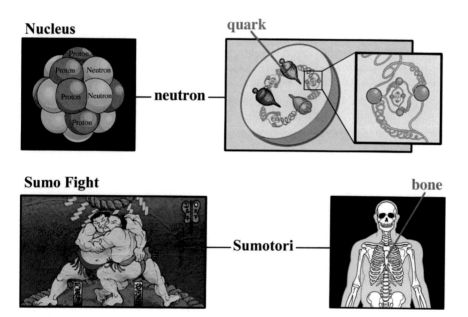

Fig. 2.4 As for a sumo fight which does not require any knowledge of the constituents of the Sumotori (Sumo wrestler), the nucleus can be described in terms of nucleons (protons and neutrons) with no explicit treatment of their internal degrees of freedom, namely quarks and gluons. One can watch, understand and analyze a sumo fight by considering the wrestlers as main entities, without any knowledge of their bones and position of organs in their body. Even though the fights are sometimes violent, fortunately the released energy during the combat is much less than the necessary energy to dismember a fighter. This very same separation of energy scales leads to the emergence of complexity and new phenomena from one scale to another

Lagrangian which incorporates all relevant symmetries of the underlying theory, in particular chiral symmetry in nuclear dynamics. Consequently, such an effective theory can only be applied for momenta and masses (the so called the "*soft*" scale), that are small compared to masses of particles not considered (the "*hard*" scale). If the theory is built by introducing a hard scale Λ in momentum, the corresponding spatial scale is $\hbar/\Lambda$, and the theory can not be applied above this scale Λ. In such a setting with the soft scale denoted Q and the hard scale Λ, any interaction or matrix element can be expanded by a power of the small parameter Q/Λ,

$$ V \propto \sum_\nu \left(\frac{Q}{\Lambda} \right)^\nu F_\nu(C_i(\Lambda)), \tag{2.1} $$

where ν is the order of the expansion and the functions F_ν depend on a series of coupling constants C_i, often called the low-energy constants.[2] The effect of unresolved high energy (short distance) scale is contained in the low-energy constants C_i which are determined by fitting experimental observables. The number of constants varies from one order to the other. By construction, observables should not depend on the cutoff value. By definition, for $Q \ll \Lambda$, the series will converge if the theory is perturbative. The procedure to modify the low-energy constant values for a given cutoff value is called renormalization. If the renormalization procedure takes place properly, the observables should not depend on the adopted value $Q \ll \Lambda$, which is called *renormalization-group (RG) invariance*. This is the paradigm of Chiral Effective Field Theory (ChEFT) that is presented in the following.

In the case of low-energy nuclear physics, as we will see below, the typical soft scale Q is the nucleon momentum and the pion mass $m_\pi \sim 140$ MeV/c^2, while the hard scale is specified by the mass of heavy mesons m_ρ or $m_\omega \sim 800$ MeV/c^2. Effective field theory for nuclear physics originates in the seminal work of Steven Weinberg in 1979 [8].

2.3 Nuclear Forces

After the discovery of quarks, physicists understood that the force binding protons and neutrons inside the nucleus is in reality a residual interaction, i.e., a "leak" of the strong force outside the nucleons. Even though the nuclear force is not fundamental, the residual from the strong force between quarks is still extremely "strong", i.e., nucleons are very closely packed together. The atomic nucleus, this tiny piece of dense matter inside the atom, was first discovered in 1901 by Ernest Rutherford in England. The physicist was so surprised by the structure of the atom, mostly an electron cloud surrounding an unexpectedly small and dense nucleus containing most of the atomic mass, that he later poetically compared the nucleus in the atom as a *fly*

---

[2] A regularization scale might be also introduced in F_ν to avoid the ultraviolet divergence appearing in some of Feynman diagrams, so called loop diagrams.

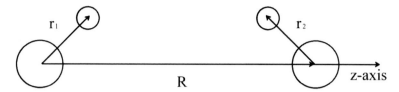

Fig. 2.5 Two hydrogen atoms separated by a fix distance R and their electrons at displacements $\mathbf{r}_i$ from their nuclei

inside a cathedral. Indeed, if one compares the relative size of a fly (one cm) with the length of a cathedral (100 m), it reaches 10000, similar to the comparison of the atom (1 Å(angstrom) $= 10^{-10}$ m $= 10^5$ fm) to the nucleus typical diameter (10 fm $= 10^{-14}$ m) with the major difference that the atomic fly has the weight of the entire cathedral.

Problem

2.1 Imagine how large the density is inside a nucleus. What is the weight of a spherical pea with a diameter of 0.4 cm at the nuclear saturation density (0.16 nucleon/fm^3)? Compare it to the weight of the Eiffel Tower ($\sim$8000 tons).

Residual interactions are encountered elsewhere in nature. For instance, they play an utmost importance in chemistry: although molecules are electrically neutral (the positive charge of ions is exactly compensated by the negative charge of electrons), the fine detail of their structure leads to a residual electric interaction at short distances. This residual interaction, called van der Waals interaction, is responsible for the global cohesion of liquid and solid states of matter such as liquid water or ice composed of H_2O molecules bound together. While the electrical potential evolves radially as $\propto 1/r$, the van der Waals potential is much weaker and evolves as $\propto 1/r^6$.

Let us have a short step aside at this stage: how can we consider nuclei as made of protons and neutrons mostly interacting through two-body and (weaker) three-body forces instead of interacting quarks and gluons? In other words, how is it possible that one can describe the nucleus made only of nucleons without considering explicitly its most fundamental constituents? The reasons lies in the concept of separation of energy scales encountered in almost all domains of physics. For example, hydrodynamics and weather forecast do not consider the chemistry of individual water molecules H_2O! Indeed, the nucleon binding energy in the nucleus is very different from the necessary energy to break a nucleon into pions and other particles.

Problem

2.2 Consider the two protons of a hydrogen molecule H_2, at a fixed distance R from each other along the z-axis. The fist electron is located at the distance $\mathbf{r}_1$ from the first proton and the second electron is placed at the distance $\mathbf{r}_2$ from the second electron (see Fig. 2.5). The Hamiltonian is written by two parts, H_0 and H_1, where H_0 is the atomic electron-proton interactions and H_1 is proton-proton, electron-electron and electron-proton interactions between the two atoms;

$$H_0 = -\frac{\hbar^2}{2m}(\nabla_1^2 + \nabla_2^2) - \frac{e^2}{r_1} - \frac{e^2}{r_2}, \tag{2.2}$$

$$H_1 = \frac{e^2}{R} + \frac{e^2}{|\mathbf{R} + \mathbf{r}_2 - \mathbf{r}_1|} - \frac{e^2}{|\mathbf{R} + \mathbf{r}_2|} - \frac{e^2}{|\mathbf{R} - \mathbf{r}_1|}. \tag{2.3}$$

1. Under the condition $R \gg r_1, r_2$, expand H_1 in powers of $\mathbf{r}_i/\mathbf{R}$ and show that the lowest order is proportional to $1/R^3$.
2. Show that the lowest order correction of the interaction H_1 to the ground state energy is proportional to the power of $1/R^6$.

While efforts are being made to describe atomic nuclei starting from dynamics of quarks and gluons (notably in lattice quantum chromodynamics approaches), such calculations are incredibly computationally expensive. Today, the best QCD calculations have difficulties to reproduce the experimental value of the proton radius and we still do not know why. New experiments to confirm the proton radius are ongoing. Lattice QCD (LQCD) calculations have made great progress during the last decades and the *NN* interaction is successfully described by the LQCD calculations even the hard core properties. On the other hand, even if full LQCD calculations for nuclei would be possible, it would not be the proper way to cope with nuclear structure for most applications. Modern nuclear structure theories do not treat explicitly the internal structure of protons and neutrons. They are all *effective* and depend on parameters adjusted on some measured quantities on finite nuclei or nucleon-nucleon scattering data. The validity of this assumption and its dependence with the resolution scale are still under debate in nuclear physics (see illustrative examples in Fig. 2.6).

Although many-body problems in nuclear physics are difficult to solve, the *NN* interaction exhibits few properties that can explain general features of nuclei. First, it is obviously attractive since nuclei are bound systems of nucleons. Nature and animals, including human beings, need this fundamental property to exist. Secondly, it is short range since it originates in the strong interaction. Nucleons feel each other only at distances smaller than few femtometers (1 fm = 10^{-15} m). Finally, the fact that the binding energy of nuclei with more than $\sim$10 nucleons has an almost constant value close to 8 MeV per nucleon (see Chap. 4) shows that the range of the nuclear interaction can not be long, as in the case of the electric interaction. More quantitatively, the binding energy per nucleon increases rapidly for light particles but only by one MeV from ^{4}He to heavier nuclei: this indicates that the range of the nuclear interaction is approximately the radius of a ^{4}He particle. This short range of the interaction explains why the nucleus is confined within the atom: it never gets influenced by the neighbouring nuclei kept at large distance by the electron screening and the nuclear charge repulsion. The last important property of the nuclear interaction is that it is repulsive at very short distances: it costs a tremendous amount of energy to two nucleons to get close at a distance of less than one femtometer, essentially because they are not point-like particles but have a finite size. This is called the *hard repulsive core* of the interaction. These three essential properties deduced from few simple observations are cartooned in Fig. 2.7 through the comparison with sumo wrestlers.

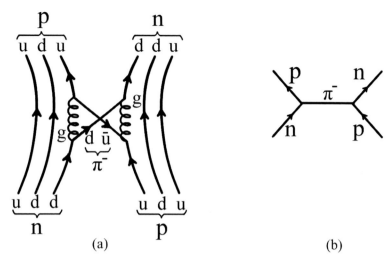

Fig. 2.6 **a** Artistic view of a QCD diagram contributing to the *NN* interaction. This high resolution scale is not considered in today's nuclear models and nuclear many-body problems can be treated, in the example of chiral effective field theories, by considering only nucleon and pion effective fields as degrees of freedom as shown in (**b**)

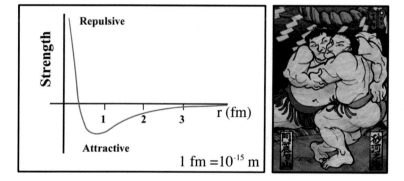

Fig. 2.7 (Left) Strength of the nuclear interaction (negative value for attraction, positive for repulsion) as a function of distance between the two interacting nucleons. (Right) Illustration of two sumotori wrestlers showing similar interaction behaviour: when at a large distance to each other, for example during the salutation, the two wrestlers do not interact with each other. When they fight, at a distance of less than a meter, they grab each other with their arms: they have a strongly attractive interaction. Whenever they are too close to each other, their bodies make a strong repulsion: due to their large volumes, the two wrestlers can obviously not interpenetrate each other

2.4 Meson Theory for Nucleon-Nucleon Interactions

The nuclear attraction at short distances was first explained by Hideki Yukawa in 1935 in terms of exchange of pions [6], a massive meson particle whose existence was postulated by Yukawa before its experimental discovery in 1947. We will introduce the basic idea of Yukawa's theory in an analogy with the electromagnetic potential.

The scalar potential A produced by a charge distribution $\rho(\mathbf{r})$, is expressed by

$$A(\mathbf{r}) = \int d\mathbf{r}' \frac{\rho(\mathbf{r}')}{|\mathbf{r} - \mathbf{r}'|}. \tag{2.4}$$

For a point charge q, the potential A reduces to the Coulomb potential $A(r) = q/r$. It is shown that the potential A satisfies the Poisson equation

$$\nabla^2 A(\mathbf{r}) = -4\pi \rho(\mathbf{r}), \tag{2.5}$$

with help of the relation,[3]

$$\nabla^2 (1/r) = -4\pi \delta(\mathbf{r}). \tag{2.6}$$

Yukawa's idea of the strong interaction was inspired by the electromagnetic interaction, but it did not fall off rapidly enough as a function of r. To make it decrease more rapidly, he introduced an equation for a scalar potential ϕ as,

$$(\nabla^2 - k^2)\phi(\mathbf{r}) = 4\pi f \rho(\mathbf{r}), \tag{2.7}$$

adding a term $k^2 \phi(\mathbf{r})$ to Eq. (2.5). This is nothing but the Klein–Gordon equation. The strength of field is determined by the hadron density $\rho(\mathbf{r})$ and the field strength f. The different signs of the right-hand-side terms between Eqs. (2.5) and (2.7) stem from the nature of electromagnetic and strong interactions. The solution of Eq. (2.7) is given by

$$\phi(\mathbf{r}) = -f \int \frac{\exp(-k|\mathbf{r} - \mathbf{r}'|)}{|\mathbf{r} - \mathbf{r}'|} \rho(\mathbf{r}') d\mathbf{r}'. \tag{2.8}$$

---

[3] Equation (2.6) can be proven by multiplying a function $f(\mathbf{r})$ on both sides of the equation and integrate by $\mathbf{r}$,

$$\int d\mathbf{r} f(\mathbf{r}) \nabla^2 \left(\frac{1}{r}\right) = -\int d\mathbf{r} \nabla f(\mathbf{r}) \cdot \nabla \left(\frac{1}{r}\right) = -4\pi \int r^2 dr \frac{df(\mathbf{r})}{dr} \frac{d}{dr} \left(\frac{1}{r}\right)$$

$$= 4\pi \int dr \frac{df(\mathbf{r})}{dr} = -4\pi f(\mathbf{r} = 0),$$

which is identical to the integral of r.h.s. of Eq. (2.6):

$$-4\pi \int d\mathbf{r} f(\mathbf{r}) \delta(\mathbf{r}) = -4\pi f(\mathbf{r} = \mathbf{0}).$$

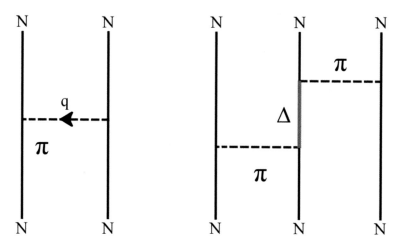

Fig. 2.8 Two-body *NN* and three-body *NNN* interactions mediated by π meson exchange processes. The latter is called the Fujita–Miyazawa three-body interaction in which the Δ particle is excited in the intermediate state, invented in 1957 [7]

For a point particle $\rho(\mathbf{r}') = \delta(\mathbf{r}')$, this solution becomes the Yukawa potential,

$$\phi(r) = -f\,\frac{\exp(-kr)}{r}. \tag{2.9}$$

From this form, we can see that the coupling $f > 0$ provides an attractive potential of hadronic interaction.

Problem

2.3 Prove that the potential of the form of (2.8) satisfies Eq. (2.7).

Because of the uncertainty principle, we can see that the range k of the interaction (2.9) is connected to the mass of the exchanged mesons. Assume two nucleons, one emits a meson of mass m_π and momentum q absorbed by the second nucleon as shown in the left panel of Fig. 2.8. Upon emission of the meson, energy conservation is violated by an energy of $\Delta E = \sqrt{m_\pi^2 c^4 + q^2 c^2}$, where c is the velocity of light. The uncertainty principle already mentioned in Chap. 1 gives the maximum time Δt, during which the meson can "fly" between the two nucleons: $\Delta t \approx \hbar/\Delta E$. If the meson travels at the velocity of light, it can travel a distance of $c\Delta t = \hbar/\Delta E \approx \hbar/m_\pi c$ (assuming $m_\pi c^2 \gg qc$), which is then the maximum range of a force mediated by a particle of mass m_π. This basic idea was formulated mathematically by Yukawa. He invented a functional form of strong interaction expressed in terms of the pion mass m_π and the distance r between the two nucleons

$$V(r) = f\,\frac{e^{-\frac{m_\pi c}{\hbar} r}}{r}. \tag{2.10}$$

This is known as the *Yukawa potential*. The nuclear force is actually more complex than the above description: the central potential, that depends mostly on the distance between the two nucleons, contains the exchange of different mesons, not only pions. The net contribution of several mesons (pions π, omega mesons ω, rho mesons ρ and so on) with different masses leads to the radial dependance of Fig. 2.7. Also, the nuclear potential contains non-central terms such as a tensor term and a spin-orbit term.

Due to the intrinsic structure of nucleon, the three-body force is needed to describe structure and reactions of many-body systems. More details will be discussed in Sect. 2.6.

2.5 *NN* Potentials and Phase Shifts

Nucleon-nucleon elastic scattering data allows one to determine *NN* potentials that can be used as inputs in nuclear structure calculations. The asymptotic radial part of the scattered wave function can be decomposed on partial waves and expressed in terms of phase shifts. The definition of phase shifts from the stationary scattering theory is given in this section.

The Hamiltonian driving the collision of two nucleons a and b is

$$H = H_a + H_b + T_{ab} + V_{ab}, \tag{2.11}$$

where H_a and H_b are the intrinsic Hamiltonians of a and b. In the following the eigenvalues and eigenstates of $H_{a(b)}$ are indicated by $E_{a(b)}$ and $\phi_{a(b)}$, respectively. The kinetic operator of the relative motion is

$$T_{ab} = -\frac{\hbar^2}{2\mu_{ab}}\Delta_{r_{ab}}, \tag{2.12}$$

which contains the reduced mass of the system composed of the two nucleons

$$\mu_{ab} = \frac{m_a m_b}{m_a + m_b}, \tag{2.13}$$

and the Laplacian operator $\Delta_{r_{ab}}$ acts on the relative motion between a and b

$$\mathbf{r}_{ab} = \mathbf{r}_a - \mathbf{r}_b. \tag{2.14}$$

In the particular case of no interaction potential V_{ab} between a and b, the Hamiltonian reduces to

$$H_0 = H_a + H_b + T_{ab} \tag{2.15}$$

and the initial state wave function that satisfies

$$H_0\Phi(\mathbf{r}_a), \mathbf{r}_b) = (E_a + E_b + E_{ab})\Phi(\mathbf{r}_a, \mathbf{r}_b) \tag{2.16}$$

is found to be a plane wave

$$\Phi(\boldsymbol{r}_a, \boldsymbol{r}_b) = \phi_a(\boldsymbol{r}_a)\phi_b(\boldsymbol{r}_b) \exp(i\boldsymbol{k} \cdot \boldsymbol{r}_{ab}), \tag{2.17}$$

where $\boldsymbol{k}$ is the relative momentum

$$\boldsymbol{k} = \mu_{ab} \left(\frac{\boldsymbol{k}_a}{m_a} - \frac{\boldsymbol{k}_b}{m_b} \right), \tag{2.18}$$

which enters the energy of relative motion

$$E_{ab} = \frac{\hbar^2 k^2}{2\mu_{ab}}. \tag{2.19}$$

The above relations and quantities are valid for any initial state i and final state f of the elastic reaction.

The full solution of the Schrödinger equation for the Hamiltonian (2.11) reads

$$H\Psi^{(\pm)} = E\Psi^{(\pm)}, \tag{2.20}$$

where the solution $\Psi^{(\pm)}$ with "+" ("−") is a wave function with asymptotic outgoing (incoming) spherical waves in the final states. The asymptotic form of these solutions can be written as (see Chap. 7 for detailed discussions on the scattering theory)

$$\Psi^{(\pm)} \rightarrow \Phi_i + \sum_f \phi_f f_{fi}^{(\pm)} \frac{\exp(\pm i\boldsymbol{k}_f \cdot \boldsymbol{r}_f)}{r_f}, \tag{2.21}$$

where Φ_i is the initial incoming wave and $f_{fi}^{\pm}$ are the so-called scattering amplitudes. In this asymptotic form of the wave function, the meaning of the $\pm$ appears clearly. The scattering amplitudes are directly connected to the reaction cross section σ_{fi} from the initial state i to any final state f,

$$\frac{d\sigma_{fi}}{d\Omega} = \frac{v_f}{v_i} |f_{fi}^{(+)}|^2, \tag{2.22}$$

where $v_i = \hbar k_i/\mu_i$ and $v_f = \hbar k_f/\mu_f$ are the velocities of relative motion in the entrance (i) and exit (f) channels, respectively.

One can write the full scattering wave function (2.21) as an expansion on partial waves with angular momenta l. In the case of an elastic scattering process $a + b \rightarrow a + b$ shown in Fig. 2.9, assuming a radial short range potential $V_{ab}(r)$ (i.e., no Coulomb potential), the outgoing scattering wave function is given by

Fig. 2.9 *NN* elastic
scattering of particles *a* and
b in the center of mass
(c.o.m.) frame

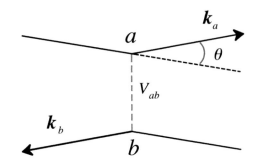

$$\Psi^{(+)}(\mathbf{r}) = \sum_{\ell=0}^{\infty} \sum_{m=-\ell}^{\ell} \frac{\psi_\ell^{(+)}(r)}{r} Y_{\ell m}(\hat{r}) \phi_a \phi_b. \tag{2.23}$$

The above solution can be extended in case of particles with arbitrary spin. The wave functions of radial motion are written by $\psi_\ell^{(+)}$, and the spherical harmonics $Y_{\ell m}$ depending on the direction $\hat{r} = \mathbf{r}/r$. The radial wave function $\psi_\ell^{(+)}$ is a solution of the radial Schrödinger equation

$$E_{ab}\psi_\ell^{(+)}(r) = \left(-\frac{\hbar^2}{2\mu_{ab}} \frac{1}{r^2} \frac{d}{dr} \left(r^2 \frac{d}{dr} \right) + \frac{\ell(\ell+1)}{r^2} + V_{ab}(r) \right) \psi_\ell^{(+)}(r). \tag{2.24}$$

The solutions $\psi_\ell^{(+)}(r)$ have to meet the correct boundary conditions at $r \sim 0$ and $r \to +\infty$. In the literatures, different ansatzes are given for short-distance boundary conditions. However, the dependence on the ansatz is demonstrated to be small for relative energies up to several hundreds of MeV. At large distances, they should be solutions of the Schrödinger equation with vanishing potential. Therefore, they must be linear combinations of Bessel functions (under the assumption of a short-range interaction). In the case of no interaction potential $V_{ab} = 0$, the solution of the Schrödinger equation is given by a plane wave (2.17) and can be expanded as

$$\Phi_{ab} = \sum_\ell i^\ell j_\ell(kr)(2\ell+1)P_\ell(\cos\theta)\phi_a\phi_b \tag{2.25}$$

$$= 4\pi \sum_{\ell,m} i^\ell j_\ell(kr) Y_{\ell m}(\hat{k}) Y_{\ell m}^*(\hat{r}) \phi_a \phi_b, \tag{2.26}$$

where j_ℓ are the regular spherical Bessel function and $\hat{k} = \mathbf{k}/k$. The two formulations (2.25) and (2.26) are equivalent: the first decomposes the plane wave on Legendre polynomials P_ℓ where θ is the angle between $\mathbf{r}$ and $\mathbf{k}$, and the second is an expansion on spherical harmonics. We use Rayleigh formula

$$e^{ikr\cos\theta} = \sum_l i^l(2l+1)j_l(kr)P_l(\cos\theta) \tag{2.27}$$

$$= \sum_{lm} i^l 4\pi j_l(kr)Y_{\ell m}(\hat{k})Y_{\ell m}^*(\hat{r}), \tag{2.28}$$

to derive Eq. (2.26). As a reminder, Bessel functions are defined as follows. For $\ell = 0$, j_0 is defined as

$$j_0(z) = \frac{\sin(z)}{z} = \frac{\exp(iz) - \exp(-iz)}{2iz}. \tag{2.29}$$

Other Bessel functions with $\ell > 0$ can be obtained from derivatives of j_0 as

$$j_\ell(z) = z^\ell \left(-\frac{1}{z}\frac{d}{dz}\right)^\ell j_0(z). \tag{2.30}$$

They can be written as a difference between two functions, analogous to Eq. (2.29);

$$j_\ell(z) = \frac{1}{2iz}(u_\ell^{(+)}(z) - u_\ell^{(-)}(z)), \tag{2.31}$$

where the functions $u_\ell^{(\pm)}(z)$ are spherical Hankel functions and have the asymptotic behavior

$$u_\ell^{(\pm)}(z) \to \exp\left(\pm i\left(z - \ell\frac{\pi}{2}\right)\right) \tag{2.32}$$

for $z \to +\infty$. In this representation, the plane wave (2.26) can be rewritten as

$$\Phi_{ab} = \frac{4\pi}{2ikr} \sum_{\ell,m} i^\ell (u_\ell^{(+)}(kr) - u_\ell^{(-)}(kr))Y_{\ell m}(\hat{r})Y_{\ell m}^*(\hat{k})\phi_a\phi_b. \tag{2.33}$$

Now, let us consider a non-zero interaction potential V_{ab} which will modify the radial wave function $\psi_\ell^{(+)}(r)$. For large distances, since the interaction potential vanishes, they can still be decomposed as a linear combination of spherical Bessel functions j_l or spherical Hankel functions $u_\ell^{(\pm)}$. Since $u_\ell^{(-)}$ correspond to incoming waves, only the outgoing waves are affected by the scattering process. The asymptotic form of the wave function can then be written as

$$\psi^{(+)} \to \frac{4\pi}{2ikr} \sum_{\ell,m} i^\ell (S_\ell(k)u_\ell^{(+)}(kr) - u_\ell^{(-)}(kr))Y_{\ell m}(\hat{r})Y_{\ell m}^*(\hat{k})\phi_a\phi_b, \tag{2.34}$$

where S_ℓ is called the S-matrix or scattering matrix and takes into account the distortion by the interaction V_{ab}. S_ℓ depends on the momentum k or the energy E_{ab}. The difference between the full solution $\psi^{(+)}$ and the plane wave Φ_{ab} gives the scattering part of the wave function

$$\psi^{(+)} - \Phi_{ab} \to \frac{4\pi}{2ikr} \sum_{\ell,m} i^{\ell} (S_{\ell}(k) - 1) u_{\ell}^{(+)}(kr) Y_{\ell m}(\hat{r}) Y_{\ell m}^{*}(\hat{k}) \phi_a \phi_b. \qquad (2.35)$$

By comparison to Eq. (2.21), in the case of elastic scattering, in which the final state is the same as the initial state denoted as $f = i$, the scattering amplitude reads

$$f_{ii}^{(+)}(\theta) = \sum_{\ell} \frac{2\ell + 1}{2ik} (S_{\ell}(k) - 1) P_{\ell}(\cos\theta). \qquad (2.36)$$

It is advantageous to introduce a new quantity: the scattering phase shift δ_{ℓ} which is defined as

$$S_{\ell} = \exp(2i\delta_{\ell}). \qquad (2.37)$$

In the case of an elastic scattering process with a real potential V_{ab}, the S-matrix is a complex number with $|S_l| = 1$ such that the incoming and outgoing fluxes in each partial wave ℓ are conserved. In this case, the phase shifts δ_{ℓ} are real quantities that can be limited to the range $\theta = [0°, 180°]$ as a convention. The asymptotic form of the wave function is then given by

$$u_{\ell}(kr) = S_{\ell} u_{\ell}^{+} - u_{i}^{-} \to \exp(i\delta_{\ell}) \sin\left(kr + \delta_{\ell} - \ell\frac{\pi}{2}\right). \qquad (2.38)$$

The meaning of the phase shift appears clearly in the above formulation. The action of the interaction potential V_{ab} leads to a shift δ_{ℓ} of the phase of the asymptotic wave function as illustrated in Fig. 2.10.[4] It depends on the momentum k in the initial state. An increase (decrease) of δ_{ℓ} as a function of k signifies an attractive (repulsive) potential as illustrated in Fig. 2.11. A sharp increase of the phase shift from $0°$ to $180°$ crossing through $90°$ indicates a resonant behavior that can be seen in the cross section, i.e., the differential cross section of the elastic scattering can be written as

$$\frac{d\sigma_{el}}{d\Omega} = \left| \sum_{\ell} \frac{2\ell + 1}{2ik} [S_{\ell}(k) - 1] P_{\ell}(\cos(\theta)) \right|^2, \qquad (2.39)$$

and the contribution of a partial wave ℓ is maximum if $S_{\ell} = -1$ or $\delta_{\ell} = 90°$.

Elastic scattering data at different incident energies then lead to phases shifts for each partial wave $\delta_{\ell}(E)$ as a function of the incident energy E. While the above description did not consider the spin of interacting nucleons, the phase shifts also depend on the total spin s to which the two nucleons couple and the total angular momentum j, which is the eigenvalue of the operator j defined by $j = \ell + s$. The

---

[4] A s-wave corresponds to an angular momentum $\ell = 0$. Indeed, the angular momenta $\ell = 0, 1, 2, 3$ are often named s,p,d,f, respectively. Historically, these terms were used to qualify groups of transitions in the atomic spectra of alkali metals and stand for *sharp, principal, diffuse* and *fundamental*. For higher values of angular momentum, the increasing angular momenta are addressed by letters taken in alphabetical order (g for $\ell = 4$, h for $\ell = 5$, etc.).

Fig. 2.10 Asymptotic behavior of the radial wave function $u_0(k_i r)$ for a *s*-wave without potential (blue) and with a short-range attractive (red). The attractive potential results in a positive scattering phase shift δ_0. Courtesy S. Typel, TU Darmstadt

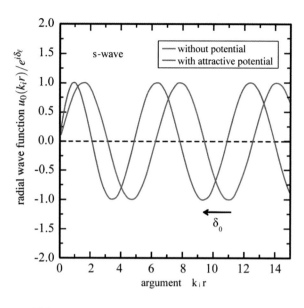

Fig. 2.11 Energy (E_{aA}) dependence of the nucleon-nucleus scattering phase shift δ_ℓ in the cases of an attractive (blue), repulsive (green) potential, and a resonance at 5 MeV (red). Courtesy S. Typel, TU Darmstadt

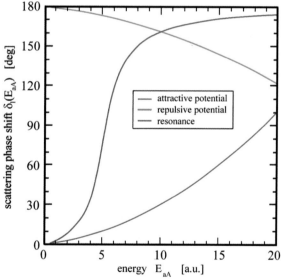

phase shift is therefore extracted for each (ℓ, s, j) combination. The usual convention uses the capital letter $S, P, D, F, G, \ldots$ for angular momenta $\ell = 0, 1, 2, 3, 4, \ldots$, respectively. The total spin and angular momentum values are given as $2s + 1$ and j in the exponent and index, respectively. For example, the partial wave corresponding to $(\ell = 0, s = 0, j = 0)$ is denoted by $^1S_0$ while the partial wave corresponding to $(\ell = 1, s = 1, j = 2)$ is denoted by $^3P_2$. Examples of phase shifts extracted from experiment are given in Fig. 2.16 at the end of this chapter.

2.5.1 Symmmetry Requirement on NN Interactions

We discuss in this subsection basic symmetry requirements on NN interactions. The NN two-body interaction $V(1, 2)$ can depend on the position $\mathbf{r}$, the momentum $\mathbf{p}$, the spin σ and the isospin τ of two interacting nucleons

$$V(1, 2) = V(\mathbf{r}_1, \mathbf{r}_2, \mathbf{p}_1, \mathbf{p}_2, \sigma_1, \sigma_2, \tau_1, \tau_2), \tag{2.40}$$

and the two-body potential is evaluated by the expectation value

$$\langle \phi_1^* \phi_2^* | V(1, 2) | \phi_{1'} \phi_{2'} \rangle, \tag{2.41}$$

where a single particle wave function ϕ is specified by the coordinate, spin and isospin. The functional form of V is restricted by the underlying invariance (or symmetry) requirements, given below.

- Translational invariance which implies the conservation of momentum; the dependence of positions $\mathbf{r}_1$ and $\mathbf{r}_2$ must be the relative distance $\mathbf{r} = \mathbf{r}_1 - \mathbf{r}_2$.
- Galilean (Lorentz) invariance. It guarantees the invariance of the interaction with respect to the transformation from one reference frame to another reference frame moving with uniform relative velocity. This invariance requires that the dependence of momenta $\mathbf{p}_1$ and $\mathbf{p}_2$ should be the relative velocity $\mathbf{p} = \mathbf{p}_1 - \mathbf{p}_2$. The local potential has a functional form $V(\mathbf{r}_1, \mathbf{r}_2) \propto \delta(\mathbf{r}_1 - \mathbf{r}_2)$, while the non-local potential is represented by the function of momentum $\mathbf{p}^2$.
- Rotational invariance. This invariance reflects the conservation of angular momentum and requests that all terms in the interaction should have the total angular momentum of zero (scalar product).
- Isospin invariance. It requires the conservation of isospin. The isospin degree of freedom is expressed by the operators τ_1 and τ_2. The isospin invariance demands that the interaction should be isoscalar under the rotation in the isospin space so that the isoscalar product $\tau_1 \cdot \tau_2$ and its power terms are only allowed in the interaction.
- Parity invariance. The parity operator P is the space reflection operator, which changes properties of operators as

$$(\mathbf{r}, \mathbf{p}, \sigma, \tau) \xrightarrow{P} (\mathbf{r}' = -\mathbf{r}, \mathbf{p}' = -\mathbf{p}, \sigma' = \sigma, \tau' = \tau). \tag{2.42}$$

Accordingly, the terms containing even power of **r** and **p** fulfill this symmetry.[5]
- Time reversal invariance. The corresponding operator T acts

$$(\mathbf{r}, \mathbf{p}, \sigma, \tau) \overset{T}{\rightarrow} (\mathbf{r}' = \mathbf{r}, \mathbf{p}' = -\mathbf{p}, \sigma' = -\sigma, \tau' = \tau), \tag{2.43}$$

so that even powers of **p** and σ are allowed in the interaction.

These invariances imply the general form of interaction with the terms

$$1, \; \sigma_1 \cdot \sigma_2, \; \tau_1 \cdot \tau_2, \; (\mathbf{r} \cdot \sigma_1)(\mathbf{r} \cdot \sigma_2), \tag{2.44}$$

$$\mathbf{L} \cdot \mathbf{S} = -i\frac{\hbar}{2}(\mathbf{r} \times \mathbf{p}) \cdot (\sigma_1 + \sigma_2), \; \mathbf{L}^2 = -\hbar^2(\mathbf{r} \times \mathbf{p}) \cdot (\mathbf{r} \times \mathbf{p}). \tag{2.45}$$

Each of these terms can be multiplied by an arbitrary function of $\mathbf{r}^2$ and $\mathbf{p}^2$. The spin enters symmetrically ($\mathbf{S} = (\sigma_1 + \sigma_2)/2$) in the spin-orbit interaction $\mathbf{L} \cdot \mathbf{S}$. The symmetric choice is required since the potential must be symmetric for the interchange of all coordinates of particles 1 and 2.

Problem

2.4 Among the following different terms, show which ones are allowed in *NN* interaction satisfying the symmetry requirements

$$(\mathbf{p}^2 + \mathbf{r}^2), \quad r^2\sigma_1 \cdot \sigma_2, \quad \mathbf{r} \cdot \mathbf{L}. \tag{2.46}$$

Charge independent interactions must be symmetric in the isospin variables τ_1 and τ_2 as well as in the space and spin variables. The symmetry in the spin variables implies that the non-central forces, such as the spin-orbit force and tensor force $S_{12} = (3(\mathbf{r} \cdot \sigma_1)(\mathbf{r} \cdot \sigma_2)/r^2 - \sigma_1 \cdot \sigma_2)$, commute with the total spin operator $\mathbf{S}^2$ and therefore conserve the total spin quantum number of two particle system which has either $S = 0$ or $S = 1$. With the charge independent interaction, the antisymmetrization condition of two-particle wave function under the exchange of space, $P_r(ij)$, spin, $P_\sigma(ij)$, and isospin, $P_\tau(ij)$, variables requires

$$P(ij) \equiv P_r(ij)P_\sigma(ij)P_\tau(ij) = -1, \tag{2.47}$$

which gives the relation

$$\pi(-1)^{S+T} = -1, \tag{2.48}$$

where π is the parity of the two-particle system, and S and T are the total spin and isospin of two-particle wave functions, respectively.

---

[5] The spin and isospin operators σ and τ are vectors, but do not change the sign under the space reflection so that these operators are called *pseudovector* operators. On the other hand, the scalar product $\mathbf{p} \cdot \sigma$ and $\mathbf{r} \cdot \sigma$ are scalars, but change the sign under the space reflection. These operators are called *pseudoscalar* operators.

2.6 Many-Body Interactions

Three-body forces exist at all scales in nature, including in classical mechanics. An often given example is the existence of three-body gravitational forces in the earth-moon-sun system when planets are considered as point-like objects. Indeed, the relative position of the sun and the earth leads to tides which modify the envelop of the earth, i.e., polarizes slightly the mass distribution of the earth. The relative motion of the earth and moon cannot be reproduced by only considering the gravitational interaction between the centers of mass of the sun, moon and earth: the tide effect should be included as an earth-moon interaction that depends on the position of the sun. This is a three-body force, i.e., a force that does not exist in a system of two objects but appears in a system with three and more objects.

The strong interaction does exhibit such a behaviour. The three-body forces are caused by a polarization of the considered interacting objects under the effect of their environment. In nuclear physics, three-body forces exist because nucleons are composite objects made of quarks and gluons which may excite inside the nucleus. The nucleons then may be polarized by the surrounding density, leading to a modification of the nucleon-nucleon interaction. If these nucleonic excitations are not described explicitly in the theory, those *intrinsic* three-body forces need to be included. In the ChEFT, medium effects lead to the emergence of many-body forces in higher order terms of chiral perturbation. In effective theories, the medium effects are often mimicked by a density dependent effective force to obtain the saturation of nuclear matter.

In nuclear physics, the three-body force can be described as mediated by two π meson exchange processes in which the Δ-isobar state is excited in the intermediate state. It was proposed by Jun-Ichiro Fujita and Hironari Miyazawa in 1957. This, so-called Fujita–Miyazawa three-body force is displayed diagrammatically in Fig. 2.8. This three-body force has been known to play an important role in the study of nuclear structure of few-body systems. It is revived in the context of radioactive drip line nuclei studies, which are discussed in Chap. 5. The inclusion of three-body forces was also recognized as essential to give proper saturation properties in the mean field model with phenomenological interactions by Tony Hilton Roy Skyrme in 1956.

After this pioneer work by Fujita–Miyazawa, more realistic three-body forces were developed to fit the nucleon-nucleus scattering phase shifts and the binding energies of three-body systems. They were the Tucson–Melbourne (TM), the Urbana IX and Illinois 2 and 4 (IL2, IL4) three-body forces. The ChEFT model gives rise to the three-body forces as the next-to-next leading order (NNLO) expansion, while the main part of two-body forces are derived by the LO and NLO terms in the perturbative expansion as shown in Fig. 2.15 and Sect. 2.7.3. This hierarchy explains why the three-body forces are relatively weaker than the two-body forces.

The first empirical evidence for the three-body forces was found in the binding energies of three-body nuclear systems ^{3}H and ^{3}He. The binding energies of these nuclei are not reproduced by the Faddeev-type calculations with any realistic two-body interactions such as Argonne V18 (AV18), CD Bonn and Nijmegen I and II

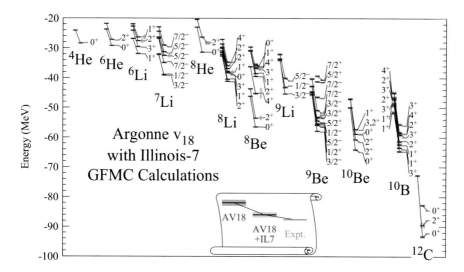

Fig. 2.12 Energies of ground and low-lying excited states of light nuclei with the mass A = (4 ∼ 12) calculated by GFMC method with the two-body AV18 and the two-body+three-body AV18+IL7 forces, respectively, compared with experimental data. The light shading shows the statistical error of Monte Carlo method. This figure is reprinted with permission from [10]. ©2021 by the American Physical Society. Courtesy R. B. Wiringa, Argonne National Laboratory

forces (the Faddeev model is introduced in Sect. 3.2 of Chap. 3). This problem was cured by the introduction of Fujita–Miyazawa-type three-body forces. The importance of three-body forces was further confirmed by the study of the binding energies of light nuclei and also by the saturation point of nuclear matter in the plot of the binding energy as a function of the nuclear density.

Ab initio model calculations such as the Green's function Monte Carlo (GFMC) and the no-core shell models show a clear evidence of the three-body forces (see Sect. 3.11 for more details about GFMC and no-core shell model). In Fig. 2.12, the calculated binding energies of light nuclei with the mass $A = 4 \sim 12$ are shown together with the experimental data. In these GFMC calculations, the two-body interaction AV18 was adopted with the three-body force IL7. It is clearly seen that the introduction of three-body forces dramatically improves the agreement between theory and experiment by increasing the binding of all considered nuclei. Quantitatively, the three-body forces have an effect of 15–20% on the total binding energies of these nuclei.

More direct evidence of three-nucleon forces might be obtained by reactions which involve three nucleons at different projectile energies. High precision measurements were carried out in proton-deuteron (pd) and neutron-deuteron (nd) scatterings at energies of 70–400 MeV/nucleon [11] and also break-up reactions at lower energies $E/A \leq 20$ MeV. In Fig. 2.13, experimental results of (pd) and (nd) elastic scatterings are compared with the Faddeev calculations with and without three-body forces. The red (blue) shaded band is the result with (without) the Tucson–Melbourne three-

Fig. 2.13 Differential cross sections of elastic (d, p) and (n, d) scatterings at the projectile energies of 70–400 MeV/nucleon (MeV/N). The red (blue) shaded bands are the calculations with (without) Tucson–Melbourne three-body forces. For the two-body forces, several modern forces (CD Bonn, AV18, Nijmegen I and II) are used to check the model dependence which gives the bands in the curves. The black solid lines for the energies 70, 135 and 250 MeV/N are the results of AV18 two-body force with Urbana IX three-body forces. The experimental data are shown with circles, squares and diamonds at different energies, respectively. Courtesy K. Sekiguchi, Tohoku University

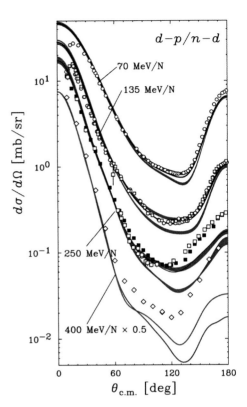

body force which is constructed being consistent with the chiral symmetry. For the two-body forces, the several modern ones (CD Bonn, AV18, Nijmegen I and II) are used to check model dependence, which is shown as shaded bands. The cross sections show specific features depending on the scattering angles in the center of mass system $\theta_{c.m.}$. At forward angle $\theta_{c.m.} \leq 80°$, theoretical calculations based on the various NN forces are well converged and the three-body forces have a minor effect since the direct processes due to the NN forces govern the scatterings. There are some discrepancies at the very forward angles which are caused by the absence of Coulomb interaction in the Faddeev calculations. At intermediate angles $80° \leq \theta_{c.m.} \leq 140°$, where the cross sections show the minimum, we can see clear discrepancies between the experimental data and the calculations with the NN forces only. The discrepancies become larger for larger incident energies of the projectiles. This discrepancies are cured largely by the Fujita–Miyazawa type two π-exchange three-body forces. At backwards angles $\theta_{c.m.} \geq 140°$, where the exchange processes by the NN forces are dominant, differences remain between the experimental data and the calculations at higher projectile energies >250 MeV/nucleon even when the three-body forces are included in the Faddeev calculations. This problem is open for future studies.

So far the isospin $T = 1/2$ channel of the three-body forces has been investigated by the three-body scattering with (p, d) and (d, n). Future experiments are planned to investigate new aspects of three-body forces, for example, the isospin $T = 3/2$ channel of the three-body forces by four-body scattering with $(p,^3\text{He})$ or $(^3\text{He},n)$ reactions.

2.7 QCD and Chiral Effective Field Theory (ChEFT)

2.7.1 QCD and NN Interaction

There have been attempts to construct NN interactions based on the low-energy limit of quantum chromodynamics (QCD). While the interaction between colored objects (quarks and gluons) is weak at short distances or high momentum transfer ("asymptotic freedom"), it is strong at long distances (≥ 1 fm) or low energies, leading to the confinement of quarks into colorless objects, the hadrons. The origin of confinement is still an open question in the field theory. Consequently, QCD allows for a perturbative analysis at large energies, whereas it is highly non-perturbative in the low-energy regime. Nuclear physics may deal with low-energy phenomena and the force between nucleons as a residual color interaction similar to van der Waals forces. Therefore, in terms of quarks and gluons, the nuclear force is a very complicated problem that, nevertheless, can be attacked with high performance computing power on a discretized space-time lattice, known as lattice QCD (LQCD). LQCD calculations have made a big step forward by using the Japanese "K" supercomputer which has 10.51 petaflops computing power (a petaflops $= 10^{15}$ flops) and ranked number one of TOP500 supercomputers in the World by the evaluation of June and November, 2011 and retired August 2019. The World number one by the evaluation of June, 2020 is "Fugaku" installed at RIKEN, which has 415.5 petaflops computing power.

An illustration of LQCD calculations is given in Fig. 10.8 in Chap. 10, where the nucleon–nucleon potential calculated from lattice QCD is shown as a function of distance between the two nucleons. Some approximations, which we will not detail here, have been used to make this calculation feasible. Nevertheless, it demonstrates that the short range repulsion, the range of the attractive part of the interaction and the asymptotic behavior as we can deduce them from nuclear structure are predicted by QCD. More discussions are given in Chap. 10.

2.7.2 Chirality

In 1990, a chiral effective theory was proposed by S. Weinberg to construct effective NN interactions based on pion and nucleon degrees of freedom instead of the quarks and gluons in QCD. The chiral invariance is required to obtain an effective

Lagrangian. The chiral invariance is one of the important symmetries imposed on QCD Lagrangian, and the chiral invariance is defined with the operator

$$\gamma_5 \equiv i\gamma_0\gamma_1\gamma_2\gamma_3 = \begin{pmatrix} 0 & I \\ I & 0 \end{pmatrix}. \tag{2.49}$$

The Dirac γ_μ matrices ($\mu = 0, 1, 2, 3$) are defined in Eq. (1.96). The chiral operator γ_5 has the eigenvalues ± 1 for any Dirac spinor field. For massless particles, the chirality is equivalent to the helicity defined by the operator $(\mathbf{p} \cdot \boldsymbol{\sigma})/|\mathbf{p}|$ (see Fig. 2.14). There are two kinds of chiral symmetry breaking. One is due to the quark mass and another is the dynamical effect of QCD. The former is named *explicit* symmetry breaking and the second one is called *spontaneous* symmetry breaking. The latter mechanism is responsible to create the masses of hadrons.

Problem

2.5 Prove that γ_5 has the eigenvalues ± 1 for any Dirac field ψ.

The eigenstates of chirality are denoted as

$$\gamma_5\psi_R = +\psi_R, \quad \gamma_5\psi_L = -\psi_L, \tag{2.50}$$

where ψ_R and ψ_L are right-handed and left-handed chiral states, respectively. We can define the projection operators

$$P_L = \frac{1}{2}(1 - \gamma_5), \quad P_R = \frac{1}{2}(1 + \gamma_5), \tag{2.51}$$

such that P_L and P_R operators project out the left-handed and right-handed chiral particle states, respectively, from any Dirac state ψ. The projection operators have the properties,

$$P_{L(R)}^2 = P_{L(R)}, \quad P_L + P_R = 1, \quad P_L P_R = P_R P_L = 0. \tag{2.52}$$

Any Dirac spinor field can be projected out as a sum of its left-handed and right-handed chiral states,

$$\psi = (P_L + P_R)\psi = P_L\psi + P_R\psi = \psi_L + \psi_R. \tag{2.53}$$

The chiral symmetry plays a key role in QCD. Let us consider the baryonic current

$$j^\mu = \overline{\psi}\gamma^\mu\psi, \quad \mu = \{0, 1, 2, 3\}, \tag{2.54}$$

where $\overline{\psi} = \psi^\dagger\gamma_0$ is a Dirac adjoint field. The four-current $j^\mu = (\rho, \mathbf{j})$ contains the normal density $\rho \equiv \psi^\dagger\psi$ and the vector current $\mathbf{j}$. The equation of continuity is written as

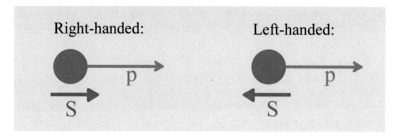

Fig. 2.14 Chirality is equivalent to helicity for massless particles. Here **p** and **s** show the momentum and the spin directions, respectively

$$\partial_\mu j^\mu = 0 \tag{2.55}$$

in the covariant form. This type of current is called a vector current, which is responsible for the electromagnetic interaction.

We can show that the current can be decomposed into

$$\overline{\psi}\gamma_\mu\psi = (\overline{\psi_L} + \overline{\psi_R})\gamma_\mu(\psi_L + \psi_R) = \overline{\psi_L}\gamma_\mu\psi_L + \overline{\psi_R}\gamma_\mu\psi_R, \tag{2.56}$$

where the cross terms $\overline{\psi_L}\gamma_\mu\psi_R$ and $\overline{\psi_R}\gamma_\mu\psi_L$ are dropped out. This property can be proved by using the formulas $\overline{\psi_L} = \psi_L^\dagger\gamma_0 = \psi^\dagger\frac{1}{2}(1-\gamma_5)\gamma_0 = \overline{\psi}\frac{1}{2}(1+\gamma_5) = \overline{\psi}P_R$ derived from the facts that $\gamma_5^\dagger = \gamma_5$ and $\gamma_5\gamma_\mu = -\gamma_\mu\gamma_5$. Using these properties of chiral states and the γ matrices, we can show

$$\overline{\psi_L}\gamma_\mu\psi_R = \overline{\psi}P_R\gamma_\mu P_R\psi = \overline{\psi}\gamma_\mu P_L P_R\psi = 0, \tag{2.57}$$

where we use the identity $P_L P_R = 0$. We can prove also in the same way $\overline{\psi_R}\gamma_\mu\psi_L = 0$. Thus, for the current (2.54), the left-handed particle can couple only to the left-handed field and the right-handed particle can couple only to the right-handed field. A condition for the conservation of chiral fields is given by the anti-commutation relation

$$\{D(\gamma_\mu, \gamma_0), \gamma_5\} = D(\gamma_\mu, \gamma_0)\gamma_5 + \gamma_5 D(\gamma_\mu, \gamma_0) = 0 \tag{2.58}$$

for any operator D written by a function of γ_μ and/or γ_0. If this anti-commutation relation holds, we can show

$$DP_L = P_R D \quad \text{and} \quad DP_R = P_L D, \tag{2.59}$$

so that the chirality is conserved with any action by the operator D (Fig. 2.14).

Let us consider next a model in which massless particles are described by a Lagrangian (see the Dirac equation (1.95))

$$L = i\overline{\psi}\gamma_\mu\partial^\mu\psi, \tag{2.60}$$

where $i\gamma_\mu \partial^\mu$ is called the vector coupling and it satisfies the anti-commutation relation (2.58). Then the Lagrangian is rewritten to be

$$L = i\overline{\psi}\gamma_\mu \partial^\mu \psi = i\overline{\psi_L}\gamma_\mu \partial^\mu \psi_L + i\overline{\psi_R}\gamma_\mu \partial^\mu \psi_R, \tag{2.61}$$

where the right-handed and the left-handed currents are conserved. We should notice that the mass term m (for example, the quark mass term in the Lagrangian) mixes the two chiral states since

$$\overline{\psi_L}m\psi_R = \overline{\psi}P_R m P_R \psi = \overline{\psi}mP_R\psi = (\overline{\psi}_R + \overline{\psi}_L)m\psi_R = \overline{\psi}_R m\psi_R \neq 0, \tag{2.62}$$

where the mass term commutes with the projection operator $P_R m = mP_R$ and $P_R^2 = P_R$. This is called an *explicit* breaking of the chiral symmetry by fermionic masses. Another breaking process is called a *spontaneous* chiral symmetry breaking (CSB) associated with Goldstone bosons (pions), which will be discussed in Chap. 9. We can show that the chirality eigenstates are equivalent to the helicity eigenstates for massless particles or in the limit of very high energy. While the helicity is conserved in a process of free propagation in space for any particles (regardless of finite mass or massless particles), the chirality is conserved only for massless particles.

2.7.3 ChEFT and NN and Many-Body Interactions

Since lattice QCD calculations are very time-consuming, they can only be used to check a few representative key issues such as *NN* interactions or the binding energy of the deuteron. For nuclear structure physics, a more efficient approach is needed. A promising approach is called the chiral effective field theory (ChEFT). ChEFT considers only pions and nucleons rather than quarks and gluons as the relevant degrees of freedom to cope with nuclear structure. Chirality for a Dirac fermion ψ (quark) is defined through the operator γ_5, which has eigenvalues ± 1. Any Dirac field can thus be projected into its left- or right-handed component by acting with the projection operators $P_L = \frac{1}{2}(1 - \gamma_5)$ or $P_R = \frac{1}{2}(1 + \gamma_5)$ on ψ, as was discussed in Sect. 2.7.2. For massless quarks, the chirality is well conserved (the chiral limit). Spontaneous chiral symmetry breaking occurs in quantum chromodynamics and creates finite masses of composite hadrons (pion) from massless quarks (or very small masses of u- and d-quarks). In some sense, ChEFT is a revival of Yukawa's meson theory under the constraints of the chiral symmetry and the broken symmetry for the dynamics. The idea of ChEFT by S. Weinberg tells how to implement chiral symmetry in the theory of pionic and nucleonic interactions. While QCD in the low energy limit is non-perturbative, one needs to implement a perturbative expansion such that only a finite number of terms contribute at a given order of the expansion. This expansion is provided by powers of a ratio between an external momentum Q or pion mass (the soft scale) and the chiral symmetry breaking scale $\Lambda \sim 1$ GeV (the hard scale), as already introduced in Eq. (2.1). The problem of nuclear force receives further constraints because of its non-perturbative nature and the existence of nuclear bound states. To overcome these difficulties, Weinberg in 1990 proposed a two-step procedure to calculate nuclear potentials [9]: first, obtain the NN interaction by ChEFT in a perturbative manner. Secondly, iterate it to all orders by using the Lippman–Schwinger equation[6] to obtain NN scattering amplitudes. This idea is now implemented with success.

As mentioned earlier, the relevant degrees of freedom of ChEFT are pions and nucleons although pion-less ChEFT theories also exist. Since the interactions of pions (Goldstone bosons) must vanish at zero momentum transfer and in the chiral limit ($m_u, m_d, m_\pi \to 0$), the low-energy expansion of the Lagrangian is arranged in powers of derivative (momentum) and pion mass. The pion mass sets the soft scale $m_\pi c^2 \approx 140$ MeV and the hard scale is the chiral-symmetry breaking scale, $\Lambda \approx 1$ GeV which may represent the masses of heavy mesons. Thus, the expansion is classified in terms of powers of $\mu \equiv Q/\Lambda$. It should be noticed that the value of Λ and Q may have some arbitrariness without any solid criterion. In the context of the chiral perturbation theory (ChPT), the graphs are analyzed by a power of ν of the ratio μ. According to Weinberg, an effective theory can be useful if one writes down all terms consistent with the chiral symmetry of QCD. In general, there is an infinite number of terms which meet this requirement. Therefore in order to make any physical prediction, one assigns a power counting scheme to the theory. To differ from other phenomenologies, ChEFT satisfies all relevant symmetries of QCD as a prerequisite.

To construct the Lagrangian of ChEFT, a chiral field must be introduced. One common choice of the field in the isospin SU(2) framework (see Sect. 1.2.3 in Chap. 1 for the isospin and its group structure SU(2)) is

$$U(x) = \exp\left(\frac{i}{f_\pi}\boldsymbol{\tau} \cdot \boldsymbol{\pi}(x)\right) = \exp\left(\frac{i}{f_\pi}\begin{pmatrix} \pi^0 & \sqrt{2}\pi^+ \\ \sqrt{2}\pi^- & -\pi^0 \end{pmatrix}\right), \qquad (2.63)$$

where f_π is the pion decay constant, $\boldsymbol{\tau} = (\tau_x, \tau_y, \tau_z)$ are the Pauli isospin matrices, and $\boldsymbol{\pi}(x) = (\pi_x, \pi_y, \pi_z)$ are the pion fields, which have three components, the same as the isospin operator discussed in Sect. 1.2.3 in Chap. 1. The pion fields are further decomposed into $\pi^\pm(x) = (\pi_x \pm i\pi_y)/\sqrt{2}$ and $\pi^0 = \pi_z$. The chiral field (2.63) can generate infinite order of pion fields by the expansion with $\boldsymbol{\tau} \cdot \boldsymbol{\pi}(x)$;

$$U(x) = \exp\left(\frac{i}{f_\pi}\boldsymbol{\tau} \cdot \boldsymbol{\pi}(x)\right) = 1 + \frac{i}{f_\pi}\boldsymbol{\tau} \cdot \boldsymbol{\pi}(x) - \frac{1}{2f_\pi^2}\boldsymbol{\pi}(x)^2 + \cdots. \qquad (2.64)$$

In the ChEFT, the QCD Lagrangian (see Chap. 9) is replaced by an effective Lagrangian with the chiral field

---

[6] The Lippman–Schwinger equation is a basic equation to solve scattering problems of photons, atoms, molecules and nuclei. This equation is derived from the Schrödinger equation, especially to handle continuum states as detailed in Chap. 8.

$$L_{QCD} \to L_{eff}(U, \partial^{\mu}U) = L_{\pi\pi} + L_{\pi N}, \tag{2.65}$$

where $L_{\pi\pi}$ deals with the dynamics of pions and $L_{\pi N}$ describes the interaction between pions and a nucleon. Lorentz invariance[7] permits only even numbers of derivatives

$$L_{\pi\pi} = L_{\pi\pi}^{(2)} + L_{\pi\pi}^{(4)} + \cdots, \tag{2.66}$$

where the superscripts refer to the numbers of derivatives or pion mass insertions. The leading order (LO) $\pi\pi$ Lagrangian now reads

$$L_{\pi\pi}^{(2)} = \frac{f_{\pi}^2}{4} \mathrm{tr} \left[\partial_{\mu} U \partial^{\mu} U^{\dagger} + m_{\pi}^2 (U + U^{\dagger}) \right], \tag{2.67}$$

where the first and second terms involve a quadratic derivative and a mass square term, respectively.

The nucleon field is represented by an isospin T $= 1/2$ doublet,

$$\Psi = \begin{pmatrix} p \\ n \end{pmatrix}. \tag{2.68}$$

The free Lagrangian for nucleons is written as

$$L_N^{free} = \overline{\Psi}(i\gamma^{\mu}\partial_{\mu} - m_N)\Psi, \tag{2.69}$$

where the first term is the kinetic term and m_N is the nucleon mass. The πN interaction can be described by the so called pseudo-vector coupling or gradient coupling to the nucleon

$$L_{\pi N} = -\frac{g_A}{2f_{\pi}} \overline{\Psi}\gamma^{\mu}\gamma_5 \Psi \boldsymbol{\tau} \cdot \partial_{\mu}\boldsymbol{\pi}, \tag{2.70}$$

which is the isospin invariant field.

Baryon fields can also be incorporated into ChEFT in a chiral consistent manner. The LO πN Lagrangian has a chiral covariant derivative D_{μ}, which is a replacement of ∂_{μ}, and an axial vector coupling field u_{μ}[8]

$$L_{\pi N} = \overline{\Psi}(i\gamma^{\mu}D_{\mu} - m_N + \frac{g_A}{2}\gamma^{\mu}\gamma_5 u_{\mu})\Psi, \tag{2.71}$$

---

[7] Physics laws should be invariant under the transformation between two coordinate systems moving at a constant velocity. In the special relativity, the transformation is called "*Lorentz transformation*" and the invariance is named "*Lorentz invariance*". In classical mechanics, it is named "*Galilean invariance*".

[8] The chiral covariant derivative is an analogous one to the covariant derivative in the electromagnetic field $\partial_{\mu} \to D_{\mu} \equiv \partial_{\mu} - iqA_{\mu}$, where q is the electric charge and A_{μ} is the photon field. This is called "*minimal coupling*". This idea is generalized in QCD introducing the minimal coupling between quarks and gluons (see Chap. 10 for more details).

where D_μ is defined by $D_\mu = \partial_\mu + \Gamma_\mu$ with

$$\Gamma_\mu = \frac{1}{2}[\xi^\dagger, \partial_\mu \xi] = \frac{i}{4f_\pi^2}\boldsymbol{\tau} \cdot (\boldsymbol{\pi} \times \partial_\mu \boldsymbol{\pi}) + \cdots , \tag{2.72}$$

where

$$\xi \equiv \sqrt{U} = 1 + \frac{i}{2f_\pi}\boldsymbol{\tau} \cdot \boldsymbol{\pi}(x) - \frac{1}{8f_\pi^2}\boldsymbol{\pi}(x)^2 + \cdots . \tag{2.73}$$

The axial vector term u_μ is also defined as

$$u_\mu = i\{\xi^\dagger, \partial_\mu \xi\} = -\frac{1}{f_\pi}\boldsymbol{\tau} \cdot \partial_\mu \boldsymbol{\pi} + \cdots . \tag{2.74}$$

As stated before and illustrated in Fig. 2.15, chiral perturbation theory and its power counting imply that nuclear forces emerge as a hierarchy controlled by the

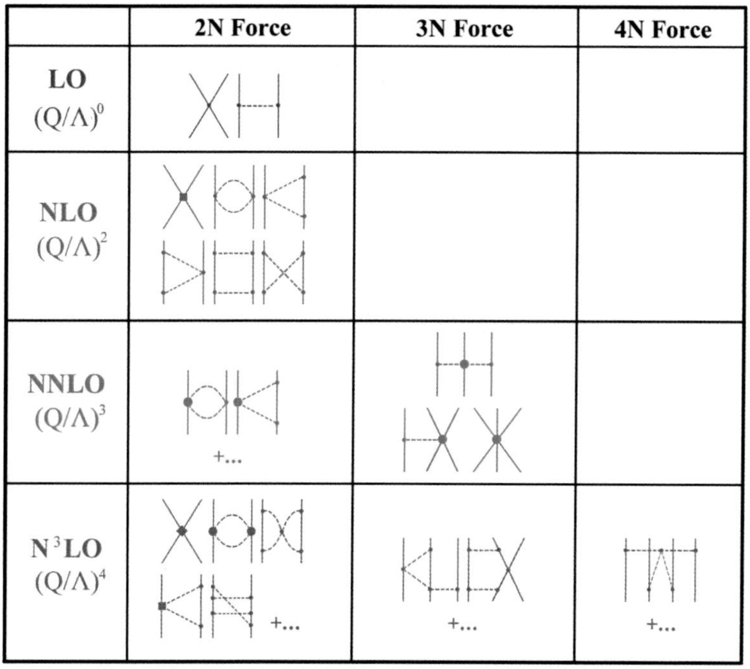

	2N Force	3N Force	4N Force
LO $(Q/\Lambda)^0$			
NLO $(Q/\Lambda)^2$			
NNLO $(Q/\Lambda)^3$			
N^{3}LO $(Q/\Lambda)^4$			

Fig. 2.15 Hierarchy of nuclear forces in ChEFT. The chiral expansion is classified in terms of powers of $\mu \equiv Q/\Lambda$, in which Q is an external momentum (the soft scale) close to the pion mass, and Λ is the chiral-symmetry breaking scale, LO, NLO, NNLO and N^{3}LO stand for the leading order, the next-to-leading order, the next-to-next-to-leading order and the next-to-next-to-next-to-leading order expansions, respectively. $\Lambda \approx 1$ GeV (the hard scale). Dashed lines represent pions and vertical solid lines are for nucleons. Filled circles show the coupling between nucleons and mesons, while filled squares stand for contact terms between nucleons

power ν of the ratio $\mu \equiv Q/\Lambda$. In the lowest order, known as leading order (LO, $\nu = 0$), the NN amplitudes are written as momentum-independent contact terms, represented by the four-nucleon-leg graph in the first row of Fig. 2.15. These contact terms take into account a short range nature of interaction mediated by heavy mesons which are not included in the theory explicitly. In the non-relativistic limit, the leading order NN Lagrangian has no derivative and reads

$$L_{NN}^{(LO)} = -C_S \overline{N}N\overline{N}N - C_T(\overline{N}\sigma N)(\overline{N}\sigma N), \tag{2.75}$$

where N is the nucleon field having only upper components of Dirac wave function, and $\overline{N} = N^\dagger$. In Eq. (2.75), σ is the Dirac spin matrix, and C_S (spin-singlet term) and C_T (spin-triplet term) are the low-energy constants, i.e., the parameters to be determined by fitting NN scattering data. The short-range part of NN interaction is determined by the Lagrangian (2.75) together with the contact terms of higher-order expansions.

Static one-pion exchange (OPE) also appears at LO as shown in the second diagram in the first row of the figure. The pseudo-vector coupling interaction in Eq. (2.71) is expressed by using Eq. (2.74) as

$$V_{pv} = \overline{\Psi}\frac{g_A}{2f_\pi}\gamma^\mu\gamma_5\boldsymbol{\tau}\cdot\partial_\mu\boldsymbol{\pi}\Psi. \tag{2.76}$$

In the non-relativistic limit, Eq. (2.76) is further rewritten as

$$V_{pv} = \overline{N}\frac{g_A}{2f_\pi}\boldsymbol{\tau}\boldsymbol{\sigma}\cdot\nabla\boldsymbol{\pi}N, \tag{2.77}$$

where we take the non-relativistic limit, $\partial_\mu \rightarrow \nabla$, and $\gamma^\mu\gamma_5 \rightarrow -\sigma$. The one-pion exchange potential can be obtained by the second-order perturbation with the pion propagator $D(q) = 1/(q^2 - m_\pi^2)$ with four component momentum $q^2 = q_\mu q^\mu = E^2 - |\mathbf{q}|^2$ (see the Feynman diagram at the left side of Fig. 2.8). The derivative on the pion field $\boldsymbol{\pi}$ with incoming momentum $\mathbf{q}$ gives $\nabla^j\pi = -iq^j\pi$. The one-pion exchange potential is then given in the momentum space as

$$\begin{aligned} V_\pi(\mathbf{q}) &= V_{pv}(1)^\dagger D(q) V_{pv}(2) \\ &= \frac{g_A}{2f_\pi}^2 \overline{N}(1)\sigma^i\tau^a iq^i N(1)\frac{\pi^{\dagger a}\pi^b}{q^2 - m_\pi^2}\overline{N}(2)\sigma^j\tau^b(-iq^j)N(2), \end{aligned}$$

where the sums of double indices such as (i, i), or (a, a) are implicit. The pion field satisfies the orthogonality condition $\pi^{\dagger a}\pi^b = \delta_{ab}$ and the four momentum q^2 is expressed as $q^2 = E^2 - |\mathbf{q}|^2 = -|\mathbf{q}|^2$ at the static limit $E = 0$. Finally the one-pion exchange potential is given in the momentum space as

$$V_\pi(\mathbf{q}) = -\frac{g_A^2}{2f_\pi}\overline{N}(1)\sigma^i\tau^aN(1)\frac{q_iq_j}{|\mathbf{q}|^2+m_\pi^2}\overline{N}(2)\sigma^j\tau^aN(2)$$

$$= -\left(\frac{g_A}{2f_\pi}\right)^2(\boldsymbol{\tau}_1\cdot\boldsymbol{\tau}_2)\frac{(\boldsymbol{\sigma}_1\cdot\mathbf{q})(\boldsymbol{\sigma}_2\cdot\mathbf{q})}{|\mathbf{q}|^2+m_\pi^2}, \qquad (2.78)$$

where the nucleon wave function $N(i)$ carries the spin, and isospin variables and satisfies the orthonormalization condition $\overline{N}(i)N(j) = \delta_{ij}$. The LO two-body interaction is then written in the coordinate space as

$$V^{(LO)}(\boldsymbol{r}) = (C_S + C_T\boldsymbol{\sigma}_1\cdot\boldsymbol{\sigma}_2)\delta(\boldsymbol{r}) - \left(\frac{g_A}{2f_\pi}\right)^2(\boldsymbol{\tau}_1\cdot\boldsymbol{\tau}_2)(\boldsymbol{\sigma}_1\cdot\boldsymbol{\nabla}_1)(\boldsymbol{\sigma}_2\cdot\boldsymbol{\nabla}_2)\frac{e^{-m_\pi r}}{4\pi r}, \qquad (2.79)$$

where g_A is the axial vector coupling constant,[9] while f_π is the pion decay constant which determines the lifetime of the pion. This is, of course, a rather crude approximation for the two-nucleon force, but accounts already for some important features. The OPE provides the tensor force, necessary to make the deuteron bound, and it explains the NN scattering amplitudes of high orbital angular momentum. At the LO, the contact interaction which contributes only in S waves provides the short-range and intermediate-range interaction.

Problem

2.6 Derive the LO two-body interaction in the coordinate space as expressed in Eq. (2.79) from a Fourier transform of the leading order ChEFT Lagrangian (2.75) and the pion exchange interaction (2.78) expressed in momentum space.

In the next order, $\nu = 1$, all contributions vanish due to parity and time-reversal invariance. Therefore, the next-to-leading order (NLO) is $\nu = 2$. Two-pion exchange (TPE) occurs for the first time as shown in Fig. 2.15 and, thus, the creation of a more sophisticated description of the intermediate-range interaction is starting here. At this order, the lowest order πNN and $\pi\pi NN$ vertices are allowed which is why the leading TPE is rather weak. Furthermore, there are several contact terms shown by the four-nucleon-leg graph with a black square, which contribute to S, P and higher-multipole waves in opposition to the LO which contributes only to S waves. The operator structure of these contact terms include a spin-orbit term besides central, spin-spin, and tensor terms. Especially the spin-orbit term plays the important role to give shell structure of single-particle orbits in the one-body potential. Thus, essentially all spin-isospin structures necessary to describe the two-nucleon force phenomenologically have been generated at this order. The force should also verify fundamental symmetries such as invariance after translation or rotation, symmetry

---

[9] The pion is a pseudoscalar meson with $J^\pi = 0^-$ so that the coupling between nucleon and pion fields would be either the pseudoscalar or the pseudovector couplings due to the Lorenz invariant condition. For the pseudoscalar coupling, the pion field π appears in the Hamiltonian density while the derivative of pion field $\partial_\mu\pi$ will appear in the pseudovector coupling case. It is known that the two couplings give the same result for the nuclear force. Here the pseudovector coupling is adopted since it guarantees the chiral symmetry of the Lagrangian.

under the exchange of two nucleons. Due to (LO+NLO) contributions of ChEFT, the interaction between two nucleons can be described in a form

$$
\begin{aligned}
V^{(\mathrm{LO+NLO})}&(\mathbf{r}, \mathbf{q}, \boldsymbol{\sigma}_1, \boldsymbol{\sigma}_2, \boldsymbol{\tau}_1, \boldsymbol{\tau}_2) \\
&= V_0(r) + V_\sigma(r)\boldsymbol{\sigma}_1 \cdot \boldsymbol{\sigma}_2 + V_\tau(r)\boldsymbol{\tau}_1 \cdot \boldsymbol{\tau}_2 + V_{\sigma\tau}(r)\boldsymbol{\sigma}_1 \cdot \boldsymbol{\sigma}_2\boldsymbol{\tau}_1 \cdot \boldsymbol{\tau}_2 \\
&+ V_T(r)(3(\boldsymbol{\sigma}_1 \cdot \mathbf{r})(\boldsymbol{\sigma}_2 \cdot \mathbf{r})/r^2 - \boldsymbol{\sigma}_1 \cdot \boldsymbol{\sigma}_2) + V_{LS}(r)(\boldsymbol{\sigma}_1 + \boldsymbol{\sigma}_2) \cdot \mathbf{L} \\
&+ V_{Ls}(r)(\boldsymbol{\sigma}_1 \cdot \mathbf{L})(\boldsymbol{\sigma}_2 \cdot \mathbf{L}) + V_{qs}(r)(\boldsymbol{\sigma}_1 \cdot \mathbf{q})(\boldsymbol{\sigma}_2 \cdot \mathbf{q})
\end{aligned}
\tag{2.80}
$$

where $\mathbf{q}$ is the relative momentum of the two interacting nucleons and $\mathbf{L}$ is the angular momentum $\mathbf{L} = (\mathbf{r_1} - \mathbf{r_2}) \times (\mathbf{P_1} - \mathbf{P_2})$. The chiral NN interactions have also non-local terms together with Eq. (2.80).

The first four terms describe the central part of the nuclear force that depends solely on the distance between the nucleons and their relative spin and isospin, including spin-spin (V_σ), isospin-isospin (V_τ) and spin-isospin ($V_{\sigma\tau}$) terms. The tensor term V_T has the particularity to mix spin and angular momentum states. The spin-orbit term of the interaction V_{LS} also depends on the relative momentum of the two nucleons. The last two terms are derived from the second-order NN Lagrangian and are believed to have less importance than the others due to a quadratic dependence on the angular or kinetic momenta in the low L and the low q limits but they are formally not excluded by symmetry arguments. All these terms play a crucial role in the nuclear structure and its evolution across the nuclear chart (see Chap. 3).

The main deficiency at this stage of the development, i.e., LO+NLO, is an insufficient intermediate-range attraction. This problem is finally fixed at the third order (ν = 3), i.e., at the next-to-next-to-leading order (N²LO). The three-body force appears for the first time in N²LO diagrams. The TPE involves the two-derivative $\pi\pi NN$ vertices denoted by a filled square-shaped vertex in Fig. 2.15. These vertices represent correlated TPE as well as intermediate Δ-isobar contributions. It is well known from the meson phenomenology of nuclear forces that these two contributions are crucial for a realistic and quantitative TPE model. Consequently, the TPE now has a realistic amplitude and describes the intermediate-range attraction of the nuclear force properly. The next-to-next-to-next-to-leading order (N³LO) will provide not only two-body and three-body forces, but also four-body forces. The ChEFT now provides the NN interaction in the functional form (2.80) with the proper coupling strength which describes well the NN scattering phase shifts up to higher angular momenta.

Phase shifts of np scatterings are calculated from NN potentials at different orders of ChEFT. The agreement becomes better when higher order of ChEFT are taken into account. The red solid line shows in Fig. 2.16 the best agreement with empirical phase shifts of all partial waves. The number of parameters of two-body interactions up to the order of N³LO is 24, while the other realistic interactions, Nijmegen and CD-Bonn potentials, have 35 and 38 parameters, respectively. For nuclear structure calculations, the ChEFT are often adapted so called ab initio type calculations (see Chap. 3 for detailed discussion) with an aid of heavy duty computing power. All the coupling constants of two-body interactions up to the order N³LO in the ChEFT

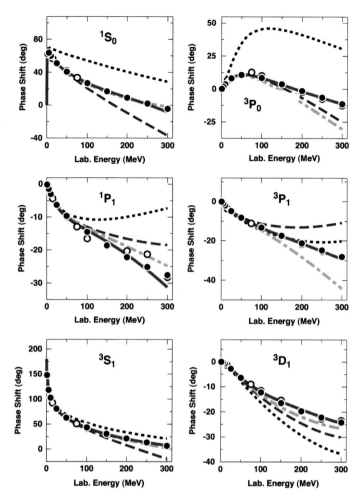

Fig. 2.16 Phase shifts of *np* scatterings calculated from *NN* potentials at different orders of ChEFT. The black dotted line is LO, the blue dashed line is NLO, the green dashed-dotted line is N^2LO and the red solid line is N^3LO. Data are indicated by dots and circles. Figure reprinted with permission from [12]. ©2011 by Elsevier

are well constrained by the phase shifts and scattering lengths of the *NN* scatterings except two parameters defining the three-body interactions illustrated by the right-most two diagrams in the N^2LO row. These two parameters are constrained by the binding energies of three-body systems ^{3}He and ^{3}H (triton).

After Yukawa's hypothesis on strong interaction, a systematic approach to the *NN* interaction was proposed by Mitsuo Taketani and his collaborators in 1951. The basic idea originated from the Yukawa's meson exchange potential and the hierarchy of meson-exchange diagrams given in Taketani diagram in Fig. 2.17. According to this concept, the long range part, region I, was determined by one-pion exchange,

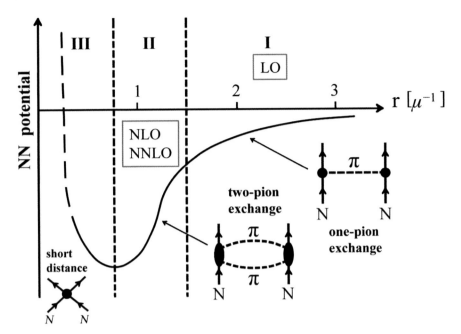

Fig. 2.17 Taketani diagram. Hierarchy of scaling in nuclear forces in pion-exchange potential and in expansion of ChEFT. The unit of distance r is the pion Compton wave length $\mu^{-1} = \hbar/m_\pi\,c \approx 1.4$ fm

while the region II at intermediate distance was dominated by the two-pion exchange. The detailed behaviour of the interaction in the short distance region III remained unclear at that time. It is surprising that Taketani's strategy for *NN* interaction is in close analogy to the QCD and ChEFT approaches to describe *NN* interactions. In the high momentum region above 1 GeV ($R < 1$ fm), QCD gives a short-range behaviour of *NN* interaction as the interaction between quarks and gluons. At low momentum scales considerably smaller than 1 GeV ($R > 1$ fm), the ChEFT becomes a useful model and describes successfully the long-range and the medium-range *NN* potential. In the ChEFT, the short-range part is determined by the low-energy constants C_S and C_T in Eq. (2.79) in the LO expansion. There are additional 7 low-energy constants in the NLO expansion. These constants appear by the regularization procedure, and should be determined by the nucleon-nucleon scattering data. In this way, Taketani's concept is revived in the modern theory of *NN* interaction based on QCD.

Summary

The nucleon-nucleon (NN) interaction is repulsive at short range (< 1 fm), and attractive in the typical range from 1 to 2 fm.

The one pion exchange potential (OPEP) is the main part of nucleon-nucleon potential which is responsible for nuclear binding. The OPEP has a long-range nature and is written in the momentum space as,

$$V_\pi(\mathbf{q}) = - \left(\frac{g_A}{2f_\pi}\right)^2 (\boldsymbol{\tau}_1 \cdot \boldsymbol{\tau}_2) \frac{(\boldsymbol{\sigma}_1 \cdot \mathbf{q})(\boldsymbol{\sigma}_2 \cdot \mathbf{q})}{|\mathbf{q}|^2 + m_\pi^2},$$

where f_π is the pion decay constant and g_A is the axial vector coupling constant. The two pion exchange potential becomes important in the intermediate range of NN potential.

Effective NN interactions can be empirically constrained from precision NN elastic scattering experiments and the determination of *phase shifts*. The elastic scattering amplitude is written as

$$f_{ii}^{(+)} = \sum_l \frac{2l+1}{2ik} (S_l(k) - 1) P_l(\theta),$$

where $S_l(k)$ is the S-matrix and $P_l(\theta)$ is the Legendre polynomials with rank l. The S-matrix is described by the phase shift δ_l as

$$S_l(k) = \exp(2i\delta_l).$$

Three-body forces stem from the *intrinsic* structure of nucleons and plays an important role on microscopic study of nuclei far from stability line. Empirical evidences of the three-body forces are found in the GFMC calculations of few-body systems (Fig. 2.12) and also the elastic three-body scatterings of (d, p) and (n, p) at the intermediate energies (see also discussions on the shell evolution in Sect. 6.3.4 in Chap. 6). Three-body forces are *induced* by low-momentum regularization of effective interactions in renormalized theories.

Chirality is an important concept in QED and QCD. Chirality is a mirror symmetry of an object and classified to be the left-handed and the right-handed states by the eigenvalue of chiral operator γ_5

$$\gamma_5 \psi_R = +\psi_R, \quad \gamma_5 \psi_L = -\psi_L.$$

For a massless particle and the object with light speed, the chirality is equivalent to the helicity.

Chiral effective field theory (ChEFT) is an effective theory based on a perturbative treatment of chiral Lagrangian of nucleons and mesons instead of quarks and gluons. ChEFT follows a *power counting* scheme where the interaction is perturbatively expanded in powers of a ratio between an external momentum Q or pion mass (soft scale) and the chiral symmetry breaking scale $\Lambda \sim 1$ GeV (the hard scale). The chiral field in the isospin SU(2) framework is introduced as

$$U(x) = \exp\left(\frac{i}{f_\pi} \boldsymbol{\tau} \cdot \boldsymbol{\pi}(x)\right) = \exp\left(\frac{i}{f_\pi}\begin{pmatrix} \pi^0 & \sqrt{2}\pi^+ \\ \sqrt{2}\pi^- & -\pi^0 \end{pmatrix}\right), \qquad (2.81)$$

where f_π is the pion decay constant, $\boldsymbol{\tau} = (\tau_x, \tau_y, \tau_z)$ are the Pauli isospin matrices, and $\boldsymbol{\pi}(x) = (\pi_x, \pi_y, \pi_z)$ are the pion fields. The phase shifts of nucleon nucleon scattering are well reproduced by NN interactions derived up to the next-to-next-to-next leading expansion order (N^3LO) of ChEFT.

In ChEFT, the renormalization procedure takes place to avoid the Q dependence of the observables, which is called renormalization-group (RG) invariance. The contact terms take into account short-range physics which is not included explicitly in the theory. These short-range interactions are mediated by heavy mesons in the meson-exchange theory.

Solutions of Problems

2.1 The volume of a spherical pea with a diameter of 4 mm is $V = 4/3\pi r^3 \approx 0.033$ cm^3. The nuclear density at saturation is $\rho = 0.16$ fm^{-3}. The weight of a nucleon is $m_N = 940$ MeV/c^2. The weight of a pea $M_{pea} = V \times \rho \times m_N$ MeV/c$^2 = 5.10^{33}$ MeV/c^2. This weight is transformed to gram divided by 6×10^{26}. Then, the mass of a pea at nuclear saturation density is $M_{pea} \sim 8 \times 10^6$ tons. The weight of Eiffel tower is approximately $M_{Eiffel} = 8000$ tons. Thus, the weight of pea with nuclear density is equivalent to (8 Megatons / 8 kilotons) = 1000 Eiffel towers.

2.2 (1) The term $1/|\mathbf{R} + (\mathbf{r_2} - \mathbf{r_1})|$ in H_1 of Eq. (2.2) is expanded up to the order of $O(1/R^3)$ as

$$\begin{aligned}
\frac{1}{|\mathbf{R} + (\mathbf{r_2} - \mathbf{r_1})|} &= \frac{1}{\sqrt{(R + z_2 - z_1)^2 + (x_2 - x_1)^2 + (y_2 - y_1)^2}} \\
&= \frac{1}{R\left[1 + \frac{2(z_2 - z_1)}{R} + \frac{(\mathbf{r_2} - \mathbf{r_1})^2}{R^2}\right]^{1/2}} \\
&\simeq \frac{1}{R}\left[1 - \frac{1}{2}\left[\frac{2(z_2 - z_1)}{R} + \frac{(\mathbf{r_2} - \mathbf{r_1})^2}{R^2}\right] + 3\frac{(z_2 - z_1)^2}{R^2}\right].
\end{aligned}$$
$$(2.82)$$

The terms $1/|\mathbf{R} + \mathbf{r_2}|$ and $1/|\mathbf{R} - \mathbf{r_1}|$ are also expanded up to the order of $O(1/R^3)$ in a similar manner. Keeping the terms of order expansion up to $O(1/R^3)$, H_1 is expressed as

$$\begin{aligned}
H_1 &\simeq \frac{e^2}{R}\left(1 + \left[1 - \frac{1}{2}\left(\frac{2(z_2 - z_1)}{R} + \frac{(\mathbf{r_2} - \mathbf{r_1})^2}{R^2}\right) + 3\frac{(z_2 - z_1)^2}{R^2}\right]\right. \\
&\quad \left. - \left[1 - \frac{1}{2}\left(\frac{z_2}{R} + \frac{r_2^2}{R^2}\right) + 3\frac{z_2^2}{R^2}\right] - \left[1 + \frac{1}{2}\left(\frac{z_1}{R} - \frac{r_1^2}{R^2}\right) + 3\frac{z_1^2}{R^2}\right]\right) \\
&\simeq \frac{e^2}{R^3}(x_1 x_2 + y_1 y_2 - 2z_1 z_2).
\end{aligned}$$
$$(2.83)$$

(2) The numerator of lowest-order term in Eq. (2.83) corresponds to the interaction of two electric dipoles er_1 and er_2 separated by the distance $\mathbf{R}$. The unperturbed Hamiltonian H_0 has the solution of products of two s-wave eigenstates,

$$u_{gs}(\mathbf{r}_1, \mathbf{r}_2) = u_{l_1=0}(\mathbf{r}_1)u_{l_2=0}(\mathbf{r}_2) \tag{2.84}$$

for the ground state. Here $u_{l=0}(\mathbf{r})$ is the lowest-energy eigenstate of Coulomb potential. Since the ground state configuration of two electrons has the total angular momentum $L = l_1 + l_2 = 0$, the first-order term $E^{(1)} = \langle u_{gs}|H_1|u_{gs}\rangle = 0$. The second-order perturbation term

$$
\begin{aligned}
E^{(2)}(R) &= \sum_k \frac{|\langle k|H_1|u_{gs}\rangle|^2}{E_{gs}-E_k} \\
&= \frac{e^4}{R^6}\sum_k \frac{|\langle k|x_1x_2+y_1y_2-2z_1z_2|u_{gs}\rangle|^2}{E_{gs}-E_k}
\end{aligned} \tag{2.85}
$$

can be non-zero for the excited states k having the angular momenta $l_1 = 1$ and $l_2 = 1$ for the two electrons. One can see that the interaction (2.85) varies as $1/R^6$ and it is attractive since $E_{gs} - E_k < 0$.

2.3 The evaluation of $\nabla^2\phi(\mathbf{r})$ by use of the potential form (2.8) gives

$$
\nabla_{\mathbf{r}}^2 \int \frac{\exp(-k|\mathbf{r}-\mathbf{r}'|)}{|\mathbf{r}-\mathbf{r}'|}\rho(\mathbf{r}')d\mathbf{r}'
$$

$$
= \int [\nabla^2(\exp(-k|\mathbf{r}-\mathbf{r}'|))\frac{1}{|\mathbf{r}-\mathbf{r}'|} + 2\nabla(\exp(-k|\mathbf{r}-\mathbf{r}'|))\cdot\nabla\frac{1}{|\mathbf{r}-\mathbf{r}'|}
$$

$$
+ \exp(-k|\mathbf{r}-\mathbf{r}'|)\nabla^2(\frac{1}{|\mathbf{r}-\mathbf{r}'|})]\rho(\mathbf{r}')d\mathbf{r}'
$$

$$
= \int [(k^2\exp(-k|\mathbf{r}-\mathbf{r}'|) - k\frac{2}{|\mathbf{r}-\mathbf{r}'|}\exp(-k|\mathbf{r}-\mathbf{r}'|))\frac{1}{|\mathbf{r}-\mathbf{r}'|}
$$

$$
+ k\frac{2}{|\mathbf{r}-\mathbf{r}'|^2}\exp(-k|\mathbf{r}-\mathbf{r}'|) - 4\pi\delta(\mathbf{r}-\mathbf{r}')\exp(-k|\mathbf{r}-\mathbf{r}'|)]\rho(\mathbf{r}')d\mathbf{r}'
$$

$$
= k^2\int \frac{\exp(-k|\mathbf{r}-\mathbf{r}'|)}{|\mathbf{r}-\mathbf{r}'|}\rho(\mathbf{r}')d\mathbf{r}' - 4\pi\rho(\mathbf{r}). \tag{2.86}
$$

This is exactly the Klein–Gordon equation (2.7).

2.4 Even powers of $\mathbf{p}$ and $\mathbf{r}$ are conserved the symmetries required for the NN interaction in Sect. 1.2 as well as $\sigma_1 \cdot \sigma_2$ term. On the other hand, $\mathbf{r}$ is not parity invariant and $\mathbf{L}$ violates the time reversal invariance. Therefore, $\mathbf{r}^2+\mathbf{p}^2$ and $r^2\sigma_1\cdot\sigma_2$ are allowed in the two-body Hamiltonian, but $\mathbf{r}\cdot\mathbf{L}$ is not allowed.

2.5 The eigenvalue equation

$$\gamma_5\psi = x\psi \tag{2.87}$$

is equivalent to the following matrix formulation

$$\begin{vmatrix} -x & 0 & 1 & 0 \\ 0 & -x & 0 & 1 \\ 1 & 0 & -x & 0 \\ 0 & 1 & 0 & x \end{vmatrix} = 0. \tag{2.88}$$

The determinant (2.88) gives

$$x^4 - 2x^2 + 1 = 0, \tag{2.89}$$

which is solved with the eigenvalues $x = \pm 1$.

2.6 The leading order NN Lagrangian has no derivative term and reads

$$L_{NN}^{(LO)} = -C_S \bar{N} N \bar{N} N - C_T (\bar{N} \sigma N)(\bar{N} \sigma N) \tag{2.90}$$

where N is the nucleon field and $\bar{N} = N^\dagger$ is its adjoint. A Fourier transform of $L_{NN}^{(LO)}$ can be performed by using the relation

$$\frac{1}{(2\pi)^3} \int e^{i\mathbf{q}\mathbf{r}} d\mathbf{q} = \delta(\mathbf{r}). \tag{2.91}$$

The contact term in the coordinate space becomes

$$V_{contact}^{(LO)}(\mathbf{r}) = -\frac{1}{(2\pi)^3} \int e^{i\mathbf{q}\mathbf{r}} d\mathbf{q} L_{NN}^{(LO)} = (C_S + C_T \boldsymbol{\sigma}_1 \cdot \boldsymbol{\sigma}_2)\delta(\mathbf{r}). \tag{2.92}$$

The coupling constants C_S and C_T are free parameters to be fixed by experimental data. Unresolved physics of ChEFT is contained in these terms. The one-pion exchange potential is written in the momentum space as

$$V_\pi(\mathbf{q}) = -\left(\frac{g_A}{\sqrt{2}f_\pi}\right)^2 (\boldsymbol{\tau}_1 \cdot \boldsymbol{\tau}_2)\frac{(\boldsymbol{\sigma}_1 \cdot \mathbf{q})(\boldsymbol{\sigma}_2 \cdot \mathbf{q})}{|\mathbf{q}|^2 + m_\pi^2}. \tag{2.93}$$

The Fourier transform of $V_\pi(\mathbf{q})$ is given by

$$\begin{aligned} V_\pi(\mathbf{r}) &= \frac{1}{(2\pi)^3} \int e^{i\mathbf{q}\mathbf{r}} V_\pi(\mathbf{q}) d\mathbf{q} \\ &= \left(\frac{g_A}{\sqrt{2}f_\pi}\right)^2 (\boldsymbol{\tau}_1 \cdot \boldsymbol{\tau}_2)(\boldsymbol{\sigma}_1 \cdot \boldsymbol{\nabla}_1)(\boldsymbol{\sigma}_2 \cdot \boldsymbol{\nabla}_2)\frac{1}{(2\pi)^3} \int e^{i\mathbf{q}\mathbf{r}} \frac{d\mathbf{q}}{|\mathbf{q}|^2 + m_\pi^2}. \end{aligned} \tag{2.94}$$

The integral on r.h.s will be performed as

$$I(r) \equiv \frac{1}{(2\pi)^3} \int e^{i\mathbf{q}\mathbf{r}} \frac{d\mathbf{q}}{|\mathbf{q}|^2 + m_\pi^2}$$

$$= \frac{1}{(2\pi)^3} \int_0^\infty dq \int_{-1}^1 d(\cos\theta) \int_0^{2\pi} d\phi q^2 e^{iqr\cos\theta} \frac{1}{q^2 + m_\pi^2}$$

$$= -\frac{i}{8\pi^2 r} \int_{-\infty}^\infty dq \frac{q}{q^2 + m_\pi^2} (e^{iqr} - e^{-iqr})$$

$$= \frac{1}{4\pi r} e^{-m_\pi r}. \qquad (2.95)$$

The integral of the q variable is performed to take the residue of upper (lower) half complex plane for the first (second) term

$$\int_{-\infty}^\infty dq \frac{q}{q^2 + m_\pi^2} (e^{iqr} - e^{-iqr}) = \frac{1}{2} \int_{-\infty}^\infty dq \left(\frac{1}{q - im_\pi} + \frac{1}{q + im_\pi} \right) (e^{iqr} - e^{-iqr})$$

$$= \frac{1}{2} \int_{-\infty}^\infty dq \left(2\pi i\delta(q - im_\pi)e^{iqr} - (-2\pi i)\delta(q + im_\pi) \right) e^{-iqr} = 2\pi i e^{-m_\pi r}. \qquad (2.96)$$

Finally the one-pion exchange potential in the coordinate space reads

$$V_\pi(\mathbf{r}) = -(\frac{g_A}{\sqrt{2}f_\pi})^2 (\boldsymbol{\tau}_1 \cdot \boldsymbol{\tau}_2)(\boldsymbol{\sigma}_1 \cdot \boldsymbol{\nabla}_1)(\boldsymbol{\sigma}_2 \cdot \boldsymbol{\nabla}_2) \frac{1}{4\pi r} e^{-m_\pi r}. \qquad (2.97)$$

Further Readings

"Nuclear effective field theory: status and perspectives" by H.-W. Hammer, S. König, and U. van Kolck, Rev. Mod. Phys. 92, 025004 (2020), gives an overview of the state of the art of nuclear effective field theory.

"Modern theory of nuclear forces" by E. Epelbaum, H.-W. Hammer, and U.-G. Meißner, Rev. Mod. Phys. 81, 1773 (2009).

"High-precision nuclear forces from chiral EFT: state-of-the-art, challenges, and outlook" by E. Epelbaum, H. Krebs and P. Reinert, Front. Phys. 17 (2020), outlines the conceptual foundations of nuclear chiral effective field theory, and reviews a new generation of nuclear forces derived in chiral effective field theory using the recently proposed semi-local regularization method.

References

1. F.G. Hasert et al., Phys. Lett. B **46**, 138 (1973)
2. G. Arnison et al., Phys. Lett. B **122**, 103 (1983)
3. G. Arnison et al., Phys. Lett. B **129**, 273 (1983)
4. B.P. Abbott et al., Phys. Rev. Lett. **116**, 061102 (2016)
5. B.P. Abbott et al., Phys. Rev. Lett. **119**, 161101 (2017)
6. H. Yukawa, Proc. Phys.-Math. Soc. Jpn. **17**, 48 (1935)
7. J.-I. Fujita, H. Miyazawa, Prog. Theo. Phys. **17**, 360 (1957)
8. S. Weinberg et al., Physica **96A**, 327 (1979)
9. S. Weinberg et al., Phys. Lett. B **251**, 288 (1990)
10. J. Carlson et al., Rev. Mod. Phys. **87**, 1067 (2015)
11. K. Sekiguchi et al., Few-Body Sys. **54**, 911 (2013)
12. R. Machleidt, D.R. Entem, Phys. Rep. **503**, 1 (2011)

Chapter 3
Nuclear Structure Theory

Abstract We describe various nuclear structure theories from mean field models to modern ab initio approaches. Firstly, mean field models are introduced to describe nuclear ground states and also collective excitations including pairing correlations for the open-shell nuclei. As beyond mean field models, cluster models, modern shell model approaches are also discussed as well as advanced ab initio models.

Keywords Mean-field and beyond-mean-field approximations · Shell models · Cluster models · Ab initio methods

3.1 Preamble

The association of several nucleons close together may lead to a bound nucleus. A given association of nucleons, for example 6 neutrons and 6 protons making a ^{12}C nucleus, can end up into different quantum wave functions, all corresponding to different states of ^{12}C. The wave function that minimizes the total energy of the nucleus is called its *ground state*. How can one describe such wave functions? It is mathematically complex and approximations are necessary. One famous approach to nuclear structure is the mean field model. The nuclear shell model was initially considered in 1930s in analogy to the model for electrons in atoms. The shell model is based on the very important assumption that nucleons in nuclei can be described fairly well as independent particles evolving in a mean attractive potential (the mean field potential) created by all other nucleons. The Hartree–Fock (HF) model implements this idea of the mean field starting from many-body interactions and called a *self-consistent* mean field model.

The mean field gives rise to the single-particle orbitals and shell structure in the single-particle energies, which is the origin of the shell model; nucleons lie on a limited number of single-particle orbitals obeying the Pauli exclusion principle. This assumption is absolutely not intuitive and first abandoned because it was difficult to imagine how nucleons could act as independent particles while the force acting between nucleons is very strong. On the other hand, observations led physicists to admit that nuclear excitations could be fairly well reproduced by an independent-

© Springer Nature Singapore Pte Ltd. 2021
A. Obertelli and H. Sagawa, *Modern Nuclear Physics*, UNITEXT for Physics,
https://doi.org/10.1007/978-981-16-2289-2_3

particle-based model. There was a dilemma that held physicists for about 60 years in the first half of the 20th century. Finally it was understood that quantum physics and the Pauli principle could explain how nucleons exhibit an independent particle behavior in a mean field.

In this chapter, we first give a bird's eye view of theoretical models for nuclei in Sect. 3.2. Section 3.3 is dedicated to the simplest nucleus, the deuteron, whose wave function can be reduced to a one-body variable problem and can be solved exactly. We introduce effective interactions in Sect. 3.4. We then proceed to a progressive approach starting from a mean field one-body potential description in Sect. 3.5, moving to energy density functional theory and non-relativistic and relativistic mean field models in Sect. 3.6. One of the most important ingredients in nuclear collective motions is pairing correlations. We discuss how to take into account the pairing correlations in nuclear dynamics in Sect. 3.6.3. The generator coordinate model (GCM) and cluster models are also studied as models beyond the mean field approach in Sects. 3.7.3 and 3.10, respectively. Modern shell model approaches are discussed including stochastic Monte Carlo approaches in Sects. 3.8 and 3.9. We finally give an overview of the ab initio techniques to address the question of the nuclear many-body problem in Sect. 3.11.

3.2   A Bird's-Eye View of Nuclear Theory

This section presents an overview of standard and modern theoretical approaches to describe the many-body nuclear problem. Details on the formalism and examples of applications are given in the next sections. For interested readers, key references for each model are given.

1. **Nuclear mean field model** The mean field is an approximation in which each particle of a system composed of A nucleons moves in an external field (*mean field*) generated by the remaining $A - 1$ nucleons. In the HF theory, the mean field is constructed self-consistently through NN interactions. The fundamental assumption of the HF theory is an anti-symmetrized product of independent particle wave functions for A-body wave function, the so-called Slater determinant

$$\Psi_{\mathrm{HF}}(\mathbf{r}_1, \mathbf{r}_2, \cdots, \mathbf{r}_A) = \mathcal{A}\{\phi_1\phi_2 \cdots \phi_A\}, \qquad (3.1)$$

where $\mathcal{A}$ is the anti-symmetrization operator and ϕ_i is a single-particle wave function. The single-particle wave function $\phi_i(\mathbf{r})$ is determined by an application of variational principle. The variational principle states that the energy expectation value of the Hamiltonian is stational for a small variation of a single-particle wave function $\phi_i(\mathbf{r})$

$$\delta\langle\Psi_{\mathrm{HF}}|H|\Psi_{\mathrm{HF}}\rangle = \langle\delta\Psi_{\mathrm{HF}}|H|\Psi_{\mathrm{HF}}\rangle = 0. \qquad (3.2)$$

The variation gives the HF equation for the wave function $\phi_i(\mathbf{r})$ and its eigenstate and eigenenergy are called HF single-particle wave function and HF single-particle energy, respectively. The Hartree–Fock (HF) theory has been quite successful to describe the ground state properties of many nuclei using effective interactions.

The HF model is extended to a time-dependent HF (TDHF) theory to study excited states of nuclei. The TDHF is equivalent to random phase approximation (RPA) in a small amplitude limit of oscillations. The self-consistent TDHF and HF+RPA models are also successful models to describes the ground state and the excited states in the same framework.

A different model for the mean-field approach was proposed by John Dirk Walecka based on the relativistic Lagrangian of σ and ω meson-nucleon couplings. This approach is a relativistic version of a phenomenological description of the nuclear many-body problem. The model is a Hartree model and called the relativistic mean field (RMF) model. The relativistic Hartree–Fock (RHF) model was also proposed introducing π meson-nucleon coupling in the relativistic Lagrangian. The theory is fully Lorentz invariant and the nucleons are treated as Dirac particles with four component spinors. The nucleons interact between themselves by the exchange of effective point-like particles, called mesons. They are characterized by their quantum numbers, spin, parity and isospin. The masses of mesons and the meson-nucleon couplings are determined in such a way to reproduce experimental data at the best. This model involves few mesons and the number of parameters is claimed to be less than other mean field models. The Walecka model was further refined by introducing the ρ meson-nucleon coupling and the non-linear σ meson field. One of the advantage of the RMF approach is the spin-orbit interaction, which comes out naturally as a relativistic effect of nucleons.

The HF theory was invented to describe the nucleus by non-interacting independent particle wave functions. In other words, the theory separates the realistic Hamiltonian into two parts

$$H = H_{\text{HF}} + V_{\text{res}}, \tag{3.3}$$

where H_{HF} is adopted for the HF calculations and used to set up a basis for more complicated calculations which will accommodate the residual interaction V_{res}. In some cases, the residual interaction V_{res} does not destroy the independent-particle picture because of a large energy gap in the single-particle spectrum. Closed shell nuclei have such a large energy gap near the Fermi surface so that the HF description is fairly good. On the other hand, in open-shell nuclei, whose energy gap is small, the independent particle structure might be largely modified. This is precisely the situation for "*superfluid*" nuclei[1] in which the pairing interaction plays a key role. Under the influence of pairing interaction, two particles with spins of

---

[1] Superfluidity, the frictionless flow, was observed in liquid helium at temperatures near absolute zero $(-273.15°C$, or $-459.67°F)$ in 1937 by a russian physicist, P. L. Kapitsa. The superfluid phase (or similarly the superconducting phase) in nuclei was claimed form the observation of the large energy gap in the energy spectra of even-even nuclei by Aage Bohr, Ben R. Mottelson and D. Pines in 1957 just after the BCS (Bardeen-Cooper-Schrieffer) model was proposed to the metallic superconductivity. The superfluid phase in nuclei was further confirmed by the values of moments of inertia for deformed nuclei.

opposite directions will couple to the total angular momentum zero. The BCS (Bardeen-Cooper-Schrieffer) approximation is commonly adopted to accommodate the pairing correlations both in the ground and the excited states. The quasiparticle is introduced in the BCS model and the pairing gap is obtained to characterize the superfluidity phase of nuclear ground states. The HF-Bogolyubov (HFB) approximation is a generalization of the BCS model and it makes possible to calculate the loosely bound states and also the effect of continuum unbound states in the HF and the RPA calculations.

Another important correlation in V_{res} is the quadrupole-quadrupole interaction, which may induce a quadrupole deformation in its intrinsic frame. Nuclear deformation is discussed in Chap. 6.

2. **Beyond mean field models** To calculate many-body correlations beyond the mean field, several sophisticated models have been proposed. Particle-vibration coupling (PVC) was introduced to take into account both particle and vibration degrees of freedom in the calculations of single-particle states and also the width of Giant resonances. A straightforward extension of the RPA is the second RPA and the extended RPA models in which higher order effects beyond RPA are taken into account. Another advanced model is the generator coordinate method (GCM), which introduces multi-Slater determinants, for example, of different deformed single-particle basis as the basic states, while the HF and RPA calculations are performed with a single-Slater determinant. The angular momentum projection is necessary to obtain excitation spectra of deformed intrinsic states, which are mixtures of various angular momentum states. To describe many-body correlations such as cluster correlations, molecular dynamical models have been developed. An advanced version is anti-symmetrized molecular dynamics (AMD) and another one is fermionic molecular dynamics (FMD). The two models employ anti-symmetrized many-body wave functions consisting of Gaussian wave packets. The AMD employs effective phenomenological interactions, while the FMD adopts the effective interaction created by the Unitary Correlation Operator Method (UCOM) from realistic interactions.

3. **Nuclear shell model**

   The nuclear shell model was proposed by Maria Goeppert Mayer and Hans Jensen in 1949 to understand the magic numbers for the stability of nuclei. The independent-particle shell model is based on a mean field central potential and a spin-orbit one-body potential. The latter plays a key role to induce the magic numbers in nuclei.

   In the interacting shell model, particles interact with each other through the residual interaction in a limited configuration space. Typical phenomenological interactions fitted to energy spectra and transition moments were proposed by Cohen and Kurath for p-shell nuclei, and by Brown and Wildenthal for sd-shell nuclei. The quantum Monte Carlo approach based on Feynman path-integral theory is adopted for shell model calculations. To avoid time-consuming diagonalization of matrix of large dimensionality, statistical models have been proposed. One of such methods with realistic interactions is the *Variational Shell Model (VSM)*. The Monte Carlo method is also adopted in the *Shell Model Monte Carlo (SMMC)* model, in which the phenomenological interactions are used to study properties of nuclei larger than p-shell ones. Statistical generations of shell model states make possible the compression of the Hilbert space to be diagonalized efficiently, introducing auxiliary field. Another approach is the *Monte Carlo Shell Model (MCSM)*, which is a combined model of the generator coordinate method and the auxiliary field. A more ambitious approach for shell model calculations is the *no-core* shell model relying on modern high-performance computers. The model adopts the full model space without any core and all particles are active in the model space. The effective interactions adopted are derived from realistic interactions by similarity transformation methods, or the chiral effective interactions based on chiral field theory for strong interactions.

4. **Ab initio models**

   The *ab initio* theory is well established in atomic and molecular physics calculations governed by electromagnetic interactions. Ab initio means "from first principles", implying that the only inputs into an ab initio calculation are physical constants: the electric charges, electron mass and Plank constant in molecular dynamics. In nuclear dynamics with strong interactions, they are the quark-gluon coupling and quark masses. The ab initio quantum chemistry methods solve the Schrödinger equation of electrons, which is specified by the positions and charges of the nuclei and the number of electrons, in order to obtain electron densities, total energies, the dispersion relation (band structure) and other properties of the system. The ab initio theory for the strong interaction is quantum chromodynamics (QCD) of quarks and gluons. There have been many attempts to obtain NN interactions and to solve few-body problems in nuclear physics based on QCD on the Lattice (LQCD). The progress during last decades is impressive, but still it is not mature enough to be applied to many-body calculations. Presently, ab initio approaches in nuclear physics are characterized by the following requirements:

   - All nucleons are active.
   - The Pauli principle is exactly taken into account in the many-body wave function.
   - Realistic interactions, which describe accurately the NN scattering data and also bound nuclei with two and three nucleons, are adopted.

   Under these requirements, the many-body wave function is obtained by solving exactly the many-body Hamiltonian constructed by realistic interactions,

   $$H|\Psi\rangle = E|\Psi\rangle, \tag{3.4}$$

   where $|\Psi\rangle$ is the exact many-body wave function. In the ab initio approach, a many-body Hamiltonian H is written as,

   $$H = T + V = \sum_{i=1}^{A} t_i + \sum_{i<j}^{A} V_{ij} + \sum_{i<j<k}^{A} V_{ijk} + ..., \tag{3.5}$$

where t_i, V_{ij} and V_{ijk} are the single-particle kinetic energy, the 2-body NN interaction, and the 3-body NNN interaction, respectively. We should notice that the reason why the Hamiltonian H contains many-body interactions is due to the fact that the nucleons are treated as a point particle in the Schrödinger equation and the effect of intrinsic degree of freedom of quarks and gluons are implemented as many-body interactions. The chiral effective field theory (ChEFT) shows that the nuclear interactions obey a hierarchy: more many-body forces appear at higher orders of the perturbative expansion of effective field theory (see Chap. 2 for more details).

There are several ways to solve many-body problems at the ab initio level:

- Faddeev-Yakubowski (FY) method,
- Gaussian expansion method (GEM),
- Variational methods,
- No-core shell model (NCSM),
- Green's function Monte Carlo (GFMC) model,
- Similarity transformation methods,
- Quantum Monte Carlo methods,

and

- Combinations of above methods.

Faddeev proposed in 1961 the exact coupled equations for three-body Hamiltonian which can be applied both to scattering and bound-state problems. This model was extended to four-body problems by Yakubowski, but systems more than four particles cannot be handled on the same footing. The Gaussian expansion method (GEM) was proposed in 1988 for bound and scattering states of few-body systems. The method has been applied to a variety of few-body systems including hypernuclei (see Chap. 5). Alternatively, last two decades, many ab initio models were proposed to study the ground state and the excited states of nuclei, such as Variational Monte Carlo method, Green's function Monte Carlo, no-core shell model (NCSM), self-consistent Green's function method, coupled cluster model, and nuclear lattice effective field theory. These ab initio models were firstly applied to study light nuclei with mass $A < 20$. Because of the rapid increase of computer power and advance on algorithm for numerical computations, some ab initio models can now handle medium-mass nuclei with $A \sim 100$, and low-energy reactions involving light nuclei.

3.3 The Deuteron

A complete general solution can be obtained for an extremely important problem, that of the motion of a system consisting of two interacting particles (the two-body problem). As a first step towards the solution of this problem, we can simplify it by separating the motion of the system into the motion of the centre of mass and the relative motion of two particles. The potential energy of the interaction of two particles depends only on the distance between them, i.e. on the magnitude of the difference in their position vectors. The Hamiltonian of such a system reads

$$H = T_1 + T_2 + V(\mathbf{r}), \tag{3.6}$$

where T_1 and T_2 are kinetic energy of particle 1 and 2, respectively, and $V(\mathbf{r})$ is the two-body interaction depending on the relative distance $\mathbf{r} = \mathbf{r}_1 - \mathbf{r}_2$. By using the relative velocity $\mathbf{v} = \mathbf{v}_1 - \mathbf{v}_2$ and the center of mass (c.o.m.) velocity $\mathbf{V}_C = (m_1\mathbf{v}_1 + m_2\mathbf{v}_2)/(m_1 + m_2)$, the kinetic energies of Eq. (3.6) can be transformed into the relative and center-of-mass kinetic energies, T_r and $T_{c.o.m.}$, and Eq. (3.6) is rewritten as

$$H = T_r + T_{c.o.m.} + V(r) \equiv \frac{1}{2}\mu v^2 + \frac{1}{2}MV_C^2 + V(r), \tag{3.7}$$

where $M = m_1 + m_2$ and μ is the reduced mass defined by

$$\mu = \frac{m_1 m_2}{m_1 + m_2}. \tag{3.8}$$

In the rest frame of c.o.m. $V_C = 0$, the c.o.m. kinetic energy $T_{c.o.m.}$ is dropped in the Hamiltonian (3.7) and it becomes formally identical with the Hamiltonian of a particle of mass μ moving in an external field $V(\mathbf{r})$, i.e., the problem of the motion of two interacting particles is equivalent to that of the motion of one particle in a given external field $V(\mathbf{r})$.

The Schrödinger equation for a two-body system is expressed by

$$H\Psi = E\Psi, \tag{3.9}$$

and can be solved exactly in the coordinate space introducing the reduced mass (3.8) and the two-body potential $V(r)$. The two-body wave function is expressed by the angular momentum, the spin and the isospin as

$$\Psi = \Psi([(l_1 l_2)L(s_1 s_2)S]J, (t_1 t_2)T), \tag{3.10}$$

where l_i, s_i, t_i are the angular momentum, the spin and the isospin of the ith single-particle state, respectively, and J, L, S, T are the total angular momentum, the angular momentum, the spin and the isospin of two-particle system, respectively. The two-body wave function is anti-symmetric for the exchange of particles 1 and 2

$$P \equiv P_r P_S P_T = (-1)^{(L+S+T)} = -1, \tag{3.11}$$

Table 3.1 Selection rule of two-body wave function for the spin, isospin and angular momentum

S	T	L
0	0	Odd
0	1	Even
1	0	Even
1	1	Odd

Fig. 3.1 Configurations for spin-triplet state $S = 1$ of deuteron: **a** prolate configuration in which the spins and the radial unit vector are parallel. **b** oblate configuration in which the spins and the radial unit vector are orthogonal

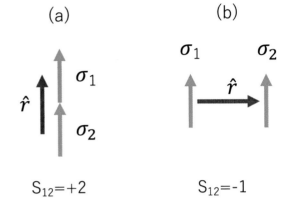

where P_r, P_S and P_T are the space, spin and isospin exchange operators, respectively. The symmetry condition (3.11) for the two-particle wave function gives the selection rule for the angular momentum, spin and isospin as tabulated in Table 3.1.

The S-wave ($L = 0$) two-body interaction is known as a strongly attractive one compared with other channels $L > 0$. Because of this strong attractive channel $L = 0$, the two-nucleon system could make a bound state in this channel. For di-proton and di-neutron systems, the selection rule (3.11) provides uniquely the $T = 1$ and $S = 0$ for $L = 0$ channel. However it was found the central S-wave interaction is not enough to make any bound state for these two nucleon systems with T=1. Among the three possible two-nucleon systems, only the n-p system has a bound state: the ground state of the deuteron. An essential difference between n-p and p-p or n-n systems is the isospin coupling. For di-neutrons and di-protons, the isospin T=1 is only allowed. However for n-p system, the isospin $T = 1$ and $T = 0$ are allowed. For $T = 0$ case, the spin coupling $S = 1$ is possible for the angular momentum L =even case as can be seen in Table 3.1. For $(T, S) = (0, 1)$ system, the tensor force can act and might create an extra binding energy and contribute for the total binding added to the central interactions. Actually this is the reason why the deuteron is bound as a two-body system, but di-proton and di-neutron systems are not bound.

The tensor interaction is written as

$$V_T(r) = f(r)S_{12}, \tag{3.12}$$

where $f(r)$ is the coupling strength, and S_{12} is the tensor operator;

$$S_{12} = \left[3(\boldsymbol{\sigma}_1 \cdot \hat{r})(\boldsymbol{\sigma}_2 \cdot \hat{r}) - (\boldsymbol{\sigma}_1 \cdot \boldsymbol{\sigma}_2)\right], \tag{3.13}$$

where $\hat{r}$ is the unit radial vector from particle 1 to particle 2. The tensor operator S_{12} gives a positive value for the case if the spins operators $\boldsymbol{\sigma}$ and the unit vector $\hat{r}$ are in a parallel configuration (case (a) in Fig. 3.1), i.e., $S_{12} = +2$. On the other hand, if the spin and the unit vector $\hat{r}$ are orthogonal, the S_{12} becomes negative $S_{12} = -1$ (configuration (b) in Fig. 3.1). Since the sign of coupling strength $f(r)$ is negative, the configuration (a) gives an attractive binding energy and makes the n-p system bound. i.e., the deuteron can be made. The configuration (a) induces a prolate deformation and makes the intrinsic Q moment positive (see Sect. 7.2.3 of Chap. 7 for details). The measured electric quadrupole-moment value of $Q = 0.2860 \pm 0.0015$ fm^2 [1] confirmed that the deformation of deuteron is prolate and consistent with the tensor coupling effect on the deuteron binding.

3.4 Effective Interactions for Nuclear Many-Body Problems

Realistic interactions are determined by fitting NN scattering data as discussed in Chap. 2. A straightforward application of the realistic interactions for nuclear many-body problems is extremely difficult because of its nature of strong interaction, i.e., one difficulty is its short-range nature and another is many channels depending on spin and isospin degrees of freedom. In practice, one has to introduce *effective interactions* for all microscopic models.

There are two categories of effective interactions: one is based on realistic interactions, which are determined by fitting NN scattering data and another is determined by nuclear matter and finite nuclei data (also referred as phenomenological). There have been many attempts to derive NN interactions from exact and reliable theory in an analogy of electromagnetic interactions for atomic and molecular dynamics. A study of the NN interactions from QCD of quarks and gluons have been advanced rapidly over the past ten years. Recent developments succeed to describe S-wave NN scattering phase shifts and to show the hard core by Lattice QCD (LQCD) calculations (see Chap. 10 for details), but still not sufficiently mature to be used in nuclear structure calculations. Thus, there are quite a few effective interactions that are used for various purposes.

The realistic interactions are quite successful in the description of NN scattering data, but do not work very well in the descriptions of nuclear matter properties and the structure of heavy nuclei with mass $A > 100$. For heavy nuclei, phenomenological interactions are derived to be used in the HF and the RPA calculations. The phenomenological interactions are determined to fit a set of experimental data such as the nuclear matter properties and also the binding energies and the radii of several closed shell nuclei. By definition, the phenomenological interactions are good to describe nuclear matter properties as well as structure of heavy nuclei, while they are not fitted on NN scattering data and do not reproduce NN scatterings satisfactory. These interactions are considered to take into account complicated correlations in complex nuclei through density-dependent terms. Recently, a model of chiral effective theory was proposed based on the chiral invariant Lagrangian of pions and nucleons. This model describes well NN scattering data and also is claimed to be applicable to nuclear-structure calculations.

Phenomenological effective interactions are commonly adopted in shell model, the HF, and microscopic cluster models. Skyrme and Gogny interactions are commonly used in the self-consistent HF and RPA calculations. Density functional theory (DFT) is a successful theory to investigate the electronic structure (principally the ground state) of many-body systems, in particular atoms, molecules, and the condensed phases. The same idea to construct energy density functionals (EDF) is applied to obtain the effective interactions for nuclear many-body dynamics. It has been a challenging issue to construct useful EDFs for describing nuclear dynamics of nuclei consistently in a wide region of the nuclear chart. These phenomenological EDFs are constrained by nuclear matter properties, and nuclear structure properties such as binding energies and radii of reference nuclei.

There are several realistic interactions which are determined to reproduce the phase shifts of nucleon-nucleon scattering and the properties of few nucleon systems such as deuteron and triton. As NN interactions, Bonn potentials were based on a meson-exchange picture of two-body interactions, while AV18 was obtained to fit phenomenologically the phase shift. The three-body interactions are also constructed in the same strategy. Commonly-adopted ones are Tucson-Melbourne, Urbana IX, and Illinois 7 three-body interactions. Idaho and Jülich groups constructed realistic potentials based on chiral effective field theory (ChEFT). The ChEFT provides NN, NNN and higher many-body interactions in the same framework according to a power counting scheme. For the ab initio many-body calculations, these realistic interactions are renormalized to obtain good convergence in numerical calculations. The commonly used renormalization schemes are

- Brückner G-matrix method,
- Lee-Suzuki-Okamoto similarity transformation method,
- truncation to V_{low-k} limit,
- Unitary Correlation Operator Method (UCOM),
- Similarity Renormalization Group (SRG) and In Medium SRG (IMSRG).

The NCSM have first adopted the Lee-Suzuki similarity transformation method to obtain effective interactions for the shell model calculations. Recently, the SRG method is commonly applied to obtain the effective interactions for the NCSM. Fermionic molecular dynamics (FMD) model uses UCOM to obtain the effective interaction. Effective interactions in momentum space were also obtained for use in the ab initio calculations by V_{low-k} approximation or SRG approach from the realistic potentials. All these interactions are invented equivalently to describe nuclear properties at low momentum regime, renormalizing the unresolved high-momentum components of realistic interactions.

3.5 Nuclear Mean Field

The mean field is an approximation in which each particle of a system composed of A particles moves in an external field (*mean field*) generated by the remaining $A - 1$ particles. It is an enormous simplification of the many-body problem neglecting a large part of the inter-nucleon forces. The neglected part of the force is called *residual interaction*. The major problem of theoretical models is how to include the residual interaction, especially to describe excited states. The idea of mean field theory can reduce a many-particle interacting system into a system of non-interacting particles in a mean field potential. In this theory, the mean field is constructed self-consistently through NN interactions. This is called the self-consistent Hartree and Hartree–Fock theory. In the former, the exchange interaction due to the anti-symmetrization of many-body wave function is neglected, while the latter takes into account the exchange interaction. The HF theory can be extended to the time-dependent HF (TDHF) theory to describe excited states taking into account the residual interactions. It is proved that the RPA is formally equivalent to TDHF theory to describe low-energy collective excitations.

3.5.1 Slater Determinant and the Second Quantization

In the Hartree or the HF theory, the fundamental assumption is to consider a A-body wave function as a direct product (Hartree) or an anti-symmetrized product (HF) of independent particle wave functions. The direct product is given by

$$\Phi_{\mathrm{H}}(\mathbf{r}_1, \mathbf{r}_2, \cdots, \mathbf{r}_A) = \phi_\alpha(\mathbf{r}_1)\phi_\beta(\mathbf{r}_2) \cdots \phi_\omega(\mathbf{r}_A), \tag{3.14}$$

where $\phi_\alpha(\mathbf{r}_i)$ is a single-particle wave function specified by the quantum number α. The anti-symmetrized product is given in Eq. (3.1). The anti-symmetrized operation generates the normalized Slater determinant,

$$\Phi_{\mathrm{HF}}(\mathbf{r}_1, \mathbf{r}_2, \cdots, \mathbf{r}_A) = \mathcal{A}\{\phi_\alpha\phi_\beta \cdots \phi_\omega\} = \frac{1}{\sqrt{A!}} \begin{vmatrix} \phi_\alpha(\mathbf{r}_1) & \phi_\beta(\mathbf{r}_1) & \cdots & \phi_\omega(\mathbf{r}_1) \\ \phi_\alpha(\mathbf{r}_2) & \phi_\beta(\mathbf{r}_2) & \cdots & \phi_\omega(\mathbf{r}_2) \\ \cdot & \cdot & \cdots & \cdot \\ \cdot & \cdot & \cdots & \cdot \\ \phi_\alpha(\mathbf{r}_A) & \phi_\beta(\mathbf{r}_A) & \cdots & \phi_\omega(\mathbf{r}_A) \end{vmatrix}. \tag{3.15}$$

Problem

3.1 Give the anti-symmetrized wave function for a three-body system.

To manipulate wave functions, to evaluate matrix elements of two-body interactions, transition strengths, etc., the second quantization representation is conveniently adopted. We first define the *particle creation operator* $a_\mu^\dagger$, which creates a particle in the state with the set of quantum numbers μ. In the second quantization form, the Slater determinant (3.15) can be expressed in a simple form,

$$|\alpha, \beta, \cdots \omega\rangle \equiv a_\alpha^\dagger a_\beta^\dagger \cdots a_\omega^\dagger |0\rangle, \tag{3.16}$$

where $|0\rangle$ is the vacuum state describing a state of zero particle. Two basic properties of fermions are the anti-symmetrization under interchange of particles,

$$a_\alpha^\dagger a_\beta^\dagger \cdots a_\omega^\dagger |0\rangle = -a_\beta^\dagger a_\alpha^\dagger \cdots a_\omega^\dagger |0\rangle, \tag{3.17}$$

and the Pauli exclusion principle, which tells two particles can not occupy the same state,

$$a_\alpha^\dagger a_\alpha^\dagger a_\beta^\dagger \cdots a_\omega^\dagger |0\rangle = 0. \tag{3.18}$$

These two requirements can be simultaneously expressed by the anti-commutation relation

$$\{a_\alpha^\dagger, a_\beta^\dagger\} \equiv a_\alpha^\dagger a_\beta^\dagger + a_\beta^\dagger a_\alpha^\dagger = 0. \tag{3.19}$$

The operator a_μ is introduced as the Hermitian conjugate of $a_\mu^\dagger$, which annihilates a particle in the state μ,

$$a_\mu |\mu\rangle = a_\mu a_\mu^\dagger |0\rangle = |0\rangle, \quad a_\mu |0\rangle = 0, \tag{3.20}$$

and called the *particle annihilation operator*. Thus, the a_μ and $a_\nu^\dagger$ satisfy the anti-commutation relation,

$$\{a_\mu, a_\nu^\dagger\} = \delta_{\mu\nu}. \tag{3.21}$$

3.5.2 The Mean-Field Potential

We take a simple one-body interaction to illustrate the role of the mean field on evaluating the single-particle spectrum. The nuclear mean field is approximated by a one-body potential with two terms,

$$V_{\mathrm{MF}}(r) = V(r) + V_{ls}(r), \tag{3.22}$$

where $V(r)$ is the central potential and $V_{ls}(r)$ is the spin-orbit potential. The central part is parametrized in several forms, for example, the harmonic oscillator potential, the square well potential, the Nilsson potential, or the Woods-Saxon potential. The profile of the central potential is determined by the following conditions. Firstly, the potential should be flat at the center ($r = 0$) where a nucleon feels the NN forces uniformly from all the directions;

$$\frac{\partial V(\mathbf{r})}{\partial r}\bigg|_{r=0} = 0. \tag{3.23}$$

The nuclear potential is largely negative at the center and becomes weaker going from the center to the surface ($r = R_0$)

$$\frac{\partial V(\mathbf{r})}{\partial r}\bigg|_{r=R_0} > 0, \tag{3.24}$$

and becomes zero at far outside of the nuclear surface, $V(\mathbf{r}) \to 0$ ($r \to \infty$). These requirements are fulfilled by the Woods-Saxon potential given below. The Woods-Saxon potential is commonly adopted since its radial profile resembles to that of nuclear density. The Woods-Saxon potential reads

$$V(r) = V_0 f(r) = V_0 \frac{1}{1 + \exp(\frac{r-R_0}{a})}, \tag{3.25}$$

where V_0 is the potential depth and $f(r)$ is a Fermi function with the radius parameter often taken as $R_0 = r_0 A^{1/3}$ and the surface diffuseness parameter a. Typical values for the potential parameters are $V_0 = -50$ MeV, r_0=1.27 fm and a=0.67 fm. The spin-orbit potential is commonly given by

$$V_{ls}(r) = V_{ls}r_0^2 \frac{1}{r}\frac{df(r)}{dr}\boldsymbol{\ell} \cdot \mathbf{s}, \tag{3.26}$$

where V_{ls} is the spin-orbit coupling strength and a typical value is V_{ls}=22 MeV. The spin-orbit potential must be a surface term since it involves a derivative operator $\boldsymbol{\ell} = \mathbf{r} \times \mathbf{p} = -i\hbar\mathbf{r} \times \nabla$ and gives no contribution in the central part of the nucleus, where the density is almost constant and its derivate is zero. In the surface region, however, the density gradient is finite so that the spin-orbit potential can be expressed as a local potential

$$V_{ls}(r) \propto \nabla\rho(r) \times \mathbf{p} \cdot \mathbf{s} = \frac{1}{\hbar}(\boldsymbol{\ell} \cdot \mathbf{s})\frac{1}{r}\frac{\partial\rho(r)}{\partial r}. \tag{3.27}$$

The spin operator $\mathbf{s}$ has a pseudovector nature and a vector product $\nabla\rho(r) \times \mathbf{p}$ is also a pseudovector so that the scalar product of the two operators satisfies the symmetry constrain of potential under the space reflection. The parameters are chosen to give a good description of the gross features of the low-energy bound state spectra and also to agree with the average potentials found from the optical model analysis of nucleon and nucleus scatterings of different nuclei at different incident energies.

The solution of the Shrödinger equation for the nuclear potential (3.22) gives a shell structure as illustrated in Fig. 3.2. In this description, assuming spherical symmetry, nucleons are placed on energy orbitals tagged by the principal quantum number, the angular momentum, and total spin ($n\ell j$), exactly as for the description

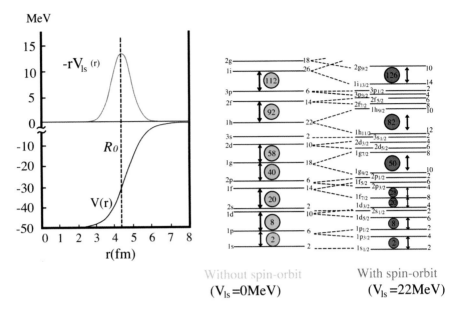

Fig. 3.2 (Left) Woods-Saxon potential. Here, R indicates the position of nuclear surface. The derivative of the potential is also drown on top of the central potential $V(r)$. (Right) Calculated single-particle energies without and with spin-orbit potential, respectively. The calculated states are classified by the number of nodes n of wave function, orbital angular momentum ($l = 0, 1, 2, 3, 4$) by the notation (s, p, d, f, g), total spin $j = l \pm 1/2$, where each orbit has the degeneracy of $(2j + 1)$. The numbers in circles indicate the magic numbers in which nuclei are very stable for external perturbations. Each level is assigned with its degeneracy $2(2l + 1)$ and $2j + 1$ for w/o and with the spin-orbit interaction, respectively

of electronic structure in atoms. For example, the orbital with the lowest energy presents an angular momentum $\ell = 0$, identified as an s wave, with a total spin of $j = 1/2$. It is quoted as the orbital $1s_{1/2}$. An orbital with a total spin j has a degeneracy of $2j + 1$. The nuclear orbitals are not homogeneously spaced in energy but are packed in *shells* separated from each other by energy gaps. A nucleus that fills a shell, either for protons or neutrons, will be stabilized against excitations and called a magic nucleus. This over-stability and shell effects are encountered in several fields of quantum mechanics. The most obvious example stands in atomic physics: noble gases have quantum orbitals filled with electrons and are much less reactive than other atoms. It is the case of He, Ne, Ar, Xe,... In the case of atomic nuclei, if one refers to Fig. 3.2, the first shell is composed of the $1s_{1/2}$ alone separated by a rather large energy gap from the second shell composed of the $1p_{3/2}$ and $1p_{1/2}$ orbitals. The third shell is constituted by the $1d_{5/2}$, $2s_{1/2}$, and $1d_{3/2}$ orbitals, therefore the first three magic numbers are 2, 8 and 20. These shell orbits are classified as s-shell, p-shell, and sd-shell orbits, respectively. When the shell structure was first observed from mass measurements in 1930s no model was able to reproduce magic numbers beyond 20. There was a very good reason for this theoretical failure: the sensitivity of the mean field to the spin orientation and orbital on which lie the nucleons, also called the spin-orbit part of the nuclear potential, was not intuited at that time. It is only in 1949 that M. Goppert-Mayer[2] and J. H. D. Jensen simultaneously and separately came up with the same conclusion: a spin-orbit term coming from the nucleon-nucleon interaction was necessary in the nuclear potential of the shell model to explain the enigma of the sequence of nuclear magic numbers. This term is essential to explain the energy splitting of orbitals with the same angular momentum but different total spin such as observed between the orbitals $1f_{7/2}$ and $1f_{5/2}$ that explains the 28 magic number as can be seen in Fig. 3.2. Both Goppert-Mayer and Jensen won the physics Nobel prize in 1963 for their contribution in understanding the nucleus and its structure. One can then determine all magic numbers of nucleons: 2, 8, 20, 28, 50, 82, 126. We are daily in contact with doubly magic nuclei: the most abundant oxygen isotope is ^{16}O made of 8 protons and 8 neutrons, milk contains a sizable amount of calcium whose most abundant isotope is ^{40}Ca with 20 protons and 20 neutrons, part of canalisations in urban dwellings are still built with lead mainly composed of the doubly magic nucleus ^{208}Pb composed of 82 protons and 126 neutrons.

3.6 Energy Density Functionals (EDF)

The mean field and the resulting shell structure are direct outcomes of the nucleon-nucleon interaction and it is legitimate to calculate the structure of a nucleus directly from effective many-body interactions. It appears that solving the dynamics of nucleus is a similar problem to untie an unsolved Gordian Knot[3]. The fact that there are up to hundreds of nucleons bound by the strong interaction in a nucleus leads to two serious challenges: (i) we do not know the force acting between particles inside the nucleus with enough precision. As discussed in Chap. 2, important progresses have been made in developing effective interactions from bare nucleon-nucleon scattering. (ii) even if we knew the nucleon-nucleon interaction to a very high degree of precision, we could not solve exactly the many-body Hamiltonian with 100 particles. We need an approach that renders the severe constraints of the problem in several stages of microscopic calculations, just like Alexander the Great untied the Gordian knot by cutting it with his sword.

There are two categories to obtain the effective interaction for nuclear many-body problems. The first one is based on either realistic potentials or the quantum chromodynamics (QCD). In this approach, the effective interactions are obtained starting from realistic potentials to reproduce the two-body scattering data. The relativistic

---

[2] The idea of the spin-orbit potential was suggested to Mayer by Enrico Fermi. In her epoch-making article of [2], she acknowledged his suggestion by saying "Thanks are due to Enrico Fermi for the remark; Is there any indication of spin-orbit coupling?" which was the origin of this paper.

[3] The Gordian Knot is a legend of Phrygian Gordium associated with Alexander the Great. It is often used as a metaphor for an intractable problem.

or non-relativistic Brückner Hartree–Fock (BHF) theory[4] or variational method with the correlation functions are used to derive the effective interaction for the calculations of observables. As modern approaches, the lattice QCD, and the chiral effective models also are applied to obtain the effective interactions. The second category is more phenomenological and pragmatic; one selects the functional forms of Hamiltonian with spin and isospin degree of freedom and determines the parameters of the functional forms under constrains of a set of experimental data. They are, for example, the binding energy of nuclear matter and finite nuclei, the rms radii, or quantities also extracted from data such as the nuclear incompressibility, the symmetry energy and the spin-orbit splitting between spin-orbit partner orbitals. One of the second approach is called "*Density functional theory* (DFT) or *Energy density functional theory* (EDF)" analogous to many-electron theories such as Hohenberg-Kohn theorem and Kohn-Sham method. This approach is not based on the interactions between nucleons, but expresses the Hamiltonian by the local one-body density.

DFT is a theory to map the many-body problem onto a one-body problem. DFT starts from the Hohenberg-Kohn (HK) theorem [3] which states that "the exact ground state energy E is a universal functional of the local density $\rho(\mathbf{r})$". Namely, the HK theorem says that the ground state energy of any interacting many-particle system with a given two-body interaction is uniquely determined by its ground state density. Furthermore the knowledge of the local density $\rho(\mathbf{r})$ determines the external potential of the system (except a trivial constant).

The many-body Hamiltonian H can be expressed as,

$$H = T + V + W, \tag{3.28}$$

where T is the kinetic energy, V is the sum of the external potentials $v(\mathbf{r}_i)$ acting on the particle i,

$$V = \sum_i v(\mathbf{r}_i), \tag{3.29}$$

and W is the two-body interparticle interaction. The expectation value for exact many-body wave function Ψ can be described as

$$E_{\mathrm{HK}}[\rho] = \langle \Psi | H | \Psi \rangle = \langle \Psi | T + V + W | \Psi \rangle, \tag{3.30}$$

where $E_{\mathrm{HK}}[\rho]$ depends only on the local density $\rho(\mathbf{r})$. In Eq. (3.30), a term containing the ground-state density and the external potential explicitly separates out as

$$\langle \Psi | V | \Psi \rangle = \int \rho(\mathbf{r})v(\mathbf{r})d\mathbf{r}. \tag{3.31}$$

and the functional $F_{\mathrm{HK}}[\rho]$ is defined as

---

[4] The BHF theory is a Hartree–Fock theory with the effective interaction derived from bare nucleon-nucleon forces.

$$F_{HK}[\rho] = \langle \Psi | T + W | \Psi \rangle. \tag{3.32}$$

The functional F_{HK} is universal in the sense that it does not depend on the external potential V. The ground state is determined by minimizing $E_{HK}(\rho)$ with respect to $\rho(\mathbf{r})$.

To apply the DFT formalism one obviously needs good approximations for the functional $F_{HK}(\rho)$. Accurate approximations have been obtained by means of the Kohn-Sham (KS) method. A main advantage of the KS scheme is that it allows a straightforward determination of a large part of the kinetic energy in a simple way. Another advantage, from a more physical point of view, is that it provides an exact one-particle picture of interacting many-body systems. In the KS scheme the functional $F_{HK}(\rho)$ is divided into three parts, kinetic energy part, the direct (Hartree) part and the exchange correlation part

$$F_{HK}(\rho) = T + E_H + E_{xc}. \tag{3.33}$$

The HK theory is quite successful to describe the total binding energy of many-electron system, but fails to reproduce the shell structure. In order to describe shell structure, Kohn and Sham [4] introduced an effective single-particle potential U_{eff} under the assumption that the potential creates the exact density as a solution of single-particle Schrödinger equation,

$$\left\{ -\frac{\hbar^2}{2m} \nabla^2 + U_{eff} \right\} \phi_i(\mathbf{r}) = \varepsilon_i \phi_i(\mathbf{r}), \tag{3.34}$$

i.e., the density obtained $\rho(\mathbf{r}) = \sum_i |\phi_i(\mathbf{r})|^2$ is the exact density. The energy density function is given by

$$E_{KS}[\rho, \tau] = \int d\mathbf{r} \left\{ \frac{\hbar^2}{2m} \tau(\mathbf{r}) + \rho(\mathbf{r}) v(\mathbf{r}) \right\} + E_H[\rho(\mathbf{r})] + E_{xc}[\rho(\mathbf{r})], \tag{3.35}$$

where τ is the kinetic energy density. The effective single-particle potential is obtained by the functional derivative of $E_{KS} - T$ with respect to the density,

$$U_{eff} = \frac{\delta}{\delta\rho} E_H[\rho] + \frac{\delta}{\delta\rho} E_{xc}[\rho] + v(\mathbf{r}). \tag{3.36}$$

The reliability of KS theory depends entirely on the knowledge of E_H and $E_{xc}(\mathbf{r})$. In atomic and molecular physics, the interaction is electromagnetic so that the theory is interpreted as ab initio method. In practical applications, the KS theory takes the following steps:

Kohn-Sham method

(1) determine the functional forms of $E_H[\rho]$ and $E_{xc}[\rho]$ at one's best,
(2) obtain an initial guess for the density $\rho_{n=0}$ ("n" is the number of iteration),

(3) calculate the effective potential U_{eff} in Eq. (3.36) with $\rho = \rho_n$,

(4) solve the Schrödinger equation (3.34) with the effective potential U_{eff} in Eq. (3.36) and obtain $\phi_i(\mathbf{r})$,

(5) calculate the density $\rho_{n+1}(\mathbf{r})$ for the next trial with the single-particle wave functions $\phi_i(\mathbf{r})$ obtained in step 4,

(6) go back to step (3) and repeat the circle steps (3)–(5) replacing ρ_n by ρ_{n+1} until convergence is achieved.

In molecular dynamics, the Hamiltonian consists of the kinetic energies, the electron-electron interactions and the external field (the Coulomb interaction between electron and nucleus). The HK theory proves one-to-one correspondence between the density $\rho(\mathbf{r})$ and an external potential; the ground state energy is a unique functional of the density $\rho(\mathbf{r})$. The HK theorem is also valid in a localized self-bound system such as a nucleus. For DFT of the KS scheme, there are many attempts to determine universal EDF, $E[\rho, \tau]$, which can be applied to the ground state and the excited states of nuclei in a wide region of mass table. These EDF give reasonable success in practical applications.

3.6.1 Hartree–Fock Models

In nuclear physics, the first attempt to establish an EDF was attacked by D. Skyrme in 1956. He proposed an effective interaction with two-body and three-body forces of the form

$$V = \sum_{i<j} V(\mathbf{r}_i, \mathbf{r}_j) + \sum_{i<j<k} V(\mathbf{r}_i, \mathbf{r}_j, \mathbf{r}_k). \tag{3.37}$$

To simplify the calculations, he adopted momentum-dependent interactions as a short-range expansion of finite range interactions

$$
\begin{aligned}
V(\mathbf{r}_1, \mathbf{r}_2) =\ & t_0(1 + x_0 P_\sigma)\delta(\mathbf{r}_1 - \mathbf{r}_2) \\
& + t_1(1 + x_1 P_\sigma)\frac{1}{2}[\delta(\mathbf{r}_1 - \mathbf{r}_2)\mathbf{k}^2 + \mathbf{k}'^2\delta(\mathbf{r}_1 - \mathbf{r}_2)] \\
& + t_2(1 + x_2 P_\sigma)\mathbf{k}' \cdot \delta(\mathbf{r}_1 - \mathbf{r}_2)\mathbf{k} \\
& + i W_0(\boldsymbol{\sigma}_1 + \boldsymbol{\sigma}_2)\mathbf{k}' \times \delta(\mathbf{r}_1 - \mathbf{r}_2)\mathbf{k},
\end{aligned} \tag{3.38}
$$

where $\mathbf{k}$ and $\mathbf{k}'$ are relative wave vector of two nucleons, and P_σ is the spin-exchange operator $P_\sigma = (1 + \boldsymbol{\sigma}_1 \cdot \boldsymbol{\sigma}_2)/2$. In Eq. (3.38), $t_0, x_0, t_1, x_1, t_2, x_2$ and W_0 are the parameters of Skyrme interaction. The first line of Eq. (3.38) is the zero-range S-wave interaction, the second and third lines are the momentum-dependent S-wave and P-wave interactions, respectively, and the fourth line is the spin-orbit interaction. The three-body interaction is also assumed to be a zero-range force of the form

$$V(\mathbf{r}_1, \mathbf{r}_2, \mathbf{r}_3) = t_3\delta(\mathbf{r}_1 - \mathbf{r}_2)\delta(\mathbf{r}_2 - \mathbf{r}_3), \tag{3.39}$$

Table 3.2 Parameter sets of the Skyrme interactions SIII [6] and SkM* [7]

	t_0 (MeVfm^3)	x_0	t_1 (MeVfm^5)	x_1	t_2 (MeVfm^5)	x_2	t_3 (MeVfm^6)	x_3	W_0 (MeVfm^5)	α
SIII	−1128.75	0.45	395.0	0.0	−95.0	0.0	14000.0	1.0	120.0	1.0
SkM*	−2645.0	0.09	410.0	0.0	−135.0	0.0	15595.0	0.0	130.0	1/6

where t_3 is another parameter. It is shown in even-even nuclei that the contact-type three-body force (3.39) is equivalent to the linearly density-dependent two-body contact interaction

$$\frac{1}{6}t_3(1 + x_3 P_\sigma)\rho((\mathbf{r}_1 + \mathbf{r}_2)/2)\delta(\mathbf{r}_1 - \mathbf{r}_2), \tag{3.40}$$

where $\rho((\mathbf{r}_1 + \mathbf{r}_2)/2)$ is the density. For the HF calculation, it can be proved that a choice of $x_3 = 1$ in Eq. (3.40) gives exactly the same HF density as that of the three-body interaction (3.39). The density-dependent term is generalized to introduce a parameter α for the power of density dependence,

$$\frac{1}{6}t_3(1 + x_3 P_\sigma)\rho^\alpha((\mathbf{r}_1 + \mathbf{r}_2)/2)\delta(\mathbf{r}_1 - \mathbf{r}_2). \tag{3.41}$$

The parameters of Skyrme interaction are adjusted to the experimental binding energies and radii and nuclear matter saturation properties. The parameter sets SIII and SkM* are given in Table 3.2.

Despite the success of the Skyrme interaction, it has been argued that the zero-range interactions might not be able to mimic all features of medium-range or long-range parts of realistic interactions. In particular, most of the Skyrme interactions do not given properly the pairing correlations in nuclei so that additional pairing interaction is introduced for applications to the HF-BCS or the HF Bogolyubov calculations. D. Gogny, in 1975 [8], introduced a finite-range two-body interactions as a sum of two Gaussians with spin-isospin exchange terms instead of t_0, t_1, and t_2 terms in (3.38)

$$\begin{aligned}
V(\mathbf{r}_1, \mathbf{r}_2) = &\sum_{i=1,2} e^{-(\mathbf{r}_1 - \mathbf{r}_2)^2/\mu_i^2}(W_i + B_i P_\sigma - H_i P_\tau - M_i P_\sigma P_\tau) \\
&+ i W_0(\sigma_1 + \sigma_2)\mathbf{k}' \times \delta(\mathbf{r}_1 - \mathbf{r}_2)\mathbf{k} \\
&+ t_3(1 + P_\sigma)\rho^{1/3}((\mathbf{r}_1 + \mathbf{r}_2)/2)\delta(\mathbf{r}_1 - \mathbf{r}_2), \tag{3.42}
\end{aligned}$$

together with the contact spin-orbit and a density-dependent two-body interaction. In Eq. (3.42), μ_i is the range parameter of interaction, W_i, B_i, H_i and M_i are the interaction strengths so called, Wigner-, Bartlett-, Heisenberg- and Majorana-types, respectively, and P_τ is the isospin-exchange operators, $P_\tau = (1 + \tau_1 \cdot \tau_2)/2$. The typical parameter sets are given in Table 3.3.

Table 3.3 Parameter sets of the Gogny interaction D1. μ_i is given in unit of fm, while interaction strength is given in unit of MeV. W_0 and t_3 are taken to be $W_0 = 115$ MeVfm^5 and $t_3 = 1350$ MeVfm^4, respectively

i	μ_i (fm)	W_i (MeV)	B_i (MeV)	H_i (MeV)	M_i (MeV)
1	0.7	-402.4	$-100.$	-496.2	-23.56
2	1.2	-21.30	-11.77	37.27	-68.81

Problem

3.2 Derive the momentum-dependent terms of the Skyrme interaction from the matrix element of a finite-range interaction, $V(\mathbf{r}) = V_0 e^{-r^2/\mu^2}$, using the plane wave functions for the initial and final states with the relative wave vectors $\mathbf{k} = \mathbf{k}_1 - \mathbf{k}_2$ and $\mathbf{k}' = \mathbf{k}'_1 - \mathbf{k}'_2$, respectively.

D. Vautherin and D. M. Brink [9] applied the Skyrme interaction to the HF calculations from light to heavy nuclei to reproduce the binding energies and nuclear radii from ^{16}O to ^{208}Pb. The HF calculations start from calculating the expectation value of an effective Hamiltonian with respect to a Slater determinant of the single-particle wave function $\phi_i(\mathbf{r}, s, t)$ in Eq. (3.15). The total energy of Skyrme Hamiltonian can be evaluated as,

$$E = \sum_{i=1}^{A} < i|\frac{p_i^2}{2m}|i > +\frac{1}{2}\sum_{i,j}^{A} < ij|\widetilde{V}(\mathbf{r}_i, \mathbf{r}_j)|ij > +\frac{1}{6}\sum_{i,j,k}^{A} < ijk|\widetilde{V}(\mathbf{r}_i, \mathbf{r}_j, \mathbf{r}_k)|ijk >$$

$$= \int h(\mathbf{r})d\mathbf{r}, \tag{3.43}$$

where $\widetilde{V}$ means both the direct and the exchange matrix elements, and "i, j" stand for the single-particle wave functions $\phi_i(\mathbf{r}, s, t)$ and $\phi_j(\mathbf{r}, s, t)$, respectively. The energy density $h(\mathbf{r})$ can be expressed by one-body densities and one-body spin-orbit current density. For $N = Z$ nuclei it can be expressed as

$$h(\mathbf{r}) = \frac{\hbar^2}{2m}\tau + A\rho^2 + B\rho^3 + C\rho\tau + D(\nabla\rho)^2 + E\rho\nabla \cdot \mathbf{J} + F\mathbf{J}^2, \tag{3.44}$$

where τ and $\mathbf{J}$ are the kinetic density and the spin-orbit density, respectively, given by

$$\tau(\mathbf{r}) = \sum_i |\nabla^2\phi_i(\mathbf{r})|^2 \tag{3.45}$$

$$\mathbf{J}(\mathbf{r}) = -i\sum_i \phi_i^*(\mathbf{r}) \cdot (\nabla\phi_i(\mathbf{r}) \times \boldsymbol{\sigma}). \tag{3.46}$$

The coefficient of each term in Eq. (3.44) can be expressed analytically by the Skyrme parameters $t_0, x_0, t_1, x_1, t_2, x_2, t_3$ and W_0. For $N \neq Z$ nuclei, the isospin dependence of density such as $\rho_\tau = (\rho_n - \rho_p)$ should be taken into account in the EDF.

The HF equation can be obtained by a variational procedure for Eq. (3.43) with respect to the single-particle wave function ϕ_i,

$$\frac{\delta}{\delta\phi_i}(E - \sum_i \varepsilon_i \int |\phi_i(\mathbf{r}_i)|^2 d\mathbf{r}_i) = 0, \tag{3.47}$$

where ε_i is the single-particle energy. We thus obtain the HF equation,

$$-\frac{\hbar^2}{2m}\nabla^2\phi_i(\mathbf{r}_i) + v_i^{HF}(\mathbf{r}_i)\phi_i(\mathbf{r}_i) = \varepsilon_i\phi_i(\mathbf{r}_i), \tag{3.48}$$

where the two-body part of the mean field potential v_i^{HF} is expressed as

$$v_i^{HF}(\mathbf{r_i}) = \sum_j \int d\mathbf{r_j}\langle ij|\widetilde{V}(\mathbf{r_i}, \mathbf{r_j})|ij\rangle + \sum_{j,k} \int d\mathbf{r_j} \int d\mathbf{r_k}\langle ijk|\widetilde{V}(\mathbf{r_i}, \mathbf{r_j}, \mathbf{r_k})|ijk\rangle.$$

$$\tag{3.49}$$

Here the interaction $\widetilde{V}$ contains both the direct and exchange terms. The HF method follows a quite similar numerical technique to that of KS theory described in this section. One essential difference is that HF needs HF potential extracted from many-body interactions, while KS theory adopts EDF (not the interactions) for the iterations.

Modern EDF theory selects the parameters in Eq. (3.44) directly to fit observables such as the binding energies and the rms radii of many nuclei over the mass table. See calculated binding energies and root mean square radii for several spherical closed shell nuclei given in Table 3.5 in comparisons with experimental data. The calculated HF density is shown in Fig. 3.3.

Problem

3.3 Calculate the matrix element of the two-body contact interaction $t_0\delta(\mathbf{r}_1 - \mathbf{r}_2)$ with anti-symmertized two-particle wave function for an even-even nucleus with $N = Z$.

3.6.2 Relativistic Mean Field (RMF) Model

A conceptually different mean field model, the so-called *relativistic mean field model* (RMF model), was introduced by Walecka in 1975 [10]. This is also a phenomenological model which includes the nucleonic and mesonic degrees of freedom in a relativistic formalism. This model is called σ-ω model (or Walecka model) since the original Lagrangian introduced two mesonic degrees of freedom, i.e., the scalar σ and the vector ω mesons with $J^\pi = 0^+$ and $J^\pi = 1^-$, respectively with the isospin $T = 0$. It is an effective theory in the sense that the coupling constants of the σ and ω mesons with nucleons are treated as free parameters which can be adjusted in such a way to reproduce the empirical saturation properties of nuclear matter, in particular at the nuclear saturation point. While the σ-ω model provides a simple form of a relativistic density functional for the nuclear equation of states (EOS), its incompressibility at the saturation point is much larger than the realistic value. More realistic model Lagrangians for the EOS include the vector-isovector ρ meson with $J^\pi = 1^-$, $T = 1$ and non-linear σ meson couplings, of importance for an accurate description of finite nuclei properties.

The most important meson is the pion for the strong interaction. It carries the spin and parity quantum numbers $J^\pi = 0^-$ and $T = 1$, so called the pseudo-scalar meson. Since its negative parity character breaks the parity symmetry of the ground state in the Hartree approximation (direct terms only in Eq. (3.49)), the pion has no contributions to the EDF. Two and any even numbers of pion exchanges can contribute to the mean field potential without violating the parity symmetry. The σ meson carrying $J^\pi = 0^+$ is considered to mimic the higher order processes of pion exchanges (even numbers) by a simplified one-meson exchange process. The contribution of one-pion exchange appears only in the exchange term (Fock term) in the relativistic HF model.

The starting point of RMF is the following classical Lagrangian density [11],

$$\mathcal{L} = \mathcal{L}_N + \mathcal{L}_M + \mathcal{L}_{int}, \tag{3.50}$$

where $\mathcal{L}_N$, $\mathcal{L}_M$ and $\mathcal{L}_{int}$ are the Lagrangians for free nucleons, free mesons, and interaction between nucleons and mesons, respectively. The Lagrangian density for free nucleons reads

$$\mathcal{L}_N = \overline{\psi}(i\gamma^\mu \partial_\mu - m)\psi, \tag{3.51}$$

where m is the bare nucleon mass, and ψ is a four component Dirac spinor, and $\overline{\psi}$ is a Dirac adjoint defined by $\overline{\psi} = \psi^\dagger \gamma^0$. The Lagrangian for the free mesons has contributions of σ, ω and ρ mesons,

$$\mathcal{L}_M = \frac{1}{2}(\partial_\mu \sigma \partial^\mu \sigma - m_\sigma^2 \sigma^2) - \frac{1}{2}(\Omega_{\mu,\nu}\Omega^{\mu,\nu} - \frac{1}{2}m_\omega^2 \omega_\mu \omega^\mu)$$
$$- \frac{1}{2}(\mathbf{R}_{\mu,\nu}\mathbf{R}^{\mu,\nu} - \frac{1}{2}m_\rho^2 \boldsymbol{\rho}_\mu \boldsymbol{\rho}^\mu), \tag{3.52}$$

where m_σ, m_ω and m_ρ are the rest masses of mesons, and σ, ω_μ, and $\boldsymbol{\rho}_\mu$ are the scalar, the vector and the vector-isovector field operators, respectively. The arrow indicates a vector in the isospin space. The field tensors of ω and ρ mesons are defined by

$$\Omega^{\mu,\nu} = \partial^\mu \omega^\nu - \partial^\nu \omega^\mu, \tag{3.53}$$
$$\mathbf{R}^{\mu,\nu} = \partial^\mu \boldsymbol{\rho}^\nu - \partial^\nu \boldsymbol{\rho}^\mu. \tag{3.54}$$

The interaction between the nucleons and the mesons is given by the ansatz of minimal coupling:

$$\mathcal{L}_{int} = -g_\sigma \overline{\psi}\sigma\psi - g_\omega \overline{\psi}\gamma_\mu \omega^\mu \psi - g_\rho \overline{\psi}\gamma_\mu \boldsymbol{\tau} \cdot \boldsymbol{\rho}^\mu \psi, \tag{3.55}$$

Table 3.4 Parameter set of NL1 for the RMF Lagrangian. The masses are given in MeV, the parameter g_2 in fm^{-1} and the other coupling constants are dimensionless

m_N	m_σ	m_ω	m_ρ	g_σ	g_ω	g_ρ	g_2	g_3
938.0	492.25	795.359	763.0	10.138	13.285	4.976	−12.172	−36.265

with the coupling constants g_σ, g_ω, and g_ρ and the Pauli isospin matrix τ. It was recognized at an early stage that essential physical properties such as nuclear incompressibility and surface properties cannot be reproduced quantitatively by a simple linear parameter sets in the Lagrangian. The model was extended further to include a non-linear self-coupling of σ mesons, replacing the mass term $\frac{1}{2}m_\sigma^2\sigma^2$ by a non-linear potential

$$U(\sigma) = \frac{1}{2}m_\sigma^2\sigma^2 + \frac{1}{3}g_2\sigma^3 + \frac{1}{4}g_3\sigma^4, \tag{3.56}$$

where g_2 and g_3 are adjustable parameters to reproduce the physical properties of finite nuclei. The equations of motion for the baryon and meson fields can be obtained by a variational principle for a Lagrangian density which yield the Euler-Lagrange equations of motion

$$\partial_\mu\left(\frac{\partial\mathcal{L}}{\partial(\partial_\mu q_j)}\right) - \frac{\partial\mathcal{L}}{\partial q_j} = 0 \tag{3.57}$$

where the variables are $(q_j = \phi_1, ..., \phi_A, \sigma, \omega, \rho)$. The wave function ψ is a direct product of single-particle wave function: $\psi = \phi_1\phi_2...., \phi_A$. For the spinor field ϕ_i, one obtains the Dirac equation

$$\left(\gamma_\mu(i\partial^\mu + g_\omega\omega^\mu(x) + g_\rho\tau\cdot\rho^\mu(x)) + m + g_\sigma\sigma(x)\right)\phi_i = 0. \tag{3.58}$$

For the meson fields, one obtains the Klein-Gordon equations for the scalar and vector mesons. A parameter set NL1 [12] is given in Table 3.4.

The calculated binding energies and the charge radii of four closed shell nuclei are shown in Table 3.5 together with experimental data. The Brückner HF (BHF) calculations are performed with the Reid soft core potential as a bare force and summing up all the two-particle scattering processes in the nuclear medium. The BHF results are not at all satisfactory giving only less than half the experimental binding energies and the rms radii are about 10–20% smaller than the empirical ones. To cure this shortcoming it was argued that three-particle correlation terms or relativistic effects should be taken into account in the calculations. This BHF problem remains still an open serious question of nuclear many-body problem for future study. The calculated results of three commonly used effective models are shown in Table 3.5. HF models with Skyrme and Gogny interactions are non-relativistic mean field models, while RMF is a relativistic Hartree model. All these models give

Table 3.5 Binding energies and rms charge radii of closed-shell nuclei. The BHF stands for Brückner-HF calculations with the Reid soft core potential. Skyrme and Gogny are HF results of effective interactions SIII and D1, respectively, while RMF is relativistic mean field (Hartree) calculations with NL1 interaction

Nuclei		BHF	Skyrme	Gogny	RMF	Exp.
$^{16}_{8}$O	E/A	−3.91	−7.96	−7.80	−7.95	−7.98
	r_c	2.50	2.69	2.74	2.78	2.73
$^{40}_{20}$Ca	E/A	−3.88	−8.54	−8.45	−8.56	−8.55
	r_c	3.04	3.48	3.44	3.50	3.49
$^{90}_{40}$Zr	E/A	—	−8.70	−8.66	−8.74	−8.71
	r_c	—	4.32	4.23	4.28	4.25
$^{208}_{82}$Pb	E/A	−2.52	−7.87	−7.86	−7.85	−7.87
	r_c	4.51	5.57	5.44	5.51	5.50

satisfactory results for both the binding energies and the rms charge radii for the three nuclei. The HF charge radii and charge densities of Gogny model are compared with empirical one extracted from electron scattering experiments. The charge density was obtained by convoluting the finite-size proton form factor

$$f_p(r) = \frac{1}{(r_0\sqrt{\pi})^3} e^{-r^2/r_0^2}, \qquad r_0 = 0.65 \text{ fm}, \tag{3.59}$$

into the proton density

$$\rho_c(\mathbf{r}) = \int f_p(\mathbf{r} - \mathbf{s})\rho_p(\mathbf{s})d\mathbf{s}. \tag{3.60}$$

The calculated HF charge density distribution of ^{208}Pb with a Gogny D1' interaction is shown in Fig. 3.3. The single-particle wave functions of three s orbits are also shown in the same figure. The empirical evidence of electron scattering data shows that independent particle approximation (mean field) is to a large extent a satisfactory approximation. More details about the empirical evidence of independent particle approximation (mean field) are discussed in Chap. 5.

3.6.3 Pairing Interactions and BCS/Bogolyubov Approximation

The Hartree–Fock (HF) method describes quantitatively the ground state properties of closed-shell nuclei such as ^{16}O, ^{40}Ca, ^{90}Zr and ^{208}Pb as shown in the previous Sect. 3.6.2. The variational principle is equivalent to the requirement to impose no interference between the ground state and the excited states. In other words, the ground state and the excited states are calculated separately in the model. The excited

Fig. 3.3 Calculated proton and charge densities of ^{208}Pb by the HF model using a Gogny interaction D1'. The wave functions of $s_{1/2}$ orbitals are also shown in the figure. The charge density is obtained by folding the finite size of the proton into the HF proton density. See Chap. 4 for details of experimental methods to determine charge densities. Figure reprinted with permission from [8]. ©2021 by the American Physical Society

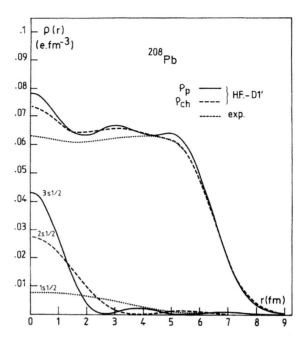

states are described by particle-hole (p-h) excitations from the HF ground state. The HF method takes into account partially the two-body interactions as the mean field approximation, but the residual interaction V_{res} is neglected in the long-range part of the nuclear force as discussed with Eq. (3.3). For non-closed-shell nuclei (the open-shell nuclei), we should take into account an important residual interaction, i.e., the pairing interaction, which causes the superfluidity phase, in which two nucleons couple to zero total angular momentum and behave like a boson with zero viscosity (very large mean free path) in the ground state. This pairing correlations appears in many experimental observables. Typical examples are given below.

1. The energy difference between the ground and the first excited states in even-even nuclei is significantly larger than that of even-odd nuclei. For example, a typical energy difference between the first excited and the ground states is more than 1 MeV in even Sn isotopes, while there are many states below a few hundreds keV in odd Sn isotopes.
2. The moment of inertia of deformed rare-earth nuclei is typically one half of the rigid rotor one because of the superfluidity.
3. The so-called pairing energy gap manifests itself in the mass staggering between even and odd mass nuclei over wide regions of the mass table.

In the liquid-drop (or the Bethe-Weizsäcker) mass formula (4.3) in Chap. 4, the odd-even effect can be described by an extra binding energy term Δ,

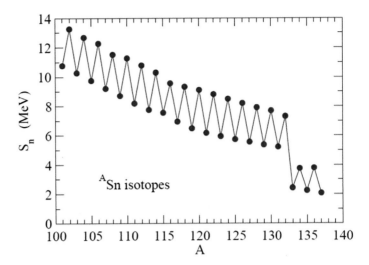

Fig. 3.4 The neutron separation energies of Sn isotopes as a function of mass number A. The separation energy is obtained by a difference of binding energies (3.62) between neighboring Sn isotopes. Since the even isotope gets an extra binding energy Δ in Eq. (3.61), the separation energy is larger than that of the odd isotope as seen in the figure. A sudden drop of S_n at $A = 132$ is due to the closure shell effect at N=82

$$\delta B = \begin{cases} \Delta & Z \text{ even } N \text{ even} \\ 0 & A \text{ odd} \\ -\Delta & Z \text{ odd } N \text{ odd} \end{cases}, \qquad (3.61)$$

where $\Delta = a_P/\sqrt{A}$ with $a_P = 12$ MeV. The term Δ is often called the pairing gap energy due to the *pairing effect* in nuclei and creates an odd-even staggering in the binding energies of isotopes and isotones. To see this staggering in the experimental data, the separation energy of one neutron from the nucleus is shown in Fig. 3.4.

The one neutron separation energy is defined as a difference of binding energies between two neighboring isotopes

$$S_n(A = N + Z) = B(N, Z) - B(N - 1, Z). \qquad (3.62)$$

In Fig. 3.4, the separation energies are plotted as a function of mass number for Sn isotopes. One can see a clear staggering effect in the separation energies which shows larger values for even-Sn isotopes than those of neighbouring odd-Sn isotopes. The pairing gap index Δ is equally expressed by a difference of S_n between neighboring isotopes

$$\Delta = (-1)^{(A+1)} \frac{1}{2} (S_n(A + 1) - S_n(A))$$

$$= (-1)^{(A+1)} \frac{1}{2} (B(N + 1, Z) - 2B(N, Z) + B(N - 1, Z)). \qquad (3.63)$$

In Sn isotopes, the pairing gap is determined empirically to be $\Delta \sim 1$ MeV.

To take into account the pairing correlations on top of the mean field HF model, the BCS or HF+Bogolyubov model are often introduced [13]. These methods do not provide an exact solution of the eigenvalue problem, but are adopted to be a useful tool based on the variational principle. The basic novelty is to introduce a new type of fermions, the "quasi-particles". The concept of quasi-particle is a general concept in many-body physics. By introducing quasi particles, the simplicity of single-particle models is preserved. However, the ground state is no longer a single-Slater determinant, but will be defined as the so-called BCS vacuum in Eqs. (3.68) and (3.69), which does not have a definite number of physical particles. The quasi-particle $\alpha_k^\dagger(\alpha_{\bar{k}}^\dagger)$ is defined by the following Bogolyubov transformation,

$$\alpha_k^\dagger = u_k a_k^\dagger - v_k a_{\bar{k}},$$
$$\alpha_{\bar{k}}^\dagger = v_k a_k^\dagger + u_k a_{\bar{k}}, \tag{3.64}$$

where $a_k^\dagger$ and a_k are the creation and the annihilation operators of bare particles defined in Sect. 3.5.1, respectively, and k stands for the angular momentum and its projection $k \equiv (j, m)$. For each state $k > 0$ ($j, m > 0$), the conjugate state $\bar{k}$ can be given by the time-reversed state

$$a_{\bar{k}}^\dagger \equiv a_{\overline{jm}}^\dagger = (-1)^{j+m} a_{j-m}^\dagger, \tag{3.65}$$

which corresponds to a creation operator of a nucleon with the same j, but its projection has the opposite sign. The coefficients u_k and v_k in Eq. (3.64) are variational parameters with the meaning of the unoccupation and the occupation amplitudes, respectively. They satisfy the normalization condition

$$|u_k|^2 + |v_k|^2 = 1. \tag{3.66}$$

The quasi-particle operators satisfy the anti-commutation relation as

$$\{\alpha_k, \alpha_{k'}^\dagger\} = \alpha_k \alpha_{k'}^\dagger + \alpha_{k'}^\dagger \alpha_k = \delta_{k,k'}. \tag{3.67}$$

The quasi-particle vacuum $|BCS\rangle$ is defined by

$$\alpha_k |BCS\rangle = 0. \tag{3.68}$$

The ground state of the many-body wave function is represented by the quasi-particle vacuum state as

$$|BCS\rangle = \prod_k \alpha_k |0\rangle, \tag{3.69}$$

where $|0\rangle$ is the HF vacuum state defined by $a_k |0\rangle = 0$ where k represents all single-particle states. The BCS state is rewritten to be

$$|\text{BCS}\rangle = \prod_{k>0}(u_k + v_k a_k^\dagger a_{\bar{k}}^\dagger)|0\rangle. \tag{3.70}$$

It is clearly seen in Eq. (3.70) that the particle number conservation is badly violated with respect to the HF vacuum.

Problem

3.4 Derive the BCS vacuum state (3.70) from Eq. (3.69).

We take the many-body Hamiltonian

$$H = \sum_{k_1,k_2} t_{k_1 k_2} a_{k1}^\dagger a_{k2} + \frac{1}{4} \sum_{k_1,k_2,k_3,k_4} V_{k_1 k_2 k_3 k_4} a_{k1}^\dagger a_{k2}^\dagger a_{k4} a_{k3}, \tag{3.71}$$

where $t_{k_1 k_2}$ and $V_{k_1 k_2 k_3 k_4}$ are the matrix elements of kinetic energy and the two-body interaction, respectively. The parameters v_k and u_k in the BCS vacuum (3.70) are determined by a variation of the energy with respect to v_k,

$$\langle \text{BCS}|H|\text{BCS}\rangle, \tag{3.72}$$

under the subsidiary condition for the particle number conservation,

$$\langle \text{BCS}|\hat{N}|\text{BCS}\rangle = \sum_k v_k^2 = 2\sum_{k>0} v_k^2 = N, \tag{3.73}$$

where $\hat{N}$ is the number operator $\hat{N} = \sum_k a_k^\dagger a_k$. Since the BCS vacuum (3.70) does not hold the particle number conservation, the constraint (3.73) for the average particle number is needed in nuclear system with mass $N = (100 \sim 200)$. For metallic superconductor, the number of electrons is of the order of the Avogadro number $\sim 6 \times 10^{23}$ or more, so that the particle number fluctuation of the BCS vacuum is not an issue to be taken care of. The subsidiary condition for the particle number conservation can be implemented by the variational Hamiltonian,

$$H' = H - \lambda\hat{N}, \tag{3.74}$$

where the Lagrange multiplier λ is fixed by the condition (3.73). The expectation value of the variational Hamiltonian H' for the BCS vacuum reads

$$\langle \text{BCS}|H'|\text{BCS}\rangle = \sum_k (t_{kk} - \lambda)v_k^2 + \frac{1}{2}\sum_{k,k'} V_{kk'kk'} v_k^2 v_{k'}^2 + \sum_{k,k'>0} V_{k\bar{k}k'\bar{k}'} u_k v_k u_{k'} v_{k'}. \tag{3.75}$$

The variation is taken for the constrained Hamiltonian $H' \equiv H - \lambda\hat{N}$ with respect to v_k:

$$\delta \langle \text{BCS}|H'|\text{BCS}\rangle = 0 \Rightarrow \left(\frac{\partial}{\partial v_k} + \frac{\partial u_k}{\partial v_k} \frac{\partial}{\partial u_k} \right) \langle \text{BCS}|H'|\text{BCS}\rangle = 0. \qquad (3.76)$$

The variation of v_k (3.76) turns out to be

$$2\varepsilon_k u_k v_k + \Delta_k (v_k^2 - u_k^2) = 0 \text{ for } k > 0, \qquad (3.77)$$

where the single-particle energy ε_k reads

$$\varepsilon_k = \frac{1}{2} \left\{ t_{kk} + t_{\bar{k}\bar{k}} + \sum_{k'} (V_{kk'kk'} + V_{\bar{k}k'\bar{k}k'}) v_{k'}^2 \right\} - \lambda, \qquad (3.78)$$

and the gap parameter Δ_k is defined by

$$\Delta_k = -\sum_{k'>0} V_{k\bar{k}k'\bar{k}'} u_{k'} v_{k'}. \qquad (3.79)$$

The BCS equation (3.77) gives relations between the gap parameter Δ_k and v_k, and also between Δ_k and u_k. The v_k and u_k are eventually determined with the use of the condition (3.66) to be

$$v_k^2 = \frac{1}{2} \left(1 - \frac{\varepsilon_k - \lambda}{\sqrt{(\varepsilon_k - \lambda)^2 + \Delta_k^2}} \right), \qquad (3.80)$$

$$u_k^2 = \frac{1}{2} \left(1 + \frac{\varepsilon_k - \lambda}{\sqrt{(\varepsilon_k - \lambda)^2 + \Delta_k^2}} \right). \qquad (3.81)$$

The gap parameter Δ_k can be expressed in a self-consistent way as

$$\Delta_k = -\frac{1}{2} \sum_{k'>0} V_{k\bar{k}k'\bar{k}'} \frac{\Delta_{k'}}{\sqrt{\varepsilon_{k'} + \Delta_{k'}^2}}. \qquad (3.82)$$

Problem

3.5 Derive the gap equation (3.77) from the variational equation (3.76).

Problem

3.6 Obtain the BCS parameters v_k and u_k of Eqs. (3.80) and (3.81) by using the normalization condition (3.66).

In the BCS model, the radial wave function is the same for both the bare and quasi-particle wave functions since the quasi-particle transformation involves only one single-particle state in Eq. (3.64). In general, the radial wave function of quasi-particles may differ from the one of bare particles, for example, in the case of loosely-bound single-particle state due to the coupling to the unbound continuum states in the HF potential. The HFB theory is a generalization of the BCS theory being able to accommodate the loosely-bound nature of single-particle wave function. The quasi-particle is defined by an unitary transformation,

$$\alpha_k^\dagger = \sum_l u_{kl} a_l^\dagger - v_{kl} a_{\bar{l}},$$

$$\alpha_{\bar{k}} = \sum_l v_{kl} a_l^\dagger + u_{kl} a_{\bar{l}}, \qquad (3.83)$$

where u_{kl} and v_{kl} satisfy an orthonormal condition,

$$\sum_l (u_{kl}^* u_{k'l} + v_{kl}^* v_{k'l}) = \delta_{k,k'}. \qquad (3.84)$$

The HFB formula (3.83) involves many-orbitals with the same jm but different nodes so that the quasi-particles contain both bound and unbound states by the HFB transformation (3.83). When active orbitals are well bound as in ordinary nuclei along the valley of stability, major orbitals involved in the BCS or the HFB calculations are well-bound and do not change appreciably their wave functions by the pairing correlations. Thus, the standard BCS treatment is entirely adequate. This may not be the case for drip-line nuclei, in which the Fermi energy is close to the particle threshold and the pairing can couple bound and unbound continuum states. Therefore, the radial profile of a quasi-particle wave function will be largely different to the bare particle state even having the same jm. Thus the HFB will be a better model for describing the pairing correlations in nuclei far from the stability line where the coupling to unbound continuum states becomes important.

3.6.4 Tamm-Dancoff and Random Phase Approximations

The HF ground state can be constructed to fill up the mean field potential with A nucleons up to a Fermi level ε_F from the bottom of the potential in order. With respect to this ground state, there is in general a set of many-particle many-hole (m-particle m-hole) excitations with $m = 0, 1, 2, \cdots, A$ from the Fermi level, which is a complete orthogonal set to be used to construct the true many-body wave functions of the ground state and the excited states for a given Hamiltonian H. The exact solution can be obtained by a diagonalization of the full model space or by a variation of the expansion coefficients X for the ground state $|0\rangle$ and the excited state $|\lambda\rangle$

$$|0\rangle = X_0^0 |\text{HF}\rangle + \sum_{mi} X_{mi}^0 a_m^\dagger a_i |\text{HF}\rangle + \frac{1}{4} \sum_{mnij} X_{mn,ij}^0 a_m^\dagger a_n^\dagger a_i a_j |\text{HF}\rangle + ..., (3.85)$$

$$|\lambda\rangle = X_0^\lambda |\text{HF}\rangle + \sum_{mi} X_{mi}^\lambda a_m^\dagger a_i |\text{HF}\rangle + \frac{1}{4} \sum_{mnij} X_{mn,ij}^\lambda a_m^\dagger a_n^\dagger a_i a_j |\text{HF}\rangle + ..., (3.86)$$

where m, n stand for the particle (unoccupied) states and i, j stand for the hole (occupied) states, respectively. The HF wave function $|\text{HF}\rangle$ is a determinant of A particles to fill up the lowest single-particle levels of the mean field potential up to the Fermi level. The exact diagonalization of H with the full model space is an extremely difficult task because of astronomical large number of configurations. Limiting the model space makes the calculations feasible for not only light nuclei but also very heavy nuclei. Some commonly adopted models are [14]:

1. The Tamm-Dancoff approximation (TDA) limits the model space within 1 particle-1 hole ($1p$-$1h$) excitations from the HF ground state (vacuum). The TDA model takes the first term of Eq. (3.85) for the ground state and the second term of Eq. (3.86) for the excited state.
2. The random phase approximation (RPA) is an extension of TDA. The excited states have $1p$-$1h$ and $1h$-$1p$ excitations in the model space from the RPA vacuum. The ground state, the RPA vacuum, is not the HF vacuum, but includes implicitly many-body correlations on top of the HF vacuum by definition. In a simple shell-model picture, the particles in the ground state occupy the single-particle states from the bottom of the single-particle potential in order. The energy of the highest occupied state is named "*the Fermi energy*". The p-h excitation means that a particle below the Fermi energy is excited to an unoccupied state above the Fermi energy and makes a hole below the Fermi energy. This is the TDA state. On the other hand, since the RPA ground state is not the single-Slater determinant as the TDA one, but also has $2p$-$2h + \cdots$ many-p-many-h components, it is possible to have h-p excitation, in which the particle above the Fermi surface is excited to the state below the Fermi energy in the TDA picture. In RPA as well as TDA, one can adopt appropriately large model space for both low-energy collective states and highly excited giant resonance states.
3. The interacting shell model limits a model space for active single-particle states and takes into account all the terms in Eqs. (3.85) and (3.86). For example, take one major shell such as sd shell or pf shell and the configuration mixings among active particles are fully taken into account in the calculations.

The TDA and RPA theories can be extended to open shell nuclei replacing the bare particles by the quasi-particles of Bogolyubov transformation for the pairing correlations. These models are also applied to study the excited states of deformed nuclei adopting the Nilsson single-particle states as the basis ones.

In the HF model, the many-body interactions (the diagonal matrix elements such as $\langle ij|\tilde{V}|ij\rangle$ in Eq. (3.49)) are partially taken into account in the single-particle energies. Because of this approximation, some symmetries guaranteed in the original Hamiltonian are badly broken, for example, the translational invariance. As was

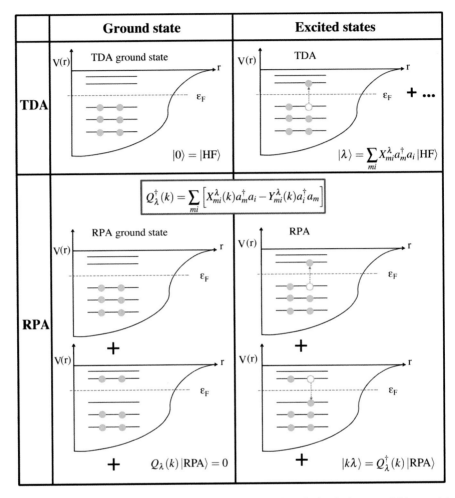

Fig. 3.5 (Upper left) TDA ground state. (upper-right): A p-h excitation in the mean field potential in TDA. The single-particle orbitals up to the Fermi energy ε_F are fully occupied by A particles from the bottom of the potential $V(r)$ in order in the HF vacuum for TDA. (Lower left) RPA ground state. (Lower right): p-h and h-p excitations from the ground state in RPA. The RPA ground state involves $0p$-$0h$+$2p$-$2h$+many-body correlations which make $h - p$ excitations possible in the excited state

discussed in Sect. 3.4, the translational invariance is required to the two-body interaction so that its position dependence of two particles should be the relative coordinate $\mathbf{r}_1 - \mathbf{r}_2$. In the mean field approximation, the Hamiltonian density does not depends on the relative coordinate, but on the coordinate $\mathbf{r}$ which is the radial coordinate of a particle measured from the center of mass of nucleus. In order to restore this symmetry, we have to take into account the full interactions in the model in due course. One possible model is a self-consistent HF+RPA model in which the residual interaction V_{res} in Eq. (3.3) is taken into account in RPA calculations so that the original full

Hamiltonian is included in the calculations. Then the translational invariance broken in the HF level will be recovered in the final RPA results. This point is particularly important to eliminate the spurious modes such as the center of mass motion from excitation spectra.

With the pairing correlations, the quasi-particle is introduced instead of real particles in the theory. Thus, quasi-particle RPA (QRPA) model on top of the HF-BCS or the HF-Bogolyubov model is adopted to describe excitation spectra of open-shell nuclei. The RPA correlations are responsible for substantial enhancement of electromagnetic and weak transition rates. Particularly the RPA theories are applied to study several different kinds of vibrational modes; isoscalar (IS) low-lying quadrupole and octupole excitations, IS monopole and quadrupole giant resonances (GR), and the isovector (IV) dipole giant resonance (GDR). It is also applied to study the charge exchange excitations; the isobaric analogue state (IAS), the Gamow-Teller giant resonances (GTGR), and the spin-dipole (SD) excitations.

Firstly, we start to discuss the Tamm-Dancoff approximation (TDA). In this model, the ground state is assumed to be a HF Slater determinant and the excited state is described by a linear combination of 1-particle 1-hole ($1p$-$1h$) states on top of the HF ground state. That is, we take the first term in Eq. (3.85) for the ground state and the second term in Eq. (3.86) for the excited states:

$$|0\rangle = |\mathrm{HF}\rangle, \tag{3.87}$$

$$|\lambda\rangle = \sum_{mi} X_{mi}^{\lambda} a_m^{\dagger} a_i |\mathrm{HF}\rangle, \tag{3.88}$$

where λ assigns the quantum number of excited state. The first term of Eq. (3.86) disappears in Eq. (3.88) to ensure the orthogonality condition $\langle 0|\lambda\rangle=0$. For TDA calculation, we consider the general Hamiltonian

$$H = \sum_k \varepsilon_k a_k^{\dagger} a_k + \frac{1}{4} \sum_{kk',ll'} \tilde{V}_{kk'll'} : a_k^{\dagger} a_{k'}^{\dagger} a_{l'} a_l :, \tag{3.89}$$

where ε_k is the HF single-particle energy and $\tilde{V}_{kk'll'} = V_{kk'll'} - V_{kk'l'l}$ is the antisymmetrized two-body matrix element. The second term, the residual interaction, in Eq. (3.89) is expressed by the normal product of creation and annihilation operators $a^{\dagger}$ and a since the HF terms of two-body interactions are taken into account in the single-particle energy ε_k as is shown in Eq. (3.49). In the normal product, the indices have the selection rules $k \neq l$, $k \neq l'$, $k' \neq l$, and $k' \neq l'$.[5] These selection rules are due to the fact that the two-body matrix elements such as $k = l$ and $k' = l'$ are already included in the HF single-particle energy. The eigenstate of the Hamiltonian (3.89) is defined by an equation

$$H|\lambda\rangle = E_{\lambda}|\lambda\rangle, \tag{3.90}$$

---

[5] For more details of the second quantization and the normal product, see D. J. Rowe, Nuclear Collective Motions: Models and Theory, Appendix C (World Scientific, 2010).

with the eigenenergy E_λ. The excited state $|\lambda\rangle$ is written as

$$|\lambda\rangle = Q_\lambda^\dagger |0\rangle \quad \text{and} \quad Q_\lambda |0\rangle = 0, \tag{3.91}$$

where $Q_\lambda^\dagger$ is the excitation operator given by

$$Q_\lambda^\dagger = \sum_{mi} X_{mi}^\lambda a_m^\dagger a_i. \tag{3.92}$$

We now obtain the TDA equations for the amplitudes X_{mi}^λ from (3.90). Multiplying a p-h state $|nj\rangle$ to Eq. (3.90) from the left, we obtain the equation, the so-called secular equation,

$$(\langle nj|H|\lambda\rangle - E_\lambda \langle nj|\lambda\rangle) = \sum_{mi} \left\{ \delta_{mn}\delta_{ij}(\varepsilon_m - \varepsilon_i) + \tilde{V}_{njmi} - E_\lambda \right\} X_{mi} = 0, \tag{3.93}$$

where the following relations of normal products are used:

$$\langle nj| \sum_k \varepsilon_k a_k^\dagger a_k |mi\rangle = \langle 0|a_j^\dagger a_n \left(\sum_k \varepsilon_k a_k^\dagger a_k \right) a_m^\dagger a_i |0\rangle = \delta_{nm}\delta_{ij}(\varepsilon_n - \varepsilon_i)$$

and

$$\langle nj| \frac{1}{4} \sum_{kk',ll'} \tilde{V}_{kk'll'} : a_k^\dagger a_{k'}^\dagger a_{l'} a_l : |mi\rangle = \langle 0|a_j^\dagger a_n \left(\frac{1}{4} \sum_{kk',ll'} \tilde{V}_{kk'll'} : a_k^\dagger a_{k'}^\dagger a_{l'} a_l : \right) a_m^\dagger a_i |0\rangle$$

$$= \frac{1}{4}(V_{nijm} - V_{injm} - V_{nimj} + V_{inmj}) = V_{nijm}.$$

The TDA equation (3.93) is solved by a diagonalization of the matrix equation for given HF single-particle energies and the residual interaction $\tilde{V}$ with the normalization condition

$$\langle \lambda|\lambda\rangle = \sum_{mi} |X_{mi}^\lambda|^2 = 1. \tag{3.94}$$

In the shell model term, TDA is equivalent to the interacting shell model with $1p$-$1h$ configuration space.

Problem

3.7 Prove that the two-body matrix element of V_{res} in Eq. (3.89) for the p-h states $|nj\rangle$ and $|mi\rangle$ reads

$$\langle nj|V_{\text{res}}|mi\rangle = \tilde{V}_{nijm}, \tag{3.95}$$

where $V_{\text{res}} = \frac{1}{4} \sum_{kk',ll'} \tilde{V}_{kk'll'} : a_k^\dagger a_{k'}^\dagger a_{l'} a_l :$.

The TDA theory assumes the ground state is the HF Slater determinant so that only $1p$-$1h$ excitations are available in the model space. In general, however, the ground state might have $2p$-$2h$, $3p$-$3h$, and many-p many-h correlations. If it is the case, the excitation operator involves not only p-h pair but also h-p pair with respect to the ground state. A straightforward generalization of the TDA operator (3.92) is the RPA operator

$$Q_\lambda^\dagger(k) = \sum_{mi} [X_{mi}^\lambda(k) a_m^\dagger a_i - Y_{mi}^\lambda(k) a_i^\dagger a_m], \tag{3.96}$$

where the second term creates a hole state destroying a particle state with respect to the Fermi energy shown in Fig. 3.5. The label k denotes kth excited state for the given multipole λ. The RPA excited and ground states are defined by analogy with the TDA ones (3.91) as

$$|k\lambda\rangle = Q_\lambda^\dagger(k)|\text{RPA}\rangle \quad \text{and} \quad Q_\lambda(k)|\text{RPA}\rangle = 0. \tag{3.97}$$

Since $a_i^\dagger a_m|\text{HF}\rangle = 0$, one can see in Eq. (3.97) that the RPA ground state is not just the HF Slater determinant, but includes the many-particle many-hole correlations such as $2p - 2h$ states as sketched in Eq. (3.85). For RPA, we have two amplitudes X_{mi}^λ and Y_{mi}^λ, therefore we need two sets of equations to be solved. The equation of motion for RPA can be expressed as

$$[H, Q_\lambda^\dagger]|\text{RPA}\rangle = H||\text{RPA}\rangle - Q_\lambda^\dagger H|\text{RPA}\rangle = (E_\lambda - E_0)||\text{RPA}\rangle,$$
$$= \hbar\omega||\text{RPA}\rangle, \tag{3.98}$$

where $\hbar\omega = (E_\lambda - E_0)$. To make the commutators of $a_j^\dagger a_n$ and $a_n^\dagger a_j$ with $[H, Q_\lambda^\dagger]$ in Eq. (3.98), we can obtain

$$\langle\text{RPA}| \left[a_j^\dagger a_n, [H, Q_\lambda^\dagger]\right] |\text{RPA}\rangle = \hbar\omega_\lambda \langle\text{RPA}|[a_j^\dagger a_n, Q_\lambda^\dagger]|\text{RPA}\rangle, \tag{3.99}$$

$$\langle\text{RPA}| \left[a_n^\dagger a_j, [H, Q_\lambda^\dagger]\right] |\text{RPA}\rangle = \hbar\omega_\lambda \langle\text{RPA}|[a_n^\dagger a_j, Q_\lambda^\dagger]|\text{RPA}\rangle. \tag{3.100}$$

To evaluate the commutators in Eqs. (3.99) and (3.100), we take an approximation known as the "*quasi-boson approximation*", in which the correlated RPA ground state is taken to be not so much different from the HF ground state so that all the expectation values of the commutators are calculated as

$$\langle RPA|[a_j^\dagger a_n, a_m^\dagger a_i]|RPA\rangle \approx \langle\text{HF}|[a_j^\dagger a_n, a_m^\dagger a_i]|\text{HF}\rangle = \delta_{mn}\delta_{ij}. \tag{3.101}$$

The name "*quasi-boson* " comes from the fact that Eq. (3.101) is exactly the same as the boson commutation relation $[B_b, B_a^\dagger] = \delta_{ab}$ for creation and annihilation boson operators $B_a^\dagger$ and B_b, namely, so that the p-h pairs behave in Eq. (3.101) like the boson operators. The r.h.s of Eqs. (3.99) and (3.100) are evaluated now for the HF vacuum in a similar way to those of TDA in Eq. (3.93);

$$A_{minj} \equiv \langle HF| \left[a_i^\dagger a_m, [H, a_n^\dagger a_j]] \right] |HF\rangle = (\varepsilon_m - \varepsilon_i)\delta_{mn}\delta_{ij} + \widetilde{V}_{mjin},$$

$$B_{minj} \equiv -\langle HF| \left[a_i^\dagger a_m, [H, a_j^\dagger a_n]] \right] |HF\rangle = \widetilde{V}_{mnij}, \tag{3.102}$$

where the matrices are calculated with respect to the HF vacuum and the matrix A is Hermitian, while B is anti-Hermitian. The l.h.s. of Eqs. (3.99) and (3.100) give

$$\langle HF|[a_j^\dagger a_n, Q_\lambda^\dagger]|HF\rangle = X_{nj}^\lambda, \tag{3.103}$$

$$\langle HF|[a_n^\dagger a_j, Q_\lambda^\dagger]|HF\rangle = Y_{nj}^\lambda. \tag{3.104}$$

The RPA equations (3.99) and (3.100) can be rewritten in a compact form:

$$\begin{pmatrix} A & B \\ B^* & A^* \end{pmatrix} \begin{pmatrix} X \\ Y \end{pmatrix} = \hbar\omega \begin{pmatrix} 1 & 0 \\ 0 & -1 \end{pmatrix} \begin{pmatrix} X \\ Y \end{pmatrix}, \tag{3.105}$$

where X, Y are column vectors as $(X)_{mi}^\lambda = X_{mi}^\lambda$, $(Y)_{mi}^\lambda = Y_{mi}^\lambda$, respectively.

For the self-consistent HF+RPA model, the HF equation (3.48) is firstly solved iteratively to obtain the single-particle energies and wave functions. Secondly, the two-body ph matrix elements $\widetilde{V}_{mjin}$ and $\widetilde{V}_{mnij}$ are calculated by using the same interaction for the HF calculations. Finally, the RPA secular equations (3.105) are diagonalized in a large model space even enough for highly excited states under the normalization condition,

$$\langle RPA|[Q_\lambda, Q_\lambda^\dagger]|RPA\rangle = \sum_{mi} \left(|X_{mi}^\lambda|^2 - |Y_{mi}^\lambda|^2 \right) = 1. \tag{3.106}$$

The transition strength for one-body transition operator O from the ground state to the excited state is expressed by the X and Y amplitudes as

$$|\langle\lambda|O|RPA\rangle|^2 = \left| \sum_{mi} (X_{mi}\langle i|O|m\rangle + Y_{mi}\langle m|O|i\rangle) \right|^2. \tag{3.107}$$

Since the RPA solution includes the ground state correlations expressed by Y_{mi} amplitudes as seen in the normalization condition for RPA (3.106), the solutions of RPA equations (3.105) give in general better descriptions of excited states providing more collectivity for the transition strength (3.107) of both low excitation collective states and highly excited giant resonance states.

3.7 Beyond the Mean Field Approaches

Mean field models such as the HF, the RMF and the RPA[6] have been successful to describe various types of collective states. There are several beyond mean-field models to take into account correlations more than $1p$-$1h$ and $1h$-$1p$ excitations. One of the sophisticated models is the particle-vibration coupling model (PVC), another is the second RPA or the extended model. These models accommodate $2p$-$2h$ excitations on top of $1p$-$1h$ excitations to describe better the phenomena mentioned earlier. The generator coordinate method (GCM) is also used to accommodate the correlations beyond mean field models.

3.7.1 Particle-Vibration Coupling (PVC) Model

The PVC model was invented to take into account the correlations beyond RPA type $1p$-$1h$ excitations (or two quasi-particle configurations in QRPA) [15]. The PVC is analogous to a specific potential employed in the analysis of condensed atomic systems, for example, electron-phonon coupling in metals and particle-phonon coupling in liquid ^{3}He-^{4}He mixtures. The PVC model was originally developed based on the Fermi liquid theory by Landau and Migdal. In this section, we present a modern PVC theory in details based on the self-consistent HF+RPA model. The self-consistent Green's function (SCGF) model is also based on the idea of the PVC formalism.

We divide the nuclear configurations into two subspaces S_1 and S_2 projected by the operators Q_1 and Q_2, respectively. The sum of two subspaces S_1 and S_2 includes all physical states. The subspace S_1 is made up with the HF ground state and all possible $1p$-$1h$ TDA or RPA excitations, while S_2 represents complex compound states, which involve highly extensive configuration mixings in the continuum and resonance states. The operators Q_1 and Q_2 satisfy

$$Q_1 + Q_2 = 1, \quad Q_1^2 \equiv Q_1, \quad Q_2^2 = Q_2, \quad Q_1 Q_2 = 0. \tag{3.108}$$

In PVC model, the coupling between the two subspaces is analyzed introducing rather simple configurations called "doorway states", representing the configurations of $2p$-$2h$ states in S_2 model space, equivalently the coupled states of p-h configurations to other vibrational degrees of freedom. The doorway states represent compound many-particle many-hole states with all degrees of freedom distributed in a statistical way in energy. The effective Hamiltonian, which includes the coupling between S_1 and S_2 subspaces, is given in the lowest order by

---

[6] RPA is the first-order approximation of the time-dependent HF theory so that RPA is included in the category of the (dynamical) mean field model.

$$H(\omega) = Q_1 H Q_1 + W^\downarrow(\omega) \equiv Q_1 H Q_1 + Q_1 H Q_2 \frac{1}{\omega - Q_2 H Q_2 + i\varepsilon} Q_2 H Q_1,$$
$$(3.109)$$

where ω is the excitation energy and H is the Hamiltonian adopted in the HF and RPA calculations. The coupling between the space S_1 and S_2 by the Hamiltonian H gives rise to the term $W^\downarrow(\omega)$.

The energy-dependent complex Hamiltonian (3.109) allows to work inside the space S_1. It has complex eigenenergy whose imaginary part originates from the coupling to more complicated configurations and gives rise to the so called spreading width.

The transition strength (3.107) for the excited states can be generalized as the *strength function* at the excitation energy $\hbar\omega$;

$$S(\omega) = \sum_\lambda \!\!\!\!\!\!\!\int |\langle \lambda | O | 0 \rangle|^2 \delta(\omega - \omega_\lambda) d\omega_\lambda, \qquad (3.110)$$

where $|\lambda\rangle$ is an excited state with energy $E_\lambda = \hbar\omega_\lambda$ and the ground state is denoted by $|0\rangle$. The symbol $\sum\!\!\!\!\int$ implies the inclusions of both the discrete and continuum states. For the discrete states, the strength function (3.110) is rewritten to be

$$S(\omega) = \sum_\lambda |\langle \lambda | O | 0 \rangle|^2,$$

which is equivalent to the sum of all excited states λ for the transition strength (3.107), assuming the ground state $|RPA\rangle = |0\rangle$. The strength function can be rewritten by using an operator equation $G(\omega)$ as

$$S(\omega) = -\frac{1}{\pi} \lim_{\varepsilon \to 0} \mathrm{Im}\langle 0 | \Omega^\dagger G(\omega) O | 0 \rangle, \qquad (3.111)$$

where $G(\omega)$ is called the "*Green's function*" and defined as

$$G(\omega) = \frac{1}{\omega - H(\omega) + i\varepsilon}. \qquad (3.112)$$

Problem

3.8 Prove the operator equation Eq. (3.109) starting from the Green's function (3.112).

By using the closure relation $\sum\!\!\!\!\int d\omega_\lambda |\lambda\rangle\langle\lambda| = 1$, the expectation value of Eq. (3.111) is expressed as

$$\langle 0\,\Omega^\dagger G(\omega)O|0\rangle = \langle 0\,\Omega^\dagger \frac{1}{\omega - H + i\varepsilon} O|0\rangle$$

$$= \sum_\lambda \!\!\!\!\!\int d\omega_\lambda \langle 0\,\Omega^\dagger|\lambda\rangle\langle\lambda|\frac{1}{\omega - H + i\varepsilon}|\lambda\rangle\langle\lambda|O|0\rangle$$

$$= \sum_\lambda \!\!\!\!\!\int d\omega_\lambda |\langle\lambda|O|0\rangle|^2 \frac{1}{\omega - \omega_\lambda + i\varepsilon}$$

$$= -i\pi \sum_\lambda \!\!\!\!\!\int d\omega_\lambda |\langle\lambda|O|0\rangle|^2 \delta(\omega - \omega_\lambda), \qquad (3.113)$$

where the residue relation

$$\frac{1}{\omega - x + i\varepsilon} = \mathrm{P}\frac{1}{\omega - x} - i\pi\delta(\omega - x) \qquad (3.114)$$

is used. Here the symbol P stands for the Cauchy principal value. We can confirm now Eq. (3.111) is equivalent to the strength function (3.110), which is an appropriate formula to accommodate the continuum effect in the strength function.

Problem

3.9 Prove Eq. (3.114) performing the integral

$$\lim_{\varepsilon \to +0} \int_{-\infty}^{\infty} \frac{f(x)}{x + i\varepsilon} dx \qquad (3.115)$$

with the relation

$$\frac{1}{x + i\varepsilon} = \frac{x}{x^2 + \varepsilon^2} - \frac{i\varepsilon}{x^2 + \varepsilon^2}. \qquad (3.116)$$

We introduce the response function

$$R(\omega) = \langle 0|\hat{O}^\dagger \frac{1}{\omega - H(\omega) + i\varepsilon} \hat{O}|0\rangle, \qquad (3.117)$$

which has real and imaginary parts. By using the response function, we can formulate the excitation spectrum for any one-body transition operator $\hat{O}$ in a compact form. From Eq. (3.111), the strength function is given by the imaginary part of the response function as

$$S(\omega) = -\frac{1}{\pi}\mathrm{Im}R(\omega). \qquad (3.118)$$

Let us discuss how to obtain the eigenstates of the PVC Hamiltonian (3.109).

- We first solve the standard RPA equation (3.105) derived in Sect. 3.6.4. The RPA equation has solutions with energies $\pm\omega_k$. The solution for $+\omega_k$ is denoted $|k\rangle = O_k^\dagger|0\rangle$, while that for $-\omega_k$ is written as $|\bar{k}\rangle = \overline{O}_k^\dagger|0\rangle$.

- Using these RPA solutions, the PVC phonon creation operator $O_\nu^\dagger$ is expressed as a linear combination of $O_k^\dagger$ and $\overline{O}_k^\dagger$

$$O_\nu^\dagger = \sum_{\omega_k > 0} F_k^{(\nu)} O_k^\dagger - \overline{F}_k^{(\nu)} \overline{O}_k^\dagger. \tag{3.119}$$

Then the eigenvalue equation for the Hamiltonian (3.109) is determined by an equation of motion for $O_\nu^\dagger$,

$$[H(\omega), O_\nu^\dagger] = (\Omega_\nu - i\frac{\Gamma_\nu}{2}) O_\nu^\dagger, \tag{3.120}$$

where Ω_ν is the eigenenergy and Γ_ν is the width of the eigenstate ν, respectively. The equation of state for PVC phonon state is an analogous equation to the RPA one (3.98). The difference stems from the second part of the Hamiltonian, $W \downarrow (\omega)$, in (3.109). The PVC secular equation becomes the energy-dependent complex matrix equations for the RPA basis states as

$$\begin{pmatrix} \mathcal{D} + \mathcal{A}_1(\omega) & \mathcal{A}_2(\omega) \\ -\mathcal{A}_3(\omega) & -\mathcal{D} - \mathcal{A}_4(\omega) \end{pmatrix} \begin{pmatrix} F^{(\nu)} \\ \overline{F}^{(\nu)} \end{pmatrix} = \left(\Omega_\nu - i\frac{\Gamma_\nu}{2} \right) \begin{pmatrix} F^{(\nu)} \\ \overline{F}^{(\nu)} \end{pmatrix}, \tag{3.121}$$

where $\mathcal{D}$ is a diagonal matrix containing the physical RPA eigenvalues $\mathcal{D}_{kl} = \delta_{kl}\omega_k$. In Eq. (3.121), $F^{(\nu)}$ and $\overline{F}^{(\nu)}$ are the amplitudes of positive and negative energy eigenvalues and satisfy the orthonormal relation

$$\sum_k F_k^{(\nu)} F_k^{(\nu')} - \overline{F}_k^{(\nu)} \overline{F}_k^{(\nu')} = \delta_{\nu,\nu'}. \tag{3.122}$$

The matrices $\mathcal{A}_i (i = 1, 2, 3, 4)$ in the PVC secular equation are the expectation values of the Hamiltonian $W \downarrow (\omega)$ for the p-h configurations of RPA basis states $O_k^\dagger$ and $\overline{O}_k^\dagger$ in Q_1 space. The explicit forms can be obtained in matrix forms,

$$(\mathcal{A}_1)kl = \langle \hat{0}|[O_l, [W^\downarrow, O_k^\dagger]]|\hat{0}\rangle, \tag{3.123}$$

$$(\mathcal{A}_2)kl = \langle \hat{0}|[\overline{O}_l, [W^\downarrow, O_k^\dagger]]|\hat{0}\rangle, \tag{3.124}$$

$$(\mathcal{A}_3)kl = \langle \hat{0}|[O_l, [W^\downarrow, \overline{O}_k^\dagger]]|\hat{0}\rangle, \tag{3.125}$$

$$(\mathcal{A}_4)kl = \langle \hat{0}|[\overline{O}_l, [W^\downarrow, \overline{O}_k^\dagger]]|\hat{0}\rangle. \tag{3.126}$$

- The electromagnetic and charge-exchange transition strength associated with RPA+PVC model is given for the transition operator $\hat{O}$ as

$$S(\omega) = -\frac{1}{\pi} \text{Im} \sum_\nu \langle 0|\hat{O}|\nu\rangle^2 \frac{1}{\omega - \Omega_\nu + i(\frac{\Gamma_\nu}{2} + \Delta)}. \tag{3.127}$$

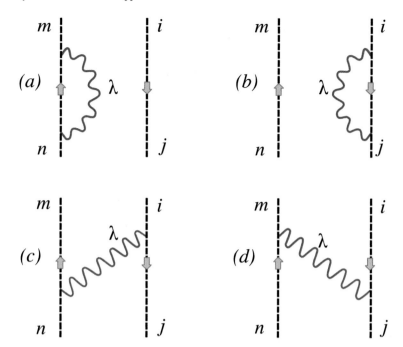

Fig. 3.6 Diagrams of Particle Vibration couplings. RPA phonon states are shown by violet solid wavy lines, while the p-h states are shown by dashed lines. The uparrow and downarrow represent particle and hole states, respectively

The coupling matrix $W^{\downarrow}_{mi,nj}(\omega)$ between p-h states (mi) and (nj) is explicitly written as

$$W^{\downarrow}_{mi,nj} = \langle mi|V\frac{1}{\omega - H}V|nj\rangle = \sum_{N}\langle mi|V|N\rangle\langle N|\frac{1}{\omega - \omega_N}|N\rangle\langle N|V|nj\rangle,$$

(3.128)

where $|N\rangle = |m'i'\rangle \otimes |k\lambda\rangle$ represents a doorway state with particle-hole states $m'i'$ and the kth phonon state with the multipolarity λ. The energy of the doorway state is $\omega_N = \omega^k_\lambda + \varepsilon_{m'} - \varepsilon_{i'}$. The matrix element of the two-body interaction V in Eq. (3.128) is expressed as

$$\langle mi|V|N\rangle = \langle mi|V|m'i'\otimes k\lambda\rangle = \langle 0|a^\dagger_i a_m V a^\dagger_{m'}a_{i'}Q^\dagger_\lambda(k)|0\rangle,$$

(3.129)

where a_i and $a^\dagger_m$ are the annihilation and creation operators, and $Q^\dagger_\lambda(k)$ is the creation operator for phonon with the multiplicity λ, defined in Eq. (3.96). The coupling matrix between $1p$-$1h$ and doorway states is rewritten to be

$$\langle mi|V|N\rangle = \delta_{ii'}\langle m|V|m', k\lambda\rangle + \delta_{mm'}\langle i|V|i', k\lambda\rangle,$$

(3.130)

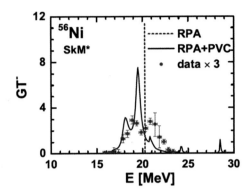

Fig. 3.7 Gamow-Teller strength distributions calculated with the Skyrme interaction SkM* for the nucleus ^{56}Ni. The blue dashed discrete line denotes the RPA strength (with dimensionless units), and the black solid line represents the distribution (with units of MeV^{-1}) calculated by the RPA+PVC model (3.127). The experimental data (with units of MeV^{-1}) are taken from [17]. Figure reprinted with permission from [16]. ©2021 by the American Physical Society

and

$$\langle N|V|mi\rangle = \langle m'i' \otimes k\lambda|V|mi\rangle,$$
$$\langle mi|V|N\rangle = \delta_{ii'}\langle m', k\lambda|V|n\rangle + \delta_{mm'}\langle i', k\lambda|V|j\rangle. \qquad (3.131)$$

The spreading width (3.128) can be diagrammatically classified by four graphs shown in Fig. 3.6. The diagonal terms of Eqs. (3.130) and (3.131) give the contributions (a) and (b);

$$\langle mi|V|N\rangle\langle N|V|nj\rangle|_{\text{diagonal}} \rightarrow \langle m|V|m', k\lambda\rangle\langle m', k\lambda|V|n\rangle + \langle i|V|i', k\lambda\rangle\langle i', k\lambda|V|j\rangle, \qquad (3.132)$$

which give the spreading effect of strength distributions. The cross terms of Eqs. (3.130) and (3.131) give those of (c) and (d),

$$\langle mi|V|N\rangle\langle N|V|nj\rangle|_{\text{cross}} \rightarrow \langle i|V|j, k\lambda\rangle\langle m, k\lambda|V|n\rangle + \langle m|V|n, k\lambda\rangle\langle i, k\lambda|V|j\rangle, \qquad (3.133)$$

which will renormalize the excitation energy of the phonon state.

Figure 3.7 shows RPA and RPA+PVC calculations for charge-exchange Gamow-Teller (GT) states in ^{56}Ni. The GT states are excited by charge exchange reactions such as (p, n) or $(^{3}\text{He}, t)$ reactions with the spin-isospin operator

$$O_-(GT) = \sum_i t_-\sigma, \qquad (3.134)$$

where t_- is the operator for charge exchange between neutron and proton $t_-|n\rangle = |p\rangle$. The GT transition is one of the simplest type of β transition process so called allowed transitions in which the decay amplitude is independent of the position of transforming nucleon[7]. Thus, the GT transition operator depends on the spin operator, but independent of the position and velocity of the nucleon. Another type of allowed transition is the Fermi transition in which the transition operator depends only on the isospin operator. More details of β decays are discussed in Chap. 6.

For RPA calculations, the available configuration for GT transition in ^{56}Ni is $1f_{7/2}(\nu) \rightarrow 1f_{5/2}(\nu)$ configuration so that only one state is shown in Fig. 3.7 as RPA results. The coupling with phonons in RPA+PVC shifts the peaks about 1–2 MeV downward and produces a spreading width. The energy shift originates from the second term $W^{\downarrow}$ in Eq. (3.109); the effect of Q_2 space on Q_1 space. The excitation energies of doorway configurations are much higher than the RPA states in Q_1 space so that the RPA states will be pushed down by the perturbation effect of Q_2 space. The doorway states have more complicated configuration than the $1p - 1h$ RPA states. The couplings to the doorway states will give more fragmentation of RPA strength in general. These are the main reasons for the lower energy and more spreading of the GT strength in PVC calculations than those of RPA.

3.7.2 Second RPA

The Second RPA (SRPA) is a direct way to include $2p$-$2h$ configurations on top of $1p$-$1h$ RPA configuration. In a standard SRPA model, firstly, the RPA equation is solved in the $1p$-$1h$ model space. Then, $2p$-$2h$ configurations are coupled with the RPA states. However, the correlations between $2p$-$2h$ configurations are not taken into account.[8]

In practice, the same interaction as that of RPA calculations will be adopted as the coupling interactions between $2p$-$2h$ and $1p$-$1h$ model space. In SRPA, the excitation operator is written as a natural extension of the RPA one

$$\tilde{Q}^{\dagger}_{\lambda}(k) = \sum_{mi}[X^{\lambda}_{mi}(k)a^{\dagger}_m a_i - Y^{\lambda}_{mi}(k)a^{\dagger}_i a_m]$$
$$+ \sum_{m<n,i<j}[X^{\lambda}_{minj}(k)a^{\dagger}_m a_i a^{\dagger}_n a_j - Y^{\lambda}_{minj}(k)a^{\dagger}_i a_m a^{\dagger}_j a_n]. \quad (3.135)$$

The X and Y are solutions of the SRPA equation

---

[7] There are other kinds of transitions, called "forbidden" transitions, whose operators involve the position and/or momentum of nucleon. The allowed transition is favored (fast), while the forbidden one is unfavored (slow).

[8] Recently, the full SRPA calculations were done in the closed-shell nuclei ^{16}O and ^{40}Ca taking into account the correlations between $2p$-$2h$ configurations [19].

$$\begin{pmatrix} \tilde{A} & \tilde{B} \\ \tilde{B}^* & \tilde{A}^* \end{pmatrix} \begin{pmatrix} \tilde{X} \\ \tilde{Y} \end{pmatrix} = \hbar\omega \begin{pmatrix} 1 & 0 \\ 0 & -1 \end{pmatrix} \begin{pmatrix} \tilde{X} \\ \tilde{Y} \end{pmatrix}, \tag{3.136}$$

where $\tilde{A}$ and $\tilde{B}$ contain the couplings between $1p\text{-}1h$ and $2p\text{-}2h$ configurations;

$$\tilde{A} = \begin{pmatrix} A_{11} & A_{12} \\ A_{21} & A_{22} \end{pmatrix}, \quad \tilde{B} = \begin{pmatrix} B_{11} & B_{12} \\ B_{21} & B_{22} \end{pmatrix},$$

$$\tilde{X}(k) = \begin{pmatrix} X_1 \\ X_2 \end{pmatrix}, \quad \tilde{Y}(k) = \begin{pmatrix} Y_1 \\ Y_2 \end{pmatrix}, \tag{3.137}$$

where indices 1 and 2 denotes $1p\text{-}1h$ and $2p\text{-}2h$ configurations, respectively. The matrices A_{11} and B_{11} are the same RPA matrices as those appeared in Eq. (3.105). The coupling matrices between $1p\text{-}1h$ and $2p\text{-}2h$ configurations are denoted by A_{12}, A_{21}, B_{12} and B_{21}, while the coupling matrices in $2p\text{-}2h$ configurations are written as A_{22} and B_{22}. Both in RPA and SRPA, the quasi-boson approximation is used to evaluate the matrix elements in Eq. (3.136).

The second RPA includes only $1p\text{-}1h$ and $2p\text{-}2h$ states and assumes the Hartree–Fock ground state in which the single-particle states up to the Fermi energy are fully occupied [18]. There is also a version of extended SRPA which includes $p\text{-}h$, $p\text{-}p$ and $h\text{-}h$ configurations as the one-body amplitudes, and $2p\text{-}2h$, $2p\text{-}2p$ and $2h\text{-}2h$ configurations as the two-body amplitudes assuming a correlated ground state where the single-particle states are fractionally occupied and the non-vanishing two-body correlation matrix. For open-shell nuclei, the pairing correlations may take an important role for nuclear spectroscopy. So far there is no theoretical model to take into account the pairing correlations in the SRPA.

The coupled cluster (CC) theory in Sect. 3.11.5 adopts the same procedure as SRPA to take into account complex configurations $1p\text{-}1p$, $2p\text{-}2h$, $\cdots$ in the model space. The CC model with the singlet and doublet clusters (CCSD) has the same configuration space as SRPA. The advantage of CC model is to construct the effective interaction to renormalize the effect of model space truncation properly by the similarity transformation method on the interaction (see more details in Sect. 3.11.5).

3.7.3 Generator Coordinate Method (GCM)

The Generator coordinate method (GCM), introduced by David Hill, John Wheeler and James Griffin, is a powerful tool to treat the collective motions in various circumstances in nuclear physics [20]. The GCM can describe configuration mixings of different shell model configurations or different shapes in the intrinsic frame. The trial wave function is constructed by a continuous superposition of generator functions $|\Phi(\alpha)\rangle$,

$$|\Psi\rangle = \int d\alpha f(\alpha)|\Phi(\alpha)\rangle, \tag{3.138}$$

where α is a set of parameters, "*the generator coordinates*", to be selected appropriately depending on the problem which should be solved. Since the coordinate α is integrated and does not appear in the wave function $|\Psi\rangle$, this method is called "*generator coordinate method*". In order to determine the amplitude $f(\alpha)$, the variational principle for $f(\alpha)$ is applied to the expectation value of the Hamiltonian:

$$\delta\left(\frac{\langle\Psi|H|\Psi\rangle}{\langle\Psi|\Psi\rangle}\right) = 0. \tag{3.139}$$

We then obtain an integral equation, the Hill-Wheeler-Griffin (HWG) equation,

$$\int d\alpha'\langle\Phi(\alpha)|H|\Phi(\alpha')\rangle f(\alpha') = E\int d\alpha'\langle\Phi(\alpha)|\Phi(\alpha')\rangle f(\alpha'), \tag{3.140}$$

choosing a set of the generator coordinates α, for example, the quadrupole deformation parameters, or the octuple deformation parameters.

Problem

3.10 Derive the Hill-Wheeler-Griffin equation (3.140) from the variation (3.139).

The trail wave function $|\Phi(\alpha)\rangle$ is constructed by the mean field wave functions and violates several basic symmetries: translational, number conservation, rotational invariance, isospin invariance and reflection symmetry. These symmetry violations are not physical, but artifact of the mean field approximation and should be recovered by the projection method in the GCM solutions. However, the symmetry violations tell us physics behind phenomena observed in experiments. For example, the spatial deformation of intrinsic mean field wave function explains rotational bands. Reflection asymmetry (pier shape) tells parity doublets and a combination with spatial deformation leads to alternating parity rotational bands. Particle number breaking implies superfluidity and justifies the difference in level densities in even-even and odd-A nuclei.

The trail wave function $|\Phi(\alpha)\rangle$ can be projected out on the basis with a definite parity, a definite particle number, and a definite angular momentum,

$$|\Phi_M^{J\pi}(\alpha)\rangle_N = P^\pi P^N P_M^J|\Phi(\alpha)\rangle, \tag{3.141}$$

where P^π, P^N and P_M^J are the parity, particle number and angular momentum projection operators, respectively. In practice, there are two strategies to perform the projection. One is the projection after variation (3.139) (PAV) and another one is the projection before the variation (VAP). The VAP includes additional correlations by the projection and considered a better approximation, while computationally more demanding.

The HWG equation is formally written as

$$\mathcal{H}f = E\mathcal{N}f, \tag{3.142}$$

with the overlap functions

$$\mathcal{H} = \langle \Phi(\alpha)|H|\Phi(\alpha')\rangle, \tag{3.143}$$

$$\mathcal{N} = \langle \Phi(\alpha)|\Phi(\alpha')\rangle, \tag{3.144}$$

which are called Hamiltonian kernel and the overlap kernel, respectively. If the parameter α is discrete, the HWG equation becomes a standard diagonalization problem of a matrix equation

$$\sum_{\alpha'} \mathcal{H}_{\alpha,\alpha'} f_{\alpha'} = E \sum_{\alpha'} \mathcal{N}_{\alpha,\alpha'} f_{\alpha'}, \tag{3.145}$$

with the non-orthogonal basis, i.e., the basis states are not orthogonal but have overlaps $\mathcal{N}_{\alpha,\alpha'}$. In numerical algorithm, we first invert the overlap kernel $\mathcal{N}$ in Eq. (3.142) and then diagonalize the equation $\mathcal{N}^{-1}\mathcal{H}f = Ef$. This is possible if the overlap kernel $\mathcal{N}$ does not have zero eigenvalues, but it happens often in real calculations. Because of this reason, it is not easy to solve HWG equation (3.142) directly and some methods were introduced to avoid this singularity problem. One example is the Gram-Schmidt orthonormalization method.

The Bohr Hamiltonian was proposed in 1962 by Aage Bohr and has been used as a unified model to describe both the vibrational and rotational degree of freedoms. It is known that the GCM provides the Bohr Hamiltonian from a microscopic approach. The Hamiltonian was firstly given by a classical Hamiltonian and then quantized introducing the dynamical deformation degree of freedom β and γ, which represent the quadrupole deformation and its axial asymmetry as will be seen in Eq. (3.148) (see also Chap. 7 for details). The quantized Hamiltonian reads

$$H_{\text{Bohr}} = -\frac{\hbar^2}{2B} \left[\frac{1}{\beta^4} \frac{\partial}{\partial\beta} \left(\beta^4 \frac{\partial}{\partial\beta} \right) + \frac{1}{\beta^2 \sin 3\gamma} \frac{\partial}{\partial\gamma} \left(\sin 3\gamma \frac{\partial}{\partial\gamma} \right) \right]$$
$$+ T_{\text{rot}} + V(\beta, \gamma), \tag{3.146}$$

where B is the mass parameter, $V(\beta, \gamma)$ is the collective potential. The rotational energy is given by

$$T_{\text{rot}} = \frac{I_1^2}{2\mathcal{I}_1} + \frac{I_2^2}{2\mathcal{I}_2} + \frac{I_3^2}{2\mathcal{I}_3}, \tag{3.147}$$

where the operators I_κ are the projections of the total angular momentum onto the intrinsic frame axis and $\hat{\mathcal{I}}_\kappa$ are the corresponding moments of inertia. The β and γ are deformation parameters which specify the increments of the three axes of the intrinsic frame

$$\delta R_\kappa = \sqrt{\frac{5}{4\pi}} \beta R_0 \cos\left(\gamma - \frac{2\pi}{3}\kappa\right), \quad \kappa = 1, 2, 3, \tag{3.148}$$

where R_0 is the radius parameter. In case of fixed deformation, $\beta = \beta_0$ and $\gamma = \gamma_0$, the Bohr Hamiltonian has only the rotational degree of freedom to describe rotational motions of deformed nuclei. When β and γ are considered as dynamical variables to describe the fluctuations around the static deformation, the rotational and vibrational degrees are coupled in the Hamiltonian.

The HWG equation can be rewritten as

$$H_{\text{coll}}\phi(\alpha) = E\phi(\alpha), \tag{3.149}$$

where $\phi(\alpha) = \mathcal{N}^{1/2}f$ and

$$H_{\text{coll}} = \mathcal{N}^{1/2}\mathcal{H}\mathcal{N}^{1/2}. \tag{3.150}$$

This equation involves the integral operator (see Eq. (3.142)), which makes practical calculations extremely difficult. However, it can be reduced to a more plausible form by introducing a Gaussian overlap approximation (GOA). Under GOA, the overlap kernel reads

$$\mathcal{N}(\alpha, \alpha') = \mathcal{N}(q, s) = e^{-\lambda(q)s^2/2}, \tag{3.151}$$

where $q = (\alpha + \alpha')/2$ and $s = \alpha - \alpha'$. With the GOA approximation, the transformation $\phi(\alpha) = \mathcal{N}^{1/2}f$ can be evaluated analytically and H_{coll} is expressed by a function of the collective variable q only. The mass parameter such as B in the Bohr-Hamiltonian (3.146) is evaluated now microscopically from the realistic Hamiltonian. The HWG equation can be thus transformed to be a Bohr collective Hamiltonian (3.146) in GOA approximation.

The GCM method has been extensively applied to the collective vibrational and rotational modes, fission problem, and shape coexistence. Especially, GCM is a technique that is well suited to describe shape coexistence with the quadrupole moment as generator coordinate. The method determines the configuration mixing in a variational way and provides excitation spectra and transition matrix elements. An illustrative example of GCM calculations with the form of the Bohr collective Hamiltonian (3.146) is given for the shape coexistence of neutron-deficient Kr isotopes in Sect. 7.7.

3.8 Many-Body Shell Model Wave Function

The shell model, in which valence nucleons influence each other through a residual interaction, is a very useful framework for quantum many-body calculations. The many-body wave functions are approximated by one or multi anti-symmetrized products of single-particle wave functions. The first step for shell model calculation is to select the active model space, called "valence space" outside of the inert core. The second step is to construct an appropriate effective Hamiltonian for the model space. The third step is to calculate matrix elements of the effective interaction for the diagonalization in the model space. The harmonic oscillator wave functions are commonly adopted for the calculations of matrix elements. Some shell models adopt phenomenological two-body matrices directly derived from selected experimental spectra without referring the two-body interactions. The fourth step is to obtain eigenstates and the eigenenergies by the diagonalisation. Finally, the observables such as electromagnetic transitions and the reaction cross sections are calculated by using the obtained wave functions with the effective operator which takes into account configurations outside the model space in the normalized procedure. A schematic view of the model spaces of standard and NCSMs is shown in Fig. 3.8 for a sd-shell nuclei ^{20}Ne.

Calculations of the matrix elements of the model Hamiltonian and the diagonalization will be performed in the model space. To classify many-body wave functions, one can use either the total angular momentum $\mathbf{J}$ or its projection M on the z-axis J_z since the many-body Hamiltonian commutes with both quantum numbers;

$$[H, \mathbf{J}^2] = 0, \quad [H, J_z] = 0. \tag{3.152}$$

These classification schemes are denoted J-scheme and M-scheme, respectively. In large scale shell model calculations, the M-scheme has an advantage to calculate the Hamiltonian matrix element and also the projection on the definite angular momentum state since it does not involve any angular momentum coupling coefficients such as Clebsch-Gordan coefficients.

There is a general rule to count the permissible states for a given configuration of N identical particles in the angular momentum j orbit. This can be done in the M-scheme representation by considering all possible m values for a given value of the sum $M = \sum_{i=1,N} m_i$ where

$$-j \leq m_i \leq j. \tag{3.153}$$

In the summation, the values m_i should be all different because of the Pauli blocking principle of Fermions. As a simple example, we consider the configuration of two identical particles in the $j = 7/2$ orbit:$(j = 7/2)^2$. The maximum M is $M = 7/2 + 5/2 = 6$. For this M value, the maximum total angular momentum J can not exceed this M value so that $J = 6$ is the maximum total angular momentum. When we act the angular momentum down operator $J_- = \sum_i j_-^i$, we have only one $(7/2,3/2)$ configuration for the $M=5$ state, which could be the projection of either $J = 6$ or $J = 5$ angular momentum. We can continue this operation up to the total $M = 0$. The summary is given in Table 3.6.

We have determined the number of M-scheme states for each M value of the two-particle configuration, $(j)^2$. For each M value, possible total angular momenta J for the two-particle states are restricted by a condition $J \geq M$. In other words, the state in the J−scheme, is classified with $|JM\rangle$ with $J \geq |M|$. Thus, if n_J is the number of states for a given value of J, the number of states in m scheme $n(M)$ is expressed as

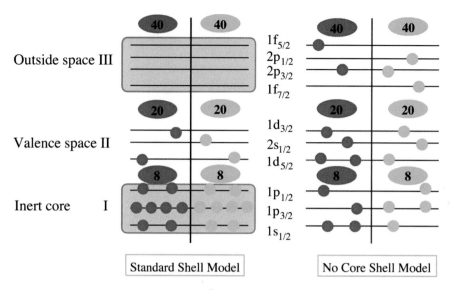

Fig. 3.8 (Left) A schematic view of shell model configurations of the sd-shell nucleus ^{20}Ne in a standard shell model calculation. Regions I, II and III correspond to the inert core, active valence single-particle states and those outside the model space, respectively. Shell model calculations are performed putting valence particles in the model space II above the inert core. In standard calculations, single-particle states in one major shell are taken for the model space. Particles in the inert core are not involved in the shell model calculations. Excitations of single-particle states above the model space are not allowed. The model space consists of three sd-shell orbitals. (Right) A schematic view of no-core shell model (NCSM) configuration. All particles are active in a model space truncated by the major quantum number of the harmonic oscillator wave functions. Particles are allowed to occupy any states within the model space. The numbers 8, 20 and 40 show the shell gaps between the major shells

$$n(M) = \sum_{J=|M|}^{J_{max}} n_J, \qquad (3.154)$$

and therefore

$$n_J = \begin{cases} n(|M| = J) & \text{if } M = J_{max} \\ n(|M| = J) - n(|M| = J + 1) & \text{if } |M| < J_{max}. \end{cases} \qquad (3.155)$$

Let us apply this rule to Table 3.6. Firstly for the $M = 6$ state, we have one $J = 6$ state. Next, for $M = 5$, J could be 5 or 6. However, we have one $J = 6$ state and $n(M = 5) = 1$ so that $J = 5$ is not possible. Continuing this process, we obtain possible J values are $J = 6, 4, 2, 0$ with one configuration $n_J = 1$ for each J value.

The number of possible configurations in a model space is calculated by using the procedure explained above. In Fig. 3.9, the number of sd-shell configurations and pf-shell configurations are shown as a function of the mass number A. In Fig. 3.9, the states are further classified with the isospin quantum numbers $|T M_T\rangle$

Table 3.6 Configurations in M-scheme wave functions for two identical particles in the valence $j = 7/2$ orbit: $(j = 7/2)^2$. These configurations correspond to those of ^{42}Ca or ^{42}Ti with the ^{40}Ca core. M is defined as $M = m_1 + m_2$ and $n(M)$ denotes the number of configurations in the M-scheme

M	(m_1, m_2)	$n(M)$	possible J
6	(7/2,5/2)	1	6
5	(7/2,3/2)	1	6, 5
4	(7/2,1/2), (5/2,3/2)	2	6, 5, 4
3	(7/2,−1/2), (5/2,1/2)	2	6, 5, 4, 3
2	(7/2,−3/2), (5/2,−1/2), (3/2,1/2)	3	6, 5, 4, 3, 2
1	(7/2,−5/2), (5/2,−3/2), (3/2,−1/2)	3	6, 5, 4, 3, 2, 1
0	(7/2,−7/2), (5/2,−5/2), (3/2,−3/2), (1/2,−1/2)	4	6, 5, 4, 3, 2, 1, 0

with $T \geq M_T = (N - Z)/2$ including the isospin degree of freedom. The number of active valence particles A_{act} is evaluated by $A_{act} = A - A_{core}$, where A_{core} is the number of particles in the inert core. A_{core} is 16 for the sd-shell configuration and 40 for the pf-shell configuration. The number of basis states is about 10^5 in nuclei at the middle of the sd-shell which is relatively easy to handle in the modern computer power. On the other hand, the number of basis states is going up to more than 1 billion in nuclei at the middle of the pf-shell, which is not so easy to handle even with modern supercomputers. It is almost impossible to handle heavier nuclei beyond the pf-shell by the standard shell model techniques. To cure this problem, the Monte Carlo method was introduced and made it feasible to perform shell model calculations with heavier mass nuclei and also with much larger model spaces. This point will be discussed in Sect. 3.9.

The shell model Hamiltonian can be divided into two parts, the monopole field H_m and the multipole correlations;

$$H = H_\mathrm{m} + H_\mathrm{M}, \tag{3.156}$$

where the monopole field is the mean field Hamiltonian consisting from the kinetic part and the monopole interaction,

$$H_\mathrm{m} = T + V_\mathrm{m}. \tag{3.157}$$

The multipole correlations represent the pairing and the quadrupole-quadrupole correlations which make a link between the collective behaviour and underlying shell structure. *Effective single-particle energies* are often adopted in the shell model cal-

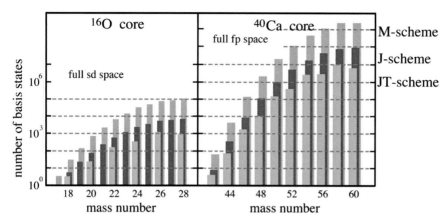

Fig. 3.9 Matrix dimensions for the diagonalization of shell model calculations with sd- and pf-shell model space. The numbers of valence particles are given subtracting the core particle A_{core} from mass number A. A_{core} is 16 for the full sd-shell model, while A_{core} is 40 for the full pf-shell model. The number of basis states depends on J–scheme, JT-scheme or M-scheme configurations. For the standard shell model calculations, the dimension of possible maximum diagonalization matrix is around 10^9 (1 billion) (in 2016) so that nuclei in the fp-shell might be the limit for standard shell model study

culations, which represent mean field effects from the other nucleons on a nucleon in a specified single-particle orbit. *The monopole interaction V_m gives an idea on how the effective single-particle energy is governed by the many-body interactions.* The two-particle wave function in j_1 and j_2 orbits is written as

$$|(j_1 j_2) J M\rangle = \sum_{m_1, m_2} \frac{1}{\sqrt{(1 + \delta_{j_1 j_2})}} \langle j_1 m_1 j_2 m_2 | J M \rangle | j_1 m_1 j_2 m_2 \rangle, \quad (3.158)$$

and the monopole interaction for two identical particles is defined by

$$V_{\text{m}}(j_1, j_2) = \frac{\sum_J (2J + 1)(1 + \delta_{j_1 j_2}) \langle (j_1 j_2) J M | \widetilde{V} | (j_1 j_2) J M \rangle}{(2j_1 + 1)(2j_2 + 1)}, \quad (3.159)$$

where $\widetilde{V}$ is the two-body interaction with direct and exchange terms (see Eq. (3.43)). A factor $(2J + 1)$ can be replaced by a sum of "M" quantum number in Eq. (3.159). In Eq. (3.158), $\langle j_1 m_1 j_2 m_2 | J M \rangle$ is the Clebsch-Gordan coefficient (see Appendix 7.10.1 of Chap. 7 for more details on the Clebsch-Gordan coefficients). The two-body matrix elements in Eq. (3.159) are taken to be independent to the m quantum number, the z- component of angular momentum, due to the rotational invariance of the two-body interaction. The monopole interaction is related with the diagonal matrix element of HF potential (3.49):

$$
\begin{aligned}
V_{\mathrm{m}}(j_1, j_2) &= \frac{\sum_{J,M}(1 + \delta_{j_1 j_2})\langle (j_1 j_2)JM|\widetilde{V}|(j_1 j_2)|JM\rangle}{(2j_1 + 1)(2j_2 + 1)} \\
&= \frac{\sum_{J,M,m_1,m_2,m_1',m_2'}\langle j_1 m_1 j_2 m_2|JM\rangle\langle j_1 m_1' j_2 m_2'|JM\rangle\langle j_1 m_1 j_2 m_2|\widetilde{V}|j_1 m_1' j_2 m_2'\rangle}{(2j_1 + 1)(2j_2 + 1)} \\
&= \frac{\sum_{m_1,m_2,m_1',m_2'}\delta_{m_1,m_1'}\delta_{m_2,m_2'}\langle j_1 m_1 j_2 m_2|\widetilde{V}|j_1 m_1' j_2 m_2'\rangle}{(2j_1 + 1)(2j_2 + 1)} \\
&= \langle j_1 m_1 j_2 m_2|\widetilde{V}|j_1 m_1 j_2 m_2\rangle, \quad\quad (3.160)
\end{aligned}
$$

where the orthonormal relation,

$$
\sum_{J,M}\langle j_1 m_1 j_2 m_2|JM\rangle\langle j_1 m_1' j_2 m_2'|JM\rangle = \delta_{m_1,m_1'}\delta_{m_2,m_2'} \quad\quad (3.161)
$$

is used. Equation (3.160) shows explicitly that the monopole interaction turns out to be nothing but the diagonal matrix element of the HF potential (3.49);

$$
\sum_{j_2}(2j_2 + 1)V_{\mathrm{m}}(j_1, j_2) = \sum_{j_2 m_2}\langle j_1 m_1 j_2 m_2|\widetilde{V}|j_1 m_1 j_2 m_2\rangle = \langle j_1 m_1|v^{\mathrm{HF}}|j_1 m_1\rangle.
$$

$$(3.162)$$

The two-body matrix element in the monopole interaction (3.159) depends on the angular momentum J coupled by two interacting nucleons in orbits j_1 and j_2. Since this is a mean field effect, the J-dependence is averaged out with a weight factor $2J + 1$, which is equivalent to the diagonal matrix elements of the HF potential. The conventional shell model has a truncation of the model space so that the effective single-particle energies are defined by adding the core and the active model space contributions. In the NCSM, there is no truncation of the model space so that all contributions of occupied orbits are taken into account in the single-particle energies through the monopole interactions.

The monopole interaction has the central and the long-range part of tensor contributions, and the tensor force is particularly related with the shell evolution of several isotopes. The monopole field H_m is constructed to reproduce precisely the energies of closed shells and single-particle (hole) energies built on them. It was also pointed out that the contributions from three-nucleon interactions in the monopole interaction are important for the shell evolution of neutron-rich nuclei.

In the shell model calculations of medium-heavy nuclei, the quadrupole interaction plays an important role to derive the deformation of the intrinsic system. The importance of the quadrupole interaction was already noticed in the pairing+quadrupole-quadrupole interaction model to describe various vibrational and rotational collective excitations in medium-heavy and heavy nuclei in 1960s. The important role of the quadrupole interaction is recently revived in modern shell model calculations to study the evolution of deformation in competition with the pairing interaction.

3.9 The Monte Carlo and No-Core Shell Models

Shell models with many-body correlations have been powerful methods describing quantitatively a spectrum of atomic nuclei as well as transition strengths among nuclear states. In order to make quantitative predictions in medium and heavy mass nuclei, a larger model space is required in order to take into account properly correlations among valence nucleons. One thus has to diagonalize a shell model Hamiltonian with a large dimension, which significantly increases the computational time. The Monte Carlo method is introduced in shell model calculations to avoid to diagonalize an astronomically large model space. In recent years, computer powers have been enhanced steadily and such large-scale shell model calculations even no-core shell models are now feasible.

3.9.1 Shell Model Monte Carlo (SMMC)

As was shown in Fig. 3.9, the dimension of the shell model basis grows astronomically depending on both the number of valence nucleons (N_V) and the size of the single-particle model space (N_S). The dimension of the diagonalization would become tremendously large and essentially the straightforward diagonalization is extremely time consuming and almost impossible to perform in heavy nuclei with $A > 80$ or with a larger model space more than one major shell even in light or medium heavy nuclei.

An alternating efficient method, the Monte Carlo method, was proposed for statistical treatment of a shell model Hamiltonian [21]. In order to deal with a large Hamiltonian matrix, the Shell Model Monte Carlo (SMMC) has been developed, in which the auxiliary field is introduced. The auxiliary field method is a technique that reduces a two-particle Hamiltonian to a form that has only one-particle fields. It is used in many areas of physics, particularly condensed matter, and it is closely related to the path integral method of quantum field theory (QFT). We refer to Chap. 10 for the application of QFT to lattice quantum chromodynamics (LQCD).

A general form for H in SMMC is given by the kinetic energy and the interaction parts. The interaction part V contains only the one-body operators and is expressed by a sum over products of one-body operators,

$$H = T + V = T + \frac{1}{2} \sum_{\alpha=1}^{N_\alpha} V_\alpha O_\alpha^2, \qquad (3.163)$$

where T is the kinetic energy, and O_α are a set of one-body operators and V_α is a residual interaction strength. The interaction V is considered as a mapping of residual interactions on the separable forms with various multipolarities. The operator O_α is either a multipole density operator $\rho_{(LM)} = [a^\dagger a]^{(LM)}$ or a pair density operator $\Delta_{(JM)}^\dagger = [a^\dagger a^\dagger]^{(JM)}$ and its adjoint.

The model is based on using the imaginary (Euclidian) time β for the many-body evolution operator, $\exp(-\beta H)$, either to define the canonical or grand canonical ensemble at a temperature $kT=\beta^{-1}$ or to converge a many-body trial state to the exact ground state for a large β.

We first consider a trial wave function $|\Psi\rangle$ to the exact ground state. The imaginary time evolution of $|\Psi\rangle$ will be done by the operator $\exp(-\beta H)$ and the imaginary time β is divided into N_t steps;

$$e^{-\beta H} = \prod_{n=1}^{N_t} e^{-\Delta\beta H}, \tag{3.164}$$

where $\Delta\beta = \beta/N_t$. This is called the Trotter expansion. Assuming $\Delta\beta$ to be small, an individual term in Eq. (3.164) can be separated into the kinetic and the two-body potential parts as

$$e^{-\Delta\beta H} \approx e^{-\Delta\beta T} e^{-\Delta\beta V}. \tag{3.165}$$

A common trick of the Monte Carlo method is *the Hubbard-Stratonovich transformation* which transforms the two-body operator in the exponent to a one-body operator with an identity,

$$e^{-\frac{1}{2}VO^2} = \sqrt{\frac{|V|}{2\pi}} \int d\sigma e^{-\frac{1}{2}|V|\sigma^2 - s\sigma VO}, \tag{3.166}$$

where σ is an auxiliary field and $s = 1$ ($s = i$) if $V < 0$ ($V > 0$).

Problem

3.11 Prove the Hubbard-Stratonovich identity (3.166).

Applying the Hubbard-Stratonovich transformation at each time step, the evolution operator $\exp(-\beta H)$ can be expressed as an integral of one-body evolution operators with respect to the auxiliary fields $\sigma_{\alpha n}$, which couple to operators O_α at each time step $n = 1, 2, \cdots, N_t$. Thus, each auxiliary field is denoted by an ensemble of auxiliary fields over all the time steps $\sigma_\alpha = \{\sigma_{\alpha 1}, \sigma_{\alpha 2}, \cdots, \sigma_{\alpha N_t}\}$:

$$e^{-\beta H} = \prod_{n=1}^{N_t} e^{-\Delta\beta H} = \int_{-\infty}^{\infty} \cdots \int_{-\infty}^{\infty} \prod_{\alpha,n} d\sigma_{\alpha n} \left(\frac{\Delta\beta|V_\alpha|}{2\pi}\right)^{1/2} G(\sigma) e^{-\Delta\beta \sum_n H(\sigma_{\alpha n})}, \tag{3.167}$$

where the Gaussian weight factor $G(\sigma)$ is defined by

$$G(\sigma) = \exp\{-\frac{\Delta\beta}{2} \sum_{\alpha,n} |V_\alpha|\sigma_{\alpha,n}^2\}, \tag{3.168}$$

and the one-body Hamiltonian $H(\sigma_{\alpha n})$ is defined by

$$H(\sigma_{\alpha n}) = T + s_\alpha V_\alpha \sigma_{\alpha n} O_\alpha, \tag{3.169}$$

where $s_\alpha = 1(s_\alpha = i)$ if $V_\alpha < 0$ ($V_\alpha > 0$). Equation (3.167) can be expressed by using a symbol for the multiple integral $D[\sigma]$ as

$$e^{-\beta H} = \int D[\sigma] e^{-\int d\beta \sum_\alpha |V_\alpha|\sigma^2/2} e^{-\int d\beta h(\sigma)}, \tag{3.170}$$

where $D[\sigma] \equiv \prod_\alpha d\sigma_\alpha (\Delta\beta |V_\alpha|/2\pi)^{1/2}$.

For any observable operator Q, we define the expectation value

$$\langle Q \rangle = \frac{\langle \Psi | e^{-\beta H/2} Q e^{-\beta H/2} | \Psi \rangle}{\langle \Psi | e^{-\beta H} | \Psi \rangle}. \tag{3.171}$$

In the limit $\beta \to \infty$, $\langle Q \rangle$ will approach the exact ground state expectation value. In the limit $\Delta\beta \to 0$, the value $\langle Q \rangle$ can be written as

$$\langle Q \rangle = \frac{\int D[\sigma] G(\sigma) \Phi(\sigma) Q(\sigma)}{\int D[\sigma] G(\sigma) \Phi(\sigma)}, \tag{3.172}$$

where

$$\Phi(\sigma) = \langle \Psi | U(N_t, 0) | \Psi \rangle, \tag{3.173}$$

and

$$Q(\sigma) = \frac{\langle \Psi | U(N_t/2, 0) Q U(N_t/2, 0) | \Psi \rangle}{\Phi(\sigma)}, \tag{3.174}$$

with

$$U(N_t, 0) = \prod_{n=1}^{N_t} e^{-\Delta\beta H(\sigma)}. \tag{3.175}$$

In numerical calculations, the integral is evaluated by discretizing the $\sigma_{\alpha n}$ variables. In the Monte Carlo integral, each set of σ is generated stochastically and many sets of auxiliary fields are generated and the corresponding expectation values are added with an equal weight. This algorithm is called "*Metropolis Algorithm*". There is a well-known problem in this method, which is called the minus-sign problem in the Hamiltonian (3.169). If the sign of the residual interaction V_α is always the same, the ($N_\alpha \times N_t$)-th dimensional integral has good convergence. However, in nuclear physics, the sign of residual interaction V_α will be positive or negative and needs a special care to have good convergence.

To avoid the computational limit of the shell model, another method so called "*Monte Carlo Shell Model (MCSM)*" is also proposed [23]. The MCSM utilizes the idea of the auxiliary field Monte Carlo method, that was also adopted in SMMC, but the basic algorithm is completely different. The MCSM introduces a set of Slater determinants with deformed single-particle states as basis states for a many-body

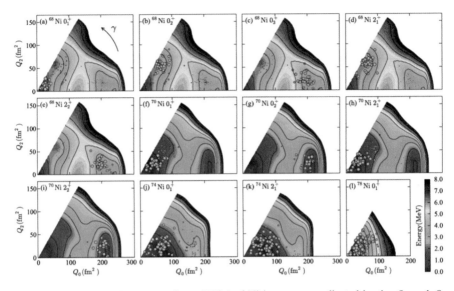

Fig. 3.10 The Potential energy surfaces (PESs) of Ni isotopes, coordinated by the Q_0 and Q_2 calculated by the constrained HF method with a Hamiltonian A3DA. The energy relative to the minimum is shown by contour plots. Circles on the PES represent shapes of MCSM deformed basis vectors. Figure reprinted with permission from [22]. ©2021 by the American Physical Society

wave function $|\Psi\rangle$. The Monte Carlo sampling method is applied to select the basis states among many candidates for the intrinsic states $|\Psi(\sigma)\rangle$ for different sets of auxiliary fields σ. For independent sets of $|\Psi(\sigma)\rangle$, the Hamiltonian H is diagonalized. The number of basis states for $|\Psi(\sigma)\rangle$ is increased until the convergences of energy is obtained. This approach is a hybrid model of the auxiliary fields and the GCM. In the MCSM, the original Hamiltonian is diagonalized for the set of Slater determinants so that the sign problem encountered in the SMMC is not an issue. Since such single-particle wave functions violate the rotational symmetry, the angular momentum and parity projections are performed on the wave function $|\Psi\rangle$ before diagonalizing the Hamiltonian.

The MCSM method has been applied from light neutron-rich nuclei up to Ni and Cr neutron-rich isotopes. Since deformed basis wave functions are used in MCSM, one can easily analyze with this method the intrinsic structure of each state of neutron-rich nuclei in the (β, γ) deformation plane. This has been carried out, e.g., for ^{68}Ni to discuss the shape coexistence in this nucleus. Figure 3.10 depicts, for selected states of ^{68,70,74,78}Ni isotopes, potential energy surfaces (PESs) in the β-γ plane calculated by the constrained Hartree–Fock (CHF) method with the constraints on the quadrupole moments Q_0 and Q_2. The energy relative to the minimum is shown by contour plots. Circles on the PES represent shapes of MCSM deformed basis vectors. The area of the circle is set to be proportional to the overlap probability of the intrinsic state in the projected shell model state.

We can see two energy minima for ^{68}Ni; a spherical minimum which extends slightly towards the oblate region, as well as a prolate local minimum. Figure 3.10a shows that the ground state configuration of ^{68}Ni is dominated by the spherical intrinsic states showed by many circles near $Q_0 = 0$. We see also some small circles spreading toward to oblate region. This reflects the effect of shape fluctuation in the shell model wave function. For the 0_2^+ and 2_1^+ states in Fig. 3.10b, d, the distributions of circles are concentrated in the oblate region ($\beta \sim -0.2$). The 0_3^+ and 2_2^+ states in Fig. 3.10c, e show many circles in a profound prolate minimum with $\beta_2 \sim 0.4$. This suggests that the shape of these states is different from the ground state, calculated to be spherical. One can see also the intrinsic structure of the shell model states for ^{70,74,78}Ni in Fig. 3.10f–l as shown by the number of dots in the β-γ plane.

3.10 Anti-symmetrized Molecular Dynamics and Fermionic Molecular Dynamics

The anti-symmetrized molecular dynamics (AMD) [25] and fermionic molecular dynamics (FMD) models [26] have been developed to study the clustering phenomenon in nuclei, especially in unstable nuclei, in which the clustering is expected to be enhanced in the isotopes toward the neutron drip line. The methods based on molecular dynamics have originally been developed for the description of the cluster structure of nuclei with masses $12 \leq A \leq 40$. In the HF and shell models, the single-particle wave function is measured from the center of the nucleus. However, the single-particle wave packet in molecular dynamics is defined from a different central position $\mathbf{Z}_i$ for each particle. This is a big advantage to describe the localization of wave functions such as cluster structure. In this method, one considers a Slater determinant

$$\Psi(\mathbf{Z}) = \frac{1}{\sqrt{A!}} \det\{\varphi_i\} \tag{3.176}$$

constructed with Gaussian wave packets for single-particle wave functions,

$$\varphi_i(\mathbf{r}) = \phi_i(\mathbf{r})\chi_i\xi_i, \tag{3.177}$$

where ξ_i is the isospin wave function and χ_i is the spin wave function. $\phi_i(\mathbf{r})$ is the radial wave function,

$$\phi_i(\mathbf{r}) \propto \exp\left[-\nu_i\left(\mathbf{r} - \frac{\mathbf{Z}_i}{\sqrt{\nu_i}}\right)\right], \tag{3.178}$$

where ν_i is the width of the wave packet and the center of the wave function for the ith particle is expressed by a complex parameter $\mathbf{Z}_i$. In the limit of $|\mathbf{Z}_i| \to 0$, the Gaussian wave packet becomes the harmonic oscillator wave function measured from the center of nucleus. In AMD, the width parameters $\{\nu_i\}$ are set to be equal for all the nucleons, while they are allowed to take a different value for each nucleon in

FMD. The spin wave function χ_i is parametrized by variational variables η_i;

$$\chi_i = \left(\frac{1}{2} + \eta_i\right) | \uparrow\rangle + \left(\frac{1}{2} - \eta_i\right) | \downarrow\rangle. \tag{3.179}$$

In AMD, the parameters $\mathbf{Z}_i$ and η_i $(i = 1, \cdots A)$ are determined by the variational procedure for energy,

$$\frac{\delta}{\delta \mathbf{Z}_i}\left(\frac{\langle \Psi | H | \Psi \rangle}{\langle \Psi | \Psi \rangle}\right) = 0, \tag{3.180}$$

$$\frac{\delta}{\delta \eta_i}\left(\frac{\langle \Psi | H | \Psi \rangle}{\langle \Psi | \Psi \rangle}\right) = 0. \tag{3.181}$$

Without any assumption for the formation of constituent clusters, multi-cluster structure can be described by the spatially localized groups of Gaussian wave packets, thanks to the flexible coordinate vectors $\mathbf{Z}_i$ which assign the different centroid of each particle. In general $|\mathbf{Z}_i| \neq 0$, the Gaussian wave packet is not labelled by the angular momentum l commonly used for mean field single-particle wave functions. If all the Gaussian centroids $\mathbf{Z}_i$ gather at the same position, the AMD wave function is equivalent to a harmonic oscillator shell model wave function. That is, the AMD wave function can describe in principle both the shell and cluster structures of nuclei in an unified way. The energy minimization will determine whether the cluster or the shell structure is favorable energetically for given quantum numbers to assign each state. This feature of AMD is common to FMD which employs also the Gaussian wave packets with multi Gaussian centroids.

The intrinsic wave function $\Psi(\mathbf{Z})$ is the mixture of parities and angular momenta because the Gaussian wave packet $\varphi_i(\mathbf{r})$ in Eq. (3.177) has no good angular momentum and parity. The wave function $\Psi(\mathbf{Z})$ must be projected to the eigenstate of parity and angular momentum as,

$$\Psi^{J^\pi}(\mathbf{Z}) = P_{MK}^J P^\pi \Psi(\mathbf{Z}) = \int d\Omega \, D_{MK}^J(\Omega) \Psi^\pi(\mathbf{Z}), \tag{3.182}$$

where P^π and P_{MK}^J are projection operators of parity and angular momentum, respectively, and D_{MK}^J is the Wigner D function. By varying the complex numbers $\{\mathbf{Z}_i\}$ and $\{\eta_i\}$ together with the real numbers $\{\nu_i\}$ for $i = 1, \cdots A$, one obtains the ground state as well as low-lying excited states of the calculated nucleus.

As many-body interactions, a realistic nucleon-nucleon interaction has been employed in FMD by introducing the unitary correlation operator method (UCOM), while effective interactions have been adopted in AMD.

Compared to atomic or molecular systems, an unique character in nuclear system is found in a self-bound nature by attractive nuclear interactions between nucleons. Another characteristic feature is the saturation properties of nuclei, where the binding energy per nucleon is rather constant from light to heavy nuclei. We list the empirical

Table 3.7 Empirical binding energies $B(A)$ and the binding energy per particle $B(A)/A$ of $A = 4 \times n$ nuclei

A	$B(A)/A$	$B(A)$
^{4}He	7.074	28.30
^{8}Be	7.072	56.50
^{12}C	7.680	92.16
^{16}O	7.976	127.62
^{20}Ne	8.032	160.64

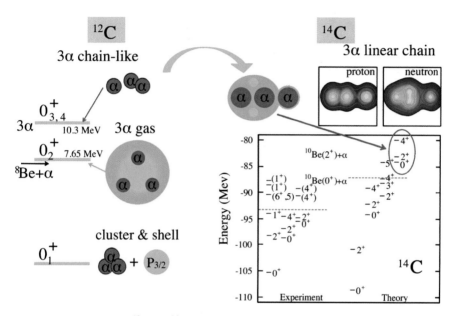

Fig. 3.11 Cluster states in ^{12}C and ^{14}C predicted by AMD calculations. In the left-hand side, the configuration of the ground state of ^{12}C is shown as a mixture of a shell-model configuration, 4 particles in the $p_{3/2}$-orbit, and the 3α clusters. A diluted 3α cluster state is predicted at the excitation energy close to the threshold energy of ^{8}Be+α in ^{12}C. This state is called the "Hoyle State" and played a key role to create nuclei beyond $A > 8$ in the nucleosynthesis. A chain like 3α cluster state is predicted in ^{12}C at E_x=10.3 MeV. On the right-hand side, the experimental and the calculated spectra of ^{14}C are shown in the large window. A chain like 3α cluster state is predicted in ^{14}C, which produces a rotational band marked by a red circle. A schematic picture of the 3α cluster state and the corresponding calculated proton and neutron densities are also shown in the upper part of the right panel. The results are taken from [29]. Courtesy Y. Kanada-En'yo, Kyoto University

binding energies of light $A = 4 \times n$ in Table 3.7, which shows a surprising constancy of the binding energies per particle $B(A)/A$ in all the listed nuclei. Because of the strong attractive interaction and saturation property of nuclei, the cluster structure may appear even in relatively small excitation energies of $A = 4 \times n$ system with $n = 2, 3, 4, \cdots$. Let us discuss possible cluster states in ^{12}C and ^{16}O in terms of the total binding energy. For each nucleus, the threshold energy of the α particle is

$$B(^{12}C) - \left(B(^4 He) + B(^8 Be)\right) = 7.36\text{MeV},$$

$$B(^{12}C) - \left(3 \times B(^4 He)\right) = 7.26\text{MeV},$$

$$B(^{16}O) - \left(B(^4 He) + B(^{12}C)\right) = 7.16\text{MeV}.$$

These values suggest possible α-cluster states, $\alpha + ^8$Be in ^{12}C and $\alpha + ^{12}$C in ^{16}O at the excitation energy around 7 MeV. Experimentally, the second 0_2^+ state is observed at $E_x = 7.65$ MeV in ^{12}C, which is identified as a cluster state and called "Hoyle state" (see Fig. 3.11). The Hoyle state plays a very important role to create nuclei heavier than $A = 8$ in the nucleosynthesis. The cluster-like second 0_2^+ state is also observed in ^{16}O at $E_x = 6.05$ MeV.

As an example of theoretical study of cluster states in nuclei, AMD calculations of cluster states in ^{12}C and ^{14}C are shown in Fig. 3.11. The configuration of the ground state of ^{12}C is shown as a mixture of a shell-model configuration, 4 particles in the $p_{3/2}$-orbit, and the 3α clusters. A diluted 3α cluster state (the Hoyle state) is predicted at the excitation energy close to the threshold energy of ^{8}Be+α in ^{12}C. A chain like 3α cluster state is also predicted in ^{12}C at $E_x \sim 10.3$ MeV. The experimental and the calculated spectra of ^{14}C are shown in the large window on the right-hand side of Fig. 3.11. A chain like 3α cluster state is predicted in ^{14}C, which produces a rotational band marked by a red circle. A schematic picture of the 3α cluster state and the corresponding calculated proton and neutron densities are also shown in the upper part of the right panel.

One disadvantage of the Gaussian wave function (3.178) is the difficulty to treat the extended nature of loosely-bound single-particle state in unstable nuclei. For example, a halo structure with an extended density distribution cannot be easily described with the Gaussian function. However, a method has recently been developed to correct the Gaussian tail of a wave function to a desired exponential tail using an idea based on the *resonating group method*.

3.11 Ab Initio Approaches

The ab initio approach is motivated by the two key requirements. (1) The Hamiltonian is diagonalized in the full model space, which has no truncation (no core) and is expanded until the results get converged. (2) The effective interactions are derived from realistic interactions which describe well the NN scatterings, and also two and three-body bound states. There are many approaches which aim to solve nuclear many-body problems satisfying these conditions. The first attempt was the Faddeev-Jakubowski method, followed by the Fermionic molecular dynamics (FMD), the Green's function Monte Carlo (GFMC), the no-core shell model (NCSM), the Variational Monte Carlo (VMC), the coupled cluster (CC) model, the self-consistent Green's function (SCGF), and the nuclear lattice effective field theory (NLEFT).

As modern realistic interactions, Argonne V18 and CD Bonn interactions are adopted commonly in the ab initio models as well as recently developed ChEFT interactions. In this section, we present basic features of the ab initio models. Firstly, we introduce microscopic models, UCOM and RSG, to obtain effective interactions from realistic interactions or QCD-based interactions. In the following subsections, we discuss various ab initio models to solve many-body nuclear structure problems.

3.11.1 Unitary Correlation Operator Method (UCOM)

The basic idea of the UCOM is the explicit treatment of the short-range central and tensor correlations as a transformation of the relative distance between particle pairs, and the generation of analytic expressions for the correlated wave functions and correlated operators [25, 27]. These are imprinted into an uncorrelated many-body state $|\Phi\rangle$ (e.g., a Slater determinant) through a state-independent unitary transformation defined by the unitary correlation operator C, resulting in a correlated state

$$|\Psi\rangle = C|\Phi\rangle. \tag{3.183}$$

The correlation operator C is written by an unitary operator

$$C = \exp(-iG) \quad \text{with} \quad G = G^\dagger, \tag{3.184}$$

where G is the hermitian generator of the correlations. The generator can be two-body, three-body and higher ones,

$$C = \sum_{i>j} g(i, j) + \text{three-body} + \text{higher}. \tag{3.185}$$

The complexity of the correlated wave function $|\Psi\rangle$ is reshuffled into the Hamiltonian,

$$\langle\Psi|H|\Psi\rangle = \langle\Phi|C^\dagger H C|\Phi\rangle = \langle\Phi|\overline{H}|\Phi\rangle. \tag{3.186}$$

The unitary transformation of the initial Hamiltonian H produces the correlated Hamiltonian $\overline{H}$,

$$\overline{H} = C^\dagger H C. \tag{3.187}$$

The construction of the two-body generators $g(i, j)$ follows the physical mechanisms by which the interaction induces correlations. The short-range central correlations, caused by the repulsive core of the interaction, are introduced by a radial distance dependent shift pushing nucleons apart from each other if they are within the range of the core.

For a given bare potential such as Argonne V_{18} and Bonn potentials or ChEFT, the corresponding correlation functions are determined by an energy minimization in the two-body system for each spin-isospin channel as will be seen later. Matrix elements of an operator O with correlated many-body states Φ can be equivalently written as matrix elements of a "correlated" (transformed) operator

$$\overline{O} \equiv C^\dagger O C = O^{[1]} + O^{[2]} + \cdots + O^{[A]}, \qquad (3.188)$$

where $O^{[n]}$ is the n-body operator.

In actual applications of the UCOM, a two-body approximation is usually employed, i.e., three-body and higher-order terms of the expansion are neglected. Starting from the Hamiltonian H_P for the A-body system, consisting of the kinetic energy operator T and a two-body potential V, the formalism of the UCOM constructs the correlated Hamiltonian in the two-body approximation;

$$H^{C2} = T^{[1]} + T^{[2]} + V^{[2]} \equiv T + V_{\text{UCOM}}, \qquad (3.189)$$

where the one-body contribution comes only from the uncorrelated kinetic energy $T^{[1]} = T$. Two-body contributions arise from the correlated kinetic energy $T^{[2]}$ and the correlated potential $V^{[2]}$, which together constitute the correlated interaction V_{UCOM}.

In contrast to the SRG approach as will be described in the next subsection, in the UCOM method, an explicit ansatz is taken for the unitary correlation operator C which is motivated by physical considerations on the structure induced by realistic nuclear interactions. First of all, two important correlations are introduced; the correlations caused by the short-range repulsion in the central part of the interaction, the so-called central correlations, and those induced by the tensor part, so called tensor correlations. The correlation operator is written as a product of two unitary operators C_Ω and C_r accounting for tensor and central correlations, respectively,

$$C = C_\Omega C_r = \exp(-i \sum_{j>k} g_\Omega(j,k)) \exp(-i \sum_{j>k} g_r(j,k)). \qquad (3.190)$$

Explicit expressions for the hermitian generators g_Ω and g_r are constructed based on the physical mechanism responsible for the correlations. The central correlations induced by the short-range repulsion are revealed through the suppression of the two-body density at short distances. Schematically speaking, the interaction pushes close-by nucleons apart and thus out of the region of the mutual repulsion. This kind of distance-dependent radial shift is described by the generator $g(1,2)$ which produces an outward radial shift depending on the uncorrelated distance of the particles. The relative distance and the relative momentum for equal mass particles are denoted. by

$$\mathbf{r} = \mathbf{r}(1) - \mathbf{r}(2), \quad \mathbf{q} = \frac{1}{2}(\mathbf{p}(1) - \mathbf{p}(2)), \qquad (3.191)$$

respectively. A generator $g(1,2) = g(\mathbf{r}, \mathbf{q})$ which creates a position-dependent shift may be written in the hermitian form

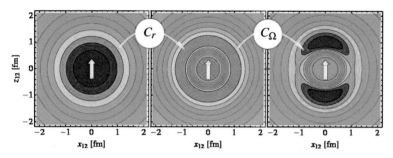

Fig. 3.12 Two-body density $\rho^{(2)}_{SM_S,TM_T}(\mathbf{r}_{12})$ in the $S = 1$, $M_S = 1$, $T = 0$, $M_T = 0$ channel of ^{4}He. Left: for the uncorrelated trial state $|\Phi\rangle$, middle: including central correlations C_r, right: including central and tensor correlations, $C_\Omega C_r$. The arrow indicates the orientation of the spin in the two-body channel. The red (blue) region shows a high (low) two-body density, while green and yellow colors show the region between two extremes. The projection operator C_r creates a low density region at the middle, and the combined operator $C_\Omega C_r$ polarizes strongly the density to the spin direction. Figure reprinted with permission from [26]. ©2021 by Elsevier

$$g(\mathbf{r}, \mathbf{q}) = \frac{1}{2}\{(\mathbf{q} \cdot \frac{\mathbf{r}}{r})s(r) + (\frac{\mathbf{r}}{r} \cdot \mathbf{q})s(r)\}, \qquad (3.192)$$

where $\mathbf{q}$ acts as an operator ∇/i. One expects the unitary operator $C(1, 2) = \exp\{-ig(\mathbf{r}, \mathbf{q})\}$ to shift a relative position r to about $\mathbf{r} + s(r)(\mathbf{r}/r)$. The idea is to find a suited function $s(r)$ such that $g(\mathbf{r}, \mathbf{q})$ moves the probability amplitude out of the classically forbidden region of the short-range repulsion core. The tensor correlations caused by the tensor part of the interaction connect the spin and angular degrees of freedom and result in the mixing of states with the orbital angular momenta L and $L \pm 2$. This can be created with the generator

$$g_\Omega = \theta(r)S_{12}(\mathbf{r}, \mathbf{q}_\Omega) = \theta(r)\frac{3}{2}\Big((\sigma_1 \cdot \mathbf{q}_\Omega)(\sigma_2 \cdot \mathbf{r}) + (\sigma_1 \cdot \mathbf{r})(\sigma_2 \cdot \mathbf{q}_\Omega)\Big), \quad (3.193)$$

where $\mathbf{q}_\Omega = \mathbf{q} - (\mathbf{q} \cdot \mathbf{r})\mathbf{r}/r^2$. The strength and distance dependences of the two transformations are described by the functions $s(r)$ and $\theta(r)$ that depend on the potential under consideration. In general, the generators are amend by projection operators on two-body spin S and isospin T in order to allow for a spin-isospin dependence of the unitary transformation.

Figure 3.12 shows the two-body density $\rho^{(2)}_{SM_S,TM_T}(\mathbf{r}_{12})$ in the deuteron channel for ^{4}He. For the uncorrelated wave function, the two-body density has a maximum at $\mathbf{r}_{12} = 0$, where the potential is strongly repulsive. This defect in the wave function is cured by the central correlator which shifts the nucleons apart. For the centrally correlated wave function we find now the largest density at $|\mathbf{r}_{12}| \sim 1$ fm. At this distance, the potential is most attractive. The tensor force provides attraction if the spins are aligned parallel with the distance vector. This is reflected in the two-body density. After applying the tensor correlations, the density is enhanced at the "poles" and reduced at the "equator". The UCOM method has been applied successfully to the calculations of FMD, SRPA, NCSM and CC models.

3.11.2 Similarity Renormalization Group (SRG)

Similarity Renormalization Group (SRG) methods, in particular the In-Medium SRG (IMSRG) approach have been applied to solve the nuclear many-body problem [34]. These methods use continuous unitary transformations to evolve the nuclear Hamiltonian to a desired shape. The IMSRG is used to decouple the ground state from all excitations and solve the many-body Schrödinger equation. SRG methods are associated with the effective field theories, because they smoothly map theories on different resolution scales and also degrees of freedom.[9]

Effective Field Theory (EFT) and Renormalization Group (RG) methods have become important tools of modern (nuclear) many-body theory. Effective Field Theories allow to systematically take into account the separation of scales when we construct theories to describe quantum many-body phenomena. ChEFT is one well-known example. The RG models are natural companions for EFTs and smoothly connect theories with different resolution scales and degrees of freedom, i.e., provide a systematic framework for formalizing many effects on the renormalization of nuclear interactions and many-body effects. The idea of decoupling energy scales in RG is used to study the nuclear many-body problem and its implementation is called the SRG and In-Medium SRG (IMSRG).

The basic idea of the Similarity Renormalization Group (SRG) method is quite general. The goal is to simplify any many-body Hamiltonian by means of a continuous unitary transformation that is parametrized by a one-dimensional flow parameter s;

$$H(s) = U(s)H(0)U^{\dagger}(s), \tag{3.194}$$

where $H(s = 0)$ is the starting original Hamiltonian and $U^{\dagger}(s)$ is a unitary operator

$$U(s)U^{\dagger}(s) = I. \tag{3.195}$$

As in any quantum mechanical problem, we are primarily interested in finding the eigenstates of H by diagonalizing its matrix representation. This task is made easier if

---

[9] The term renormalization group (RG) refers to a formal apparatus that allows systematic investigation of the changes of a physical system as viewed at different scales. A change in scale is called a scale transformation. The renormalization group is intimately related to scale invariance, a symmetry, in which a system appears the same at all scales (so-called self-similarity). The renormalization group emerges from the renormalization of the quantum field variables, which address the problem of infinities in a quantum field theory. This problem of systematically handling the infinities of quantum field theory to obtain finite physical quantities was solved for quantum electrodynamics (QED) by Richard Feynman, Julian Schwinger and Shin'ichiro Tomonaga, who received the Nobel prize in 1965.

we can construct a unitary transformation that renders the Hamiltonian to be diagonal as s increases. The SRG consists of a continuous sequence of unitary transformations that suppress off-diagonal matrix elements, driving the Hamiltonian towards a band- or block-diagonal form. Mathematically, we want to split the Hamiltonian into suitably defined diagonal and off-diagonal parts,

$$H(s) = H_d(s) + H_{off}(s), \tag{3.196}$$

and find $U(s)$ to satisfy the conditions

$$H(s) \xrightarrow{s \to \infty} H_d(s), \qquad H_{off}(s) \xrightarrow{s \to \infty} 0. \tag{3.197}$$

To implement the continuous unitary transformation, we take the derivative of Eq. (3.194) with respect to s;

$$\begin{aligned}
\frac{dH(s)}{ds} &= \frac{dU(s)}{ds} H(0) U^\dagger(s) + U(s) H(0) \frac{dU^\dagger(s)}{ds} \\
&= \frac{dU(s)}{ds} U^\dagger(s) H(s) + H(s) U(s) \frac{dU^\dagger(s)}{ds}.
\end{aligned} \tag{3.198}$$

This equation is further reformulated as a commutator form with $\eta(s) \equiv \frac{dU(s)}{ds} U^\dagger(s)$,

$$\frac{dH(s)}{ds} = [\eta(s), H(s)]. \tag{3.199}$$

Here we use the identity

$$\frac{d\left(U(s)U^\dagger(s)\right)}{ds} = \frac{d\mathbf{I}}{ds} = 0 \Rightarrow \frac{dU(s)}{ds} U^\dagger(s) = -U(s) \frac{dU^\dagger(s)}{ds}. \tag{3.200}$$

Eq. (3.199) is called the SRG flow equation for the Hamiltonian which describes the evolution of $H(s)$ under the action of a dynamical generator $\eta(s)$. To specify the generator which will transform the Hamiltonian to the desired structure, one possible choice is so-called the double-brackent flow generator,

$$\eta(s) = [H_d(s), H_{off}(s)]. \tag{3.201}$$

The renormalization is performed by splitting the Hamiltonian into diagonal and off-diagonal parts in Eq. (3.196). The idea is also imposed on a block or band-diagonal structure on the representation of the Hamiltonian in bases that are organized by momentum or energy. We can obtain $H(s)$ by integrating Eq. (3.199) numerically, without explicitly constructing the unitary transformation itself. Through different choices for $H_d(s)$ and $H_{off}(s)$, one can tailor the SRG evolution to transform the initial Hamiltonian to a form that is most convenient for a particular problem. This flexibility makes the SRG a powerful alternative to conventional effective interaction methods

such as Lee-Suzuki-Okamoto similarity transformations which was discussed in
Sect. 3.9.1.

In Fig. 3.13, we show examples for two types of RG evolutions that decouple the
low- and high-momentum pieces of NN interactions. The first example, Fig. 3.13a,
is a RG decimation, in which the interaction is evolved to decreasing cutoff scales
$\Lambda_0 > \Lambda_1 > \Lambda_2$, and high-momentum modes are integrated out. This is the so-called
$V_{\text{low}-k}$ approach, which was first used in nuclear physics in the early 2000s. Note that
the resulting low momentum interaction is entirely confined to states with relative
momentum $q \leq \Lambda$. In contrast, Fig. 3.13b shows the SRG evolution of the NN
interaction to a band-diagonal shape via the flow equation (3.199), using a generator
built from the relative kinetic energy in the two-nucleon system:

$$\eta(\lambda) = [\frac{\mathbf{q}^2}{2\mu}, H(\lambda)], \qquad (3.202)$$

where μ is the reduced mass $\mu = m/2$ and $H(\lambda)$ is the Hamiltonian (3.194) with
the flow parameter $\lambda = s^{-1/4}$ of a dimension of momentum. When the SRG-evolved
Hamiltonian $H(\lambda)$ is decomposed into the relative kinetic energy and the two-body
interaction $V(\lambda)$, the flow equation (3.199) is expressed in a simple form

$$\frac{dH(\lambda)}{d\lambda} = \frac{dV(\lambda)}{d\lambda} = [[\frac{\mathbf{q}^2}{2\mu}, V(\lambda)], \frac{\mathbf{q}^2}{2\mu} + V(\lambda)]. \qquad (3.203)$$

Even in this simple form (3.203), the direct solution of the operator equation is far
from trivial and is solved practically under the matrix representation of the equation.

As suggested by Fig. 3.13(b), λ is a measure for the width of the band in momentum
space. Thus, momentum transfers between nucleons are limited according to

$$Q = |\mathbf{q}_1 - \mathbf{q}_2| \leq \lambda, \qquad (3.204)$$

and low- and high-lying momenta are decoupled as λ is decreased as $\lambda_0 > \lambda_1 > \lambda_2$.
Equation (3.204) implies that the spatial resolution scale of a SRG-evolved inter-
action (or $V_{\text{low}-k}$ if we determine the maximum momentum transfer in the low-
momentum block) is $1/Q = 1/\lambda$; only long-ranged components of the NN interaction are resolved explicitly and short-range components of the interaction can be
replaced by contact interactions in the same way as ChEFT. This is the reason why
the realistic NN interactions that accurately describe NN scattering data collapse
into a universal long-range interaction when RG is evolved; that is the one-pion
exchange potential (OPEP). As a theoretical framework, SRG evolutions or ChEFT
are preferred over $V_{\text{low}-k}$ models in nuclear many-body theory, because they can be
naturally extended to $3N$ interaction, $4N$ interaction, also and to many-body observables. Namely, the SRG method to nuclear forces has been carried out in free space
to construct effective nucleon-nucleon (NN) and three-nucleon (3N) interactions to
be used as input in ab initio calculations. An alternative is to perform the SRG evolution directly in the A-body system of interest. Unlike the free-space evolution, the

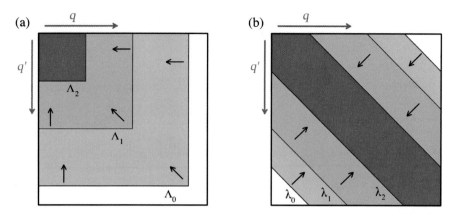

Fig. 3.13 Schematic illustration of two types of RG evolution for NN interactions in momentum space: **a** $V_{\text{low}-k}$ running in Λ, and **b** SRG running in λ. Here, q and q' denote the relative momenta of the initial and final state, respectively. At each Λ_i or λ_i, the matrix elements outside of the corresponding blocks or bands are negligible, implying that high- and low-momentum states are decoupled

in-medium SRG (IM-SRG) has the appealing feature that one can approximately evolve 3-, 4-,$\cdots A$-body operators using only one- and two-body interactions and operators. The IM-SRG was applied thus to generate effective interactions of modern shell model calculations of closed-shell and open-shell nuclei to calculate wave functions and observables not only the ground state but also for excited states.

In Reference [35], a complete N^2LO analysis of p-shell nuclei was presented using the NN and $3N$ interactions generated by the SRG method. In Fig. 3.14, the NLO and N^2LO results are shown for nuclei up to A = 16. It should be noticed that since the Hamiltonian has been completely determined in the NN and $3N$ systems, the ground-state energies are parameter-free predictions. In Fig. 3.14, the truncation errors with the expansion parameter $Q = m_\pi/\Lambda$ have been estimated. The m_π is the soft scale and the Λ is the hard scale parameters, respectively. The inclusion of the three-nucleon force is found to improve the agreement with the data for most of the considered observables.

3.11.3 No-Core Shell Model (NCSM)

The no-core shell model (NCSM) is a straightforward but computationally demanding method to solve quantum many-body problems [28]. This method is a challenge for new algorithms and advances of new super-computers for book-keeping and diagonalization of a large sparse matrix eigenvalue problem. The model adopts ab initio realistic 2-body and 3-body interactions

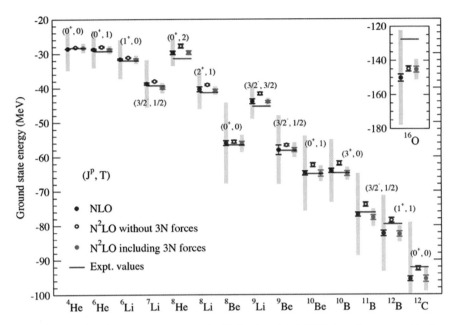

Fig. 3.14 Calculated ground state energies in MeV using chiral NN interactions in combination with the chiral $3N$ force generated by the SRG method (open and solid dots) in comparison with experimental values (red levels). For each nucleus the NLO and N²LO results are the left and right symbols and bars, respectively. The open blue symbols correspond to incomplete calculations at N²LO using NN-only interactions. Blue and green error bars indicate the uncertainty of the SRG model. The shaded bars indicate the truncation error with the expansion parameter $Q = m_\pi/\Lambda$. Figure reprinted from [36]

$$H = T + V_{2N} + V_{3N}, \tag{3.205}$$

and renormalizes the Hamiltonian into the adopted model space. The renormalization procedure is adopted commonly to obtain the effective Hamiltonian and the effective operator for the adopted model space by the appropriate projection method. The matrix element of H is calculated by using the harmonic oscillator Slater determinants in the m-scheme basis

$$|\psi_n\rangle = [a_\alpha^\dagger a_\beta^\dagger \cdots]_n |0\rangle, \tag{3.206}$$

where the bracket $[\cdots]$ means the anti-symmetrized wave function and $\alpha, \beta, \cdots$ denote the angular momentum and isospin quantum numbers $\alpha \equiv (j_\alpha m_\alpha t_{z,\alpha})$. The number of basis states n becomes larger for a larger model space which is truncated by the major quantum number of harmonic oscillator. Diagonalizing the matrix $\langle \psi_m | H | \psi_n \rangle$, we obtain the exact eigenstate

$$|\Phi\rangle = \sum_n C_n |\psi_n\rangle, \tag{3.207}$$

where real coefficients C_n are obtained by the diagonalization.

The effective Hamiltonian for NCSM is obtained by the Lee-Suzuki-Okamoto similarity renormalization scheme from realistic many-body interactions such as the Chiral EFT interaction. In this renormalization scheme, the full Hilbert space is divided into the active model space P and a complement space Q with a condition $Q = 1 - P$. In solving low-lying states of the many-body system, the fundamental assumption is that the main components of these states can be constructed from configurations of a few active particles and holes occupying a few active orbits. The set of states with the configurations of active orbits is called model space, which is referred to as the P space. The complement of the P space is called the Q space.

Consider $|\Phi_\nu\rangle$ is a complete set of eigenstates of H,

$$H|\Phi_\nu\rangle = E_\nu |\Phi_\nu\rangle. \tag{3.208}$$

Each state $|\Phi_\nu\rangle$ is decomposed into the P and Q spaces as

$$|\Phi_\nu\rangle = (P + Q)|\Phi_\nu\rangle = |\nu\rangle + Q|\Phi_\nu\rangle \tag{3.209}$$

where $|\nu\rangle \equiv P|\Phi_\nu\rangle$. Define the operator S which maps the state in Q space into P space

$$S|\nu\rangle = Q|\Phi_\nu\rangle. \tag{3.210}$$

Then, we have

$$|\Phi_\nu\rangle = (I + S)|\nu\rangle \tag{3.211}$$

where I is the identity operator. The operator S satisfies the relations,

$$QSP = S, \quad PSP = QSQ = 0, \quad S^2 = 0. \tag{3.212}$$

Problem

3.12 Prove that $QSP = S$ from Eq. (3.210).

Since $S^2 = 0$, we have

$$e^{\pm S} = (I \pm S). \tag{3.213}$$

Thus, the operator e^{+S} has the inverse e^{-S} being a unitary operator, which makes possible the mapping,

$$|\Phi_\nu\rangle = e^S |\nu\rangle, \quad |\nu\rangle = e^{-S}|\Phi_\nu\rangle. \tag{3.214}$$

These mappings are called "similarity transformations". Then, $|\nu\rangle$ is an eigenstate of a Hamiltonian

$$H_S = e^{-S} H e^S, \tag{3.215}$$

with the eigenvalue E_ν since

P   Q

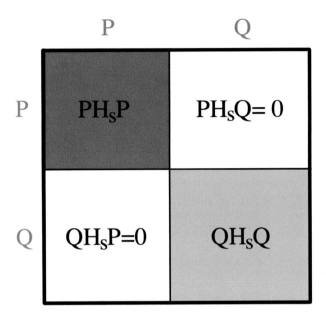

Fig. 3.15 A schematic representation of the effective Hamiltonian in the Lee-Suzuki-Okamoto renormalization scheme. The Hamiltonian $H_P = PH_S P$ in the active model space is renormalized to take into account the Hamiltonian $H_Q = QH_S Q$ in the complimentary space $Q = 1 - P$ by using the similarity transformation operator S defined in Eq. (3.210). There is no coupling between P and Q spaces in the model Hamiltonian

$$\langle \Phi_\nu | H | \Phi_\nu \rangle = \langle \Phi_\nu | e^S e^{-S} H e^S e^{-S} | \Phi_\nu \rangle = \langle \nu | H_S | \nu \rangle = E_\nu. \qquad (3.216)$$

The operator S has remarkable properties. This operator maps the complete basis $|\Phi_\nu\rangle$ to the basis $|\nu\rangle$ in a model space P. The same operator, being the similarity transformation, maps a Hamiltonian H for the complete basis to H_S in a new model space. The important condition of the Lee-Suzuki-Okamoto approach is a decoupling condition for the effective Hamiltonian,

$$QH_S P = Qe^{-S} H e^S P = 0, \qquad (3.217)$$

so that the Hamiltonian is block diagonal and there is no matrix element between P and Q spaces $PH_S Q=0$. and $QH_S P=0$ as seen in Fig. 3.15.

The effective Hamiltonian H_P in the model space P is expanded to be

$$H_S = e^{-S} H e^S = H + [H, S] + \frac{1}{2}[[H, S], S] + \cdots. \qquad (3.218)$$

In general, the commutator terms contain two-, three-, $\cdots$, A-body operators, i.e., the Hamiltonian H_S can be written in a cluster expansion form as

$$H_S = \mathcal{H}^{(1)} + \mathcal{H}^{(2)} + \mathcal{H}^{(3)} + \cdots, \qquad (3.219)$$

where $\mathcal{H}^{(i)}$ is a i-body Hamiltonian. While both S and H_S are A-body operators, the simplest, but non-trivial approximation of H_S is a two-body effective Hamiltonian which is a truncation of the full Hamiltonian H. To determine the operator S is an essential problem in the Lee-Suzuki-Okamoto approach. Taking the two-body terms in H in Eq. (3.205), the operator S is expressed by the two-body operator S_{12} in order to map the Q space into the P space.

With the condition of Eq. (3.217), the operator S_{12} can be expressed as

$$S_{12} = \tanh^{-1}(\omega - \omega^{\dagger}), \tag{3.220}$$

with the operator

$$\omega = \sum_{k=1}^{d} Q|\Phi_k\rangle\langle\tilde{\phi}_k|P, \tag{3.221}$$

where d is the dimension of the P space, $|\Phi_k\rangle$ is the eigenstate of H and $|\phi_k\rangle = P|\Phi_k\rangle$ which satisfies the normalization condition $\langle\tilde{\phi}_i|\phi_k\rangle = \delta_{ik}$ [10]. With the effective Hamiltonian H_P, we can take into account effectively the full model space in NCSM calculations. The mapping method adopted is similar to the one adopted in the PVC model discussed in Sect. 3.7.1.

The next improvement is to develop a three-body Hamiltonian. This approximation can be expanded in principle to include higher order correlations of n-body clusterings with $n \leq A$ although the computational burden becomes heavier.

In the standard parmetrization of ChEFT, the "low-energy constants" (LECs) are adjusted in the NN scattering data, and the binding energies and charge radii of ^{3}H, and ^{3,4}He. In the NNLO$_{\text{sat}}$ model in [37], binding energies and charge radii for ^{3}H, ^{3,4}He, ^{14}C, and ^{16,22,23,24,25}O are employed in the optimization of the parameter sets. In Fig. 3.16, the results for ^{4}He are computed with the NCSM. For nuclei with A > 4, the coupled-cluster method in its singles and doubles approximation (CCSD) are employed. The triplet contribution (CCSDT) is also included in a perturbative method (named Λ-CCSD(T)). The previous models predict too-small radii and too-large binding energies of ^{16}O and ^{40}Ca. The NNLO$_{\text{sat}}$ cures this shortcomings of the previous models and gives fairly agreements with the empirical data as shown in Fig. 3.16. This model is further applied for the spectra for ^{40}Ca and selected isotopes of lithium, nitrogen, oxygen and fluorine isotopes and reproduces well the experimental data.

3.11.4 Variational Monte Carlo (VMC) and Green's Function Monte Carlo (GFMC) Approaches

Quantum Monte Carlo methods are based on Feynman's path integral formalism. Variational Monte Carlo (VMC) [31] and Green's function Monte Carlo (GFMC) [32] approaches make samplings of the configuration space and give access to important

---

[10] The derivation of Eq. (3.220) can be found in [30].

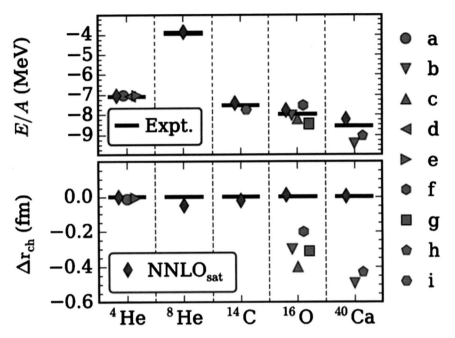

Fig. 3.16 Ground-state energy per nucleon (top), and differences between computed and experimental values of charge radii (bottom) for selected nuclei computed with chiral interactions. In most cases of the previous models labeled [a] and [i], the low-energy constants are determined without introducing the many-body data of medium-heavy nuclei, and predicts too-small radii and too-large binding energies of ^{16}O and ^{40}Ca. The red diamonds are NNLO$_\mathrm{sat}$ results. In the figure, the results for ^{4}He are computed with the NCSM. For nuclei with $A > 4$, the coupled-cluster method in its singles and doubles approximation (CCSD) are employed. See text and [37] for details. Figure reprinted with permission from [37]. ©2021 by the American Physical Society

properties of nuclear structure in light nuclei. The VMC finds an upper bound to an eigenenergy of the Hamiltonian H by evaluating the expectation value with a trial wave function $|\Psi_T\rangle$. The trial wave function can be expressed in the separable form

$$|\Psi_T\rangle = \mathcal{F}|\Psi\rangle, \tag{3.222}$$

where the factorization of the wave function is performed into two parts: the long-range low-momentum components is separated into $|\Psi\rangle$, and the short-range high-momentum components is taken care by $\mathcal{F}$. The operator $\mathcal{F}$ includes the so-called Jastrow-like correlations between pairs and triplets of particles

$$\mathcal{F} = \left(\mathcal{S} \prod_{i<j<k} (1 + F_{ijk}) \right) \left(\mathcal{S} \prod_{i<j} F_{ij} \right), \tag{3.223}$$

where S is the symmetrization operator, F_{ij} is a two-body correlation and F_{ijk} is a three-body correlation. This correlation operator is called *Jastrow operator*. The two-body correlation operator has a strong spin and isospin dependence and is expressed by a combination of solutions of Schrödinger-type equations for different spin-isospin channels. The three-body correlation operator appears naturally when the three-body forces are taken into account in the calculations. For the long-range part of the trial wave function $|\Psi\rangle$ can be expressed by the sum of few Slater determinants obtained by modest scale shell model calculations. The wave function $|\Psi_T\rangle$ in Eq. (3.222) is then called the *Slater-Jastrow* wave function. The BCS type pair correlation or the α cluster correlation is also used as the trial wave function $|\Psi\rangle$ if it is necessary.

The wave function $|\Psi_T\rangle$ is varied to minimize the expectation value,

$$E_T = \frac{\langle \Psi_T | H | \Psi_T \rangle}{\langle \Psi_T | \Psi_T \rangle} \geq E_0, \tag{3.224}$$

with respect to the parameters to describe two-body and three-body correlations in Eq. (3.223).

Monte Carlo methods can be used to calculate E_T and to minimize the energy with respect to changes in the variational parameters. The spacial integral of Eq. (3.224) is evaluated using the Metropolis Monte Carlo technique, in which a weighting factor $W(\mathbf{R})$ is introduced to sample points in the $3A$ coordinate space, $\mathbf{R} = (\mathbf{r}_1, \mathbf{r}_2, \cdots, \mathbf{r}_A)$. The simplest choice of $W(\mathbf{R})$ is $\langle \Psi(\mathbf{R}) | \Psi(\mathbf{R}) \rangle$, in which the trial wave function will be the Slater-Jastrow determinants, or the BCS state+Jastrow operator depending on physical circumstances. The energy can be computed as the average over the N points in the random walk sampling

$$E_V = \frac{1}{N} \sum_{i=1}^{N} \frac{\langle \Psi_T(\mathbf{R}_i) | H | \Psi_T(\mathbf{R}_i) \rangle}{W(\mathbf{R}_i)}. \tag{3.225}$$

In VMC, the Monte Carlo technique is used to evaluate the many-dimensional energy functional integrals (3.225). On the other hand, GFMC is based on the path integral formulation of quantum many-body problems in which the evolution of the system is described by the imaginary time (Euclidian time). The GFMC typically starts from a trial wave function $|\Psi_T\rangle$ which is evolved by the imaginary time τ in the form

$$|\Psi(\tau)\rangle = \exp\left(-(H - E_0)\tau\right)|\Psi_T\rangle = \sum_n \exp\left(-(E_n - E_0)\tau\right) a_n \psi_n, \tag{3.226}$$

where E_0 is the ground state energy and ψ_n is a set of Slater determinants. The wave function will turn out to be the ground state in the limit $\tau \to \infty$;

$$|\Psi(\tau \to \infty)\rangle = a_0 \psi_0. \tag{3.227}$$

For strongly-interacting systems one cannot compute $\exp[-(H - E_0)\tau]$ directly, but one can compute the high-temperature propagator, $\beta = (kT)^{-1} << 1/H$, or the short-time propagator, $\Delta\tau << 1/H$, and insert complete sets of states between each short-time propagator,

$$|\Psi(\mathbf{R}_N, \tau)\rangle \equiv \langle\mathbf{R}_N|\Psi(\tau)\rangle = \prod_{1,2,---,N} \langle\mathbf{R}_N|\exp\left(-(H - E_0)\Delta\tau\right)|\mathbf{R}_{N-1}\rangle$$
$$\cdots \langle\mathbf{R}_1|\exp\left(-(H - E_0)\Delta\tau\right)|\mathbf{R}_0\rangle\langle\mathbf{R}_0|\Psi_T\rangle, \qquad (3.228)$$

where the Monte Carlo technique is used to sample the paths of space coordinates $\mathbf{R} \equiv (\mathbf{r}_1, \mathbf{r}_2, \cdots, \mathbf{r}_A)$ in the propagator. In a simple case where the interaction V has no spin and isospin dependence, the propagator can be expressed assuming very small time slice $\Delta\tau$ as

$$G_{\Delta\tau}(\mathbf{R}, \mathbf{R}') \equiv \langle\mathbf{R}|\exp\left(-H\Delta\tau\right)|\mathbf{R}'\rangle$$
$$\approx \langle\mathbf{R}|\exp\left(-V\Delta\tau/2\right)\exp\left(-T\Delta\tau\right)\exp\left(-V\Delta\tau/2\right)|\mathbf{R}'\rangle, \qquad (3.229)$$

where T is the kinetic energy and V is a many-body interaction. The matrix for the interaction V is expressed as

$$\langle\mathbf{R}|\exp\left(-V\Delta\tau\right)|\mathbf{R}\rangle \approx S \prod_{i<j} \exp[-V_{ij}(\mathbf{r}_{ij})\Delta\tau], \qquad (3.230)$$

where S is the symmetrization operator over orders of pairs. The energy is then calculated as a function of imaginary time τ as

$$E(\tau) = \frac{\langle\Psi_T|H|\Psi(\tau)\rangle}{\langle\Psi_T|\Psi(\tau)\rangle}$$
$$\approx \frac{\sum_i \langle\Psi_T(i)|H|\Psi(\tau, i)\rangle / W_i}{\sum_i \langle\Psi_T(i)|\Psi(\tau, i)\rangle / W_i}, \qquad (3.231)$$

where the sum over i indicates the sum over samples of the wave functions and W_i is its weight. The imaginary time τ should be taken large enough to ensure the convergence of results. GFMC was applied to study the effect of the three-body force in the energy spectra of p-shell nuclei (see Fig. 2.12). The GFMC is often adopted in atoms, molecules and solid state physics and called also *Diffusion Monte Carlo* method.

3.11.5 Coupled Cluster Method

The coupled cluster (CC) model is based on the *similarity transformation theory* of many-body wave function and the many-body Hamiltonian [33]. The CC model was proved to be extremely useful in quantum chemistry. Applications to nuclear physics struggled with the repulsive nature of NN interactions and highly demanding computer power of CC calculations for handling an extremely large model space. These problems have been cured during the last decade and the CC model is now applied to nuclear structure calculations in light to medium-mass nuclei.

In this model, the many-body ground state $|\Psi_0\rangle$ is given by the exponential ansatz,

$$|\Psi_0\rangle = |\Psi_{CC}\rangle = e^T |\Phi_0\rangle = \left(\sum_{n=1}^{\infty} \frac{1}{n!} T^n \right) |\Phi_0\rangle, \tag{3.232}$$

where $|\Phi_0\rangle$ is a reference Slater determinants and T is the cluster operator that generates correlations. The operator T is expanded as a linear combination of particle-hole excitation operators

$$T = T_1 + T_2 + \dots + T_A, \tag{3.233}$$

where T_n is the n-particle n-hole (n-p n-h) excitation operator. For CCSD approximation (S and D stand for single and double p-h excitations, respectively), the operator T is given by

$$T = T_1 + T_2, \tag{3.234}$$

with

$$T_1 = \sum_{ia} t_i^a a_a^\dagger a_i, \tag{3.235}$$

$$T_2 = \sum_{ijab} t_{ij}^{ab} a_a^\dagger a_b^\dagger a_i a_j. \tag{3.236}$$

Here, the indices $(a, b, \cdots)$ denote particle states, while the indices $(i, j, \cdots)$ denote hole states. The cluster amplitudes t play a similar role to the shell model coefficient of each configuration after the diagonalization. The cluster amplitudes t are determined by the solutions of the equations to the amplitude equations

$$\langle \Phi_i^a | \overline{H} | \Phi_0 \rangle = 0, \quad \langle \Phi_{ij}^{ab} | \overline{H} | \Phi_0 \rangle = 0, \quad \cdots, \tag{3.237}$$

where $\Phi_i^a = t_i^a a_a^\dagger a_i |\Phi_0\rangle$ and $\Phi_{ij}^{ab} = t_{ij}^{ab} a_a^\dagger a_b^\dagger a_i a_j |\Phi_0\rangle$. In order to set up these equations, the similarity transformed Hamiltonian $\overline{H}$ is expanded using the so-called Baker-Campbell-Hausdorff expression,

$$\overline{H} = e^{-T} H e^T = H + [H, T] + \frac{1}{2}[[H, T], T] + \cdots + \frac{1}{n!}[\cdots[H, T], \cdots T] + \cdots, \tag{3.238}$$

up to the maximum order required by an adopted approximation (the second order for CCS and the fourth order for CCSD). Using the expansion (3.238), the ground state energy is expressed by a sum of commutators

$$E_{CC} = \langle \Psi_0 | \overline{H} | \Psi_0 \rangle$$
$$= \langle \Psi_0 | H + [H, T] + \frac{1}{2}[[H, T], T] + \cdots + \frac{1}{n!}[\cdots[H, T], \cdots T] + \cdots | \Psi_0 \rangle,$$

$$(3.239)$$

with the subsidiary conditions for the cluster amplitudes (3.237). The CC model is extensively applied to study not only light nuclei but also medium-heavy nuclei up to the $A \sim 100$ region in CCSD model space with additional effects of triplet clusters in a perturbative manner. The CCSD is equivalent to Second RPA in the model space, but the advantage of CCSD is to renormalize the effective interactions consistently with the adopted model space.

The CCSD calculations of O isotopes with chiral EFT interactions are shown in Fig. 3.17. The two-body interaction NN is the chiral NN interaction developed at next-to-next-to-next-to-leading order (N^{3}LO) within chiral EFT. The potential $3NF_{eff}$ is the in-medium NN interaction from the leading order chiral three-body forces ($3NF$) by integrating one nucleon over the Fermi sea up to the Fermi momentum k_F. Calculations with NN interactions yield the correct level ordering but very compressed spectra when compared to data, and all the computed excited states are well bound with respect to the neutron emission thresholds. However, the inclusion of $3NF$s increases the level spacing and significantly improves the agreement with experiment. Several of the excited states are resonances in the continuum, and the proximity of the continuum is particularly relevant for the dripline nucleus ^{24}O. The

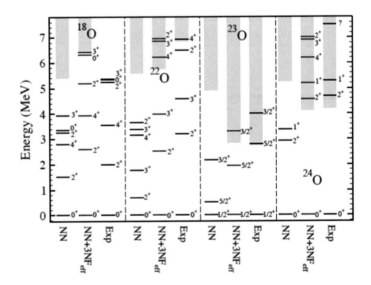

Fig. 3.17 Excitation spectra of oxygen isotopes computed from chiral nucleon-nucleon interactions, without and with inclusion of the effects of three-nucleon forces denoted by NN and $NN+3NF_{eff}$, respectively. The particle continua above the scattering thresholds are shown as gray bands. Figure reprinted with permission from [41]. ©2021 by the American Physical Society

Gamow basis states are used to take into account a description of the excited resonant states.

3.11.6 Self-consistent Green's Function (SCGF) Approach

The self-consistent calculations of the Green's function is a model to calculate the ground state and the excited states by using the Green's function theory [38]. The term "self-consistency" means in the SCGF that the input information about the ground state and excitations of the systems no longer depends on any reference state but instead it is taken directly from the computed correlated wave function (or propagator function). To achieve this, the computed spectral function is fed back into the working equations and calculations are repeated until a consistency between input and output is obtained. This approach is referred to as *self-consistent Green's function* (SCGF) method and it is implemented for nuclear structure studies.

The SCGF method is formulated based on the many-body Green's function which is also referred to as propagators. These are defined in the second quantization formalism by assuming the true ground state $|\Psi_0^A\rangle$ of a system of A nucleons, which is taken to be a vacuum of excitations. The one-body Green's function (or propagator) is then defined as

$$i\hbar g_{\alpha\beta}(t - t') = \langle \Psi_0^A | \mathcal{T}[a_\alpha(t)a_\beta^\dagger(t')] | \Psi_0^A \rangle, \tag{3.240}$$

where $\mathcal{T}$ is the time ordering operator, $a_\beta^\dagger(t')(a_\alpha(t))$ are the time-dependent creation (annihilation) operators in the so-called Heisenberg picture, and greek indices $\alpha, \beta, \cdots$ label a complete single-particle basis in the model space. These can be the continuum momentum or coordinate spaces or any discrete set of single-particle states. Note that $g_{\alpha\beta}(t - t')$ depends only on the time difference $t - t'$. For $t > t'$, Eq. (3.240) gives the probability amplitude to add a particle to the ground state $|\Psi_0^A\rangle$ in a single-particle state β at time t' and then to let it propagate to another single-particle state α at a later time t. Alternatively, for $t < t'$ a particle is removed from a state α at t and added to β at t'. The one-body Green's function contains useful information regarding single-particle behavior inside the many-body system; one-body observables, the total binding energy, and even elastic nucleon-nucleus scattering. The propagation of a particle or a hole excitation corresponds to the time evolution of an intermediate many-body system with $A + 1$ or $A - 1$ particles. One can better understand the physics information included in Eq. (3.240) from considering the eigenstates $|\Psi_n^{A+1}\rangle$, $|\Psi_k^{A-1}\rangle$ and eigenvalues E_n^{A+1}, E_k^{A-1} of these intermediate systems. The suffices n and k label the states of $A + 1$ and $A - 1$ nuclei, respectively. By expanding on these eigenstates and by Fourier transforming from time to frequency, one arrives at the spectral representation of the one-body Green's function

$$g_{\alpha\beta}(\omega) = \int d\tau e^{i\omega\tau} g_{\alpha\beta}(\tau = t - t')$$

$$= \sum_n \frac{\langle \Psi_0^A | a_\alpha | \Psi_n^{A+1} \rangle \langle \Psi_n^{A+1} | a_\beta^\dagger | \Psi_0^A \rangle}{\hbar\omega - (E_n^{A+1} - E_0^A) + i\eta} + \sum_k \frac{\langle \Psi_0^A | a_\beta^\dagger | \Psi_k^{A-1} \rangle \langle \Psi_k^{A-1} | a_\alpha | \Psi_0^A \rangle}{\hbar\omega - (E_n^{A-1} - E_0^A) - i\eta},$$

$$\equiv g_{\alpha\beta}^{(+)}(\omega) + g_{\alpha\beta}^{(-)}(\omega), \tag{3.241}$$

where η is an infinitesimal value and the creation (annihilation) operator $a_\beta^\dagger(a_\alpha)$ has no time dependence in the Schrödinger picture.[11]

Problem

3.13 Obtain the spectral representation of one-body Green's function (3.241) from Eq. (3.240).

The one-body Green's function (3.241) is completely determined by solving the Dyson equation

$$g_{\alpha\beta}(\omega) = g_{\alpha\beta}^{(0)}(\omega) + \sum_{\gamma,\delta} g_{\alpha\gamma}^{(0)}(\omega) \Sigma_{\gamma\delta}^*(\omega) g_{\delta\beta}(\omega). \tag{3.242}$$

In Eq. (3.242), the unperturbed propagator $g_{\alpha\beta}^{(0)}(\omega)$ is the initial reference state (usually a mean-field or Hartree–Fock state), while $g_{\alpha\beta}(\omega)$ is called the correlated or dressed propagator. The quantity $\Sigma_{\gamma\delta}^*(\omega)$ is a self-energy term and it is often referred to as the mass operator. This operator plays a central role in the GF formalism and can be interpreted as the non-local and energy-dependent potential that each Fermion feels due to the interactions with the medium. The medium effect is fully taken into account in the self-energy term $\Sigma_{\gamma\delta}^*(\omega)$, so called the irreducible self-energy, in which the correlations are included up to infinite order by using the Green's function perturbation technique. For frequencies $\omega > 0$ the solution of Eq. (3.242) yields a continuum spectrum with $E_n^{A+1} > E_0^A$ and the state $|\Psi_n^{A+1}\rangle$ describes the elastic scattering of nucleon on the $|\Psi_0^A\rangle$ target. It can be shown that $\Sigma_{\gamma\delta}^*(\omega)$ is an optical potential for scattering of a particle from the target with mass A. The Dyson equation is nonlinear in its solution, $g_{\alpha\beta}(\omega)$, and thus it corresponds to an all-orders summation of diagrams involving the self-energy. The Feynman diagram corresponding to the Dyson equation is shown in Fig. 3.18.

The one-body Green's function can be used to calculate the total binding energy and the expectation values of all one-body operators. The Green's function can be recast in the particle and hole spectral functions, which contain the separate responses for the attachment and removal of a nucleon. They can be obtained directly from Eq. (3.241) by using the residue formula (3.114) and the response function (3.118):

----

[11] The Heisenberg operator $a_H^\dagger(t)$ is defined by the Schrödinger operator $a_S^\dagger$ through the relation $a_H^\dagger(t) = e^{-iHt} a_S^\dagger e^{iHt}$.

Fig. 3.18 Diagrammatic representations of the Dyson equation (3.242). Single lines with an arrow represent the unperturbed propagator $g^{(0)}(\omega)$ and double lines are the fully correlated or dressed propagator $g(\omega)$ with the self-energy term Σ^*

Dyson equation

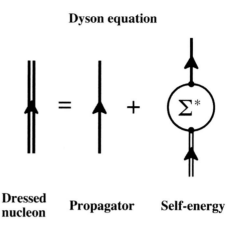

Dressed nucleon   **Propagator**   **Self-energy**

$$S^p_{\alpha\beta}(\omega) = -\frac{1}{\pi}\text{Im}g^{(+)}_{\alpha\beta}(\omega) = \sum_n \langle\Psi^A_0|a_\alpha|\Psi^{A+1}_n\rangle\langle\Psi^{A+1}_n|a^\dagger_\beta|\Psi^A_0\rangle \text{ for } \omega > E^{A+1}_n - E^A_0,$$

$$(3.243)$$

$$S^h_{\alpha\beta}(\omega) = -\frac{1}{\pi}\text{Im}g^{(-)}_{\alpha\beta}(\omega) = \sum_k \langle\Psi^A_0|a^\dagger_\beta|\Psi^{A-1}_k\rangle\langle\Psi^{A-1}_k|a_\alpha|\Psi^A_0\rangle \text{ for } \omega < E^{A-1}_k - E^A_0.$$

$$(3.244)$$

The diagonal parts of Eqs. (3.243) and (3.244) have a straightforward physical interpretation. The particle part, $S^p_{\alpha\alpha}(\omega)$, is the probability of adding a nucleon with quantum numbers α to the A-body ground state $|\Psi^A_0\rangle$, and to find the system in a final state with energy $E^{A+1}_n = E^A_0 + \omega$. Likewise, $S^h_{\alpha\alpha}(\omega)$ gives the probability of removing a particle from state α leaving the nucleus in an eigenstate of energy $E^{A-1}_k = E^A_0 - \omega$. These are demonstrated in coordinate space in Fig. 3.19 for neutrons around ^{56}Ni. Below the Fermi energy, $E_F = (E^{A+1}_n + E^{A-1}_k)/2$, one can see a single dominant hole peak corresponding to the $f_{7/2}$ orbit. The states from the sd-shell are at lower energies and are very fragmented. Just above E_F there are sharp peaks corresponding to the attachment of a neutron to the remaining pf-shell orbits. For $\omega > 0$, one has neutron-^{56}Ni elastic scattering states. One can see that isolated peaks of p- and f-orbits persist around the Fermi surface, which confirm the underlying shell structure outside the ^{40}Ca core for this nucleus. The existence of these isolated dominant peaks shown in Fig. 3.19 indicates that the eigenstates $|\Psi^{A+1}_n\rangle$ and $|\Psi^{A-1}_k\rangle$ are a good approximation constructed of a nucleon or a hole independently orbiting the ground state $|\Psi^A_0\rangle$. How much a real nucleus deviates from this assumption can be shown by the deviations between the experimental spectroscopic factors and the calculated values from Eqs. (3.243) and (3.244).

The SCGF method has a large overlap with the PVC model to take into account the correlations beyond the mean field to study the spectroscopic information of single-particle states. The SCGF is also suitable to the study of infinite systems at finite temperature. There are many challenging problems for possible application of SCGF for

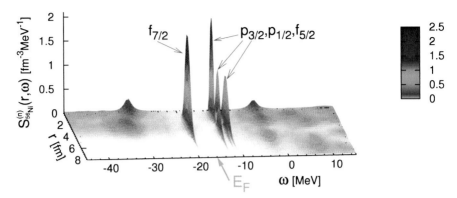

Fig. 3.19 Calculated single-particle spectral function for the addition and removal of a neutron to and form ^{56}Ni. The diagonal part, $S_{\alpha\alpha}(\omega)$, is shown in coordinate space. Energies below the Fermi level E_F correspond to the one-hole spectral function $S^h_{\alpha\alpha}(\omega)$ which describes the distribution of nucleons in energy and coordinate space. Energies above E_F are for the one-particle spectral function $S^p_{\alpha\alpha}(\omega)$. The color code shows the strength of the spectral function. Figure reprinted with permission from [42]. ©2021 by the American Physical Society

open shell nuclei; describing excited spectra, accessing deformed nuclei and describing pairing and superfluidity at finite temperatures. The recently-developed Gorkov formalism allows to calculate spherical open-shell nuclei [43]. These challenges will be crucial to the study of exotic nuclei at developing radioactive beam facilities.

3.11.7   Nuclear Lattice Effective Field Theory (NLEFT) Simulation

One ambitious ab initio approach was proposed to combine the idea of Lattice QCD simulation with ChEFT interactions[39].[12] This approach is named *"nuclear lattice EFT (NLEFT) simulations"* in which nucleons and pions are interacting on the lattice, while Lattice QCD handles quarks and gluons with the QCD Lagrangian. In the NLEFT approach, space-time is discretized in Euclidean time (imaginary time) on a hypercubic volume $L_s^3 \times L_t$, with $L_s(L_t)$ being the length in the spatial (temporal) direction. The minimal distance on the lattice, the so-called lattice spacing, is $a(a_t)$ in space (time). In most NLEFT simulations, the analysis makes use of a periodic cubic lattice with a lattice spacing of $a = \hbar c/(100 \text{ MeV}) = 1.97$ fm and the size $L_s = (10 - 12)$fm. In the time direction, a step size of $ca_t = \hbar c/(150 \text{ MeV}) = 1.32$ fm is taken and the propagation time L_t is varied to extrapolate to the limit $L_t \to \infty$. An unavoidable feature of the finite lattice spacing is the ultra-violet divergence (UV) regulator so that the largest possible momentum is taken as $p_{\max} = \hbar c \pi/a = 314$

---

[12] Details on Lattice QCD can be found in Chap. 10, while ChEFT is described in Sect. 2.7 of Chap. 2.

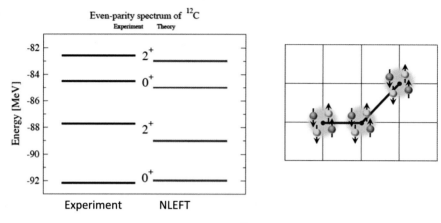

Fig. 3.20 (Left) Lowest-lying even-parity states of ^{12}C from NLEFT with NNLO interaction of ChEFT in comparison with the experimental values. (Right) A topological configuration of the Hoyle state with $J^\pi = 0^+$ at $E = -85$MeV. The data are taken from [44]. Right figure reprinted with permission from [44]. ©2021 by the American Physical Society

MeV. Thus, the interaction is very soft and therefore most higher order corrections, including also the Coulomb effects, are treated in perturbation theory.

In the actual calculations, the nucleons are treated as point-like particles residing on the lattice sites, whereas the nuclear interactions (pion exchanges and contact terms) are represented as insertions on the nucleon world lines using standard auxiliary field representations, which makes the approach particularly suited for highly parallel computation. In the auxiliary field approach, each nucleon evolves in time from the starting at $t = t_i$ up to the final time t_f (see Sect. 3.9.1 for an analogous simulation approach in SMMC). The value of t_f has to be large enough so that the asymptotic behavior of any observable for the A-nucleon state can be extracted. The nuclear forces are selected to have an approximate spin-isospin $SU(4)$ symmetry (Wigner symmetry) that is of fundamental importance in suppressing the sign oscillations that plague any Monte Carlo simulation of strongly interacting fermion systems at finite density. The $SU(4)$ symmetry implies that the combined spin-isospin symmetry of four nucleons having the total isospin and spin, $T = 0$ and $S = 0$, respectively, which leaves the nuclear forces in the S-wave invariant. Under this restriction on the nuclear force, nuclear lattice simulations allow access to problems of nuclear structure and dynamics such as the ground states of $A = 4n$ nuclei, Hoyle state in ^{12}C and α-α scattering.

As an application of nuclear lattice simulations, the 0^+ and 2^+ states in ^{12}C are studied with NNLO of ChEFT interaction including three-body forces. The first excited state of the ^{12}C nucleus with $J^\pi = 0^+$ known as the "*Hoyle state*" constitutes one of the most interesting states in nuclear physics, as it plays a key role in the production of carbon via fusion of three alpha particles in red giant stars. The results of even-parity states are shown in Fig. 3.20 together with experimental energies. The

NLEFT calculations show not only the Hoyle state but also the low-lying second 2_2^+ rotational like state above the Hoyle state. For the ^{12}C ground state and the first excited 2_1^+ state, a compact triangular configuration of alpha clusters is found to be dominant. For the Hoyle state and the second excited 2_2^+ state, a large contribution of "bent-arm" or obtuse triangular configuration of alpha clusters is pointed out as shown in right-hand side of Fig. 3.20 (see also Fig. 3.11 for the pictorial representations of cluster configurations by AMD calculations). The electromagnetic transition rates between the low-lying states of ^{12}C have also been obtained at LO level of ChEFT interaction with reasonable success. The basic features of calculated results of ^{12}C in NLEFT simulations are consistent with those of AMD calculations in Sect. 3.10.

Summary

Main characteristics of the nuclear-structure models reviewed in this chapter are summarised in Table 3.8. The mean-field is based on a one-body potential which is used to determine the single-particle nuclear shell structure. Main terms are a central potential and a spin-orbit potential, which gives magic numbers of nucleons as was shown in Fig. 3.2. Magic numbers in the shell structure indicate the number of nucleon at which nuclei are very stable with respect to external excitations.

The Hartree–Fock (HF) model is a self-consistent mean field model to calculate the ground state of many-body system. The HF equation (3.48) reads

$$-\frac{\hbar^2}{2m}\nabla^2\phi_i(\mathbf{r}_i) + v_i^{HF}(\mathbf{r_i})\phi_i(\mathbf{r}_i) = \varepsilon_i\phi_i(\mathbf{r}_i),$$

where v_i^{HF} is the mean field potential with the direct and exchange terms obtained from the effective many-body interactions.

The BCS model takes into account the pairing correlations in open-shell nuclei within the mean field approximation. To take into account the pairing correlations in the mean field, the quasi-particles $\alpha_k^\dagger$ and $\alpha_{\overline{k}}$ are introduced by the Bogolyubov unitary transformation (3.64);

$$\alpha_k^\dagger = u_k a_k^\dagger - v_k a_{\overline{k}},$$
$$\alpha_{\overline{k}} = v_k a_k^\dagger + u_k a_{\overline{k}},$$

where $a_k^\dagger$ and $a_{\overline{k}}$ with $k \equiv (j_k m_k)$ are the creation and the annihilation operators of bare particles, respectively. The HFB is an extended version of the BCS model and can accommodate loosely-bound nature of single-particle wave function in nuclei close to the drip lines.

RPA and QRPA models describe excited states of closed-shell nuclei and open shell nuclei, respectively. The RPA equation is given in Eq. (3.105);

Table 3.8 Characteristic points of theoretical models. The model is distinguished by the ab initio type or non ab initio type one. The applicability of the model is specified by the mass region A, nuclei with the closed shell structure or open shell structure, and for the ground states or the excited states. Refer the text and the summary of this chapter for abbreviations of the models. The abbreviation GS and EX denote the ground states and the excited states, respectively

Model	Section	Ab initio	Mass region (A)	Closed- or open-shell	GS or EX
HF	3.4.1	no	$12 \le A < 300$	closed	GS
RMF, RHF	3.4.2	no	$12 \le A < 300$	closed	GS
BCS, HFB	3.4.3	no	$16 < A < 300$	open	GS
TDA, RPA	3.4.4	no	$12 \le A < 300$	closed	EX
QRPA	3.4.4	no	$16 < A < 300$	open	EX
PVC+RPA	3.5.1	no	$16 \le A < 300$	closed	GS, EX
SRPA	3.5.2	no, yes	$16 \le A \le 48$	closed	EX
QPVC, QRPA	3.5.1	no	$16 < A < 300$	open	GS, EX
GCM	3.5.3	no	$16 < A < 200$	open	EX
ISM	3.6	no	$4 < A < 200$	open	EX
SMMC	3.7.1	no	$10 < A < 160$	open, closed	GS, EX
MCSM	3.7.1	no	$4 \le A < 100$	open, closed	EX
AMD	3.8	no	$4 < A \le 40$	open, closed	GS, EX
FMD	3.8	yes	$4 < A < 30$	open, closed	GS, EX
NCSM	3.9.1	yes	$4 \le A \le 30$	open, closed	GS, EX
VMC	3.9,2	yes	$4 \le A \le 30$	open, closed	GS, EX
GFMC	3.9.2	yes	$4 \le A \le 16$	open, closed	GS,EX
CC	3.9.3	yes	$4 \le A \le 100$	closed, closed ±2	GS, EX
SCGF	3.9.5	yes	$16 \le A < 60$	open, closed	GS, EX
NLEFT	3.9.6	yes	$A = 4 \times n \le 16$	open, closed	$0^+, 2^+$

$$\begin{pmatrix} A & B \\ B^* & A^* \end{pmatrix} \begin{pmatrix} X \\ Y \end{pmatrix} = \hbar\omega \begin{pmatrix} 1 & 0 \\ 0 & -1 \end{pmatrix} \begin{pmatrix} X \\ Y \end{pmatrix},$$

where the RPA amplitudes X, Y are column vectors as $(X)^\lambda_{mi} = X^\lambda_{mi}$, $(Y)^\lambda_{mi} = Y^\lambda_{mi}$, respectively, A and B are p-h matrix elements, and $\hbar\omega$ is the excitation energy.

The GCM is a powerful beyond mean field model formulated in the Hill-Wheeler equation with a multi-Slater determinant basis. This model can take into account the effect of shape coexistence and shape fluctuation properly under the angular momentum, parity, and isospin projections on the intrinsic deformed basis.

The Interacting Shell Model (ISM) is the most used model for detailed spectroscopy, and for which phenomenological or ChEFT-based interactions are used.

AMD and FMD adopt the multi-center Gaussian wave packets for single-particle states, which were invented originally to describe dynamics of atoms and molecules, and are applied successfully to describe cluster states of light nuclei such as the Hoyle state in ^{12}C.

Shell Model Monte Carlo and Monte Carlo Shell Model adopt statistical technique of Quantum Monte Carlo approach for quantum many-body dynamics to avoid the diagonalization of astronomical dimension of shell model matrix and can thus be applied to study medium-heave nuclei.

Green's function Monte Carlo and NCSMs accommodate ab initio realistic many-body interactions and describe spectroscopy from light nuclei to medium-heavy nuclei.

Unitary Correlation Operator Method (UCOM) and Similarity Renormalization Group (SRG) provide the correlated Hamiltonian and correlated operators with the aid of the flow equation and propagator, which can be applied to ab initio type calculations, including three-body and higher order correlations. In Medium SRG (IMSRG) takes in to account further the in medium corrections on top of free space SRG.

Several ab initio approaches are also proposed recently to solve many-body wave functions with effective interactions derived from the realistic interactions or the QCD-based interactions. They are, for example, the coupled cluster (CC) theory, self-consistent Green's function (SCGF) model with the similarity transformed effective interactions, and also nuclear lattice effective field theory (NLEFT) with ChEFT interaction up to NNLO expansion.

The coupled cluster (CC) approach includes the singlet and doublet clusters (CCSD) with ChEFT interactions and additional triplet effect is treated perturbatively (CCSDT) to avoid heavy burden on computing power. The nuclear structure of not only light nuclei but also medium-heavy nuclei with mass $A \sim 100$ is studied with the realistic effective interactions.

The self-consistent Green's function (SCGF) model includes self-consistently the effect of particle-vibration coupling in the single-particle spectra to solve the Dyson equation.

Nuclear lattice effective field theory (NLEFT) is the approach to solve the nuclear many-body problem on the Lattice with ChEFT interactions. This model was applied to several nuclei with the mass $A = 4n$. The ground state and the Hoyle state in ^{12}C are reproduced not only in the absolute binding energies but also electric transition rates between $J^\pi = 0^+$ and 2^+ states.

Solutions of Problems

3.1 For a three-body system, the anti-symmetrized wave function reads,

$$\Phi_{\rm HF}(\mathbf{r}_1, \mathbf{r}_2, \mathbf{r}_3) = \mathcal{A}\{\phi_\alpha \phi_\beta \phi_\gamma\}$$

$$= \frac{1}{\sqrt{6}}\Big[\phi_\alpha(\mathbf{r}_1)\phi_\beta(\mathbf{r}_2)\phi_\gamma(\mathbf{r}_3) - \phi_\beta(\mathbf{r}_1)\phi_\alpha(\mathbf{r}_2)\phi_\gamma(\mathbf{r}_3) - \phi_\alpha(\mathbf{r}_1)\phi_\gamma(\mathbf{r}_2)\phi_\beta(\mathbf{r}_3))$$

$$- \phi_\gamma(\mathbf{r}_1)\phi_\beta(\mathbf{r}_2)\phi_\alpha(\mathbf{r}_3) + \phi_\gamma(\mathbf{r}_1)\phi_\alpha(\mathbf{r}_2)\phi_\beta(\mathbf{r}_3) + \phi_\beta(\mathbf{r}_1)\phi_\gamma(\mathbf{r}_\beta)\phi_\alpha(\mathbf{r}_3)\Big]. \quad (3.245)$$

3.2 Let us derive the momentum-dependent terms of Skyrme interaction from the finite-range interaction by using the plane wave functions for the initial and final states. Suppose the initial and final two particle state have the relative wave vectors $\mathbf{k}$ and $\mathbf{k}'$, the two-body matrix element of Gaussian interaction $V(\mathbf{r}) = V_0 e^{-r^2/\mu^2}$ can be calculated as

$$\langle \mathbf{k}'|V_0 e^{-r^2/\mu^2}|\mathbf{k}\rangle = \int d\mathbf{r} e^{i(\mathbf{k}-\mathbf{k}')\cdot\mathbf{r}} V_0 e^{-r^2/\mu^2} = 4\pi V_0 \int_0^\infty r^2 dr \frac{\sin|\mathbf{k}-\mathbf{k}'|r}{|\mathbf{k}-\mathbf{k}'|} e^{-r^2/\mu^2}$$

$$= V_0 (\mu\sqrt{\pi})^3 e^{-|\mathbf{k}-\mathbf{k}'|^2\mu^2/4}. \quad (3.246)$$

The first order expansion of the last expression gives

$$\langle \mathbf{k}'|V_0 e^{-r^2/\mu^2}|\mathbf{k}\rangle \approx V_0(\mu\sqrt{\pi})^3 \left[1 - \frac{|\mathbf{k}-\mathbf{k}'|^2\mu^2}{4}\right]$$

$$= V_0(\mu\sqrt{\pi})^3 \left[1 - \frac{\mu^2}{4}(\mathbf{k}^2 + \mathbf{k}'^2) + \frac{\mu^2}{2}\mathbf{k}' \cdot \mathbf{k}\right]. \quad (3.247)$$

The functional forms of $\mathbf{k}$ and $\mathbf{k}'$ in Eq. (3.247) are the exactly same as those in Skyrme interaction (3.38).

3.3 The anti-symmetrized matrix element is expressed as

$$\langle \widetilde{V}\rangle \equiv \frac{1}{2}\sum_{i,j}^A \langle ij|\widetilde{V}(\mathbf{r}_i, \mathbf{r}_j)|ij\rangle$$

$$= \frac{1}{2}t_0 \sum_{i,j}^A \int\int \phi_i(\mathbf{r}_1)^* \phi_j(\mathbf{r}_2)^* \delta(\mathbf{r}_1 - \mathbf{r}_2)(1 - P_r P_\sigma P_\tau)\phi_i(\mathbf{r}_1)\phi_j(\mathbf{r}_2)d\mathbf{r}_1 d\mathbf{r}_2,$$

where P_r, P_σ and P_τ are the space, the spin and the isospin exchange operators, respectively. Since the δ interaction acts on the relative S-wave between two particles, the space exchange operator has the sign, $P_r = (-)^l = +1$. The spin and isospin operators can be expressed as $P_\sigma = (1 + \boldsymbol{\sigma}_1 \cdot \boldsymbol{\sigma}_2)/2$ and $P_\tau = (1 + \boldsymbol{\tau}_1 \cdot \boldsymbol{\tau}_2)/2$, the matrix element is evaluated as

$$\langle \widetilde{V} \rangle = \frac{1}{2} \sum_{i,j}^{A} \langle ij | \widetilde{V}(\mathbf{r}_i, \mathbf{r}_j) | ij \rangle$$

$$= \frac{1}{2} t_0 \sum_{i,j}^{A} \int \int \phi_i(\mathbf{r}_1)^* \phi_j(\mathbf{r}_2)^* \delta(\mathbf{r}_1 - \mathbf{r}_2)$$

$$\times (\frac{3}{4} - \frac{1}{4}\boldsymbol{\sigma}_1 \cdot \boldsymbol{\sigma}_2 - \frac{1}{4}\boldsymbol{\tau}_1 \cdot \boldsymbol{\tau}_2 - \frac{1}{4}\boldsymbol{\sigma}_1 \cdot \boldsymbol{\sigma}_2 \boldsymbol{\tau}_1 \cdot \boldsymbol{\tau}_2) \phi_i(\mathbf{r}_1) \phi_j(\mathbf{r}_2) d\mathbf{r}_1 d\mathbf{r}_2.$$

$$(3.248)$$

For even-even $N = Z$ nucleus, the spin and isospin dependent terms have no contributions to the matrix elements since the sum of the diagonal matrix elements cancels out. Consequently, we have only a finite contribution from the first term in the equation which is given as

$$\langle \widetilde{V} \rangle = \frac{3}{8} t_0 \int \rho(\mathbf{r}_1)\rho(\mathbf{r}_2)\delta(\mathbf{r}_1 - \mathbf{r}_2) d\mathbf{r}_1 d\mathbf{r}_2$$

$$= \frac{3}{8} t_0 \int \rho(\mathbf{r})^2 d\mathbf{r}.$$

$$(3.249)$$

3.4 Equation (3.69) reads

$$|\mathrm{BCS}\rangle = \prod_{k>0} \alpha_k \alpha_{\bar{k}} |0\rangle = \prod_{k>0} (u_k a_k - v_k a_{\bar{k}}^\dagger)(v_k a_k^\dagger + u_k a_{\bar{k}}) |0\rangle, \qquad (3.250)$$

where $|0\rangle$ is the HF vacuum state defined by $a_k |0\rangle = 0$ and k represents all single-particle states. Since $a_k |0\rangle = 0$, Eq. (3.250) is further rewritten to be,

$$|\mathrm{BCS}\rangle = \prod_{k>0} (u_k v_k a_k a_k^\dagger - v_k^2 a_{\bar{k}}^\dagger a_k^\dagger) |0\rangle$$

$$= \mathcal{N} \prod_{k>0} (u_k + v_k a_k^\dagger a_{\bar{k}}^\dagger) |0\rangle, \qquad (3.251)$$

where $\mathcal{N}$ is a normalization factor coming from a product of $\prod_{k>0} v_k$.

3.5 The variation of u_k with respect of v_k is obtained from the Eq. (3.66) as

$$2u_k \frac{\partial u_k}{\partial v_k} + 2v_k = 0$$

$$\frac{\partial u_k}{\partial v_k} = -\frac{v_k}{u_k}. \qquad (3.252)$$

Using Eq. (3.252), the variation (3.76) gives

$$(t_{kk} + t_{\bar{k}\bar{k}} - 2\lambda)2v_k + \sum_{k'}(V_{kk'kk'} + V_{\bar{k}k'\bar{k}k'})2v_k v_{k'}^2$$

$$+ 2\sum_{k'>0} V_{k\bar{k}k'\bar{k}'}(-\frac{v_k}{u_k}v_k + u_k)u_{k'}v_{k'}) = 0, \quad (3.253)$$

where the symmetry of two-body matrix element V between k and k' gives a factor 2 for the second and the third terms. Multiplying $u_k/2$ on Eq. (3.253), we obtain

$$(t_{kk} + t_{\bar{k}\bar{k}} - 2\lambda)u_k v_k + \sum_{k'}(V_{kk'kk'} + V_{\bar{k}k'\bar{k}k'})v_{k'}^2 u_k v_k$$

$$+ \sum_{k'>0} V_{k\bar{k}k'\bar{k}'}u_{k'}v_{k'}(u_k^2 - v_k^2) = 0. \quad (3.254)$$

Defining the single-particle energy (3.78) and the gap parameter (3.79), we finally obtain

$$2\varepsilon_k u_k v_k + \Delta_k(v_k^2 - u_k^2) = 0 \text{ for } k > 0. \quad (3.255)$$

3.6 The gap equation (3.77) gives the equation

$$2\varepsilon_k u_k v_k = \Delta_k(u_k^2 - v_k^2). \quad (3.256)$$

Taking a square of both side of this Eq. (3.256) and inserting $u_k^2 = 1 - v_k^2$, we obtain the quadratic equation of v_k^2,

$$(\Delta_k^2 + \varepsilon_k^2)v_k^4 - (\Delta_k^2 + \varepsilon_k^2))v_k^2 + \frac{1}{4}\Delta_k^2 = 0. \quad (3.257)$$

We can get the same equation for u_k^2 inserting $v_k^2 = 1 - u_k^2$ in Eq. (3.256). Equation (3.257) is easily solved and v_k^2 and u_k^2 are obtained as

$$v_k^2 = \frac{1}{2}\left\{1 \pm \frac{\varepsilon_k}{\sqrt{\Delta_k^2 + \varepsilon_k^2}}\right\}$$

$$u_k^2 = \frac{1}{2}\left\{1 \mp \frac{\varepsilon_k}{\sqrt{\Delta_k^2 + \varepsilon_k^2}}\right\}. \quad (3.258)$$

In the limit $\Delta_k \to 0$, the occupation probability $v_k = 1$ and the unoccupation probability $u_k = 0$ for the occupied orbit $\varepsilon_k < 0$. This asymptotic limit determines the v_k, u_k coefficients as

$$v_k^2 = \frac{1}{2} \left\{ 1 - \frac{\varepsilon_k}{\sqrt{\Delta_k^2 + \varepsilon_k^2}} \right\}$$

$$u_k^2 = \frac{1}{2} \left\{ 1 + \frac{\varepsilon_k}{\sqrt{\Delta_k^2 + \varepsilon_k^2}} \right\}. \tag{3.259}$$

3.7 To calculate two-body matrix elements in the second quantized form, we use the anti-commutation relations

$$\{a_\beta, a_\alpha^\dagger\} = a_\beta a_\alpha^\dagger + a_\alpha^\dagger a_\beta = \delta_{\alpha\beta}, \tag{3.260}$$

and

$$\{a_\alpha^\dagger, a_\beta^\dagger\} = \{a_\alpha, a_\beta\} = 0. \tag{3.261}$$

The two-body p-h matrix element is expressed as

$$\langle nj|V_{res}|mi\rangle = \frac{1}{4} \sum_{kk',ll'} \tilde{V}_{kk'll'} \langle HF|a_j^\dagger a_n : a_k^\dagger a_{k'}^\dagger a_{l'} a_l : a_m^\dagger a_i|HF\rangle. \tag{3.262}$$

Since the suffices have the selection rules $k \neq l$, $k \neq l'$, $k' \neq l$, and $k' \neq l'$ for the normal product, $k(k')$ should be the same as $i(n)$ or $n(i)$, and at the same time $l(l')$ should be the same as $m(j)$ or $j(m)$ to obtain non-zero value of the two-body matrix element. We use also the identity $a_n|HF\rangle = 0$ and $a_j^\dagger|HF\rangle = 0$ for the p-h excitation $|nj\rangle$. Then we have four combinations of indices of p-h states for the matrix elements;

$$\langle nj|V_{res}|mi\rangle = \frac{1}{4}(-\tilde{V}_{nimj} + \tilde{V}_{nijm} + \tilde{V}_{inmj} - \tilde{V}_{injm}). \tag{3.263}$$

Since the wave function is anti-symmetric, we have relations,

$$-\tilde{V}_{nimj} = \tilde{V}_{nijm} = \tilde{V}_{inmj} = -\tilde{V}_{injm}. \tag{3.264}$$

Thus, we can prove

$$\langle nj|V_{res}|mi\rangle = \tilde{V}_{nijm}. \tag{3.265}$$

3.8 The Green's function (3.112) is rewritten to be

$$(\omega - H + i\varepsilon)G = 1. \tag{3.266}$$

Sandwiching this Green's function by Q_1 from the right and from the left, one obtains

$$Q_1(\omega - H + i\varepsilon)Q_1 Q_1 G Q_1 - Q_1 H Q_2 Q_2 G Q_1 = Q_1. \tag{3.267}$$

Next one gets the following equation sandwiching Eq. (3.266) by Q_2 from the left and by Q_1 from the right,

$$Q_2(\omega - H + i\varepsilon)Q_2Q_2GQ_1 - Q_2HQ_1Q_1GQ_1 = 0. \tag{3.268}$$

From Eq. (3.268), Q_2GQ_1 is expressed as

$$Q_2GQ_1 = [Q_2(\omega - H + i\varepsilon)Q_2]^{-1}Q_2HQ_1Q_1GQ_1. \tag{3.269}$$

Inserting Eq. (3.269) into Eq. (3.267), Q_1GQ_1 is obtained as

$$Q_1GQ_1 = \frac{1}{\omega - H(\omega) + i\varepsilon} \tag{3.270}$$

where

$$H(\omega) = Q_1HQ_1 + Q_1HQ_2\frac{1}{\omega - H + i\varepsilon}Q_2HQ_1. \tag{3.271}$$

3.9 We will prove in an intuitive way the Sokhotski-Plemelj Formula (3.114),

$$\frac{1}{x + i\varepsilon} = P\frac{1}{x} - i\pi\delta(x), \tag{3.272}$$

where $\varepsilon > 0$ is an infinitesimal quantity. The l.h.s. of Eq. (3.272) is transformed as

$$\frac{1}{x + i\varepsilon} = \frac{x}{x^2 + \varepsilon^2} - \frac{i\varepsilon}{x^2 + \varepsilon^2}. \tag{3.273}$$

The first term of Eq. (3.273) is

$$\lim_{\varepsilon \to 0} \frac{x}{x^2 + \varepsilon^2} = \begin{cases} \frac{1}{x} & \text{if } x \neq 0 \\ 0 & \text{if } x = 0. \end{cases} \tag{3.274}$$

For any function $f(x)$ without singularity, we have an integral,

$$\lim_{\varepsilon \to 0} \int_{-\infty}^{\infty} \frac{x}{x^2 + \varepsilon^2}f(x)dx = P\int_{-\infty}^{\infty} \frac{f(x)}{x}dx, \tag{3.275}$$

where the symbol P stands for the Principal value of Cauchy's integral. The second term of Eq. (3.273) has a property

$$\lim_{\varepsilon \to 0} \frac{\varepsilon}{x^2 + \varepsilon^2} = \begin{cases} 0 & \text{if } x \neq 0 \\ \frac{1}{\varepsilon} = \infty & \text{if } x = 0, \end{cases} \tag{3.276}$$

which shows a divergency at $x = 0$. This term shows a similar behaviour to Dirac's δ-function. Thus we can express the second term of Eq. (3.273) as

$$\lim_{\varepsilon \to 0} \int_{-\infty}^{\infty} \frac{\varepsilon}{x^2 + \varepsilon^2} f(x) dx = \lim_{\varepsilon \to 0} f(0)\varepsilon \int_{-\infty}^{\infty} \frac{1}{x^2 + \varepsilon^2} dx = \pi f(0) = \pi \int_{-\infty}^{\infty} \delta(x) f(x) dx,$$

(3.277)

where we use the formula for the integral

$$\int_{-\infty}^{\infty} \frac{1}{x^2 + \varepsilon^2} dx = \frac{\pi}{\varepsilon}.$$

(3.278)

In a more mathematically rigorous way, we can prove Eq. (3.114) by using the technique of contour integral in the complex plane.

3.10 The variational equation (3.139) is equivalent to the ratio $\langle \Psi | H | \Psi \rangle / \langle \Psi | \Psi \rangle$ being constant. Therefore, the variational equation (3.139) can be rewritten as

$$\frac{\delta \langle \Psi | H | \Psi \rangle}{\delta f(\alpha)} = E \frac{\delta \langle \Psi | \Psi \rangle}{\delta f(\alpha)},$$

(3.279)

where E is a constant. This equation turns out to be the Hill-Wheeler equation (3.140).

3.11

- $V > 0$ (repulsive interaction)

$$\int_{-\infty}^{\infty} d\sigma e^{-\frac{1}{2} V \sigma^2 - i \sigma V O} = \int_{-\infty}^{\infty} d\sigma e^{-\frac{1}{2} V \sigma^2} \cos(\sigma V O)$$

$$= \sqrt{\frac{2\pi}{V}} e^{-\frac{1}{2} V O^2}.$$

(3.280)

- $V < 0$ (attractive interaction)

$$\int_{-\infty}^{\infty} d\sigma e^{\frac{1}{2} V \sigma^2 - \sigma V O} = \int_{-\infty}^{\infty} d\sigma e^{\frac{1}{2} V (\sigma - O)^2} e^{-\frac{1}{2} V O^2}$$

$$= \int_{-\infty}^{\infty} d\sigma' e^{\frac{1}{2} V (\sigma')^2} e^{-\frac{1}{2} V O^2} = \sqrt{\frac{2\pi}{|V|}} e^{-\frac{1}{2} V O^2}$$

(3.281)

where $\sigma' = \sigma - O$.

3.12 Equation (3.210) reads

$$S|\nu\rangle = SP|\Phi_\nu\rangle = Q|\Phi_\nu\rangle.$$

(3.282)

Multiplying the operator Q from the left, one gets

$$QSP|\Phi_\nu\rangle = Q|\Phi_\nu\rangle.$$

(3.283)

Other relations $PSP = QSQ = 0$ and $S^2 = 0$ are also obtained from $QSP = Q$ and $QP = 0$.

3.13 In the Heisenberg picture, the creation and annihilation operators are given by,

$$a_\alpha^\dagger(t) = e^{iHt/\hbar} a_\alpha^\dagger e^{-iHt/\hbar}, \tag{3.284}$$

$$a_\alpha(t) = e^{iHt/\hbar} a_\alpha e^{-iHt/\hbar}.$$

Let us note E_0^A as the eigenenergy of the unperturbed state $|\psi_0^A\rangle$, and E_n^A as that of the excited state $|\psi_n^A\rangle$. The Green's function is given by Eq. (3.240),

$$g_{\alpha\beta}(t - t') = -\frac{i}{\hbar} \langle \psi_0^A | T \left[a_\alpha(t) a_\beta^\dagger(t') \right] | \psi_0^A \rangle, \tag{3.285}$$

with the time ordering operator T. Since we are dealing with fermions, there is a change in the sign for exchange of operators such that

$$T \left[a_\alpha(t) a_\beta^\dagger(t') \right] = \begin{cases} a_\alpha(t) a_\beta^\dagger(t') & t > t' \\ -a_\beta^\dagger(t') a_\alpha(t) & t < t' \end{cases}. \tag{3.286}$$

If the Hamiltonian does not depend on time, we get by substituting (3.284) and (3.286) into (3.285),

$$g_{\alpha\beta}(t - t') = -\frac{i}{\hbar} \theta(t - t') \langle \psi_0^A | a_\alpha e^{-i(H - E_o^A)(t - t')/\hbar} a_\beta^\dagger | \psi_0^A \rangle$$
$$+ \frac{i}{\hbar} \theta(t' - t) \langle \psi_0^A | a_\beta^\dagger e^{i(H - E_o^A)(t - t')/\hbar} a_\alpha | \psi_0^A \rangle. \tag{3.287}$$

Let us perform the Fourier transform with respect to time as,

$$g_{\alpha\beta}(\omega) = \lim_{\to +0} \int_{-\infty}^{\infty} d\tau e^{i\omega\tau \mp \eta\tau} g_{\alpha\beta}(\tau = t - t'), \quad \text{for} \begin{cases} \tau > 0 \\ \tau < 0 \end{cases}, \tag{3.288}$$

where $\mp\eta\tau$ is added to avoid the divergence of integral at infinity. By performing the time integral, we obtain

$$g_{\alpha\beta}(\omega) = \langle \psi_0^A | a_\alpha \frac{1}{\hbar\omega - (H - E_0^A) + i\eta} a_\beta^\dagger | \psi_0^A \rangle$$
$$+ \langle \psi_0^A | a_\beta^\dagger \frac{1}{\hbar\omega + (H - E_0^A) - i\eta} a_\alpha | \psi_0^A \rangle. \tag{3.289}$$

Using the completeness relation $\sum_n |\psi_n\rangle \langle \psi_n| = 1$ for there $(A + 1)$ system and for the $(A - 1)$ system in the first and second terms, respectively, in Eq. (3.289) and replacing H with its eigenenergies, we obtain

$$g_{\alpha\beta}(\omega) = \sum_n \frac{\langle\psi_0^A | a_\alpha | \psi_n^{A+1}\rangle \langle\psi_n^{A+1} | a_\beta^\dagger | \psi_0^A\rangle}{\hbar\omega - (E_n^{A+1} - E_0^A) + i\eta} + \sum_k \frac{\langle\psi_0^A | a_\beta^\dagger | \psi_n^{A-1}\rangle \langle\psi_n^{A-1} | a_\alpha | \psi_0^A\rangle}{\hbar\omega + (E_k^{A-1} - E_0^A) - i\eta}.$$

$$(3.290)$$

Books for Further Readings

"The Nuclear Many-Body Problem" by P. Ring and P. Schuck (Springer, 2013) is a reference textbook that introduce the concepts of effective interactions, mean field and collective models towards a description of the nuclear many-body problem.

"An advanced course in computational nuclear physics" edited by M. Hjorth-Jensen, M. P. Lombardo, and U. van Kolck (Springer, 2017) gives an overview of ab initio models including LQCD.

The above reference list contains several reviews which provide further details about ab initio methods, in particular [32–34, 38].

References

1. D. M. Bishop and L. M. Cheung, Phys. Rev. A **20**, 381 (1979)
2. M. Goppert-Mayer, Phys. Rev. **75**, 1969 (1949)
3. P. Hohenberg, W. Kohn, Phys. Rev. **136**, B864 (1964)
4. W. Kohn, L.J. Sham, Phys. Rev. **140**, A1133 (1965)
5. T.H.R. Skyrme, Phil. Mag. **1**, 1043 (1956); Nucl. Phys. **9**, 615 (1959)
6. M. Beiner, H. Flocard, N. Van Giai, P. Quentin, Nucl. Phys. A **238**, 29 (1975)
7. J. Bartel, P. Quentin, M. Brack, C. Guet, H.-B. Håkansson, Nucl. Phys. A **386**, 79 (1982)
8. J. Dechargé, D. Gogny, Phys. Rev. C **21**, 1568 (1980); J.F. Berger, M. Girod, D. Gogny, Comp. Phys. Comm. **63**, 365 (1991)
9. D. Vautherin, D.M. Brink, Phys. Rev. C **5**, 626 (1972)
10. J.D. Walecka, Ann. Phys. **83**, 491 (1974)
11. P. Ring, Prog. Part. Nucl. Phys. **37**, 193 (1996)
12. P.G. Reinhard, Rep. Prog. Phys. **52**, 439 (1989)
13. D.M. Brink, R. Broglia, *Nuclear Superfluidity (Cambridge Monographs on Particle Physics, Nuclear Physics and Cosmology* (Cambridge University Press, Cambridge, 2005)
14. D.J. Rowe, *Nuclear Collective Motion: Models and Theory* (World Scientific Pub Co Inc, Revised version, 2010/11/30)
15. C. Mahaux, P.F. Bortignon, R.A. Broglia, C.H. Dasso, Phys. Rep. **120**, 1 (1985)
16. Y.F. Niu, G. Colò, M. Brenna, P.F. Bortignon, J. Meng. Phys. Rev. C **85**, 034314 (2012)
17. M. Sasano et al., Phys. Rev. Lett. **107**, 202501 (2011)
18. D. Drozdz et al., Phys. Rep. **197**, 1 (1990)
19. P. Papakonstantinou, R. Roth, Phys. Lett. **B 671**, 356 (2009); P. Papakonstantinou, R. Roth, Phys. Rev. C **81**, 024317 (2010)
20. D.L. Hill, J.A. Wheeler, Phys Rev. **89**, 1102 (1953); P.G. Reinhard, K. Goeke, Rep. Prog. Phys. **50**, 1 (1987)
21. G.H. Lang, C.W. Johnson, S.E. Koonin, W.E. Ormand, Phys. Rev. C **48**, 1518 (1993)

22. Y. Tsunoda, T. Otsuka, N. Shimizu, M. Honma, Y. Utsuno, Phys. Rev. C **89**, 031301(R) (2014)
23. M. Honma et al., Phys. Rev. Lett. **77**, 3315 (1996)
24. M. Freer, H. Horiuchi, Y. Kanada-En'yo, D. Lee, U.-G. Meißner, Rev. Mod. Phys. **90**, 035004 (2018)
25. H. Horiuchi, K. Ikeda, K. Kato, Prog. Theo. Phys. Sup. **192**, 1 (2012); Y. Kanada-En'yo, M. Kimura, A. Ono, Prog. Theor. Exp. Phys. **2012**, 01A202 (2012)
26. R. Roth, T. Neff, H. Feldmeier, Prog. Part. Nucl. Phys. **65**, 50 (2010)
27. T. Neff, H. Feldmeier, Nucl. Phys. A **713**, 311 (2003)
28. P. Navratil, J.P. Vary, B.R. Barrett, Phys. Rev. Lett. **84**, 5728 (2000); Phys. Rev. C **62**, 054311 (2000)
29. T. Suhara, Y. Kanada-En'yo, Phys. Rev. C **82**, 044301 (2010)
30. K. Suzuki, R. Okamoto, Prog. Theor. Phys. **92**, 1045 (1994)
31. J. Lomnitz Adler, V.R. Pandharipande, R.A. Smith, Nucl. Phys. A **315**, 399 (1981); V.R. Pandharipande, S.C. Pieper, R.B. Wiringa, Phys. Rev. B **34**, 4571 (1986); S.C. Pieper, R.B. Wiringa, V.R. Pandharipande, Phys. Rev. Lett. **64**, 364 (1990)
32. J. Carlson, S. Gandolf, F. Pederiva, Steven C. Pieper, R. Schiavilla, K.E. Schmidt, R.B. Wiringa, Rev. Mod. Phys. **87**, 1067 (2015)
33. G. Hagen, T. Papenbrock, M. Hjorth-Jensen, D.J. Dean, Rep. Prog. Phys. **77**, 096302 (2014); G. R. Jansen, J. Engel, G. Hagen, P. Navratil, A. Signoracci, Phys. Rev. Lett. **113**, 142502 (2014)
34. S.K. Bogner, R.J. Furnstahl, A. Schwenk, Prog. Part. Nucl. Phys. **65**, 94 (2010)
35. E. Epelbaum et al., Phys. Rev. C **99**, 024313 (2019)
36. E. Epelbaum, H. Krebs, P. Reinert, Front. Phys. **8**, 98 (2020)
37. A. Ekström, G.R. Jansen, K.A. Wendt, G. Hagen, T. Papenbrock, B.D. Carlsson, C. Forssén, M. Hjorth-Jensen, P. Navrátil, W. Nazarewicz, Phys. Rev. C **91**, 051301(R) (2015)
38. W.H. Dickhoff, C. Barbieri, Prog. Part. Nucl. Phys. **52**, 377 (2004); Vittorio Somà, Front. Phys. **8**, 340 (2020)
39. E. Epelbaum, H. Krebs, T.A. Lähde, D. Lee, Ulf-G. Meißner, Phys. Rev. Lett. **109**, 252501 (2012)
40. J. Dechargé, M. Girod, D. Gogny, B. Grammaticos, Nucl. Phys. A **358**, 203c (1981)
41. G. Hagen, Hjorth-Jensen, Jansen, Machleidt, T. Papenbrock, Phys. Rev. Lett. **108**, 242501 (2012)
42. C. Barbieri, Phys. Rev. Lett. **103**, 202502 (2009)
43. V. Somà, C. Barbieri, T. Duguet, Phys. Rev. C **87**, 011303(R) (2013)
44. E. Epelbaum, H. Krebs, T. A. Lähde, D. Lee, and U.-G. Meisner, Phys. Rev. Lett. **109**, 252501 (2012)

Chapter 4
Nuclear Observables and Measurement Techniques

Abstract The description of atomic nuclei in terms of energy shells is at the origin of our representation of nuclear structure. This representation has been built on experimental findings. In this chapter we review the most important observables and the corresponding experimental methods. The measurements of masses, charge and matter radii, β-decay half-lifes, electric and magnetic moments and spectroscopy are addressed.

Keywords Masses $\cdot$ Radii $\cdot$ β decay $\cdot$ Nuclear moments $\cdot$ Spectroscopy

4.1 Masses

4.1.1 Definitions

The mass of a nucleus reflects the sum of all interactions at play among its nucleons. Following the Einstein's relation $E = mc^2$, c being the speed of light in vacuum, the mass $m(Z, N)$ of a nucleus is expressed as

$$m(N, Z)c^2 = N\,m_n c^2 + Z\,m_p c^2 - B(N, Z), \tag{4.1}$$

where N, Z are the numbers of neutrons and protons with masses $m_n = 939.57$ MeV/c^2 and $m_p = 938.28$ MeV/c^2, respectively, and $B(N, Z)$ is the total binding energy of the nucleus. In the above identity, the binding energy is given in MeV and, by definition, represents the energy gain by the nucleus compared to an ensemble of non-interacting N and Z nucleons. Mostly used to characterize ground states, the binding energy can be trivially extended to characterize excited states. $B(N, Z)$ contains both bulk information on the nuclear interactions and, although less pronounced, Coulomb and weak forces in the nucleus, as well as nuclear structure effects such as shells, pairing and deformation.

The atomic mass unit (u) is also routinely used to quantify nuclear masses. It is defined relative to the mass of a ^{12}C atom so that $m(^{12}\text{C}) = 12$ u, and therefore

© Springer Nature Singapore Pte Ltd. 2021

A. Obertelli and H. Sagawa, *Modern Nuclear Physics*, UNITEXT for Physics,

https://doi.org/10.1007/978-981-16-2289-2_4

$1 \text{ u} = 931.49 \text{ MeV/c}^2$. If one further considers the example of ^{12}C, its binding energy in MeV can be recovered from the definition of the atomic mass unit

$$B = 6 m_n c^2 + 6 m_p c^2 - m(N = 6, Z = 6)c^2 = 6 m_n c^2 + 6 m_p c^2 - (12 \text{ u} - 6 m_e)c^2,$$

where $m_e = 511 \text{ keV/c}^2$ is the electron mass. From the above translation of 1 u in MeV/c^2, one gets a binding energy of 92.28 MeV for the ground state of the ^{12}C nucleus, i.e., 7.69 MeV/nucleon.

Relative masses are often discussed in terms of the neutron (proton) separation energy S_n (S_p) defined as the difference of binding energy between two nuclei which differ by one neutron (proton):

$$S_n(N, Z) = B(N, Z) - B(N - 1, Z) = m(N - 1, Z)c^2 + m_n c^2 - m(N, Z)c^2. \tag{4.2}$$

The relation for S_p is obtained by exchanging protons and neutrons in Eq. (4.2).

4.1.2 The Liquid-Drop Model

The short range of the nuclear interaction (see Chap. 2) is at the origin of the saturation density encountered inside nuclei: the inside of nuclei composed of $A \gtrsim 10$ nucleons reaches a density of $\rho_0 \sim 0.16 \text{ fm}^{-3}$. Saturation density leads to a binding energy scaling nearly with the number of nucleons A. For medium-mass and heavy nuclei, the binding energy per nucleon is ~ 8 MeV/nucleon, as seen for ^{12}C. This feature is grasped by the liquid-drop model which describes the binding energy of the nucleus in a macroscopic way, assuming it is a liquid drop of matter. It is expressed as

$$B_{LD} = a_V A - a_S A^{2/3} - a_C \frac{Z(Z - 1)}{a^{1/3}} - a_A \frac{(N - Z)^2}{A} + a_P \frac{1}{\sqrt{A}}, \tag{4.3}$$

where the first term (a_V) is the volume term, and which reflects the short range of the nucleon-nucleon interaction: the binding between two nucleons inside the nucleus depends on their mean distance $\propto A^{1/3}$ which gives an A-dependence when integrated over the volume of the nucleus. The second term (a_S) is a correction to the volume term taking into account that saturation is not reached at the surface of the liquid drop. The third term (a_C) reflects the Coulomb repulsion among all proton pairs $\propto Z(Z - 1)$. The last two terms take into account the Pauli blocking which penalizes energetically asymmetric nuclei (a_A) and the pairing effect which favours even-even nuclei (a_P). The roles of each term are illustrated in Fig. 4.1. A typical set of empirical values for the parameters of the liquid-drop model are provided in the caption of Fig. 4.2 and the resulting binding energy as a function of number of nucleons for stable nuclei is shown. The maximum binding energy encountered in ^{56}Fe is driven by the competition between the surface and Coulomb terms which counterbalance the volume term for low and large mass nuclei, respectively.

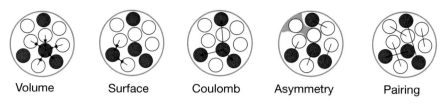

| Volume | Surface | Coulomb | Asymmetry | Pairing |

Fig. 4.1 Illustration of the individual contributions to the binding energy in the liquid-drop model. The volume, surface, Coulomb, proton-neutron asymmetry and pairing terms are quantified by the a_V, a_S, a_C, a_A and a_P parameters, respectively. See text for details

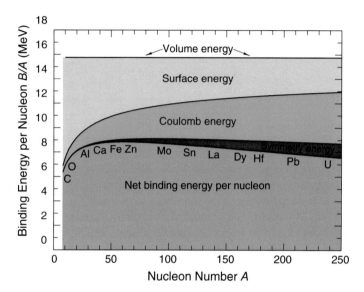

Fig. 4.2 Liquid-drop-model binding energy as a function of the number of nucleons for stable nuclei. The parameters of the liquid-drop model are: $a_V = 15.85$ MeV, $a_S = 18.34$ MeV, $a_C = 0.71$ MeV, $a_A = 23.21$ MeV and $a_P = +12, 0, -12$ MeV for even-even, even-odd and odd-odd nuclei, respectively. The contribution of the volume, surface, Coulomb, asymmetry and pairing terms are shown. Courtesy P. Moller

4.1.3 Overview of Measurement Techniques

Masses are a pillar of our understanding of the shell structure as illustrated in Chaps. 5 and 6. Masses are most often determined for ground states but mass measurement techniques also provide an equally efficient tool for isomers,[1] if their half life is long

---

[1] An isomer is a nuclear excited state whose half life is significantly longer than the typical half life of prompt electromagnetic decays, i.e., $\sim$ fs to ps. As a rule of thumb, states with an half life larger than 1 ns are usually called isomers.

enough to allow the measurement. Accurate mass measurements are used to test the weak interaction in nuclei, quantum electrodynamics and the standard model as illustrated in Chap. 9. Finally, masses and nucleon separation energies are essential ingredients of the different astrophysical nucleosynthesis processes as discussed in Chap. 8.

The key concept behind most mass measurements is the motion of a particle with electric charge q and (rest) mass m in a magnetic field. This motion is driven by the cyclotron frequency, which is the frequency of the particle in the plane perpendicular to a uniform magnetic field B. The relativistic cyclotron frequency is given by

$$\omega_c = \frac{qB}{\gamma m}, \tag{4.4}$$

where $\gamma = 1/\sqrt{1 - \beta^2}$ is the relativistic Lorentz factor, and β is the velocity of the particle in the laboratory frame in units of speed of light. Any precise mass measurement is in fact a determination of the ratio m/q. m is extracted once q is known. The measurements are then based on a frequency determination or, its Fourier equivalent, a time of flight.

Different mass-measurement techniques have been developed in the past century with a major improvement in precision from traps and storage rings over the past decades. They can be decomposed into two main families: (i) time of flight and momentum measurements over a known distance of flight. Originally the method was developed to be used for one-path trajectories and a magnetic spectrometer, but later it was largely improved by extension to multiple-turn closed-path trajectories with storage rings. (ii) cyclotron frequency determination of trapped ions. The latest method was made possible once ion trapping techniques were developed. Since the 90s, mass measurements have been extensively applied to radioactive isotopes. Mass measurements and their precision face limitations due to the short lifetime and the low production yield of the radioactive nuclei to be studied. Depending on the case of interest, some techniques cannot be applied or are more suited than others.

Figure 4.3 compares the domain of applicability in terms of accuracy and reachable lifetimes for the most common techniques which are described in the following. While time of flight measurements with fast beams allow mass measurements for very short-lived nuclei down to 100 ns with a relative mass precision $\delta m/m$ of typically 10^{-4} down to 10^{-6}, rings and traps can reach a precision of 10^{-8} but for nuclei with a lifetime not shorter than 10 μs to 1 s, depending on the technique. The recently developed phase-imaging ion-cyclotron-resonance technique allows to reach the 10^{-9} precision level.

Fig. 4.3 Schematic comparison of relative mass uncertainties $\delta m/m$ as a function of the half life of the measured nucleus for the different mass-measurement techniques detailed in this chapter: $B\rho$ - time of flight, multi-reflection time of flight (MR-ToF), frequency measurement with storage rings, cyclotron frequency measurements with Penning traps

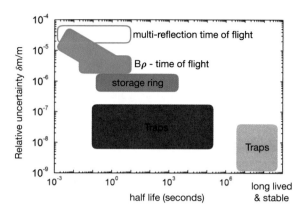

4.1.4 Time of Flight

4.1.4.1 The $B\rho$ Time-of-flight Technique

The Lorentz force acting on a charge q moving at the velocity v in a magnetic field B is

$$F = qv \times B. \tag{4.5}$$

In a uniform magnetic field the trajectory of a charged particle is circular on a trajectory of radius ρ in a plane perpendicular to the magnetic field, following

$$B\rho = \frac{p}{q} = \frac{\gamma mv}{q}, \tag{4.6}$$

where $B\rho$ is called the magnetic rigidity of the particle in the magnetic field. For a momentum p given in MeV/c, the magnetic rigidity in Tesla meter (Tm) is $B\rho = 0.0033356 \times p/q$. This leads to

$$\frac{m}{q} = \frac{B\rho}{c}\sqrt{\left(\frac{ct}{L}\right)^2 - 1}. \tag{4.7}$$

By measuring precisely the magnetic rigidity and the velocity, i.e., the time of flight t over a known distance L, the m/q ratio of a nucleus can be extracted. The measurements are usually performed at intermediate energy with fully stripped ions. The absolute mass of the nucleus can then be obtained from a combined measurement of the time of flight and an energy loss measurement in a thin detector along the path of flight, allowing the determination of Z, and therefore q. The relative precision on (m/q) can be easily estimated from Eq. (4.7) by

$$\left(\frac{\delta(m/q)}{m/q}\right)^2 = \left(\frac{\delta(B\rho)}{B\rho}\right)^2 + \gamma^2\left(\frac{\delta L}{L}\right)^2 + \gamma^2\left(\frac{\delta T}{T}\right)^2. \tag{4.8}$$

In realistic cases, the $B\rho$ of individual particles are measured event by event with a high resolution spectrometer of typical resolution $\delta(B\rho)/B\rho = 10^{-4}$. The distance of flight from the production target to the focal plane of the spectrometer is typically in the order of 30 m, leading to a time of flight of about 300 ps. The length of the trajectory can be determined with an accuracy of a few millimeters while the time of flight is measured with a resolution down to 10 ps. From the relation (4.8), a characteristic absolute mass precision is therefore 10^{-4}. Additional corrections can be performed when reference masses are measured at the same time. Relative precision for masses can then reach 10^{-5} as illustrated in Fig. 4.3.

4.1.4.2   The Multi-reflection Time-of-Flight (MR-ToF) Technique

The time-of-flight technique, without the use of any magnetic field, was also applied to low-energy ions reflected by electrostatic mirrors. Historically, the first of its kind was developed in the 1960s by Werner Tretner and known as the "Farvitron". The multi-reflection time-of-flight (MR-ToF) technique as it is used today was developed by Hermann Wollnik and Martin Przewloka in 1990. An MR-ToF mass analyser is composed of two ion-optical mirrors made of electrostatic electrodes. The mirrors are facing each other and separated by a field-free drift section defined by the available space in the setup, in the order of fifty centimeters to one meter. Low-energy ions travel freely between the two mirrors. When approaching one mirror, the ions are slowed down by a repulsive electric potential and reflected back, comparable to light in a spherical mirror. The multi-reflection allows long flight paths in a limited space region. Although it does not compete with the most precise mass measurement techniques (see below), it offers a robust way of achieving a good mass resolution of 10^{-5} in a reasonable time ($\sim 10-50$ ms) within a compact geometry. MR-ToF spectrometers can be used as preselectors for high-precision mass spectrometry. The principle of a MR-ToF is shown in Fig. 4.4.

At injection at a given potential V, ions of different species j with charge q_j and mass m_j gain the kinetic energy $q_j V = m_j v_j^2/2$, where v_j is the velocity of ions j after acceleration. After one revolution between the two mirrors, the time of flight for each species will depend on its mass-over-charge ration: $T_j \propto v_j^{-1} \propto \sqrt{m_j/q_j}$. After n revolutions, two species i, j will be separated by

$$\Delta T_{ij} \propto n \left| \sqrt{\frac{m_j}{q_j}} - \sqrt{\frac{m_i}{q_i}} \right|. \tag{4.9}$$

All species are separated in time of flight, depending on their mass-over-charge ratio. The ions of interest can then be selected by lowering the potential of an ion mirror at the time they approach the mirror, for a time short enough so that only these ions are transmitted out of the trap while contaminants with different q/m, and therefore with a different timing, are reflected by the mirror and not transmitted. Alternatively, one can use the in-trap lift technique which gives the ions a high enough kinetic energy

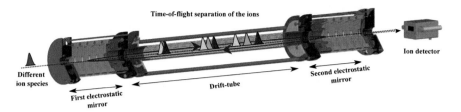

Fig. 4.4 Principle of the multi-reflection time-of-flight (MR-ToF) technique. It can be used directly for mass measurements via precise determination of the time-of-flight of the ions or as a separator to transmit the masses of interest to a precision trap for a more precise mass measurement. Courtesy F. Wienholtz, TU Darmstadt

to allow them to overcome the mirror potential well. Up to now, the transmission of unwanted species can be reduced up to four orders of magnitude.

4.1.5 Storage Rings

Storage rings use the Lorentz force of Eq. (4.5) to "store" particles in a magnetic field in closed trajectories. Typically, storage rings allow re-circulation orbits of individual ions in high vacuum 10^6 times, which allow a better mass precision for a given number of produced ions. The storage is limited by the lifetime of the ions and the method is competitive for rather long-lived isotopes (at least several ms). The period of an ion with velocity v in an orbit with length L is given by

$$T = \frac{L}{v}.$$

The variation from one period to another and frequency $f = 1/T$ depends on both the variation of the particle orbit length and the velocity spread among identical nuclei

$$\frac{\Delta T}{T} = -\frac{\Delta f}{f} = \frac{\Delta L}{L} - \frac{\Delta v}{v}. \tag{4.10}$$

Ions with different mass-to-charge ratios have different bending radii in the dipole magnetic field sections of the storage ring, leading to different orbit lengths. The ratio of the relative change of orbit length to the relative change in $B\rho$ of stored ions is characterized by the so-called compaction factor α_p. It is an ion-optical quantity that describes the extra distance that particles of an increased rigidity with respect to the central trajectory have to fly:

$$\alpha_p = \frac{dL/L}{d(B\rho)/(B\rho)}. \tag{4.11}$$

α_p can be obtained from the dispersion function of the considered storage ring beam optics resulting from the magnet configuration of the ring. The optics of the storage ring is defined by the design of the optical elements (electric dipoles, quadrupoles and higher-order multipoles) and their layout. This connection between orbital length and mass-over-charge ratio explains why different ion species have different revolution times and therefore give separate peaks in measured revolution frequency spectra.

Substituting Eq. (4.11) into Eq. (4.10), one gets

$$\frac{\Delta T}{T} = \alpha_p \frac{\Delta(B\rho)}{B\rho} - \frac{\Delta v}{v}. \tag{4.12}$$

To calculate the right-hand side (r.h.s) term of Eq. (4.12), one uses the definition of the magnetic rigidity (Eq. (4.6)) to obtain after derivation

$$\frac{d(B\rho)}{B\rho} = \frac{d(m/q)}{m/q} + \frac{d_v(v\gamma)}{v\gamma}, \tag{4.13}$$

where the derivative $d_v(v\gamma)$ can be expressed as

$$d_v(v\gamma) = d_\beta(\beta\gamma) = \gamma d\beta + \beta(1 - \beta^2)^{-3/2}\beta d\beta = \gamma^3 d\beta. \tag{4.14}$$

Substituting into Eq. (4.13), one gets

$$\frac{d(B\rho)}{B\rho} = \frac{d(m/q)}{m/q} + \gamma^2 \frac{dv}{v}, \tag{4.15}$$

which gives

$$\frac{\Delta T}{T} = \alpha_p \left(\frac{\Delta(m/q)}{m/q} + \gamma^2 \frac{\Delta v}{v} \right) - \frac{\Delta v}{v}, \tag{4.16}$$

and can be re-written as

$$\frac{\Delta T}{T} = \frac{1}{\gamma_t^2} \frac{\Delta(m/q)}{m/q} - \left(1 - \frac{\gamma^2}{\gamma_t^2} \right) \frac{\Delta v}{v}, \tag{4.17}$$

where the quantity $\gamma_t = 1/\sqrt{\alpha_p}$, called the *transition energy*, is used. Equation (4.17) is the key equation for mass measurements with storage rings which provides the relative time difference between ions of different m/q and different velocity v. By minimising the second term of the r.h.s. of the equation, a time-of-flight (or frequency) measurements directly determines the m/q ratio of a particle, with resolution $\gamma_t^2 \Delta T/T$. There are two ways to minimise this term: by reducing the velocity $\Delta v/v$ fluctuation from one particle to another by beam cooling (the method is called Schottky Mass Spectrometry (SMS)), or to ensure that $\gamma^2 = \gamma_t^2$ by a specific tuning of the ring optics (known as Isochronous Mass Spectrometry (IMS)). The two methods are illustrated in Fig. 4.5.

SCHOTTKY MASS SPECTROMETRY   **ISOCHRONOUS MASS SPECTROMETRY**

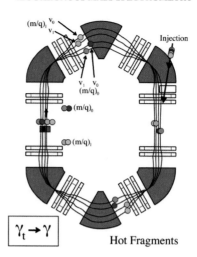

Fig. 4.5 (Left) Schottky mass spectrometry. The ions are cooled to reduce their velocity spread. Their revolution frequencies are measured by so-called Schottky pick up. This technique can be applied to long-lived nuclei. (Right) Isochronous mass-spectrometry of uncooled ions. Their revolution times are measured by a time-of-flight technique. This method is well suited to short-lived nuclei

The time of flight in a ring can be measured turn by turn, in a non-destructive way, i.e., the ions are not lost after the measurement. The measurement relies on the Schottky pickup effect during which the moving charges in the ring induce a small voltage change on the electrodes located in the ring, which can then "pick up" this signal. Schottky pickup measurements actually measure frequencies and not the time of flight directly.

4.1.5.1   Beam Cooling and Schottky Mass Spectrometry

In this section, we introduce the basics of the beam cooling required by the SMS technique. Cooling a beam means reducing its *emittance*. The emittance of a beam defines its quality in terms of size, angular divergence and velocity spread. The emittance of a beam is given for the three space directions. In the beam direction, the emittance is called *longitudinal*, while the two other emittances are called *transverse*. The beam emittance in a storage ring comes mostly from the velocity spread and straggling caused by the production reaction mechanism upstream the storage ring, and caused by detection systems inserted in the beam line and ring.

The Liouville's theorem plays a central role in beam optics. It states that the density ρ in phase space, i.e., positions and momenta, of a system of particles is constant when only conservative forces[2] are applied to it. For a fixed number of particles, it is equivalent to say that the volume in the phase space and emittances are maintained. The Liouville's theorem is a general physics law and is not restricted to accelerators. In the case of a storage ring, it means that the velocity spread of a beam can be reduced via beam cooling only if non-conservative forces are applied to the ions forming the beam bunches. Two main methods exist for the beam cooling of fast beams: stochastic and electron cooling.

Stochastic cooling relies on measuring the signal induced by a particle on electrodes (Schottky pickup) at each turn and on correcting the perpendicular momentum of particles that deviate from the centroid momentum of the bunch with a electric kicker. This negative feedback is used to reduce the momentum spread of the beam, and therefore is equivalent to lowering the temperature (cooling) of the bunch in its center of mass. Within the same turn, the signal is used to correct the motion of the ion bunch. Although the correction is valid for some of the ions, it may not be valid for all. Nevertheless, it is shown that after some time, typically a second to a minute, the transverse emittance of the beam is reduced. Stochastic cooling is effective for beams with a rather high velocity spread ($\Delta p / p \geq 10^{-3}$). It is often used as a pre-cooling technique, followed by electron cooling. An illustration of stochastic cooling followed by electron cooling is given in Fig. 4.6.

Electron cooling uses the collision of ion bunches with electron bunches at the same velocity of the beam, i.e., with zero relative mean velocity. The electron bunches are prepared so that they have a much better emittance than the ion bunches. Usually, the overlap zone for electron cooling in the ring runs over few meters. Collisions between ions and electrons result in a reduction of the ion beam emittance. The electron bunches are renewed after each turn. Electron cooling takes about one second. As an example of typical conditions at the ESR storage ring at the GSI[3] accelerator complex, Germany, the beam velocity spread of a $^{238}U^{92+}$ beam at 1 GeV with 10^7 ions per bunch can be reduced to 10^{-5}. In the case of low beam intensities where the ion-ion scattering within the bunch (space charge effects) can be neglected, the velocity spread can be reduced down to 10^{-7}.

4.1.5.2 Isochronous Mass Spectrometry (IMS)

The Isochronous Mass Spectrometry technique is a second way to minimize the influence of velocity spread on the frequency, and therefore mass resolution. The main idea is to modify the optical setting of the ring so that the velocity difference of

---

[2] A force $\mathbf{F}$ is called *conservative* when its resulting work on a particle moving from two points does not depend on the path taken. It is verified if any of the following conditions is met: (i) $W \equiv \int \mathbf{F} \cdot d\mathbf{r} = 0$, (ii) $\nabla \times \mathbf{F} = \mathbf{0}$, (iii) the force can be written as the negative gradient of a potential Φ: $\mathbf{F} = -\nabla \Phi$. This is the case for a charged particle in a static magnetic field and with central electric forces, while friction forces are not conservative.

[3] Gesellschaft für Schwerionenforschung.

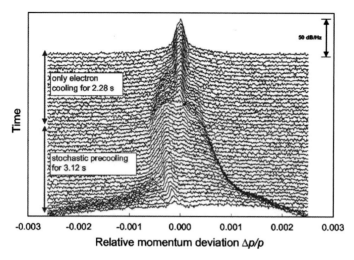

Fig. 4.6 Stochastic cooling followed by electron cooling of a primary Uranium beam ($^{238}U^{92+}$) at 400 MeV/u in the ESR ring of GSI. The measurements shown by lines are separated by 120 ms each. The vertical axis is proportional to the number of particles for a given deviation in momentum $\Delta p/p$ compared to the central value. Figure reprinted with permission from [1]. ©2021 by Elsevier

two identical particles is counterbalanced by the length difference of their respective orbits. If the *isochronicity condition*

$$\gamma = \gamma_t \tag{4.18}$$

is reached, the revolution frequency becomes independent of the velocity. In IMS, no electron cooling is necessary and the conditions for mass measurement are reached from the injection in the ring. The method is suited for the most exotic species with a short lifetime not compatible with the time necessary for cooling. The isochronous condition is reached for one particular m/q ratio. The IMS is therefore suited for one mass measurement at a time. The modification and tuning of γ_t is made by the modification of fields in quadrupoles and sextupoles in the case of the ESR at GSI, Germany. At the R3 storage ring of RIKEN, Japan, composed only of magnetic dipoles, the isochronous condition is reached by modifying the magnetic field by extra concentric coils, called *trim coils*, added to some of the magnet dipoles. A drawback of the IMS is that the optics setting implies a large dispersion (several meters) in some sections of the storage ring, which limits the momentum acceptance of the isochronous mode. The momentum acceptance of the ESR and R3 ring in the IMS mode are typically ±0.5%. Precision better than 10^{-6} can be reached in this mode.

4.1.6   Time-of-flight Ion-Cyclotron-Resonance (ICR)

The motion of a charged particle trapped in a three-dimension (3D) harmonic potential well have three characteristic frequencies of motion which depend, among other terms, on its mass. The precision measurements of "resonant" frequencies give therefore access to the mass of the trapped particle.

Charged particles can be trapped, in principle indefinitely, by use of electric fields or a combination of electric and and magnetic fields. Static electric fields alone are not sufficient to create a harmonic well for ion trapping (Earnshaw's theorem). There are two main families of apparatus to produce a harmonic potential well to trap ions: Penning and Paul traps. Penning traps are based on the combination of a magnetic field for radial confinement and a static electric field produced by electrodes on which electric potentials are applied for a longitudinal confinement. The electric field can either be produced by hyperbolic or cylindrical electrodes, as illustrated in Fig. 4.7. The electrodes of Paul traps are supplied by a high frequency potential. The AC potential leads to a time-varying electric field which generates the potential well.

The most precise mass measurements are performed with Penning traps nowadays.

4.1.6.1   Motion in a Penning Trap

In the following, we describe the basics of a Penning trap measurement. Apart from how the trapping potential is designed, the equations of motion of a charged particle in a Paul trap can be derived in a similar way. We assume an ideal Penning trap composed of a constant magnetic field B along the z-axis and a quadrupole electrostatic potential

$$U(\rho, z) = \frac{U_0}{4d^2}(2z^2 - \rho^2), \qquad (4.19)$$

where $\rho^2 = x^2 + y^2$, U_0 is the electric potential of the central ring relative to the end caps and d is a characteristic dimension of the trap. In the case of a hyperbolic trap, as shown on the l.h.s. of Fig. 4.7, $4d^2 = 2z_0^2 + \rho_0^2$ where ρ_0^2 is the inner-ring radius and z_0 is the closest distance of the two endcaps. If a cylindrical trap is used (sketch on the r.h.s. of Fig. 4.7), a 5-electrode geometry defines a quadrupole potential of the same form in the vicinity of the central electrode.

Under the assumption of an ideal quadrupole potential with radial symmetry as expressed in relation (4.19), the longitudinal motion and the radial motion of a trapped ion are decoupled and both follow harmonic oscillations driven by one longitudinal ω_z and two radial $\omega_\pm$ eigenfrequencies. The longitudinal frequency describes the back and forth oscillations of the charged particle between the repulsive potential by the two endcaps of the trap

$$\omega_z = \sqrt{\frac{qU_0}{md^2}}. \qquad (4.20)$$

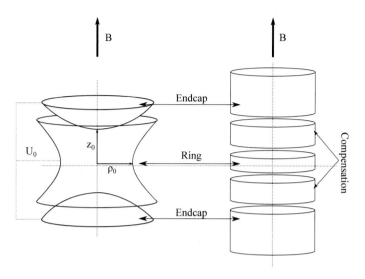

Fig. 4.7 Design of traps with (left) hyperbolic electrodes, (right) cylindrical geometry. Both designs can provide a first-order harmonic electric potential for ion trapping

As seen earlier, a charged particle in a magnetic field has a bent trajectory due to the Lorentz force. In the case of closed trajectories, the motion due to the Lorentz force is driven by the *cyclotron* frequency

$$\omega_c = \frac{q}{m} B. \tag{4.21}$$

The equations of motion of a charged particle in Cartesian coordinates can therefore be written as

$$\ddot{x} = \omega_c \dot{y} + \frac{\omega_z^2}{2} x \ ; \ \ \ddot{y} = -\omega_c \dot{x} + \frac{\omega_z^2}{2} y \ ; \ \ \ddot{z} = -\omega_z^2 z. \tag{4.22}$$

For a charged particle in a Penning trap, the radial motion is decomposed into two motions due to the interplay of the magnetic field and the quadrupole electric field: the *modified cyclotron frequency* ω_+ and the *magnetron frequency* ω_-. These radial frequencies are expressed as

$$\omega_\pm = \frac{\omega_c}{2} \pm \sqrt{\frac{\omega_c^2}{4} - \frac{\omega_z^2}{2}}, \tag{4.23}$$

and the time-dependent solutions of the equation of motion are expressed as

Fig. 4.8 Eigenfrequencies $\omega_z, \omega_\pm$ of a charged-particle motion in a Penning trap. Courtesy F. Wienholtz, TU Darmstadt

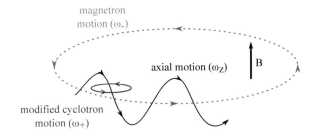

$$x(t) = \rho_+ \sin(\omega_+ t + \phi_+) + \rho_- \sin(\omega_- t + \phi_-)$$
$$y(t) = \rho_+ \cos(\omega_+ t + \phi_+) + \rho_- \cos(\omega_- t + \phi_-) \qquad (4.24)$$
$$z(t) = \alpha_z \cos(\omega_z t + \phi_z).$$

The motional amplitudes ρ_+, ρ_- and α_z are illustrated in Fig. 4.8, where a typical trajectory of a trapped particle with longitudinal and cyclotron frequencies is shown. In the limit of $\omega_z/\omega_c \ll 1$, we have $\omega_+ \sim \omega_c$ and $\omega_- \sim 0$, showing that $\omega_+ \gg \omega_-$ is often reached.

Problem

4.1 Solve the equation of motion for a particle with charge q and mass m in a Penning trap with an ideal quadrupole potential and constant magnetic field. Deduce the eigenfrequencies as expressed in Eqs. (4.20) and (4.23).

The motion can be modified by changing the potential of the trapping electrodes or adding high-frequency multipolar excitations, for example, quadrupole, to the trapping electrodes. In the later case of radio-frequency excitations, the energy of one motion mode can be transferred to another. From Eq. (4.23), the following identities are always verified

$$\omega_c = \omega_+ + \omega_-,$$
$$\omega_c^2 = \omega_z^2 + \omega_+^2 + \omega_-^2, \qquad (4.25)$$
$$\omega_z^2 = 2\omega_+\omega_-.$$

The relations (4.25) are key for the mass determination: by measuring precisely both ω_+ and ω_-, one gets the cyclotron frequency ω_c which gives access to the mass of the trapped ion assuming that the magnetic field B and the charge q of the ions are known. Eigenfrequencies depend on the trap settings and vary from trap to trap. As a numerical example, for a precision trap with typical settings $\rho_0 = 5$ mm, $z_0 = 5.5$ mm, $U_0 = 10$ V, and $B = 5$ T, the motional frequencies for an ion with $A/q = 50$ are $\omega_+ \sim 10$ MHz, $\omega_z \sim 1$ MHz, and $\omega_- \sim 10$ kHz.

The eigenfrequencies can be measured either in a non-destructive way, i.e., the particle is not lost after the measurement, or in a destructive manner, i.e., the particle is lost after the measurement. Non-destructive techniques are based on the measurement of very small signals, typically of the order of μV, induced on the electrodes by the motion of the trapped ions. In this scenario, the ions are conserved during the measurement. The precision of the measurement scales with the number of times the ions pass the electrodes, i.e., with the storage time. In the case of short lived nuclei, the destructive way is preferred. Then, each ion used for the measurement is adequately manipulated via high frequency excitations (dipole or higher multipole excitations) in the trap and subsequently ejected and detected using a suitable time-of-flight detector. It is then lost and thus used only once. Both methods reach the sensitivity to a single trapped ion.

4.1.6.2 Limits of Accuracy

The accuracy of a mass measurement is limited by several factors:

(1) The time and spatial stability of the magnetic field is currently the limit. The trapping region is usually of a size of few millimeters for precision traps. The spatial homogeneity that can be reached in such a volume is in the order of ± 0.01 ppm. The best time stability of a solenoid magnetic field reaches values better than $(\delta B/\delta t)/B \sim 10^{-10}$ per hour.

(2) The accuracy scales with the time of measurement. A high vacuum is therefore important, especially for highly charged ions which can easily pick up electrons from the atoms of the residual gas. The mass measurement for radioactive nuclei is particularly challenging in the case of the most exotic nuclei with lifetimes below a hundred of milliseconds.

(3) Short-lived nuclei are produced with limited yields. Traps have been developed for mass measurements with very few ions down to single ions. The mass accuracy limit is then a combination of the storage time and the number of ions available for the measurement.

4.1.6.3 An ICR-ToF Measurement at ISOLTRAP

In the following, we detail how a mass measurement with a Penning trap is performed in practice. We consider the particular example of the ISOLTRAP mass spectrometer, developed at CERN over the past thirty-five years which has been the first to perform high precision Penning-trap mass measurements of short-lived ions. In this particular example, the cyclotron frequency is determined in a destructive way.

1. After production, low-energy ions are slowed down, cooled and bunched by a RFQ[4] cooler and buncher, and then transmitted to a set of preparation traps. This step allows to separate isobars from the beam and to select the species to be transmitted to the precision trap and measured.

---

[4] Radio-frequency quadrupole.

2. The ions of interest are injected into the precision trap. In the case of ISOLTRAP, it is composed of parabolic electrodes embedded in a magnetic field of 5.9 T provided by a superconducting solenoid.

3. Ions stored in the center of the trap are excited by an azimuthal dipole excitation at their magnetron frequency. The ions transfer to centered circular orbits with the magnetron radius. The magnetron radius depends on the initial magnetron radius of the ion cloud, the length of the dipole excitation and its amplitude.

4. The trajectory of the ions is then modified by a quadrupole radiofrequency excitation at the cyclotron frequency $\omega_{rf} = \omega_c$ produced by oscillatory potentials applied to the segmented trapping electrodes. In this way, the two radial motions are converted into each other by quadrupole excitation on a periodic basis, as illustrated in Fig. 4.9. If the excitation frequency is different from the cyclotron frequency, the conversion is partial. Along with the transfer from the magnetron to the cyclotron mode, the ions acquire off centered orbits with a cyclotron radius. The velocity component $v_\perp$ perpendicular to the magnetic field is the source of a magnetic moment μ of the trapped charged particle

$$\mu = \frac{v_\perp^2}{2mB} = \frac{E_r}{B}, \qquad (4.26)$$

where E_r is the radial kinetic energy. The magnetic moment can be written as a function of the magnetron and modified cyclotron frequencies $\omega_\pm$ and radii $\rho_\pm$ from the solutions of the equation of motion as expressed in Eq. (4.26). One gets

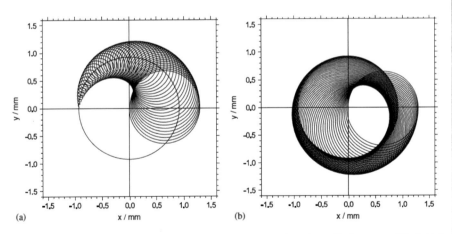

Fig. 4.9 Radial motion of a charged particle excited by a quadrupolar radio frequency. The initial motion is a pure magnetron motion (solid line circle in panel (**a**)), converted into a modified cyclotron motion. Panel **a** shows the first conversion from pure magnetron to pure cyclotron motion from $t = 0$ to $t = T_{conv}/2$. Panel **b** shows the inverse conversion from $t = T_{conv}/2$ to $t = T_{conv}$. Figure reprinted with permission from [2]. ©2021 by Elsevier

$$\mu = \frac{\omega_+^2 \rho_+^2 + \omega_-^2 \rho_-^2}{B} \sim \frac{\omega_+^2 \rho_+^2}{B}. \tag{4.27}$$

The modified cyclotron frequency is usually much larger than the magnetron frequency, $\omega_+ >> \omega_-$, and the magnetic moment is proportional to ω_+^2.

5. Ions are released from the trap by lowering the trapping electric potential. They are detected by a micro channel plate (MCP) detector which gives the time of flight of the ions from the trap to the detector with a time of flight resolution better than few nanoseconds. When exiting the trapping region, and therefore the constant magnetic field region, the ions experience a field gradient flying towards the time-of-flight detector, located outside the trap. They are accelerated by the interaction of their magnetic moment μ with the gradient of the magnetic field through the force $\mu \partial B / \partial z$. The ions with the largest magnetic moment are faster than the others and will arrive first at the detector. The time of flight is then directly related to the radial kinetic energy of the ions when trapped, therefore depends on their mass.

To be more quantitative, the time-of-flight spectrum is measured as a function of the frequency of the quadrupole excitation ω_{rf}, for a fixed excitation time. Assuming an excitation time T_{rf}, the radial kinetic energy E_r gained during the excitation, between $t = t_0$ and $t = t_0 + T_{rf}$, can be derived as

$$E_r \propto \frac{\sin^2(\omega_B T_{rf})}{\omega_B^2}, \tag{4.28}$$

where

$$\omega_B = \frac{1}{2} \sqrt{(\omega_{rf} - \omega_c)^2 + k_0^2}$$

and

$$k_0 = \frac{V_{rf}}{2a^2} \frac{q}{m} \frac{1}{\omega_+ - \omega_-},$$

V_{rf} being the maximum potential of the quadrupole radiofrequency field on a circle with radius a. As stated above, the ion exiting the trap experiences an axial force $F_z(\omega_{rf}, z) = -\mu_z \nabla_z B(z)$ in the gradient of the magnetic field cause by its magnetic moment (Eq. (4.27)). This force leads to a reduction of the time of flight from the ion to the detector. The time of flight is given by

$$T = \int_{z_0}^{z_1} \frac{dz}{v(z)}, \tag{4.29}$$

where $v(z)$ is the velocity of the ion along its path from the trap z_0 to the detector z_1. By using the non relativistic relation between kinetic energy $E(z)$ and velocity $v(z)$, the time of flight can be written as

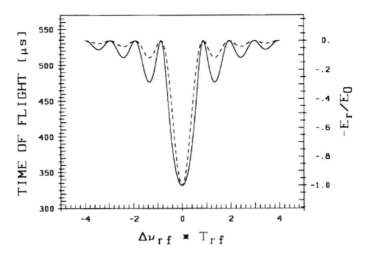

Fig. 4.10 Radial kinetic energy E_r normalized to its maximum value at the cyclotron frequency (dotted line), as a function of the detuning frequency ($\Delta\nu_{rf} = \omega_c - \omega_{rf}$) times the excitation time (T_{rf}). The corresponding time-of-flight spectrum is shown with the continuous line. At $\Delta\nu_{rf} = 0$, i.e., $\omega_{rf} = \omega_c$, the time of flight is smaller thanks to the acceleration produced by the interaction of the magnetic moment of the ion and the field gradient along the flight path. Figure reprinted with permission from [3]. ©2021 by Elsevier

$$T = \int_{z_0}^{z_1} \sqrt{\frac{m}{2E(z)}} dz = \int_{z_0}^{z_1} \sqrt{\frac{m}{2[E_0 - qV(z) - \boldsymbol{\mu}(\omega_{rf}).\mathbf{B}(z)]}} dz, \qquad (4.30)$$

where $V(z)$ and $\mathbf{B}(z)$ are the electric potential and magnetic field along the ion path.

Equation (4.30) shows that the measured spectrum is directly connected to the radial kinetic energy spectrum, via the ion magnetic moment, as a function of the excitation frequency around the resonant cyclotron frequency. This is illustrated in Fig. 4.10. Note the damped oscillation pattern which originates from the convolution of the distribution of scanned frequencies and the time-of-light expression for a given excitation frequency. Note that in the case of a rectangular time excitation, as in Fig. 4.10, the experimental signal follows rather closely the theoretical line shape

$$f(\omega) = \frac{\sin(\alpha\omega)}{\alpha\omega}. \qquad (4.31)$$

An experimental time-of-flight spectrum measured by this technique is shown in the panel (a) of Fig. 4.11. The data show an excellent agreement with the line shape as expressed in Eq. 4.31. In this case the quadrupole excitation time was 1.2 s at each detuning frequency ν_c.

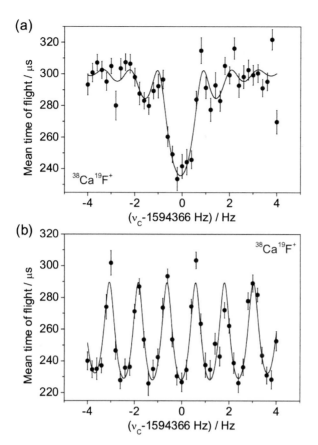

Fig. 4.11 Time of flight for $^{38}\text{Ca}^{19}\text{F}^+$ as a function of the frequency of the quadrupole excitation. Data were measured at ISOLTRAP in CERN. A molecule composed of ^{38}Ca and ^{19}F is chosen to measure the mass of ^{38}Ca to avoid A=38 isobars in the trap, the mass of the stable ^{19}F being known with high precision. (Top) The quadrupole excitation in the trap was constant for a 1.2 second duration for each detuning frequency. The total number of collected molecules is about 2500. (Bottom) Same but with a so-called Ramsey quadrupole excitation scheme composed of two 100 ms pulses separated by a 1 s waiting period, for the same number of collected molecules. An improvement by a factor three in the mass uncertainty is reached in this case. Solid curves are expected line shape from theory. Figure reprinted from [4]. ©2021 by the American Physical Society

The above described ICR-ToF method can be used with a different quadrupole excitation function scheme. If the excitation function is composed of several pulses separated by waiting periods, it is called the *Ramsey method*. Since the excitation function is modified, the line shape of the time-of-flight response as a function of the detuning frequency will also be modified. This can be used and optimized to gain in precision for a given number of collected ions as illustrated in Fig. 4.11.

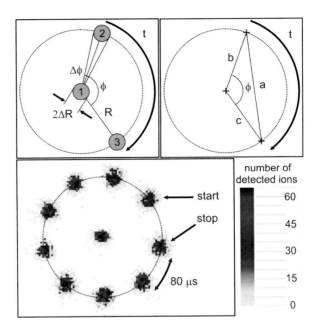

Fig. 4.12 Principle of the Phase-Imaging mass-measurement technique. (Top) Ions are trapped in the center of the trap (position 1). After a dipole excitation, they are prepared at a radius R (position 2). After a free evolution time t, they acquire a phase ϕ. (Bottom) Experimental data for ^{130}Xe$^+$ and different free evolution times t are shown. The method is applied to determine the magnetron and the modified cyclotron frequencies separately. Figure reprinted with permission from [5]. ©2021 from the American Physical Society

4.1.7 Phase-Imaging Method

The phase shift ϕ of the trapped-ion radial motion after a free evolution time t is connected to the radial angular frequency ω, either the magnetron or the modified cyclotron frequency depending on the excitation scheme of the ions, by

$$\phi + 2\pi n = \omega t, \tag{4.32}$$

where n is the number of full revolutions the ion performs during the time period of t. A new method has been recently developed to measure the phase shift ϕ. It is called phase-imaging ion-cyclotron-resonance (PI-ICR) technique. It is based on the determination of the cyclotron frequency via the projection of the ion motion in the trap onto a high-resolution position-sensitive micro-channel plate detector. This is the main technical difference from the above TOF-ICR and Ramsey methods where the time-of-flight detector outside the magnetic field does not have to be position sensitive.

The ions are first excited in the trap with a dipole excitation so that their trajectory has a mean radius R. The magnetron (ω_-) and modified cyclotron (ω_+) are measured separately with the same technique. The cyclotron frequency $\omega = \omega_+ + \omega_-$ can then be obtained for mass determination. In the case of the modified cyclotron frequency an additional step is needed before the ions are released from the trap: their magnetron motion is converted into the cyclotron one with a quadrupole excitation, then released. In the case of an ideal axially symmetric magnetic field, the phase is conserved after the distance of flight to the detector. Figure 4.12 shows an illustration of a measurement with $^{130}Xe^+$ ions performed at GSI with the SHIPTRAP apparatus. The frequency precision that can be obtained is driven by statistics and the radial spread ΔR of the ions. For very short-lived nuclei, the Phase-Imaging technique has been shown to be 25 times faster and provides a 40-fold increased resolving power compared to the Ramsey technique.

4.2 Beta Decay Strength and Half-Life

4.2.1 The Discovery of the Weak Interaction

Radioactivity was discovered in 1896 by Henri Becquerel, and it became clear within a few years that the decaying nuclei emitted three types of radiation. In 1899, Ernest Rutherford separated radioactive emissions into two types: alpha and beta, based on penetration of objects and ability to cause ionization. Alpha rays could be stopped by thin sheets of paper or aluminium, whereas beta rays could penetrate several millimeters of aluminium. In 1900, Paul Villard evidenced a more penetrating type of radiation, which Rutherford identified as a new type in 1903 and named them *gamma rays*.

In both alpha and gamma decay, the resulting spectrum has a narrow energy distribution, since the particle carries the energy from the difference between the initial and final nuclear states. However, the kinetic energy distribution, or spectrum of beta decays have a continuous spectrum. If beta decay were simply electron emission, then the energy of the emitted electron should have a particular, well-defined value. However, the observed broad distribution of energies suggested that energy is lost in the beta decay process, i.e., that there is violation of the energy conservation. This spectrum was puzzling for many years.

Eventually in 1930, Wolfgang Pauli resolved the beta-particle energy puzzle by suggesting that, in addition to electrons and protons, atomic nuclei also contained an extremely light neutral particle, which was named "neutrino"[5] by Enrico Fermi.

---

[5] It means a small neutral object in Italian.

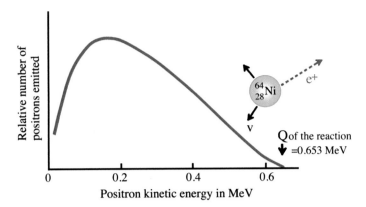

Fig. 4.13 Positron energy spectrum of beta-decay of ^{64}Co. In the process of beta decay $\beta^-(\beta^+)$, both an electron (a positron) and an antineutrino (neutrino) are emitted. Because an electron (positron) and an antineutrino (neutrino) both carry the energies, a spectrum of energies for the electron (positron) is continuous, depending upon what fraction of the reaction energy Q_β is carried by the massive particle. The shape of this energy curve is predicted from the Fermi theory of beta decay

The β disintegration of a nucleus can follow three processes:

$$(Z, N) \rightarrow (Z + 1, N - 1) + e^- + \bar{\nu}_e : \beta^- \text{desintegration}$$
$$(Z, N) \rightarrow (Z - 1, N + 1) + e^+ + \nu_e : \beta^+ \text{desintegration}$$
$$(Z, N) + e^- \rightarrow (Z - 1, N + 1) + \nu_e : \text{electron capture}$$

corresponding to the following processes

$$n \rightarrow p + e^- + \bar{\nu}_e$$
$$p \rightarrow n + e^+ + \nu_e$$
$$p + e^- \rightarrow n + \nu_e,$$

where ν_e and $\bar{\nu}_e$ are the electron neutrino and anti-neutrino, respectively. To occur, these processes have to be energetically possible, i.e., the sum of the initial masses (M_i) has to be larger than the sum of the final masses (M_f):

$$Q_\beta = \sum_f M_f - \sum_i M_i \leq 0. \tag{4.33}$$

In free space, since the mass of the neutron is larger than the proton mass, the neutron disintegration is allowed, but the proton cannot decay via β decay. On the other hand, all processes can occur in nuclei.

Enrico Fermi geniusly considered the β decay to be an analogous process to the electromagnetic process in 1934, although there was not enough experimental

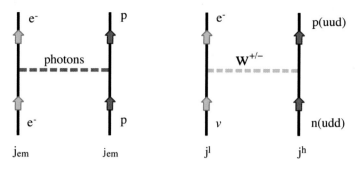

Fig. 4.14 Comparison of the electromagnetic and the weak interactions. The superscripts *l* and *h* of weak currents denote leptons and hadrons, respectively

evidence at that time. He considered that the creation of an electron and a neutrino in β decay is similar to the photon emission of the electromagnetic process. Fermi introduced a hypothetical interaction Hamiltonian, so called "the weak interaction" for β decay. The basic idea is drawn in Fig. 4.14, i.e., the electromagnetic interaction is transmitted by a virtual photon[6], while the weak interaction is mediated by the so called "hypothetical weak boson W". The Fermi β-decay theory was incredibly successful and explained all experimental data for nearly 25 years. However, in 1957, parity violation was found in the beta decay and the Fermi theory was finally modified to accommodate it.

4.2.2 The Fermi β-decay Theory

The analogy between the two interactions is implemented in a form of "current-current" interaction. The electromagnetic interaction is written as (see also Sect. 4.6.1 in this chapter),

$$H_{\text{em}} = - \int j_\mu(\mathbf{r}) A^\dagger_\mu(\mathbf{r}) d\mathbf{r} \tag{4.34}$$

where $j_\mu(\mathbf{r})$ is the four-vector charge current,

$$j_\mu(\mathbf{r}) = \left(\rho, \frac{1}{c} \mathbf{j}(\mathbf{r}) \right), \tag{4.35}$$

and $A_\mu(\mathbf{r})$ represents the four-vector potential,

---

[6] The concept of "*virtual*" particles arises in perturbation theory of quantum mechanics where the interaction between "*real*" particles is described in terms of exchanges of virtual particles. A process involving virtual particles is expressed by a schematic representation known as a Feynman diagram, in which virtual particles are represented by internal lines. As an illustration, see Fig. 4.14.

$$A_\mu(\mathbf{r}) = (\psi, \mathbf{A}) = \int j_\mu(\mathbf{r}') \frac{1}{|\mathbf{r} - \mathbf{r}'|} d\mathbf{r}'. \tag{4.36}$$

Inserting $A_\mu(\mathbf{r})$ in the expression of $H_{\rm em}$, the electromagnetic interaction is expressed in a current-current interaction as

$$H_{\rm em} = -\int \int j_\mu(\mathbf{r}) j_\mu^\dagger(\mathbf{r}') \frac{1}{|\mathbf{r} - \mathbf{r}'|} d\mathbf{r} d\mathbf{r}', \tag{4.37}$$

where the range of interaction $\frac{1}{|\mathbf{r}-\mathbf{r}'|}$ comes from the fact that photons are massless and mediate a long-range interaction. The current-current interaction (4.37) for the electromagnetic field gives a helpful guide in elucidating the weak interaction.

In an analogy to the electromagnetic interaction, Fermi introduced the current-current type weak interaction as

$$H_{\rm weak} = g \int \int j_\mu^{w\dagger}(\mathbf{r}) j_\mu^w(\mathbf{r}') f(|\mathbf{r} - \mathbf{r}'|) d\mathbf{r} d\mathbf{r}', \tag{4.38}$$

where g is the coupling constant, $j_\mu^w(\mathbf{r})$ is a hypothetical four-vector weak current and $f(r)$ is the range of weak interaction mediated by the weak bosons. The meson-exchange force is customary described by a Yukawa-type radial extension (see Chap. 2)

$$f(r) = \frac{e(-r/r_W)}{r}, \tag{4.39}$$

where the range of the weak interaction r_W is determined by the mass of weak boson m_W. The mass of the weak boson was found experimentally much later in 1983 at CERN to be $m_W \simeq 80 \,\mathrm{GeV}/c^2$ so that the interaction range is very short, even compared with the strong interaction mediated by π mesons. The typical range r_w of weak interaction is

$$r_W = \frac{\hbar}{m_W c} \simeq 2.5 \times 10^{-3} \mathrm{fm}. \tag{4.40}$$

The range $f(r)$ is so short that it can be approximated by a delta interaction $\delta(\mathbf{r})$. Inserting the delta interaction in Eq. (4.38), we get

$$H_{\rm weak} = g \int \int j_\mu^{w\dagger}(\mathbf{r}) j_\mu^w(\mathbf{r}) d\mathbf{r}. \tag{4.41}$$

The coupling constant g governs the strength of the weak interaction, playing the same role as the electric charge of electromagnetic interaction, and is related to the vector current coupling constant g_V, which will appear later for the Fermi-type beta decay current.

The weak current has two parts, the lepton and the hadron parts

$$j_\mu^w(\mathbf{r}) = j_\mu^l(\mathbf{r}) + j_\mu^h(\mathbf{r}), \tag{4.42}$$

which create three types of current-current interaction: lepton-lepton, lepton-hadron and hadron-hadron interactions

$$H_{\text{weak}} = g \left[\int \int j_\mu^{l\dagger}(\mathbf{r}) j_\mu^l(\mathbf{r}) d\mathbf{r} + \left(\int \int j_\mu^{l\dagger}(\mathbf{r}) j_\mu^h(\mathbf{r}) d\mathbf{r} + \text{H.C.} \right) \right.$$
$$\left. + \int \int j_\mu^{h\dagger}(\mathbf{r}) j_\mu^h(\mathbf{r}) d\mathbf{r} \right],$$
(4.43)

where H.C. denotes the Hermitian conjugate term. Fermi introduced originally the lepton-hadron current interaction only, which governs the beta decays in all nuclear processes. The pure leptonic and pure-hadronic processes predicted by the general form of the weak Hamiltonian (4.43) were observed later, for example, a lepton muon (denoted by the Greek character μ) decays as

$$\mu \rightarrow e \bar{\nu} \nu \qquad \text{pure leptonic process,} \qquad (4.44)$$

and a positively charged meson kaon denoted K^+ also decays as

$$K^+ \rightarrow \begin{cases} \pi^+ \pi^+ \pi^- \\ \pi^+ \pi^0 \pi^0 \end{cases} \qquad \text{pure hadronic process.}$$

The scattering of neutrinos with charged leptons,

$$\nu_e e^- \rightarrow \nu_e e^-, \quad \nu_e e^- \rightarrow \nu_e \mu^-, \qquad (4.45)$$

also involves only leptons.

The hadronic part of the original weak interaction (4.43) has only the vector current term. Since it was discovered that the weak interaction violates parity conservation, the hadronic currents have been generalized to a combination of a vector term (V) and an axial-vector (A) term

$$j_\mu^h(r) = j_\mu^{h,V}(r) + j_\mu^{h,A}(r). \qquad (4.46)$$

With this formulation, the vector term changes sign under parity transformation, while the axial-vector term does not. The current j_μ^h therefore violates parity. The above effective Hamiltonian (4.43) can describe properly weak processes which involve a small momentum transfer. It is a low-energy approximation of the unified electroweak theory developed by Glashow, Weinberg and Salam (GWS) where weak processes are implemented as the exchange of $W^\pm$, Z^0 bosons.

4.2.3 Conservation Laws

The conservation of angular momentum in the β decay process implies that

$$\mathbf{J}_m = \mathbf{J}_d + \mathbf{L} + \mathbf{S}, \tag{4.47}$$

where $\mathbf{J}_m$ and $\mathbf{J}_d$ are the total angular momentum of the states of mother and daughter nuclei involved in the decay, $\mathbf{L}$ and $\mathbf{S}$ are the orbital angular momentum and spin carried by the leptons in the exit channel and defined as

$$\mathbf{L} = \mathbf{L}_e + \mathbf{L}_\nu, \quad \mathbf{S} = \mathbf{S}_e + \mathbf{S}_\nu, \tag{4.48}$$

with $\mathbf{L}_e$, $\mathbf{S}_e$ the orbital angular momentum and the spin carried by the electron (positron in case of β^+), respectively. Identical notations are used for the antineutrino $\bar{\nu}_e$ (neutrino ν_e in the case of β^+). One distinguishes two modes of β decay: the so-called Fermi (F) decay with $S = 0$ and the so-called Gamow-Teller mode with $S = 1$, as illustrated in Fig. 4.15.

The small mass of the leptons in the exit channel imply that $L = 0$ decays are favored, while processes with $L = 1, 2, \ldots$ are possible but have much smaller probabilities. In practice, $L > 0$ decays are only observed when $L = 0$ is not possible. Intuitively, one can obtain this conclusion by estimating the kinetic energy E carried by the leptons for a decay with angular momentum L occurring at a radius r

$$E \sim pc = \frac{Lc}{r}.$$

Fig. 4.15 Illustration of the two β-decay modes: in Fermi decays (top) the spin of the leptons in the exit channel are combined to $S = 0$ while in the Gamow-Teller decays (bottom) the spin of the leptons are combined to $S = 1$. The arrows indicate the direction of the spin projections on the vertical axis

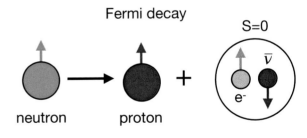

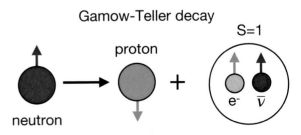

Considering a decay with angular momentum $L = 1\,\hbar$ and a decay occurring at the nuclear surface of $r = 3$ fm, one gets

$$E = \frac{\hbar c}{r} = \frac{197\ [\text{MeV fm}]}{3\ [\text{fm}]} = 66\ \text{MeV},$$

larger than most of the Q_β values. Therefore, the lepton orbital angular momentum can differ from zero only if they are emitted at high energy, which cannot exceed Q_β, or if the leptons are emitted at large radii, although the emission at large radius is suppressed.

The conservation of angular momentum imposes the following condition for the initial (π_m) and final (π_d) parity states

$$\pi_m = \pi_d(-1)^L \Rightarrow \Delta\pi = \pi_m \pi_d = (-1)^L. \tag{4.49}$$

Based on the conservation laws, the different β decays are classified into groups which correspond to different transferred angular momentum $L=|\mathbf{L}|$. The L=0 decay is called the allowed decay. There are two types of allowed decays; Fermi-type and Gamow-Teller (GT)-type. The parity of allowed decay is not changed between the mother and daughter states $\pi_m = \pi_d$ in Eq. (4.49) since $L = 0$. Among the allowed decays, the Fermi decay between $J_m^\pi = J_d^\pi = 0^+$ is called the "super-allowed" Fermi decay, while the GT decay between $J_m^\pi = 0^+$ and $J_d^\pi = 1^+$ is called "unique" GT decay. The beta decays with $L = 1, 2, 3, \cdots$ are called the first-, second-, third- $\cdots$ forbidden decays according to the involved transfer angular momentum. The parities of decays change (do not change) in the cases of $L =$ odd ($L =$ even). The decays with higher L are slower and have smaller decay probabilities characterized by the so-called ft value, as described in the following.

4.2.4 Decay Probabilities and ft Values

The β decay is characterized by its half-life $t_{1/2}$ related to the transition probability per time unit $\lambda_{i \to f}$ as follows

$$t_{1/2} = \frac{\ln(2)}{\lambda_{i \to f}}, \tag{4.50}$$

where i is the initial state of the mother nucleus and f is the final state of the daughter nucleus, ground state or excited state. The transition probability is given by the so-called *Fermi's golden rule*

$$d\lambda_{i \to f} = \frac{2\pi}{\hbar} |\langle f; e^-, \hat{\nu}_e | H_{\text{weak}} | i \rangle|^2 d\rho_f, \tag{4.51}$$

where $d\rho_f$ is the final state density in the daughter nucleus and H is the weak interaction Hamiltonian. According to Eq. (4.51), $d\lambda_{i\to f}$ can be factorized

$$d\lambda_{i\to f} = \frac{2\pi}{\hbar} g^2 |\langle f|j_\mu^{h,\dagger}|i\rangle|^2 |\langle e^-, \hat{\nu}_e|j_\mu^l|0\rangle|^2 d\rho_f, \tag{4.52}$$

where $\langle f|j_\mu^{h\dagger}|i\rangle$ and $\langle e^-, \hat{\nu}_e|j_\mu^l|0\rangle$ are the matrix elements of hadronic and leptonic currents, respectively. In principle, the hadronic and leptonic matrix elements are connected to each other due to momentum and energy conservation. In the case of β decay, the momentum transfer to the nucleus is very small so that the two matrices can be evaluated separately.

The final state density $d\rho_f$ is expressed by the electron distribution, which depends on the nuclear charge Z, the nuclear radius R and the electron energy E_e. Then the transition probability $\lambda_{i\to f}$ is proportional to the nuclear matrix element and the integral of electron phase space $f(Z, R, E_e)$ called f-factor,

$$\lambda_{i\to f} \propto |\langle f|j_\mu^{h,\dagger}|i\rangle|^2 f(Z, R, E_e). \tag{4.53}$$

The charge and current densities of beta-decay $\beta^\pm$ are formulated in analogy to the electromagnetic current. In the non-relativistic limit, the velocity-independent currents read,

$$\rho_V(L=0) = g_V \sum_i t_\pm(i)\delta(\mathbf{r} - \mathbf{r}_i), \tag{4.54}$$

$$\mathbf{j}_A(L=0) = g_A \sum_i t_\pm(i)\sigma_i\delta(\mathbf{r} - \mathbf{r}_i), \tag{4.55}$$

where g_V and g_A are the vector and axial vector coupling constants of beta decays, respectively. The isospin raising (lowering) operator $t_\pm = t_x \pm it_y = \tau_\pm/2$ transforms a neutron (proton) to a proton (neutron) with the matrix element $\langle p|t_-|n\rangle$ ($\langle n|t_+|p\rangle$)=1. Next-order terms depending on the velocity or the gradient ∇ are expressed as

$$\rho_A(L=1) = g_A \sum_i t_\pm(i)\frac{1}{2c} (\sigma_i \cdot \mathbf{v}_i\delta(\mathbf{r} - \mathbf{r}_i) + \delta(\mathbf{r} - \mathbf{r}_i)\sigma_i \cdot \mathbf{v}_i), \tag{4.56}$$

$$\mathbf{j}_V(L=1) = g_V \sum_i t_\pm(i)\left[\frac{1}{2c} (\mathbf{v}_i\delta(\mathbf{r} - \mathbf{r}_i) + \delta(\mathbf{r} - \mathbf{r}_i)\mathbf{v}_i) + \frac{\hbar}{2mc}\mu_\beta\nabla \times \sigma_i\delta(\mathbf{r} - \mathbf{r}_i)\right],$$
$$\tag{4.57}$$

where the term proportional to μ_β is the analog of the magnetic moment in the electromagnetic field (see Eq. (4.269) in Sect. 4.6.4).

The multipole moments for beta decay are written as

$$M(\rho_{V,A}, \lambda, \mu) = \int r^{\lambda} Y_{\lambda\mu}(\hat{r}) \rho_{V,A} d\mathbf{r}, \tag{4.58}$$

$$M(\mathbf{j}_{V,A}, \lambda, \mu) = \int r^{\lambda} \left[Y_{\kappa}(\hat{r}) \mathbf{j}_{V,A}(\mathbf{r}) \right]_{(\kappa 1)\lambda\mu} d\mathbf{r}, \tag{4.59}$$

where $Y_{\lambda\mu}(\hat{r})$ are the spherical harmonics, and $[Y_{\kappa} \mathbf{j}]_{(\kappa 1)\lambda\mu}$ is the vector spherical harmonics with the rank $\lambda\mu$ determined by the coupling of the spherical harmonics and the vector current operator $\mathbf{j}$ such as Eqs. (4.55) and (4.57).

For the allowed transitions, the contributions come from two operators

$$M(\rho_V, \lambda = 0) = \frac{g_V}{(4\pi)^{1/2}} \sum_I t_-(i), \tag{4.60}$$

$$M(\mathbf{j}_A, \kappa = 0, \lambda = 1, \mu) = \frac{g_A}{(4\pi)^{1/2}} \sum_I t_-(i)\sigma_\mu(i), \tag{4.61}$$

which are referred as Fermi (F) and Gamow-Teller (GT) transitions, respectively.

As is shown in Fig. 4.13, the β decay spectrum shows a broad energy spectrum, while the electromagnetic transition is always a discrete spectrum. Because of this broad spectrum of β decay, it is not straightforward to relate the nuclear matrix element with the decay life time. In order to cure this problem, it is common to express the transition probability for the β decay in terms of the ft value, where $t \equiv t_{1/2}$ is the half life, and f is a quantity depending on the nuclear charge and the electron energy and the multipolarity of transitions. Essentially, the f value represents the phase space for the leptons and is analogous to the factor $(E_\gamma)^{2\lambda+1}$ in the electromagnetic transition rate in Table 4.6. The f value can be evaluated by integrals over the electron spectra.

For the allowed transition $L=0$, the ft is evaluated to be

$$ft(B(F) + B(GT)) = \frac{\pi^2 \hbar^7 \ln 2}{2m_e^5 c^4} \equiv D \frac{g_V^2}{4\pi} \tag{4.62}$$

with

$$D \equiv \frac{2\pi^3 \hbar^7 \ln 2}{g_V^2 m_e^5 c^4} = 5974 \text{ sec},$$

when both Fermi and GT decays are possible. By the muon life time measurement, the vector coupling constant g_V is empirically determined as

$$\frac{g_V}{(\hbar c)^3} = 1.166 \times 10^{-5} \text{GeV}^{-2}. \tag{4.63}$$

The reduced transition probability for the Fermi operator in Eq. (4.62) is given by

$$B(F; T M_T \rightarrow T M_T \pm 1) = \frac{g_V^2}{4\pi} \frac{1}{2I + 1} |\langle T | T_\pm | T \rangle|^2 = \frac{g_V^2}{4\pi} (T \mp M_T)(T \pm M_t + 1)$$

(4.64)

for the initial and final states with the same total spins $I = I_i = I_f$. The GT transition is defined as

$$B(GT) = \frac{g_A^2}{4\pi} \frac{1}{2I_i + 1} |\langle I_f || \sum_i t_-(i)\sigma(i) || I_i \rangle|^2.$$

(4.65)

The allowed transitions are independent of the positions of nucleons.

The first forbidden transitions are governed by the matrix elements of the various operators $M(\rho_A, \lambda = 0)$, $M(\rho_V, \lambda = 1)$, $M(\mathbf{j}_A, \kappa = 1, \lambda = 0)$, $M(\mathbf{j}_A, \kappa = 1, \lambda = 1)$, $M(\mathbf{j}_A, \kappa = 1, \lambda = 2)$, and $M(\mathbf{j}_V, \kappa = 0, \lambda = 1)$. The transitions by three operators $M(\mathbf{j}_A, \kappa = 1, \lambda = 0, 1, 2)$ are called the spin-dipole transitions, which depend on the position of nucleons in the linear order. Higher forbidden transitions are defined by using velocity-dependent operators as expressed above.

4.2.5 Examples of β-decay Measurements

Considering the quark substructure of the nucleons, the neutron is composed of two "down" type quarks and one "up" type quark, while the proton is composed of two "up" type quarks and one "down" type. As shown in Fig. 4.14, the beta decay process thus corresponds to the conversion of a down quark to an up quark, converting the neutron into a proton, with the emission of a virtual W^- particle that decays into an electron and an anti-electron neutrino. This process is governed by the weak nuclear interaction, which is unified with electromagnetism in the standard model of particle physics to form the electroweak force.

For the superallowed beta decays, the entire wave functions of the initial and final nuclei are unchanged essentially with the exception of one proton being converted into a neutron. These special decays are particularly suited to theoretical analysis and precise experimental measurements for superallowed beta decays thus provide demanding tests of Conserved Vector Current hypothesis (CVC) and the Standard Model description of electroweak interactions (see detailed discussion in Sect. 10.3). The ft value of superallowed beta decay process from ^{14}O,

$$^{14}\text{O}(J^\pi = 0^+, T = 1, T_z = -1) \xrightarrow[\beta_+ (\text{Fermi})]{} {}^{14}\text{F}(J^\pi = 0^+, T = 1, T_z = 0) + \nu_e + e^+,$$

(4.66)

can be evaluated by using Eq. (4.62),

$$ft = 5974 \frac{g_V^2}{4\pi} / B(F) = 2987 \text{ s},$$

(4.67)

Fig. 4.16 Beta decay of
^{11}C. Since the spin-parities
of mother and daughter
nuclei are the same, the
decay process is induced by
both allowed Fermi and
allowed GT transitions

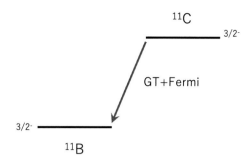

with use of the Fermi transition matrix

$$B(F; T = 1, M_T = 1 \rightarrow T = 1, M_T = 0) = \frac{g_V^2}{4\pi} 2. \qquad (4.68)$$

This value can be compared with the empirical value $ft(\exp) = (3041.2 \pm 2.7)$ s.

The superallowed Fermi beta decays have been measured precisely for 20 nuclei so far. The calculated value (4.67) gives a good account of experimental value of all nuclei. These superallowed beta decay data currently provide the most precise determination of the "up-down" element of Cabibbo-Kobayashi-Maskawa (CKM) quark mixing matrix that describes the transformation between the mass eigenstates and the weak interaction eigenstates of the Standard Model quarks.

Let us estimate an allowed GT decay rate from the ground state ^{11}C to the ground state ^{11}B shown in Fig. 4.16;

$$^{11}\text{C} \xrightarrow[\beta_+(\text{GT+Fermi})]{} {}^{11}\text{B} + \nu_e + e^+. \qquad (4.69)$$

The ground states of the two nuclei have the same spin and parity $J^\pi = \frac{3}{2}^-$ and the shell model configurations might be assigned as $1p_{3/2}^\pi$ and $1p_{3/2}^\nu$ for ^{11}C and ^{11}B, respectively on top of the ^{10}C core. We will estimate the β decay rate by using a formula for the single-particle GT transition

$$B(GT; lj \rightarrow lj') = \frac{g_A^2}{4\pi} \begin{cases} \left(\frac{j+1}{j}\right)^{\pm 1} & j = j' = l \pm 1/2 \\ \frac{2j'+1}{(l+1)/l} & |j - j'| = 1. \end{cases} \qquad (4.70)$$

In the ^{11}C decay, $B(GT; j = 3/2 \rightarrow j = 3/2) = \frac{g_A^2}{4\pi} 5/3$ from Eq. (4.70) and $B(F; T = 1/2, T_z = 1/2 \rightarrow T = 1/2, T_z = -1/2) = \frac{g_V^2}{4\pi}$ from Eq. (4.64). The ft value is then obtained as

Table 4.1 ft values and extracted $B(GT)$ for allowed β decays. The $B(GT)$ values are obtained from Eq. (4.62) taking $B(\text{Fermi}) = g_V^2/4\pi$ for $T_m = T_d = 1/2$. The $B(GT)$ value is given in unit of $g_A^2/4\pi$

nuclei	nlj	Q_β (MeV)	$\log(ft)$	ft(sec)	$B_{exp}(GT)$	$B_{sp}(GT)$
$^3\text{H} \rightarrow \,^3\text{He}$	$1s_{1/2}$	18.59	3.0524	1128.	2.72	3.0
$^{15}\text{O} \rightarrow \,^{15}\text{N}$	$1p_{1/2}^{-1}$	2.754	3.6377	4342.	0.24	0.333
$^{17}\text{F} \rightarrow \,^{17}\text{O}$	$1d_{5/2}$	3.761	3.3573	2277.	1.03	1.40
$^{39}\text{Ca} \rightarrow \,^{39}\text{K}$	$1d_{3/2}^{-1}$	6.524	3.63	4266.	0.25	0.60
$^{41}\text{Sc} \rightarrow \,^{41}\text{Ca}$	$1f_{7/2}$	6.495	3.4529	2873.	0.70	1.286

$$ft = \frac{D}{B(F) + \left(\frac{g_V}{g_A}\right)^2 B(GT)} = 1644 \text{ s},$$

$$\log(ft) = 3.22, \tag{4.71}$$

where the empirical ratio g_A/g_V is taken as

$$\frac{g_A}{g_V} = -1.257. \tag{4.72}$$

The empirical ft value of ^{11}C is $\log(ft) = 3.592$.

Experimental data of β decay between mirror nuclei with $T_m = T_d = 1/2$ are listed in Table 4.1. Main single-particle configurations are also listed in this table. The $B(GT)$ values are extracted by using Eq. (4.62) assuming $B(\text{Fermi}) = g_V^2/4\pi$. The beta decay of $n \rightarrow p$ is the reference decay and normalized to be $B(GT)/(g_V^2/4\pi) = 3$. The single particle $B_{sp}(GT)$ value is calculated from Eq. (4.70) for each configuration. We can see the experimental $B(GT)$ values are always smaller than the calculated ones. This is called a "*quenching effect*" on the GT transitions. The quenching effect on $B(GT)$ can be seen for the GT transitions in Table 4.1. The quenching of the GT matrix and also the M1 transition matrix have been discussed intensively in the last few decades. It turns out that the configuration mixings are the main effects on the quenching, while other effects such as the Δ-hole coupling will also contribute.

The observation of spectra for unstable nuclei have bee carried out by β decays as an efficient experimental tool to study unstable nuclei near the drip line. As an example, the level scheme of ^{31}Mg is shown in Fig. 4.17 observed by β_- decay from ^{31}Na [6]. Many new states were found through this β decay combining the successive γ decay rates. The spin assignment of each level is feasible combining the anisotropy of the β decay, the observed ft values and the successive γ transitions. Thus, the β decay is a promising tool to find new levels and related structure of deformation and shape coexistence in exotic nuclei.

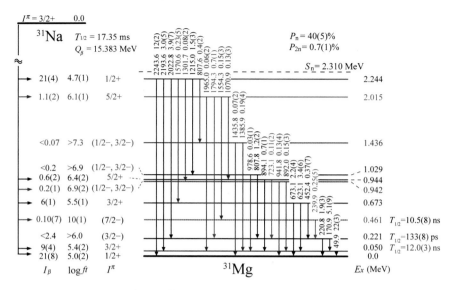

Fig. 4.17 Beta decay from ^{31}Na to ^{31}Mg and following γ transitions. The ft value and branching ratio I_β are given next to each level. The spins and parities in parenthesis are tentative assignments. The β and γ transitions, levels, spins, and parities labeled displayed in red are newly found. The levels and transitions labeled in blue are those observed in the β decay for the first time. Figure reprinted with permission from [6]. ©2021 from the American Physical Society. Courtesy H. Nishibata and A. Odahara, Osaka University

4.3 Charge and Matter Radii

The spatial density distribution $\rho(\mathbf{r})$ of a nucleus is the closest observable to the spatial wave functions $|\phi(\mathbf{r})|^2$ of the constituent nucleons. Indeed, in the Hartree-Fock description of the nucleus (see Chap. 3), the matter density ρ is expressed as the sum of the presence probabilities of each individual nucleon of the N neutrons and Z protons

$$\rho(\mathbf{r}) = \rho_N(\mathbf{r}) + \rho_Z(\mathbf{r}) = \sum_i^{N,Z} |\phi_i(\mathbf{r})|^2, \tag{4.73}$$

where ρ_N and ρ_Z are the neutron and proton densities, respectively. The charge density distribution $\rho_c(r)$ can be obtained from electron elastic scattering. The method is so far restricted to stable nuclei and not easily applicable to radioactive nuclei (see Sect. 4.3.1.3), while several other methods aim at determining the matter radii, as discussed further in this section. Since a proton is not a point-like particle, the charge radius $\langle r^2 \rangle_c$ and the proton-density radius $\langle r^2 \rangle_Z$ are slightly different. If one considers the density distribution of a proton ρ_p, the charge and point-like proton densities are connected by

$$\rho_c(\mathbf{r}) = \int \rho_Z(\mathbf{r} - \mathbf{r}')\rho_p(\mathbf{r}')d\mathbf{r}'. \qquad (4.74)$$

Note that the two densities are very close to each other since the root mean square charge radius of the proton[7] is 0.84 fm.

Accessible information on the charge distributions for short lived nuclei is mostly reduced to the root mean square radius noted as $\langle r^2 \rangle_c$ and given by

$$Z\langle r^2 \rangle_c = \int r^2 \rho_c(\mathbf{r})d\mathbf{r}. \qquad (4.75)$$

This quantity is a many-body observable, sensitive to deformation and shell effects. It can be extended to the neutron radius $\langle r^2 \rangle_N$ and to the matter root mean square radius $\langle r^2 \rangle_m$, related together by

$$A\langle r^2 \rangle_m = \int r^2 \rho(r)d\mathbf{r} = N\langle r^2 \rangle_N + Z\langle r^2 \rangle_Z. \qquad (4.76)$$

4.3.1 Electron Elastic Scattering

Electron scattering provides one of the most objective probe for hadronic structure. This is because the electromagnetic forces are well known and relatively weak compared with the forces responsible for the structure. This feature allows one to obtain fairly precise theoretical descriptions of the scattering. Compared with weak-interaction probes (e.g., neutrinos), electrons have the advantage of a larger cross section. The results of electron scattering experiments are often compared to photon absorption experiments. Indeed for light nuclei where $\alpha Z = Z/137 \ll 1$, electron scattering can be described as the exchange of only one virtual photon.[8] In electron scattering, the photon is virtual, meaning that the transfer of momentum q can be varied independently of the transfer of energy ω. This makes the electron scattering a much more versatile tool than the equivalent photon absorption, where ω and q are rigidly coupled by $q^2 \equiv |q^2 - \omega^2| = 0$, where natural units ($c = 1$) are used.

4.3.1.1 Sensitivity

The first electron scattering on nuclei was performed in 1951 at the Illinois Betatron, where "high"-energy electron beams of 15.7 MeV were reached for the first time.

---

[7] Values for the protons root mean square charge radius from fine structure and electron elastic scattering measurements, reported in the literature, are inconsistent with each other. Two values of 0.84 fm and 0.87 fm were claimed. The most recent studies conclude for a root mean square charge radius of 0.84 fm for the proton.

[8] This approximation corresponds to the so-called *first order Born approximation*, detailed in Chap. 8.

Soon after, in 1953, the first results from Stanford, where electrons from 100 to 500 MeV were available, were published by Robert Hofstadter and collaborators. The results showed a clear deviation of the measured differential cross section with predictions assuming a point-like nucleus for electron scattering. These were the first nuclear charge radius measurements ever. Hofstadter got the Nobel prize from this series of measurements "for his pioneering study of electron scattering in atomic nuclei and the structure of the nucleons[9]".

Since then, in nuclear physics, electrons have been used for elastic and inelastic scattering, quasi-elastic scattering, photonuclear experiments and for the production of secondary beams of pions, muons and high-energy gamma rays. Soon after the results from Stanford, several electron machines were developed worldwide. Examples of locations where such facilities were developed for nuclear physics are Darmstadt and Mainz in Germany, Tohoku in Japan, Kharkov in Russia, Bates in the USA and Saclay in France.

How does electron scattering work exactly? When light from a point source passes through a small circular aperture, it does not produce a bright dot as an image, but rather a diffuse image, known as Airy's disc for a circular aperture, surrounded by regular, much fainter bright zones. When the light source can be approximated by a plane wave and the detection plane is located far away from the object, the phenomenon is called *Fraunhofer diffraction*, or far-field diffraction. In the case of a square aperture (or an opaque square), the diffraction minima are located at angles θ in both the horizontal and vertical directions from the center,

$$\sin \theta = \frac{m\lambda}{d}, \quad m = 1, 2, 3, \ldots \tag{4.77}$$

where λ is the wavelength of the considered light source and d is the dimension of the aperture (Fig. 4.18). The diffraction pattern is sensitive to the size and shape (a square in this example) of the diffractive object. Similarly, the pattern of angular distributions of an elastically scattered particle off a nucleus is sensitive to the size of the nucleus.

Electron scattering can be used to learn about charge distribution in nuclei. Electrons indeed constitute optimal probes for the study of atomic nuclei. Their point-like nature, and the fact that the electromagnetic interaction is weak (implying low re-scattering rates) and well understood (QED) make the reaction mechanism well under control. Their sensitivity to nuclear structure depends on their incident energy. The electron-probe resolution can be deduced from the de Broglie relation between the electron's momentum p and its wavelength

$$\lambda = \frac{\hbar}{p}. \tag{4.78}$$

---

[9] Indeed, similar deviations were also observed from the scattering of protons, showing in the same way that the proton is not a point-like charged particle. This was the starting point of hadronic physics. Another (captivating) story.

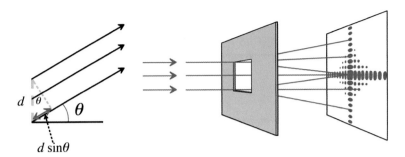

d θ

θ

$d \sin\theta$

Fig. 4.18 Illustration of a diffraction pattern from a light beam going through a rectangular aperture. The diffraction pattern reflects the size and shape of the aperture. The angles θ at diffraction minima are given by Eq. (4.77) at which all the contributions of waves from the square aperture cancel out

An electron of momentum $p = 200$ MeV/c will then have a "resolution" of about $\Delta r = \hbar c / p = 1$ fm, perfect for nuclear structure, while an electron of 2 GeV/c with wavelength of 0.1 fm will probe the internal structure of the nucleon.

4.3.1.2 Cross Section and Coulomb Form Factor

In the case of a spinless point-charge target, with no internal constituents nor spatial extension, the electron scattering cross section can be calculated exactly and is known as the Mott cross section σ_{Mott}

$$\sigma_{Mott} = \frac{Z^2 \alpha^2}{4 p^2 \beta^2} \frac{1 - \beta^2 \sin^2(\theta/2)}{\sin^4(\theta/2)}, \tag{4.79}$$

where Z is the charge of the target nucleus, θ is the scattering angle, α is the fine-structure constant, $p^2 = m_e^2 \gamma^2 \beta^2$ is the momentum of the incoming electron, and β is its velocity in light velocity units ($\beta = v/c$). The term $\beta^2 \sin^2(\theta/2)$ in the Mott cross section (4.79) comes from the interaction of the electron's magnetic moment with the magnetic field of the target. This term manifests a characteristic feature of spin 1/2 particles, i.e., it disappears as $\beta \to 0$, but becomes as important as the ordinary electric interaction as $\beta \to 1$ since the magnetic and electric fields are of equal strength in the relativistic limit. In the case of the non relativistic limit ($\beta \to 0$), Eq. (4.79) can be reduced to the well-known formula of the so-called Rutherford cross section.

The differential cross section on a realistic nucleus will be modified by its internal structure and spatial extension through its *form factors* which contain all relevant structure information. For electron scattering on a nucleus in the case of $q \gg \omega$, the differential cross section as a function of the transferred momentum q reads

$$\frac{d\sigma}{d\Omega}(q) = \sigma_{Mott}(q) \left[|F_L(q)|^2 + \left(\frac{1}{2} + \tan^2 \left(\frac{\theta}{2} \right) \right) |F_T(q)|^2 \right], \tag{4.80}$$

where $q = |\mathbf{p}_f - \mathbf{p}_i|$ is the momentum of the exchanged virtual photon, $\mathbf{p}_{i(f)}$ the momentum of the incoming (outgoing) electron, and the F_L and F_T are longitudinal and transverse form factors, respectively. The longitudinal and transverse form factors are associated with the fields along and perpendicular to the momentum-transfer direction, respectively. Both magnetic and electric currents contribute to $|F_T(q)|^2$, whereas $|F_L(q)|^2$ is purely due to the electric Coulomb field. It is seen that the transverse part of the cross section tends to become more important at backward angles. For a given scattering angle θ, q can be calculated from the incident (exit) electron energy E (E') as $q^2 = 4EE' \sin^2(\theta/2)$, neglecting the electron mass relative to the kinetic energy, which is quite a reasonable approximation for a high-energy electron with $E > 100$ MeV.

Let us consider the elastic scattering off spin-zero nuclei. In this case, the formula (4.80) simplifies considerably since only the $J=0$, or the monopole moment of the charge density in the $|F_L|$ can contribute by angular momentum considerations. The Coulomb field of the nucleus reads

$$\phi(r) = e \int \frac{\rho(r')}{|r - r'|} dr', \tag{4.81}$$

where ρ is the charge density of the nucleus normalized to be $\int \rho(r')dr' = Z$. The Coulomb field of an electron at position r is then given by

$$V_C(r) = -e\phi(r) = -e^2 \int \frac{\rho(r')}{|r - r'|} dr'. \tag{4.82}$$

The scattering amplitude in the first-order Born approximation is given for spinless particle as (see section 8.2.5 in Chap. 8)

$$f(\mathbf{p}_f, \mathbf{p}_i) = -\frac{m}{2\pi\hbar^2} \int e^{iqr} V_C(r) dr. \tag{4.83}$$

Inserting an identity, $e^{iqr} = -\frac{1}{q^2}\nabla^2 e^{iqr}$ in Eq. (4.83), the amplitude $f(q = \mathbf{p}_f - \mathbf{p}_i)$ can be expressed in terms of the charge density $\rho(r)$ by integrating by parts twice and making use of the Poisson equation for the Coulomb field $\phi(r)$

$$f(q) = \frac{2me^2}{\hbar^2 q^2} \int e^{iqr} \rho(r) dr = \frac{2mc^2\alpha}{\hbar c q^2} \int e^{iqr} \rho(r) dr. \tag{4.84}$$

Then the differential cross section for spin zero particle is given by a product of the Rutherford cross section and the Coulomb form factor,

$$\frac{d\sigma}{d\Omega} = |f(q)|^2 = \left(\frac{d\sigma}{d\Omega}\right)_R |F_C(q)|^2, \tag{4.85}$$

where the Rutherford cross section and the Coulomb form factor are defined, respectively, as

$$\left(\frac{d\sigma}{d\Omega}\right)_R = \frac{4(mc^2)^2 Z^2 \alpha^2}{(\hbar c)^2 q^4}, \tag{4.86}$$

and

$$F_C(q) = \frac{1}{Z} \int e^{iqr} \rho(r) dr = \frac{4\pi}{Z} \int_0^\infty \rho(r) \frac{\sin(qr)}{qr} r^2 dr. \tag{4.87}$$

For elastic scattering with high-energy electrons with $E = E'$ and $q^2 = 4E^2 \sin^2(\theta/2)$, the cross section (4.85) reads

$$\frac{d\sigma}{d\Omega} = |f(q)|^2 = \left(\frac{mc^2 Z\alpha\hbar c}{2E^2}\right)^2 \frac{1}{\sin^4(\theta/2)} |F_C(q)|^2. \tag{4.88}$$

The Rutherford cross section (4.86) is derived under the following assumptions.

- The Born approximation is valid.
- The projectile and target particles have spin 0 and no structure; they are assumed to be point-like particles.
- The target is very heavy and does not take up energy (no recoil effect).

For the scattering of spin 1/2 charged particles, the Rutherford cross section in Eq. (4.88) is replaced by the Mott cross section (4.79), getting an extra factor $1 - \beta^2 \sin(\theta/2)^2$. Then Eq. (4.88) becomes the product of Mott cross section (4.79) and the Coulomb form factor (4.87) as

$$\frac{d\sigma}{d\Omega} = \left(\frac{d\sigma}{d\Omega}\right)_{\text{Mott}} |F_C(q)|^2. \tag{4.89}$$

Problem

4.2 Derive Eq. (4.84) by using the Poisson relation for the Coulomb field,

$$\Delta\phi(r) = -4\pi\rho(r)$$

and the partial integration twice.

We can have a direct access to the charge density from its Fourier transform, i.e., the form factor (4.87), through the elastic cross sections at various momentum transfer q. Cross sections at low momentum transfer give information about the gross features of the charge distribution while, at high momentum transfer, they disclose details about the full distribution, including the interior of the nucleus. From cross sections at low transferred momenta (in the order of 1 fm^{-1}) the charge radius can be deduced. The state-of-the-art of electron scattering is illustrated in Fig. 4.19 where the elastic scattering of electrons off ^{208}Pb is shown. The elastic channel is selected by the missing mass technique (see Sect. 4.5.1). Required luminosities are in the order of 10^{26} cm^{-2}s^{-1} for light nuclei and lower for heavier nuclei. Charge densities are typically obtained by fitting the scattering data with cross sections based on a parameterized charge form factor. The higher the momentum, the higher the number of parameters that can be constrained and thus the details on the density profile that can be obtained.

Problem

4.3 Considering the form factor of Eq. (4.87) and assuming a constant charge distribution

$$\rho(r) = \begin{cases} \frac{3Ze}{4\pi R^3} & r < R_c \\ 0 & r > R_c, \end{cases} \tag{4.90}$$

(i) determine the analytical form differential cross section as function of the transferred momentum for a given scattering angle θ. (ii) Determine at which q, as a function of R_C, E, θ, is the first minimum of the calculated cross section. (iii) Extract from Fig. 4.19 the charge radius of ^{208}Pb and compare to the experimental value of $R_c = 5.51$ fm.

4.3.1.3 Electron Scattering from Radioactive Nuclei

Electron elastic-scattering measurements exist for many isotopes but remain currently confined to the valley of stability. Although electrons have been shown to be one of the most precise tools to examine nuclear structure, it is extremely difficult to perform collisions of electrons with unstable nuclei. The proper energy of electrons in the center of mass should be one of the order of several hundreds of MeV, implying a high-energy electron machine at the same location as a radioactive-ion beam (RIB) facility. Furthermore, electron-ion cross sections are small. Therefore, a large luminosity is required, which is not easy for a collider: the beam size at the collision vertex should be less than a millimeter and intensities should be pushed at the maximum authorized by the space-charge limit beyond which the charge density and the Coulomb repulsion among charged particles are too high for stable beam-optics conditions. The very first facility, called SCRIT (Self-Confining Radioactive Ion Target), dedicated to collisions of unstable ions and electrons has been commissioned recently in RIKEN, Japan. The concept of SCRIT is not a collider properly speaking: the unstable ions are trapped and at rest in the electron ring as illustrated in Fig. 4.20. Unstable ions produced from the fission of Uranium are transmitted at very low energy inside the electron beam and trapped by the electron beam itself which, by its negative charge, creates an attractive radial potential for the positive ions. A set

Fig. 4.19 Elastic scattering cross sections of electrons at $E = 502$ MeV as a function of momentum transfer from ^{208}Pb. In the lower panel, the calculated HF densities by Gogny and Negele interactions are shown together with empirical charge density distribution obtained from the electron scattering experiment. Figure reprinted with permission from [7]. ©2021 by the American Physical Society

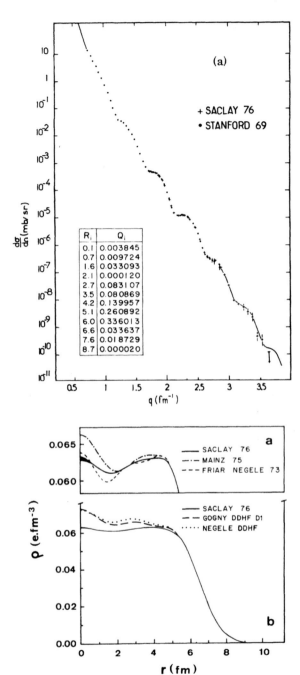

of electrodes confine the positive charges longitudinally. While the state-of-the-art stable beam experiments could reach luminosities up to $\sim 10^{33}$ cm^{-2}s^{-1}, the SCRIT facility should reach luminosities up to 10^{27} cm^{-2}s^{-1} and be mainly dedicated to elastic scattering of medium-mass fission fragments. The luminosity will allow to determine the charge radius of nuclei, as well as the diffusiveness of the charge density but will be insufficient to explore further the charge distribution of short lived nuclei. The first "real" collider for electrons with exotic nuclei has been proposed at FAIR, in Europe. The facility, called ELISe, aims at luminosities up to 10^{28}cm^{-2}s^{-1} with a very large detection efficiency for the Lorentz focusing of the heavy ions and an in-beam magnetic spectrometer for high-resolution spectroscopy from electron detection. ELISe is not expected to be realised during the coming decade(s).

4.3.2 Laser Spectroscopy

4.3.2.1 The Atomic Fine Structure

The atomic *fine structure* is caused by the interaction of the electron spin with atomic fields at the electron position. This interaction leads to shifts in an electron's atomic energy levels.

The interaction between the magnetic field induced by the electron orbital and the nuclear spin gives rise to the hyperfine level splitting. Sometimes, relativistic effects are also called hyperfine structure. Reporting here the definition of Gordon Drake given in his handbook of atomic, molecular and optics physics [10], the "hyperfine structure in atomic and molecular spectra is a result of the interaction between electronic degrees of freedom and nuclear properties other than the dominant one, the central nuclear Coulomb field." In other words, everything going beyond the Dirac expectation for a infinite heavy, point-like spin-zero nucleus is called hyperfine structure.

In this section related to charge radii, only the fine-structure splitting is relevant while the hyperfine splitting is sensitive to other nuclear properties, such as the nuclear electric quadrupole and magnetic dipole moments described in Sect. 4.4.3. Still, both splittings contribute to the so-called hyperfine structure.

The hyperfine structure can be measured from laser spectroscopy: the "sample" is irradiated with a laser of a given frequency ν_1. If this frequency matches a transition ν_0 in an atom of the sample, a resonant absorption may occur. The populated excited state can either decay (typical lifetime of $\tau = 10$ ns) or be the object of a subsequent excitation. Note that the natural width of a state is given by the Heisenberg's principle

$$\Delta\nu \sim \frac{1}{2\pi\tau} \sim 16 \, \text{MHz}. \tag{4.91}$$

Typical resolutions reached by commercial lasers are today below 1 MHz, which is smaller than the intrinsic width of $\sim$16 MHz. The atomic hyperfine structure can

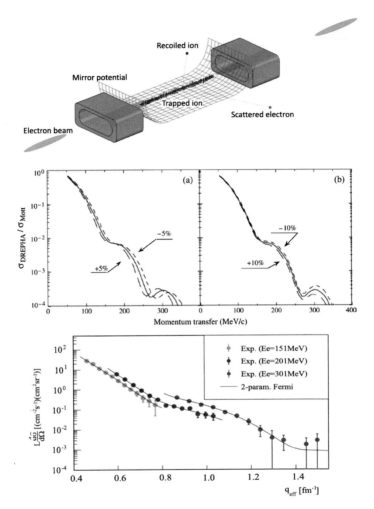

Fig. 4.20 (Top) View of the SCRIT concept. Electron bunches (pale pink) result in a transverse confinement of ions (blue dots). By placing electrodes (green) for longitudinal confinement, one can form a localized cloud of trapped ions. Courtesy W. Wakasugi, Kyoto University. (Middle) Result of a distorted-wave calculation of the elastic-scattering cross section for Sn. The ratio to the Mott cross section is shown as a function of momentum transfer. The parameters for the size and diffuseness, assuming a two-parameter Fermi distribution $\rho_c(r) = \rho_0/\{1 + \exp(4 \ln 3(r - c)/t)\}$ for the nuclear charge distribution, are changed for $\pm 5\%$ and $\pm 10\%$, respectively (dashed lines). Courtesy T. Suda, Tohoku University. (Bottom) Electron elastic scattering from ^{132}Xe measured with SCRIT. This measurement cannot be performed with standard setups because Xe is a noble gas and cannot be prepared as a solid target. The panel shows the elastic cross section multiplied by the luminosity as a function of the momentum transfer in inverse-Fermi units ($1\text{fm}^{-1} \equiv \hbar c = 197 \text{ MeV/c}$) for the incident energies of 151 MeV (green), 201 MeV (blue) and 301 MeV (purple). The elastic scattering was selected from the missing-mass excitation energy spectrum from the measurement of the scattered electrons. Assuming a two-parameter Fermi distribution, the shape parameters were determined as $c = 5.42^{+0.11}_{-0.08}$ fm and $t = 2.71^{+0.29}_{-0.38}$ fm, leading to a root-mean-square radius of $\langle r^2 \rangle^{1/2} = 4.79^{+0.12}_{-0.10}$ fm. The data are published in [9]. Figure reprinted from [11]

therefore be measured. Typical energies of principal transitions in atoms are of the order of 10^{15} Hz while the hyperfine splitting are of the order of 1 GHz, as indicated in Fig. 4.21.

In practice, there are two ways to carry out laser-spectroscopy studies. The first method consists in detecting the de-excitation photons following the excitation. A second technique relies on the ionization of the atom after the excitation of the state of interest. The latter method as the advantage to rely on ion detection (instead of photon detection) and therefore does not suffer from any background from the excitation laser. On the other hand, the ionization technique requires a high-intensity laser for a probable second-photon absorption before decay. The two techniques are illustrated in Fig. 4.22.

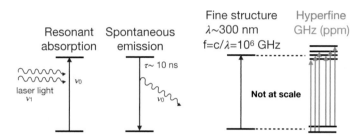

Fig. 4.21 (Left) Resonant absorption of photons is at the heart of the principle of laser spectroscopy: when the frequency of a laser matches the energy difference between two atomic levels, the resonant absorption of a photon might take place. (Right) The fine structure splitting of atoms is sensitive to the charge radius of the atomic nucleus. At a much finer resolution (GHz), the hyperfine splitting is sensitive to the magnetic dipole moment and the electric quadrupole moment of the nucleus, at lowest order

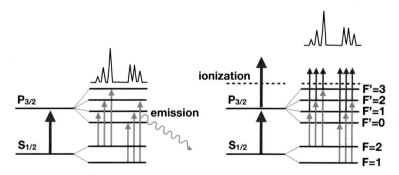

Fig. 4.22 Two experimental methods for laser spectroscopy: (left) detection of de-excitation photons following the excitation, (right) detection of ions after a laser ionization (red) following the excitation (green) of a hyperfine-structure state

4.3.2.2 Isotopic Shifts

The origin of the dependence of the hyperfine structure to the charge radius of the nucleus is that the electronic wave functions have a non zero probability of presence at the origin, i.e., inside the nuclei. The finite size of the nucleus leads to a shift in energy of electronic transitions compared to expectations from a point-like nucleus. The absolute values of transition energies are usually not used to study charge radii. Instead the difference of a transition energy from an isotope to another, called *isotopic shift* is a key observable that allows to determine precise relative charge radii from a reference of a stable nucleus whose charge radius is known from another technique such as electron elastic scattering. Considering a transition of energy ν^A and $\nu^{A'}$ for two isotopes A and A', the isotopic shift reads

$$\delta\nu^{AA'} = \nu^A - \nu^{A'}. \tag{4.92}$$

The isotopic shift can be decomposed into two components

$$\delta\nu^{AA'} = \delta\nu_{MS}^{AA'} + \delta\nu_{FS}^{AA'}. \tag{4.93}$$

The *field shift* (FS), or nuclear volume shift, comes from the change in the charge radius from isotope A to isotopes A'. Indeed, the strong attraction between protons and neutrons induces an increase of the charge radius when neutrons are added to a nucleus. This component of the field shift is the one of interest to determine charge radii. In addition to it, the *finite nuclear mass shift* (MS) also contributes to the isotopic shift. It originates in the change of mass of the nucleus from isotope A to isotope A'.

4.3.2.3 The Field Shift

Electrons with a finite probability $|\Psi(r=0)|^2$ inside the (finite size) nucleus are less bound than the estimate for a point-like nucleus, where Ψ is the electron wave function. Indeed, in the case of a point-like nucleus, the attractive potential of the nucleus on electrons evolves as Ze^2/r, while a finite nuclear size makes it shallower at small distances.

The Coulomb energy shift E_{FNS} induced by the finite nuclear size (FNS) of the nucleus is given by the difference between the Coulomb energy caused by a finite size nucleus of charge density ρ_c and a point-like nucleus of total electric charge Ze

$$E_{FNS} = -e^2 \int d\mathbf{r} \int d\mathbf{r}' \frac{\rho_c(\mathbf{r})\rho_e(\mathbf{r}')}{|\mathbf{r}-\mathbf{r}'|} - \left(-Ze^2 \int d\mathbf{r} \int d\mathbf{r}' \frac{\delta(\mathbf{r})\rho_e(\mathbf{r}')}{|\mathbf{r}-\mathbf{r}'|}\right). \tag{4.94}$$

If one considers a sphere of radius R inside which the density of the electron $\rho_e(\mathbf{r}') = |\Psi_e(\mathbf{r}')|^2 \sim |\Psi_e(0)|^2$ and for R larger than the nuclear radius, all contributions for $r \geq R$ will diminish and one gets

$$E_{FNS} = -e^2|\Psi_e(0)|^2 \int_0^R d\mathbf{r} \int_0^R d\mathbf{r}' \frac{\rho_c(\mathbf{r})}{|\mathbf{r}'-\mathbf{r}|} + Ze^2|\Psi_e(0)|^2 \int_0^R d\mathbf{r} \int_0^R d\mathbf{r}' \frac{\delta(\mathbf{r})}{|\mathbf{r}-\mathbf{r}'|}.$$
(4.95)

At leading order, the integrals can be approximated in the following way

$$\int_0^R d\mathbf{r}' \frac{1}{|\mathbf{r}-\mathbf{r}'|} \sim 4\pi \int_0^r r'^2 dr' \frac{1}{r} + 4\pi \int_r^R r'^2 dr' \frac{1}{r'}.$$
(4.96)

Equation (4.95) is then simplified to be

$$E_{FNS} = -e^2|\Psi_e(0)|^2 \int_0^R d\mathbf{r}\rho_c(\mathbf{r}) \left[4\pi\frac{r^2}{3} + 4\pi(\frac{R^2}{2} - \frac{r^2}{2})\right] + Ze^2|\Psi_e(0)|^2 4\pi\frac{R^2}{2}.$$
(4.97)

The terms in R^2 cancel and the result does not depend on the arbitrary choice of R. By using the definition of the mean square charge radius

$$\langle r_c^2 \rangle = \int_0^\infty r^2 \rho_c(\mathbf{r}) d\mathbf{r},$$
(4.98)

one then gets for the Coulomb energy shift

$$E_{FNS} = \frac{4\pi Ze^2}{6} \langle r_c^2 \rangle |\Psi_e(0)|^2.$$
(4.99)

Remember that the above formulas adopt the standard units, where $4\pi\epsilon_0 \equiv 1$. The energy shift of the electron orbits caused by the finite size of the nuclear charge distribution is positive: in the interior of the nucleus, the electron is less attracted than for a point-like nucleus. The absolute E_{FNS} is accessible experimentally only for hydrogen-like while work on helium-like atoms is ongoing. Otherwise, the many-body system composed of the nucleus and several electrons is too difficult to compute and reach the exact value for $|\Psi_e(0)|^2$. This is the reason why relative charge radii from isotopic shifts are preferred. In the case of two isotopes, the field shift for a given transition from a state i of electron wave function $\Psi_e^i(r)$ to a state f with a wave function $\Psi_e^f(r)$ is

$$\delta\nu_{FS}^{AA'} = \frac{2\pi Ze^2}{3}(|\Psi^f(0)|^2 - |\Psi^i(0)|^2)(\langle r_c^2 \rangle^{A'} - \langle r_c^2 \rangle^A)$$

$$= \frac{2\pi Ze^2}{3}\Delta|\Psi(0)|^2\delta\langle r_c^2 \rangle^{AA'}$$

$$\equiv F\delta\langle r_c^2 \rangle^{AA'},$$
(4.100)

where $\Delta |\Psi(0)|^2 = |\Psi^f(0)|^2 - |\Psi^i(0)|^2$ is the difference of the presence probability of the electron at the location of the nucleus for the two levels defining the considered electronic transition and F is called the field shift constant. It does not vary from one isotope to another at first order. We see clearly from Eq. (4.100) that the field shift component of the isotopic shift depends linearly on the charge radius difference from an isotope to another.

4.3.2.4   The Mass Shift

The mass shift contribution to the isotopic shift comes from the nucleus motion in the center of mass (c.m.) frame of the atom. By definition of the center of mass, the motion of the nucleus in the c.m. frame is given by its momentum P_A, with A being the number of nucleons composing the nucleus

$$\mathbf{P}_A = - \sum_{\text{electrons } i} \mathbf{p}_i, \tag{4.101}$$

where $\mathbf{p}_i$ are the momenta of the individual electrons i. Therefore its kinetic energy in the center-of-mass frame can be expressed as a function of the electrons' momenta

$$E_{kin} = \frac{P_A^2}{2M_A} = \frac{1}{2M_A} \left(\sum_i p_i^2 + \sum_{ij, i \neq j} \mathbf{p}_i \cdot \mathbf{p}_j \right). \tag{4.102}$$

The above Eq. (4.102) shows that the addition of neutrons to a nucleus of an atom leads to a change of nuclear motion and therefore implies an effect on the electronic levels and transition energies as a recoil effect. In the case of two isotopes with A and $A' = A + 1$ nucleons

$$\delta\nu_{MS}^{AA'} \propto \frac{1}{M_A} - \frac{1}{M_{A'}} = \frac{M_{A'} - M_A}{M_A M_{A'}} \sim \frac{A - (A + 1)}{A(A + 1)} \sim -A^{-2}. \tag{4.103}$$

From Eq. (4.102), one sees that there are two origins for the mass shift component:

- An effect on single-particle orbits, i.e., the $\sum p_i^2$ term of Eq. (4.102). It is called the *Normal Mass Shift* (NMS). At first order, its amplitude is proportional to the energy of the transition.
- An effect on the change of correlations among the electrons, i.e., the $\sum \mathbf{p}_i \cdot \mathbf{p}_j$ term of Eq. (4.102). It is called the *Specific Mass Shift* (SMS). Today, the SMS cannot be computed for systems with more than three electrons.

The mass shift cannot be computed with high accuracy for systems with more than 5 electrons. For heavier atoms, the calculations can be typically performed at a 10–20% level. The mass shift can be seen as an unwanted contribution to the isotopic shift since it does not contain any information on the charge radius. Its dependence with

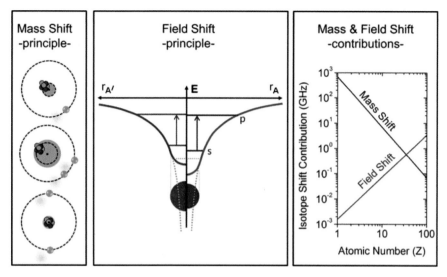

Fig. 4.23 Illustration of the principle of the mass and field shifts. (Left) the *mass shift* is a nuisance to charge radii measurements: it contributes to the isotopic shift but does not depend on the charge radius but on the mass of the nuclei considered, at first order. The moment distribution of electrons in the center of mass of the atom implies, by definition, a momentum of the nucleus in the same mass frame. The mass shift depends on correlations between the electrons. (Middle) The *field shift* is what gives to isotopic shifts their dependence on the charge radius. The larger is the spatial distributions of the protons in the nucleus, the less attractive will be the potential for the electrons. A change of charge radius from one isotope to another will impact the energy of fine-structure transitions. (Right) The impact of the mass and field shifts on the isotopic shift have an opposite dependence with mass. Relative charge radii are easier to obtain from heavy nuclei than light nuclei. Reprinted with permission from [12]. ©2021 by Elsevier

the nuclear mass indicated that it will be larger for light nuclei, for which a relative charge radius will then be more difficult than for heavier masses, as illustrated on the r.h.s. panel of Fig. 4.23.

4.3.2.5   General Form

The above description of the terms of the isotopic shift leads to its most general form

$$\delta\nu^{AA'} = (K_{NMS} + K_{SMS}) \times \frac{M_{A'} - M_A}{M_{A'}M_A} + F \times \delta\langle r_c^2\rangle^{AA'}, \tag{4.104}$$

where K_{NMS} and K_{SMS} are coefficients, at first order independent on the isotope, corresponding to the Normal Mass Shift and Specific Mass Shift, respectively. One needs to measure the same fine-structure transition of at least three isotopes to extract empirically the two unknown constants F and $(K_{NMS} + K_{SMS})$. Among the considered isotopes, at least one should have one known r.m.s charge radius $\langle r_c^2\rangle$. The charge radius of other isotopes can then be obtained.

4.3.3 Reaction Cross Sections

4.3.3.1 Definitions

The essential quantity in nuclear reactions is the cross section which quantifies the probability of a specific reaction to occur. We refer the reader to Chap. 8 for an introduction to nuclear reactions. Assuming a beam of particles of current J propagating along the z axis and impinging on a target of uniform density of scattering centers ρ, the beam current is reduced by reactions. The reduction of current over an infinitesimally small length dz of target material of atomic density ρ is driven by the first-order differential equation

$$\frac{dJ}{dz} = -\sigma_I J(z)\rho, \qquad (4.105)$$

where σ_I is the interaction cross section. The solution is given by

$$J(z) = J(0)e^{-\sigma_I \rho z}. \qquad (4.106)$$

The nuclear interaction is strong and of short range: the interaction of a beam particle and a target nucleus is driven by the nucleus-nucleus potential strongly dependent on the matter density profiles of the colliding particles. This feature is used to estimate the matter radius of nuclei from interaction cross sections.

The interaction cross section σ_I gathers all reaction events for which the final nucleus has a change of nucleons compared to the projectile. It is related to the reaction cross section σ_R as follows

$$\sigma_R = \sigma_I + \sigma_{inelastic}, \qquad (4.107)$$

where $\sigma_{inelastic}$ is the inelastic cross section which leads to an excitation of the projectile, the target nucleus or both but leaving the projectile with the same proton and neutron number, i.e., any excitation of the projectile which is below the nucleon separation energy.[10] If $\sigma_{inelastic}$ is small enough, one can assume that $\sigma_I \sim \sigma_R$, depending on the targeted accuracy. As an example, at the relativistic energy of 950 MeV/nucleon, the interaction cross section for ^{12}C with ^{34}Al is measured to be $\sigma_I = 1334(28)$ mb, while the inelastic cross section is measured to be $\sigma_{inelastic} = 10$ mb [13].

4.3.3.2 The Interaction Radius

One can define an *interaction radius* R_I of a nucleus B with a target A, following the black disk approximation

---

[10] Note that the terms interaction, reaction and inelastic cross sections can be encountered in the literature with different definition, which sometimes lead to confusions, unfortunately.

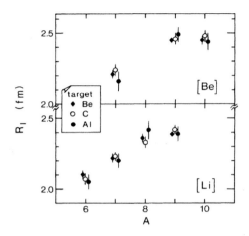

Fig. 4.24 Interaction radii of Be (top panel) and Li (bottom panel) nuclei obtained from interaction cross sections with different targets. Figure reprinted with permission from [14]. ©2021 by the American Physical Society

$$\sigma_I(A, B) = \pi[R_I(A) + R_I(B)]^2 \tag{4.108}$$

The target interaction radius $R_I(A)$ can be extracted from a symmetric reaction

$$\sigma_I(A, A) = 4\pi[R_I(A)]^2, \tag{4.109}$$

leading to the interaction radius of the projectile

$$R_I(B) = \sqrt{\frac{\sigma_I(A, B)}{\pi}} - \sqrt{\frac{\sigma_I(A, A)}{4\pi}}. \tag{4.110}$$

An example of interaction radii extracted from heavy-ion collisions at relativistic energies of several 100 MeV/nucleon is shown in Fig. 4.24.

Problem

4.4 Using the transmission method, interaction cross sections at 790 MeV/nucleon were measured by all the known He isotopes (^{3}He, ^{4}He, ^{6}He, and ^{8}He) on Be, C, and Al targets [15]. Based on the measured interaction cross sections summarized in Table 4.2, calculate the difference of radii between ^{4}He and ^{3}He

$$R_I\left(^4\text{He}\right) - R_I\left(^3\text{He}\right) \tag{4.111}$$

as a function of A, fit them to a power law A^α, and compare it to the $A^{1/3}$ dependence.

4.3.3.3 Microscopic Approaches

The above approach can be refined by using a microscopic reaction model in which microscopic density profiles can be used as inputs. Reaction cross sections at relativistic energies can be predicted employing the eikonal approximation which assumes that the projectile is not deflected before the reaction vertex, neither the reaction products after the reaction. Details on the eikonal approximation are given in Chap. 8. The reaction cross section can be expressed as a summation of the interaction probability of the projectile over the impact parameter b

$$\sigma_R = 2\pi \int_0^{+\infty} [1 - T(b)]b\,db, \tag{4.112}$$

where $T(b)$ is the transmission function at impact parameter b. $T(b)$ expresses the probability for projectiles impinging at impact parameter b not to interact with the target. $T(b)$ is the square modulus of the projectile-target S matrix (see Chap. 8). It depends on both the structure of both the projectile and the target nuclei, and the reaction mechanism. We explicit here one expression for $T(b)$ in the static density limit, assuming that the reaction time is much faster than the rearrangement of the nucleons

$$T(b) = \exp[-\overline{\sigma} \int \rho_t^{(z)}(b + \mathbf{x})\rho_p^{(z)}(\mathbf{x})\,d^2\mathbf{x}], \tag{4.113}$$

where $\overline{\sigma}$ is the effective NN cross section, taking into account the isospin asymmetry of the target and projectile. The integral is a convolution of the projectile and target densities over the directions perpendicular to the beam axis. The target and projectile matter densities $\rho_{t,p}^{(z)}$ are integrated over the beam propagation axis z following

$$\rho_{t,p}^{(z)}(b) = \int_{-\infty}^{+\infty} \rho_{t,p}(\sqrt{b^2 + z^2})\,dz.$$

Equation (4.113) shows the sensitivity of the reaction cross section to the projectile density ρ_p. For a measured reaction cross section, the extracted matter radii $\langle r^2 \rangle_m$ would be taken as those of microscopic density profiles that reproduce the experimental reaction cross section within the exposed formalism.

It is important to underline that such an approach to determine matter radii from σ_R is considered as very model dependent and to be taken with caution. The systematic uncertainties in such analysis are difficult to determine. We give here a qualitative

Table 4.2 Measured interaction cross sections σ_I of He isotopes using different target nuclei. Cross sections are given in mb

Target	^{3}He	^{4}He	^{6}He	^{8}He
Be	498 ± 4	485 ± 4	672 ± 7	757 ± 4
C	550 ± 5	503 ± 5	722 ± 6	817 ± 6
Al	850 ± 9	780 ± 13	1063 ± 8	1197 ± 9

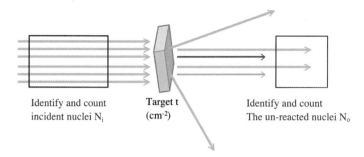

Fig. 4.25 Transmission method for interaction cross section measurement: by detecting and counting (N_0) the unreacted beam particles going through the target, as well as counting the incident beam particle (N_i), the interaction cross section can be obtained from Eq. (4.114)

example to illustrate difficulties: the inelastic collective excitation of a loosely-bound nucleus to its continuum might be a non-negligible part of the reaction cross section. If the inclusive reaction cross section is compared to prediction to a model that does not take into account these collective excitation but only nucleon removal from individual nucleon-nucleon collisions, the theory will underestimate the reaction cross section. The interpretation of the experimental cross section in such formalism might then lead to an overestimation of the matter radius, especially for loosely-bound nuclei.

One technique to measure interaction cross sections is the so called *transmission method* and it consists in measuring the unreacted beam particles after transmission through a target of thickness t and scattering center density ρ. The method is illustrated in Fig. 4.25. By integrating the relation of Eq. (4.106) over the measurement time and for the total target length L_t the target length, one gets

$$\sigma_I = \frac{1}{L_t \rho} \ln \left(\frac{N_i}{N_0} \right), \tag{4.114}$$

with N_0 the number of unreacted particles and N_i the initial number of beam particle impinging on the target. As seen before, the number of unreacted particles is the number of beam particles measured downstream the reaction target and corrected from the inelastic scattering to bound excited states. The inelastic cross section to bound excited states can either be obtained from another measurement, or approximated to a theoretical estimation.

4.3.4 Proton Elastic Scattering

4.3.4.1 Diffraction Pattern

Similarly to the diffraction of light in an aperture and electron scattering, the angular distribution obtained from proton elastic scattering is also sensitive to the nuclear size. Unlike electron scattering, proton elastic scattering offers the advantage to be applicable to short-lived nuclei in inverse kinematics, i.e., the nucleus to study is sent as a beam onto a proton target. Among all hadronic probes, proton elastic scattering might also be preferred because the proton can be considered with no internal structure at the relevant incident energies, reducing the uncertainties in the reaction mechanism. Still, the treatment of the proton-nucleus reaction mechanism requires approximations which are the main source of uncertainties in extracting matter radii. We refer the reader to Chap. 8 where nuclear reactions and optical potentials are introduced. In the following, we provide the essential information relevant for the determination of matter radii.

The transition amplitude $T_{\beta\alpha}$ for elastic scattering is given in its prior form by

$$T_{\beta\alpha} = \langle \phi_\beta | V | \Psi_\alpha \rangle, \tag{4.115}$$

where the entrance and exit channels α and β are composed of the same nuclei A and a. The differential elastic scattering cross section is directly related to the transition amplitude by

$$\frac{d\sigma}{d\Omega}(\theta) = \frac{m_\alpha^2}{(2\pi\hbar^2)^2} \frac{k_\beta}{k_\alpha} |T_{\beta\alpha}|^2, \tag{4.116}$$

where m_α is the reduced mass of the system composed of the two scattering nuclei, $k_{\alpha,\beta}$ the momenta of the relative motion in the entrance and exit channels. If one makes the further approximation that the solution of the Schrödinger equation is not distorted by the optical potential, i.e., that the relative motion reduces to the solution of the homogeneous equation, i.e., $\Psi_\alpha = e^{i\mathbf{k}_\alpha.\mathbf{r}_\alpha} \Phi_\alpha$ where Φ_α is the intrinsic wave function of the nucleus, one gets

$$T_{\beta\alpha} = \langle \phi_\beta | V | \phi_\alpha \rangle = \int e^{i\frac{(\mathbf{k}_\beta - \mathbf{k}_\alpha).\mathbf{r}_\alpha}{\hbar}} V(\mathbf{r}_\alpha) d\mathbf{r}_\alpha. \tag{4.117}$$

The transition amplitude for the elastic scattering is the exact Fourier transform of the optical potential. Since the nuclear interaction is short range, one expects that the cross section reflects the size and shape of the nuclear matter density, in a similar way that the electron scattering is sensitive to the charge spatial distribution.

This can be seen in Fig. 4.26 where the differential cross sections for Ca, Sn and Pb isotopes are plotted. The heavier the nucleus (i.e., the larger radius it has), the smaller is the angle between two diffraction minima. This correlation can be understood in classical physics considering that elastic scattering is a process confined to the surface because at the interior of the nucleus the potential is highly absorptive, as would be the scattering of light by a black disk. In that analogy, the angular distance between two diffractive minima is given by

$$\Delta\theta = \frac{\hbar}{pR}, \tag{4.118}$$

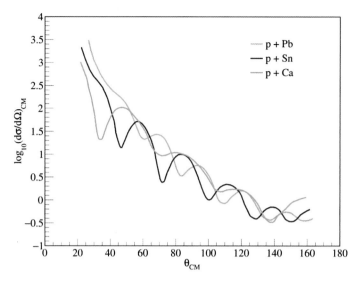

Fig. 4.26 Calculated cross sections of proton elastic scattering from Ca, Sn and Pb isotopes at an incident energy of 30 MeV. The angular distance between two diffraction minima is correlated to the nuclear radius of the target. Values are taken from [16]

where R is the radius of the target nucleus and p is the momentum of the incoming proton.

Elastic scattering is then an effective tool to extract the matter radius of a nucleus. When the statistics of the measurement is good enough (achieved in experiments of few days with a beam intensity of about 10^4 pps), it is shown that uncertainties down to 0.1 fm can be reached, mostly due to the uncertainties in the interaction potential and in the reaction mechanism itself. To limit the effect of the latter couplings, intermediate energies above 100 MeV/nucleon are believed to be most reliable to extract nuclear radii from elastic scattering. Such elastic scattering data, when combined to measurements of the charge distribution of the nucleus, can be used to determine the neutron distribution and, at the same time, the neutron skin thickness in neutron rich nuclei. The latter is an information to understand the nuclear symmetry energy (see Chap. 9).

In inverse kinematics, where the heavy ion is impinging on a proton target, a proper angular resolution can be obtained by measuring the recoil proton. The sensitivity to the matter radius is used at incident energies from few tens of MeV/nucleon to several hundreds of MeV/nucleon. The angular distribution as illustrated in Fig. 4.26 is employed as a sensitive observable of the matter radius. Equivalently, the differential cross section as a function of momentum transfer is also used. The angular distribution should be measured with precision to at least the first diffraction minimum to be sensitive to the root mean square radius of the nucleus.

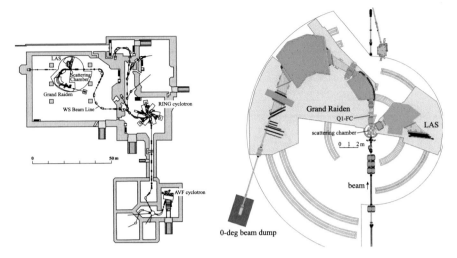

Fig. 4.27 (Left) Footprint of the RCNP facility in Osaka, Japan. Heavy ions are accelerated by two coupled acceleration stages: the AVF cyclotron and the RING cyclotron. The principle of cyclotrons is detailed in Chap. 6. At RCNP, protons can be accelerated up to 295 MeV. (Right) Bird view of the Gran Raiden spectrometer. The proton beam arrives from the bottom of the figure and hit the reaction target. Scattered protons which are transmitted through the spectrometer are bent by the Gran Raiden set of two dipoles. The magnetic field of the two dipoles are set such that the scattered protons of interest reach the detectors in the focal plane. Compared to elastically scattered protons, inelastically scattered protons have less momentum after the interaction with a nucleus from the target and are therefore more bent and detected at a smaller radius, also referred as the *magnetic rigidity* $B\rho$, where B is the magnetic field and ρ is the curvature radius of the proton trajectory in this field. The spectrometer is mounted on circular rails and can be rotated around the target. Specific scattering angles can be therefore be selected. In the show picture, protons emitted at and close to zero degree are selected. In this particular configuration, the direct high-intensity beam is also transmitted through the dipole. It is magnetically separated from scattering events. Beam particles are collected in a *beam dump*. Courtesy A. Tami, RCNP

4.3.4.2 Proton Elastic Scattering in Direct Kinematics at RCNP

We present in the following a state of the art setup for elastic scattering measurements in direct kinematics (a proton beam impinging on a heavy-ion target) at RCNP, Japan. The experimental setup is composed of a magnetic spectrometer located downstream the target. The aim of the spectrometer is to measure the scattered protons after elastic scattering and determine the scattering angle and momentum with enough resolutions to (i) select events which correspond to elastic scattering and separate them from inelastic reactions, and (ii) to separate the scattered protons from the high intensity beam protons which traveled through the target without nuclear interaction.

A magnetic spectrometer is mainly composed of a (set of) magnetic dipole(s), generally with a field of ~ 1 Tesla, which will bend the protons proportionally to their momentum. Downstream of the dipole magnet, position sensitive detectors such as drift chambers are place to determine the position of the transmitted protons

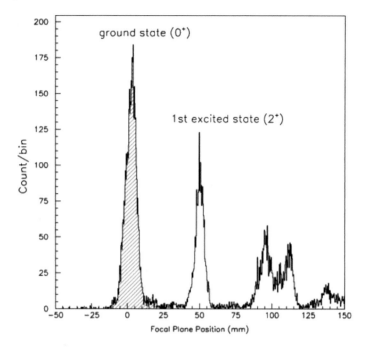

Fig. 4.28 Excitation energy spectrum obtained from missing-mass reconstruction after the interaction of a 295 MeV proton beam with a ^{120}Sn target at RCNP with the Gran Raiden magnetic spectrometer. The spectrum is measured at a laboratory angle of 35°, i.e., the Gran Raiden spectrometer was rotated by 35° relative to the incident beam direction for this measurement. Note that the spectrometer, thanks to the narrow angular acceptance and the large magnetic deviation, leads to an excitation spectrum with almost no background. Figure reprinted with permission from [17]. ©2021 from the American Physical Society

and their angles relative to the central trajectory of the direct beam. The Gran Raiden magnetic spectrometer of RCNP is shown in Fig. 4.27.

The Gran Raiden spectrometer is designed to achieve an excellent energy resolution of ~15 keV (σ)[11] for the excitation energy reconstructed by missing mass (see Sect. 4.5.1) from the detection of the proton after elastic or inelastic scattering.

A measured excitation energy spectrum obtained from missing-mass reconstruction after the interaction of a 295 MeV proton beam with a ^{120}Sn target at RCNP is shown in Fig. 4.28. The energy resolution of the spectrometer allows to clearly separate the population of the ground state of ^{120}Sn (main peak) from inelastic events where excited states of ^{120}Sn are populated.

The measurements are performed at different angles of the Gran Raiden spectrometer in the laboratory frame. This is performed by rotating around the target the full spectrometer on a circular rail system, as visible in the left panel of Fig. 4.27.

---

[11] The quote resolution is for low excitation energies. For an excitation energy of 10 MeV, the energy resolution is ~100 keV.

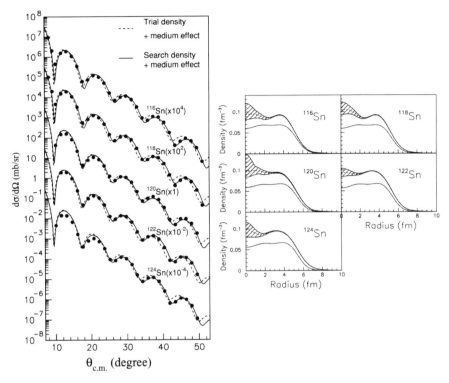

Fig. 4.29 (Left) Differential cross sections for proton elastic scattering from tin isotopes. The measurements were performed in direct kinematics at RCNP with incident proton beam of 295 MeV with the Gran Raiden spectrometer. Angular distributions as a function of the center of mass scattering angle ($\theta^\circ_{c.m.}$) are shown for the stable even-even isotopes ^{116}Sn, ^{118}Sn, ^{120}Sn, ^{122}Sn and ^{124}Sn. The dashed lines show calculations based on an effective interaction proton-nucleon interaction using trial density distributions. The solid lines are best-fit calculations based on the same effective proton-nucleon interaction. The best fit is obtained from adjusting parameters to describe the neutron distribution. (Right) Point-like proton (solid line) and neutron (dashed line) density distributions for tin isotopes. The neutron distributions are deduced from the best fit of the elastic scattering data. Hatched areas represent the error envelopes encompassing trial neutron density distributions. Figures reprinted with permission from [17]. ©2021 from the American Physical Society

For each measurement, the elastic events are selected in the focal plane as illustrated in Fig. 4.28, and the elastic cross section at the specific angle of the measurement is determined from their number. Such angular distributions performed with incident protons at 295 MeV are shown for several even-even tin isotopes in the left panel of Fig. 4.29. The data show the characteristic diffraction pattern due to the size of the nucleus. Comparing carefully the diffraction patterns from one isotope to another, one observes the effect of increasing matter radius as qualitatively expressed in (4.118).

Such differential cross sections can be interpreted quantitatively within a theoretical reaction framework by use of an *optical potential* which describes effectively the interaction of the incident proton with the target nucleus. If this optical potential is defined as a convolution of an effective nucleon-nucleon interaction and a parameterized density distribution for the target nucleus, the parameters of the matter distribution can be obtained by fitting the data. Such an analysis was performed in the tin data measured at RCNP. Calculated differential cross sections are compared to the data in the left panel of Fig. 4.29. The obtained matter density distributions are shown in the right panel of the figure. Similarly to the use of electron scattering to determine the charge density distributions, high-precision proton elastic scattering can be used to determine the matter density distribution with the difference that the effective nucleon-nucleon interaction and the *ansatz* chosen to derive the optical potential are model dependent. The systematic uncertainties from such analysis are difficult to assess.

4.3.5 Parity Violation Scattering

Where does the parity violation come from and how can such asymmetry be observed experimentally? This is an intriguing question and many smart ideas have been proposed and implemented to answer it. Let us summarize first what is parity. The parity transformation is defined with the following set of rules on space and momentum coordinates, angular momentum and spin

$$\mathbf{x} \rightarrow -\mathbf{x} \qquad (4.119)$$
$$\mathbf{p} \rightarrow -\mathbf{p}$$
$$\mathbf{L} \rightarrow \mathbf{L}$$
$$\mathbf{s} \rightarrow \mathbf{s}.$$

Angular momentum and spin are not modified by a parity transformation and are therefore qualified as *pseudo vectors*.

4.3.5.1 Parity Violation of the Weak Interaction: Wu's Experiment

A first signature of parity violation in nature was observed in the β-decay of ^{60}Co by Chien-Shiung Wu in 1957 from a concept proposed by Tsung-Dao Lee and Chen-Ning Yang in 1956. The experiment aimed at demonstrating that the weak interaction indeed violates parity. The experiment was based on the measurement of electrons produced in the β decay of ^{60}Co. A ^{60}Co radioactive source was located in a strong magnetic field so that the angular momentum of the nucleus was aligned to the B-field direction. Because the total angular momentum of ^{60}Co is $I = 5^+$, and the state of the daughter nucleus to which most of the decay process has a total angular momentum of

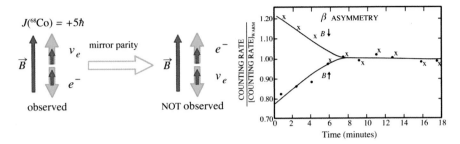

Fig. 4.30 (Left) Sketch of Wu's experiment and expectations in case of parity violation. (Right) Result obtained from the β electron asymmetry measurement from the original paper by Wu. The symbol "cross (x)" ("filled circle") is the counting rate of electrons with anti-parallel (parallel) momenta to the magnetic field. The loss of asymmetry with time is related to the experiment method: to detect the electrons, the magnetic field was stopped at $t = 0$. After few minutes, the ^{60}Co nuclei disalign and there is no preferred direction anymore, therefore no asymmetry can be measured. Figure reprinted with permission from [20]. ©2021 by the American Physical Society

$I = 4^+$, the neutrino and the electron ejected back to back (momentum conservation) need to have their spin $1/2$ in the same direction to conserve angular momentum. Parity symmetry would imply that the two mirror-parity-symmetry decays sketched on the left of Fig. 4.30 have the same probability to occur. It is not the case: parity symmetry is violated in that electrons emitted with a spin in the same direction of the magnetic field have an asymmetric emission distribution. The obtained historical result is shown on the r.h.s. of Fig. 4.30: a clear asymmetry in the emission direction with respect to the magnetic field direction is seen, showing that β-decay leads to the spin of the electron always antiparallel with its emission momentum, and therefore violates the parity symmetry.

4.3.5.2 Electroweak Interaction and Electron Scattering

In the electroweak unified model, the electron interacts with a nucleus by exchanging either a photon, the well known Coulomb interaction, or a Z_0 boson mediating the weak interaction. The general form of the propagators[12] involved in the interaction, as illustrated in Fig. 4.31, is

$$\frac{1}{q^2 + m_B^2},\tag{4.120}$$

where m_B is the mass of the exchanged boson. For the photon $m_B = 0$, whereas for the Z_0 the mass is very heavy, $m_Z = 91.2$ GeV/c^2, and dominates the propagator. Since for elastic scattering from nuclei, $m_Z \gg q^2$, the photon propagator is much larger

---

[12] The propagator is a function that specifies the probability amplitude for a particle to travel from one place to another in a given period of time. In Feynman diagrams, virtual particles (photons or weak bosons) contribute the propagator to the rate of the scattering by the respective diagram.

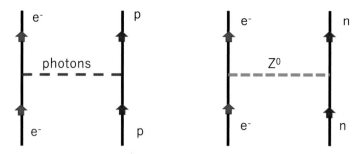

Fig. 4.31 (left) Feynman diagrams of the electromagnetic process (left), and weak process (right) of photon and Z_0 exchange, respectively, in the electron scattering experiment. The weak process is experimentally difficult to observe because it is much weaker than the electromagnetic scattering

Fig. 4.32 Calculated proton and neutron densities of ^{208}Pb compared to the electromagnetic charge density and the weak charge density (4.123)

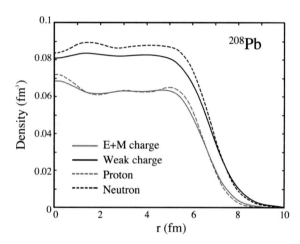

than the Z_0 one, leading to a much higher cross section. Another difference between the exchange of the photon and the Z_0 is the couplings to both the electron and the nucleons. The photon has purely vector couplings, and couples only to protons at $q^2 = 0$ limit for the $J = 0$ nuclei. The Z_0 boson has both vector and axial vector couplings (Fig. 4.32).

The potential between an electron and a nucleus to a good approximation is written as

$$V(r) = V_C(r) + \gamma_5 V_W(r), \qquad (4.121)$$

where V_C is the standard Coulomb potential dominated by the charge density. On the other hand, the axial potential V_W is described by the so-called "*weak charge, q_W*", which can be evaluated by the standard model of weak interaction as

$$q^W = -2t_z - 4q^{em}\sin^2\theta_W = \begin{cases} 1 - 4\sin^2\theta_W & \text{for protons} \\ -1 & \text{for neutrons,} \end{cases} \qquad (4.122)$$

where q^{em} is the charge of electromagnetic interaction, θ is the weak mixing angle (Weinberg angle) connecting the masses of W and Z bosons, $\cos\theta_W = m_W/m_Z$. The mixing angle is observed to be $\sin^2\theta_W = 0.23$ and gives a very small proton weak charge, $q_p^W \simeq +0.07$, compared with the neutron weak charge $q_n^W = -1$. The weak charge density of nucleus with (N, Z) is then written by using the neutron and proton densities as

$$\rho_W(r) = (1 - 4\sin^2\theta_W)Z\rho_p(r) - N\rho_n(r), \qquad (4.123)$$

where $\int \rho_p(\mathbf{r})d\mathbf{r} = \int \rho_n(\mathbf{r})d\mathbf{r} = 1$. The weak density (4.123) is dominated by the neutron density as shown in Fig. 4.32, and offers an opportunity to measure the neutron density distribution by the parity-violating experiment as is discussed below. The axial potential $V_W(r)$ is expressed by the weak charge density as

$$V_W(r) = \frac{g_V}{2^{3/2}}\left[(1 - 4\sin^2\theta_W)Z\rho_p(r) - N\rho_n(r)\right], \qquad (4.124)$$

where g_V is the vector coupling constant (called often the weak interaction Fermi coupling constant). The axial potential is in the order of 1 eV and much smaller than the vector potential V_C, which is in the order of MeV. The weak interaction violates parity, and the axial potential can be measured only in the parity-violating cross section (4.127).

4.3.5.3   Parity-Violating Scattering Experiments

From Wu's experiment, it was not straightforward to conclude that this parity violation should also be seen in electron scattering. This was proved experimentally in 1959 by Yakov Zel'dovich. The parity violation in electron scattering is extremely small since the weak interaction is much weaker than the electromagnetic one. Nevertheless, the effect can be measured via an asymmetry measurement. The concept of an asymmetry measurement and its link to the parity violation is illustrated in Fig. 4.33. Suppose the elastic scattering measurement as illustrated on the l.h.s. of the figure. After parity transformation, all momentum are reflected, while the spin direction remains the same as for a pseudo vector. Since physics is invariant under rotation, the parity-symmetric experiment is then also equivalent to the r.h.s. of the figure. A change of the sign of the longitudinal polarization of the electron beam is equivalent to realize the conditions for a parity-symmetric experiment. Any difference between the two measurements would sign a parity violation. In these experiments, the main cross section comes from the electromagnetic interaction and therefore the asymmetry to be measured is typically in the order of 10^{-6}, and corresponds to an interference term between electromagnetic and weak amplitudes.

The helicity is the projection of the spin onto the direction of momentum. In the zero-limit of electron mass, the Dirac equation (1.107) in Chap. 1 is solved for helicity eigenstates $\Psi_{R(L)} = \frac{1}{2}(1 \pm \gamma_5)\Psi$,

$$(c\boldsymbol{\alpha} \cdot \mathbf{p} + V_{\pm}(r)) \, \Psi_{R(L)} = E\Psi_{R(L)}, \tag{4.125}$$

where

$$V_{\pm}(r) = V_C(r) \pm V_W(r). \tag{4.126}$$

Since $\gamma_5 \Psi_{R(L)} = \pm \Psi_{R(L)}$. As sketched in Fig. 4.33, the spin of state $\Psi_{R(L)}$ is parallel (anti-parallel) to its momentum and the helicity is $\hat{h} = (\mathbf{p} \cdot \boldsymbol{\sigma})/|\mathbf{p}| = \pm 1$. Thus, the state Ψ_R scatters according to a potential $V_C + V_W(r)$, while the state Ψ_L scatters according to a potential $V_C - V_W(r)$. To observe the parity violating asymmetry

$$A_{PV}(q) = \frac{d\sigma_R(q)/d\Omega - d\sigma_L(q)/d\Omega}{d\sigma_R(q)/d\Omega + d\sigma_L(q)/d\Omega}, \tag{4.127}$$

the effect of the parity-violating part of the weak interaction may be isolated in the interference term between $V_C + V_W$ and $V_C - V_W$. In Eq. (4.127), $d\sigma_{R(L)}/d\Omega$ is the cross section for the scattering of right- (left-) handed electrons given by

$$\frac{d\sigma_{R(L)}(q)}{d\Omega} = |f_C(q) \pm f_W(q)|^2$$

$$= \left| \frac{m}{2\pi\hbar^2} \left(\int V_C(\mathbf{r})e^{i\mathbf{q}\mathbf{r}} d\mathbf{r} \pm \int V_W(\mathbf{r})e^{i\mathbf{q}\mathbf{r}} d\mathbf{r} \right) \right|^2. \tag{4.128}$$

Then the parity violating asymmetry is expressed as

$$A_{PV}(q) \simeq \frac{f_C(q)^* f_W(q) + f_C(q) f_W(q)^*}{|f_C(q)|^2} \tag{4.129}$$

since $|f_C(q)| >> |f_W(q)|$. Here $f_C(q)$ is the Coulomb form factor (4.87) and $f_W(q)$ is the weak form factor defined by

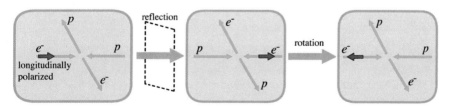

Fig. 4.33 Sketch of the elastic scattering of an electron with longitudinal polarization on a proton (left). Parity-symmetric experiment verifies the rules of parity transformation as expressed in Eq. (4.120) (center). Since physics laws are conserved by rotation in space, the first experiment with reverse polarization of the electron spin (right) is equivalent to its parity symmetry

$$f_W(q) = -\frac{m}{2\pi\hbar^2}\frac{g_V}{2^{3/2}}\int e^{i\mathbf{q}\mathbf{r}}\rho_W(\mathbf{r})d\mathbf{r}$$

$$= -\frac{m}{4\pi\hbar^2}\frac{g_V\,Q_W}{\sqrt{2}}F_W(q) \qquad (4.130)$$

with

$$F_W(q) = \frac{1}{Q_W}\int \rho_W(\mathbf{r})e^{i\mathbf{q}\mathbf{r}}d\mathbf{r}, \qquad (4.131)$$

and the weak charge of nucleus,

$$Q_W = \int d\mathbf{r}\rho_W(r). \qquad (4.132)$$

The form factors $F_W(q)$ and $F_C(q)$ are normalized to be unity $F_W(0) = F_C(0) = 1$ at $q \to 0$, respectively. The asymmetry A_{PV} is then given by the interference term between the Coulomb and weak form factors in Eqs. (4.87) and (4.131) as

$$A_{PV}(q) = \frac{g_V q^2}{\sqrt{2}4\pi\alpha\hbar c}\frac{Q_W F_W(q)}{Z F_{ch}(q)}. \qquad (4.133)$$

Since the parity violation asymmetry $A_{PV}(q)$ is proportional to the ratio of form factors $F_W(q)/F_{ch}(q)$, if one measures the asymmetry A_{PV} as a function of q and knows the charge form factor from electron elastic scattering, the neutron density, in principle, can be accessed from the weak form factor since the first term of the r.h.s. $1 - 4\sin^2(\theta_W)$ in the weak density (4.123) is very small. Unfortunately, this is "only theory" since, in practice, the parity-violating asymmetry A_{PV} is extremely difficult to measure because of strong statistics limitations, even at one given q^2. Therefore, the followed strategy is to choose a momentum transfer at which A_{PV} strongly depends on the neutron radius and at which the measurement will lead to sufficient statistics. Once A_{PV} is obtained, it can be compared to different theoretical predictions, and constrain the neutron density and at the same time the neutron radius.

4.3.5.4   The PREX Experiment and the Neutron Radius of ^{208}Pb

The PREX experiment at Jefferson National Laboratory (JLab) initiated a program to determine the neutron radius of ^{208}Pb with controlled errors using parity-violation asymmetry measurements of electron elastic scattering. Two measurements were performed: PREX-1 with a 1.06 GeV electron beam and a measurement at a momentum transfer of $Q^2 = 0.0088\,\mathrm{GeV}^2/c^2$ [21], and PREX-2 with a 0.953 GeV electron beam and a measurement at $Q^2 = 0.00616\,\mathrm{GeV}^2/c^2$ [18]. The longitudinal polarization reaches 90% at JLab and, during the measurement, the spin was reversed at frequencies 120 Hz and 240 Hz so that any bias from changes in external conditions cancels out. The asymmetry was measured in quadruplets of helicities with quasi-random

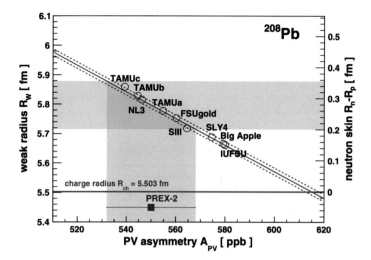

Fig. 4.34 Theoretical correlation between the parity-violation asymmetry A_{PV} (horizontal axis) and the weak rms radius R_W (ordinate, left scale). From this correlation, the measured asymmetry and its uncertainties can be translated in a value for R_W and corresponding uncertainties, which leads to a value of 0.283(71) fm for the neutron skin thickness in ^{208}Pb (ordinate, right scale). Figure reprinted with permission from [21]. ©2021 by the American Physical Society

patterns. The accumulated data correspond in total to 7.0×10^7 helicity quadruplets. The sources of uncertainty of the measurement are mostly coming from statistics and the obtained asymmetry was $A_{PV} = 657 \pm 60$ (stat) ± 13 (syst) ppb for the first measurement, and $A_{PV} = 550 \pm 16$ (stat) ± 8 (syst) ppb for the second measurement. The correlation between the asymmetry A_{PV} the weak radius R_W as illustrated in Fig. 4.34 allows the transition from to the measured asymmetry to the extracted neutron rms radius. The extracted value of the weak rms radius of ^{208}Pb is 5.800(75) fm taking into account both measurements. Combined to the precisely measured charge rms radius, this value leads to a 0.283(71) fm neutron skin thickness in ^{208}Pb.

4.3.6 Coherent π_0 Photoproduction

The coherent neutral pion photo production aims at determining the matter radius of the nucleus via the interaction with high-energy real photons of few hundreds of MeV. The method focuses on the very specific final state where the target nucleus remains in its ground state. This implies that the photon interacts *coherently* with the target nucleus, i.e., with all nucleons and not with a single nucleon. In the interpretation of a one-step direct reaction, the photon interaction results in the excitation of a delta resonance, the first excited state of the nucleon with an excitation energy of ~300 MeV, which decays via a pion emission. Once produced the π_0 decays onto two photons with a branching ratio of 98.8%. The measurement of the energy of the two photons allows, from invariant mass reconstruction (details given later in this chapter), to select coherent events.

The differential cross section (angular distribution) of the reconstructed π_0 is then analysed to constrain the nuclear matter form factor, sensitive to matter radius and diffusiveness. In the Plane Wave Impulse Approximation (PWIA), the cross section is expressed as

$$\frac{d\sigma}{d\Omega}(PWIA) = \frac{s}{m_N^2} A^2 \left(\frac{q_\pi^*}{2k_\gamma}\right) F_2(E_\gamma^2, \theta_\gamma^*)^2 |F_m(q)|^2 \sin^2(\theta_{\pi_0}^*), \qquad (4.134)$$

where s is the square of the total energy of the γ-nucleus pair, q is the transferred momentum in the reaction, F_2 is a spin independent amplitude and $|F_m(q)|^2$ is the nuclear-matter form factor. Asterisks denote quantities in the γ-nucleus center of mass.

Such an experiment was performed at the MAMI facility located in Mainz, Germany. High-energy bremsstrahlung gamma rays were produced from the interaction of electrons with a copper target, used as a *radiator*, and their energy was tagged event by event. Bremsstrahlung tagging is an established technique for providing large fluxes of energy-labelled photons. The principle is as follows: when an electron of incident energy E_0 strikes a radiator, generating a bremsstrahlung photon of energy E_γ, the energy of the outgoing scattered electron is $E_e = E_0 - E_\gamma$. The energy of the radiated photon is therefore completely determined by the knowledge of E_0 and E_e. At MAMI and in most of the other bremsstrahlung applications, the scattered electron's momentum is analysed in the field of a magnetic spectrometer and correlated in time with some subsequent reaction product of the bremsstrahlung photon.

The obtained results for ^{208}Pb at four different incident energies ranging from 180 to 240 MeV are shown in Fig. 4.35. The data are compared to Plane-Wave Impulse Approximation (PWIA) calculations, as described in Eq. (4.134), and full calculations taking into account the final state interaction of the π_0 with ^{208}Pb. Distortions are higher at higher incident energies but it is worth mentioning that photons probe the entire volume of the nucleus and the relatively close agreement of PWIA predictions with the data demonstrates that the final state interactions of the produced pion with the recoil ^{208}Pb nucleus are rather small. These final state interactions are taken into account with the full calculation in Fig. 4.35. The shown calculations shown are the result of a best fit procedure with minimization on two parameters c, a related to the radius and diffusiveness of the neutron density, respectively, assuming a two-parameter Fermi density distribution

$$\rho(r) = \frac{\rho_0}{1 + exp(\frac{r-c}{a})}. \qquad (4.135)$$

Table 4.3 Summary of methods to extract charge, matter or neutron radii, together with their accuracies. Some of the techniques are only applicable to stable nuclei and require high statistics, others can be applied to low-yield radioactive isotopes. The accuracy of each experimental method is qualitatively indicated (++: very accurate, +: believed accurate but some model dependences can not be assessed simply, -: model dependent)

Method	Observable	Extracted quantity	Accuracy
Fine structure	Isotopic shift	Relative charge radii	++
e^- elastic scattering	Differential cross section	Charge density	++
Parity-violation scattering	Asymmetry	Neutron radius	++
Coherent π_0 photoproduction	Differential cross section	Matter radius	+
Proton elastic scattering	Differential cross section	Matter radius	+
Interaction cross section	Inclusive cross section	Matter radius	-

The proton distribution was assumed of the same shape with $c_p = 6.68$ fm and $a_p = 0.447$ fm set from electron elastic scattering data. The analysis led to a neutron skin thickness for ^{208}Pb evaluated to $\Delta r_{np} = 0.15 \pm 0.03(\text{stat.})^{+0.01}_{-0.03}(\text{sys.})$ fm.

4.3.7 Comparison of Radii Extraction Techniques

The different methods described above to extract charge and matter radii are summarised in Table 4.3.

4.4 Nuclear Moments

Nuclear moments provide unique information on nuclear structure. As detailed in the following, they provide information on the nucleonic distribution inside the nucleus: magnetic moments give information on the angular momentum and spin distribution in the nucleus, related to the single-particle distribution of valence nucleons, while electric quadrupole moments are the key observable for nuclear deformation. Several experimental techniques exist to measure these moments, depending on the lifetime of the state of interest.

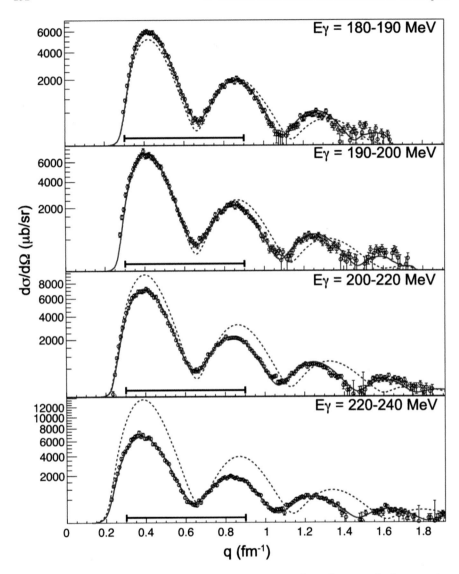

Fig. 4.35 Differential cross sections for the coherent pion photo production reaction $^{208}Pb(\gamma, \pi_0)^{208}Pb$ (black circles) for four different ranges of the photon energy E_γ. The red solid line shows the fit of the theoretical model to the data. The transferred momentum q range of the fit is shown by horizontal black bar. The dashed blue line shows the model predictions without including the pion-nucleus final-state interaction. Figure reprinted with permission from [22]. ©2021 by the American Physical Society

4.4.1 The Electric Quadrupole Moment

Solving the Schrödinger equation for electrons in the electric field of a point-like nucleus with a positive charge Ze,

$$V(r) = \frac{Ze^2}{r}, \tag{4.136}$$

results in the fine structure of the atom where each electron orbital energy depends on the principal quantization number n and the total angular momentum $J = \ell \pm 1/2$, ℓ being the angular momentum of the electron. All magnetic substates m are degenerate in energy in spherical symmetry.

In a realistic situation, the nucleus has a spatial extension and the potential can be extended in multiple orders of r^{-1}. Considering the nucleus as an ensemble of point-like charges q_i located at positions $\mathbf{r}_i$, the electric potential created at a point $\mathbf{r}$ is given by

$$V(\mathbf{r}) = \sum_i \frac{q_i}{|\mathbf{r} - \mathbf{r}_i|}. \tag{4.137}$$

It can be formulated into a multipole expansion as follows

$$V(\mathbf{r}) = \sum_i \frac{q_i}{|\mathbf{r}|} + \frac{\hat{r} \cdot \mathbf{p}}{r^2} + \frac{1}{2} \frac{\sum_{ij} Q_{ij} \hat{n}_i \hat{n}_j}{r^3} + \cdots, \tag{4.138}$$

where $\hat{r}$ is a unit vector oriented as $\mathbf{r}$, $\hat{n}_i$ is the unit vector starting from the point where the potential is evaluated to the charges q_i. $\mathbf{p}$ and Q_{ij} are the electric dipole and quadrupole electric moments, respectively. The term with multipole ℓ of the expansion scales as $1/r^{\ell+1}$. A classical illustration of charge distributions having a dipole and quadrupole electrical moment are shown in Fig. 4.36.

The first term of the r.h.s. expansion in Eq. (4.138), the monopole moment, is nothing else than the Coulomb potential created by a point-like particle with a charge $\sum_i q_i$. An electric dipole moment $\mathbf{p}$ emerges when the barycenters of positive and negative charges are not identical and is defined as

$$\mathbf{p} = \sum_i q_i \mathbf{r}_i. \tag{4.139}$$

This dipole moment appears at the order $\ell = 1$ of the multipole expansion. It does not exist in nuclei since they are composed of only one type of charges. Therefore, in nuclei, the first term of the expansion of the electric potential after the monopole corresponds to the quadrupole moment ($\ell = 2$)

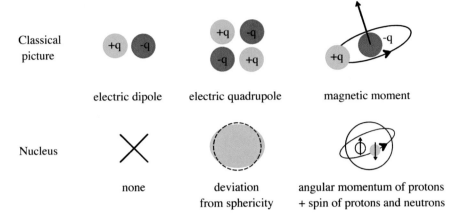

Fig. 4.36 Classical view of electric dipole and quadrupole moments and magnetic moment. Since the nucleus has only positive charges, it does not have any electric dipole moment. The nuclear magnetic moment has contributions from both the proton orbital angular momentum and the intrinsic spins of both protons and neutrons

$$\mathbf{Q}_{mn}(\mathbf{R}) = \sum_i q_i (3r_{mi}r_{ni} - |\mathbf{r}_i|^2\delta_{mn}), \qquad (4.140)$$

where m, n are indices representing the cartesian coordinates (x,y,z). In three dimensions, the quadrupole moment is represented by a tensor of rank 2 (a 3×3 matrix). It is traceless and symmetric and therefore is defined by five independent quantities: $Q_{xx}, Q_{zz}, Q_{xy}, Q_{yz}, Q_{zx}$. It represents a deviation from sphericity in nuclei. Its determination is central to determine the nuclear shape, as detailed in Chap. 7. For a continuous distribution of charges with density $\rho_c(\mathbf{r})$, the electric quadrupole moment is defined as

$$Q_{mn} = \int \rho_c(\mathbf{r})(3r_m r_n - |\mathbf{r}|^2\delta_{mn})d\mathbf{r}, \qquad (4.141)$$

where $\rho_c(\mathbf{r})$ is the charge density. The electric Q-moment is commonly defined by the operator Q_{zz} in Eq. (4.141),

$$Q_{zz} = \int \rho_c(\mathbf{r})(3z^2 - |\mathbf{r}|^2)d\mathbf{r} = \sqrt{\frac{16\pi}{5}} \int \rho_c(\mathbf{r})r^2 Y_{20}(\hat{r})d\mathbf{r}. \qquad (4.142)$$

The above definitions are valid in an *intrinsic* frame attached to the nucleus. Measured quadrupole moments in the laboratory frame are called *spectroscopic* Q moments and can be related to the intrinsic Q moments through a model assumption (see Chap. 7). The Q-moment for a single-particle state $|ljm\rangle$ can be expressed by using the reduced matrix element as

Table 4.4 Q-moments of single-particle (hole) states near closed shell nuclei ^{40}Ca and ^{16}O

Nucleus	lj	Q_{exp} $(b=10^{-24}cm^2)$
^{17}O	$(d_{5/2})_\nu$	$-0.026e$
^{17}F	$(d_{5/2})_\pi$	$|0.058|e$
^{39}K	$(d_{3/2}^{-1})_\pi$	$+0.060e$
^{39}Ca	$(d_{3/2}^{-1})_\nu$	$|0.038|e$
^{41}Ca	$(f_{7/2})_\nu$	$-0.090e$
^{41}Sc	$(f_{7/2})_\pi$	$-0.156e$

$$Q_{s.-p.} = \langle j, m = j|Q_{zz}|j, m = j\rangle = e\sqrt{\frac{16\pi}{5}} \frac{\langle jj20|jj\rangle}{\sqrt{2j+1}} \langle j||Y_2||j\rangle \langle r^2\rangle_j$$

$$= 2e\langle jj20|jj\rangle \langle j\frac{1}{2}20|j\frac{1}{2}\rangle \langle r^2\rangle_j$$

$$= -e\frac{2j-1}{2j+2} \langle r^2\rangle_j, \tag{4.143}$$

where $\langle r^2\rangle_j$ is the mean square radius of the single-particle state $|ljm\rangle$.

Problem

4.5 Derive the Q-moment (4.143) for a single-particle state $|ljm\rangle$.

Problem

4.6 Calculate the Q-moments (4.143) for single-particle states $|ljm\rangle$ in Table 4.4 and obtain effective charges defined by

$$e_{eff}/e = Q_{exp}/Q_{s.-p.}, \tag{4.144}$$

where Q_{exp} is the experimentally observed (measured) value. Use the mean square radius of the harmonic oscillator wave function:

$$\langle r^2\rangle_j = \frac{\hbar}{m\omega_0}(N + \frac{3}{2}) \sim A^{1/3}(N + \frac{3}{2}), \tag{4.145}$$

where N is the major quantum number. Notice the particle-hole conjugation of the Q-moment;

$$\langle (j, m = j)^{-1}|Q_{zz}|(j, m = j)^{-1}\rangle = -\langle j, m = j|Q_{zz}|j, m = j\rangle. \tag{4.146}$$

4.4.2 The Magnetic Dipole Moment

In addition to electric moments, the nucleus also has magnetic moments caused by the motion of the (charged) protons within the nucleus: the orbiting of protons gives rise to an orbital magnetic field. The definition of the magnetic moment of two charges with relative motion was already introduced in Chap. 1 in the atomic case of the Stern-Gerlach experiment. The intrinsic spin of nucleons (protons and neutrons) also induces an intrinsic magnetic field. The magnetic dipole operator is expressed as a sum of these two contributions

$$\mu = \sum_{i=1}^{A} g_l^i \mathbf{l}^i + \sum_{i=1}^{A} g_s^i \mathbf{s}^i, \tag{4.147}$$

where the free-nucleon gyromagnetic factors for protons and neutrons are

$$g_l = \begin{cases} 1 \\ 0 \end{cases}, \quad g_s = \begin{cases} +5.587 & \text{for protons} \\ -3.826 & \text{for neutrons} \end{cases}, \tag{4.148}$$

in units of nuclear magnetron $\mu_N = e\hbar/2mc$. The magnetic moment of a state with total angular momentum J is the expectation value of μ on the axis defined by the total angular momentum

$$\mu(J) = \langle J, m = J | \mu_z | J, m = J \rangle. \tag{4.149}$$

The magnetic moment of the nuclear state under study is often written in terms of the g factor, $\mu = g\mu_N/\hbar$ with $\mu_N = e\hbar/(2m_N c) = 3.144 10^{-8}$ eV·T^{-1} being the nuclear magnetron. Note that the nuclear magnetron is much smaller than the so-called Bohr magnetron for the electron $\mu_e = e\hbar/(2m_e c) = 5.788 10^{-5}$ eV·T^{-1}.

The nuclear magnetic moment provides a sensitive test of the nuclear coupling scheme. The magnetic moment for a single-particle configuration of shell model orbit can be evaluated by using the relation

$$\mu_z = j_z \frac{\langle \mu \cdot \mathbf{j} \rangle}{\langle \mathbf{j} \cdot \mathbf{j} \rangle}, \tag{4.150}$$

where the ratio between two values μ_z and j_z is expressed by expectation values of two inner products $\langle \mu \cdot \mathbf{j} \rangle$ and $\langle \mathbf{j} \cdot \mathbf{j} \rangle$, which are the projections of μ and $\mathbf{j}$ on the vector $\mathbf{j}$. Inserting Eq. (4.150) into Eq. (4.149), the magnetic moment is expressed as

Table 4.5 Magnetic moments of single-particle (hole) states near closed shell nucleus ^{16}O

Nucleus	μ_{exp}
^{15}N	−0.283
^{15}O	+0.719
^{17}O	−1.894
^{17}F	+4.721

$$\mu_{sp} = \langle j, j = m | \mu_z | j, j = j_z \rangle = \frac{\langle \boldsymbol{\mu} \cdot \mathbf{j} \rangle}{j(j+1)} \langle j, , j = m | j_z | j, j = j_z \rangle$$

$$= \frac{1}{j+1} [g_l j(j+1) + (g_s - g_l)\langle \mathbf{s} \cdot \mathbf{j} \rangle]$$

$$= j \left[g_l \pm (g_s - g_l) \frac{1}{2l+1} \right] \quad \text{for} \quad \begin{cases} j = l + 1/2 \\ j = l - 1/2 \end{cases} , \qquad (4.151)$$

where the quantity $\langle \mathbf{s} \cdot \mathbf{j} \rangle$ can be evaluated by squaring the identity $\mathbf{j} - \mathbf{s} = \mathbf{l}$. The values of magnetic moments in Eq. (4.151) are called *Schmidt values*.

Electric or magnetic static moments of nuclei are measured via the interaction of the nuclear charge distribution and magnetism with electromagnetic fields. The fields used for this purpose can be of different origins, either induced (i) by the atomic electrons bound to the nucleus, (ii) by bulk electrons from a crystal in which the nucleus under study has been implanted or (iii) by the electromagnetic field created by the moving charge of a colliding nucleus. This last technique called *Coulomb reorientation* gives access to the quadrupole moment of short lived states. It is detailed in Chap. 7. The two former methods are restricted to ground states or long-lived isomers, and are detailed in this section.

Problem

4.7 Observed magnetic moments of nuclei $A = 16 \pm 1$ are listed in Table 4.5. Predict the single-particle (hole) configuration of the last particle (hole) by comparing the Schmidt magnetic moment (4.151) with experimental data. Notice the particle-hole conjugation of magnetic moment:

$$\langle (j, m = j)^{-1} | \mu_z | (j, m = j)^{-1} \rangle = \langle j, , m = j | \mu_z | j, m = j \rangle. \qquad (4.152)$$

4.4.3 Hyperfine Structure

The electric quadrupole and magnetic dipole moments modify the hyperfine structure of the atomic spectrum. The measurement of the hyperfine splittings on top of the fine structure leads to unambiguous determination of the quadrupole and magnetic moment of the nuclear state(s) under study. As seen above, the atomic hyperfine structure originates from the anisotropy of the electron and nucleus fields and their interactions. For this reason, each fine-structure energy level will be split into several states depending on the coupling of the nuclear and electronic total angular momenta.

In the following, we denote with I the total angular momentum of the nucleus and J the total angular momentum of the electrons which is the total angular momentum of the unpaired electron in the case of an odd number of bound electrons. The total angular moment $\mathbf{F}$ of the atomic system composed of the nucleus and bound electrons is given by

$$\mathbf{F} = \mathbf{I} + \mathbf{J} \tag{4.153}$$

implying

$$|I - J| \leq F \leq I + J. \tag{4.154}$$

$\mathbf{F}$ drives the hyperfine splitting of the atomic energy levels.

4.4.3.1 Magnetic Hyperfine Splitting

In the case of a non-zero angular momentum, i.e., with a total angular momentum $I > 1/2$, a nucleus has a magnetic moment μ, which interacts with the magnetic field $\mathbf{B}$ produced by the electrons. The energy gained by this magnetic interaction is given by the Hamiltonian

$$H = \mu \cdot \mathbf{B} \tag{4.155}$$

with

$$\mathbf{B} = B\frac{\mathbf{J}}{J} \quad \text{and} \quad \mu = \mu\frac{\mathbf{I}}{I}. \tag{4.156}$$

The resulting shift in energy of the atomic orbital can be expressed as

$$\Delta E_D = \mu B \cos\theta, \tag{4.157}$$

where θ is the precession angle between the nuclear total spin $\mathbf{I}$ and the electron total spin $\mathbf{J}$. Classically, it can be expressed as

$$\cos\theta = \frac{\mathbf{I} \cdot \mathbf{J}}{IJ} = \frac{(\mathbf{I}+\mathbf{J})^2 - I^2 - J^2}{2IJ}. \tag{4.158}$$

In terms of eigenvalues of quantum angular momentum operators, this converts into

$$\Delta E_D = \frac{\mu B}{2IJ}(F(F+1) - I(I+1) - J(J+1)), \tag{4.159}$$

where B is the magnetic field created at the nucleus site by the orbiting of electrons and μ is the projection along the magnetic field direction of the magnetic moment

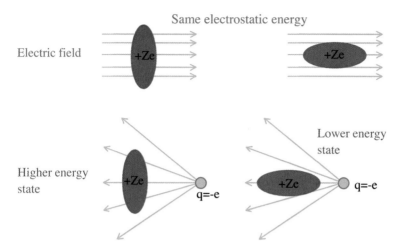

Fig. 4.37 Quadrupole interaction of a quadrupole moment in an electric field

of the nuclear state of total angular momentum I. The following notation is often encountered in the literature

$$\Delta E_D = \frac{A}{2} C, \qquad (4.160)$$

with $C = F(F+1) - I(I+1) - J(J+1)$ and $A = \frac{\mu B}{IJ}$. Note that the magnetic field created by an electron on a specific orbital (with total angular momentum J) can be assumed identical along an isotopic chain.

4.4.3.2 Quadrupole Hyperfine Splitting

Similarly, in the case of a deformed nucleus with a spectroscopic quadrupole moment Q_s, the electric potential seen by the electrons will be shifted by the deviation from sphericity of the charge density. This is illustrated in Fig. 4.37. In a uniform electric field (top part of Fig. 4.37), the energy of a quadrupole moment is independent of the orientation since the system is invariant under rotation. In an electric field gradient, there is an angle dependence of the quadrupole interaction and therefore the energy of the system will vary depending on the orientation of the nucleus in the field gradient. The quadrupole interaction between a nucleus of total angular momentum I and electrons with total angular momentum J is given by

$$E = \frac{e}{4} Q_s \langle \frac{\partial^2 V}{\partial z^2} \rangle P_2(\cos\theta), \qquad (4.161)$$

where P_2 is the second-order Legendre polynomial, and $\langle \partial^2 V / \partial z^2 \rangle$ is the mean gradient of the electric field caused by the electron at the site of the nucleus.

The shift in energy is calculated to be

$$\Delta E_Q = B \times C(C + 1),\qquad(4.162)$$

with

$$B = \frac{3Q_s \frac{\partial^2 V}{\partial z^2}}{8IJ(2I - 1)(2J - 1)},\qquad(4.163)$$

and $C = F(F + 1) - I(I + 1) - J(J + 1)$, as defined previously. The electric quadrupole perturbation is much smaller than the magnetic dipole term, as expressed in the perturbation expansion of the potential since they appear at different order of the multipole decomposition: the energy splitting caused by the nuclear magnetic moment is in the order of 100 MHz to 10 GHz, the shift in energy caused by the quadrupole moment is typically in the order of 1 to 100 MHz.

Problem

4.8 Determine the order of magnitude of ΔE_D.

4.4.3.3   Example of a Hyperfine Spectrum

Experimentally, the excitation of electronic states corresponding to electron spin-flip partners (fine structure) from laser frequency scans gives access to the energy difference between hyperfine multiplets (n, ℓ, J, I, F) and (n, ℓ, J', I, F')

$$\nu(J, F, J', F', I) = \Delta_D(J, F, J', F', I) + \Delta_Q(J, F, J', F', I),\qquad(4.164)$$

where $\Delta_D = \Delta E_D(I, J, F) - \Delta E_D(I, J', F')$ and $\Delta_Q = \Delta E_Q(I, J, F) - \Delta E_Q(I, J', F')$ with the above definition of ΔE_D and ΔE_Q. The magnitude of the above-mentioned transition energies ν are illustrated in Fig. 4.38 in the case of excitations from $J = 3/2$ to $J' = 1/2$ atomic partners assuming a nuclear spin 3/2. By use of Eqs. (4.160) and (4.162), the hyperfine splittings caused by the nuclear magnetic dipole moment and the electric quadrupole moment of the nucleus, respectively, can be determined as a function of the quantities A, A' and B. Note that there is no splitting ($B' = 0$) caused by the quadrupole moment for the electronic orbital with $J' = 1/2$. A and A' depend on the magnetic moment, while B depends on the electric quadrupole moment.

Experimental data for the hyperfine spectroscopy of the ground state of ^{59}Cu are shown in Fig. 4.39, for the $^2S_{1/2} -^2 P_{3/2}$ atomic transitions. The spectrum shows that the total angular momentum of ^{59}Cu is larger than $1/2$. Indeed, the coupling of the $J = 1/2$ angular momentum of electrons and I should give two groups of transitions to hyperfine partners of the coupling of I and $J = 3/2$. The two groups are well separated in energy and appear as triplets. The photon decay transitions from total spin F' to F allow only $\Delta F = 0, \pm 1$, therefore leading to the first-glance conclusion that $I > 1/2$. The fit of the spectrum with Eq. (4.160), (4.162) and (4.164),

Fig. 4.38 Hyperfine structure of an atomic system assuming a nuclear spin $I = 3/2$ and atomic transitions from an orbital with $J = 3/2$ to an orbital with $J' = 1/2$

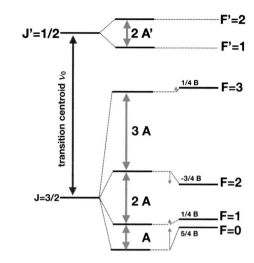

Fig. 4.39 Hyperfine spectrum of the $^2S_{1/2} - ^2P_{3/2}$ transition (324.74 nm) in ^{59}Cu. Figure reprinted with permission from [23]. ©2021 by the American Physical Society

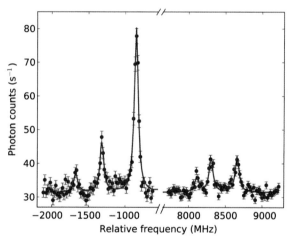

for several hypothesis of nuclear spin, gives a best χ^2 for $I = 3/2$. Note that the A' and A splitting parameters, as illustrated in Fig. 4.38, are constant along an isotopic chain (after correction of the I factor in case of change). The obtained values after a fit assuming a certain spin can be compared to the known values of the corresponding stable isotopes, as a further consistency check for the spin assignment. The detailed study of the energies of the transitions gives an electric quadrupole moment of $-19.3(19)$ e fm^2 from the value of B giving the best agreement with experimental energies.

4.4.4 Magnetic and Quadrupole Nuclear Resonance

Nuclear moments can also be measured from the interaction of the nuclear magnetic substates with (strong) external fields. When a nucleus is implanted into a material, the interaction of the nucleus with the medium is governed by electric fields induced by the medium with a much stronger magnitude than those caused by the atomic electrons bound to the nucleus. The method is restricted to ground states and long lived states with non-zero nuclear spin.

The magnetic splitting of nuclear states is governed by the Zeeman Hamiltonian

$$H_B = -\boldsymbol{\mu} \cdot \mathbf{B} = -\omega_L I_z, \tag{4.165}$$

where I_z is the projection of the nuclear spin onto the symmetry axis of the magnetic field and

$$\omega_L = \mu B \tag{4.166}$$

is the Larmor frequency. The method is called Nuclear Magnetic Resonance (NMR).

Similarly, the Quadrupole Nuclear Resonance (QNR) is based on the interaction of a nuclear quadrupole moment with a strong external electric field gradient $V_{zz} = \partial^2 V/\partial z^2$. Very high electrical field gradients can be reached in crystals where the ion of interest is implanted. The quadrupole Hamiltonian resulting from the interaction of the field gradient with the quadrupole moment of the nuclear state with spin I is

$$H_Q = \hbar \omega_Q (3I_z^2 - I^2), \tag{4.167}$$

where the coupling frequency ω_Q is given by

$$\omega_Q = Q \frac{e\langle \frac{\partial^2 V}{\partial z^2} \rangle}{h} \frac{2\pi}{4I(2I-1)}. \tag{4.168}$$

The eigenvalues of H_Q for each projection m of the spin onto the field gradient direction are

$$E_m = \hbar \omega_Q [3\,m^2 - I(I+1)]. \tag{4.169}$$

The later eigenvalues give the quadrupole splitting of so-called "nuclear hyperfine states" in the electric field gradient. A typical energy splitting of nuclear levels is of the order of nano electronvolts which corresponds to a MHz frequency range. Such a small splitting, sensitive to the nuclear quadrupole and magnetic moments, cannot be measured with the standard spectroscopy techniques. In the case of radioactive isotopes, the preferential technique is based on the anisotropy of decaying particles after implantation, either photon, beta or alpha decay. The direction in which the decay radiation is emitted depends on the direction of the nuclear spin (Fig. 4.40). The technique requires that the initial nucleus of interest has a total angular momen-

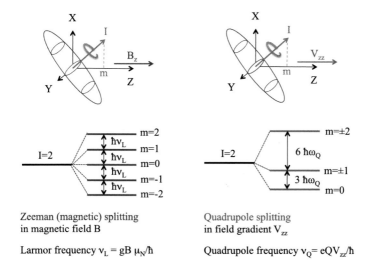

Zeeman (magnetic) splitting in a magnetic field B

Larmor frequency $v_L = gB\,\mu_N/\hbar$

Quadrupole splitting in field gradient V_{zz}

Quadrupole frequency $v_Q = eQV_{zz}/\hbar$

Fig. 4.40 Magnetic (Zeeman) splitting in a magnetic field (left) and quadrupole splitting in an electric field gradient (right) in the case of a deformed nucleus with spin $I = 2$

tum oriented in space so that a reference direction exists in the laboratory. Several techniques to produce oriented ion beams have been developed.

4.4.5 Time-Differential Perturbed Angular Distributions (TDPAD)

The magnetic and quadrupole moments of excited states can also be measured with precision by measuring the angular distribution of the γ-ray from their decay. Indeed the direction of the γ emission is correlated with the direction of the total angular momentum of the initial state of the emitting nucleus. The angular distribution of γ radiations emitted from an axially symmetric oriented nucleus is given by

$$W(\theta) = \sum_{\lambda=\text{even}} B_\lambda(I_i) A_\lambda(\gamma) P_\lambda(\cos\theta), \qquad (4.170)$$

where the orientation parameters $B_\lambda(I_i)$ describe the orientation of the axially symmetric oriented states, $A_\lambda(\gamma)$ are tabulated directional distribution coefficients for γ transitions, θ is the relative angle between the emitted γ direction and the initial total angular momentum, and P_λ are Legendre polynomials. In most practical applications, $\lambda \leq 4$ and it is often assumed that $\lambda > 2$ are negligible. An angular distribution can be measured in the laboratory only if the emitting nucleus is spin oriented. This is usually done by selecting the reaction mechanism to produce the state of interest which should lead to a certain spin alignment of the products.

An external magnetic field or an external electric field gradient will induce a spin precession of the total angular momentum at the Larmor frequency ω_L or the coupling frequency ω_Q, respectively, as defined in Eqs. (4.166) and (4.168). This precession produces a perturbation in the angular distribution of Eq. (4.170) or, equivalently, a time dependent perturbation of the rate at a given angle θ in the laboratory. The measurement of these time perturbations allows to access the magnetic moment and quadrupole moment of excited states. In order to measure the time dependence at different angles, the lifetime of the state of interest τ should be at least few tens of nanoseconds, i.e., long enough so that the condition

$$\omega_{L,Q}\tau > 1 \tag{4.171}$$

is verified.

In the case of a dipole magnetic moment measurement, the effect can be detected in a plane perpendicular to the magnetic field. The time-dependent intensity of the emitted γ ray observed in a detector at angle θ relative to the beam axis is given by

$$I(\theta, t) = I_0 e^{-t/\tau} W(\theta, t), \tag{4.172}$$

where τ is the lifetime of the decaying state and W is the angular distribution perturbed by the presence of the magnetic field $\mathbf{B}$. Considering only the dominant terms $\lambda = 2$ in Eq. (4.170), one gets

$$W(\theta, t) = B_2(I_i, t) A_2(\gamma) P_2(\cos(\theta - \omega_L t)). \tag{4.173}$$

If the detectors are placed at an angle of 90° from each other, one can define the time-dependent asymmetry

$$R(t) = \frac{W(\theta, t) - W(\theta + \pi/2, t)}{W(\theta, t) + W(\theta + \pi/2, t)}, \tag{4.174}$$

which can be shown, from Eq. (4.173), to be equal to

$$R(t) = \frac{3 A_2 B_2(I_i, t = 0)}{4 + A_2 B_2(I_i, t = 0)} \sin(2\omega_L t + \alpha). \tag{4.175}$$

The measurement of the asymmetry gives direct access to ω_L. A typical setup is sketched in Fig. 4.41. The method is illustrated by a measurement performed at RIKEN for the 4_1^+ state ($\tau \sim 200$ ns) of ^{32}Al.

The TDPAD method can also be applied to the measurement of quadrupole moment of isomeric sates by implanting spin-aligned nuclei in a crystal whose struc-

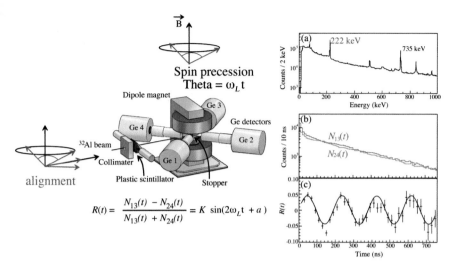

Fig. 4.41 Time-dependent perturbed angular distribution (TDPAD) method for the magnetic dipole moment measurement of excited states. Spin-aligned projectiles in the isomeric state under study are implanted in a region where an external magnetic field is applied. The asymmetry $R(t)$ of the measured transition at different angles as a function of time evolves as a sine wave depending solely of the Larmor frequency ω_L. The case of the 4_1^+ state ($\tau \sim 200$ns) of ^{32}Al is shown. The transition used for the asymmetry is the 222 keV transition from the 4_1^+ state at 956.6 keV to the 2_1^+ state. Data taken from [24]

ture generates strong field gradients within its lattice. As an example, a crystal of Zirconium metal has an hexagonal close-packed lattice with a typical mean electric field gradient of 50×10^{20} V/m^2. In a quadrupole moment measurement, the asymmetry $R(t)$ depends on multiples of the coupling constant

$$\omega_0 = \frac{\epsilon \omega_Q}{4 I_i (2 I_i - 1)}, \tag{4.176}$$

where $\epsilon = 3$ for an integer I_i and $\epsilon = 6$ for a half-integer I_i. The general form of the asymmetry for detectors placed at 90° from each other, in a plane perpendicular to the electric field gradient direction is given by

$$R(t) = \frac{3 A_2 B_2 (I_i, t = 0)}{4 + A_2 B_2 (I_i, t = 0)} \sum_n \alpha_n \cos(n \omega_0 t), \tag{4.177}$$

where α_n are tabulated coefficients with n running over integer values depending on I_i.

4.5   Particle Spectroscopy

The spectroscopy of a nucleus measures the response of a system after excitation. The ways to produce a nucleus in an excited state are numerous: Coulomb excitation, nuclear inelastic scattering, addition or removal of a nucleon at low energy via transfer, the knockout of a nucleon with high-momentum transfer, multi-nucleon transfer via deep inelastic scattering, fragmentation, fission, as well as nucleon annihilation from the interaction with an antiproton. All these reactions will lead to the population of specific eigenstates in the final nucleus. The mechanism and sensitivity of the different reaction processes are detailed in Chap. 8. In the following, the most used techniques to access the spectroscopy of nuclei are described. Particle spectroscopy via *missing mass* allows to determine the excitation energy of bound and unbound states with a sensitivity of the transferred angular momentum in the reaction. The *invariant-mass spectroscopy* allows the spectroscopy of unbound states decaying via the emission of a particle. This technique is abundantly used for neutron spectroscopy of neutron-rich nuclei. Finally, it can also be combined to gamma spectroscopy since (most) bound excited states decay to lower states via gamma-ray emission.

4.5.1   Nucleon Transfer and the Missing-Mass Technique

One interest of nucleon transfer reactions lies in their kinematics. A schematics of a two-body kinematics of two nuclei (1) and (2) interacting together and leading in the laboratory two nuclei (3) and (4) is illustrated in Fig. 4.42. From energy and momentum conservation, all information about the residue (4) can be obtained by measuring the momentum p_3 and using the known mass m_3 of the second outgoing particle (3). Indeed, the excitation energy $E_4^\star$ of particle (4), its spectroscopy, is given by:

$$E_4^\star = \sqrt{E_4^2 - p_4^2 c^2} - m_{4,gs}, \qquad (4.178)$$

where $m_{4,gs}$ is the ground-state mass of particle (4), while the total energy E_4 and the momentum vector p_4 of particle (4) can be obtained from

$$E_4 = T_1 + m_1 + m_2 - (T_3 + m_3) \qquad (4.179)$$

$$p_4^2 = p_1^2 + p_3^2 - 2p_1 p_3 \cos(\theta_3), \qquad (4.180)$$

where p_1 and $m_{1,2}$ are known and $T_{1,3}$, the kinetic energy of particles (1) and (3) can be obtained from their momenta following the well known identity $pc = \sqrt{T(T + 2mc^2)}$, all quantities given in MeV (with the convention $c = 1$ for the velocity of light).

This technique of reconstructing the mass, and therefore the excitation energy spectrum, of one of the two ejectiles without measuring it is denominated as *missing*

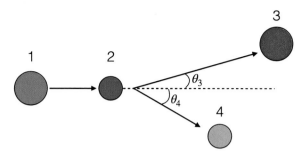

Fig. 4.42 Schematics of a two-body kinematics. In this illustration, particle (1) reacts with particle (2) at rest in the laboratory frame. The reaction ends up into two products: the particles (3) and (4) scattered at angles θ_3 and θ_4 relative to the particle (1) incident direct, respectively

mass technique. Since the nucleus of interest is not measured, all bound and unbound rates can be measured in the same way. The missing mass technique is one of the few possible (and most used) techniques for the spectroscopy of unbound states. The kinematics of a (p, d) reaction in inverse kinematics is shown for the case of ^{8}He+p at 15.7 MeV/nucleon measured at GANIL in Fig. 4.43. It is clear from the data that other reaction channels take place at the same time with rather strong cross section, namely elastic and inelastic scatterings and (p, t) transfer reaction. From the analysis of the data, it was shown that the DWBA approximation is not sufficient and that the coherent coupling to other reaction channels should be taken into account (see Chap. 8).

Fig. 4.43 Kinematics of ^{8}He+p at 15.7 AMeV measured at the SPIRAL facility, GANIL. Figure reprinted with permission from [25]. ©2021 by Springer Nature

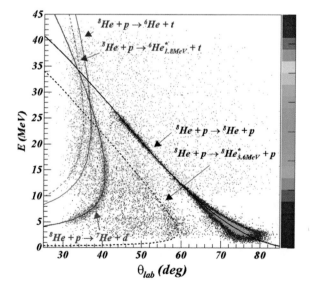

A nucleon transfer reaction is a quantum process during which a quantized angular momentum is transferred. The differential cross section to a given final state, as a function of the scattering angle, shows an oscillation pattern whose structure (position of maxima and minima) depends on the transferred angular momentum. In the following, we provide a classical correspondence between the position of the first angular maximum after a transfer reaction and the transferred angular momentum.

For simplicity, we assume a reaction in direct kinematics where a nucleon is removed from a target whose mass is considered infinite compared to the projectile and light residue, so that all the transferred angular momentum is taken away by the light particle. In a quantum system for which the angular momentum is a good quantum number, the operator L^2 commutes with the Hamiltonian and its application to the nucleus wave function gives

$$L^2|\Phi\rangle = \ell(\ell+1)\hbar^2|\Phi\rangle. \tag{4.181}$$

In the classical limit of a transfer reaction occurring at the surface of the nucleus, the transferred momentum in the reaction is given by $L = p_\perp R$, where $p_\perp$ is the transverse momentum of the scattered particle and R is the distance between the two nuclei at grazing, leading to the classical limit approximation

$$p_\perp R = \sqrt{\ell(\ell+1)}\hbar. \tag{4.182}$$

In the infinite mass target approximation, the perpendicular momentum of the light recoil can be approximated to $p_\perp = p\sin(\theta_\circ)$ with the scattering angle $\theta_\circ$. Under these approximations, the scattering angle (the first maximum of the cross section) is given by

$$\theta_\circ = \arcsin\left(\frac{\sqrt{\ell(\ell+1)}\hbar}{pR}\right). \tag{4.183}$$

One can verify the validity of this classical estimate of the first maximum of the differential cross section on various examples. Consider the example $^{52}\mathrm{Cr}(d,p)^{53}\mathrm{Cr}$ at 10 MeV/nucleon, as shown in Fig. 4.44. The projectile energy 10 MeV/nucleon for a deuteron is equivalent to a momentum $p = 193$ MeV/c. If one assumes the standard parameterization of the nuclear radius $R = 1.2\,A^{1/3}$ fm, one gets $R = 4.5$ fm for $^{52}\mathrm{Cr}$. Following Eq. (4.183), for $\Delta\ell = 0, 1, 2$, one calculates $\theta_\circ = 0°, 19°, 34°$, respectively. The $\Delta\ell = 1$ is in good agreement with the data. The values estimated for $\Delta\ell = 0, 1, 2$ are in good agreement with the microscopic calculations shown in Fig. 4.44.

Intuitively, the beam velocity will impact strongly the population of states depending on the transferred angular momentum they imply. Low incident energy will favour the population of states with low $\Delta\ell$ momentum transfer, for a given Q-value of the reaction. The optimum population of given states requires the *momentum matching* imposed by Eq. (4.182). Note that nucleon transfer at very high incident energy can

Fig. 4.44 Measured angular distribution for ^{52}Cr(d,p)^{53}Cr$_{gs}$ at 10 MeV/nucleon. The data points are compared to calculations supposing different transfer angular momenta $\Delta\ell$ from 0 to 2. The location of the first maximum shows a $\Delta\ell = 1$ transfer. Figure reprinted with permission from [26]. ©2021 by Elsevier

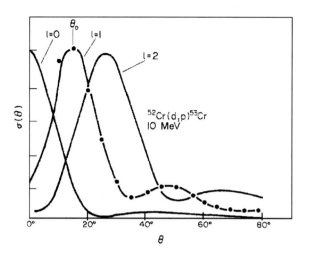

be used to probe the high momentum component of the momentum distribution of nucleons in nuclei.

A large part of our understanding of nuclear structure is based on transfer reactions. Transfer is a key tool and a work horse for low-energy ion-beam facilities worldwide. Transfer reaction experiments with radioactive beams are mostly based on the detection of recoil light charged particles. Several recent or ongoing developments of detectors aim at improving the energy or angular resolution as well as the detection efficiency, compared to the previous generation of detectors. Semiconductor detectors to measure light reaction products, hydrogen and helium isotopes in most transfer reactions, give today the best resolution when thin targets are used with large efficiency. The recent development of large-volume time-projection-chambers for nuclear physics have opened new opportunities, in particular for low-intensity radioactive beams.

4.5.2 Charged-Particle Detectors

Any charged particle passing through a material at velocity v loses energy from electromagnetic interaction between the nucleus and the electrons of the material. In the case of nuclear physics energies, ions and atoms experience the same energy losses: the nucleus interacts with the electrons of the material, nearly independently on the amount of electrons bound to it when it impinges the material. The energy loss process frees charge carriers in the material of the detector which, when an electric field is present in the volume of the detector, can drift towards electrodes located at the edge of the detector. The movement of charge carriers within the detector induce an electric signal on the electrode, leading to the detection of the event.

The basic steps of charged-particle detection in most nuclear and particle physics starts with an energy loss in the sensitive region of a detectors, where the produced charges are separated. The energy loss of particles with matter is described hereafter. The principle of two important types of particle detectors, semiconductors (solid-state detectors) and time-projection chambers (most often gas detectors), is presented. Note that there are other detection mechanisms (for example, crystal calorimeters which rely on the measurement of the rise of temperature of the detectors from the crystal vibrations, called phonons, induced by the particle and not based on the ionization), which are not covered in this textbook.

4.5.2.1 Energy Loss and the Bethe-Bloch formula

The energy loss of charged particles through matter has interested many prominent physicists: Bohr provided the first classical derivation, Bethe and Bloch gave a quantum mechanical description while Fermi added density corrections to the quantum formulation. We first describe here the classical derivation.

Let us consider the interaction of a charged particle with mass M and electric charge Ze moving at velocity v through a medium with a density n_e of electrons assumed to be at rest (Fig. 4.45).

The momentum transfer to the electron by the charged particle travelling along a straight line at impact parameter b can be calculated as

$$\Delta \mathbf{p} = \int_{-\infty}^{+\infty} \mathbf{F}_{\text{Coulomb}} dt = \int_{-\infty}^{+\infty} Ze^2 \frac{\mathbf{r}}{r^3} dt. \tag{4.184}$$

The longitudinal forces cancel since $F_\parallel(-x) = -F_\parallel(x)$ and only the perpendicular component remains: the kinetic energy acquired by an electron corresponds to an impulse perpendicular to the nucleus trajectory;

$$\Delta p_\perp = \frac{2Ze^2}{bv}, \quad \Delta p_\parallel = 0. \tag{4.185}$$

For a path length dx in a slice of impact parameter between b and $b + db$, the total number of electrons found is $n_e 2\pi b db dx$. The corresponding energy loss by the charged particle is then

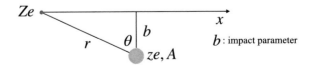

Fig. 4.45 Assumed trajectory to calculate the momentum transferred to a nucleus of charge Ze from collisions with individual electrons through a material of electron density n_e. In this illustration, the material is composed of atoms of atomic number z and a nucleus of A nucleons

$$dE(b) = -\frac{4\pi Z^2 e^4}{m_e v^2} n_e \frac{db}{b} dx, \tag{4.186}$$

which diverges for $b \to 0$. The integration of Eq. (4.186) requires the choice of adequate limits for integration over the impact parameter. The integration over collision parameters b_{min} and b_{max} gives

$$-\frac{dE}{dx} = \frac{4\pi Z^2 e^4}{m_e c^2 \beta^2} n_e \ln \frac{b_{max}}{b_{min}}, \tag{4.187}$$

where $\beta = v/c$. The lower impact parameter corresponds, in the classical picture, to a maximum energy transfer to the electron. This maximum transfer corresponds to a head-on collision. In the case of a head-on collision, the momentum and energy conservations lead to an electron velocity v_e gained after a collision with an incident ion of mass M at velocity v:

$$v_e = \frac{2v}{1 + \frac{m}{M}} \sim 2v. \tag{4.188}$$

Using Eq. (4.185), in a non-relativistic approximation, the energy loss is given by

$$\Delta E(b) = \frac{p_e^2}{2m_e} = \frac{(\gamma m_e (2v))^2}{2m_e} = \frac{4Z^2 e^4}{2m_e b_{min}^2 v^2}, \tag{4.189}$$

where γ is the Lorenz factor. Equation (4.189) gives a relation,

$$b_{min} = \frac{Ze^2}{m_e v^2}. \tag{4.190}$$

Note that this classical derivation is kinematically inconsistent since Eq. (4.185) assumes an electron at rest while a head-on collision implies momentum transfer to the electron during the collision. The upper limit for integration is the impact parameter which corresponds to an interaction time shorter than one period of the electron ($b/v \leq 1/\omega$), where ω is the electron orbiting frequency. We then obtain

$$b_{max} = \frac{v}{\omega}. \tag{4.191}$$

Integrating over these limits, one gets the classical energy loss formula as derived by Bohr

$$-\frac{dE}{dx} = \frac{4\pi Z^2 e^4}{m_e c^2 \beta^2} n_e \ln \frac{m_e c^3 \beta^3}{\omega Z e^2}. \tag{4.192}$$

Bethe derived the energy loss formula using quantum mechanics, including relativistic effects still under the assumption that the ion velocity is larger than the orbital electron velocity. The result is the so-called Bethe-Bloch formula for charged-particle energy loss in medium:

$$- \frac{dE}{dx} = K Z^2 \frac{z}{A} \rho \frac{1}{\beta^2} \left[\frac{1}{2} \ln \left(\frac{2 m_e c^2 \beta^2 \gamma^2 T_{max}}{I^2} \right) - \beta^2 - \frac{\delta}{2} \right], \qquad (4.193)$$

where z/A is the ratio of atomic number and mass number of the considered material (most stable materials have the same ratio z/A) and $\rho = n_e \times (A/z)$ the atomic density in mol/m^3, $I = \hbar \omega$, $K = 4 \pi N_A r_e^2 m_e c^2$ with N_A being the Avogadro number and with the classical electron radius defined as $r_e = e^2/(m_e c^2)$. T_{max} is the maximum energy transferred during a single collision given by $T_{max} \sim 2 m_e c^2 \gamma^2 \beta^2$, for $M \gg m_e$. The mean excitation energy for elements beyond oxygen is given by $I \sim 10 \times Z$ eV. The last term $\delta/2$ of Eq. (4.193) is a density correction term for high-energy particles due to a Lorentz contraction of the electric field. This correction term originates in a medium polarisation effect absente from the above derivation. The effect flattens the logarithmic increase of the relativistic term of the above derived Bethe-Bloch equation and leads to the so-called *Fermi plateau*. The term $\delta/2$ depends on $\beta \gamma$.

The calculated energy loss for charged particles as a function of their mass-normalized momentum $\beta \gamma$ is shown for different media in Fig. 4.46. All curves show the same feature: at low energy, energy loss per unit length is larger and decreases down to a broad minimum. Particles around this minimum are called minimum-ionizing particles (MIP). If a detector is built to be sensitive to MIP particles, it will therefore be sensitive to all (within the range of the readout electronics, not discussed here). The energy loss then increases slowly with energy. The Bethe-Bloch formula is precise at the level of few % within the momentum range of $0.1 < \beta \gamma < 1000$ in the case of intermediate-Z materials. For lower momenta, the ion velocity is comparable to the orbital velocity of electrons in the material, and for higher momenta the radiative losses become non negligible.

In the case of a compound material, the energy loss per distance unit can be determined from the energy loss for individual constituents as given by the Bethe-Bloch equation. Assuming that the stopping power per atom of the compound is additive (assumption also called the Bragg-Kleeman rule), the energy loss per distance unit in a compound material composed of atoms of mass number A_i in stoechiometric fraction ω_i is

$$- \frac{1}{\rho} \frac{dE}{dx} = - \sum_i \frac{\omega_i}{\rho_i} \left(\frac{dE}{dx} \right)_i, \qquad (4.194)$$

where r_i is the atomic density of the constituent i, approximated by

$$\rho_i = \frac{A_i}{A_{\text{eff}}} \rho, \qquad (4.195)$$

where we define $A_{\text{eff}} = \sum_i \omega_i A_i$. The stoechiometric ratios are normalized to unity: $\sum_i A_i \omega_i = 1$. In the case of a thick material, a charged particle will loose energy

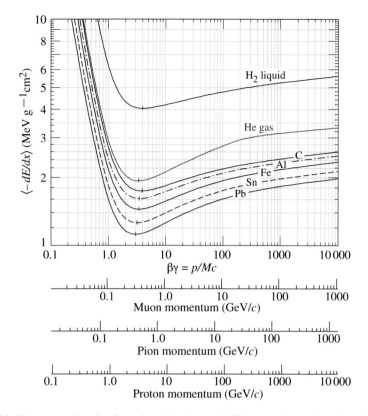

Fig. 4.46 Mean energy loss by charged particles through different media as a function of incident momentum. Except light-atom media such as hydrogen and helium, all media exhibit a similar curve with a minimum energy loss per unit distance of ~1.5 MeV.g^{-1}.cm^2. A particle having a momentum corresponding to this minimum is called a *MIP*, standing for Minimum Ionizing Particle. Figure from [27]

inside the material before exiting. The calculation of its energy loss might then require to integrate Eq. (4.193) to take into account the energy-dependence of the energy loss in the material. In the extreme case, the charge particle might stop inside the material. One defines the range R of a charged particle at a given incident kinetic energy in a specific medium as the mean distance it should travel inside the medium until it is stopped. Under the assumption that the kinetic energy of the particle is lost continuously inside the medium, one can calculate the range of a particle by integrating the inverse of Eq. (4.193). The range is defined as

$$R = \int_0^{E_0} \frac{dx}{dE} dE,$$ (4.196)

where E_0 denotes the initial kinetic energy at $x = 0$.

Problem

4.9 Calculate from Eq. (4.193) the energy loss of a 200-MeV/nucleon carbon ion in 1 cm of water at rom temperature. By integrating the inverse of the Bethe-Bloch relation, calculate the range of a 10 MeV, 200 MeV and 400 MeV ^{12}C ion in water. Note that by assuming that the shape of the dE/dx curve as a function of material depth is independent of the material, it can be shown [34] that the range in a compound medium can be obtained from the range in each of the constituents by

$$R = \frac{\rho}{\sum_i \rho_i / R_i}, \tag{4.197}$$

where ρ_i is the atomic density of the constituent i. Conclude about the most suitable energy to build an accelerator for carbon-ion therapy?

For a non-relativistic charged particle, in a given thin detector of thickness Δx, the Bohr formula of Eq. (4.192) relates the energy loss ΔE of the nucleus to its mass M and atomic number Z following

$$\Delta E \propto \frac{M Z^2}{E} \Delta x, \tag{4.198}$$

where $E = M \beta^2 c^2 / 2$ is the kinetic energy of the particle. The energy loss contains therefore information on the mass, atomic number and kinetic energy. Its measurement is at the basis of most of identification techniques for light recoil charged particles in transfer reactions.

4.5.2.2 Semiconductor Detectors

Semiconductors made of silicon are a wide spread technology for detecting charged particles with high position and energy resolution. We recall here the principle of a semiconductor with the objective to quantify the energy resolution one can expect from such detectors.

What is a semiconductor detector and how does it work? Solids can be classified as a function of their resistivity or conductivity, the inverse of the resistivity. The conductivity of a material quantifies its capability to conduct a current when an electrical potential is applied to it. The conductivity of a metal is high, while low for an insulator, by definition. Semi-conductors stand in between these two extremes and are usually the choice for solid-state detectors. They usually have a crystal structure. The most used in nuclear and particle physics are the single-atom crystals Si and Ge although other composite crystals, such as GaAs, CdTe and HgI$_2$, can be used. Such semiconductors composed of pure crystal without any impurity are called *intrinsic* semiconductors.

The macroscopic properties of conductors can be related to their quantum structure: the energy levels accessible to electrons in crystals are separated in energy

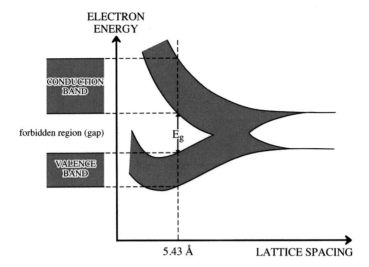

Fig. 4.47 Energy levels of Si atoms in a two-level model. When the atoms interact with each other, the degeneracy of identical levels is broken and transforms into high level-density energy bands. For a semiconductor, the inter-atom spacing leads to a full valence band separated in energy from the a conduction band by a so-called forbidden region or gap E_g

bands. The electrons are indeed distributed over groups of energy levels, themselves separated by energy gaps. This can be illustrated in a simple two-level scheme. Let us assume a crystal composed of one single kind of atoms, each presenting two energy levels: one filled with electrons and one empty. If the inter-atom distance is high enough so that they do not interact, the energy of such a crystal will be the sum of individual-atom energies. As soon as the atoms get closer with each other and interact, the levels will mixed and, up to the very large number of atoms involved, will lead to regions of high density levels. For specific inter-atom spacing, separate energy bands might appear: this is the case for semiconductors, as illustrated in Fig. 4.47.

The last occupied band is called the *valence* band. In this energy band, the electrons are bound to the atoms of the crystal. The next band is called the *conduction* band where electrons can move freely from the atoms of the crystal. These two bands are separated by a *forbidden* band (energy gap).

The size of the energy gap is what characterizes a semiconductor. In an insulator, the valence band is fully occupied and is separated by a large energy gap from the conduction band so that the electrons can not move inside the conduction band. In a metal there is an overlap between the conduction and the valence band (no energy gap). The main characteristics of conductors, insulators and semiconductors are compared in Fig. 4.48. The electrons can therefore move freely whenever a potential difference is applied to the metal. In an intrinsic semiconductor, the valence band is fully occupied as in an insulator but the energy gap separating the electrons from the conduction band is small, typically 1eV (1.12 eV for Si, 0.66 eV for Ge at 300 K). At room temperature, thermal fluctuations are enough to promote some electrons in the

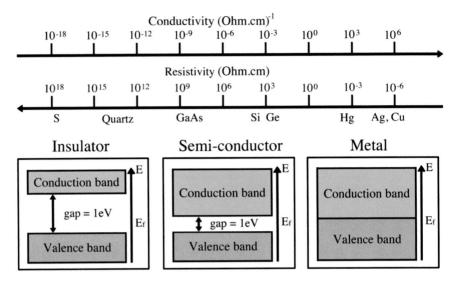

Fig. 4.48 Classification of solids as a function of their resistivity and conductivity. The energy-band structures for insulators, semi-conductors and metals are illustrated

conduction band and semiconductors are therefore slightly conducting. Applying a potential difference across a semi-conductor then leads to a current: the promotion of an electron to the conduction band leaves a vacancy in the crystal. An electron from a neighbouring atom in the valence band will fill this vacancy and free another vacancy as if a positive charge, called *hole*, moves in the opposite direction compared to electrons. The migration of electron-hole pairs across the crystal after the energy loss a particle going through the semi-conductor is at the heart of the detection principle. In the case of silicon, the size of the forbidden band is 1.12 eV at the room temperature 300 K and, because not all energy loss goes into the promotion of electrons from the valence to the conduction band, the energy cost to create an electron-hole pair is $\epsilon = 3.61$ eV at 300 K. The corresponding mean energy to create an electron-hole pair in germanium is $\epsilon = 2.96$ eV. The mean number of electron-hole pairs created is

$$N = \frac{E}{\epsilon}. \tag{4.199}$$

When an ionizing radiation interacts with the crystal, not all the energy loss contributes in creating electron-hole pairs, indeed a significant part of the energy loss does not lead to ionization but can end up into crystal collective excitations which dissipate then into heat. The Fano factor F defines the mean fluctuation of charge carriers of electron or holes $\langle \Delta N^2 \rangle$, created from an energy loss E

$$\langle \Delta N^2 \rangle = FN = F\frac{E}{\epsilon}. \tag{4.200}$$

The Fano factor can be seen as a correction to the Poisson statistics. The Fano factor is measured $F = 0.115$ for Si, $F = 0.13$ for Ge. The full width at half maximum for a given energy loss is then given by

$$\sigma_{\text{FWHM}} = 2.35\sqrt{F\epsilon E}. \tag{4.201}$$

For a 1 MeV energy loss in a germanium detector, one gets a theoretical resolution of

$$\sigma_{\text{FWHM}} = \sqrt{0.13 \times 10^6 \text{ eV} \times 3.61 \text{ eV}} = 1.5 \text{ keV}, \tag{4.202}$$

from statistical fluctuations only.

The use of an intrinsic semiconductor material as discussed so far requires to produce a very pure crystal with little impurities, at the level lower than 10^7 cm^{-3}, which is extremely difficult to achieve technically. Even with high purity, when a potential difference is applied across the detector, a current will flow through the detector and this might prevent from measuring the charges steaming from the energy loss of a charged particle if the signal over noise ratio is too low. In a silicon crystal with a surface of 1 cm^2 and a thickness of 1 mm, an applied voltage of 500 V will result in a 0.1 A current, much higher than the typical 1 μA generated by a charged particle punching through the silicon detector! One option to overcome the problem is to cool down the detector so that thermal fluctuations prevent valence electrons to "jump" into the conduction band. There is also the possibility to use a semiconductor with a large gap, such as diamond (carbon) crystals, whose gap is 5.47 eV and large enough to prevent thermal electrons to promote to the conduction band at the cost of resolution.

The most spread solution is to dope the semiconductor with low-density impurities in the crystal and rather low voltages. These impurities are other types of atoms which replace atoms of the semiconductor crystal at some sites. These impurities will change the energy level structure of the semiconductor and modify its properties. One can dope a crystal with *donor* (n doped) or an *acceptor* (p doped) atoms. A donor has more valence electrons than the atoms of the crystal where it is implanted. The implantation of a donor results in the creation of an energy level in the forbidden band, close to the conduction band. An acceptor is an atom that has less valence electrons than the atoms of the crystal where it is implanted. Its implantation results in the creation of an energy level in the forbidden region close to the valence band.

Most of the semi-conductor detectors used in nuclear physics are based on *p-n junctions*. Such a junction is illustrated in Fig. 4.49. It is composed of a n-doped region in contact with a p-doped region. The impurity densities are of the order of 10^{12} to 10^{18} cm^{-3}, i.e., still at the level of 10^{-12} to 10^{-6} compared to the majority atoms composing the crystal. At thermodynamical equilibrium, the Fermi energy should be the same over the crystal: this leads to a gradient of density of charge carriers in the crystal around the junction. Due to the difference of electron and hole densities in the two regions, charges move: holes from the p region diffuse towards the n region and electrons from the n region diffuse towards the p zone. Two space charge zones appear in the bulk of the crystal as illustrated in Fig. 4.49. These space

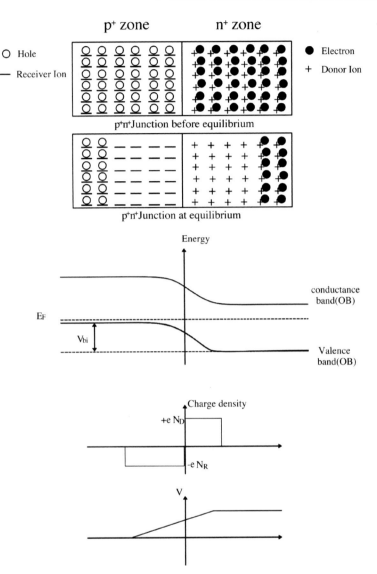

Fig. 4.49 (Top) *pn* junction before contact and at thermodynamical equilibrium. The contact of the doped *p* and *n* semiconductors implies a drift of charge carriers across the interface and lead to a depleted region with no charge carrier inside. The energy levels and electric potential are modified across the junction (middle). The Fermi level is constant across to the material. It is close to the valence band in the *p* type region, while it is close to the conduction band in the *n* type material. The junction leads to a built-in voltage bias V_{bi}. (Bottom) A space charge density develops in the depleted region. It is shown under the *abrupt change approximation*

Fig. 4.50 Example of a realistic design of a $p^+ - n - n^+$ junction used for charged particle detection

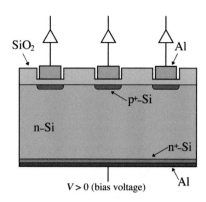

charges lead to the creation of an electric field across the non neutral region. This electric field opposes to the natural diffusion of charges and an equilibrium phase is reached. The zone located in between the p and the n zones is said *depleted*, i.e., without any free charge carrier. It has therefore a very large resistance, and this is the feature that is used for the detection. The charges created by an ionising particle are not anymore diluted in a large number of diffusing charge (a current). When the created electrons and holes move towards the edges of the depleted region, where the electrodes are located, they induce a signal on these electrodes which can be detected. In practice, the depleted region as described here is small and the field is too weak for an effective collection of charges. An additional bias voltage is applied through the junction to increase the size of the depleted region, i.e., with the cathode to p and the anode to n so that electrons and holes are pulled out of the depletion zone. In practice, many extrinsic semiconductor detectors are composed of a n-type crystal sandwiched between p-type and n-type regions with much higher impurity densities, therefore called p^+ and n^+, respectively. These high-impurity density regions are used (i) to facilitate the electrical contact of the crystal to be used as a detector, (ii) to create a sensitive volume mostly composed of the n-type region. The larger the external potential, the larger is the depletion zone. The applied voltage is most of the time chosen so that the n-type region is fully depleted, i.e., charge carriers are confined to the p^+ (holes) and n^+ (electrons) regions. There is an upper limit to the increase of the voltage beyond which sparks across the junction may appear, leading to irreversible damages to the detector. Typically, depleted regions in silicon can reach up to few millimeters. The principle of the $p^+ - n - n^+$ junction has been extensively applied with various geometries to satisfy the requirements of specific experiments. In particular, as illustrated in Fig. 4.50, the implantation of impurities can be arranged in a strip pattern to give a position resolution from millimeters to $\sim 50\,\mu m$ depending on the pitch size of the strips.

Energy measurements and particle identification with Si semiconductor detectors are illustrated here. With two detectors assembled in a telescope, i.e., in two detection layers one after the other, a charged particle with the right amount of energy may lose a part E_1 of its total kinetic energy E in the first detector and may stop in the second

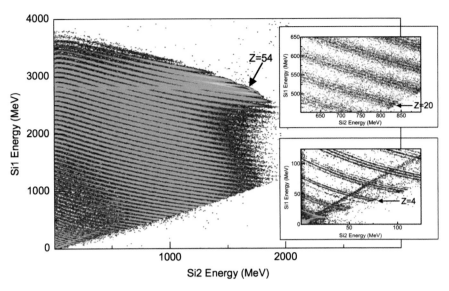

Fig. 4.51 Illustration of a charged-particle identification with the $\Delta E - E$ technique. Two $300\mu m$-thick high-homogeneity silicon detectors are used for this measurement. The separation in Z is achieved for all produced nuclei. On the main plot, nuclei with $Z = 54$ (Xe) are indicated. The inserts are enlargements of the main plot to show the isotopic separation: one can distinguish four Ca $(Z = 20)$ isotopes and two Be $(Z = 4)$ isotopes produced in deep inelastic collisions of a ^{129}Xe beam at 35 MeV/nucleon with an heavy-ion target. Figure reprinted with permission from [28]. ©2021 by Elsevier

layer while depositing its residual kinetic energy $E_2 = E - E_1$. One can apply the so-called $\Delta E - E$ technique to identify the mass and charge of the particle. From Eq. (4.198), one gets the relation between E_1 and E_2:

$$(E_1 + E_2)E_1 \propto MZ^2dx. \tag{4.203}$$

It is here clear that the measurement of the two energies leads to M and Z if the energy resolution is sufficient. In Fig. 4.51 is illustrated a charged-particle identification performed with two high-homogeneity and high-planarity semiconductor silicon detectors with the $\Delta E - E$ technique. Each of them are 300 μm thick and were placed against each other. Charged particles are produced from deep inelastic collisions of a ^{129}Xe beam at 35 MeV/nucleon with an heavy-ion target. The isotopic and mass separation is visible. The atomic number is identified for all produced particles. Zooms in the low mass region shows that isotopes up to $Z\sim22$ can be separated in mass.

Thin stripped Si semiconductors are natural choice of detector for low-energy transfer reactions and charged-particle spectroscopy. Indeed, the requirements for a light charged-particle detector used for transfer reactions can be summarized as: (i) a good position resolution to determine the scattering angle of recoil particles.

In inverse kinematics, angular resolution of less than a degree is required in most application for a reasonable missing-mass reconstruction as defined in Eq. (4.180). (ii) A good energy resolution for the determination of the kinetic energy of the particle, and (iii) particle identification capabilities which also lead to requirements in energy and time resolution. An example of such an array for charged-particle spectroscopy is the MUST2 detector showed in Fig. 4.52. The array is composed of several telescopes which can be arranged in space depending on the kinematics of the reaction of interest. Each module is composed of three detection layers: a first 300 μm thick silicon detector (p^+n), stripped on both sides with a pitch of 700 μm followed by a 3-mm Si(Li) detector and a 30-mm Cs(I) detector. These different layers offer a variety of options for particle identification depending on the energy of the particle. Low-energy particles that do not punch through the first detection stage, and that can not be identified by the above-described $\Delta E - E$ method, are identified via the correlation of their time of flight from the target to the first stage of the detector and the total kinetic energy measured by this first stage driven by the definition of kinetic energy

$$E = \frac{1}{2}Mv^2 = \frac{1}{2}M\frac{d^2}{t^2}, \tag{4.204}$$

where v is the velocity of the charged particle, d its flight distance between the target and the detector and t the corresponding time of flight. The above non-relativistic limit is valid for low-energy ions.

4.5.2.3 Active Targets and Time Projection Chambers

Radioactive-ion beams can be produced at rather low intensity and may require the use of a thick secondary target to reach a sufficient luminosity. Experiments based on charged-particle detection usually favour the use of thin targets to preserve the invariant-mass energy resolution. Indeed, the thicker the target, the more uncertainty there is on the not-measured energy loss inside the target which impacts directly the invariant-mass energy resolution. Statistics and resolution are usually competing objectives. Active targets, i.e., devices where the target medium is also the sensitive region of the detection, offer a solution to this dilemma. A time projection chamber is a well-suited device for such purpose. It is a gas detector whose volume acts as a target and as a medium to reveal the tracks of charged particles emitted from a reaction occurring in the active volume.

We describe here the main characteristics of a time projection chamber, as illustrated in Fig. 4.53. A strong electric field of several 100 V/cm is applied over the gas volume of the TPC. The homogeneity of the field across the gas volume is usually guaranteed by a field cage, a set of equidistant strips or wires between the cathode and the anode which are separated from each other by adequate resistors. Charges produced from ionization of gas molecules will drift along the electric field lines within the detector. The electrons, at the end of their drift, reach an amplification

Fig. 4.52 Picture of four
MUST2 telescopes arranged
in a compact forward angle
geometry for an experiment
at GANIL. Courtesy F.
Flavigny, LPC Caen

region where the produced charges induce a signal on detection pads connected to electronics for readout. The drift time gives the position of the ionization in the gas volume.

A charged particle passing through the gas of the detector losses energy by ionizing and exciting molecules of the gas in similar amounts. The energy loss of the charged particle is not uniform along its path but occurs randomly. The number k of collisions along a segment of length ℓ follows a Poisson statistics

$$P(k) = \frac{(\ell/\lambda)^k}{k!} \exp(-\ell/k),   \qquad (4.205)$$

where $\lambda = 1/(n_e\sigma_I)$ is the mean distance between clusters, n_e the electron density of the gas and σ_I the ionization cross section per electron. Most ionization collisions produce a single electron-ion pair but in some occurrences the electron receives enough kinetic energy to ionise one or more neighbouring molecules. In rare collisions, the electron issued from the primary collision receives enough kinetic energy

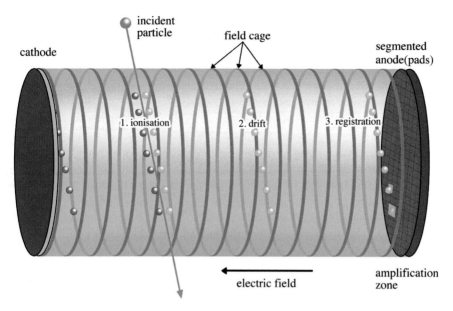

Fig. 4.53 Principle of a time projection chamber. A charged particle traversing the active region of the detector will ionize atoms along its trajectory. Under the electric field imposed across the so-called drift region of the TPC, the electrons (green) and ions (purple) from the ionization do not recombine and drift apart, the positive ions towards the cathode and the electrons towards the anode. On the anode side, the electrodes are amplified in a zone of high field. The movement of the produced charges induces a signal on the electrodes of the anode. The induced signal on the anode electrodes and the corresponding drift time allow a 3D reconstruction of the track

to produce its own ionization track in the gas of the detector. These energetic electrons are called δ *electrons* and represent a nuisance in most active-target measurements.

The energy loss following the Bethe-Bloch formula of Eq. (4.193) translates into ionization by introducing the mean energy necessary to produce an electron-ion pair

$$w = \frac{1}{n_T} \langle \frac{dE}{dx} \rangle, \tag{4.206}$$

where n_T is the total number of electrons per length unit along the path of the charged particle inside the gas volume of the detector. For rare gases, $w \sim 1.5 E_I$ and for common molecular gases $w = 2$ to $2.5 E_I$, where E_I is the ionization energy of the gas. The difference between w and E_I results from energy losses leading to no ionization, i.e., excitation of gas molecules below the ionization energy. w is typically 30 eV.

Electrons produced via ionization, when an electric field E is present in the gas volume of the detector, are accelerated in between two electron-electron collisions and acquire a net drift velocity $\langle u \rangle$ which depends on the gas properties, the field applied and its pressure P. This average velocity is a slow motion compared to the instant velocity. At a given pressure and for a specific gas, it is the result of the competition between the acceleration caused by the electric force between two collisions and the friction force caused by the collisions with gas molecules. It is therefore proportional to the electric field. By considering the mean time τ between two collisions of drifting electrons with molecules of the gas, one can get an intuitive interpretation of the relation between the drift velocity and the electric field:

$$\langle \mathbf{u} \rangle = - \left(\frac{\tau e}{m} \right) \mathbf{E} = -\mu_e \mathbf{E}, \tag{4.207}$$

which defines the drift mobility μ_e as a proportional factor. In the same way, a drift mobility can also be defined for ions, with the difference that $\langle \mathbf{u} \rangle$ is in the same direction than $\mathbf{E}$ for ions. Electrons are typically 1000 times faster than ions in gas since they are much lighter.

With typical gases used for gas detectors and drift electric fields of about 200 V/cm, the drift velocity of electrons is of the order of 1–10 cm/μs at atmospheric pressure, as illustrated in Fig. 4.54. Because of the random collisions of the drifting electrons with gas molecules, any electron will present a dispersion compared to the drift trajectory solely dictated by the drift velocity. The mean square deviation of the electron is driven by a transverse and a longitudinal (along the electric field direction) diffusion coefficients D_T and D_L, respectively. After a drift distance d, the transverse (σ_T) and the longitudinal (σ_L) dispersions are given by

$$\sigma_T^2 = \sigma_x^2 + \sigma_y^2 = 4D_T d/u, \tag{4.208}$$
$$\sigma_L^2 = \sigma_z^2 = 2D_L d/u. \tag{4.209}$$

Typical values are $\sigma_{L,T} \sim 0.1 - 1\text{mm} \times \sqrt{d\,[\text{cm}]}$.

The drifting electrons are amplified when reaching the anode so that they can induce a detectable signal on the pads of the TPC. The induced signal can be estimated by use of the Shockley-Ramo theorem, as detailed in the following Sect. 4.5.2.4. The set of induced charges collected on the pads gives an image of the track, projected in the (x, y) transverse plane. The drift time for each pad gives the longitudinal component of the track.

The amplification of the electrons is made by creating a small region of a high electric field (several 10 kV/cm) above the position sensitive plane. In this high-field region, the electrons are accelerated and acquire sufficient kinetic energy to ionize the gas along their path. A proper choice of accelerating field leads to a regime of controlled amplification with typical gains from 10^3 to 10^6. Micro Pattern Gas Detectors (MPGD) are commonly used for the amplification stage of time projection chambers. There are two main technologies for such amplification at the anode: (i) Micromegas and (ii) Gas Electron Multiplier (GEM).

The principle of a Micromegas is illustrated in the top panel of Fig. 4.55. The high electric field is obtained by a mesh at about 100 μm above the detection plane. This mesh, typically biased at few hundred volts, is transparent to the transmission of

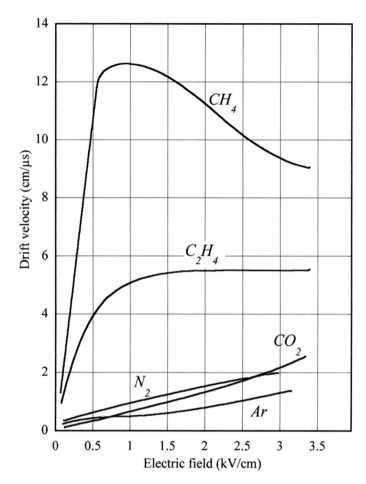

Fig. 4.54 Electron drift velocity in different gases as a function of the electric field strength at atmospheric pressure. The figure is reprinted with permission from [8]. ©2021 CERN

drifting electrons. The mesh is located very close to the anode so that the electric field between the mesh and the anode amplifies the electrons through multiple collisions after acceleration. The motions of charges (primary and secondary) in the amplification region induce a signal on the pad plane composing the anode. The signal is composed of a fast component due to the electron's motion and a slower component due to the ions flowing back to the mesh. A critical aspect of Micromegas detectors is to achieve an homogeneous width of the amplification region. Micromegas detectors are robust and can be easily produced in large dimensions.

GEM detectors are another popular amplification technology. They are illustrated in the bottom panel of Fig. 4.55. A GEM is produced by chemical etching of holes at high density into a metal coated polymer foil with a typical thickness of 50 μm.

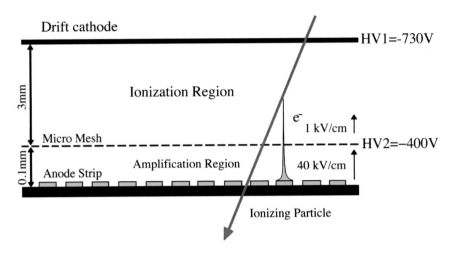

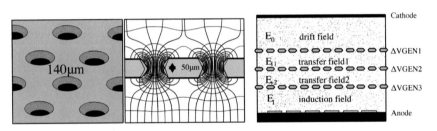

Fig. 4.55 (Top) Principle of a Micromegas detector. (Bottom) View of a GEM detector and its principle. Stacks of GEM detetors can be used for a better amplification and lower the ion back flow into the active volume of the detector

Amplification occurs inside the holes by applying a voltage across the foil, typically few hundred volts. The high mobility of drifting electrons make the GEM transparent since the electrons are conducted towards the holes. Ions produced during the amplification are mostly directed towards the metal coating of the GEM. GEMs are usually used in series to obtain sufficient gain with low discharge probability. And can be used in combination with Micromegas amplification as a pre-amplification stage. A stack of GEMS allows to reduce significantly the ion back flow in the TPC and therefore, by reducing the space charge of slow drifting ions in the active volume of the detector, allows the use of TPCs with relatively high beam intensity.

TPCs are used in many different types of experiments in nuclear and particle physics. They can be combined with other detectors or be operated inside a magnetic field. In that case, the trajectories of charged particles are curved in the volume of the gas. In case of a constant magnetic field B, the projection of the trajectory onto a plane perpendicular to $\mathbf{B}$ is given by

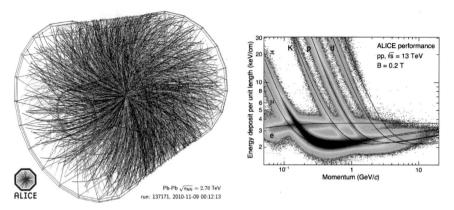

Fig. 4.56 (Left) Three-dimension representation of tracks recorded by the ALICE TPC after one central collision event of two lead ions at an energy of 2.76 TeV. ©2021 by CERN (Right) Energy loss versus momentum as measured by the ALICE TPC. Figure from [27]

$$B\rho_\perp = \frac{p_\perp}{q}, \tag{4.210}$$

where $p_\perp$ is the momentum of the particle perpendicular to the magnetic field direction, q its charge and $\rho_\perp$ the curvature radius of the trajectory in the perpendicular plane. As an illustration of the diversity of applications for TPCs, a large TPC is the main detector of the ALICE experiment at CERN where thousands of tracks are recorded after central heavy ion collisions at relativistic energies. The active volume of the ALICE TPC is a 5-meter diameter cylinder with a length of 5 meters. The TPC is embedded inside a magnetic field of 0.6 T. Figure 4.56 shows the energy loss versus momentum recorded in the ALICE TPC.

On the small-size side, the AT-TPC (Active Target Time-Projection Chamber) at the National Superconducting Cyclotron Laboratory, USA, is used for low-energy nuclear physics. It is located inside a solenoid that produces a magnetic field of 4 T. The AT-TPC, shown in Fig. 4.57, has approximately a diameter of 70 cm and a length of 120 cm.

4.5.2.4 Signal Genesis: The Shockley-Ramo Theorem

As seen in the detector examples above, the electrodes (wires, strips, pads,...) are connected to readout electronics and interconnected by discrete elements. The voltage induced on all electrodes by a charge q moving along a trajectory $x(t)$ can be estimated following a two-step procedure:

(i) calculate the induced current on grounded electrodes. The Shockley-Ramo theorem gives a prescription to estimate this induced signal.

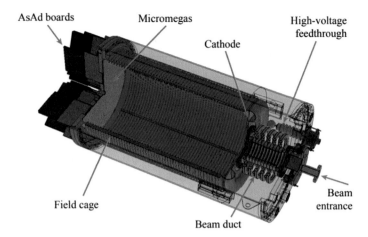

Fig. 4.57 Schematic view of the AT-TPC. The field cage is made of a non-conductive material. The beam enters the field cage through a thin window which seals the 250-litre volume of the TPC. The front-end cards (AsAd boards) of the digital electronics are directly connected to the back of the Micromegas sensor plane. Figure from [31]

(ii) use an analog circuit simulation after having considered the calculate currents (first step) as ideal current sources.

Here we focus on the induced currents. The brute-force calculation would imply to calculate the potential generated by the moving charge on the electrodes. From this potential, one could deduce the electric field and, by application of the Gauss theorem, the induced charge density on the electrodes. This way is cumbersome, and much simplified by the Shockley-Ramo theorem.

Let us consider a generic system of three electrodes as described in Fig. 4.58. The Shockley-Ramo theorem states that the induced current $I_i(t)$ on electrode i is given by

$$I_i(t) = -q\mathbf{E}_i^w(x(t), y(t), z(t)) \cdot \mathbf{v}(x(t), y(t), z(t)), \qquad (4.211)$$

where $\mathbf{v}(x(t), y(t), z(t))$ is the velocity of the charge q in the detector with all electrodes at their nominal bias. $\mathbf{v}$ is calculated by considering the full detector and electrodes at the potential set in the experiment. $\mathbf{E}_i^w$ is the so-called *weighting field* which is the electric field at $(x(t), y(t), z(t))$ generated by electrode i when put at voltage V=1 V in the absence of the charge q and all other electrodes at V=0 V. E^w depends only on the detector topology. The above formulation of the theorem assumes the following convention for the current sign: a negative current is created by a negative charge approaching the electrode.

For illustration, we consider the example of the left panel of Fig. 4.59: a charge $q = -e$ moves in the electric field produced by two electrodes of infinite dimensions in the (x, y) directions placed at $z = 0$ and biased at $V_2 = 0$ V and at $z = D$ and biased at $V_1 = 100$ V, respectively. The volume in between the two plates is filled with

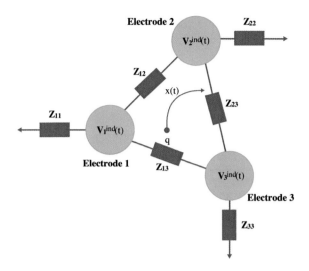

Fig. 4.58 Sketch of a detector system composed of three electrodes. A moving charge will induce currents and potentials V_i^{ind} on each electrode i. The electrodes are interconnected by impedances Z_{ij} and connected to the outside by impedances Z_{ii}

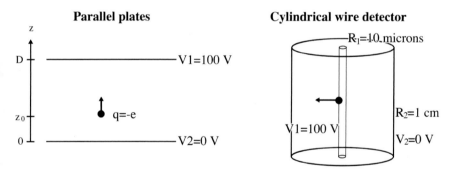

Fig. 4.59 (Left) Sketch of two parallel plates at potentials V_1 and V_2. An electron created at $z=z_0$ moves towards the electrode 1 and induces currents on both electrodes. (Right) Sketch of a cylindrical detector composed of a wire of radius r_1 at a potential V_1 inside a grounded cylinder of radius r_2. Ions of charge q created at $r=r_1$ move towards the outer electrode and induce currents on both electrodes

a gas in which the charge moves at velocity $\mathbf{v}(z) = \mu \mathbf{E}(z)$, where $\mathbf{E}(z)$ is the electric field in the drift region and $\mu(<0)$ the drift mobility of electrons in the conditions of this example. The charge is produced at z_0 at $t = 0$. In order to predict the current induced on each electrode, we apply the Shockley-Ramo theorem as expressed in Eq. (4.211).

The electric field field between the two electrodes is given by

$$\mathbf{E} = -\frac{\Delta V}{D}\mathbf{u}_z. \tag{4.212}$$

The induction will therefore induce a signal as long as the charge moves in between the two electrodes, i.e., up to the time T for which $z(T) = D$,

$$T = \sqrt{\frac{D(D - z_0)}{\mu}}\Delta V, \tag{4.213}$$

where, in this example, $\mu < 0$ ($q = -e$) and the charge moves towards the first electrode. From $t = 0$ to $t = T$, the induced current on the first electrode is

$$I_1(t) = +e(-\frac{1}{D}\mathbf{u}_z) \cdot (-\mu\frac{\Delta V}{D}\mathbf{u}_z) = +e\mu\frac{\Delta V}{D^2}, \tag{4.214}$$

for $0 \leq t \leq T$. Similarly, when one applies the Shockley-Ramo theorem for the second electrode, one gets

$$I_2(t) = +e(+\frac{1}{D}\mathbf{u}_z) \cdot (-\mu\frac{\Delta V}{D}\mathbf{u}_z) = -e\mu\frac{\Delta V}{D^2}. \tag{4.215}$$

By integration, one can also determine the induced charge on each electrode. The induced charges on the two electrodes are opposite, at all times, as expected from charge conservation. Once the charge reaches the first electrode, the induction stops: the induced currents vanish and the induced charges remain constant at the value $Q_1 = e\frac{D-z_0}{D}$ for the first electrode (1), i.e., of opposite sign than the moving charge towards the electrode, and $Q_2 = -Q_1$ on the electrode (2).

Problem

4.10 Consider a cylindrical detector composed of a conducting wire of radius $r_1=10$ μm in a grounded cylinder of internal radius $r_2=1$ cm, as illustrated in the r.h.s. of Fig. 4.59. Apply the Shockley-Ramo theorem to determine the current induced on the central wire by a positive charge q created at rest at t=0 at r=r_1.

4.5.3 Neutron Spectroscopy

Neutrons do not have an electric charge and therefore do not interact via the electromagnetic interaction with the electrons of the detector material, which is the main process for energy loss of charged particles. The detection of neutrons requires a two step process: (i) transfer of energy from the neutron to a nucleus of the detector material, (ii) detection of recoils after the nuclear reaction of the first step.

The interaction of neutrons with matter strongly depends on its energy, and therefore the detection technique to be applied also strongly depends on the energies of the neutrons to be detected. One distinguishes

- thermal and slow neutrons with an energy between 0.01 eV and 1 keV,
- intermediate-energy neutrons between 1 keV and $\sim$1 MeV,
- fast neutrons with an energy greater than $\sim$1 MeV.

4.5.3.1 Slow Neutrons

Elastic scattering of neutrons with nuclei from the detector material is not very efficient to detect neutrons since the nucleus recoil energy is often very low. Capture reactions where the neutron is absorbed by a nucleus are favoured, especially because their cross sections often increase when the neutron energy is lower. For thermal neutrons, the capture is the main reaction process. The detection of low-energy neutrons is often a sequence of thermalization from elastic scattering ended by the neutron capture.

Reactions for detecting low-energy neutrons are exothermic. The most used are the following:

- $^{10}B + n \rightarrow {}^{7}Li + \alpha$ or $^{10}B + n \rightarrow {}^{7}Li^* + \alpha$ with a reaction Q-value of 2.729 MeV and 2.310 MeV, respectively. The cross section is 3840 barns for thermal neutrons.
- $^{6}Li + n \rightarrow {}^{3}H + \alpha$ with a reaction Q-value of 4.78 MeV and a cross section of 940 barns for thermal neutrons.
- $^{3}He + n \rightarrow {}^{3}H + p$, with $Q=0.764$ MeV and $\sigma=5330$ barns,
- $^{235}U + n \rightarrow$ fission, with $Q \sim 160$ MeV and $\sigma = 600$ barns,
- $^{155,157}Gd + n \rightarrow {}^{156,158}Gd \rightarrow {}^{156,158}Gd + \gamma$, with $Q=8$ MeV and $\sigma= 56$ kb and 242 kb, respectively.

Since these reactions are very exothermic, the energy taken away by the reaction products do not reflect the energy of the incident neutron. The detection of low-energy neutrons by these reactions allows to measure neutron fluxes, or count the neutrons individually, but not to measure their energy.

4.5.3.2 Fast Neutrons from Invariant Mass

Reaction cross sections for fast neutrons are much smaller and elastic scattering with nuclei from the detector material is the main detection process. The energy transfer during the collision is more efficient with light nuclei. For this reason, materials containing hydrogen are often considered as a good choice to detect or slow down fast neutrons. Fast incident neutrons might transfer only a part of their kinetic energy in the detector: the energy of the recoils do not necessarily reflect the incident energy, and the energy of fast neutrons is most often determined from their time of flight.

The detection of neutrons is of importance for the spectroscopy of unbound states, in particular for neutron-rich nuclei. In inverse kinematics, the *invariant mass* technique is often used to measure the decay energy of neutron-decaying states relative to the neutron separation energy. The invariant mass of a system is a Lorentz invariant and is calculated as the Minkowski norm of the energy quadrivector. The method aims at determining it from quantities measured in the laboratory. It relies on the detection of all in-decay products emitted in flight. The energy E_{rel} of the decaying system relative to the separation threshold follows

$$E_{rel} = M_{inv} - \sum_{i=1}^{n} m_i = \sqrt{(\sum_{i=1}^{n} E_i)^2 - (\sum_{i=1}^{n} \mathbf{p}_i)^2} - \sum_{i=1}^{n} m_i, \qquad (4.216)$$

where E_i, m_i and $\mathbf{p}_i$ are the total energy, mass and momentum of the ith decay product, respectively. The quantity M_{inv} is called the invariant mass of the system.

The invariant mass spectrum of ^{26}O from its two-neutron decay into ^{24}O and two neutrons was measured at the RIBF and is shown in Fig. 4.60. An excellent energy resolution was achieved and the ground state energy of ^{26}O was determined to be 18 ± 3 (statistics) ± 4 (systematics) keV above the two-neutron separation energy, meaning that ^{26}O is extremely weakly unbound consistently with a neutron dripline located at ^{24}O along the oxygen isotopic chain.

4.5.3.3   Intermediate-Energy Neutrons

Neutrons with energy below 1 MeV are usually emitted isotropically in the laboratory frame and the techniques for fast neutrons are usually not suited. On the other hand, the capture cross section drops quickly with increasing incident energy making the capture-reaction based technique inefficient for energies above 1 keV. The detection of intermediate-energy neutrons, 1 keV $\leq$ E $\leq$ 1 MeV, often combines a

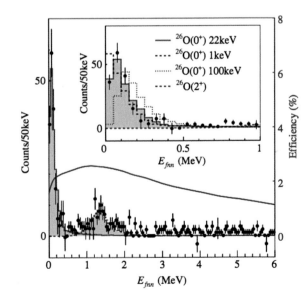

Fig. 4.60 Example of invariant mass measurement performed at the RIBF, RIKEN, with the NEBULA neutron detector. The relative-energy $E_{fnn} \equiv E_{rel}$ spectrum of ^{26}O reconstructed from the momentum vectors of the in-flight decay products, ^{24}O and two neutrons, is shown. The spectrum is background subtracted. Figure reprinted with permission from [32]. ©2021 by the American Physical Society

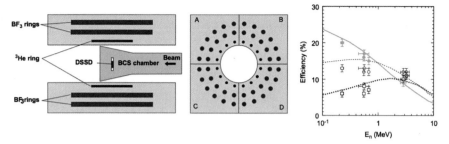

Fig. 4.61 (Left) Schematic view of the NERO detector. The side view shows the β counting station (BCS) composed of a double-sided-silicon detector (DSSD) for radioactive ion implantation. The detector is composed of a matrix of polyethylene in which ^{3}He and BF$_3$ cylindrical detectors are inserted in concentric rings, as shown on the front view. The ^{3}He detectors constitute the inner ring while the two outer rings are composed of BF$_3$ detectors. (Right) Measured and simulated detection efficiency for the three detection rings as a function of the neutron energy. Efficiencies are decomposed for the innermost ring (green), intermediate ring (red) and outer ring (blue) of detectors. The figure is reprinted with permission from [33]. ©2021 by Elsevier

moderator to slow them down and a detection based on capture reaction mechanism, as exposed above. The moderator is usually composed of a material that contains a large number of hydrogen atoms to slow efficiently the neutrons from elastic collisions. Polyethylene and paraffin are common choices. The detection efficiency of a moderator detector combination depends on the neutron energy and the detector geometry, in particular the thickness of moderator between the source of neutrons and the detection. A schematic view of such the neutron detection system NERO for β-decay studies at NSCL is shown in Fig. 4.61.

4.6 Gamma-Ray Spectroscopy

Excited states below the particle separation energy can only decay via electromagnetic decay. During such a decay, radiations are emitted, while satisfying energy and momentum conservation. The emitted radiations are most often a photon, although conversion electron, e^+e^- pair (if energetically allowed) and two-photon decay are also possible. For nuclear excitations, de-excitation photons have typical energies of 10 keV to 10 MeV. The measurement of the photon energy gives access to the relative energy between the two states involved in the transition. The lifetime of the transition is also an important information on the overlap of the initial and final wave functions. Finally, in some experiments, the angular distribution of the emitted gamma rays and their scattering properties within the detector can provide information on the electric or magnetic nature of the transition and its multipolarity. Gamma detection allows to reach high energy resolutions, an advantage compared to other spectroscopy techniques, and is the most used tool for nuclear spectroscopy.

4.6.1 General Propoerties of Electromagnetic Transitions in Nuclei

A nucleus in a bound excited state may proceed to an electromagnetic decay, i.e., via gamma-ray emission. The typical lifetime of states decaying via gamma emission ranges from a femtosecond to nanoseconds although some particular states known as isomers can have longer lifetimes. Gamma spectroscopy is essential to nuclear structure since it offers excellent energy resolution in experiments. The achieved resolutions originate from the interaction of photons with matter: there is no process which will blur the energy of the photon around its initial value. A photon is neither interacting nor losing a significant portion of its energy. This feature implies that the total energy of photons in experiments can be measured with precision. Today, the typical energy resolution of high-resolution gamma detectors based on high-purity Ge semiconductors is of the order of $\delta E / E = 0.2\%$.

During the electromagnetic decay of a state to another, energy and momentum are conserved. The large mass of the nucleus compared to the gamma ray implies that its recoil energy after the decay can be neglected in most applications and the energy E_γ of emitted photon in the center of mass of the initial nucleus is equal to the difference of energies E_i and E_f of the initial and final states

$$E_\gamma = E_f - E_i. \tag{4.217}$$

Photons carry an angular momentum $\mathbf{L}$ and a parity π. Conservation rules imply

$$\mathbf{I}_f = \mathbf{L} + \mathbf{I}_i \tag{4.218}$$

$$\pi_f = \pi \times \pi_i, \tag{4.219}$$

where notations are given in Fig. 4.62. For given initial and final states with total angular momenta I_i and I_f, respectively, the only possible angular momenta for any decaying photon are restricted to

$$|I_f - I_i| \leq L \leq I_f + I_i. \tag{4.220}$$

Since a photon carries at least one quantum of angular momentum ($L > 0$), there is no γ ray transition of type $0^+ \rightarrow 0^+$.

There are two types of radiations, *electric* or *magnetic*, depending on the angular momentum L and parity π carried by the photon. Both the electric and magnetic fields are transverse to the direction of propagation of the photon (also called Transverse Magnetic (TM) and Transverse Electric (TE) modes). At the same time, the two fields are orthogonal to each other. If the parity does not change from π_i to π_f (i.e., $\pi=+1$), the transition is either an even-electric or odd-magnetic multipole transition. In case of parity change, the transition is either an odd-electric or even-magnetic multipole transition. A decay between two states can occur in general via several multipole transitions in competition.

Fig. 4.62 A γ decay from an initial state of energy E_i to a final state of energy E_f with total angular momentum and parity (I_i, π_i) and (I_f, π_f), respectively, scale by factor 0.8 carrying an angular momentum L and a parity π

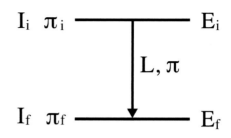

4.6.2 Electric and Magnetic Multipole Transitions

In the following, we derive the general form of electromagnetic radiations from the Maxwell equations. The transition probability between two single-particle states is then derived following a model proposed by V. Weisskopf.

All electromagnetic transitions are fully characterized by the electric and magnetic fields $\mathbf{E}(\mathbf{r}, t)$ and $\mathbf{B}(\mathbf{r}, t)$. In vacuum, they satisfy the Maxwell equations,

$$\nabla \times \mathbf{E} + \frac{\partial \mathbf{B}}{\partial t} = 0, \qquad \nabla \cdot \mathbf{E} = \rho/\varepsilon_0 \tag{4.221}$$

$$\nabla \times \mathbf{B} - \varepsilon_0 \mu_0 \frac{\partial \mathbf{E}}{\partial t} = \mu_0 \mathbf{j}, \qquad \nabla \cdot \mathbf{B} = 0, \tag{4.222}$$

where ρ and $\mathbf{j}$ are the charge and current densities, respectively. The conservation of electric charge is expressed by the continuity equation

$$\nabla \cdot \mathbf{j}(\mathbf{r}, t) + \frac{\partial}{\partial t} \rho(\mathbf{r}, t) = 0. \tag{4.223}$$

The four Maxwell equations can be reduced to two when the vector potential $\mathbf{A}(\mathbf{r}, t)$ and the scalar potential $\Phi(\mathbf{r}, t)$ are introduced. The electric and magnetic fields are obtained from the potentials as follows:

$$\mathbf{B} = \nabla \times \mathbf{A}, \quad \mathbf{E} = -\frac{\partial \mathbf{A}}{\partial t} - \nabla \phi. \tag{4.224}$$

The two potentials are not uniquely defined since, for example, any field gradient can be added to $\mathbf{A}$ and will still verify Eq. (4.224). The gauge conditions can be chosen in such a way that the two potentials decouple. This simplification can be obtained with the so-called *Lorentz gauge* set by the condition

$$\nabla \cdot \mathbf{A} + \frac{\partial \phi}{\partial t} = 0. \tag{4.225}$$

The quantization of the electromagnetic field in this gauge is difficult and one prefers the so-called *Coulomb gauge*

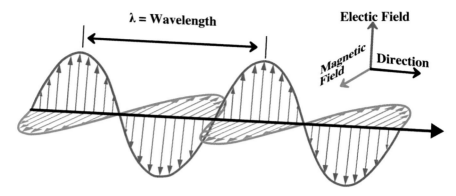

Fig. 4.63 Illustration of electromagnetic fields propagating orthogonally to the momentum direction. They are also orthogonal to each other

$$\nabla \cdot \mathbf{A} = 0. \tag{4.226}$$

This gauge is not covariant. It selects only those solutions of the Maxwell equations that correspond to transverse vector field, as illustrated in Fig. 4.63. Therefore, solutions in the Coulomb gauge form a complete set for the expansion of transverse fields only. Within this gauge, by combining Eqs. (4.221), (4.222) and (4.224), the equations for the vector and scalar fields can be written as

$$\nabla^2 \mathbf{A} - \frac{1}{c^2} \frac{\partial^2 \mathbf{A}}{\partial t^2} = \mu_0 \mathbf{j}, \tag{4.227}$$

$$\nabla^2 \Phi = -\rho/\varepsilon_0, \tag{4.228}$$

where the light velocity is defined by $c^2 = 1/\varepsilon_0\mu_0$. In free space, the charge density $\rho(\mathbf{r}, t)$ and the current density $\mathbf{j}(\mathbf{r}, t)$ vanish. Hence the equation for the vector potential is simplified to be

$$\nabla^2 \mathbf{A} - \frac{1}{c^2} \frac{\partial^2 \mathbf{A}}{\partial t^2} = 0. \tag{4.229}$$

Depending on the boundary conditions and choice of coordinates, simple solutions of the homogeneous wave equation (4.229) can be found. The plane wave and the spherical wave solutions are common choices in many applications. We focus in the following on the spherical wave solutions which are of two types, with different reflection (parity) properties.

An intermediate step before solving Eq. (4.229) is to consider the homogeneous scalar equation

$$\left(\nabla^2 - \frac{1}{c^2} \frac{\partial^2}{\partial t^2}\right)\Phi(\mathbf{r}, t) = 0. \tag{4.230}$$

The solution of the vector field can be obtained by generalizing the scalar field $\Phi(\mathbf{r}, t)$ to the vector one as will be shown below. By assuming a time dependence for $\Phi(\mathbf{r}, t)$ with the momentum transfer q,

$$\Phi(\mathbf{r}, t) = \Phi(\mathbf{r})e^{-iqct}, \tag{4.231}$$

one gets a time-independent differential equation for $\Phi(\mathbf{r})$, so called Helmholtz equation

$$(\nabla^2 + q^2)\Phi(\mathbf{r}) = 0, \tag{4.232}$$

which can be solved easily in spherical coordinates (r, θ, ϕ) by separating the radial and angular coordinates

$$\Phi_{LM}(\mathbf{r}) = j_L(qr)Y_{LM}(\theta, \phi), \tag{4.233}$$

where Y_{LM} is a spherical harmonic function and j_L is a spherical Bessel function being regular at the origin. The scalar solution (4.233) will serve as a basis to build up the vector solution of Eq. (4.229). In view of the transversality of the Coulomb gauge condition (4.226), only two vector fields $\mathbf{L}\Phi_{LM}(\mathbf{r})$ and $\nabla \times \mathbf{L}\Phi_{LM}(\mathbf{r})$ satisfy (4.229) and the Coulomb gauge. The angular momentum operator is defined by

$$\mathbf{L} = \mathbf{r} \times \mathbf{p} = -i\mathbf{r} \times \nabla \tag{4.234}$$

and acts on the spherical harmonics as

$$\begin{aligned}\mathbf{L}^2 Y_{LM}(\theta, \phi) &= L(L+1)Y_{LM}(\theta, \phi) \\ L_z Y_{LM}(\theta, \phi) &= M Y_{LM}(\theta, \phi).\end{aligned} \tag{4.235}$$

The vector fields $\mathbf{L}\Phi$ and $\nabla \times \mathbf{L}\Phi$ are indeed "transverse" since they are perpendicular to the direction of propagation, specified by the operator $\mathbf{p} \to \nabla/i$, and also they are themselves perpendicular to each other. They are distinguished by the parity, and called *transverse electric* and *transverse magnetic* multipole vector fields,

$$A_{LM}^{(E)}(\mathbf{r}) = \frac{1}{q\sqrt{L(L+1)}}(\nabla \times \mathbf{L})\Phi_{LM}(\mathbf{r}), \tag{4.236}$$

$$A_{LM}^{(M)}(\mathbf{r}) = \frac{1}{\sqrt{L(L+1)}}\mathbf{L}\Phi_{LM}(\mathbf{r}). \tag{4.237}$$

Each multipole vector field $A_{LM}^{E(M)}(\mathbf{r})$ carries an angular momentum L with its z-component M along the propagation axis and, in the case of an electromagnetic de-excitation of a nuclear state, imposes a certain parity change between the initial and final state, as already seen in Eq. (4.219). The two transverse solutions differ in their parity properties. Since one has the following parity properties

$$\mathcal{P}Y_{LM}(\theta, \phi)\mathcal{P}^{-1} = Y_{LM}(\pi - \theta, \phi + \pi) = (-1)^L Y_{LM}(\theta, \phi), \qquad (4.238)$$

$$\mathcal{P}\mathbf{L}\mathcal{P}^{-1} = \mathbf{L}, \qquad (4.239)$$

$$\mathcal{P}\mathbf{p}\mathcal{P}^{-1} = -\mathbf{p} \quad (\text{or } \mathcal{P}\nabla\mathcal{P}^{-1} = -\nabla), \qquad (4.240)$$

the parities π of the electromagnetic operators (4.252) and (4.253) are

$$\pi = (-1)^L \quad \text{for electric multipole operators}, \qquad (4.241)$$

$$\pi = (-1)^{L+1} \quad \text{for magnetic multipole operators}. \qquad (4.242)$$

The electromagnetic transitions are classified as electric or magnetic based on whether the transition is due to a motion of the charge or a motion of the spin. These points will be shown explicitly in relation with the electromagnetic field operators (4.236) and (4.237) in Sect. 4.6.3.

Problem

4.11 Show that the vector fields of Eqs. (4.236) and (4.237) are solutions of the Helmholtz equation Eq. (4.232) and of the Coulomb gauge (4.226).

4.6.3 Transitions Probabilities

In this subsection, we introduce the radiation emission mechanism in a nucleus. The electromagnetic field is a quantum entity that interacts with the nucleus. The total Hamiltonian can be decomposed into the nucleus Hamiltonian H_{nuc}, the electromagnetic Hamiltonian H_{em} and the interaction of the electromagnetic field with the nucleus H_{int}

$$H = H_{nuc} + H_{em} + H_{int}. \qquad (4.243)$$

Let us consider two eigenstates of H_{nuc}, $|i\rangle$ and $|j\rangle$, which will be the initial and final states of an electromagnetic transition, respectively. The electromagnetic transition is driven by the coupling of the vector field $\mathbf{A}(\mathbf{r}, t)$ and the current density operator $\mathbf{j}(\mathbf{r})$ as follows

$$H_{int} = -\frac{1}{c} \int \mathbf{j}(\mathbf{r}) \cdot \mathbf{A}(\mathbf{r}, t) d\mathbf{r}. \qquad (4.244)$$

The time-dependence of the vector field $\mathbf{A}(\mathbf{r}, t)$ is taken the same as that of the scalar field in Eq. (4.231). Note that the above formulation of H_{int} is exact and relativistic but not trivial.

We will derive the Hamiltonian H_{int} by the second-kind gauge transformation[13] $\mathbf{p} \to \mathbf{p} - e\mathbf{A}$ for the non-relativistic Hamiltonian of free particles

$$H = \sum_i \frac{1}{2m_i}(\mathbf{p}_i - \frac{e_i}{c}\mathbf{A}(\mathbf{r}_i))^2 \tag{4.245}$$

which leads to

$$H = \sum_i \frac{1}{2m_i}\mathbf{p}_i^2 - \sum_i \frac{e_i}{2m_i c}(\mathbf{p}_i \cdot \mathbf{A}(\mathbf{r}_i) + \mathbf{A}(\mathbf{r}_i) \cdot \mathbf{p}_i) + \sum_i \frac{e_i^2}{2m_i c^2}\mathbf{A}(\mathbf{r}_i)^2. \tag{4.246}$$

The first r.h.s. term is the kinetic energy of nucleons and the second term is the interaction term called the "*minimal interaction*" between electromagnetic fields and particles. This term can be re-written in the form of Eq. (4.244)

$$-\frac{1}{c}\int \mathbf{j}_c(\mathbf{r}) \cdot \mathbf{A}(\mathbf{r}, t)d\mathbf{r} \tag{4.247}$$

with a current operator

$$\mathbf{j}_c(\mathbf{r}) = \sum_i e\left(\frac{1}{2} - t_z(i)\right)\frac{1}{2}(\mathbf{v}_i\delta(\mathbf{r} - \mathbf{r}_i) + \delta(\mathbf{r} - \mathbf{r}_i)\mathbf{v}_i), \tag{4.248}$$

where $t_z = +1/2$ for neutrons and $-1/2$ for protons. The first term on the r.h.s. of Eq. (4.248) acts on the left and the second one acts on the right when the expectation value is evaluated. This is called the "*convection current*" arising from the motion of protons. There is another type of current which is associated with the spin of nucleons. This is called the magnetization current and expressed as

$$\mathbf{j}_m(\mathbf{r}) = \sum_i \frac{e\hbar}{2m_i c}g_s(i)\nabla \times \mathbf{s}_s\delta(\mathbf{r} - \mathbf{r}_i). \tag{4.249}$$

The total current is expressed by the sum of two currents

$$\mathbf{j}(\mathbf{r}) = \mathbf{j}_c(\mathbf{r}) + \mathbf{j}_m(\mathbf{r}). \tag{4.250}$$

The third term of Eq. (4.246) is of higher order and can be neglected here.

---

[13] The first-kind gauge transformation is the phase change of wave function $\psi_i(\mathbf{r}) \to \exp(i\lambda Q_i)\psi_i(\mathbf{r})$ where Q_i is the charge of a particle and λ is a constant. The first-kind gauge transformation implies the conservation of total charge before and after the reaction. On the other hand, the second-kind gauge transformation gives rise to the interaction between a particle and a field. The interaction raised by the second-kind gauge transformation is called the "*minimal interaction*" between a particle and electromagnetic fields.

The transition probability for emission (or absorption) of a photon of a given multipole type depends on the matrix element

$$\langle f | \int \mathbf{j}(\mathbf{r}) \cdot \mathbf{A}(\mathbf{r}) d\mathbf{r} | i \rangle. \tag{4.251}$$

One defines the so-called electric and magnetic multipole operators

$$\mathcal{M}(EL, M) = \frac{(2L+1)!!}{q^L(L+1)}[L(L+1)]^{1/2} \int \mathbf{j}(\mathbf{r}) \cdot \mathbf{A}_{LM}^E(\mathbf{r}) d\mathbf{r}, \tag{4.252}$$

$$\mathcal{M}(ML, M) = -i \frac{(2L+1)!!}{q^L(L+1)}[L(L+1)]^{1/2} \int \mathbf{j}(\mathbf{r}) \cdot \mathbf{A}_{LM}^M(\mathbf{r}) d\mathbf{r}. \tag{4.253}$$

These multipole operators are spherical tensors of rank L. The double factorial $n!!$, with n being an integer, is an extension to the normal factorial $n!$. It is defined recursively as follows: $n!! = 1$ if $n = 0$ or $n = 1$, and $n!! = n(n-2)!!$ if $n \geq 2$.

In the case of a nuclear photoprocess, the wavelength of photons $\bar{\lambda} \equiv \lambda/2\pi$ is usually large compared to the nuclear radius

$$\frac{R}{\bar{\lambda}} = qR = 6.1 \times 10^{-3} A^{1/3} E_\gamma \text{ (MeV)} \tag{4.254}$$

$$(R = 1.2 \times A^{1/3} \text{fm}),$$

where $E_\gamma = \hbar c q$ is typically of the order of a few MeV. For the long wavelength limit

$$qr << 1, \tag{4.255}$$

one can employ the expansion of the Bessel function

$$j_L(qr) = \frac{(qr)^L}{(2L+1)!!} \left(1 - \frac{1}{2}\frac{(qr)^2}{2L+3} + \dots\right). \tag{4.256}$$

In most cases, the lowest order is a good approximation and the multipole moments can then be expressed as

$$\mathcal{M}(EL, M) = \frac{1}{q(L+1)} \int \mathbf{j}(\mathbf{r}) \cdot \nabla \times \mathbf{L} r^L Y_{LM}(\theta, \phi) dr^3, \tag{4.257}$$

$$\mathcal{M}(ML, M) = \frac{-i}{c(L+1)} \int \mathbf{j}(\mathbf{r}) \cdot \mathbf{L} r^L Y_{LM}(\theta, \phi) dr^3. \tag{4.258}$$

It is shown that the above electric multipole operator can be rewritten in a simple form by using the identity

$$\nabla \times \mathbf{L} r^L Y_{LM}(\theta, \phi) = -i \nabla \times (\mathbf{r} \times \nabla) r^L Y_{LM}(\theta, \phi) = i \nabla (L+1) r^L Y_{LM}(\theta, \phi), \tag{4.259}$$

where the angular momentum operator is replaced by $-i(\mathbf{r} \times \nabla)$ in Eq. (4.234), and the continuity equation

$$\nabla \cdot \mathbf{j}(\mathbf{r}, t) = -\frac{\partial \rho(\mathbf{r}, t)}{\partial t} = iqc\rho(\mathbf{r}, t), \tag{4.260}$$

where the time dependence of $\rho(\mathbf{r}, t)$ is taken the same as the scalar field in Eq. (4.231). With the help of Eqs. (4.259) and (4.260), the electric multipole moment is expressed as

$$\mathcal{M}(EL, M) = \int \rho(\mathbf{r}) r^L Y_{LM}(\theta, \phi) dr^3. \tag{4.261}$$

This simpler expression is the one used in most of nuclear physics estimates for electric transition probabilities. The magnetic multipole operator is expressed in a different form as

$$\mathcal{M}(ML, M) = \frac{-1}{c(L + 1)} \int \mathbf{j}(\mathbf{r}) \cdot (\mathbf{r} \times \nabla) r^L Y_{LM}(\theta, \phi) dr^3. \tag{4.262}$$

The transition amplitude for emission (or absorption) of a photon is proportional to the matrix element of the multipole operator. The decay rate by the emission of a single photon can be calculated by a second-order perturbation theory

$$T_{fi} = \frac{2\pi}{\hbar} |\langle f | H_{int} | i \rangle|^2 g(E_f), \tag{4.263}$$

where $g(E_f)$ is the final state level density. This is called the *"Fermi's golden rule"* for the decay probability for the initial state *"i"* to the final state *"f"*. In the standard detector environment, experiments adopt un-oriented radiation source and also the orientation of the angular momentum of the final state is averaged. This is implemented by expressing the decay rate by summing over the magnetic states of the photon and of the final nuclear state[14]

$$T(EL(ML); i \to f) = \frac{8\pi}{\hbar} \frac{(L + 1)q^{2L+1}}{L[(2L + 1)!!]^2} B(EL(ML); I_i \to I_f), \tag{4.264}$$

where the reduced transition probability is defined as

$$B(EL(ML); I_i \to I_f) = \sum_{M, M_f} |\langle I_f M_f | \mathcal{M}(EL(ML), M) | I_i M_i \rangle|^2$$

$$= \frac{|\langle I_f || \mathcal{M}(EL(ML)) || I_i \rangle|^2}{(2I_i + 1)}, \tag{4.265}$$

where the double-bar symbol denotes the reduced matrix element for the angular momentum.

---

[14] Peter Ring and Peter Schuck, The nuclear many-body problem, Appendix B. See also A. DeShalit and H. Feshbach, Theoretical nuclear Physics I, Chap. VIII.

Table 4.6 Electromagnetic decay rates $T(EL(Ml))$ per second. The transition energy $E_\gamma = \hbar c q$ is expressed by the unit of MeV. The Weisskopf estimates $B_W(EL)$ and $B_W(ML)$ are expressed in units of $e^2\text{fm}^{2L}$ and $e\hbar/(2mc)^2\text{fm}^{(2L-2)}$, respectively. The mean life time is given by $\tau = 1/T$ and the half life time is expressed by $t_{1/2} = 0.693/T$

Multipole	Parity	$T(1/s)$	B_W
$E1$	$-$	$1.59 \times 10^{15}(E_\gamma)^3 B(E1)$	$6.45 \times 10^{-2} A^{2/3}$
$E2$	$+$	$1.22 \times 10^9(E_\gamma)^5 B(E2)$	$5.94 \times 10^{-2} A^{4/3}$
$E3$	$-$	$5.70 \times 10^2(E_\gamma)^7 B(E3)$	$5.94 \times 10^{-2} A^2$
$E4$	$+$	$1.69 \times 10^{-4}(E_\gamma)^9 B(E4)$	$6.28 \times 10^{-2} A^{8/3}$
$M1$	$+$	$1.76 \times 10^{13}(E_\gamma)^3 B(M1)$	1.79
$M2$	$-$	$1.36 \times 10^7(E_\gamma)^5 B(M2)$	$1.65 A^{2/3}$
$M3$	$+$	$6.31 \times 10^0(E_\gamma)^7 B(M3)$	$1.65 A^{4/3}$
$M4$	$-$	$1,87 \times 10^{-6}(E_\gamma)^9 B(M4)$	$1.74 A^2$

It is useful to note that

$$B(EL(ML); I_f \to I_i) = \frac{2I_i + 1}{2I_f + 1} B(EL(ML), I_i \to I_f). \qquad (4.266)$$

The relation (4.266) expresses the detailed balance for reaction rates averaged over polarizations. The reduced transition probability (4.265) is commonly adopted to express the total transition rate, since many experiments directly measure this value by the Coulomb excitations and γ-ray spectroscopy. The decay rate of some $EL(ML)$ transitions are listed in Table 4.6. The total decay rate from $I_i \to I_f$ can be obtained summing all possible multipole transition rates (4.264) with allowed parity by the initial and final states,

$$T_t(\pi; i \to f) = \sum_{EL,ML} T(I_i \to I_f : EL(ML)), \qquad (4.267)$$
$$\text{with} \quad \pi_i \pi_f \pi = 1 \quad \text{and} \quad |I_i - I_f| \le L \le I_i + I_f.$$

4.6.4 Weisskopf Estimates of Transition Rates

In the following, we derive the lifetime of nuclear states under point-like nucleon and the single-particle state approximations. To a first-order approximation, we regard the nucleon as a point particle having a charge and a magnetic moment. Finite-size corrections of nucleon as well as correlations and exchange currents may give rise to significant effects. These effects are not discussed here.

Under the point-like approximation, the charge density $\rho(\mathbf{r})$ takes the form

$$\rho(\mathbf{r}) = \sum_i e_i \delta(\mathbf{r} - \mathbf{r}_i), \tag{4.268}$$

where the sum runs over all nucleons i of electric charge $e_i = +e$ for protons and $e_i = 0$ for neutrons. In some truncated models such as the valence-space shell model (see Chap. 3), these charges are sometimes replaced by effective charges to take into account correlation effects which are not treated explicitly in the valence space of the model. The current density operator is given by

$$\mathbf{j}(\mathbf{r}) = \sum_i e(\frac{1}{2} - t_z(i))\frac{1}{2}(\mathbf{v}_i\delta(\mathbf{r} - \mathbf{r}_i) + \delta(\mathbf{r} - \mathbf{r}_i)\mathbf{v}_i) + \sum_i \frac{e\hbar}{2mc}g_s(i)\nabla \times \mathbf{s}_i\delta(\mathbf{r} - \mathbf{r}_i), \tag{4.269}$$

where $\mu_N \equiv \hbar/2mc$ is the Bohr nuclear magnetic moment.

In the long wavelength limit, by inserting Eqs. (4.268) and (4.269) into Eqs. (4.261) and (4.262) respectively, the multipole operators in the point-nucleon approximation with magnetic moments are evaluated in the following. Inserting Eq. (4.269) into Eq. (4.262), we can rewrite the spin part of the moment by using an identity

$$\nabla \times (\nabla \times \mathbf{r})r^L Y_{LM} = (L + 1)\nabla r^L Y_{LM}, \tag{4.270}$$

together applying the partial integration on $\nabla \times \mathbf{s}$ and changing it into $\mathbf{s} \cdot \nabla \times$. The convection current provides the orbital angular momentum $\hbar\mathbf{l} = m(\mathbf{r} \times \mathbf{v})$ with the orbital g factor $g_l = 1(0)$ for protons (neutrons). Eventually, we can obtain the electromagnetic transition operators as

$$\mathcal{M}(EL, M) = \sum_i e(\frac{1}{2} - t_z(i))r_i^L Y_{LM}(\theta_i, \phi_i)), \tag{4.271}$$

$$\mathcal{M}(ML, M) = \frac{e\hbar}{2mc}\sum_i (g_s(i)\mathbf{s}_i + \frac{2g_l(i)}{L + 1}\mathbf{l}_i) \cdot \nabla r_i^L Y_{LM}(\theta_i, \phi_i). \tag{4.272}$$

We hereafter consider a single-particle transition. The single-particle wave function can be expressed as

$$|\ell j m\rangle = R_\ell(r)\sum_{m,s_z}\langle \ell m \frac{1}{2}s_z|jm\rangle Y_{\ell m}(\mathbf{r})\chi_{\frac{1}{2}s_z}, \tag{4.273}$$

where ℓ is the orbital angular momentum and $\mathbf{j} = \mathbf{l} + \mathbf{s}$ is the total angular momentum. The intrinsic spin of the nucleon is written by the spinor wave function $\chi_{\frac{1}{2}s_z}$, $R_\ell(r)$ is the radial wave function, and $\langle \ell m \frac{1}{2}s_z|jm\rangle$ is the Clebsch-Gordan coefficient. The single-particle value gives a useful quantitative estimate of nuclear properties. It provides a convenient unit in which experimental observables can be expressed. For example, a collective state may have a transition probability much larger than the single-particle estimate. Within the single-particle framework, the electric moment operator for a proton excitation of Eq. (4.271) is simplified as

$$\mathcal{M}(EL, M) = er^L Y_{LM}(\theta, \phi), \qquad (4.274)$$

and the reduced transition probability is given by

$$B_{sp}(EL, i \to f) = \frac{e^2}{2I_i + 1} |\langle i || r^L Y_L || f \rangle|^2. \qquad (4.275)$$

We consider that the proton has an angular momentum ℓ_i in the initial state and ℓ_f in the final state. The reduced transition probability can be divided into a radial part and an angular part

$$B_{sp}(EL, i \to f) = \frac{e^2}{4\pi} \langle r^L \rangle^2 S(j_i, j_f, L), \qquad (4.276)$$

where the radial part is

$$\langle r^L \rangle = \int_0^\infty R_{\ell_f}(r) r^{L+2} R_{\ell_i}(r) dr, \qquad (4.277)$$

and the angular part is given by

$$S(j_i, j_f, L) = \frac{4\pi}{2I_i + 1} \left| \langle (\ell_f \frac{1}{2}) j_f || Y_L || (\ell_i \frac{1}{2}) j_i \rangle \right|^2$$

$$= (2L + 1) \langle j_i \frac{1}{2} L0 | j_f \frac{1}{2} \rangle^2. \qquad (4.278)$$

The spin dependent factor (4.278) does not depend on the orbital angular momenta ℓ_i and ℓ_f, but only on the multipole order L of the transition and on the total spin of the initial (j_i) and final (j_f) states. For the case $j_i = L + 1/2$ and $j_f = 1/2$, the value S becomes the maximum as $S(j_i = L + 1/2, j_f = 1/2, L) = 1$.

To estimate the radial integral, one needs the radial wave functions R_{ℓ_i} and R_{ℓ_f}. Following the historical estimate of Weisskopf, we assume here that R_ℓ has a constant value over the volume of the nucleus, being a sphere of radius $R = r_0 A^{1/3}$. This approximation leads to a maximum overlap between the initial and final wave functions. The constant wave function gives the mean L-multipole radius

$$\langle r^L \rangle = \int_0^R |R_\ell(r)|^2 r^{L+2} dr = \frac{3R^L}{L + 3}, \qquad (4.279)$$

in which in which the constant renormalized wave function $|R_\ell|^2 = 3/R^3$ is inserted. Assuming $S = 1$ for the spin factor, the single-particle estimate for the reduced electric multipole transition probability becomes

$$B_W(EL, i \to f) = \frac{e^2}{4\pi} \left(\frac{3}{L + 3} \right)^2 R^{2L}. \qquad (4.280)$$

Considering the bulk parameterization of the nuclear radius as $R = r_0 A^{1/3}$, Eq. (4.280) becomes

$$B_W(EL, i \to f) = \frac{r_0^{2L}}{4\pi} (\frac{3}{L+3})^2 A^{2L/3} \ [e^2(\text{fm})^{2L}] \text{ with } j_i = j_f + L, \quad (4.281)$$

and $B_W(EL, i \to f)$ is called Weisskopf unit of electric multipole transition. The EL transition is possible under the parity selection rule $l_f + L + l_i =$ even. The radius parameter r_0 is usually taken to be $r_0 = 1.2$ fm.

The magnetic transition operator (4.272) can be rewritten with the spin $\mathbf{s}$ and total angular momentum $\mathbf{j}$ as

$$\mathcal{M}(ML, M) = \frac{e\hbar}{2mc}(L(2L+1))^{1/2} \sum_i r_i^{L-1} \left[(g_s - \frac{2g_l}{L+1})(Y_{L-1}\mathbf{s})^L + \frac{2g_l(i)}{L+1}(Y_{L-1}\mathbf{j})^L \right]_i,$$
$$(4.282)$$

where $(Y_{L-1}\mathbf{s})^L ((Y_{L-1}\mathbf{j})^L)$ is the tensor product of Y_{L-1} and the spin operator $\mathbf{s}$ (the total angular momentum $\mathbf{j}$). The matrix element of (4.282) can be calculated for the single-particle wave function (4.273) in a similar way to the electric multipole operator (4.271) and given by

$$B_{sp}(ML; j_i \to j_f) = \left(\frac{e\hbar}{2mc}\right)^2 \left(g_s - \frac{2g_l}{L+1}\right)^2 L^2 \frac{2L+1}{4\pi} \langle j_i \frac{1}{2} L0|j_f \frac{1}{2}\rangle^2 \langle l_f|r^{L-1}|l_i\rangle^2,$$
$$(4.283)$$

for the transition $|j_i - j_f| = L$ with the parity selection rule $l_f + L + l_i =$ odd. The second term of (4.282) does not contribute to the transition probability (4.283) since the operator $\mathbf{j}$ has only the diagonal matrix element and does not contribute to the transition matrix element.

The Weisskopf unit for magnetic transitions was derived following somewhat different considerations from the electric transitions. To compare the transition operator (4.262) with (4.261), the magnetic transition operator has an extra term $r \times \nabla$ which is the angular momentum operator and considered to cancels with a factor $(L+1)$. The convection current $\mathbf{j}_c$ is proportional to the velocity which acts as the $\mathbf{v} = \hbar\nabla/im$. The momentum is effectively replaced by the radius R by using the uncertainty relation $pR \sim \hbar$, and the velocity is then expressed as $\hbar/mR$. The difference between the magnetic and electric transitions is then given by

$$B_W(ML)/B_W(EL) = g'^2(\hbar/mcR)^2, \quad (4.284)$$

where g' is a factor including also the magnetic current contribution, expressed by $(g_s - 2g_l/(L+1))$ in Eq. (4.283). In the magnetic transition operator in Eq. (4.272), the $\sigma = 2\mathbf{s}$ plays a similar role to $\mathbf{l}/(L+1)$. Taking into account the fact that the spin g-factor $g_s > 1$, $g_s\mathbf{s} = g_s\sigma/2$ has a factor $2 \sim 3$ larger contribution than $g_l(p) = 1$ for the magnetic transition amplitude so that an average value for the g'-factor is taken to be $g'^2 \sim 10$ in the Weisskopf unit estimation. Under these considerations, the Weisskopf estimate for ML transition rate is expressed as

$$B_{sp}(ML; j_i \rightarrow j_f) = \frac{10}{\pi} r_0^{2L-2} \left(\frac{3}{L+3}\right)^2 A^{(2L-2)/3} \left(\frac{e\hbar}{2mc}\right)^2 \left[(\text{fm})^{2L-2}\right],$$

$$(4.285)$$

for $j_i = L + 1/2$ and $j_f = 1/2$ transition. One Weisskopf unit is a rough estimate of single-particle electromagnetic transition, but often useful to find out the enhancement or hindrance of the transition strength due to nuclear structure effects.

4.6.5 Photon-Matter Interactions

A gamma ray interacts mainly with the electrons of the detector material. It can interact via three processes: photoelectric effect, Compton scattering, and e^+e^- pair production for gamma rays with energy larger than $m_{e^+} + m_{e^-} = 1022$ keV. Depending on the energy of the gamma ray and the material in which it may interact, the probability for each process to occur is different. At low energies below $\sim$100 keV, the photoelectric effect is dominant. Between $\sim$100 keV and $\sim$5 MeV, Compton scattering is the process with the largest probability, while high-energy photons mostly interact via pair production. The relative importance of each process as a function of the photon energy is shown in Fig. 4.64 in the case of Aluminium material. The three process mechanisms are sketched in Fig. 4.65.

In a photoelectric absorption, the full energy of the photon is transmitted to an electron. The photon disappears in the process and the electron acquires a kinetic energy

$$E_{e^-} = h\nu - E_b, \qquad (4.286)$$

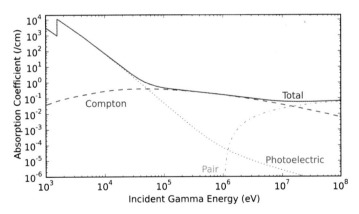

Fig. 4.64 The total absorption coefficient of aluminium (atomic number 13) for gamma rays as a function of the gamma-ray energy. The photoelectric effect is the largest at low energies, Compton scattering dominates at intermediate energies, and pair production dominates at high energies

Fig. 4.65 Photons interact with matter through three main mechanisms: (1) the photoelectric effect in which photons are absorbed and which produces an energetic electron, (2) the Compton scattering in which both the photon and an electron are scattered from an atom, (3) the e^+e^- pair production, in which the photon in the electric field of the nucleus converts into an energetic electron-positron pair

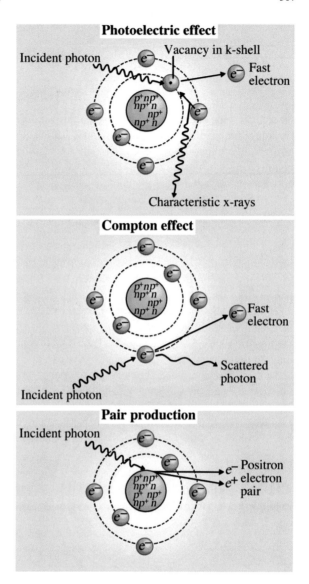

where $h\nu$ is the incident photon energy and E_b is the initial binding energy of the electron in the material. For gamma rays of energy larger than few 100 keV, this second term is usually negligible and the photo-electron carries most of the photon energy. The photoelectric process is enhanced for absorber material with high atomic number Z. Empirically, the photoelectric absorption coefficient varies as $\propto Z^{4-5}/E_\gamma^{3.5}$.

Via Compton scattering the photon transfers part of its energy to an electron. The final state comprises then a recoil electron and a scattered photon of energy $h\nu'$.

Assuming that the electron has initially no binding energy, the kinematics of the process is driven by

$$h\nu' = h\nu \frac{1}{1 + (h\nu/m_e c^2)(1 - \cos\theta)}, \tag{4.287}$$

where $h\nu'$ is the energy of the scattered photon, θ its scattering angle with respect to the initial propagation and $m_e c^2$ is the mass of the electron. The angular distribution of the scattered photon and differential cross section are given in the lowest order of quantum electrodynamics (QED) by the Klein-Nishina formula

$$\frac{d\sigma}{d\Omega} = \frac{1}{2} r_e^2 \left(\frac{\nu'}{\nu}\right)^2 \left[\frac{\nu'}{\nu} + \frac{\nu}{\nu'} - \sin^2(\theta)\right], \tag{4.288}$$

where $r_e = \alpha\hbar/m_e c \sim 0.386$ pm (picometer) is the so-called reduced Compton wavelength of the electron and θ the scattering angle of the Compton photon relative to the propagation of the incident photon. The higher is the energy of the incident photon, the more forward focused is the angular distribution of the scattered photon.

If the incident energy of the photon exceeds 1022 keV, i.e., the rest mass of a $e^+ e^-$ pair, a pair production is energetically possible from the interaction of the photon with the electric field of the absorber nucleus. The probability of the process is sizeable only for high-energy gamma rays. In the process, the incident photon is fully absorbed and a $e^+ e^-$ pair is created. The remaining energy $h\nu - 1022$ keV is shared between the electron and the positron in kinetic energy. Both the electron and the positron lose energy in the material through elastic collisions with surrounding electrons. The positron, once at rest, annihilates with an electron of the material, leading to two 511 keV photons, emitted back to back to satisfy energy and momentum conservation.

Problem

4.12 Demonstrate that the minimum kinetic energy T_{th} for an incident particle 1 impinging on a particle 2 at rest in the laboratory frame to create a final state composed of n particles is given by

$$T_{th} = \frac{\left(\sum_{k=1}^{n} M_{f,k}\right)^2 - \left(\sum_{k=1,2} M_k\right)^2}{2 M_2 c^2}, \tag{4.289}$$

where M_1 and M_2 are the masses of particle 1 and 2, respectively, and $M_{f,k}$ are the masses of the n particles in the final state. By using Eq. (4.289), determine the energy threshold for pair production from the interaction of a photon with a nucleus and show that it can be approximated to $2 m_e c^2 = 1022$ keV. What is the energy threshold for pair production from the interaction of a photon with an electron. Why the pair production from the interaction with electrons is generally negligible?

Figure 4.66 shows in logarithmic scale a typical spectrum obtained with a Ge semiconductor detector facing a radioactive source emitting one transition of energy

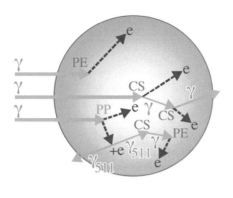

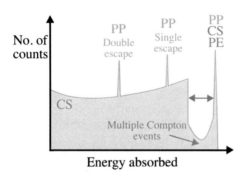

Energy absorbed

Fig. 4.66 (Top) Illustration of typical events contributing to a measured gamma spectrum. (Bottom) Typical gamma spectrum measured by a Ge detector from a radioactive source emitting one single photon energy E_0 larger than $2m_ec^2 = 1022$ keV. The total energy peak at E_0 contains photoelectric events (PE), as well as multiple Compton (CS) and pair-production (PP) events for which the full energy was eventually collected in the detector. Compton events for which only a partial energy is deposited in the detector lead to a continuum of energies with a maximum energy of $E_0/(1 + 2E_0/m_ec^2)$, the so-called Compton edge. Multiple Compton events for might lead to a collected energy between the Compton edge and the total energy peak. In the case of e^+e^- pair production, only possible for $E_0 > 2m_ec^2$, one of the two 511 keV photons resulting from the annihilation of the positron might escape the detector, leading to a shift of 511 keV in the measured spectrum. The cases where all the energy is measured except one escaped 511 keV photon leads to a peak called the single-escape peak. Events where both annihilation photons escape the detector, lead to the so-called second escape peak in the spectrum. The ratio of the different components depend on the geometry of the detector and the relative position of the source

E_0 larger than $2m_ec^2 = 1022$ keV, and therefore allowing pair production from the interaction with atomic nuclei. An example of such a source would be ^{22}Na, a β^+ emitter which decays to the first 2^+ state of ^{22}Ne (1274 keV excitation energy) with a branching ratio of 99.9%. In between the total energy peak and the corresponding Compton edge, multi-Compton events are visible.

4.6.6 Gamma Detection and Resolving Power

In most γ-ray spectroscopy experiments, the signal of interest is embedded into a large background coming from other uncorrelated transitions. A large set of research topics focuses on weakly populated transitions. The ability of a γ-ray detector to resolve the sequence of such transitions is quantified by the *Resolving Power* (RP). The resolving power depends on what one means by resolving a transition, i.e., on the peak-over-background to be achieved and the total number of counts N in the photopeak to claim that the transition is observed. It also depends on the experimental conditions. There is therefore no unique definition of the resolving power. In the following, we consider that a transition is evidenced if the corresponding peak-over-background is at least equal to unity and that the searched transition requires at least N counts in the photopeak to be claimed. We follow the definition and notations of the RP from [35].

An important technique to resolve weak transitions is to perform many-fold $\gamma - \gamma$ coincidences to increase the peak-over-background for the transition of interest. This is illustrated in Fig. 4.67 for a cascade of four gamma rays. The mean energy spacing between two transitions is $S(E)$ and the measurement is performed with an array which allows an energy resolution of $\delta(E) < S(E)$. A γ spectrum with no coincidence, called one-fold spectrum, will provide a certain peak-over-background ratio. A coincidence with an energy cut of the size of $S(E)$ will not improve the peak-over-background ratio since both the photopeak statistics and the background will be reduced by the same amount ($1/3$ in the example of Fig. 4.67). If the energy slice of the coincidence is smaller than $S(E)$, the lowest possible value not to lose a significant amount of statistics being $\delta(E)$, the background is further reduced while the photopeak events are conserved. In addition to this, only a part of the total number of γ events of interest results in the photopeak, the ratio being called the peak-to-total (P/T). Often, the gate around a transition does not comprise all the photopeak. A realistic value being 76% for a 1 σ gate around a gaussian-shape peak. Each time the fold is increased by one, the statistics is lowered by $0.76 \times P/T$. In a two-fold coincidence, the improvement in peak-over-background is then

$$R = 0.76 \frac{S(E)}{\delta(E)} P/T. \tag{4.290}$$

We consider here that the decay branch of interest has a relative intensity of α compared to all other transitions produced in the reaction of interest. We further assume that the γ transitions of interest are weak, and therefore that the peak-over-background is small for any fold. Under these conditions, the peak-over-background ratio for such a γ branch of interest is αR and it is αR^f for a f-fold coincidence spectrum where f transitions are detected.

Considering a total number of events N_0 of the studied reaction channel in the experiment, the number of counts n in a photopeak at energy E for the branch of interest with branching ratio α in a f-fold coincidence spectrum is

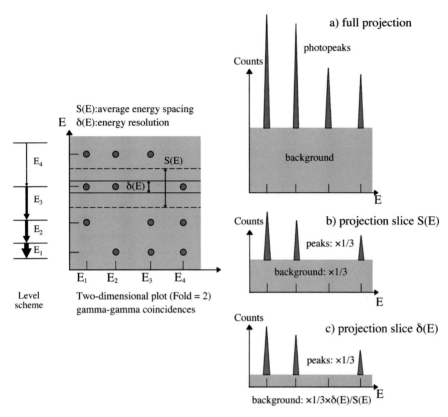

Fig. 4.67 (Left) Sketch of a two-fold gamma spectrum for a weak γ-decay chain of four transitions in coincidence with energies E_1, E_2, E_3 and E_4. (Right) Projected 1-dimensional spectra without $\gamma - \gamma$ coincidence (a), in coincidence with an energy cut equal to the mean energy spacing $S(E)$ between transitions (b), and in coincidence with an energy cut corresponding to the energy resolution $\delta(E)$ of the array (c). The gain in terms of peak-over-total (P/T) is proportional to $S(E)/\delta(E)$. Sketch from a lecture by D. Weißhaar, NSCL

$$n = \alpha N_0 \epsilon(E)^f, \tag{4.291}$$

where $\epsilon(E)$ is the photopeak efficiency of the considered γ-ray array for a transition of energy E.

With the above definitions, one can deduce the minimum branching ratio α_0 of a cascade of transitions to be resolved. One gets $N = \alpha_0 N_0 \epsilon^f$ and $\alpha_0 R^f = 1$. The minimum branching α_0 at which the cascade can be observed defines the resolving power

$$RP = \frac{1}{\alpha_0} = R^F, \tag{4.292}$$

Fig. 4.68 Progresses of gamma spectroscopy techniques as a function of time. In the case of high-spin spectroscopy with a high gamma multiplicity, the resolving power of new generation Germanium arrays based on tracking, AGATA and GRETA, represents a gain of three order of magnitudes compared to the previous state-of-the-art

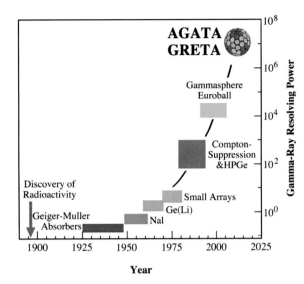

where F is the optimum fold. By combining relations (4.291) and (4.292), one can eliminate F and obtain the resolving power as a function of N_0, ϵ, R as inputs from the experiment and array, and N as the statistics condition one considers that the transitions can be claimed

$$RP = \exp[\frac{\ln(N_0/N)}{1 - \ln \epsilon / \ln R}]. \tag{4.293}$$

The evolution in time of the resolving power in the history of state-of-the-art γ-detector developments is illustrated in Fig. 4.68. The first γ detectors with a resolution suitable for detailed spectroscopy were scintillator detectors. The most commonly used scintillating material is Sodium Iodide (Na(I)) which reached an intrinsic energy resolution of about 8%. The first high-resolution detectors were developed in the early 60s and consisted of semiconductors made of Li-drifted Ge. Soon after, large volume Li-drifted Ge detectors could be produced and first $\gamma - \gamma$ coincidence measurements were performed in the 70s. An important step in high-resolution spectroscopy was reached with the development of intrinsic semiconductors made of high-purity Ge (HPGe). Since the Li drift was no longer necessary, large-volume detectors could be obtained. In the 80s, large arrays of HPGe detectors were assembled with the addition of detectors for Compton suppression. Each HPGe detector was surrounded with high-efficiency active shielding which allowed to reject Compton events where the scattered photon exit the Ge volume. The discovery of superdeformation and the systematic study of nuclear shapes have been made possible by the development of such HPGe arrays (see Chap. 7). The required purity of these crystals reaches 10^{14}, one of the most demanding purity ever produced. When the bulk volume of the detector is fully depleted by an electrical potential of several thousands of volts, any pair of electron-hole created by the interaction of a photon in the detector will drift apart towards the electrodes. The small energy necessary to create one electron-hole pair (about 3 eV) and the very high collection in the detector leads to an excellent energy resolution of typically 2 keV full-width at half maximum for the full photoelectric conversion of a 1 MeV gamma ray.

The US and Europe have a longstanding tradition in developing gamma arrays for nuclear studies. Up to the 2010s, high-purity Ge arrays consist of Ge detectors individually Compton shielded. These active shields have the drawback of lowering the photopeak efficiency. Two photons from the same event hitting the same detector could not be disentangled leading to very low efficiency for high multiplicity events. To overcome this limitation and provide a new level of sensitivity, a new concept has been developed: a full shell of germanium whose treatment is based on gamma tracking inside the bulk of the semiconductor. Two of such new generation arrays are under completion today: AGATA in Europe and GRETA in the United States. These new generation detectors are called *tracking arrays*. They are segmented electrically and the pulse shape of the charge signal induced on every segment is registered with a digital electronics. The charge deposited in a segment is recorded as well. The neighbouring segments also experience induced charges from capacitive coupling. As quantified by the Shockley-Ramo theorem, the shape of the wave forms depends on the position of the charge deposition inside the segment, leading to a position resolution of few millimeters better than the size of the segment. An individual photon that interacts with the detector may induce several charge depositions if Compton scattering occurs. Every cluster of deposited charges is analysed and time ordered to be consistent with the Compton scattering conservation law of Eq. (4.287). The initial interaction point of the photon and its total energy can then be reconstructed. The concepts of pulse-shape and tracking are illustrated in Fig. 4.69.

With the gamma-tracking technique, Compton events from one crystal to another are not rejected by treated fully. The efficiency of the array is then greatly enhanced, in particular for events with large multiplicity. The resolving power of such arrays, when available in a 4π geometry, will be three orders of magnitude larger than the current state-of-the-art Compton suppressed HPGe arrays for high multiplicity events, allowing new nuclear-physics phenomena to be investigated (see Chap. 7). GRETA and AGATA follow the same concept and have similar characteristics. When fully completed, AGATA will be composed of 180 HPGe crystals 37-fold segmented (36 segments and 1 core electrical contact) grouped into triplets of crystals, while GRETA will be composed of 120 crystals assembled into 30 quadruplets. The typical intrinsic energy resolution of each segment is about 2.1 keV FWHM for a 1332 keV transition. The full AGATA and GRETA arrays are expected to lead to a resolving power three orders of magnitude higher than Gammasphere, the former state-of-the-art Compton-suppression HPGe array.

High- purity Ge crystals

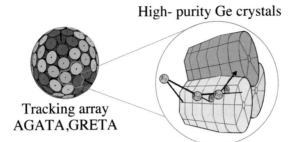

Tracking array
AGATA,GRETA

Digital electronics

Segment induced-signal waveforms

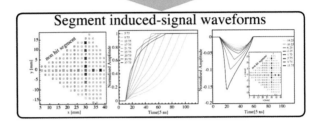

Pulse shape analysis

Identified interaction points
$(E,x,y,z,t)_i$

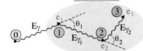

Tracking

Reconstructed gamma rays

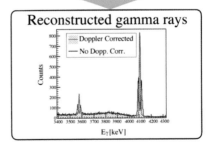

Fig. 4.69   (Continued)

◀**Fig. 4.69** (Continued) Schematic view of the data processing from a tracking array such as AGATA and GRETA. The time-dependence of the amplitude of the signals induced on the segmented electrodes of the HPGe crystals, recorded by digital electronics, are used to determine the individual sites of energy loss in the detector by pulse-shape analysis. The resulting set of identified interaction points is analysed to reconstruct the event: multiplicity of initial photons and history of interactions. In the example illustrated on the top of the figure, an event of multiplicity 1 is shown. Two interaction points occur: a Compton event in one crystal, followed by a photoelectric event of energy $E_{\gamma,2}$. The dependence of the wave form with the position of the hit inside the crystal is also shown. The induced signal on neighbouring electrodes are sensitive to the azimuthal direction around the crystal axis, while the pulse shape of the hit signal is sensitive to radial direction, i.e., the distance between the hit and the central-core electrode located on the symmetry axis of the crystal

4.6.7 In-Flight Spectroscopy

In-flight in-beam gamma spectroscopy in inverse kinematics is a powerful tool for the spectroscopy of radioactive isotopes. At intermediate energies, it allows the use of rather thick targets to counterbalance the low intensity of nuclei far from stability. The final energy resolution, and therefore the resolving power that can be reached in the experiment, is limited by several factors which can be obtained from the Doppler shift formula

$$E_0 = E_\gamma \frac{1 - \beta \cos(\theta)}{\sqrt{1 - \beta^2}}, \tag{4.294}$$

where E_γ is the measured energy of the gamma ray in the laboratory, β is the velocity in units of c of the nucleus when it emits the photon and θ is the scattering angle of the photon relative to the propagation of the emitter. E_0 is the energy of the gamma ray in the center of mass of the emitter, i.e., the energy of interest. The energy resolution δE_0 depends on uncertainties in the measurement of β, θ and E_γ. By derivating Eq. (4.294), on gets

$$\left(\frac{\delta E_0}{E_0}\right)^2 = \left(\frac{\delta E_\gamma}{E_\gamma}\right)^2 + \left(\frac{\beta \sin(\theta)}{1 - \beta \cos(\theta)}\right)^2 (\delta\theta)^2 + \frac{(\beta - \cos(\theta))^2}{(1 - \beta^2)(1 - \beta \cos(\theta))^2}(\delta\beta)^2. \tag{4.295}$$

The relative weight of these contributions to the energy resolution are illustrated in Fig. 4.70 for the two realistic cases of DALI2 and AGATA. In the case of a high-resolution detector with very good granularity, most of the energy resolution originates from the uncertainty of the energy loss in the target resulting in an uncertainty on β. Experimental spectra of Kr isotopes measured with AGATA and DALI2 are shown in Fig. 4.71.

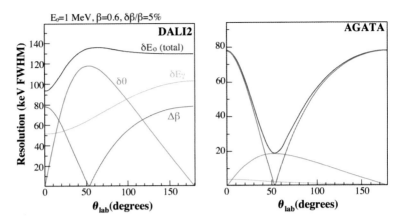

Fig. 4.70 Energy resolution (black line) as a function of the gamma scattering angle for a transition of $E_0 = 1$ MeV emitter at $\beta = 0.6$, with an uncertainty of the velocity at the emission site of $\delta\beta/\beta = 5\%$. The cases of the DALI2 scintillator array ($\delta E/E = 9\%$, $\delta\theta = 8°$) and AGATA ($\delta E/E = 0.2\%$, $\delta\theta = 1.5°$) are shown. The contribution to the energy resolution from the granularity of the detector (blue), its intrinsic energy resolution (green) and the uncertainty on the velocity at emision site (red) are shown separately

4.6.8 Short Lifetime Measurements

The lifetime of an excited bound state that decays via γ emission to another is an important observable to understand the structure of nuclear states and constrain the connection of the two states involved in the decay. At the exception of few isomeric states, most of nuclear lifetimes range between few nanoseconds and few femtoseconds (10^{-15} s).

Several methods can be used, depending on the lifetime range to be measured. A summary of the main methods and their domains of application is illustrated in Fig. 4.72. Lifetimes greater than ~ 1 nanosecond can be determined from a direct time measurement between the $t = 0$ excitation moment to the time of gamma detection. The lifetime can also be measured from the detection of a transition that populates the state of interest ($t = 0$), and inversely a transition that depopulates the state of interest. This direct method can also be applied to shorter lifetimes, down to a few picoseconds, if one uses fast timing detectors such as those made of LaBr$_3$ scintillating material and detect the decay transition in coincidence with a feeding transition. Shorter lifetimes, below 100 ps, require other techniques, based on the Doppler effect, of which the main ones are the Doppler Shift Attenuation Method (DSAM) and the Recoil Differential Doppler Shift method (RDDS).

The fast timing technique requires the measurement of two transitions in coincidence: a transition feeding the state of interest and the targeted transition from which the partial decay lifetime is to be measured. In the following, the two transitions will be named as the *feeder* and *decay* transitions. Fast scintillators and fine-tuned constant-fraction discriminators allow to reach a time resolution of few 100 ps

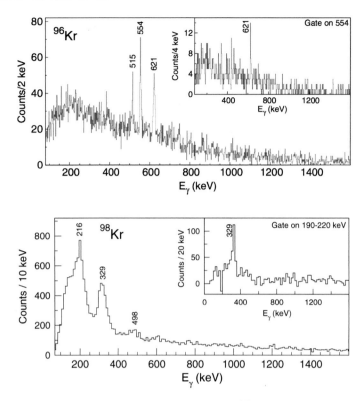

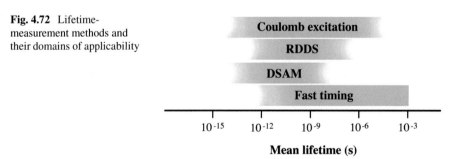

Fig. 4.71 Doppler reconstructed in-flight gamma spectra of ^{96}Kr produced from in-flight fission at $\beta = 0.1$ and measured with the high-resolution HPGe tracking array AGATA at GANIL (top) and ^{98}Kr produced from $(p, 2p)$ quasi-elastic scattering at $\beta = 0.6$ and measured with the NaI scintillator array DALI2 at the RIBF (bottom). The ^{96}Kr spectrum is reprinted from [36]. The ^{98}Kr spectrum was provided by F. Flavigny, LPC Caen

Fig. 4.72 Lifetime-measurement methods and their domains of applicability

FWHM. The measured decay time $D(t)$ between the two transitions is a convolution of the time response of the detectors $r(t)$ and the exponential decay of the state and transition of mean lifetime τ under study

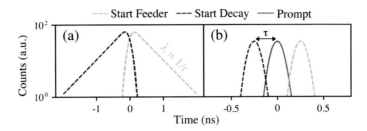

Fig. 4.73 (Left) Decay time spectrum measured between two transitions in cascade, one feeding a state and one originating from the decay of this same state. The tail of the distribution corresponds to the mean lifetime of the decay of the state via the measured transition. (Right) Illustration of the lifetime effect on the spectrum obtained from the centroid definition of Eq. (4.297)

$$D_\tau(t) = \frac{K}{\tau} \int_{-\infty}^{t} e^{-(t-t')/\tau} r(t')dt', \qquad (4.296)$$

with K being a normalization constant. A schematic representation of $D(t)$ is shown on the left panel of Fig. 4.73. The spectrum shows a tail which can be fitted by use of (4.296) and, in the absence of influence of other transitions, give access to the mean partial lifetime corresponding of the decay through the measured decay transition. If the start and stop time are exchanged between the detection of the feeder and the decay transition, the spectra are mirrored around $t = 0$. In case the lifetime is comparable to, or shorter than, the time response of the detector, the slope cannot be used anymore to extract the lifetime, but a shift of the centroid of the mean decay time distribution due to the lifetime of the state can be measured. This centroid Δt compared to the case of a prompt distribution ($\tau = 0$) is given by

$$\Delta t = \frac{\int t D_\tau(t)dt}{\int D_\tau(t)dt} - \frac{\int t D_{\tau=0}(t)dt}{\int D_{\tau=0}(t)dt} = 2\tau. \qquad (4.297)$$

For typical fast timing resolutions, lifetimes down values and precision of few picoseconds can be extracted this way. The centroid shift method is illustrated in the right panel of Fig. 4.73. In the case of lifetimes lower than ~ 10 ps, other methods are either preferred or necessary.

The RDDS technique uses a *plunger* device composed of a production target where the studied nucleus is populated in the excited state of interest and a foil located downstream the target at a known distance d. The second foil is often named as the *degrader* since it is used to slow down the nuclei produced in the production target. Any γ-decaying excited state produced in the target can decay in two distinct regions as it flies away from the target: between the target and degrader of the plunger if it decays before reaching the degrader, or downstream the degrader if the decay time is longer. In the laboratory frame, the velocity of the recoiling nucleus when emitting the photon is different in the two cases because of the significant energy loss in the degrader. This change of velocity, between before and after the degrader, leads to a measurable change in the Lorentz boost to the photon energy. The total-energy peak of the gamma ray of interest is therefore detected at two different energies corresponding to the two regions and the associated different velocities. The choice of the degrader material and thickness requires that the amount of reactions in the degrader do not overcome those in the target and that it is thick enough so that the velocity change leads to two distinct components in the measured gamma spectra. The principle of the technique and an example of measured gamma spectra are shown in Fig. 4.74. In the case of precise measurements and very short lifetimes, the distance between the target and degrader has to be kept constant over time. This is done by fixing the degrader foil on a movable frame controlled by a piezo-electric motor which, via a feedback loop, maintains the target-foil distance from the measured capacity between the target and degrader. Indeed, during the experiment the beam can heat the target which may get deformed: the distance between the production target and the degrader requires to be continuously adjusted. The influence of the lifetime of feeder states that feed the state of interest can be controlled by applied

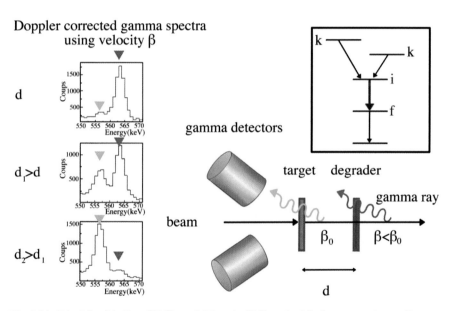

Fig. 4.74 Principle of the Recoil Differential Doppler Shift method. In the presented case, distances are of the order of $100\,\mu$m for a lifetime of about 10 ps. The mean velocity of the recoiling nucleus is β_0 after the target and $\beta < \beta_0$ after the degrader. In the laboratory frame, at an angle θ from the beam direction, the photon energy is $E_0\gamma(1 - \beta_0\cos(\theta))$ if emitted between the target and the degrader (green) and $E_0\gamma(1 - \beta\cos(\theta))$ if emitted after the degrader (pink). In the shown example, $\theta > 90°$ and the photopeak energy is shifted to lower energies compared to the center of mass energy E_0. On the l.h.s. of the figure, typical gamma energy spectra are shown for three different distances between the target and the degrader. For distances small compared to $\beta\tau c$, few decays occur between the target and the degrader (top). When $d/\beta c \sim \tau$ (middle), the two shifted total energy peaks have a similar statistics. In the case of a large distance, so that $d/\beta c > \tau$ (bottom), most of the decays take place at β_0, before the nucleus reaches the degrader

the technique to events with a coincidence on the fast component of a gamma ray from a feeding branch.

In the cases of very short lifetimes, below a picosecond, the RDDS technique cannot be applied. One may then consider the DSAM method based on the use of a production target immediately followed by a stopper material in which a nucleus produced in the production target is slowed down. A thick target can also be used and play both the role of the production target and the stopper. The method has been extensively used with low-incident production mechanisms such as fusion evaporation where the nucleus is fully stopped in the stopper material. Once produced, the nucleus quickly slows down from $\beta = \beta_0$ to $\beta = 0$. Its lifetime will impact the shape of the measured photopeak from its decay which occurred at one extreme mostly at β_0 for very short lifetime or at $\beta = 0$ for a long lifetime on the other extreme. The method is sensitive for lifetimes comparable to the slowing time of the ion inside the target, typically few 100 femtoseconds. In the DSAM technique, the influence of feeding is also to be considered with care.

Summary

The binding energy $B(N, Z)$ of a nuclear state in a nucleus composed of N neutrons and Z protons contains both bulk information on the nuclear interactions and, although less pronounced, Coulomb and weak forces in the nucleus, as well as nuclear structure effects such as shells, pairing and deformation. It is related to the mass m of the nuclear state by

$$m(N, Z)c^2 = N m_n c^2 + Z m_p c^2 - B(N, Z).$$

Nuclear masses can be determined from different techniques. (i) The $B\rho$-time-of-flight method is applicable to short lived nuclei (lifetime down to several 100 ns) and allows typical mass precision of 10^{-5}. (ii) The revolution frequency of ions in storage rings is directly proportional to the m/q ratio of the stored nuclei. A precision of 10^{-7} can be achieved. (iii) Precision ion traps are the reference apparatus for mass measurements. Penning traps allow a relative mass precision of 10^{-12} with stable nuclei, while 10^{-8} is often reached with radioactive ions. In a Penning trap, the mass is obtained from the determination of the cyclotron frequency

$$\omega_c = \frac{q B}{\gamma m},$$

where B is the magnetic field seen by the trapped ion.

The β decay of a nucleus is driven by the weak interaction and can follow three processes: (i) the change of a neutron in a proton, accompanied with the emission of an electron and an electronic anti-neutrino (β^- disintegration), (ii) the change of a proton into a neutron, accompanied with the emission of a positron and an electronic neutrino (β^+ disintegration), and (iii) electron capture which is equivalent to a β^+ decay without positron emission. A radioactive nucleus can beta decay only if the process is energetically allowed, i.e. $Q_\beta \leq 0$. β decays can be of Fermi type when the spins of the leptons in the final state couple to $S = 0$, or of Gamow-Teller type when the spins of the leptons in the final state couple to $S = 1$. The transition probability from an initial state to a final state is driven by the so-called ft value. Beta decay favours low momentum transfer.

The charge density distribution $\rho_c(r)$ can be measured for stable nuclei from electron elastic scattering. For a spin zero nucleus, the cross section reduces to

$$\frac{d\sigma}{d\Omega} = \frac{d\sigma}{d\Omega}\Big|_{\text{Mott}} |F_C(q)|^2,$$

where $(d\sigma/d\Omega)|_{\text{Mott}}$ is the Mott cross section for a point-like nucleus and $F_C(q)$ is the Coulomb form factor depending on the transferred momentum. The form factor in the sudden approximation is the Fourier transform of the charge density distribution

$$F_C(q) = \frac{1}{Z} \int e^{iqr} \rho(r) dr.$$

In the case of radioactive nuclei, most of the accessible information about the charge distribution reduces to the root mean square charge radius

$$Z\langle r^2 \rangle_c = \int r^2 \rho_c(\mathbf{r}) d\mathbf{r}.$$

The most precise relative determination of $\langle r^2 \rangle_c$ is obtained by the atomic fine structure of atoms along isotopic chains. The electromagnetic interaction of the electrons with the finite-size nucleus leads to an isotopic shift of the energy of fine structure transitions between two isotopes A and A' following

$$\delta \nu^{AA'} = K_{MS} \times \frac{M_{A'} - M_A}{M_{A'} M_A} + F \times \delta \langle r_c^2 \rangle^{AA'},$$

where K_{MS} and F are constants along an isotopic chain.

Root mean square matter or neutron radii are more difficult to determine. The parity-violating terms of electron elastic scattering, caused by the weak interaction, allows to determine the neutron root mean square radius of stable nuclei from asymmetry measurements. This method suffers mostly from the statistical uncertainties. Proton elastic scattering and coherent pion photo-production are hadronic probes used to extract the matter radius of nuclei. These methods imply systematic uncertainties from the reaction mechanism which are difficult to assess.

In nuclei, the first term of the expansion of the electric potential after the monopole corresponds to the quadrupole moment. For a continuous distribution of charges with density $\rho(\mathbf{r})$, the electric quadrupole moment is defined as

$$\mathbf{Q}_{mn}(\mathbf{R}) = \int \rho(3r_m r_n - |\mathbf{r}|^2 \delta_{mn}) d\mathbf{r},$$

and it quantifies the deviation from a spherical charge distribution. In addition to electric moments, the nucleus also has magnetic moments caused by the motion of the (charged) protons within the nucleus : the orbiting of protons gives rise to an orbital magnetic field. The intrinsic spin of nucleons (protons and neutrons) also induces an intrinsic magnetic field. The magnetic dipole operator is expressed as a sum of these two contributions

$$\mu = \sum_{i=1}^{A} g_l^i \mathbf{l}^i + \sum_{i=1}^{A} g_s^i \mathbf{s}^i,$$

where the free-nucleon gyromagnetic factors for protons (neutrons) are $g_l = 1$ (0) and $g_s = +5.587$ (-3.826). Both the electric quadrupole moment and the magnetic dipole moment can be determined from the atomic hyperfine structure.

Particle spectroscopy via missing mass allows to determine the excitation energy of bound and unbound states with a sensitivity of the transferred angular momentum in the reaction. The invariant-mass spectroscopy allows the spectroscopy of unbound states decaying via the emission of a particle. This technique is abundantly used for neutron spectroscopy of neutron-rich nuclei. Finally, it can also be combined to gamma spectroscopy since (most) bound excited states decay to lower states via gamma-ray emission. One-nucleon transfer is the most used technique for particle spectroscopy. It is performed at relatively low energy, typically from few to 50 MeV/nucleon, depending on the momentum matching required to populate the target final states. The angular distribution of the recoil nuclei can be used to determine the transferred angular momentum in the transfer.

Charged-particle spectroscopy is most often performed with semiconductor (solid) stripped silicon detectors or time projection chambers. The identification of light recoils is made possible by energy loss measurements, time of flight and, in the case of magnetic field, curvature of tracks. Typical energy resolutions achieved in inverse kinematics with radioactive beams are 0.1 - 1 MeV FWHM.

Excited nuclear bound states most often decay via electromagnetic transitions to lower states. The energy of the electromagnetic transitions is the energy difference between the initial and final state. They carry an angular momentum and can be of electric (E) or magnetic type (M).

High-resolution spectroscopy can be best obtained from the measurement of gamma transitions from the decay of nuclear excited states. Photons interact with matter via three main processes: (i) the photoelectric effect, dominant at low energy below $\sim$500 keV, (ii) the Compton effect and (iii) the pair production from the interaction with the electric field of the nucleus, with a production threshold at 1022 keV.

The partial lifetime from an initial state decaying via electromagnetic transition to a final state is an essential information to probe the overlap of nuclear wave functions. In the particular case of an E2 transition, the lifetime τ is related to the reduced transition probability $B(E2)$. It is given in $e^2\text{fm}^4$, $B(E2)$ by

$$\tau = 1.22 10^9 E_\gamma^5 B(E2),$$

with E_γ the energy of the transition in MeV. The lifetime of short lived states decaying via electromagnetic transitions can be determined from the fast timing technique (down to few ps), the recoil distance Doppler shift (RDDS) method by use of a plunger device for lifetimes down to 1 ps and from the Doppler shift attenuation method (DSAM) for lifetimes as short as 100 fs.

Solutions of Problems

4.1 The Lorentz force acting on a particle in an electric field is

$$m\frac{d^2\mathbf{u}}{dt^2} = q\mathbf{E} + q\mathbf{v} \times \mathbf{B}. \tag{4.298}$$

The electric field is given by the potential from Eq. (4.19) as

$$\mathbf{E} = -\nabla U = \frac{U_0}{4d^2}\nabla\left(\rho^2, -2z^2\right) = \frac{U_0}{d^2}\left(\frac{x}{2}, \frac{y}{2}, -z\right). \tag{4.299}$$

The magnetic field is $\mathbf{B} = B\hat{z}$ such that

$$\mathbf{v} \times \mathbf{B} = (\dot{y}B, -\dot{x}B, 0). \tag{4.300}$$

Substituting Eqs. (4.299), (4.300) into Eq. (4.298) the equations of motion are

$$\ddot{x} = \frac{x}{2}\omega_z^2 + \omega_c\dot{y}, \tag{4.301}$$

$$\ddot{y} = \frac{y}{2}\omega_z^2 - \omega_c\dot{x}, \tag{4.302}$$

$$\ddot{z} = -z\omega_z^2, \tag{4.303}$$

where we used $\omega_z = \sqrt{\frac{qU_0}{md^2}}$ (the axial motion frequency), $\omega_c = \frac{qB}{m}$ (the cyclotron frequency).

For the z component, we can write the solution as

$$z(t) = \alpha_z\cos\left(\omega_z t + \phi_z\right), \tag{4.304}$$

with α_z being the amplitude of the longitudinal motion. The longitudinal frequency ω_z describes the back and forth oscillation by the two endcaps of the trap.

For x, y components we define s such that $s = x + iy$. Then we combine x and y components in Eqs. (4.301), (4.302) to

$$\ddot{s} = \frac{x}{2}\omega_z^2 + \omega_c\dot{y} + i\frac{y}{2}\omega_z^2 - i\omega_c\dot{x} = \tag{4.305}$$

$$= \frac{\omega_z^2}{2}s + \omega_c(\dot{y} - i\dot{x}). \tag{4.306}$$

We multiply the second term in the r.h.s. by $1 = -i^2$

$$\ddot{s} = \frac{\omega_z^2}{2}s - i^2\omega_c(\dot{y} - i\dot{x}) = \frac{\omega_z^2}{2}s - i(i\dot{y} + \dot{x})\omega_c, \tag{4.307}$$

$$\ddot{s} = \frac{\omega_z^2}{2}s - i\omega_c\dot{s}. \tag{4.308}$$

The general solution to the equation can be given by a sum of sin and cos

$$x(t) = \rho_+\sin(\omega_+t + \phi_+) + \rho_-\sin(\omega_-t + \phi_-), \tag{4.309}$$
$$y(t) = \rho_+\cos(\omega_+t + \phi_+) + \rho_-\cos(\omega_-t + \phi_-), \tag{4.310}$$

where $\rho_\pm$ are the amplitudes of the motions, and the frequencies are defined as

$$\omega_\pm = \frac{\omega_c}{2} \pm \sqrt{\frac{\omega_c^2}{4} - \frac{\omega_z^2}{2}}. \tag{4.311}$$

ω_+ is called the modified cyclotron frequency, and ω_- the magnetron frequency. From Eq. (4.311) we can get the following identities between the different frequencies

$$\omega_c = \omega_+ + \omega_-, \tag{4.312}$$
$$\omega_c^2 = \omega_z^2 + \omega_+^2 + \omega_-^2, \tag{4.313}$$
$$\omega_z^2 = 2\omega_+\omega_-. \tag{4.314}$$

In the case of a hyperbolic trap

$$\rho^2 - 2z^2 = \rho_0^2. \tag{4.315}$$

This is true for every t, and we can get the relation between the amplitudes

$$\rho^2 - 2z^2 = x^2 + y^2 - 2z^2 = \rho_+^2 \sin^2(\phi_+) + \rho_-^2 \sin^2(\phi_-) + \rho_+^2 \cos^2(\phi_+)$$
$$+\rho_-^2 \cos^2(\phi_-) - 2\alpha_z^2 \cos^2(\phi_z) + 2\rho_+ \rho_-(\sin(\phi_+)\sin(\phi_-) + \cos(\phi_+)\cos(\phi_-)) = \rho_0^2.$$

$$(4.316)$$

Using the identities $\cos^2(\alpha) + \sin^2(\alpha) = 1$, and $\sin(\alpha)\sin(\beta) + \cos(\alpha)\cos(\beta) = \cos(\alpha - \beta)$ gives

$$\rho_+^2 + \rho_-^2 + 2\rho_+ \rho_- \cos(\phi_+ - \phi_-) - 2\alpha_z^2 \cos^2(\phi_z) = \rho_0^2. \qquad (4.317)$$

4.2 Let us consider the integral part of Eq. (4.83):

$$f(q) = -\frac{m}{2\pi\hbar^2} \int e^{+iqr} V_C(r) dr = \frac{2me^2}{\hbar^2 q^2} \int \int e^{+iqr} \frac{\rho(r')}{|r - r'|} dr dr'$$

We introduce q^2 in the integral and replace by the operator Δ and perform the partial integration twice;

$$\frac{1}{q^2} \int \int q^2 e^{iqr} \frac{\rho(r')}{|r - r'|} dr dr' = -\frac{1}{q^2} \int \int \Delta(e^{iqr}) \frac{\rho(r')}{|r - r'|} dr dr'$$
$$= -\frac{1}{q^2} \int \int e^{iqr} \Delta\left(\frac{\rho(r')}{|r - r'|}\right) dr dr'. \qquad (4.318)$$

We then apply an identity

$$\Delta \frac{1}{|r - r'|} = -4\pi\delta(r - r') \qquad (4.319)$$

to Eq. (4.318) and obtains

$$-\frac{1}{q^2} \int e^{iqr} \int \Delta\left(\frac{\rho(r')}{|r - r'|}\right) dr dr' = \frac{4\pi}{q^2} \int e^{iqr} \rho(r) dr \qquad (4.320)$$

which gives Eq. (4.84).

4.3 We calculate the form factor according to Eq. (4.87) using the charge distribution from Eq. (4.90)

$$F_C(q) = \frac{4\pi}{Z} \int \frac{3Ze}{4\pi R^3} \frac{\sin(qr)}{qr} r^2 dr = \frac{3e}{qR^3} \int \sin(qr) r dr. \qquad (4.321)$$

To solve the integral we use the integration in parts formula

$$\int F'G = -\int FG' + FG, \qquad (4.322)$$

where in our case $F = -\cos(qr)/q$ and $G = r$.

$$F_C(q) = \frac{3e}{qR^3} \left(\int \frac{\cos(qr)}{q} dr - \frac{\cos(qr)}{q} r \Big|_0^R \right) = \frac{3e}{qR^3} \left(\frac{\sin(qr)}{q^2} \Big|_0^R - \frac{\cos(qR)}{q} R \right)$$

$$= \frac{3e}{(qR)^3} (\sin(qR) - qR\cos(qR)). \tag{4.323}$$

Substituting Eq. (4.323) into Eqs. (4.89), we get

$$\frac{d\sigma}{d\Omega} = \frac{Z^2 \alpha^2}{4E^2} \frac{1 - \beta^2 \sin^2(\theta/2)}{\sin^4(\theta/2)} \left(\frac{3e}{(qR)^3} \right)^2 (\sin(qR) - qR\cos(qR))^2. \tag{4.324}$$

For a given E, θ we see that the first minimum depends on

$$\alpha^{-3} (\sin(\alpha) - \alpha\cos(\alpha)), \tag{4.325}$$

with $\alpha = qR$. Solving the equation numerically, we get that the first minimum appears at

$$\alpha = qR = 4.5. \tag{4.326}$$

From Fig. 4.19, the first minimum is measured at $q = 1.25$ fm^{-1}.
Therefore we get the charge radius

$$R = \frac{4.5}{1.25} = 3.6 \text{ fm.} \tag{4.327}$$

This value is smaller than the experimental value of 5.51 fm. That can be expected since the nuclei do not have a sharp edge like we assumed at this question, but have a more diffuse surfaces. Common charge distributions make use in a uniform distribution with a smoothed surface, such as Fermi distribution.

4.4 One can write the difference in radii as

$$\Delta R(^4\text{He}) = R_I(^4\text{He}) - R_I(^3\text{He}) = \frac{1}{\sqrt{\pi}} \left[\sqrt{\sigma_I(^4\text{He}, T)} - \sqrt{\sigma_I(^3\text{He}, T)} \right]. \tag{4.328}$$

The cross sections in Table 4.2 are given in mb. The conversion to fm^2 is given by 1 mb $= 10^{-31}$ m$^2 = 0.1$ fm^2. The differences in radii are listed in Table 4.7.

Figure 4.75 shows the radii differences using different target nuclei. By averaging on the target nuclei we fitted the data to A^α, which resulted in $\alpha = 0.392 \pm 0.003$ (solid line). Also shown for comparison the $A^{1/3}$ dependence (black dashed line).

It can be seen that the radii differences are essentially independent of the target nuclei (except for the Be case in ^{4}He which deviates more than the experimental uncertainty). This supports that the target and projectile parts are separable. The radius of ^{3}He is larger than that of ^{4}He, this is in agreement with known data from electron scattering. This can also be expected since for the case of ^{3}He we have asymmetric nucleus (2 protons and one neutron) and ^{4}He is a symmetric nucleus (2 protons and two neutrons) with a large binding energy. The A dependence of the radii differences is larger than $1/3$, as can be seen from the figure, ^{6}He and ^{8}He show larger increase from ^{3}He. The large excess of neutrons in these two isotopes and their very low binding energy favor a large neutron radius and therefore a large matter radius: ^{6}He and ^{8}He present a neutron halo and thick neutron skin, respectively. These nuclear physics quantum-mechanical phenomena are described in Chap. 5.

4.5 The Q-moment for a single-particle configuration is given by

$$Q_{sp} = \langle jm = j|Q_{zz}|jm = j\rangle = e\sqrt{\frac{16\pi}{5}}\frac{\langle jj20|jj\rangle}{\sqrt{2j+1}}\langle j||Y_2||j\rangle\langle r^2\rangle_j, \qquad (4.329)$$

and the reduced matrix element is evaluated to be

$$\langle j||Y_2||j\rangle = \left(\frac{5(2j+1)}{4\pi}\right)^{1/2}\langle j\frac{1}{2}20|j\frac{1}{2}\rangle.$$

The Clebsch-Gordan coefficient is given by

Table 4.7 Measured radii differences of He isotopes using different target nuclei

target	$\Delta R(^4\text{He})$ (fm)	$\Delta R(^6\text{He})$ (fm)	$\Delta R(^8\text{He})$ (fm)
Be	-0.0526 ± 0.0228	0.6437 ± 0.0289	0.9276 ± 0.0206
C	-0.1828 ± 0.0275	0.6100 ± 0.0275	0.9157 ± 0.0267
Al	-0.2188 ± 0.0498	0.6155 ± 0.0352	0.9713 ± 0.0360

Fig. 4.75 Radii differences of He isotopes using different target nuclei. Also shown a fit to a power law A^α (black line), and the $A^{1/3}$ dependence (black dashed line)

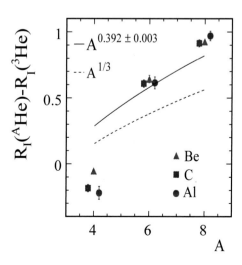

Table 4.8 Q-moment of single-particle (hole) states near closed shell nuclei ^{40}Ca and ^{16}O and effective charges of proton and neutron

Nucleus	l_j	$Q_{s.-p.}$(b=10^{-24}cm^2)	Q_{exp} (b)	e_{eff}/e
^{17}O	$(d_{5/2})_\nu$	$-0.0514e^n_{eff}$	$-0.026e$	0.51
^{17}F	$(d_{5/2})_\pi$	$-0.0514e^p_{eff}$	$|0.058|e$	1.13
^{39}K	$(d_{3/2}^{-1})_\pi$	$0.0475e^p_{eff}$	$+0.060e$	1.26
^{39}Ca	$(d_{3/2}^{-1})_\nu$	$0.0475e^n_{eff}$	$|0.038|e$	0.80
^{41}Ca	$(f_{7/2})_\nu$	$-0.103e^n_{eff}$	$-0.090e$	0.87
^{41}Sc	$(f_{7/2})_\pi$	$-0.103e^p_{eff}$	$-0.156e$	1.51

Table 4.9 Magnetic moments of p- and sd-shell single-particle states

Configuration	A=15			A=17		
	$p_{3/2}$	$p_{1/2}$	$d_{5/2}$	$s_{1/2}$	$d_{3/2}$	
π	$+3.794$	-0.265	$+4.794$	$+2.793$	$+0.123$	
ν	-1.913	$+0.638$	-1.913	-1.913	$+1.148$	

Table 4.10 Magnetic moments of single-particle (hole) states near closed shell nucleus ^{16}O, in μ_N units

Nucleus	μ_{exp}	$\mu_{s.-p.}$	(l_j)
^{15}N	-0.283	-0.265	$(p_{1/2}^{-1})_\pi$
^{15}O	$+0.719$	$+0.638$	$(p_{1/2}^{-1})_\nu$
^{17}O	-1.894	-1.913	$(d_{5/2})_\nu$ or $(s_{1/2})_\nu$
^{17}F	$+4.721$	$+4.794$	$(d_{5/2})_\pi$

$$\langle jm20|jm\rangle = \frac{3m^2 - j(j+1)}{\sqrt{(2j-1)j(j+1)(2j+3)}}. \tag{4.330}$$

Applying a formula (4.330) twice with $m = j$ and $m = 1/2$ for the Q-moment (4.329), we get

$$Q_{s.-p.} = -e\frac{2j-1}{2j+2}\langle r^2\rangle_j.$$

4.6 See Table 4.8.

4.7 The magnetic moments of p- and sd-shell orbits are calculated and listed in Table 4.9.

The most possible configurations from Table 4.9 are $(p_{1/2})_\pi^{-1}$ for ^{15}N and $(p_{1/2})_\nu^{-1}$ for ^{15}O. The single-particle state is determined unanimously to be $(d_{5/2})_\pi$ for ^{17}F, while the two configurations $(d_{5/2})_\nu$ and $(s_{1/2})_\nu$ are both possible for ^{17}O. The most plausible configurations are listed in Table 4.10.

4.8 The hyperfine structure splitting cause by the magnetic dipole moment of the nucleus μ is driven by the factor μB, where B is the magnetic field created by the electron orbital at the site of the nucleus. The magnetic dipole of a nucleus is typically of the order of

$$\mu_N = \frac{e\hbar}{2m_N c} = 0.105 \text{ e} \cdot \text{fm}.$$

Approximating the electron orbital trajectory as a closed loop with the Bohr radius a_0 and a frequency ν_0. The magnetic field created by the electron motion at the nucleus site, i.e. at the center of the closed loop in classical mechanics, is given by

$$B = \frac{\mu_0 e \nu_0}{2a_0},$$

where $\mu_0 = 4\pi \times 10^{-7} \text{ T}^{-1}$ m/A is the permeability of free space. The frequency corresponding to the electron's orbit can be related to its angular momentum, in the order of $\hbar$. Still in a classical image, one gets the angular momentum as

$$\hbar \sim a_0 m_e \nu_0,$$

where ν_0 is the velocity of the electron. The frequency is therefore given by $\nu_0 \sim \frac{\hbar}{2\pi m_e a_0}$.

The order of magnitude of the hyperfine structure splitting due to the magnetic moment of the nucleus can then be estimated as $\mu_N B \sim \frac{\mu_0 e^2 \hbar^2 \Delta E_Q}{8\pi m_N m_e c a_0}$.

4.9 Water is a compound material composed of oxygen and hydrogen atoms with stoechiometric ratios $\omega_O = 1/3$ and $\omega_H = 2/3$, respectively. The mean energy loss by distance unit is given by the weighted energy loses in oxygen and water, as expressed in Eq. (4.197),

$$-\frac{1}{\rho}\langle\frac{dE}{dx}\rangle_{\text{water}} = -\frac{\omega_O}{\rho_O}\langle\frac{dE}{dx}\rangle_O - \frac{\omega_H}{\rho_H}\langle\frac{dE}{dx}\rangle_H. \tag{4.331}$$

By using the relativistic relations $p = \sqrt{E_0(E_0 + 2Mc^2)}/c = \gamma M \beta c$, where c is the speed of light, a 200-MeV/nucleon ^{12}C ion has a momentum $p = 7.708$ GeV/c, leading to $\beta = 0.568$ and $\gamma = 1.215$. Using Eq. (4.331), with $z/A = 1$ for hydrogen and $z/A = 0.5$ for oxygen, and assuming $I \sim z \times 10$ eV for the ionization energy, one gets

$$-\frac{1}{\rho}\langle\frac{dE}{dx}\rangle_{\text{water}} = [\frac{1}{3}\frac{0.3071 \times 0.5 \times 6^2}{0.568^2} + \frac{2}{3}\frac{0.3071 \times 1 \times 6^2}{0.568^2}]$$
$$\times(\frac{1}{2}\ln(\frac{2 \times 511000 \text{ [eV]} \times 1.215^2 0.568^2}{8 \times 10 \text{ [eV]}}) - 0.568^2)$$
$$= 147 \text{ MeV mol}^{-1}\text{cm}^2. \tag{4.332}$$

In the above case, the density ρ of the l.h.s. is the molar density. At room temperature, water is a liquid with mass density $\rho_m = 1$ g $\cdot$ cm^{-3}, while its molar mass is approximately given by

$$m_{mol}(H_2O) = 2m_{mol}(H) + m_{mol}(O) \tag{4.333}$$
$$\sim 2 \times 1 \, g \cdot mol^{-1} + 1 \times 16 \, g \cdot mol^{-1} \tag{4.334}$$
$$\sim 18 \, g \cdot mol^{-1}. \tag{4.335}$$

The molar density of water can therefore be estimated as $\rho = 1/18 = 0.055 \, mol.cm^{-3}$. The energy loss in $\delta x = 1$ cm is then

$$\delta E = \Big(-\frac{1}{\rho} \langle \frac{dE}{dx} \rangle_{water} \Big) \rho \delta x = 8.1 \, MeV. \tag{4.336}$$

This energy loss is indeed significantly lower than the incident energy of the beam and justifies that we considered the mean energy loss per distance unit constant across the water layer.

We now aim at calculating the range of ions in water in the non-relativistic limit. In the case of ^{12}C, verifying $m >> m_e$, T_{max} can be taken as $2m_e(c\beta\gamma)^2$. For ions at incident kinetic energies of ~100 MeV/nucleon, the non-relativistic limit of the Bethe Bloch equation can be considered. In the particular case of this problem, Eq. (4.193) can be simplified as

$$-\frac{dE}{dx} = K Z^2 \frac{z}{A} \rho \frac{1}{\beta^2} \ln \Big(\frac{2m_e c^2 \beta^2}{I} \Big). \tag{4.337}$$

Using the non relativistic definition of the kinetic energy

$$T = \frac{1}{2} m \beta^2 c^2, \tag{4.338}$$

where m is the mass of a ^{12}C ion here, we can express the mean energy loss as a function of the velocity $v(x)$ of the ion inside water

$$\frac{dE}{dx} \equiv \frac{dT}{dx} = \frac{dT}{dv} \frac{dv}{dx} = mv(x) \frac{dv}{dx}. \tag{4.339}$$

The range is given by

$$R = \int_0^{v_0} \frac{dx}{dv} dv. \tag{4.340}$$

Substituting Eq. (4.337) into Eq. (4.339), and the result into Eq. (4.340), we get

$$R = \int_0^{v_0} \frac{v(x)^3}{K \frac{Z^2 z}{mA} c^2 \rho \ln \Big(\frac{2m_e v^2}{I} \Big)} dv, \tag{4.341}$$

which can be analytically integrated by use of the exponential integral function $E_i(x)$ defined as

$$E_i(x) = \int_{-\infty}^{+x} \frac{e^{+t}}{t}\,dt.\tag{4.342}$$

The range is expressed as

$$R = \frac{E_i(\ln(\Phi^2 v_0^4))}{2\Phi^2\Omega},\tag{4.343}$$

where Ω and Φ are defined as

$$\Omega = K\frac{Z^2 z}{mA}\rho, \quad \Phi = \frac{2m_e}{I}.\tag{4.344}$$

Note that Ω has the units of $m^3 s^{-4}$ and Φ the units of $m^{-2}.s^2$. The expression (4.343) for the range is therefore consistent with a length. In the case of ^{12}C ions losing energy inside water, one needs to express the range in water R_{water} as the function of the range in oxygen and hydrogen. Assuming here that the shape of the dE/dx curve as a function of material depth is independent of the material, the range in a compound medium is given by

$$R = \frac{\rho}{\sum_i \rho_i/R_i}.\tag{4.345}$$

Using the above expression, the range in water reads

$$R_{water} = \frac{1}{\frac{16/18}{R_O} + \frac{2/18}{R_H}}.\tag{4.346}$$

By using Eqs. (4.341) and (4.346), one obtains ranges of 5 cm, 13 cm and 49 cm in water for ^{12}C ions of 100 MeV, 200 MeV and 400 MeV, respectively. Note that these values overestimate the experimental data because of the approximate Bethe-Bloch formula and the assumed non-relativistic relation between the kinetic energy and the velocity. Experimentally, the range of a 400 MeV/nucleon carbon ion in water is $\sim$ 30 cm.

As a human body is mostly composed of water, the above ranges in water are relevant to decide the appropriate ^{12}C beam energy for carbon therapy. A typical human body does not exceed a thickness of $\sim$ 50 cm. Any location inside the body could be then accessible with such a ^{12}C beam, while 100 and 200 MeV would not be sufficient. 400 MeV is therefore a proper energy for a heavy-ion accelerator dedicated to carbon therapy. As a matter of fact, 400 MeV/nucleon carbon beams are used at the HIMAC facility in Japan, for hadrontherapy, while the University hospital of Heidelberg uses carbon beams of 425 MeV/nucleon.

4.10 By application of the Shockley-Ramo theorem, the induced current on the central wire (electrode 1) is given by

$$i_1(t) = -q\mathbf{E}^w \cdot \mathbf{v}(t),\tag{4.347}$$

where $\mathbf{E}^w$ is the weighting field at the position $x(t)$ of the moving charged particle at a velocity $v(t)$. Let us first calculate the motion of the charge particle. In between the wire and the external cylindrical tube (electrode 2), the solution of the Poisson equation

$$\Delta V = 0, \tag{4.348}$$

has a cylindrical symmetry and the Poisson equation, restricted to the radial direction r, becomes

$$\frac{1}{r}\frac{\partial}{\partial r}\left(r\frac{\partial}{\partial r}V\right) = 0. \tag{4.349}$$

The general form of the solution is

$$V(r) = a\ln(r) + b, \tag{4.350}$$

where a and b are two constant which are set by the boundary conditions $V(r_1) = V_1$ and $V(r_2) = 0$. The potential is therefore

$$V(r) = \frac{V_1}{\ln(r_1/r_2)}\ln(r/r_1). \tag{4.351}$$

The electric field is provided by

$$\mathbf{E} = E_r\mathbf{e_r} = -\frac{\partial V}{\partial r}\mathbf{e_r} = \frac{V_1}{\ln(r_2/r_1)}\frac{1}{r}\mathbf{e_r}, \tag{4.352}$$

where $\mathbf{e_r}$ is the unit radial vector pointing towards large radii.

We assume that the positive charge moves from the instant $t = 0$ at the drift velocity

$$\mathbf{v} = \mu\mathbf{E} = \mu\frac{V_1}{\ln(r_2/r_1)}\frac{1}{r}\mathbf{e_r}, \tag{4.353}$$

which leads to the motion

$$\frac{dr}{dt} = \mu\frac{V_1}{\ln(r_2/r_1)} \Rightarrow r(t) = \sqrt{\frac{t}{t_0} + r_1^2}, \tag{4.354}$$

respecting the boundary condition $r(t = 0) = r_1$ and introducing $t_0 = \ln(r_2/r_1)/2\mu V_1$. The charge reaches the second electrode at $T = t_0(r_2^2 - r_1^2)$. The weighting field is obtained by setting the potential to electrode 1 to $V_1 = 1$ V. From Eq. (4.352), we get

$$\mathbf{E}^w = \frac{1}{\ln(r_2/r_1)}\frac{1}{r}\mathbf{e_r} = \frac{1}{\ln(r_2/r_1)}\frac{1}{\sqrt{\frac{t}{t_0} + r_1^2}}\mathbf{e_r}. \tag{4.355}$$

The induced current on the wire is then given by

$$i_1(t) = -q\mathbf{E}^w \cdot \mathbf{v}(t) = \frac{-q\mu}{\ln^2(r_2/r_1)} \frac{V_1}{\frac{t}{t_0} + r_1^2}, \tag{4.356}$$

from $t = 0$ to $t = T$. The induced current is $i_1(t) = 0$ for $t \geq T$.

4.11 First, we probe that the vector fields (4.236) and (4.237) satisfy the Coulomb gauge,

$$\nabla \cdot \mathbf{A}^{E,M}(\mathbf{r}) = 0. \tag{4.357}$$

For the electric vector field, we can use identities of the vector algebra,

$$\nabla \cdot (\nabla \times \mathbf{L}) = (\nabla \times \nabla) \cdot \mathbf{L} = 0, \tag{4.358}$$

since $\nabla \times \nabla = 0$. For the magnetic one, we can prove also in a similar way,

$$\nabla \cdot \mathbf{L} = \nabla \cdot (\mathbf{r} \times \mathbf{p}) = -i\nabla \cdot (\mathbf{r} \times \nabla) = -i(\nabla \times \mathbf{r}) \cdot \nabla$$
$$= i(\mathbf{r} \times \nabla) \cdot \nabla = i\mathbf{r} \cdot (\nabla \times \nabla) = 0. \tag{4.359}$$

To prove that the vector fields satisfy the Helmholtz equation, it is enough to show the commutation relations,

$$[\nabla^2, \mathbf{L}] = 0, \tag{4.360}$$

and

$$[\nabla^2, \nabla \times \mathbf{L}] = 0, \tag{4.361}$$

since $\Phi_{LM}(\mathbf{r})$ is the solution of the Helmholtz equation. We take L_x to prove Eq. (4.360) as

$$[\frac{\partial^2}{\partial x^2} + \frac{\partial^2}{\partial y^2} + \frac{\partial^2}{\partial z^2}, L_x] = [\frac{\partial^2}{\partial x^2} + \frac{\partial^2}{\partial y^2} + \frac{\partial^2}{\partial z^2}, y\frac{\partial}{\partial z} - z\frac{\partial}{\partial y}]$$
$$= [\frac{\partial^2}{\partial y^2}, y\frac{\partial}{\partial z}] - [\frac{\partial^2}{\partial z^2}, z\frac{\partial}{\partial y}] = 2\frac{\partial}{\partial y}\frac{\partial}{\partial z} - 2\frac{\partial}{\partial z}\frac{\partial}{\partial y} = 0. \tag{4.362}$$

We can prove the commutation relations for other components, L_y and L_z, in the same way. We take also the x-component of $\nabla \times \mathbf{L}$ for the commutator (4.361). The commutator is evaluated as

$$[\nabla^2, (\nabla \times \mathbf{L})_x] = [\nabla^2, \frac{\partial}{\partial y}L_z - \frac{\partial}{\partial z}L_y] = 0, \tag{4.363}$$

since L_y and L_z commute with ∇^2 as was proved in Eq. (4.362).

4.12 In the laboratory frame, the quadrivectors for particles 1 and 2 are

$$(E_1, \mathbf{p}_1 c) \text{ and } (M_2 c^2, \mathbf{0}). \tag{4.364}$$

At the threshold kinetic energy $T_{\text{th}} = T_1 = E_1 - M_1 c^2$, all particles in the final state are at rest in the center-of-mass frame

$$E_{CM} = \sum_{k=1}^{n} M_{f,k} c^2. \tag{4.365}$$

Quoting β and γ the velocity and Lorentz factor of the center of mass in the laboratory frame, one can estimate the total energy before the reaction

$$E_{CM} = \gamma(E_1 + M_2 c^2) - \beta\gamma p_1 c = \gamma(E_1 + M_2 c^2 - \beta p_1 c). \tag{4.366}$$

We now determine γ and β. The inverse Lorentz boost from the center of mass frame to the laboratory frame gives

$$\gamma E_{CM} = (E_1 + M_2 c^2), \tag{4.367}$$

leading to

$$\gamma = \frac{E_1 + M_2 c^2}{E_{CM}}. \tag{4.368}$$

By definition, the total momentum in the center of mass is zero

$$0 = -\beta\gamma(E_1 + M_2 c^2) + \gamma p_1 c, \tag{4.369}$$

leading to

$$\beta = \frac{p_1 c}{E_1 + M_2 c^2}. \tag{4.370}$$

By inserting the above relations for γ and β in Eq. (4.366), one gets

$$E_{CM} = \frac{E_1 + M_2 c^2}{E_{CM}} \left(E_1 + M_2 c^2 - \frac{(p_1 c)^2}{E_1 + M_2 c^2} \right). \tag{4.371}$$

By using $E_1 = T_1 + M_1 c^2$, we obtain

$$T_1 = \frac{E_{CM}^2 - \left(\sum_{k=1,2} M_k c^2 \right)^2}{2 M_2 c^2}. \tag{4.372}$$

At the threshold energy, $T_1 = T_{\text{th}}$ and the center-of-mass total energy is given by Eq. (4.365). We therefore get

$$T_{\text{th}} = \frac{\left(\sum_{k=1}^{n} M_{f,k} \right)^2 - \left(\sum_{1,2} M_k \right)^2}{2 M_2} c^2. \tag{4.373}$$

In the case of a e^+e^- pair production by interaction of a photon of mass $M_1 = 0$ onto a nucleus of mass $M_2 = M$, the final state is composed of three particles, namely the nucleus, a positron and an electron, the last two having identical masses m_e. By using the general Eq. (4.373) to this case, one gets the pair production threshold for the incident photon

$$T_{\text{th}} = \frac{(M + 2m_e)^2 - M^2}{2M} c^2 = 2m_e c^2 (1 + \frac{m_e}{M}). \tag{4.374}$$

The mass of a nucleus is much larger than the mass of an electron. The threshold photon energy for the production of a e^+e^- can therefore be approximated to $2m_e c^2 = 1022$ keV.

A e^+e^- pair can also be produced from the interaction of a photon with an electron bound in an atom. The relation (4.373) also holds in this case, and the corresponding energy threshold is then obtained by replacing the mass of the nucleus by the mass of an electron

$$T_{\text{th}} = \frac{(m_e + 2m_e)^2 - m_e^2}{2m_e} c^2 = 4m_e c^2 = 2044 \text{ keV}. \tag{4.375}$$

The production threshold from the interaction with an electron is therefore twice higher than from the interaction with a nucleus. In addition, the cross section for a pair production is proportional to $(Ze)^2$ in the case of an interaction with a nucleus of atomic number Z, while it is proportional to e^2 in the case of an interaction with an electron. In case of a material of atomic number Z, the e^+e^- pair production will only occur from the interaction with nuclei for photons of energy between $2m_e c^2$ and $4m_e c^2$, while for photons above $4m_e c^2$, the cross section will be Z times more in favor of pair production from nuclei. The pair production from electrons is most often negligible.

Books for Further Readings

"Radiation detection and measurement" (Wiley, 4th edition, 2010) by G. F. Knoll is a classical textbook that provides an introduction to the particle-matter interaction processes and principle of detectors for most types of radiations relevant to low-energy nuclear physics.

"Gaseous radiation detectors" (Cambridge University Press, 2014) by F. Sauli is a reference monograph on the principles of gas detectors for charged particles. It details all relevant aspects of these detectors including drift properties, amplification, working principle of several types of gas detectors from drift chambers to MPGD micro-pattern gaseous detectors (MPGD).

"Semiconductor radiation detectors" (Springer, 2nd edition, 2007) by G. Lutz pro-vides a pedagogical and rather exhaustive introduction to semiconductor physics and detectors. The structure of semiconductors, their Working principle and the most modern designs and variations are detailed.

"The electromagnetic interaction in nuclear spectroscopy" (North-Holland Publishing Company, 1975) edited by W. D. Hamilton covers most aspects of electromagnetic radiations and interactions relevant to nuclear spectroscopy. This collection of articles has remained a reference since its publication.

References

1. B. Franze et al., Nucl. Instr. Meth. A **532**, 97–104 (2004)
2. K. Blaum, Phys. Rep. **425**, 1 (2006)
3. M. König, G. Bollen, H.-J. Kluge, T. Otto, J. Szerypo, Int. Jour. Mass. Spec. Ion Proc. **142**, 95–116 (1995)
4. S. George et al., Phys. Rev. Lett. **98**(2007)
5. S. Eliseev et al., Phys. Rev. Lett. **110**, 082501 (2013)
6. H. Nishibata et al., Phy. Rev. C **99**(2019)
7. B. Frois et al., Phys. Rev. Lett. **38**, 152 (1977)
8. F. Sauli, Principles of operation of multiwire proportional and drift chambers, CERN Academic Training Lectures, 1975–1976, 10.5170/CERN-1977-009
9. K. Tsukada et al., Phys. Rev. Lett. **118**(2017)
10. G. Drake, *Hand Book of Atomic, Molecular, and Optical Physics* (Springer, 1996)
11. T. Suda, J. Phys. Conf. Ser. **1643**(2020)
12. W. Nörtershäuser W. C. Geppert, *The Euroschool on Exotic Beams*, Vol. IV. Lecture Notes in Physics, vol. 879 (Springer, Berlin, Heidelberg, 2014), pp. 233–292
13. A. Ozawa et al., Nucl. Phys. A **709**, 60–72 (2002)
14. I. Tanihata et al., Phys. Rev. Lett. **55**, 2676 (1985)
15. I. Tanihata et al., Phys. Lett. B **160**, 380 (1985)
16. B.W. Ridley, J.F. Turner, Nucl. Phys. **58**, 497 (1964)
17. S. Terashima et al., Phys. Rev. C **77**(2008)
18. D. Adhikari et al., Phys. Rev. Lett. **126**, 172502 (2021)
19. D.R. Grimes, D.R. Warren, M. Partridge, Sci. Rep. **7**, 9781 (2017)
20. C.S. Wu, E. Ambler, R.W. Hayward, D.D. Hoppes, R.P. Hudson, Phys. Rev. **105**, 1413 (1957)
21. S. Abrahamyan et al., Phys. Rev. Lett. **108**(2012)
22. C.M. Tarbert et al., Phys. Rev. Lett. **112**(2014)
23. P. Vingerhoets et al., Phys. Lett. B **703**, 34 (2011)
24. Y. Ichikawa et al., Nat. Phys. **8**, 918–922 (2012)
25. V. Lapoux, N. Alamanos, Eur. Phys. J. A **51**, 91 (2015)
26. L.D. Knutson, W. Haeberli, Prog. Part. Nucl. Phys. **3**, 127 (1980)
27. P.A. Zyla et al. (Particle Data Group), Prog. Theor. Exp. Phys. **2020**, 083C01 (2020)
28. S. Carboni et al., Nucl. Instr. Meth. Phys. Res. A **664**, 251 (2012)
29. A. Brezkin et al., Nucl. Instr. Meth. **119**, 9 (1974)
30. F. Sauli, *Gaseous Radiation Detectors: Fundamentals and Applications* (Cambridge University Press, 2014)
31. D. Bazin et al., EPJ Web of Conferences **163**, 00004 (2017)
32. Y. Kondo et al., Phys. Rev. Lett. **116**(2016)
33. J. Pereira et al., Nucl. Inst. Meth. Phys. Res. A **618**, 275 (2010)
34. G. Cesini, G. Lucarini, F. Rustichelli, Nucl. Inst. Meth. **127**, 579 (1975)
35. M.A. Delaplanque et al., Nucl. Inst. Meth. Res. A **430**, 292 (1999)
36. J. Dudouet et al., Phys. Rev. Lett. **118**, 162501 (2017)

Chapter 5
Nuclear Shells

Abstract The experimental evidence for nuclear shells and the magic numbers of nucleons which correspond to nuclear shell closures are presented. The problematics of nuclear shells is extended to superheavy elements and hypernuclei. Eventually, the model-dependence of nuclear shells is discussed, as well as the limits of a low-energy scale description of nuclear structure.

Keywords Shell structure · Superheavy elements · Hypernuclei · Short-range correlations

5.1 Experimental Evidence in Stable Nuclei

The shell model has encountered many successes in reproducing and interpreting nuclear structure and it is today the most used description of the nucleus. At the heart of the shell model and its independent-particle picture, the description of some nuclear states as a superposition of single-particle wave functions, as we introduce the shell structure in this chapter, is widely spread. All of the observables detailed in Chap. 4 can be successfully interpreted within the shell model and have been used to validate the existence of a shell structure in nuclei, namely

1. masses,
2. charge density distributions,
3. charge radii
4. quadrupole and magnetic moments,
5. spectroscopy and electromagnetic transition probabilities,
6. direct reaction cross sections.

In the following we review the most illustrative observables upon which was built our understanding of the nuclear shell structure within the shell model framework. As we shall see in more details in Chap. 6, several of the above-mentioned observables are extensively applied today to investigate the nuclear structure of radioactive nuclei.

© Springer Nature Singapore Pte Ltd. 2021   337
A. Obertelli and H. Sagawa, *Modern Nuclear Physics*, UNITEXT for Physics,
https://doi.org/10.1007/978-981-16-2289-2_5

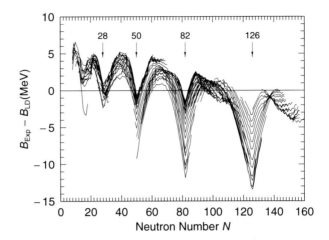

Fig. 5.1 Deviation of experimental masses from theoretical predictions based on the liquid-drop mass formula which does not contain any quantum information on the nuclear shell structure. The deviation is plotted as a function of the neutron number and shows a clear over binding at magic numbers 28, 50, 82 and 126. The modern version of the macroscopic model FRDM2012 is adopted as the liquid-drop binding energy B_{LD} [1]. Courtesy P. Möller

5.1.1 Masses

Nucleon separation energies obtained from mass measurements were the first observables that allowed the hypothesis of a shell structure in atomic nuclei. In 1933, W. M. Elsasser published his seminal work in french "about the Pauli principle in nuclei" and hinted from the large binding energy of helium and oxygen isotopes that nuclei may have a shell structure as electrons in atoms do, since nucleons also experience the Pauli principle. Remarkably, this observation of $Z = N = 2$ or 8 as filled shells (then qualified as "envelopes") pushed him to suggest that nucleons move in a flat mean-field potential, which was further demonstrated later and is nowadays the shared picture of the nuclear potential. The sensitivity of masses to the shell structure over the full range of accessible nuclei is illustrated in Fig. 5.1. This sensitivity has a direct impact on the abundance of medium-mass and heavy elements in the universe. The pattern of measured solar abundances can be seen as a further evidence of the nuclear shell structure (see Chap. 8).

5.1.2 Charge Density Distributions

The single-particle picture was further benchmarked with the measurement of charge density distributions from electron elastic scattering. Indeed, the assumption of independent particles entails that the difference of the charge density of a $^{A+1}Z + 1$ nucleus and those of the the one-proton-less nucleus ^{A}Z should be a measure of the proton single-particle density probability. The elastic scattering data for ^{206}Pb (82

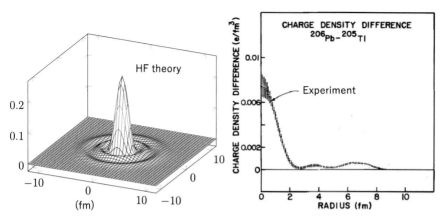

Fig. 5.2 (Left) Radial density probability of the 3s proton of ^{206}Pb from mean field theory. Courtesy N. Yoshida, Kansai University. (Right) Difference of the charge density radial distribution of the ground states of ^{206}Pb and ^{205}Tl extracted from electron elastic scattering. The distribution extracted from the data is very similar to the mean field prediction, justifying a single-particle description for nucleons at the Fermi energy in ^{206}Pb and ^{205}Tl. Figure reprinted with permission from [2]. ©2021 from Elsevier

protons and 124 neutrons) and ^{205}Tl (81 protons and 124 neutrons) measured in the 80s is considered as one of the cleanest proof for the validity of the single-particle picture. The subtraction of the charge distribution of ^{206}Pb and ^{205}Tl obtained from the experiment was in good agreement with the radial probability distribution of a 3s single-particle orbital, as illustrated in Fig. 5.2 where the characteristic maximum density at the origin and the two nodes of a 3s orbital are visible.

5.1.3 Charge Radii

As a general behaviour, it is expected that the root mean square charge radius of nuclei evolves in the same way as the matter radius with the nucleon number along an isotopic chain, i.e., it increases as $r_0 A^{1/3}$. In the case of nuclei with a large neutron excess, where thick neutron skins or halos may develop, this approximation does not stand anymore. In addition to this volume contribution to the charge radius, shell effects are visible through the onset of deformation in between closed shells. Consequently, one observes a "*kink*" at neutron shell closures in the evolution of charge radii along isotopic chains. It is illustrated in Fig. 5.3 where a strong increase of the slope of the root mean square charge radius is observed at $N = 50$ for krypton, rubidium and strontium isotopes, while a similar pattern is seen at $N = 82$ for xenon, cesium and barium isotopes.

The root mean square charge radius can be related to the axial deformation β_2 parameter (see Chap. 7) as

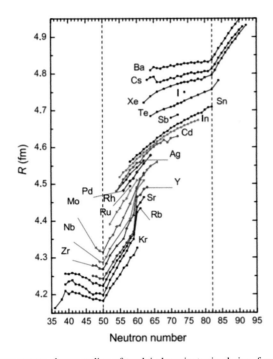

Fig. 5.3 Root mean square charge radius of nuclei along isotopic chains, from $Z = 36$ (Kr) to $Z = 56$ (Ba), as a function of the neutron number. A kink in the slope of the root mean square charge radius is observed at $N = 50$ for krypton, rubidium and strontium isotopes, while a similar pattern is seen at $N = 82$ for xenon, cesium and barium isotopes. They are manifestations of the shell closures at $N = 50$ and $N = 82$, respectively. An abrupt increase at $N \simeq 60$ in the evolution of the charge radii is caused by the rapid onset of deformation between the shell closures at $N = 50$ and $N = 82$. Figure reprinted with permission from [3]. ©2021 from Elsevier

$$\langle r_c^2 \rangle_\beta^{1/2} = \sqrt{\langle r_c^2 \rangle_0 \left(1 + \frac{5}{4\pi}\beta^2\right)} \simeq \sqrt{\frac{3}{5}} r_0 A^{1/3} \left(1 + \frac{1}{2}\frac{5}{4\pi}\beta^2\right), \qquad (5.1)$$

where the mean square charge radius of spherical shape $\langle r_c^2 \rangle_0 = 3R_0^2/5$ is estimated with a constant density of the radius $R = r_0 A^{1/3}$. The last equation of (5.1) is evaluated with an approximation $\beta^2 << 1$. Note that a sudden increase of root mean square charge radius at $N \simeq 60$ along an isotopic chain can also be the sign of a local shape transition, not connected to a spherical shell closure.

Problem

5.1 For axially deformed nuclei, evaluate the effect on the quadrupole deformation on the mean square radius, Eq. (5.1), up to the second order of β. We assume that the equidensity surface specified by a radius parameter r is deformed as,

$$\mathbf{r}_\beta = r(1 + \beta Y_{20}(\theta, \phi) + \eta). \qquad (5.2)$$

That is, the deformed density ρ at the radius $\mathbf{r}_\beta$ is identified by the spherical density $\rho_0(r)$ at the radius r,

$$\rho(\mathbf{r}_\beta) = \rho_0(r). \tag{5.3}$$

The value η is determined by the condition of volume conservation. Notice that the deformed density at the radius r is given by

$$\rho(r) = \rho(\mathbf{r}_\beta/(1 + \beta Y_{20}(\theta, \phi) + \eta)) = \rho_0(r/(1 + \beta Y_{20}(\theta, \phi) + \eta)). \tag{5.4}$$

1. Determine the value η by the volume conservation condition,

$$\int \rho_0(\mathbf{r}) d\mathbf{r} = \int \rho(\mathbf{r}) d\mathbf{r} = A, \tag{5.5}$$

up to the second order of β.
2. Evaluate the effect of deformation β on the mean square radius

$$\langle r_c^2 \rangle_\beta = \int \rho(\mathbf{r}) r^2 d\mathbf{r}. \tag{5.6}$$

5.1.4 Electric Quadrupole and Magnetic Dipole Moments

Electric quadrupole moments, as a measure of quadrupole deformation, are minimal at shell closures. Along an isotopic chain, i.e., with a constant number of protons, one can observe this minimum at neutron magic numbers. Indeed, nuclei are strongly interacting systems: correlations among neutrons will impact the correlations for protons. In particular, a neutron shell closure will quench quadrupole correlations for protons as well.

The dipole magnetic moment of even-odd nuclei plays an important role in proving the robustness of the shell model. In her Nobel lecture, Maria Goeppert-Mayer in 1963 uses magnetic moments (see Fig. 5.4) to demonstrate the role of the spin-orbit one-body potential in explaining the nuclear shells: "For example, take the magnetic moment of nuclei. For a nucleus with odd proton, even neutron number, the magnetic moments, according to the shell model, should depend only on the site of the last odd proton The lines at the two extremes are the calculated ones for the single-particle configuration (See Chap. 4). The middle lines are what one would obtain if the proton were a simple Dirac particle, and are added merely to emphasise the division into two groups. For any value of the spin, we calculate two different values of the magnetic moment, for the two different values of ℓ, $\ell = j - 1/2$ and $\ell = j + 1/2$. The nuclei in the upper group, nearer to the line $j = \ell + 1/2$ are indeed those for which we found that spin and orbital angular momentum are parallel, those in the lower group were assigned anti-parallel orientation". The magnetic moments of nuclei with even number of neutron and odd number of protons follow very well the predictions of the

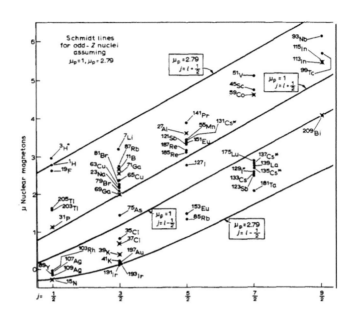

Fig. 5.4 Magnetic moments of nuclei with odd proton, even neutron number as shown by M. Goeppert Mayer in her lecture in 1963 when she received the Nobel prize in physics, together with J. H. Hansen and E. Wigner. Figure reprinted with permission from [4]

independent-particle shell model based on a strong spin-orbit splitting of harmonic oscillator orbits (see Chap. 3 for more details).

5.1.5   Spectroscopy and Transition Probabilities

In the shell model picture, the excitation of closed shell nuclei to a state higher than the ground state costs an energy at least equal to the size of the shell gap. The first excited state in even-even nuclei is mostly a 2^+ state. Observations, as illustrated in Fig. 5.5, support this picture where nuclei with a magic number of protons or neutrons systematically show a large excitation energy for the first 2^+ state compared to neighbouring nuclei. Following the same arguments, the electric transition probability from the ground state to the first 2^+ in even-even nuclei measures how easy it is to excite protons to form a 2^+ state. The transition probability is minimal for closed shell nuclei consistently with the excitation energy systematics.

The shell closured evidence for the study of stable nuclei are not universal and change across the nuclear landscape, i.e., for neutron-rich and neutron-deficient nuclei. The shell evolution with the number of protons and neutrons is discussed in Chap. 6.

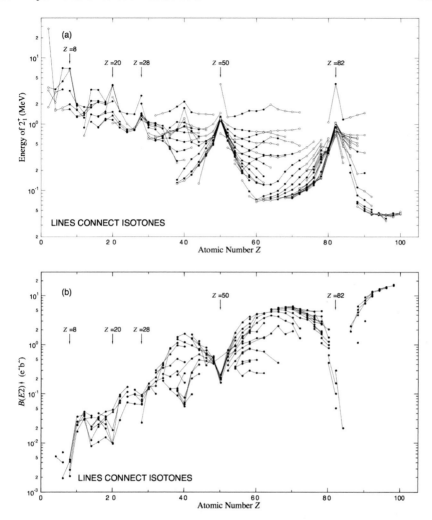

Fig. 5.5 Systematics of measured first 2^+ excitation energies (top) and transition probabilities $B(E2)$ from ground state to first 2^+ state (bottom) for even-even nuclei. Magic numbers are characterized by a high first 2^+ excitation energy and low transition probability along an isotonic chain. Similar patterns are encountered when the data are plotted as a function of the neutron number. Figure reprinted with permission from [5]. ©2021 from Elsevier

5.1.6   Direct-Reaction Cross Sections

Direct reactions from closed-shell nuclei, such as one nucleon transfer or one-nucleon knockout, populate states which are consistent with the independent-particle picture. Low-energy nucleon transfer reactions such as (d, p) or (p, d) are the basis of our quantitative understanding of the nuclear shells: so-called single-particle states are populated with most of the spectroscopic strength. The transferred angular momenta ℓ to these states reveal corresponding the shell model picture. Transfer with polarized beams or targets, i.e., with a specific spin orientation, allow to populate spin-orbit partners also in agreement with the shell model picture.

An excitation energy spectrum for the reaction $^{40}Ca(d, p)^{41}Ca$ measured at 12 MeV and the angular distributions of the recoil protons for the most populated states are illustrated in Fig. 5.6. Some states are highly populated and correspond to states which can be represented well in the independent-particle shell model, i.e., a close-shell nucleus ^{40}Ca coupled to a single neutron lying on one of the shell model orbitals $1f_{7/2}$, $2p_{3/2}$ or $2p_{1/2}$, consistently with the measured angular distributions. The cross section to populate a specific state quantifies if this state can be considered as a *single-particle* shell model state. We refer the reader to Chap. 8 for a detailed discussion on the connection between transfer reaction cross sections and the nature of populated states.

5.2 Superheavy Elements

The next proton or neutron magic numbers, beyond $Z = 82$ and $N = 126$, are not known yet. Their existence is even questionable. The existence of elements with a nucleus with more than about hundred protons is not straightforward. Indeed, the more protons a nucleus contains, the more Coulomb repulsion it experiences. Nuclei with a large number of protons are more inclined to decay via particle emission, either alpha decay or fission. The heaviest stable nuclei are ^{209}Bi and ^{208}Pb with 126 neutrons each and 83 and 82 protons, respectively. Heavier nuclei exist on earth but are radioactive. For example, ^{238}U, used as a fuel in nuclear power plants, is radioactive and decays via alpha emission but its half-life of 4.5 billion years is about the age of our planet (4.3 billion years). It is why it can be found and extracted from the earth. On the other hand, the nucleus ^{212}Po, already discussed in Chap. 1, is also radioactive: it decays via alpha emission with a short lifetime of 0.3 µs. All observed nuclei heavier than U have been synthesized in laboratories and have too short lifetimes to be observed in nature.

How heavy a nucleus can be and what would be the next doubly-magic nucleus beyond ^{208}Pb? Let us first forget for a minute quantum mechanics and consider the nucleus as a droplet of nuclear matter. In this macroscopic description, one can expect two competing nuclear shapes to minimize the energy of a considered nucleus: a spherical shape minimizes the surface tension, exactly in the same way as water droplets, while an elongated shape would increase the mean distance between protons inside the nucleus and therefore minimize the Coulomb repulsion among them. In the case of heavy nuclei, with a large number of protons, the elongated shape wins. In this liquid-drop model, one can calculate which nuclei favour an elongated shape and experience fission. It is found that beyond Nobelium isotopes ($Z = 102$) no nuclei can exist because of a too strong Coulomb repulsion among protons. In the model, all nuclei with more than 102 protons fission (Fig. 5.7).

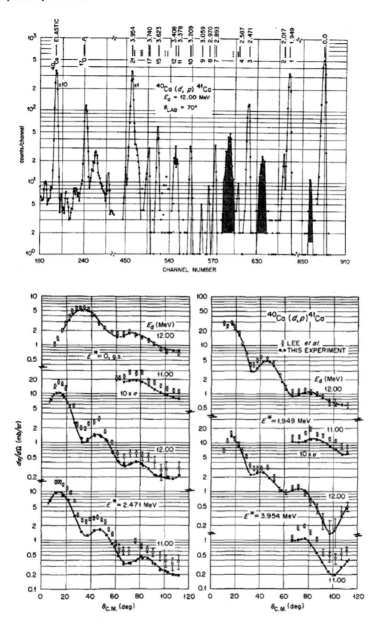

Fig. 5.6 (Top) Missing-mass energy spectrum of recoil protons from the reaction ^{40}Ca$(d, p)^{41}$Ca performed with a deuteron beam at 12 MeV. (Bottom) Angular distribution of the recoil proton in coincidence with the strongly populated physical states from missing mass. The angular distribution is given in the center of mass of the interacting system. The shape of the distributions depends on the transferred angular momentum ℓ, which informs on the angular momentum quantum number of each populated state. Figure reprinted with permission from [6]. ©2021 by Elsevier

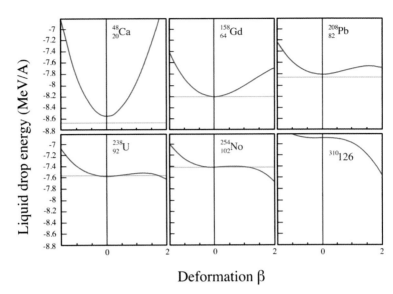

Fig. 5.7 Liquid-drop energy for different nuclei. Coulomb repulsion among protons favours deformation which leads to fission for heavy isotopes with $Z \geq 102$. The green lines indicate the experimental binding energies, when experimentally known

Reality appears to deviate from this liquid-drop model prediction, and many nuclei heavier than Nobelium have been produced in the laboratory and demonstrate sufficient lifetime to exist as chemical elements. The existence of an element is claimed when at least one of its isotopes has sufficient lifetime for a neutral atom to be formed. Such a critical lifetime is of the order of 10^{-10} s. The heaviest nuclei ever created in a laboratory contain 118 protons. The explication of their existence lies in the nuclear shell structure, which stabilizes the nucleus and counterbalance with the Coulomb repulsion: their quantum nature is essential for them to survive as superheavy nuclei against fission, even with $Z \geq 102$ protons.

The shell model predicts a region of the nuclear landscape for superheavy nuclei, where nuclei may be stabilized by shell effects and present rather long lifetimes from few seconds to hundred of years depending on the models. This region is called the "island of stability" (see Fig. 5.8) for superheavy elements, that have not been accessed so far. Several teams are dedicated to its search worldwide. The quest for the island of stability has been launched several decades ago by American and Russian research groups from Berkeley and Dubna, respectively. Nuclei with atomic number from $Z = 103$ to $Z = 106$ have been discovered between 1961 and 1974. These nuclei were produced by impinging light ions on heavy targets.

The main difficulty in the search for superheavy elements is the very low production cross sections of the order of 1–10 pb for the heaviest nuclei produced so far, as illustrated in Fig. 5.11, which corresponds to a few events detected per year. These low cross sections come from the extremely low survival probability of

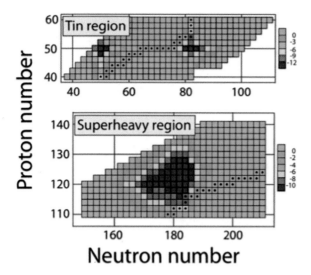

Fig. 5.8 Total shell correction in MeV for spherical even-even nuclei calculated with the SLy6 Skyrme interaction in a configuration mixing approach. White color indicates nuclei with positive shell corrections. Black dots denote nuclei which are calculated to be stable with respect to beta decay. (Top) In the Tin region, the doubly magic nuclei ^{100}Sn and ^{132}Sn show strong shell corrections. (Bottom) In the superheavy region and island of stabilisation is predicted at $Z \sim 120$ and $N \sim 180$. Different from Sn isotopes, the superheavy region does not show isolated magic nuclei but a broad region of stabilization. Figure reprinted from [7]

a superheavy element when formed at high excitation energy after formation of a compound nucleus made of the target and projectile. The high compound excitation energy favours quasi-fission or fission as a de-dexcitation mode compared to neutron emission.

At a given incident energy E, the fusion cross section of a projectile and a target nucleus can be decomposed into three components as

$$\sigma(E) = \sigma_{capture} \cdot P_{CN} \cdot P_{survival}, \tag{5.7}$$

where the capture cross section $\sigma_{capture}$ quantifies the probability to form a binary system inside the Coulomb barrier, P_{CN} is the probability that the formed binary system evolves as a compound nucleus, and $P_{survival}$ is the probability that the compound nucleus leads to a cold residue after neutron evaporation instead of fission. These three steps can be visualised in Fig. 5.9.

Once the compound nucleus is formed, the branching ratio between the fission decay channel and the formation of an heavy evaporation residue is driven by excitation energy at which the compound is formed as illustrated in Fig. 5.10. One can link this excitation energy to an effective temperature of the compound nucleus as $T = \sqrt{8E^*/A_{CN}}$, where A_{CN} is the total number of nucleons in the compound nucleus, i.e. the sum of the number of nucleons in the target and projectile. The

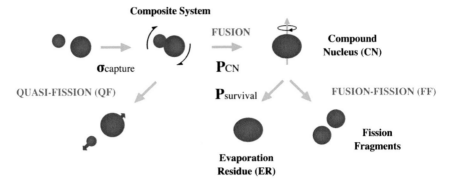

Fig. 5.9 Formation mechanism for evaporation residues after the collision of two heavy ions around the Coulomb barrier

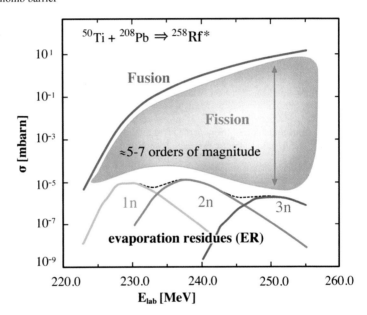

Fig. 5.10 Decay channels of the compound nucleus ^{258}Rf* from the fusion of a ^{50}Ti projectile with a ^{208}Pb target nucleus as a function of the incident energy. The decay to fission is dominant and only a minority of events lead to a cold evaporation residue after neutron emission

branching ratio of the neutron and fission partial decay width, Γ_n and Γ_f, then becomes

$$\frac{\Gamma_n}{\Gamma_f} = \frac{2T A_{CN}^{2/3}}{K_0} e^{[B_f - B_n]/T}, \qquad (5.8)$$

where B_f and B_n are the fission barrier and neutron separation energy, respectively, and $K_0 = \hbar^2/2mr_0^2 = 11.4$ MeV is a constant.

Latter on, nuclear physicists from Dubna initiated a new type of production: the fusion of heavier beam particles with heavy target nuclei. The technique was named "cold fusion" since the resulting nucleus from the fusion was populated with rather small excitation energy. The technique was extensively used by German physicists at the GSI facility in the 90s. Elements from $Z = 107$ to $Z = 112$ were discovered. In 2012, the element with $Z = 112$ was recognized and named Copernicium. Japanese teams used the same technique at RIKEN. In 2015, $Z = 113$ and $Z = 115$ were recognized as new elements.

Cold fusion does not allow the production of very heavy elements since the corresponding production cross section are too small as illustrated in Fig. 5.11. The collaboration between the US and Russia initiated the fusion of heavy projectiles such as ^{48}Ca with targets heavier than lead, such as plutonium ($Z = 94$). The method is called hot fusion because a rather large excitation energy is left in the fusion product and result in the evaporation of a significant number of neutrons. It led to the synthesis and discovery of several new elements. In 2016, the discovery of the new elements 113, 115, 117 and 118 was officially declared. For the element 113 the discoverers at RIKEN Nishina Center for Accelerator-Based Science (Japan) proposed the name Nihonium and the symbol Nh. Nihon means "Japan" in Japanese. For the element 115 the name proposed is Moscovium with the symbol Mc and for element with atomic number 117, the name proposed is Tennessine with the symbol Ts. These two names honour the locations of the Dubna laboratory in Moscow, Russia, and American laboratories in the state of Tennessee which strongly contribute to the science of super-heavy elements. For the element 118 the name Oganesson and

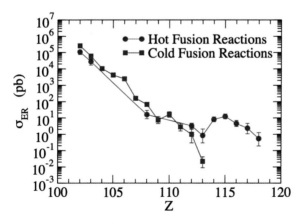

Fig. 5.11 The measured evaporation residue cross section as a function of the atomic number Z of a compound nucleus. The filled red circles denote the results of the hot fusion reactions, in which ^{48}Ca nucleus is used as a projectile. The maximum of a sum of the $3n$ and $4n$ cross sections are shown for each Z. The filled blue squares show the results of the cold fusion reactions, in which the ^{208}Pb or ^{209}Bi nuclei are used as a target. Here, the maximum of the $1n$ cross section is shown for each Z. The experimental data are taken from [8, 9]. Courtesy K. Hagino, Kyoto University

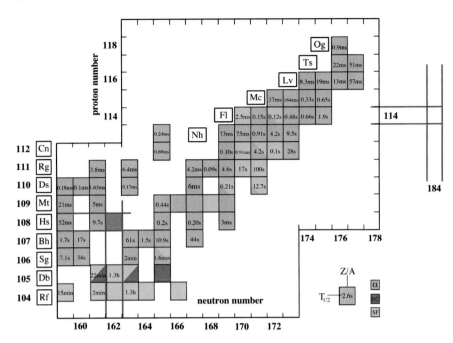

Fig. 5.12 Top part of the nuclear chart showing the current status of the search of superheavy elements. The heaviest atomic number produced experimentally is $Z = 118$ at Dubna, Russia

symbol Og were proposed in recognition to Professor Yuri Oganessian for his active work on the synthesis of super-heavy elements at DUBNA (see Fig. 5.12).

The search for new superheavy elements is being actively pursued. Physicists continue to explore the upper part of the nuclear landscape and extend the atomic periodic table first introduced in 1869 by the Russian chemist Dimitri Mendeleiev. To produce new elements with an atomic number equal or larger than 119, Dubna and RIKEN are upgrading their facilities to provide more intense primary beams and more efficient detection devices. In France, a new facility named SPIRAL2 was recently built to study in details the structure of superheavy elements and has been operational since 2021.

5.2.1 Relativistic Limit to the Existence of Chemical Elements

The existence of the Fermi sea provides a strong constrain to the existence of heavy elements. Indeed, the energy gap between the Dirac sea and the particle threshold is $2m_ec^2$ in the absence of external field (see Fig. 1.15 of Chap. 1). In the case of strong electric fields, as in the vicinity of a high-Z nucleus, this gap is reduced. Let us consider the inner s-wave electron orbit in an atom formed by a point-like nucleus with charge Ze. Obviously, the binding energy of the s-orbit increases with Z. The necessary energy for a photon to create an electron-positron pair is given by

$$E_\gamma = 2m_e c^2 + \epsilon_n. \tag{5.9}$$

At the critical point $\epsilon_n = 2m_e c^2$, the energy of the bare nucleus is equal to the energy of the atom with an electron on the inner s-wave orbit and a positron emitted with zero kinetic energy. The fully-ionized atom is not anymore energetically favored. At this point, there is no stable electronic state anymore. At which point is this limit hit? This can be obtained by solving the Dirac equation for the lowest-energy eingenvalue (inner s-wave orbit). The solution contains a factor $\sqrt{1 - \alpha^2 Z^2}$ which becomes complex for $\geq 1/\alpha \sim 137$. In a time-dependent representation, this means that the nucleus wave-function is not stable, which is consistent with the production of a $e^+ e^-$ pair. More realistic calculations taking into account the finite size of the nucleus, which smoothes down the Coulomb field, set the limit of existence of chemical elements to $Z \sim 170$. This limit is far from being reachable experimentally and, although not yet proven, it is expected that nuclear physics sets the limit of existence of superheavy elements to a lower atomic number than the above-described relativistic limit.

5.3 Hypernuclei

Nuclear structure extends to different composite baryonic systems than normal nuclei which are composed of protons and neutrons. Hypernuclei are such systems that contain at least one hyperon. Hyperons are baryons containing strange quarks. Hyperons in nuclei also exhibit a shell structure but different from the one for protons and neutrons, since the hyperon-nucleon and hyperon-hyperon interactions differ from nucleon-nucleon interactions. The study of hypernuclei therefore greatly enlarges our understanding of nuclear structure. There are two particular aspects, which make hypernuclei so appealing. Firstly, hyperons inside a nucleus do not experience Pauli blocking with nucleons. This feature makes hyperons embedded inside a hypernucleus, and opens a unique opportunity to study nuclear structure with potentially unexplored capabilities. For example, they can explore "freely" orbitals down to the deeply bound $1s$, and therefore be a probe of the interior of nucleus. Secondly, hyperons have a different structure than nucleons which has a strong impact on the amplitude of three-body forces in medium. For example, the lightest hyperon, the Λ particle, has its first excited state, the Σ baryon composed of the same constitutent quarks usd, at the excitation energy of $m(\Sigma) - m(\Lambda) = 76$ MeV, much lower than the Δ excitation of nucleons, which lies at the excitation energy of $m(\Delta) - m(N) \sim 293$ MeV. Miyazawa–Fujita type three-body forces (see Chap. 2) are then magnified for Λs compared to nucleons. This is a very interesting feature of hypernuclei to investigate new aspects of strong interactions.

5.3.1   Classification of Hyperons

In the following, we give a classification of baryons composed of the lightest three quarks u, d, s, i.e., nucleons and hyperons. The up, down and strange quarks are distinct to the three others (c, b, t: charm, bottom and top) because their masses ($m_u = 1-4$ MeV/c^2, $m_d = 4-8$ MeV/c^2, $m_s = 80-130$ MeV/c^2) are inferior to the energy scale $\Lambda_{QCD} \sim 1$ GeV of QCD, at which QCD becomes strongly coupled. A chiral expansion in powers of $m_{u,d,s}/\Lambda_{QCD}$ is therefore possible, while the masses of the charm, bottom and top quarks are too large ($m_c = 1.1 - 1.4$ GeV/c^2, $m_b = 4.1 - 4.4$ GeV/c^2, $m_t = 173 \pm 1$ GeV/c^2), which make difficult to do the chiral expansion. The ensemble of baryons built from u, d, s quarks can be labeled by the electric charge Q, the baryon number B, the strangeness S, the projection of isospin I_3,[1] and projection of spin S_z. The four conserved quantum numbers Q, B, S and I_3 are related to each other by the Gell-Mann–Nishijima relation

$$Q = I_3 + (B + S)/2 = I_3 + Y/2, \tag{5.10}$$

where $Y = B + S$ is defined as the hypercharge. Figure 5.13 represents the ground state of baryonic octet of spin 1/2 and the baryonic decuplet of spin 3/2, as a function of their hypercharge Y and their projection of isospin T_z.

When generalized to baryons composed of (u, d, s) quarks, the study of nuclear interactions extends to twenty-eight vertices for interactions between spin 1/2 particles of which twenty-five contain the barely known or unknown YN or YY interac-

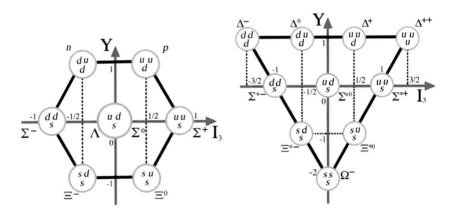

Fig. 5.13 Octet of baryons with spin 1/2 and decuplet of baryons with spin 3/2 classified as a function of their quantum numbers I_3, projection of isospin, and hypercharge Y. The constituent quarks for each baryon are indicated

---

[1] The isospin of elementary particles is denoted I_3, while it is commonly written as $T_z = (N - Z)/2$ in nuclear physics. The sign of T_z is $+1/2$ for neutrons and $-1/2$ for protons, but opposite for I_3, i.e., $I_3 = (-T_z) = +1/2$ for protons and $I_3 = (-T_z) = -1/2$ for neutrons.

tions, where Y represents hyperons, while the pp, pn and nn interactions represent only 3 of them. Due to the very short lifetime of hyperons, the study of YN interactions directly from YN scattering is therefore extremely difficult, while no YY scattering is accessible experimentally via traditional scattering experiments. The Λ particle with quark constituents usd is the lightest hyperon, with mass 1115 MeV/c^2, zero charge, isospin $I = 0$ and strangeness $S = -1$, a new quantum number not contained normally inside nuclei.

5.3.2 Production of Hypernuclei

The Λ hyperon is unstable and decays with lifetime 263(2) ps, typical of the weak interaction that does not conserve the strangeness and makes a free Λ mainly disintegrating into a nucleon-pion system. However, since the strangeness is conserved in the strong interaction and the Λ particle is the lightest particle in the family of hyperons, it can stay interacting with nucleons inside nuclei and forms short-lived hypernuclei. The existence and structure of hypernuclei is of strong interest. Hypernuclei studies shed a new light on the world of traditional nuclei by revealing new symmetries and new phenomena produced by the additional strangeness dimension. Hypernuclei have become an important means to explore the hyperon-nucleon (YN) interactions.

Hypernuclei were discovered in 1953 by M. Danysz and J. Pniewski from the analysis of emulsion plates where their productions and decays have been detected and identified after the interaction of high energy cosmic rays of protons with nuclei from the emulsion (see Fig. 5.14).

The production of hypernuclei close to the stability line in the nuclear chart has been successfully achieved by the interaction of high-energy beams (kaons, pions, electrons) with stable nuclei from the 70s, and has led to the spectroscopy of hypernuclei via gamma spectroscopy of missing mass spectroscopy. The production and spectroscopy techniques applied to hypernuclei are summarized in Table 5.1.

5.3.3 Extending the Study of Baryon-Baryon Interactions to Hyperons

Only few hypernuclei have been synthesized so far, and the hypernuclear data are still scarce as illustrated in the *hyperchart* of experimentally-produced Λ hypernuclei in Fig. 5.15. The lifetime of light hypernuclei, their binding energy and first spectroscopy could be measured. From this information, empirical one-body potentials for a Λ hyperon in a nucleus were extracted, allowing to reproduce fairly well the data. The Λ hypernuclear potential can be expressed by use of the Woods–Saxon potential as follows

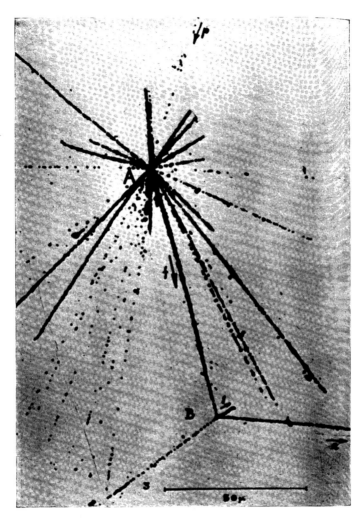

Fig. 5.14 Historical emulsion image of the first-observed hypernucleus decay. A cosmic ray inter-acts with a nucleus of the emulsion (point A), leading to a so-called star event. One of the tracks from the reaction decays after a ∼50 μm propagation (Point B), signing the weak decay of an hypernucleus. Figure reprinted with permission from [10]. ©2021 by Taylor & Francis

$$U_\Lambda = V_0^\Lambda f(r) + V_{LS}^\Lambda \left(\frac{\hbar}{m_\pi c}\right)^2 \frac{1}{r}\frac{df(r)}{dr}\mathbf{l.s},\qquad(5.11)$$

where

$$f(r) = \frac{1}{1 + e^{-\frac{r-R_0}{a}}}\qquad(5.12)$$

Table 5.1 Production and spectroscopy techniques for hypernuclei studies. BE stands for Binding Energy. HI stands for Heavy Ions

Production	Observables	Years	Experiment	Laboratory	Country
$pN \rightarrow \Lambda + K^{\pm}$	Lifetime, BE	50s–70s	Cosmic rays	–	–
$K^- + n \rightarrow \Lambda + \pi^-$ $K^- + p \rightarrow \Lambda + \pi^0$	BE, γ spectroscopy			J-Parc	Japan
$\pi^+ + n \rightarrow \Lambda + K^+$	Lifetime, spectroscopy	2003–2007	FINUDA	DAϕNE	Italy
$e + p \rightarrow e' + \Lambda + K^+$	Spectroscopy			JLab	USA
HI collisions	Lifetime, BE	Since 2006	HyPHI	GSI	Germany
HI collisions			STAR	RHIC	USA
HI collisions			CMS	CERN	Switzerland
$\bar{p}$-HI collisions	BE, γ spectroscopy	From 2025	PANDA	FAIR	Germany

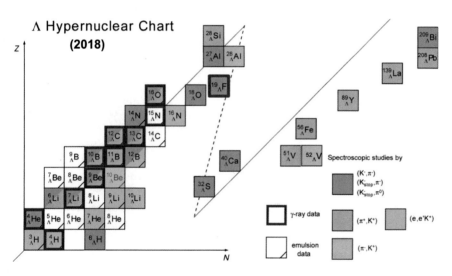

Fig. 5.15 Hypernuclear chart (2018) for hypernuclei with S = 1. Courtesy H. Tamura, Tohoku University

with R_0 and a the radius and the surface diffuseness parameter, respectively. It was deduced that the depth of the nuclear potential for Λ is $V_0^\Lambda = -30$ MeV, about 2/3 of the potential for nucleons. The spin-orbit potential for Λ was found to be $V_{LS}^\Lambda = 2$ MeV, much weaker than for nucleons. The weak spin orbit is visible in the missing-mass spectroscopy of ^{89}Y studied with the (π^+, K^+) reaction at KEK, Japan, shown in Fig. 5.16. Since the Λ does not experience any Pauli principle with protons and neutrons inside the nucleus, it can take any quantum number and populate all hyperon orbitals. The experiment revealed clearly separated s, p, d, f and g orbitals without

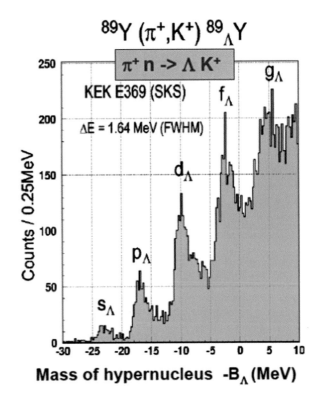

Mass of hypernucleus $-B_\Lambda$ (MeV)

significant energy separation of spin-orbit partners due to a comparable value with the energy resolution.

The gluelike role of a hyperon could shrink the size of the core nucleus and shift the neutron and proton drip lines from their normal limits. Furthermore, the hyperon's decay from the inside of a nucleus makes it a promising probe for high-density features of nuclear matter. Hyperons are expected to be an important building block of high-density matter such as neutron stars. Indeed, neutron stars are created from the collapse of the stars following supernovae explosions. These compact and massive objects have a typical radius of 10 km with masses from 1 to 2 solar masses. The relation between the mass and radius of neutron stars is highly dependent on the nuclear equation of state and the yet-unknown state of nuclear matter inside the star. Nuclear matter at the centre of the neutron star is expected to have very high densities, several times larger than the saturation density. Free hyperons are unstable and decay via the weak interaction but this process can be reversed at high densities and the internal zone of neutron stars may contain a significant amount of hyperons. The presence of hyperons in neutron stars would soften the equation of state, seemingly in disagreement with recent measurements of neutron stars with two solar masses. This important astrophysics question is under active research and urges for a better

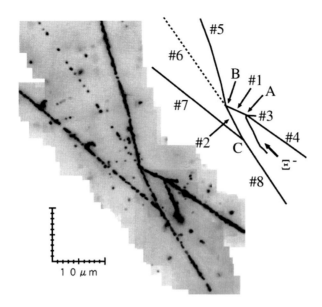

Fig. 5.17 Nagara event measured with the emulsion technique showing the production and decay of $^{6}_{\Lambda\Lambda}$He. The history of the recorded event is sketched on the top right corner of the figure: after a Ξ^{-} capture by a nucleus on the emulsion substrate in A, several fragments are produced and recoil in the emulsion from the kinetic energy gained in the reaction. The candidate hypernucleus $^{6}_{\Lambda\Lambda}$He travels until B where it experiences a first weak-interaction decay producing the following fragments: $^{5}_{\Lambda}$He, p, and π^{-}. The hypernucleus $^{5}_{\Lambda}$He decays into two charged particles at C. Figure reprinted with permission from [13]. ©2021 by the American Physical Society

understanding of YN and YY interactions. The structure of neutron stars and the presence of hyperons in their interior is further discussed in Chap. 9.

YY forces could be better understood by studying $\Lambda\Lambda$ hypernuclei, i.e., nuclei that contain two hyperons. Very few of them have been synthesized: $^{6}_{\Lambda\Lambda}$He, $^{10,11,12}_{\Lambda\Lambda}$Be, and possibly $^{12}_{\Lambda\Lambda}$B or $^{13}_{\Lambda\Lambda}$B. They are so few that each event is given a name. For example, the recorded event of show $^{6}_{\Lambda\Lambda}$He in Fig. 5.17 is called the *Nagara* event,[2] unambiguously observed at KEK after a first claim in the 60s [12]. In this particular event, the double hypernucleus $^{6}_{\Lambda\Lambda}$He is created by the reaction

$$\Xi^{-} + {}^{12}C \rightarrow {}^{6}_{\Lambda\Lambda}He + {}^{3}H + {}^{4}He, \tag{5.13}$$

and decays

$$^{6}_{\Lambda\Lambda}He \rightarrow {}^{5}_{\Lambda}He + p + \pi^{-}, \tag{5.14}$$

followed by the later weak decay of $^{5}_{\Lambda}$He, as seen in Fig. 5.17.

---

[2] This event was named Nagara event after the Nagara River which flows through Gifu city hosting the Gifu University, where the double Λ nucleus was discovered.

In total, only seven events have been interpreted as the decay of di-Λ hypernuclei. In all mentioned cases, the formation process is understood as resulting from the following steps: (i) creation of a Ξ (cascade) hyperon which has a strangeness quantum number of $S = -2$, (ii) capture of the Ξ hyperon by another nucleus in the target, (iii) decay of the Ξ hyperon towards the nucleus via X-ray and conversion electron emission, (iv) $\Xi + p \rightarrow \Lambda\Lambda$ and (v) formation of a di-Λ hypernucleus. The probability of such succession of events is very small, which explains why so few have been observed so far. The PANDA experiment at FAIR foresees an intensive program to produce and study hypernuclei, including multi-strangeness hypernuclei.

5.3.4   Femtoscopy: A Tool to Investigate YN and YY Interactions

The study and spectroscopy of hypernuclei allow to constrain YN interactions but do not represent a measurement of the bare interaction as achieved with scattering experiments for NN interactions (see Chap. 2). The possibility to perform scattering experiments with hyperons is limited due to their low production rates and short lifetimes. So far only $p\Lambda$ and $p\Sigma^{\pm}$ scattering have been performed with limited statistics and restricted to relative momentum above 100 MeV/c. Traditional YY scattering experiments cannot be performed in the laboratory, while another method can lead to unique information about YY interactions: femtoscopy.

The idea of femtoscopy is to produce the two baryons of interest in a collision of ions at a center-of-mass energy well above the particle production threshold. From the energy of the collision, particles are produced in the collision at short distances, from 1 to 5 fm relative distance depending on the colliding systems and colliding velocity. During the time span from the moment produced to the moment detected, they can interact with each other. Femtoscopy aims at determining the interaction between such pairs of particles detected at a measured relative momentum in the final state. The technique allows to study interactions among very short-lived baryons which are not accessible via usual scattering methods.

The central observable in femtoscopy is the two-particle correlation function, which is defined as the probability to find simultaneously two particles with momenta p_1 and p_2 divided by the product of the corresponding uncorrelated probability

$$C(\mathbf{p}_1, \mathbf{p}_2) = \frac{P(\mathbf{p}_1, \mathbf{p}_2)}{P(\mathbf{p}_1) \cdot P(\mathbf{p}_2)}. \tag{5.15}$$

The uncorrelated probability is constructed by considering two particles from different (and therefore uncorrelated) events.

The above-mentioned probabilities are directly related to the inclusive Lorentz invariant spectra $P(\mathbf{p}_1, \mathbf{p}_2) = E_1 E_2 \frac{d^6 N}{d^3 p_1 d^3 p_2}$ and $P(\mathbf{p}_{1,2}) = E_{1,2} \frac{d^3 N}{d^3 p_{1,2}}$. In absence of correlations, the value of the correlation function equals unity. The correlations as a function of the relative momentum $\mathbf{k} = \mathbf{p}_1 - \mathbf{p}_2$ between two baryons can be determined from experiments where the two particles are produced from the high-energy collision of two nucleons or nuclei. The correlation function is then expressed by a function of both the relative wave function $\Psi(\mathbf{r}, \mathbf{k})$ of the two baryons, containing information on their interaction, and the distribution of pairs at the production moment as a function of their relative distance $\mathbf{r}$, called the source function $S(\mathbf{r})$, as follows

$$C(\mathbf{k}) = \int d^3 r \, S(\mathbf{r}) |\Psi(\mathbf{r}, \mathbf{k})|^2. \tag{5.16}$$

Correlation functions have been used in many experimental conditions to learn about the source from the measurement of correlations between emitted particles, whose interaction is known. For example, the correlation of two neutrons in the two-neutron halo nuclei ^{6}He and ^{11}Li have been investigated successfully with femtoscopy, assuming the knowledge of the neutron-neutron interaction determined independently. Inversely, by assuming or extracting the source function from experiment, for example from a measurement of the correlation function for a system whose interaction is known, theoretical interactions can be tested against data via correlation functions. In particular, by using $p + p$ and $p + Pb$ collisions at the LHC, small distances of about 1–1.5 fm can be probed, enhancing the sensitivity to the short range strong interaction. Examples of correlation functions for $p - p$ and $p - \Sigma^-$ are shown in Fig. 5.18 obtained after $p + p$ collisions at 7 TeV measured with the ALICE detector at CERN. Attractive interactions show a positive correlation function while repulsive interactions tend to deplete the correlation function. The correlation

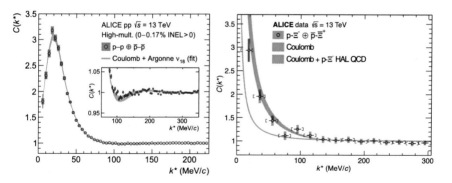

Fig. 5.18 Femtoscopy correlation functions from proton-proton (pp) collisions at 13 TeV (center of mass energy) with the ALICE detector at CERN. (Left) pp correlations compared to predictions with the Coulomg and AV18 nuclear potentials. The agreement to data is excellent. (Right) $p\Xi^-$, and $\bar{p}\bar{\Xi}^+$ antimatter equivalent, correlations compared to a theoretical correlation function assuming only Coulomb interaction between the two baryons (green) and predictions taking into account the lattice QCD-based $p\Xi^-$ potential calculated by the HAL QCD collaboration (pink). The predictions based on lattice QCD are in good agreement with the data, confirming an attractive nuclear potential between a proton and Ξ^-. Data published in [14, 15]. Courtesy B. Holhweger and L. Fabbietti, TU Munich

function of Fig. 5.18 for pp shows the interplay of the repulsive Coulomb and the attractive strong forces. The right panel of Fig. 5.18 shows the first measurement of the $p\Xi^-$ correlation function and the attractive character of the $p\Xi^-$ interaction. So far, $p\Lambda$, $\Lambda\Lambda$ and $p\Xi$ have been measured by femtoscopy from ALICE data.

5.4 Non Observability of Nuclear Shells

Despite the successes of the shell model, it is of utmost importance to realise that the very concept of a single-particle state is model dependent, and single-particle energies at the basis of the shell model should be discussed with caution. They should be considered as interim calculation quantities inside the considered model for understanding complex phenomena in order to establish an elucidate picture of nuclear structure. A unitary transform of the Hamiltonian can change the extent to which a given state is single particle or correlated.

Some quantities are discussed to interpret nuclear structure but are not strictly-speaking observables. They are related to operator matrix elements which are not invariant under unitary transformation. The nuclear Hamiltonian H is not uniquely defined. It can be transformed by any unitary transform U, exactly the same way as a gauge transformation does for a Lagrangian in particle physics. The physics, i.e., the observables, are not modified by such a transformation since

$$H_{ik} = \langle \psi_i | H | \psi_k \rangle = ((\langle \psi_i | U^\dagger) U H U^\dagger (U | \psi_k \rangle) = \langle \widetilde{\psi}_i | \widetilde{H} | \widetilde{\psi}_k \rangle = \widetilde{H}_{ik}, \qquad (5.17)$$

where $|\psi_{i,k}\rangle$ are eigenvalues of the Hamiltonian H and $|\widetilde{\psi}_{i,k}\rangle = U | \psi_{i,k}\rangle$ are eigenvalues of the Hamiltonian $\widetilde{H} = U H U^\dagger$. In this specific case of the Hamiltonian, the eigenenergies of the system are indeed observables and invariant under unitary transforms. On the other hand, the nucleonic occupancy of an orbital with quantum numbers "$n\ell j$" is driven by the operator $\hat{N}(n\ell j) = a_{n\ell j}^\dagger a_{n\ell j}$. The matrix elements of the operator $\hat{N}(n\ell j)$ are changed by the unitary transform such as the Bogolyubov transformation for the superfluidity, since the definition of the operator $\hat{N}(n\ell j)$ is not modified by the unitary transform

$$\widetilde{N}(n\ell j)_{ik} = \langle \widetilde{\psi}_i | \hat{N}(n\ell j) | \widetilde{\psi}_k \rangle \neq N(n\ell j)_{ik}, \qquad (5.18)$$

which shows that shell occupancies are model dependent and not observables [16]. It is also the case for single-particle energies and momentum distributions.

The non-observable character of single particle energies does not mean that their usefulness is weaken: the model, by the introduction of unobservable quantities and the reduction of degrees of freedom allows an interpretation, which is the essence of the process of understanding of various complex phenomena from a simplified picture. Nevertheless, the theoretical conventions should be kept under control and specified every time, when non-observables are discussed since they are valid only within the specific framework they have been derived from. The dependance of non-observable quantities with the parameters of the theory are a focus in modern theories. The problematics goes well beyond single-particle energies and shell occupancies. Nuclear reactions, the definition of direct reaction or the interpretation of cross sections (observables), are very much concerned by the same limitations. In hadronic physics, the concept of partons, or number of quark and gluons lead to similar ambiguities.

5.5 Correlations Beyond the Shell-Model Picture

The shell model and mean-field based theories describe the nucleus in a limited energy valence space. Although measurements from the 70s showed that such mean-field calculations provide successful descriptions of the energy and momentum distribution of nucleons around the Fermi surface, they do not provide a consistent picture with experiments about the occupancy of the nuclear orbitals and the overall momentum distribution of nucleons inside the nucleus. In high-quality $A(e, e'p)X$ quasi-free scattering experiments on various nuclei, several observations suggested significant strength of the nuclear spectral function beyond the in-shell correlations. More comprehensive measurements of the proton knock-out reactions show that the spectroscopic factor, basically the ratio of the observed strength within a given shell to the expected strength, is less than one. The observed strength is typically 30% to 40% below the independent particle model expectation in many nuclei in the valence shell, as summarized in Fig. 5.19. Even long range correlations such as pairing and quadrupole correlations are taken into account in the theory, the discrepancy still remains at values close to 25–30%. This discrepancy is understood as stemming from two types of correlations not included in most of nuclear structure models: (i) the coupling to highly collective states such as giant resonances in the ground state of nuclei and (ii) short range correlations that scatter nucleons at very high momentum.

The physical wave-function of a nucleus can be decomposed into components of different excitation energy or momentum domains:

$$|\Psi\rangle = \underbrace{\alpha|\Phi_{valence}\rangle + \beta|\Phi\rangle \bigotimes |GR\rangle}_{\text{long-range correlations}} + \underbrace{\gamma \ |\Phi\rangle \bigotimes |SRC\rangle}_{\text{short-range correlations}} , \qquad (5.19)$$

where $\bigotimes |GR\rangle$ denotes the coupling to high energy excitations not included in the considered valence space, such as for example the coupling to giant resonances at typically $E_x \sim 10 - 20 \text{MeV}$, and $\bigotimes |SRC\rangle$ symbolizes high momentum short range correlations stemming from the repulsive hard-core part of the nucleon-nucleon interaction. The effect of these last two terms is usually not included in low-energy nuclear structure theories and their effect on observables, such as energies and transitions probabilities, are included effectively by renormalizing the interaction and the

Fig. 5.19 Spectroscopic
strength for knocked out
valence protons measured by
quasi free scattering $(e, e'p)$,
relative to the
independent-particle shell
model (IPSM) prediction,
$(2j+1)$, as a function of
target mass. Figure adapted
from [17]. Courtesy L.
Lapikas

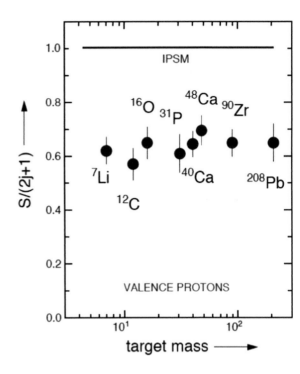

operators in the reduced valence space. The weight α, β and γ verify the normalisation: $\alpha^2 + \beta^2 + \gamma^2 = 1$. The (e, e'p) studies in stable nuclei lead to $\beta^2 \sim 10 - 15\%$ and $\gamma^2 \sim 10 - 15\%$.

The most advanced beyond-the-mean-field calculations including long-range correlations up to high excitation energy significantly overestimate the strength observed in the nuclear shells. This feature is explained by the presence of strong short-range NN interactions in the nucleus which are not be computed in most approaches. Indeed, as discussed in Chap. 2, the nucleon-nucleon interaction has a strong repulsive core at short distances which, together with tensor components of the NN force, yield hard interactions that can scatter nucleons outside of the low-energy valence space, associated with the nuclear shells, to higher energies and momenta.

The existence of short-range correlations in nuclei is evidenced by the momentum distributions of nucleons inside the nucleus. Mean-field-based calculations significantly underestimate the strength at momenta higher than the Fermi momentum. A nucleon propagates in nuclear matter with a constant momentum except when it interacts with another nucleon. When non-interacting nucleons are confined by infinite potential walls, they form an ideal Fermi gas. The Hamiltonian H of a Fermi gas system can be written as

$$H = -\sum_i \frac{\hbar^2 \nabla_i^2}{2m}, \tag{5.20}$$

where each individual term of the sum is the kinetic energy of the nucleon i, and m is the nucleon mass. This Hamiltonian describes uncorrelated particles. The ground-state of symmetric nuclear matter can be described by a Slater determinant, i.e. a product of anti-symmetrized nucleon wave functions according to the $(n\ell m)$ quantum numbers, with n the principal quantum number, ℓ the angular momentum and m the projection of the angular momentum on a reference axis. Due to the Pauli exclusion principle, the ground state is obtained by filling each momentum level with two neutrons with opposite spins and two protons with opposite spins. For symmetric nuclear matter, all momentum states are fully occupied up to the highest energy state. The momentum associated with the latter state is called the Fermi momentum k_F and is equal to

$$k_F = \left(\frac{3\pi^2 \rho}{2} \right)^{1/3} \tag{5.21}$$

where ρ is the density of nucleons.

Problem

5.2 Prove Eq. (5.21).

The obtained momentum distribution for the Fermi gas is a step function with value unity for $k \leq k_F$ and zero for $k \geq k_F$. This gas of nucleons is unbound when the infinite potential walls are removed. Let us now consider interacting particles with a realistic attractive NN potential, and therefore remove the external infinite potential wall. The interaction leads to a mean field potential U composed of direct and exchange terms. The exchange term depends on the momentum k of the nucleon. The so-called separation energy of the nucleons is then given by

$$E(k) = \frac{\hbar^2 k^2}{2m} + U(k). \tag{5.22}$$

The distribution surface of removal energy E and momentum k of nucleons inside the nucleus is called the spectral function $S(E, k)$. An example of measured spectral distribution for ^{16}O is shown in Fig. 5.20. The momentum distribution of nucleons inside a nucleus is given by integrating the measured spectral function over the energy:

$$n(k) = \int S(E, k) dE. \tag{5.23}$$

The momentum distribution can be decomposed into two components

$$n(k) = n_0(k) + n_1(k), \tag{5.24}$$

Fig. 5.20 Measured spectral distributions from $^{16}O(e, e'p)$ at 600 MeV. Figure reprinted from [18]

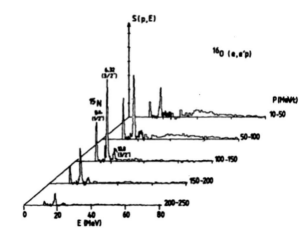

where n_0 is the momentum distribution of a bound state in the residual nucleus $(A - 1)$ after the removal of a nucleon with momentum k. $n_1(k)$ is the distribution of an unbound state in a daughter nucleus. Calculations including high-momentum correlations are compared to data from inclusive (e, e') and $(e, e'p)$ in Fig. 5.21, where the calculated distributions n and n_0 are shown by continuous and dashed lines, respectively. Data clearly show high momentum components in the momentum distributions, which account for part of the discrepancy with the independent-particle description of the nucleus. More quantitatively, it is shown that the effect of short range correlations depletes the nuclear shells by 15%.

The nucleon-nucleon short range correlations represent a pair of nucleons where each nucleon has a large momentum compare to the Fermi momentum, but the total momentum of the pair is very small, i.e. a pair of nucleons with large, back-to-back momenta. The large relative momentum of the two nucleons originates in the repulsive core of the NN interaction at a short range below 1 fm. The amount of such short range correlations leading to back-to-back high momentum pairs of nucleons was measured at the Jefferson laboratory, USA, by electron induced nucleon knockout reaction $^{12}C(e, e'pN)$ at 5 GeV. The incident electron beam couples to a nucleon-nucleon pair via a virtual photon. In the final state, the scattered electron is detected along with the two nucleons that are ejected from the nucleus, as illustrated on the left panel of Fig. 5.22. The measurements show that the correlations mostly occur for neutron-proton pairs, resulting in the decomposition of Fig. 5.22. From this measurement, it was concluded that 20% of the nucleons experience high-momentum correlations, among which 90% are composed of neutron-proton correlations. The dominance of proton-neutron correlations compared to identical-particle pairs reveals the importance of the tensor part of the interaction when high-momentum correlations are concerned. These conclusions are verified for heavier stable nuclei. Similar studies with unstable nuclei would allow to determine how the ratio of np versus pp

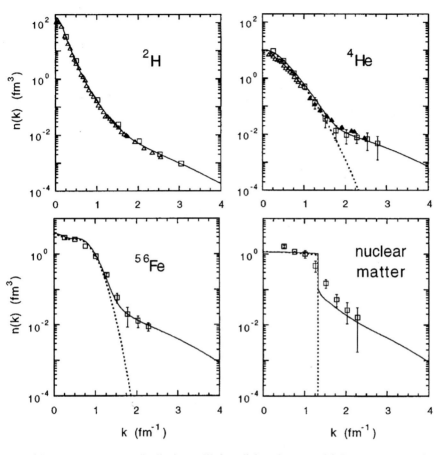

Fig. 5.21 Proton momentum distributions $n(k)$ from light to heavy nuclei. Data are compared to a model for $n(k)$ (continuous line). The component $n_0(k)$ (dashed line) that leads to bound states in the $A - 1$ residue is also shown. The data points are extracted from $(e, e'p)$ experiments. The data points for nuclear matter come from a model-dependent extraction from finite-size-nuclei data. Modified from the original figures in [19]. Reprinted with permission. ©2021 by the American Physical Society

and nn pairs is modified in nuclei with large neutron excess. The finding of np pair dominance is consistent with ab initio calculations from light nuclei, as illustrated in Fig. 5.23, where a much larger amount of np pair is found than pp pairs with high relative momentum at vanishing pair momentum geometry (the back-to-back momenta).

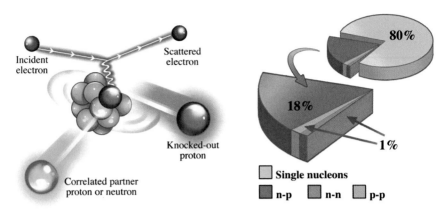

Fig. 5.22 Sketch of the ^{12}C$(e, e'pN)$ reaction at 5 GeV performed at Jefferson Laboratory, USA, to study nucleon-nucleon short range correlations in ^{12}C. Figures adapted from [20]

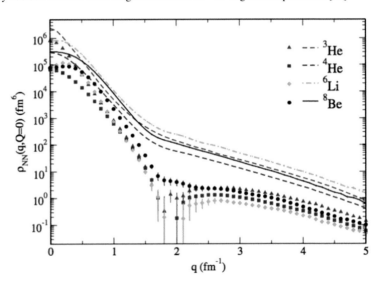

Fig. 5.23 Ab initio calculations of the relative momentum distribution of np (lines) and pp (markers) pairs at vanishing total momentum of the pair in the rest frame of the nucleus. Calculations are shown for various light nuclei and predict a dominance of np short-range correlations. Figure reprinted with permission from [21]. ©2021 by the American Physical Society

Summary

The nuclear shell structure impacts all low-energy nuclear observables. Masses, charge radii and density distributions, electromagnetic moments, spectroscopy and exclusive direct-reaction cross sections are the most commonly used observables to explore nuclear structure across the nuclear landscape. The shell structure of stable nuclei is characterized by the shell closures at the number of nucleons (protons and neutrons): 2, 8, 20, 28, 50, 82, 126. These magic numbers are not universal across the whole nuclear chart (see Chap. 6).

Despite the successes of the shell model, the experimentally-deduced shell occupancy is typically 30%–40% below the independent particle model expectation in many nuclei and looking at several nuclear shells. About 20% of this depletion comes from long-range correlations while other $\sim 20\%$ originate in short-range correlations and tensor interactions caused by the short range repulsive core of the nucleon-nucleon interaction. The latter are not included in low-energy nuclear models. The effect of nucleon-nucleon short range correlations can be represented as making pairs of nucleons where each nucleon has a large momentum compared to the Fermi momentum, but the total momentum of the pair is very small, i.e., a pair of nucleons with large, back-to-back momenta.

Nuclear structure is successfully described by models based on the assumption of mean-field and single particle states. Despite these successes, the concept of a single-particle state is model dependent. Shell occupancies, spectroscopic factors and single-particle energies are not observable strictly speaking, but extremely useful quantities to interpret various complex phenomena within a given model and for a given renormalization scheme.

Nuclei with atomic number larger than $Z = 102$ can only exist because of shell effects. A so-called "island of stability", where superheavy elements should have a large lifetime, is predicted at $Z \sim 120$ and $N \sim 180$ but not reached experimentally yet. Superheavy elements are produced via fusion which have extremely small cross sections. The current limit of measurement is ~ 10 fb $= 10^{-38}$ cm^2. The heaviest nucleus produced so far has an atomic number $Z = 118$ and named Oganesson Og in recognition to Yuri Oganessian for his active work on the synthesis of super-heavy elements.

Hypernuclei are such systems that contain at least one hyperon. Hyperons are baryons with at least one strange quark. Hypernuclei studies shed a new light on the world of traditional nuclei by revealing new symmetries and new phenomena produced by the additional strangeness dimension. Characteristics of Λ hyperons in hypernuclei are: (i) a shallower mean-field potential than nucleons and weaker spin orbit splitting ($V_0 = -30$ MeV, $W = 2$ MeV), (ii) a larger Miyazawa–Fujita type three-body forces (first excited state of Λ is 70 MeV much smaller than the first nucleonic excitation (Δ_{33} resonance) at 300 MeV), (iii) no Pauli blocking of hyperons with nucleons.

Solutions of Problems

5.1

1. The integral of the volume conservation can be performed up to the second order of β using a relation $r' \equiv r/(1 + \beta Y_{20}(\theta, \phi) + \eta)$,

$$\int \rho(\mathbf{r}) d\mathbf{r} = \int \int \rho_0(r/(1 + \beta Y_{20}(\theta, \phi) + \eta) r^2 dr d\Omega$$

$$= \int \int \rho_0(r')(1 + \beta Y_{20}(\theta, \phi) + \eta)^3 r'^2 dr' d\Omega$$

$$= 4\pi \int \rho_0(r') r'^2 dr' (1 + 12\pi\eta + 3\beta^2),$$

where we use the formulas $\int Y_{lm}^* d\Omega = 0$, $\int Y_{lm}^* Y_{lm}^* d\Omega = 1$. Since $4\pi \int \rho_0(r) r^2 dr = A$, the η is determined as

$$\eta = -\frac{\beta^2}{4\pi}. \tag{5.25}$$

2. The mean square radius for the deformed density is evaluated as

$$\langle r^2 \rangle_\beta = \int \rho(\mathbf{r}) r^2 d\mathbf{r} = \int \int \rho_0(r')(1 + \beta Y_{20}(\theta, \phi) + \eta)^5 r'^4 dr' d\Omega$$

$$= 4\pi \int \rho_0(r') r'^4 dr' \left(1 + 5\eta + 10\frac{\beta^2}{4\pi}\right) = \langle r^2 \rangle_0 \left(1 + \frac{5\beta^2}{4\pi}\right), \tag{5.26}$$

where $\langle r^2 \rangle_0$ is the mean square radius of spherical nucleus.

5.2 The nuclear density ρ is defined by the single-particle wave function $\varphi_i(\mathbf{r}, s, t)$ as,

$$\rho(\mathbf{r}) = \sum_{i,s,t} |\varphi_i(\mathbf{r}, s, t)|^2, \tag{5.27}$$

where s, t specify the spin and isospin quantum numbers, respectively. For the plane wave, the single-particle wave function is given by

$$\varphi_i(\mathbf{r}, s, t) = \frac{1}{(2\pi)^{3/2}} e^{i\mathbf{k}\mathbf{r}} \chi_s \chi_t, \tag{5.28}$$

where the factor $\frac{1}{(2\pi)^{3/2}}$ is the plane wave normalization factor to a δ-function, χ_s and χ_t are spin and isospin wave functions, respectively. We consider the nuclear matter occupied the momentum space until the Fermi momentum k_F with the spin and isospin degree of freedom. Then the sum of Eq. (5.27) is replaced by

$$\sum_{i,s,t} \rightarrow 4 \int_0^{k_F} d\mathbf{k} \tag{5.29}$$

which leads

$$\rho = \frac{4}{(2\pi)^3} \frac{4\pi}{3} k_F^3 = \frac{2}{3\pi^2} k_F^3, \tag{5.30}$$

and the Fermi momentum is given by

$$k_F = \left(\frac{3\pi^2 \rho}{2} \right). \tag{5.31}$$

Inserting the empirical nuclear matter density $\rho = 0.16 \, \text{fm}^{-3}$, the Fermi momentum is determined as

$$k_F = 1.33 \, \text{fm}^{-1}. \tag{5.32}$$

Books for Further Readings

"Many-body theory exposed" (World Scientific, 2nd edition, 2008) by W. H. Dickhoff and D. Van Neck gives a unified view of different many-body systems using the approach of self-consistent Green's functions. It provides an excellent framework for a detailed discussion on short-range correlations.

References

1. P. Moller, A. J. Sierk, T. Ichikawa, H. Sagawa, Atomic Data Nucl. Data Tables **109–110**, 1–204 (2016)
2. B. Frois et al., Nucl. Phys. A **396**, 409 (1983)
3. I. Angeli, K.P. Marinova, Atomic Data Nucl. Data Tables **99**, 69 (2013)
4. M. Goeppert Mayer, Nobel Lecture. NobelPrize.org. Nobel Media AB 2021. Mon. 18 Jan 2021, https://www.nobelprize.org/prizes/physics/1963/mayer/lecture/
5. S. Raman, C.W. Nestor, P. Tikkanen, Atomic Data Nucl. Data Tables **78**, 18 (2001)
6. K.K. Seth, J. Picard, G.R. Satchler, Nucl. Phys. A **140**, 577 (1970)
7. M. Bender, J.-H. Heenen, J. Phys.: Conf. Ser. **420**, 012002 (2013)
8. Yu. Ts. Oganessian et al., Phys. Rev. C **87**, 034605 (2013)
9. K. Morita et al., J. Phys. Soc. Jpn. **73**, 2593 (2004); **76**, 5001 (2007); **81**, 3201 (2012)
10. M. Danysz, J. Pniewski, Phil. Mag. **44**, 350 (1953)
11. H. Hotchi et al., Phys. Rev. C **64**, 044302 (2001)
12. D.J. Prose, Phys. Rev. Lett. **17**, 782 (1966)
13. H. Takahashi et al., Phys. Rev. Lett. **87**, 212502 (2001)
14. ALICE collaboration, Phys. Lett. B **811**, 135849 (2020)
15. ALICE collaboration, Nature **588**, 232 (2020)
16. R. J. Furnsthal, H.-W. Hammer, Phys. Lett. B **531**, 203 (2002)
17. L. Lapikas, Nucl. Phys. A **553**, 297c (1993)

18. J. Mougey, Nucl. Phys. A **335**, 35 (1980)
19. C. Ciofi degli Atti, S. Simula, Phys. Rev. C **53**, 1689 (1996)
20. R. Subedi et al., Science **32**, 1476–1478 (2008)
21. R. Schiavilla et al., Phys. Rev. Lett. **98**, 132501 (2007)

Chapter 6
Radioactive-Ion-Beam Physics

Abstract One main challenge of nuclear physics today is to understand the structure of radioactive nuclei since they exhibit phenomena not observed in stable nuclei. The appearance of new phenomena is triggered by the proton-to-neutron asymmetry accessible in these nuclei or their low binding energy. These radioactive nuclei have been accessible experimentally thanks to the development of production techniques introduced in this chapter. Exotic phenomena such as halos and neutron skins are presented. Shell evolution with the proton-to-neutron asymmetry is detailed, in particular for sd- and pf-shell nuclei through the examples of oxygen and calcium neutron-rich isotopes. Di-neutron correlations and molecular orbits of α-cluster states are also discussed experimentally and theoretically.

Keywords Accelerators · Neutron-rich nuclei · Halos · Neutron skins · Shell evolution · Di-neutron and cluster-chain structures

6.1 Radioactive-ion-beam Production

There are 293 isotopes which are stable or with a lifetime longer than the age of the earth, ∼4 billion years, i.e., isotopes that can be found on earth. Other radioactive isotopes need to be synthesized to be studied. About 3500 have been discovered so far [1], while theoretical models predict ∼5000–8000 isotopes to be bound by the strong force [2].

As short-lived nuclei do not exist on earth (nuclei in the green and purple regions of Fig. 6.1), they have to be synthesized from stable nuclei, accelerated in an ion state and studied shortly after production, before their radioactive decay. The development of dedicated and new-generation radioactive-ion-beam facilities has given to nuclear physicists the precious possibility to produce a broad range of nuclei with the particular number of neutrons and protons of interested for specific studies. Today, not all existing combinations of nucleons can be accessed since the nuclei further from stability are still too challenging to produce in sufficient quantities. About half of existing nuclei are still in the *Terra Incognita* (Unknown Region). Several accelerators are leading this research worldwide. The most important are GANIL in France,

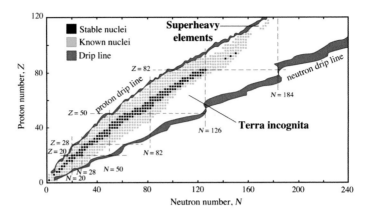

Fig. 6.1 Nuclear landscape. Black squares show stable nuclei along the stability line, the so-called "Heisenberg Valley". The green region indicates unstable nuclei so far observed by terrestrial experiments, while the green region is still unknown experimentally, denoted "Terra Incognita" (unknown region). The predicted neutron and proton drip lines are also shown by purple color in the figure. Beyond these lines, atomic nuclei decay by the emission of a proton or a neutron and can not exist to be neither bound nor resonant states. The astrophysical nucleosynthesis are held in the region "Terra Incognita"; $r-$process is held near the neutron drip line, while the $rp-$process occurs near the proton drip line. The black dashed lines show the magic numbers where nuclei get the maximum stability

ISOLDE at CERN in France and Switzerland, TRIUMF in Canada, the NSCL facility, stopped in 2021, and the new-generation FRIB facility, started in 2021, in the United States of America, GSI in Germany and the RIBF of RIKEN in Japan. A study of new elements have been performed successfully over the past 60 years at JINR, Dubna in Russia and discovered ten new elements. All above-mentioned facilities host upgrade projects under construction while the new generation level has already been reached at the RIBF operational since 2007. New accelerator complexes dedicated to radioactive isotopes are planned in China and Korea.

There are two main techniques to produce accelerated radioactive beams:

- the in-flight method,
- the Isotopic Separation On Line (ISOL) method.

The in-flight method is illustrated on the right-hand side of Fig. 6.2. An accelerated beam impinges onto a thin production target. Radioactive nuclei originate from the fragmentation or induced-fission of projectiles. The production target being thin, the beam and reaction products lose little energy in the target and the produced radioactive nuclei have about the same velocity as the beam at the exit of the target. This production method has three main advantages:

1. the production method is not sensitive to the chemical properties of the produced nuclei. The only relevant quantity for the production of a given nucleus is its production cross section,

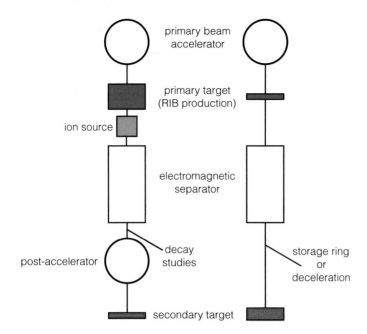

primary beam
accelerator

primary target
(RIB production)

ion source

electromagnetic
separator

post-accelerator   decay
                        studies

storage ring
or
deceleration

secondary target

Fig. 6.2 Schematics of an ISOL (left) and in-flight (right) radioactive-ion-beam production

2. radioactive nuclei are produced in flight, meaning that they can be directed right after production to the experimental zone with no further delay than the time of flight. The in-flight method is particularly suited for the production and study of very short lived nuclei down to half lives of a hundred of nanoseconds,
3. fast beams are naturally produced with the in-flight technique, requiring no further acceleration. Fast beams (above 30 MeV/nucleon) are today preferentially produced with the in-flight method.

The in-flight method is described in more details in Sect. 6.1.3.

The ISOL method is sketched on the left-hand side of Fig. 6.2. In this method, a thick production target is used. The reaction products are stopped inside the target material and only those which diffuse (in the form of neutral atoms) outside the target can contribute to produce a radioactive beam. Some of the reaction products reach an ion source where they are ionized and transported further. They are often reaccelerated. The extraction from the production target and ionization process take time, typically 100 ms. Radioactive nuclei with a half life lower than few tens of ms cannot be efficiently produced via the ISOL method.

The in-flight and ISOL methods can be considered as complementary to some extend. The ISOL production has several advantages.

1. Beams can be produced from very low energy (kV) to high energy, at the (expensive) cost of a post-accelerator for reaction products.

2. The beams produced via ISOL and reaccelerated present a high quality emittance, i.e., ensuring a small beam size on target and small energy dispersion.
3. The transmission and rates are highly dependent on the isotopes due to chemical selectivity. For each element (isotope) an ionization scheme has to be developed before the beam can be produced. On the other hand, this selectivity can ensure pure isotopic beams.

The ISOL method is described in more details in Sect. 6.1.2.

6.1.1 The Primary Beam: Ion Sources and Acceleration Systems

Almost[1] all radioactive-ion-beam production starts with the interaction of an accelerated stable beam with a production target.

6.1.1.1 Ion Sources

The first element of a heavy-ion accelerator is the ion source. Ions have to be produced at high intensity, in a charged state so that it can then be accelerated by strong electric fields. The ion source consists of two parts: an ion generator and an extraction system.

Ions are produced from the creation of a plasma, the warm state of matter reached when ions and electrons are unbound and can be represented as two fluids interacting with each other by the electromagnetic interaction. There are several techniques to create a plasma and of ionized atoms: by electric discharges in a low pressure gas volume, by heating, using lasers or beams of other particles. There are different types of ion sources. The choice for the design of an ion source are: (i) the ions to be produced, (ii) the ion intensity to be reached, (iii) the stability in time of the beam, (iv) the optical quality of the beam, quantified by the so-called *emittance* (see Sect. 4.1.5.1 in Chap. 4).

A supply of atoms must be provided to the plasma to compensate the loss of material. The material can be introduced in a gas phase inside the plasma via a needle valve or solid material can be heated and ionized inside the ion source. As an illustration of how an ion source works, we take here the example of an electron bombardment source, whose principle is sketched in Fig. 6.3. This type of source is at the origin of all ion sources. It was first developed by A. Dempster at the University of Chicago in 1916. It has a very simple design by contains all features of ion sources. In this specific case, the plasma is created by accelerated electrons created from the cathode filament. The cathode is heated by a strong current and thermionic electrons are emitted. The electrons are accelerated by the discharge power voltage between the

---

[1] One exception is the CARIBU facility at Argonne National Laboratory, USA, which relies on the production of radioactive ions from a high-activity radioactive source of Californium which releases fission fragments. Therefore, CARIBU does not require any primary beam.

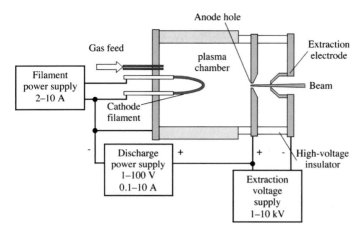

Fig. 6.3 Schematic design of an electron bombardment ion source

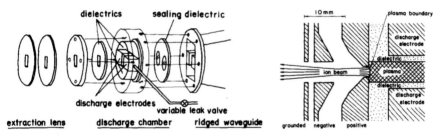

Fig. 6.4 (Left) Exploded view of a microwave ion source. (Right) Extraction of the ion beam from the plasma region. Sketches from the first article reporting a high-intensity microwave ion source from the research center of Hitachi Limited [3]

filament (cathode) and the anode where an extraction hole is drilled. If the discharge voltage is higher than the ionization energy of the gas introduced in the ion source chamber, the electrons will ionize atoms from the gas by impact ionization. The positive ions produced in the vicinity of the anode hole are extracted from the source through the hole by electric field lines created by the extraction electrode to which a negative voltage is applied.

Electron bombardment sources are easy to produce but the reached intensities are too low for the requirements of modern accelerators. Modern sources are based on other techniques to generate the plasma and the geometry of the source is made in such a way that the positive-ion density close to the extraction is as high as possible. An important type of ion source is called ECR (Electron Cyclotron Resonance) where the plasma is formed by a high-power microwave radio-frequency applied to the source. The microwave energy is used to accelerate electrons in a discharge chamber. The accelerated electrons collide with atoms in the chamber and ionize or excited them. The dissipated energy expends to the wall chamber via the heating of the walls and the emission of photons from the excited particles. The principle of microwave ion sources is ilustrated in Fig. 6.4.

The source is surrounded by a solenoid to generate a magnetic field which confines the plasma. Since the plasma in an ECR source has a relatively long lifetime, it has also the advantage to produce efficiently multi-charged ions from multi-collisions of electrons with atoms and ions.

As examples of intense ion beams produced from ion sources, the SILIS light ion source of the french research institute CEA reaches stable intensities of 100 mA for protons. The Berkeley ECR ion source produces beams of Bi^{29+} at 0.25 mA. In modern ion sources, the extraction voltage is typically 50 kV, and ranges from few kV to 400 kV.

6.1.1.2   Ion Acceleration

Ions are accelerated via electric fields. There are different technologies to accelerate ions:

- a linear accelerator composed of a unique accelerating voltage, as a Van de Graaf accelerator, or using twice an accelerating voltage by using a charge change of the acceleration ions from $q = -1$ to $q = +1$ as in a Tandem accelerator. These accelerators have an excellent energy resolution but are limited by the maximum voltage that can be imposed between the electrodes before sparks occur.
- A linear accelerator composed of a series of cavities excited in a stationary mode with a radio-frequency. Ions are accelerated in the linear accelerator in bunches. This is the way modern linear accelerators are conceived.
- A cyclotron in which low energy ions are introduced at the center. A cyclotron works with a permanent magnetic field which curves the trajectories of ions. The ions are accelerated in between two regions of magnetic field by an electric field in phase with the ions. The direction of the electric field changes at each half turn of the ions, leading to an acceleration phase and an increase of the trajectory radius every half turn. Ions are extracted in the vicinity of the outer radius of the cyclotron.
- A synchrotron aims at accelerating ions in closed orbits like a cyclotron, with the difference that it works with a fixed trajectory radius: along the acceleration, the magnetic field of the coils in the synchrotron is increased to compensate the increase of velocity of the accelerated ions and maintain the same rigidity.

Heavy-ion beam energies are given in different units. The most common is given in *energy per nucleon*. Assuming that the total kinetic energy of ions composed of $A = N + Z$ nucleons is E in units of MeV, its energy is also referred as (E/A) MeV/nucleon. Sometimes, the energy is given in mass units: $(E/(Au)) \rightarrow (Eu/m_A)$ MeV / u, where m_A is the mass of beam particle and u $= m_{^{12}C}/12$ is the atomic mass unit (see Chap. 4), where $u = 931.5$ MeV is the standard nucleon mass unit.

Table 6.1 gives an overview of the accelerators of several Radioactive-Ion-Beam (RIB) facilities worldwide. In the following subsections, we give three examples of state of the art RIB facilities which are based on different acceleration schemes for the primary beam: FRIB in the US accelerates primary beams with a linear accelerator

Table 6.1 Acceleration schemes used for primary beams at radioactive-ion beam facilities. HI stands for heavy ions

RIB facility	Type	Maximum energy
TRIUMF (Canada)	linear	520 MeV (protons)
FRIB (USA)	linear	200 MeV/nucleon (HI)
GANIL (France)	cyclotron	95 MeV/nucleon (K = 380, HI)
NSCL (USA) (up to 2021)	cyclotron	120 MeV/nucleon (K = 1200, HI)
RIBF (Japan)	cyclotron	345 MeV/nucleon (K = 2600, HI)
SPES (Italy)	cyclotron	30–70 MeV (K = 130, protons)
CERN (Switzerland)	synchrotron (PSB)	1.4 GeV (protons)
GSI (Germany)	synchrotron (SIS18)	18 Tm (400–2000 MeV/nucleon)
FAIR (Germany)	synchrotron (SIS100)	100 Tm
Langzhou (China)	Accelerating ring (CSRm)	400 MeV/nucleon (U^{72+})

up to 200 MeV/nucleon with a foreseen upgrade to 400 MeV/nucleon, the RIBF in Japan accelerate heavy ions up to 345 MeV/nucleon with a set of coupled cyclotrons and GSI accelerates fully stripped heavy ions at relativistic energies (up to about 2 GeV/nucleon) with a synchrotron. In the future, the last stage of the FAIR acceleration scheme will be composed of the SIS100 synchrotron which will enable to accelerate beams up to the magnetic rigidity, $B\rho = 100$ tesla.meter (Tm). The long term future of FAIR foresees an additional acceleration stage to 300 Tm with the SIS300 ring.

6.1.1.3 RF LINAC at FRIB

A linear accelerator is a single pass device that accelerates charged particles. Modern linear accelerators are based on sinusoidal time varying electric fields obtained by a radio frequency (RF). They are called *RF linacs*, where the term *linac* is the well-spread compressed version of LINear ACcelerator. Although a detailed description of the principle of linacs is beyond the scope of this book, we give bellow the driving concepts of RF accelerating cavities and radiofrequency quadrupoles (RFQ).

RF linacs usually operate at frequencies in the range of 100 MHz to 10 GHz. They can be used to accelerate all type of charged particles, including heavy ions. A feature of a RF linac is that its design defines the accelerating electric field inside the *cavities*, the volume whose geometry defines the boundary conditions of the electric and magnetic field. Linacs can be divided into two categories: travelling- and standing-wave linacs, depending how the electromagnetic field propagates in the cavity. The RF of the linac has to be adapted to the velocity of the particles so that the charged particles always "see" an accelerating field as they pass through the cavity. For a given cavity, the faster the charged particles, the higher the frequency. Often, a linac is composed of cavities of different lengths: the shortest for the low velocity portion and the longest at the end of the accelerator. The intensity in LINACs

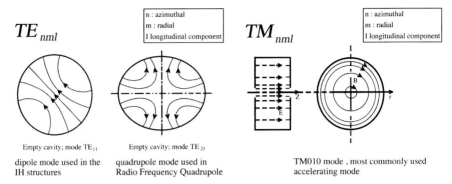

Fig. 6.5 Examples of the most common Transverse Electric and Magnetic modes used in accelerating cavities. (Left) Transverse Electric (TE) modes TE_{11} and TE_{21}. Note that in this specific case, the index of the not shown longitudinal component is dropped and the TE modes are denoted as TE_{nm}. (Right) Illustration of the TM_{010} mode, also quoted $TM010$

is limited, as all accelerating systems, by space charge effects: the beam cannot be compressed in a too small volume.

There are different ways to impose a field inside a cavity. At the surface of the cavity, made of a perfect conductor, the following boundary conditions must be verified

$$E_{\parallel} = 0 \text{ and } B_{\perp} = 0. \tag{6.1}$$

Indeed, inside a perfect conductor all fields vanish and the Maxwell equations lead to the above equalities. The choice of RF allows a much more compact device compared to static field acceleration devices. The cavity resonates with a specified frequency introduced by a wave guide from a RF source. The field pattern imposed inside the cavity will accelerate the particles flying through it.

In a cavity there are Transverse Electric (TE) or Transverse Magnetic (TM) modes. In an transverse electric (magnetic) mode, the electric (magnetic) field is perpendicular to the direction of propagation of the wave. Because of the boundary conditions, a given wave type (TE or TM) can only have certain modes which relate the dimensions of the cavity to the wave lengths in the transverse and longitudinal directions. The various modes are usually refered as TE_{nml} and TM_{nml}, where n, m qualify the transverse mode and l characterises the longitudinal component of the wave. Most common examples are illustrated in Fig. 6.5.

Cavities for linear accelerators are shaped to maintain an electric field and zero magnetic field along the particle trajectory. The particle is "surfing" on the electric wave, the driving process of the acceleration. Cavities can have a length that contains half the wavelength of a field oscillation ($L = \beta\lambda/2$) or a quarter, where β is the velocity of particles in unit of the light velocity. These cavities are also called *resonators*. Typical accelerating fields are 0.5–10 MV/m. Such cavities have often ellipsoidal shapes to reduce the dissipated energy in the cavity walls and therefore to optimise the stored energy inside the cavity. The energy stored is much higher

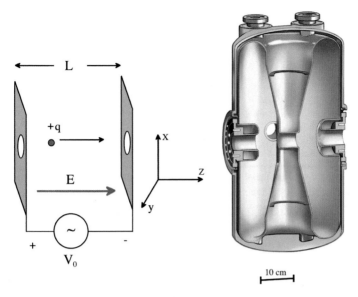

Fig. 6.6 (Left) Simplified model for a half wave resonator (HWR) cavity. (Right) Three-dimensional profile of a $\beta = 0.53$ half wave resonator (HWR) of FRIB

for cavities whose internal surface is made of a superconducting material, such as Nobium. Superconducting cavities have the technological difficulty that they should be inserted in a cryostat since superconductivity is reached at cryogenic temperature of few Kelvin. The profile of a $\beta = 0.53$ half wave resonator is shown in Fig. 6.6.

We propose now a simple model to show how the acceleration in such a cavity works. Consider infinite parallel plates separated by a distance L (the size of the cavity) with a sinusoidal voltage applied and assume an uniform electric field $E_z(t)$ in the gap, i.e., neglect the entrance and exit apertures. One gets

$$E_z(t) = E_0 \cos(\omega t + \phi), \tag{6.2}$$

where at $t = 0$ the particle is in the center of the gap ($z = 0$) and the phase of the field relative to the crest is ϕ. For a particle going through the gap, the time t and the position z are connected via the velocity v of the particle

$$t(z) = \int_0^z \frac{dz}{v(z)}. \tag{6.3}$$

The energy ΔW gained by the particle inside the gap is therefore

$$\Delta W = q \int_{-L/2}^{+L/2} E_z dz = q E_0 \int_{-L/2}^{+L/2} \cos(\omega t + \phi) dz. \tag{6.4}$$

If in addition we assume that the velocity change through the gap is small, i.e., $t(z) = z/v$, one gets

$$\omega t \sim \omega \frac{z}{v} = \frac{2\pi z}{\beta \lambda}, \tag{6.5}$$

and the energy gain reads

$$\Delta W = q E_0 cos(\phi) \int_{-L/2}^{+L/2} \cos(\frac{2\pi z}{\beta \lambda}) dz = q E_0 \frac{\sin(\pi L/\beta \lambda)}{\pi L/\beta \lambda} L \cos(\phi). \tag{6.6}$$

Usually one defines the transit time factor T that takes into account the time variation of the field during particle transit through the gap

$$T = \frac{\sin(\pi L/\beta \lambda)}{\pi L/\beta \lambda}, \tag{6.7}$$

and the energy gain by a particle of charge q in the accelerating gap at a potential $V_0 = E_0 L$ with a relative phase ϕ of the particle and the crest of the field is

$$\Delta W = q V_0 T \cos(\phi). \tag{6.8}$$

This means that there is maximum acceleration when the particle and the crest of the field are in phase. The transit time factor drives the efficiency of the acceleration: it is important to match properly the gap length L to the distance that the particle travels in one RF wavelength $\beta \lambda$. This is the reason why ion linacs are often composed of several types of cavities optimised for different β values.

A Radio Frequency Quadrupole (RFQ) device is based on the so-called TE21 mode (see Fig. 6.5). Along the beam axis, the electric mode imposes an alternating gradient with period length $\beta \lambda$, where λ is the period and β is the velocity of the ions (in half a RF period, the particles have travelled a length $\beta \lambda/2$). This has a focusing effect on the beam, as illustrated in Fig. 6.7. In addition, if the electrodes are modulated in the longitudinal direction, a longitudinal component (i.e. accelerating) can be created in the TE mode. A RFQ is not considered a very efficient accelerator since the same radio frequency is used to focus and to accelerate: a proper focusing requires limited acceleration efficiency. The typical accelerating capabilities of a RFQ is 1 MeV/m (for a charge $q = 1$). Nevertheless, a RFQ has the advantage that it can bunch the ions around the central velocity. Particles which are too slow (fast) will be accelerated (decelerated) because they are off phase. RFQ are then used as a first step to transform continuous (direct current, DC) beams into bunched beams. They are then usually combined to other types of linacs.

At FRIB, the low energy bunched beams produced by a RFQ are accelerated via a set of superconducting cavities arranged in modules. The different modules are optimised for a given energy and follow different resonating field designs. The FRIB layout is illustrated in Fig. 6.8 and shows four types of sets of cavities optimised for $\beta = 0.041, 0.085, 0.29$ and 0.53.

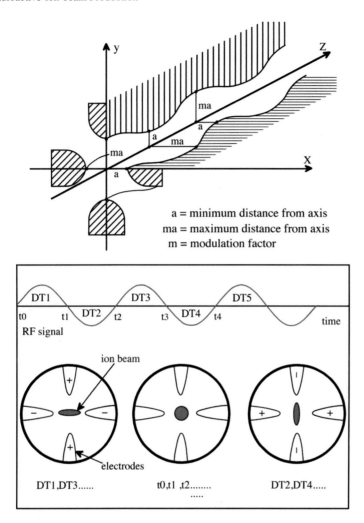

Fig. 6.7 (Top) Profile of a RFQ accelerator. It is often composed of pure copper. The geometry with four tips around the beam axis allows to impose a TE21 focusing quadrupole mode (Bottom). A longitudinal modulation of the profile of the electrodes allows an accelerating component

6.1.1.4 Cyclotrons at the RIBF

A cyclotron is composed of an accelerating device inside a dipole magnetic field, as illustrated in Fig. 6.9. The magnetic field bends the trajectory of particles while the acceleration is provided by alternative high voltage potential. In the simple example of Fig. 6.9 where the cyclotron is composed of two "Dees" (named after their half disc shape similar to the letter "D"), the acceleration occurs at each half turn. This is the main advantage of a cyclotron: only one electrode (one Dee is at high voltage

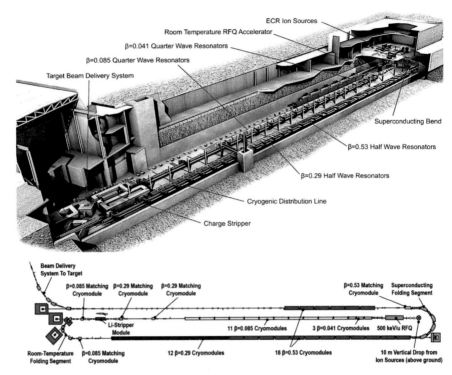

Fig. 6.8 The FRIB accelerator complex layout. For compactness reasons, the design of the accelerator follows a double-folded geometry. Ions will be produced at medium- or high-charge states from high-performance ECR ion sources. The first acceleration section is composed of $\lambda/4$ cavities and accelerates ions up to 17 MeV/nucleon. It is followed by a stripping section and a 180° magnetic bend. The second straight section, composed of $\lambda/2$ cavities, accelerates the ions to 108 MeV/nucleon and is followed by a second 180° bend. The last section accelerate the ions up to 200 MeV/nucleon. The layout is made such that additional cavities can be added at a later stage to the last section for an acceleration up to 400 MeV/nucleon. Reprinted from [4]

the other at ground) can be used to accelerate the ions to high energy, thanks to the magnetic field. The ions are introduced first at low energy in the center of the cyclotron. At each half turn, the ions are slightly accelerated and the radius of their trajectory, i.e., the magnetic rigidity $B\rho$, increases. For non-relativistic velocities, the revolution time is constant during the acceleration, and the ion takes the same time to make one turn (the so-called *isochronism condition*). When the ions have reached maximum kinetic energy, i.e., when they reached the most outer radius inside the cyclotron, they are ejected towards a beam line for use.

Modern cyclotrons are not anymore composed of Dees but of more accelerating cavities, called sectors. Usually, RF cavities have a fixed and reduced frequency range. In order to extend the energy range of the cyclotron, various harmonics are used. An harmonic "H" is the number of radio frequency oscillations per ion revolution. Maximum acceleration is obtained for $H = 1$, but any higher harmonics is

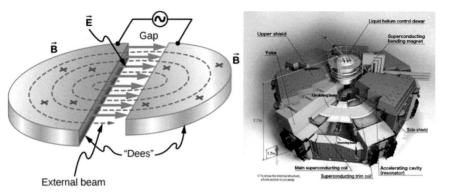

Fig. 6.9   (Left) Charged-particle acceleration principle of a cyclotron. (Right) Cut view of the SRC superconducting cyclotron in RIKEN

in principle possible. In practice, harmonics up to $\sim$7 are used for stability reasons. The principle of harmonics is indicated in Fig. 6.10.

For accelerators based on a magnetic field, such as cyclotrons and synchrotrons, it is of general use to qualify the energy of the beam by its magnetic rigidity $B\rho$, given by

$$B\rho = \frac{p}{Q}, \tag{6.9}$$

where $p = \sqrt{E(E + 2mc^2)}/c$ is the momentum of the ion, m is its mass, and Q is its electric charge. The maximum kinetic energy that a cyclotron can provide for ions of charge Q and number of nucleons A is a characteristic of the cyclotron and is qualified by the so-called K *value*. K is the kinetic energy for protons for maximum bending strength in a non-relativistic approximation. It is used to characterise the accelerating capabilities of a cyclotron and to scale the maximum kinetic energy for different charge-to-mass ratios by

$$\frac{E_{max}}{A} = K \times (\frac{Q}{A})^2, \tag{6.10}$$

where E_{max} and K are given in MeV. For example, the SRS cyclotron of RIKEN has $K = 2600$. It can accelerated $^{238}U^{92+}$ up to 345 MeV/nucleon (the non-relativistic relation (6.10) gives 388 MeV/nucleon).

The Superconducting Ring Cyclotron (SRC) of RIKEN is the largest heavy-ion cyclotron for research worldwide. The magnetic field of maximum 8 T is provided by superconducting coils. The SRC accelerates heavy-ions beams, from deuteron to uranium ions, up to 70% of the speed of light at the highest intensities of the world. The SRC, consists of six superconducting sector magnets and five radio frequency

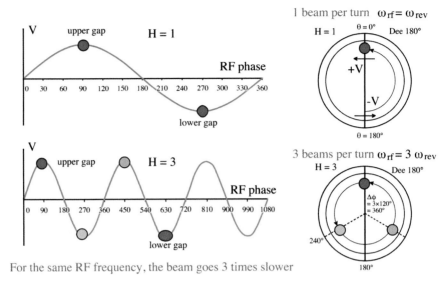

Fig. 6.10 Working principle of a cyclotron and acceleration scheme for harmonic $H = 1$ and $H = 3$. In a cyclotron, a particle bunch lays on an accelerating wave. (Top) With a cyclotron composed of two 180° Dees, $H = 1$ corresponds to a RF frequency (ω_{rf}) identical to the revolution frequency (ω_{rev}) of the ions. Usually, one Dee is at ground while the second is connected to the RF power supply. In this configuration and in $H = 1$ operation, only one ion bunch (in purple) can be accelerated by turn. The top left panel shows when and where the ion bunch is accelerated along one 360° turn. The bunched ions are accelerated twice at each passage between the two Dees: once at the upper gap from right to left by a positive potential, and once at the lower gap from left to right by the opposite phase. (Bottom) With the same magnetic field and configuration, the next possible harmonic is $H = 3$ where $\omega_{rf} = \omega_{rev}$. For the same RF, the ions are three times slower than in the $H = 1$ case. Three ion bunches (indicated in purple, yellow and green) can be accelerated in one turn. The bottom left panel shows where and when the particle bunches are accelerated. After being accelerated at the upper gap, the purple bunch will reach the lower gap and be accelerated after 3/2 periods of RF. In the mean time, the green bunch would have been accelerated at the lower gap, and the yellow bunch at the upper gap

cavities. It weighs 8,300 tons. The power consumption would have been 100 times larger if all the coils were based on conventional (non superconducting) technology.

6.1.1.5 Synchrotrons at GSI/FAIR

Heavy ion beams at GSI are accelerated first by a linear accelerator to the typical energy of 11.4 MeV/u (case of $^{238}U^{72+}$ corresponding to 1.6 Tm) and then injected into the SIS18 synchrotron (SchwerIonen[2] Synchrotron), illustrated in Fig. 6.11. Contrary to a cyclotron which works at constant magnetic field and leads to trajectories with increasing radius, a synchrotron synchronises the increase of the magnetic field

---

[2] German word for "heavy ions".

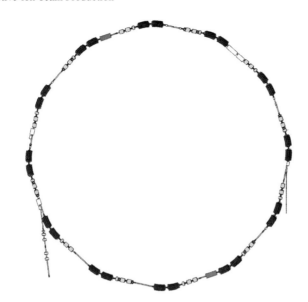

Fig. 6.11 Layout of the SIS18 synchrotron at GSI, Germany, as initially installed in 1990. Today's configuration shows slight differences. The accelerator is composed of 24 bending dipole magnets (red), 12 focusing quadrupole triplets (yellow) and 12 sextupole magnets. The ions are accelerated at each half turn when passing through accelerating cavities (blue). The maximum magnetic field of the dipoles (1.8 Tesla) and their bending radius of 10 m define the maximum bending power of the synchrotron ($B\rho = 18$Tm), and therefore its maximum accelerating capabilities. The typical magnetic field ramp is about 10 T/s and is driven by the power supplies. The total circumference of the SIS18 is 216 m. The SIS18 will be used to inject accelerated heavy-ion bunched into the new SIS100 synchrotron (100 Tm) at the upcoming FAIR facility. Credit: GSI

with the acceleration so that the particles keep a constant trajectory radius during the acceleration. The SIS18 has a circumference of 264 m and a maximum bending power of 18 Tm. At any time, the revolution frequency is given by

$$\omega = \frac{qB}{\gamma m}. \tag{6.11}$$

The bending of the ions is performed by a set of 24 dipoles along the circumference of the synchrotron, each of them with a bending radius of 10 m, while the acceleration is performed by two accelerating cavities located at two radially opposite points of the synchrotron. The cavities are at 16 kV and cover a frequency span from 0.8 to 5.6 MHz. After injection, the acceleration is combined with a linear increase of the magnetic field. Different extraction modes can be used: either a fast extraction, i.e., the full bunch is extracted in one turn, or a slow extraction, i.e., the bunch is extracted over several turns up to seconds to provide a quasi-continues beam to the experiment. A full cycle takes typically one second in case of fast extraction to several seconds if a slow extraction is chosen.

In a circular accelerator, the trajectory of particles is defined by the so-called β function. In case of the on-momentum particle, i.e., with the momentum of the reference trajectory, the equation of motion of particles in the accelerator is given by the Hill equation

$$x'' + K(s)x = 0, \tag{6.12}$$

where s is the longitudinal position along the circular orbit, x is a transverse position and K is a function of s which depends on the design of the synchrotron. The derivative is taken with respect to a variable s, i.e., $x'' = d^2x/d^2s$. It can be shown that the solution of the Hill equation is given by

$$x(s) = a\sqrt{\beta(s)}\,\exp^{\pm i\Phi(s)} \tag{6.13}$$

with $\Phi'(s) = 1/\beta(s)$ and a is a constant. Thus, the most general solution to the Hill equation is a type of harmonic oscillation whose amplitude and wavelength are both given in terms of the β function. The β function is a characteristic of the design of the accelerator. The so-called tune or Q value defines the number of oscillations per revolution in a circular accelerator

$$Q = \frac{1}{2\pi}\int \frac{ds}{\beta(s)}, \tag{6.14}$$

where the integral is performed over the circumference.

The maximum beam intensity that can be reached by a cyclotron is mainly driven by the space charge limit, i.e., the maximum number of charges that can be accelerated in one bunch. Indeed, the Coulomb forces between the charges constituting a bunch of a high intensity beam create a self field which tends to defocus the beam in the two transverse directions. These forces lead to a de-tuning of the beam after one turn. This detuning is called the space charge tune shift and is quantified as $\Delta Q_{x,y}$. Empirically, it is shown that the space-charge tune shift should be lower than 0.5, otherwise significant beam losses cannot be avoided. It is given by

$$\Delta Q_{x,y} = -\frac{q^2 R N}{2\pi m_A \epsilon_{x,y}\beta^2\gamma^3}, \tag{6.15}$$

where q is the charge of the accelerated particles, m_A is their mass, R is the radius of the accelerator, N is the number of particles per bunch in the accelerator, $\epsilon_{x,y}$ is the transverse emittance containing 100% of particles, β and γ are the velocity in units of the speed of light and Lorentz factor of the accelerated particles, respectively. In the case of the SIS18, the space charge limit is about 2×10^{11} ions per cycle and experimentally reached.

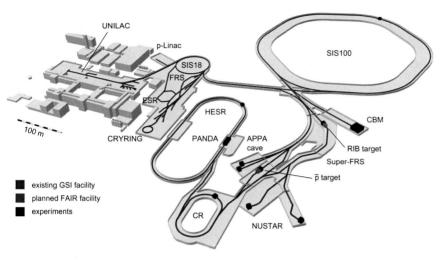

Fig. 6.12 Overview of the FAIR facility. It is composed of a new accelerator stage (red) coupled to the existing GSI facility (blue). At FAIR heavy ions are first accelerated with a LINAC then fully stripped and sent to the SIS18. At the future FAIR facility, not fully stripped ions will be accelerated by the SIS18, then fully stripped and eventually accelerated by the SIS100. At the first stage of its development, the last acceleration stage will be limited to 100 Tm. The facility is meant to reach 300 Tm (SIS300, not shown) on a longer term. Credit: FAIR

The FAIR facility shown in Fig. 6.12 aims at an increase of radioactive beam intensity of a factor 1000 to 10000 compared to GSI. This increase will come from (i) an upgrade of the SIS18 to faster cycles, (ii) an acceleration of ions with lower charges (SIS100) reducing the space charge tune shift of Eq. 6.15 and therefore allowing more ions per bunch, (iii) an increase of the angular and momentum acceptance of the spectrometer following the production target (Super FRS).

6.1.2 The ISOL Method

6.1.2.1 Overview

Radioactive nuclei produced by the ISOL method are stopped in the production target. After extraction from the target, they are ionized and injected after separation into an accelerator to bring them to the desired energy. The final radioactive-ion beam (RIB) intensity is the product of the particular nuclear production cross section σ of a desired isotope, by the interaction of a particle of a certain energy E and the target nucleus, the number of target nuclei, the driver beam intensity $I_{primary}$ and the overall efficiency ϵ. In the ISOL technique, the beam energy can vary significantly inside the target, and therefore the production cross section as well. The beam intensity I_{RIB} is given by

Fig. 6.13 Overview of the radioactive-beam production from the ISOL technique. The combination of the release from the target and the transfer to the ion source often leads to an efficiency $\epsilon_{rel}\epsilon_{trans}$ as low as $10^{-4}\%$. The ionization efficiency ϵ_{ion} can range from 0.1 to 100% depending on the ion and ionization method. The combination of the ion optical transport through the beam line (ϵ_{transp}), the electromagnetic separation (ϵ_{sep}), the storage, cooling and bunching (ϵ_{cool}) and eventual post-acceleration (ϵ_{reacc}) leads to an efficiency ranging from 10 to 100%

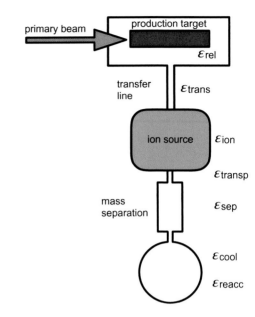

$$I_{RIB} = \left(\int \sigma(E(z)) I_{primary}(z) dz \right)_{target} \cdot N_{target} \cdot \epsilon, \qquad (6.16)$$

where N_{target} is the target density in atoms per volume and ϵ is the total transmission efficiency ranging from 10^{-8} to 10^{-4} %, as illustrated in Fig. 6.13.

6.1.2.2   Release from Target

The release time is determined by the diffusion from the target material, the desorption from the material surface and the effusion to the ion source exit aperture (or any other hole in the target ion source system). Once an ion beam is formed the mass separation and transport goes fast compared to the release time and half lives of the nuclei and no significant losses are expected. The release efficiency ϵ_{rel} strongly depends on the half life of the exotic nucleus of interest. The release time is typically 100 ms $\sim$ 1 s. This is the main limitation of the ISOL production technique. An example of the release time is illustrated in Fig. 6.14 where the release curve of ^{8}Li ions ($T_{1/2} = 840$ ms) from a tantalum foil target is shown.

ISOL target have a design to minimize the diffusion and effusion time toward the transfer line. The ionization stage should be as close as possible from the production to minimize losses.

Fig. 6.14 Release curve of ^{8}Li ions ($T_{1/2} = 840$ ms) from a tantalum foil target. The data represent a measure of the production rate as function of time after proton impact. The line represents a fitted analytical expression taking into account a release time due to diffusion ($\tau = 1$ s), effusion through the target container ($\tau = 83$ ms) and through the ion source ($\tau = 13$ ms). Data from [5] and figure from [6]

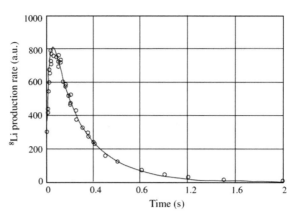

6.1.2.3 Ionization Methods

There is no universal ion source. The method to be used depends on the chemical properties of the element of interest (ionization energy). Indeed, the energy necessary to remove an electron from a neutral atom, the so-called first-ionization energy, varies from ~ 5 eV for Alkali metals to more than 10 eV for rare gases, with a maximum at 24.6 eV for He. The main ionization methods are:

- surface ionization for low ionization energy up to 5–6 eV,
- laser ionization, selective and applicable for ionization energies up to 10 eV,
- electron impact (mostly via plasma), the only solution for an ionization energy larger than 10 eV.

The methods are schematically illustrated in Fig. 6.15 together with the elements to which each method can be applied.

Surface ionization produces positive ions from atoms with a low ionization potential from their interaction with a heated metal surface with a high work function such as tungsten. The work function is the energy needed to remove an electron from a surface into the vacuum outside the solid. The work function W is therefore the difference between the Fermi energy of electrons in the material E_F and the energy of the electron in the electric potential Φ in the vicinity of the surface

$$W = -e\Phi - E_F. \tag{6.17}$$

The surface needs to be heated to increase desorption. This method works only for a small number of elements, typically alkali metals (Li, Na, K, Rb, Cs, Fr) which easily lose their outermost electron. The ionization energy of alkali metals is typically 5–6 eV.

The next is a method of stepwise *resonant laser ionization* via one or more intermediate levels. The method can be applied for most metals (ionization potential of 4–9 eV). The ionization efficiency is typically from 1 to 30%. The laser ionization has

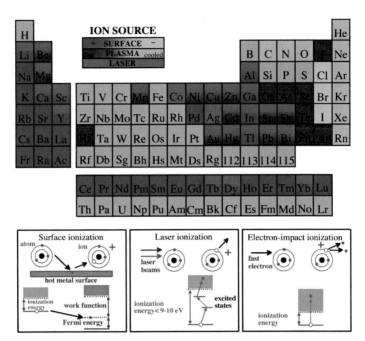

Fig. 6.15 (Top) Ion beams produced at ISOLDE from different ionization techniques. (Bottom) Sketch of the laser (left), laser (middle) and electron-impact (right) ionization techniques

the strong advantage to be chemically selective and minimizes the isobaric contaminants. Mass separation is usually performed by a magnetic spectrometer. In some cases, an isotope selection can be performed (strong isotope shift for light elements) and the method can be used to select isomers using the hyperfine structure. The laser ionization methods can be used to reduce the contamination from alkali elements by (i) reducing the cavity temperature, (ii) choosing low work-function cavity material, (iii) polarize in invert-bias the transfer line.

For chemical elements with ionization energy higher than 10 eV, the ionization can be performed efficiently via the collision of atoms with energetic electrons. This *electron-impact* ionization method requires either an intense electron beam or a hot plasma where the mean kinetic energy of the electrons is high enough to allow the ionization of the elements of interest.

6.1.2.4 Production of a ^{8}He Beam with SPIRAL at GANIL

As an illustration, we describe here the specific example of the production of a ^{8}He^{2+} beam produced at the SPIRAL-GANIL facility, France. A ^{13}C primary beam is produced in an ECR source and accelerated through a set of cyclotrons up to 95 MeV/nucleon after a series of stripping stages. The primary beam is sent onto a carbon

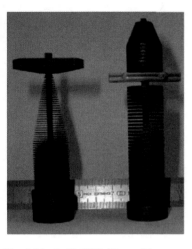
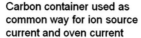
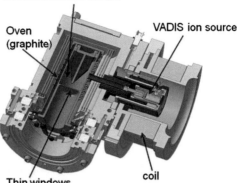

Carbon container used as common way for ion source current and oven current

Oven (graphite)

VADIS ion source

Thin windows

coil

Fig. 6.16 (Left) SPIRAL graphite targets. The target geometry is defined for a shorter diffusion path from the production position to the vacuum of the source chamber. It is optimized on the primary beam and energy to be used, as well as the isotopes to be extracted. The picture shows two targets optimized for the production of neutron-rich He isotopes (left) and Ne isotopes (right). Figure reprinted with permission from [8]. ©2021 by Elsevier. (Right) View of the FEBIAD production ion source where the production target is installed. Figure reprinted with permission from [9]. ©2021 by Elsevier

target shown in Fig. 6.16. Carbon has excellent release properties and can withstand high temperatures thanks to its high sublimation temperature of 3900 K. The design of the target guarantees the production of noble gases with reasonable yields and can be used with high power primary beams. The target temperature should be as high and uniform as possible, in order to minimize the delay time between production and release of atoms. The temperature profile is related to the properties of the Bragg peak. A specific geometry divided into two parts was developed for the ^{6,8}He production due to the long range of He in carbon. The ^{13}C primary beam only heats the first part (production target), while the second one stops the fragmentation products, while being heated by an electric current through the target axis. The target chamber is the production stage of radioactive nuclei in the form of neutral atoms. It has therefore to be followed by an ionization stage which is based on the same techniques as for ion sources. In the case of the SPIRAL target, the radioactive isotopes are transferred via a transfer line to an ECR-type ionization section. The transfer tube from the target chamber to the ECR source is heated to minimize adsorption of atoms on the surface of the tube.

6.1.2.5 Radioactive Beam Production at ISOLDE, CERN

At CERN, low-energy nuclear physics is performed at ISOLDE. Radioactive nuclei are produced from the interaction of the 1.4 GeV proton beam (2 microA, 2.4-

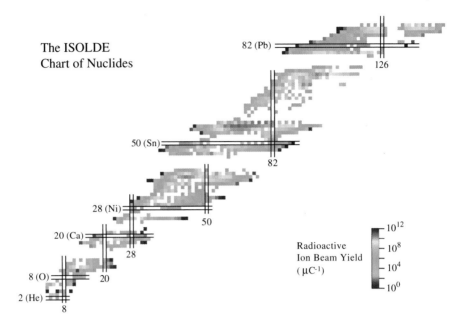

Fig. 6.17 Radioactive beams produced at ISOLDE. All targets and ionization techniques are included. Courtesy Jochen Ballof, CERN

microsecond long bunches of 3×10^{13} protons) from the PS Booster with a production target (for example, UCx for medium-mass neutron rich nuclei). The RIB are mass separated by magnetic spectrometers. They can be cooled and bunched (RFQ cooler-buncher) for low-energy measurements of re-accelerated up to 10 MeV/nucleon (REX-ISOLDE accelerating cavities) for reaction studies. A summary of the beam produced at ISOLDE and the achieved intensities is shown in Fig. 6.17.

6.1.3 The In-Flight Method

6.1.3.1 Production and Momentum Distribution

The in-flight method requires a primary beam momentum per nucleon larger than the typical Fermi momentum of $P_F \sim 200$ MeV/c. When this condition is verified, the nuclear reaction occurring between the projectile and the target can be described as a sudden ablation of nucleons from the projectile at a first step. The velocity of the fragment remains close to the projectile's velocity,

$$\beta_{fragment} \simeq \beta_{projectile}. \tag{6.18}$$

The reaction leads to a momentum distribution of the fragment after ablation of nucleons from a projectile of mass A. Under the sudden approximation, the momentum of the fragment, from which ΔA nucleons are removed, is given by

$$\mathbf{P}_{frag} = -\sum_{i=1}^{\Delta A} \mathbf{P}_i, \tag{6.19}$$

where $\mathbf{P}_i$ is the intrinsic momentum of the removed nucleon i. Under a good approximation, one can assume that the mean intrinsic momentum of the removed nucleons is equal to the Fermi momentum P_F. The width of the momentum distribution of the fragment after ablation is then given by [10]

$$\sigma_{P_{frag}}^2 = \frac{P_F^2}{5} \Delta A \frac{A_{frag}}{A-1}. \tag{6.20}$$

The fragment can be produced in an highly-excited state above the nucleon separation energy, in which case it will decay via either evaporation of nucleons or fission, depending on its structure and excitation energy. The momentum spread will then be increased in comparison to the initial fragment momentum.

Problem

6.1 Derive (6.20) from the sudden approximation of Eq. (6.19). By using Eq. (6.20), calculate the width of the momentum distribution of ^{10}Be produced from two proton removal from ^{12}C at 2.1 GeV/nucleon. Compare to the experimental distribution of Fig. 6.18.

In the case of in-flight fission of the projectile of mass A into two fragments of masses A_1 and A_2 verifying $A = A_1 + A_2$, one can assume that the kinetic energies of the two fragments in the center of mass of the projectile are equal to the Coulomb repulsion of the two fragments at the scission point. This kinetic energy can be approximated as

$$E = \frac{Z_1 Z_2 e^2}{r_0 \times (A_1^{1/3} + A_2^{1/3})} \text{ MeV}, \tag{6.21}$$

with Z_1, Z_2 the atomic number of both fragments and $r_0 \sim 1.8$ fm. Note that in this particular case, r_0 is larger than the standard $r_0 = 1.2$ fm value since at the scission point, the two fragments are generally in a deformed configuration. Numerically, the total kinetic energy to be shared by the two fragments is about 200 MeV, leading to a large momentum spread of the fragments.

The production of radioactive nuclei strongly depends on the chosen combination of projectile and target, and the incident energy. More details on reaction mechanisms are given in Chap. 8. As an illustration, the production of Sn isotopes from the in-flight method is illustrated in Fig. 6.19. The production from fragmentation, i.e., ablation followed by evaporation, leads to a parabola-shaped production of isotopes centered on the isotope with the same A/Z ratios as the projectiles, ^{124}Xe and ^{129}Xe in the cases shown in Fig. 6.19. In-flight fission from ^{238}U favours the production of neutron rich Sn isotopes.

Fig. 6.18 Momentum
distribution of ^{10}Be produced
from the interaction of ^{12}C at
2.1 GeV/nucleon with a ^{9}Be
target

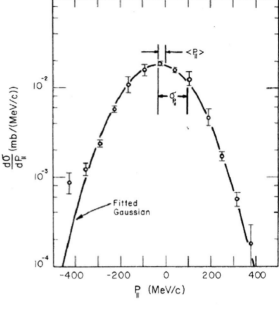

Fig. 6.19 Production cross
section (in barn) of Sn
isotopes from in-flight
fragmentation of Xe primary
beams (circles) and in-flight
fission of ^{238}U (triangles).
Figure reprinted from [11]

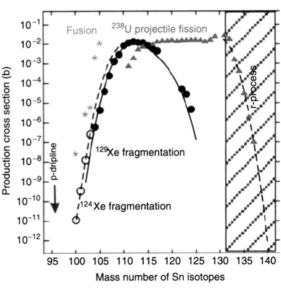

6.1.3.2 Isotope Separation

The above example of fragmentation illustrates that isotopes produced in-flight can have a broad momentum distribution, requiring a large momentum acceptance of the spectrometer in between the production target and the experimental zone to maximize the transmission of the RI of interest. An important challenge of in-flight facilities is the fragment separation. Indeed, after the production target the fragments exit the target together with the primary beam. Most often, the radioactive nuclei of interest are often the most exotic ones produced at very low intensity. The required rejection, the number of ions of interest transmitted to the experimental zone compared to the primary beam intensity, can reach 10^{12}. This is generally made possible by the use of a *fragment separator*, a multi-stage spectrometer. For energies above 100 MeV/nucleon, the separator is composed of a set of magnetic dipoles, quadrupoles and, often, higher multipoles for higher-order trajectory corrections.

The selection of isotopes with a magnetic spectrometer is usually made in three steps: (i) an initial bend for an A/Q separation from the magnetic rigidity and acceptance of the spectrometer, (ii) an energy loss degrader (wedge) for atomic number separation, (iii) a second magnetic bend for an A/Q selection for the isotope of interest.

A degrader is a piece of matter in the beam. Because the atomic slowing down of the ions is roughly proportional to Z^2/v^2 (valid at high energies), different isotopes with different Z will have different velocities after passing through the degrader. This change in velocity causes the isotopes with the same A/Z ratio to be separated in the second dispersive stage.

In practice, degraders often have a non-uniform thickness in the dispersive direction. The degrader can be wedge shaped to keep the achromatic condition for the selected isotopes, i.e., position independent on momentum in the focal plane of the spectrometer. Such an achromatic case is illustrated on the left panel of Fig. 6.20. Because the energy loss in the degrader also depends on the particle velocity, a thicker degrader at the higher $B\rho$ side and a thinner degrader at for lower $B\rho$ values can lead to the same momentum at the focal plane. A degrader that is shaped accordingly is called the achromatic degrader.

The selection in the (N, Z) plane depends on the beam energy because the energy loss in the degrader depends on the velocity. At high beam energies, such as encountered at the FRS magnetic separator at GSI, the use of a wedge leads to a selection in Z. At lower primary beam energies, such as encountered at the BigRIPS separator at the RIBF, the use of a wedge leads to a selection of isotones, i.e., in N. The wedge selection as a function of the energy of fragments is illustrated in the right panel of Fig. 6.20.

The simulated example of the selection of ^{54}Ca after the fragmentation of a ^{70}Zn beam at the RIBF is shown in Fig. 6.21. The first section of the BigRIPS spectrometer, between the production target and the first degrader, provides a first $B\rho$ selection (middle chart). A wedge-shaped degrader at the F1 focal plane of the spectrometer allows to select ^{54}Ca and neighbouring nuclei by the use of slits in the dispersive

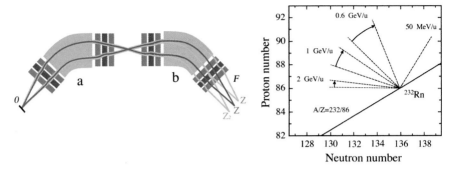

Fig. 6.20 (Left) Sketch of a magnetic spectrometer composed of two dipole (a and b) and a degrader in the dispersive focal plane. (d) The degrader is wedge shaped so that the spectrometer is achromatic: the loci of residues in the focal plane (F) do not depend on the momentum of the fragments. (Right) Selection of a wedge as a function of the fragment energy range. The particular case of ^{232}Rn ($Z = 86$) is shown

direction at the focal plane F7. A second degrader located at the focal plane F5 is used to remove from the beam the fragments produced in the first degrader.

6.1.3.3 Particle Identification

The atomic number (Z) of ions can be determined by measuring their energy loss in sensitive detectors located in the beam path such as ionization chambers, silicon detectors, and plastic scintillator. From the Bethe-Bloch formula, it can be shown (see Sect. 4.5.2.1 in Chap. 4) that the energy loss ΔE in a thin material of thickness Δx is given by

$$E \, \Delta E \propto M Z^2 \Delta x \Rightarrow \Delta E = \frac{Z^2}{\beta^2} \Delta x, \qquad (6.22)$$

when the energy loss is negligible compared to E, and where β is the velocity of the ion in units of speed of light.

The time of flight and position of ions in the dispersive direction allow to determine the magnetic rigidity and velocity of ions, therefore their mass-to-charge ratio (see Fig. 6.22)

$$B\rho = \gamma \beta c \frac{M}{Q}, \qquad (6.23)$$

where M and Q are the mass and charge of the ion, respectively. $\gamma = 1/\sqrt{1 - \beta^2}$ is the Lorentz factor.

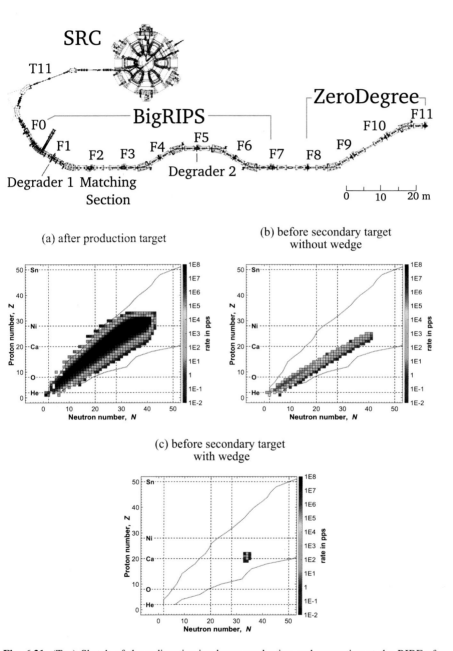

Fig. 6.21 (Top) Sketch of the radioactive-ion-beam production and separation at the RIBF of RIKEN, Japan. The primary beams are accelerated by a LINAC (first stage) and a set of coupled cyclotrons. The last acceleration stage is made by the SRC cyclotron. The beam is then sent onto a production target located at the focal plane F0. The reaction residues are separated from the beam and selected with the BigRIPS fragment separator. Figure modified with permission from [12]. ©2021 by Elsevier. (Bottom) Example of separation of ^{54}Ca ions after the fragmentation of a ^{70}Zn primary beam at 345 MeV/nucleon onto a ^{9}Be target. See text for details. The rates were calculated with the LISE++ simulation program [13]. Courtesy C. Klink, TU Darmstadt

Fig. 6.22 Particle
identification in the A1900
fragment separator at the
NSCL after fragmentation of
a ^{48}Ca primary beam at 140
MeV/nucleon onto a ^{9}Be
primary target. Figure
reprinted with permission
from [15]. ©2021 by the
American Physical Society

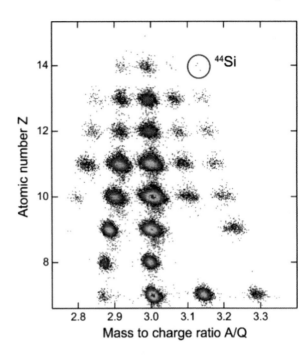

6.2   Density Distributions of Nuclei Near Drip Lines

The nuclear structure of exotic nuclei away from the stability line is far from being
well known today, and nuclear excitations spread over a wide variety that go well
beyond the simple single-particle excitation. Many new aspects of nuclear structure
have been found recently such as halos and skins, di-neutron correlations, and new
magic numbers in nuclei near the drip lines. The richness of the nucleus is such that
most models today can only predict specific types of excitations or observables and
the unified understanding of new observations is still widely open. In addition, it will
be discussed in Chap. 7 that the nuclei can exhibit different geometrical shapes, not
only spherical but also quadrupole deformed, sometimes with extreme deformation
or pear shaped.

6.2.1   Halo Nuclei

6.2.1.1   Introduction

By adding neutrons to (or equivalently removing protons from) stable nuclei, one
can reach the limit of existence of neutron-rich nuclei, the neutron drip-line beyond
which the nuclear interaction is not strong enough to bind additional neutrons to the

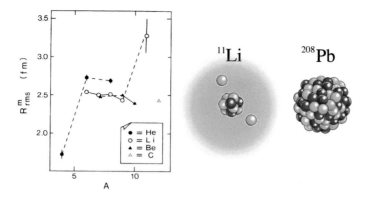

Fig. 6.23 (Left) Interaction radii of light He, Li, Be and C isotopes extracted from reaction cross sections. This historical measurement of interaction cross sections (see Chap. 4) evidenced a large increase of the matter radius of ^{11}Li compared to less exotic lithium isotopes. Figure reprinted with permission from [16]. ©2021 by the American Physical Society. (Right) Artist representation of the halo nucleus ^{11}Li compared to ^{208}Pb

nucleus. At the edge of the neutron drip-line, the neutron radial density can extend at rather large distances from the expected radius of the nucleus in a neutron cloud which was historically named *halo*. This phenomenon is one astonishing illustration of quantum mechanics in nuclei. Imagine a bag filled with balls. Intuitively, the more balls are removed, the smaller the bag becomes. It is indeed what happens when one removes a proton (ball) and two protons from ^{14}C (the bag) to lead to ^{13}Na and ^{12}Be, respectively. However, if one removes a third proton, going from ^{12}Be to ^{11}Li, the "bag" does not get smaller but inflates strongly! Here one reaches at the limit of intuition and enters the subtleties of quantum mechanics. This quantum surprise was first revealed by Isao Tanihata in 1985 at Berkeley, USA, with the first measurement of radii from reaction cross sections of light unstable nuclei (see Sect. 8.2.4 in Chap. 8 for details of the reaction cross section). As shown in Fig. 6.23, ^{11}Li was then revealed to present an extended neutron halo, pioneering the research on unstable nuclei.

6.2.1.2 Neutron-Density Profile of Light Halo Nuclei from Proton Elastic Scattering with the IKAR Active Target at GSI

The density profile of neutron halos can be determined from the differential proton elastic scattering cross section, as described in Chap. 4. Despite its sensitivity to the matter radius and matter density distribution, the proton elastic scattering in inverse kinematics has the kinematical disadvantage that the small center-of-mass angles correspond to close to 90° in the laboratory frame, and leads to low-energy recoil protons. An illustration for ^{6}He on proton at 717 MeV/nucleon is given in Fig. 6.24. In the case a solid target is used, either solid hydrogen or a composite material containing protons such as CH_2, the target has to be thin enough to let the proton

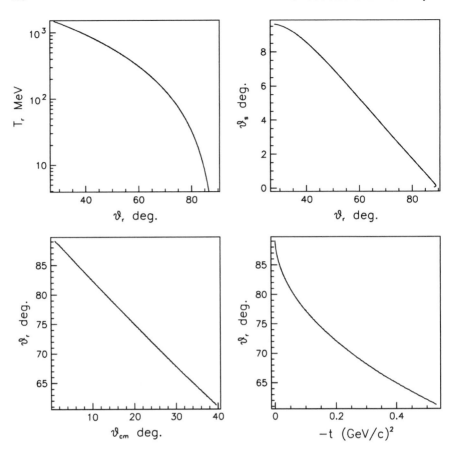

Fig. 6.24 Kinematics of proton elastic scattering on ^{6}He in inverse kinematics with at a ^{6}He beam energy of 717 MeV/nucleon. The elastic scattering cross section is the highest at 0° in the center of mass, which corresponds to 90° in the laboratory (bottom left panel). Close to 90° in the laboratory, the recoil proton energy is small (top left panel), making the measurement difficult. In inverse kinematics, at such high incident energy, the ^{6}He is focused at forward laboratory angles for the full kinematics (top right panel). The recoil proton scattering angle is uniquely connected to the transferred momentum $-t$ during the scattering (bottom right panel). Figure reprinted with permission from [17]. ©2021 by Elsevier

get out with reduced energy loss. This is in most cases at contradiction with the low intensities at which halo nuclei, in the vicinity of the dripline, can be produced.

An elegant experimental solution is to use a time projection chamber (TPC) filled with hydrogen as a drift gas and used as an active target (the drift gas is as well the target for the elastic scattering) for the low momentum transfer part of the kinematics. After an elastic collision, the proton recoils in the hydrogen gas of the TPC and ionize it over the length of its range. Its measured track is then used to determine the scattering angle and, equivalently, transfer momentum. The elastic scattering is

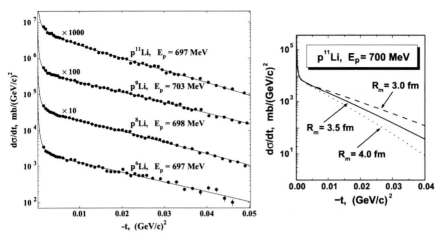

Fig. 6.25 (Left) Proton elastic differential cross sections for ^{6,8,9,11}Li isotopes in inverse kinematics at ~700 MeV/nucleon. The measurements were performed at GSI with the IKAR active target. The differential cross sections are sensitive to the matter distribution of the projectile. (Right) Illustration of the sensitivity to the matter radium R_m: calculated cross sections assuming three different density distributions with $R_m = 3.0$, 3.5 and 4.0 fm are shown. A larger R_m leads to a steeper reduction of the cross section with $-t$. Figures reprinted with permission from [18]. ©2021 by World Scientific

selected from the missing-mass spectrum and by identifying the scattered nucleus at forward angles. In the case of the very exotic nuclei ^{6}He, ^{8}He and ^{11}Li, which have no bound excited state, the detection of the nuclei downstream the reaction target is enough to select the ground sate and therefore exclude inelastic excitations. This method was successfully applied at GSI on several light nuclei with the IKAR TPC, including He and Li isotopes. The differential elastic cross sections, as a function of the transferred momentum $-t$ (Madelstam variable), for Li isotopes from ^{6}Li to ^{11}Li are shown in Fig. 6.25. The sensitivity of the method to the matter radius is illustrated in the right panel of the figure.

When measured with high statistics and over a broad range of momentum transfer, the experimental data can be used to extract a matter density distribution. The result of such an analysis is shown on in Fig. 6.26, where the extracted distributions for ^{6}Li (left panel) and ^{11}Li (right panel). The extracted matter density profiles show a neutron halo for ^{11}Li while the matter density distribution for ^{6}Li is consistent with the charge density distribution extracted from electron elastic scattering (see Sect. 6.2.2).

6.2.1.3   Theoretical Description

We now discuss how the halo wave function originates from the mean field potential. Let us consider one-neutron halo nuclei, for which a weakly-bound valence nucleon moves around the core. A typical example is ^{11}Be, in which the reaction cross section was found significantly large, similar to that of ^{11}Li. The one-neutron separation

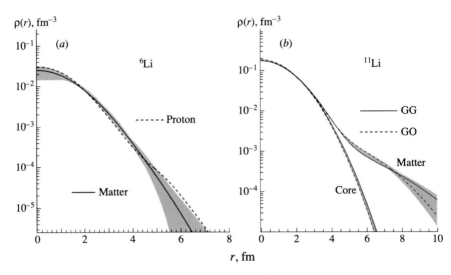

Fig. 6.26 Mass density distributions for ^{6}Li (left) and ^{11}Li (right) extracted from proton elastic scattering measurements in inverse kinematics at $\sim$700 MeV/nucleon at GSI with IKAR. The width of the grey-shaded area represents the statistical and model uncertainties in the extracted density. The matter density distribution extracted for ^{6}Li is consistent with the charge density distribution extracted from electron elastic scattering, providing a benchmark of the method. In the case of ^{11}Li, ^{11}Li is described a ^{9}Li core with two neutrons. The core and whole nucleus matter densities where parameterized. Two parameterizations of the nuclear densities were used. Both the ^{9}Li core and ^{11}Li nucleus densities were extracted from the proton elastic scattering data. The density matter distribution of ^{11}Li presents a halo. Figure reprinted with permission from [19]. ©2021 by Elsevier

energy of ^{11}Be is found to be 500 keV. This small separation energy suggests the halo structure for ^{11}Be and, in fact, the large reaction cross sections at high and medium energies can be uniquely understood by a halo nature of the last neutron wave function shown in Fig. 6.27; the density distribution is extracted by analyzing the reaction cross section data of intermediate energy $E_{lab} = 33A$ MeV, ^{11}Be+^{12}C and ^{11}Be+^{27}Al reactions. The high energy reaction cross sections of ^{11}Be projectile at $E_{lab} = 790A$ MeV are also consistently described by the density shown in Fig. 6.27. Deformation could also lead to an extended size, in a different context to the halo wave function. The density distributions calculated by the deformed Woods-Saxon potential (see Sect. 3.4 for details) are also shown in Fig. 6.27 with the quadrupole deformations $\beta = 0.0, 0.5$, and 0.7 (see Sect. 7.2 in Chap. 7 for detailed discussions on the nuclear deformation). The larger deformation gives the more extended density tail. However, the asymptotic behavior of density at $r > 6$ fm is largely different between the empirical and deformed ones. The HF density of ^{11}Be with a loosely-bound neutron of $l = 0$ having a very small neutron separation energy $S_n = 500$ keV, describes well the asymptotic behavior of the empirical halo density.

Let us estimate how a loosely-bound state gains a large radius solving the Schrödinger equation in the cartesian coordinate, which reads,

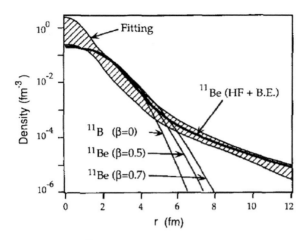

Fig. 6.27 Density distribution of ^{11}Be. The shaded area is the empirical matter density distribution of ^{11}Be extracted from the interaction cross sections of different projectile energies. The thin solid curves are density distributions calculated by the deformed Woods-Saxon potential with different quadrupole deformation $\beta = 0, 0.5$ and 0.7, respectively. The density denoted by (HF+B.E.) is calculated by the HF density of ^{10}Be plus a loosely-bound neutron with $l = 0$ having the neutron separation energy $S_n = 500$ keV. Figure reprinted with permission from [22]. ©2021 by Elsevier

$$\left[-\frac{\hbar^2}{2m}\frac{1}{r}\frac{\partial^2}{\partial r^2}r + \frac{\hbar^2}{2m}\frac{\ell(\ell+1)}{r^2} + V(r) \right]\psi(\mathbf{r}) = \varepsilon\psi(\mathbf{r}) \qquad (6.24)$$

where the wave function $\psi(\mathbf{r})$ is decomposed into the radial and angular momentum parts, $\psi(\mathbf{r}) = R_l(r)Y_{lm}(\hat{r})$. Figure 6.28 shows the radial wave function in logarithmic scale. A Woods-Saxon potential is adopted for $V(r)$ with the radius and diffuseness parameters of $R_0 = 2.74$ fm and $a = 0.75$ fm, respectively. The potential depth is adjusted for each angular momentum l to reproduce the single-particle energy $\varepsilon = -0.5$ MeV. The solid, the dashed, and the dotted lines correspond to the wave functions for $2s$ and $1p$ and $1d$ states, respectively. It is seen clearly that the wave function for $2s$ state is largely extended, while that for $1d$ state is rather compact spatially. The root-mean-square (rms) radii are 7.17, 5.17 and 4.15 fm for $2s$ and $1p$ and $1d$ states, respectively. On the right panel of Fig. 6.28, the rms radii are shown as a function of the single-particle energy ε. The rms radii are diverge for $2s$ and $1p$ states in the limit of $|\varepsilon| \to 0$, while it converges for $1d$ state being almost a constant value as a function of binding energy.

In the region far outside of the nuclear surface $r >> R_0$, the potential becomes $V(r) \to 0$ and the asymptotic behavior of loosely bound wave function of s-state can be obtained from Eq. (6.24) to be

$$R_{l=0}(r) \propto \exp(-\kappa r)/r \qquad (6.25)$$

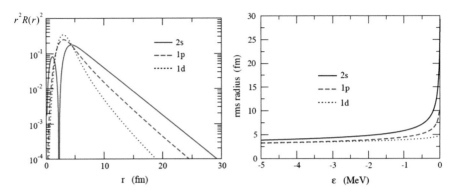

Fig. 6.28 (Left) The radial wave function $r^2 R(r)^2$ of valence neutron and (Right) rms radii for $2s$ and $1p$ and $1d$ states. The solid, the dashed, and the dotted lines correspond to the wave functions for $2s$ and $1p$ and $1d$ states, respectively. A Woods-Saxon potential is assumed for the mean field potential $V(r)$, whose depth is adjusted for each angular momentum l to reproduce the single-particle energy $\varepsilon = -0.5$ MeV

where $\kappa = \sqrt{2m|\varepsilon|/\hbar^2}$ with the single-particle energy ε. The mean square radius of wave function (6.25) will be dominated by the region $r >> R_0$ and given by

$$\langle r^2 \rangle = \frac{\int r^2 dr |R_{l=0}(r)|^2 r^2}{\int r^2 dr |R_{l=0}(r)|^2} \propto \frac{1}{\kappa^2} = \frac{\hbar^2}{2m|\varepsilon|}. \tag{6.26}$$

In the limit of $|\varepsilon| \to 0$, the mean square radius of s-wave state will diverge as a function of $1/|\varepsilon|$. This is also the case of $l = 1$ p-state wave functions since the centrifugal barrier is still very low.[3] For larger angular momentum state, $l > 1$ case, the wave function does not show any divergent behavior because of higher centrifugal barrier. This argument explains the diverging and non-diverging behavior of rms radii shown in the right panel of Fig. 6.28. Thus it is suggested that the halo phenomenon occurs only for nuclei which have the loosely bound wave function with lower centrifugal barrier $l \le 1$, for example, ^{6}He, ^{11}Be and ^{11}Li.

Problem

6.2 Estimate the rms radius of loosely bound halo particle with $\varepsilon = -400$ keV in one-dimensional square well potential

$$V(x) = \begin{cases} \infty & x < 0, \quad \text{(region I)}, \\ V_0(< 0) & 0 \le x \le L, \quad \text{(region II)}, \\ 0 & L < x, \quad \text{(region III)}. \end{cases} \tag{6.27}$$

---

[3] The centrifugal barrier stems from the three-dimensional Schrödinger equation (6.24) as a repulsive term proportional to $\ell(\ell + 1)$.

Fig. 6.29 Transverse-momentum distributions of **a** He fragments from reaction He+C and **b** Li fragments from reaction ^{11}Li+C. The solid lines are fitted Gaussian distributions. The dotted line is a contribution of the wide component in the Li distribution. Figure reprinted with permission from [14]. ©2021 by the American Physical Society

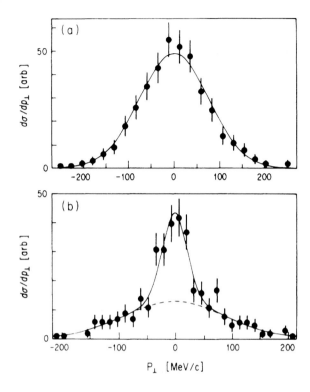

Assume that the wave function in region III determines the rms radius of the halo particle, which can be obtained by solving the one-dimensional Schrödinger equation (1.42) for the bound state.

Projectile fragmentations of ^{11}Li, ^{6}He, and ^{8}He have been measured at 0.79 GeV/nucleon at Berkeley. The momentum distribution of the projectile fragment in the one-nucleon-removal channel provides the momentum distribution of the nucleon in a projectile nucleus. The measured momentum distribution of neutrons from ^{11}Li is shown in Fig. 6.29b. From the quantum mechanical argument of Heisenberg, the width of momentum and radial distributions may be characterized by the uncertainty relation $\Delta p \cdot \Delta x = \hbar$, which means the extended wave function shows a narrow momentum distribution and vice versa. The transverse-momentum distribution of Li shows a two-Gaussian-peak structure as seen in Fig. 6.29b. The width of the wide component $\sigma = 95 \pm 12$ MeV/c gives similar value to that of C fragmentation, whereas the other component shows an extremely narrow width: $\sigma = 23 \pm 5$ MeV/c. The narrow momentum distributions of neutrons from ^{11}Li validate the large extension of two neutron wave functions, i.e., the existence of a halo structure.

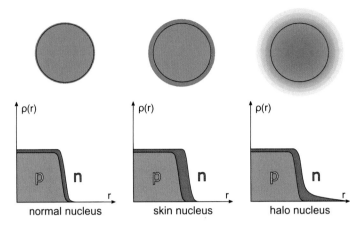

Fig. 6.30 Schematic representation of the proton (purple) and neutron (green) radial densities in a well-bound stable nuclei (left), and neutron-rich nuclei presenting a thick neutron skin (middle) and a halo (right)

6.2.2   Neutron Skins

Neutron skins, corresponding to a neutron density higher than the proton density at the surface of the nucleus, have been evidenced in stable nuclei. The development of neutron skin is illustrated in Fig. 6.30 as well as that of neutron halo in nuclei close to the drip line. The neutron skin is characterized by its thickness, defined by the difference between neutron and proton r.m.s. radii,

$$\Delta r_{\mathrm{np}} = \sqrt{\langle r_n^2 \rangle} - \sqrt{\langle r_p^2 \rangle}, \tag{6.28}$$

where $\langle r_{n(p)}^2 \rangle$ is the neutron (proton) mean square radius. Thick neutron skins, i.e., thicker than 0.2 fm, were observed in light neutron-rich nuclei.

Microscopic models predict the development of thick neutron skins in very neutron-rich medium mass nuclei. Experimental evidence of neutron skin in Sn isotopes are shown in Fig. 6.31 together with RMF and SHF model calculations. Such thick neutron skins would be a unique occurrence of low-density pure neutron matter.

The nature of the neutron skin in nuclei may not be as simple as described by mean field approaches. Spatial correlations including alpha clustering are predicted to take place at sub-saturation densities and therefore at the nuclear surface (see Chaps. 3 and 6 for discussions on the clustering phenomenon). It is expected that clustering and the formation of inhomogeneous matter at low densities modifies the proton density in the tail of the nuclear density and therefore the symmetry energy as compared to calculations assuming an uniform uncorrelated spatial distribution of constituents.

Fig. 6.31 Difference between neutron and proton rms radii Δr_{np} of Sn isotopes calculated with RMF model with NL3 Lagrangian and Skyrme model with SLy4 force. The experimental data for Sn isotopes are measured in (p, p) reaction (open stars), antiproton atoms (full stars) giant dipole resonance method (full circles), and spin-dipole resonance method (full and open squares). Figure reprinted with permission from [33]. ©2021 by the American Physical Society

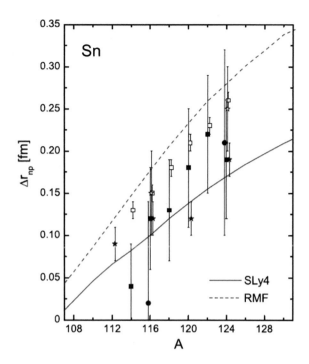

In heavy stable nuclei, the difference between the neutron and proton radii is of the order of few percent. The determination of the neutron skin thickness is indeed precious to learn about the equation of state of neutron-rich nuclear matter at low density, give information on the isospin dependence of the nuclear force and constrain our description of neutron stars. The larger is the pressure of neutron matter at the saturation density, the thicker is the skin as neutrons are pushed out against the surface tension. The nuclear pressure at higher density is responsible to support neutron stars against gravity, while the pressure at lower density below the saturation density will govern the growth of neutron skin thickness.

A strong correlation between the slope of the density-dependent symmetry energy term of nuclear equation of state and the neutron-skin thickness is predicted by mean field models. According to this correlation, one can constrain the symmetry-energy term of the equation of state. In addition, the magnitude of neutron skins may give insight in the nature of three-body forces and also the tensor forces in nuclei. At the same time, the neutron skin may have the influence on the position of nuclear drip lines, and collective nuclear excitations related to the isospin degree of freedom. A more detailed discussion on the symmetry energy and neutron skin is given in Chap. 9.

Several methods have been tested to extract the thickness of the neutron skins. A typical set of methods is based on the combination of two measurements: the measurement of the proton root mean square radius and the measurement of the nuclear matter radius. The proton radius can be measured by proton and electron elastic scatterings. The nuclear matter radius is obtained by heavy-ion reaction cross section measurements. We refer the reader to Chap. 4 for detailed measurement methods.

Indirect methods have also been considered such as the measurement of giant resonances interpreted as mutual oscillations of protons against neutrons like the giant dipole resonance (GDR), the spin-dipole resonance (SDR) or the pygmy dipole resonance (PDR). Although consistent results have been obtained from these resonance experiments and scattering experiments, they appear to be quite model dependent. Finally, the use of antiprotons and study of anti-protonic atoms were shown to be very sensitive to determine the ratio of neutron to proton densities at annihilation site, i.e., at the very surface of the nucleus at about $\rho = \rho_0/1000$, typically at a radius $r = r_{rms} + 2.5$ fm. Up to now, the former technique has been restricted to studies of stable nuclei only.

Both the X rays from the antiprotonic atom decay and the annihilation products have been used to extract matter radii. Despite the acknowledged model dependence of the past analyses, antiprotonic atoms lead to consistent results when compared to other methods. Although not feasible today, the use of this technique makes possible to assess and quantify halos in neutron rich Ne and Mg isotopes and measure neutron skin thicknesses in heavy unstable nuclei such as neutron rich tin isotopes for which a kink in the neutron-skin thickness beyond ^{132}Sn is predicted. The measurement of neutron skins in unstable heavy nuclei at and above Z=100 would offer a new insight into the behaviour of nuclear matter. A new experiment at CERN, PUMA, aims at forming and studying antiprotonic atoms built with short-lived nuclei.

6.2.3   Neutron Droplets

At the extreme of the neutron drip line of the nuclear chart, in the light mass region, the possible existence of a cluster state of four neutrons has been discussed. It has been the *tetraneutron*. A first experimental signature of its existence was reported based on a 2001 experiment at GANIL using a neutron detection after the disintegration of beryllium and lithium nuclei. From this measurement, few events were concluded to be consistent with a four neutron system. However, subsequent attempts to replicate this observation have failed. In 2016 at RIKEN, four events consistent with a tetraneutron as a sharp resonance close to threshold were reported. The experiments was performed using a beam of neutron-rich ^{8}He nuclei impinging onto a ^{4}He liquid target. The objective of the experiment was to produce tetraneutrons from low-momentum transfer and characterize its existence from a missing-mass measurement from ^{4}He nuclei produced from the decay of ^{8}Be. The low statistics of the experiment requires experimental confirmation of the result, while experiments with better statistics are being analyzed. However, current theoretical models with realistic nuclear forces do not support the existence of a tetra neutron as a bound state or a sharp resonance with a width significantly smaller than 1 MeV.

6.3 Evolution of Nuclear Shells

6.3.1 Observations

The shell structure of unstable nuclei with many additional neutrons to stable nuclei has revealed surprises. First of all, the shell structure and known magic numbers are not universal for nuclei beyond nuclear stability. Radioactive nuclei with a different balance of neutrons and protons appears to have a different structure than previously determined from stable nuclei: magic numbers of nucleons may disappear and new ones emerge in different regions of the nuclear landscape.

The shell evolution of neutron-rich nuclei was firstly noticed by Igal Talmi and Issachar Unna, in 1960, for $N = 7$ and $N = 9$ isotones. They showed clear evidence of the inversion of $s_{1/2}$- and $p_{1/2}$-orbits in $N = 7$ isotones and also the inversion of $s_{1/2}$- and $d_{5/2}$-orbits in $N = 9$ isotones as seen in Fig. 6.32. The left panel shows a melting of the shell gap at the magic number $N = 8$, while the right panel indicates the further re-ordering of sd-shell orbits in neutron-rich nuclei. They pointed out that the shell evolution is caused by the monopole interaction (see Sect. 6.3.3), which is still considered as an important driving force.

The one-neutron separation energy S_n is sensitive to the shell structure and may indicate the change of shell closures. The value is large just before a shell closure and decreases suddenly after a shell closure. Thus increasing the neutron number for a given Z, we can see a dip of S_n at the neutron shell closure. A new magic number $N = 16$ was shown in the neutron number dependence of one-neutron separation energies of nuclei with odd N even Z, and also with odd N and odd Z in Fig. 6.33. The known magic numbers $N = 8$, 20 are clearly seen close to the stable nuclei as dips in the small T_z lines. However, the dips at $N = 8$ and $N = 20$ disappear at

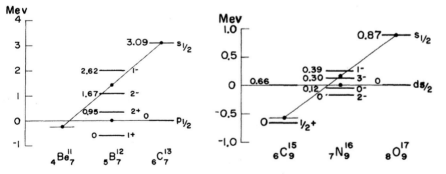

Fig. 6.32 (Right) Inversion of $2s_{1/2}$ and $1d_{5/2}$ orbits along $N = 9$ isotones. (Left) The $2s_{1/2}$ orbit in the sd-shell, which has a higher energy than that of he $1p_{1/2}$ orbit in ^{13}C, goes down below the $1p_{1/2}$ orbit in the proton deficient $N = 7$ isotone ^{11}Be. This change in the ordering of the single-particle states shows clearly a disappearance of the shell gap between the p- and sd-shells. Figure reprinted with permission from [34]. ©2021 by the American Physical Society

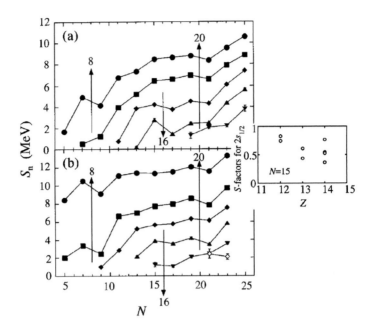

Fig. 6.33 a Neutron N number dependence for experimentally observed one-neutron separation energies S_n for nuclei with odd N and even Z. The closed circles, closed squares, closed diamonds, closed triangles, and closed inverse triangles show different isospin numbers T_z from 1/2 to 9/2. **b** Neutron N number dependence for experimentally observed S_n for nuclei with odd N and odd Z. The closed circles, closed squares, closed diamonds, closed triangles, closed inverse triangles, and open diamonds show different T_z from 0 to 5. The solid lines are guides to the eye. The breaks correspond to magic numbers, as shown by arrows. The inset shows empirical spectroscopic factors (S factors) for nuclei with N =15. Figure reprinted with permission from [35]. ©2021 by the American Physical Society

$T_z = 3/2$ and $T_z = 4$, respectively. On the other hand, a new dip in the S_n line appears at $N = 16$ for $T_z \geq 3$ in Fig. 6.33. This is an indication of a new magic number at N =16 near the neutron drip line. This magic number was further confirmed by the interaction cross section measurements [35] and spectroscopy [36].

Recently, the oxygen isotopes have been studied intensively in both experiments and theories. Along the oxygen isotopes (nuclei with 8 protons), the very neutron rich ^{22}O and ^{24}O with 14 and 16 neutrons, respectively, present features consistent with neutron magic numbers whereas the magic numbers at stability are 2, 8, 20. The shell evolution has been interpreted to originate in specific terms of the nucleon-nucleon interaction, especially those which have a strong attractive nature between protons and neutrons.

Neutron-rich pf shell nuclei also provide an interesting region in the nuclear chart to explore the variations of shell structure. In fact, a possible new magic number at $N = 32$ has been investigated abundantly over the past decades: experimental indications were found for Ca isotopes. The excitation energies of 2_1^+ states of Ca

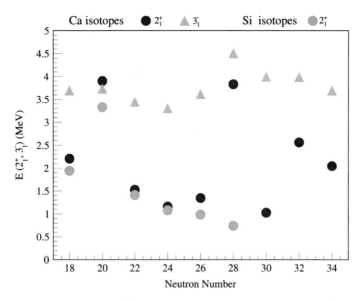

Fig. 6.34 Excitation energies of 2_1^+ and 3_1^- states of even Ca isotopes and 2_1^+ states in even Si isotopes. The doubly-magic stable isotopes ^{40}Ca and ^{48}Ca present a local maximum in the 2^+ systematics. The neutron-rich isotopes ^{52}Ca and ^{54}Ca also present rather high 2^+ excitation energies compared to non-double-magic isotopes. Shell closures at $N = 32$ and $N = 34$ in Ca isotopes have been suggested and are under debate. The excitation energies of 3_1^- states along the Ca isotopic chain is rather constant. Due to parity change, it is expected that this excitation energy is driven by single-proton excitation from the sd shell to the $f_{7/2}$ orbital: the large and stable excitation energy shows that the $Z = 20$ proton shell gap is not strongly affected by the addition of neutrons for the considered isotopes. While $N = 20$ persists as a shell closure in ^{34}Si, ^{42}Si is a well deformed nucleus where quadrupole correlations overcome the $N = 28$ shell gap. $N = 28$ disappears as a shell closure in very neutron-rich nuclei

isotopes are shown in Fig. 6.34. The nature of shell closures manifests itself as a larger excitation energy of 2_1^+ state. We can see evidence of shell closures at $N = 20$ for ^{40}Ca and at $N = 28$ for ^{48}Ca. For ^{52}Ca, the higher excitation energy of 2_1^+ state than the neighboring ^{50}Ca suggests a sub-shell closure of $N = 32$. A shell closure at $N = 34$ along the Ca isotopic chain is also claimed, motivated by a high excitation energy $E_x = 2.043$ MeV for ^{54}Ca and by one-nucleon removal cross sections consistent with a single-particle description of the ground state of ^{54}Ca. These sub-shell structures can be understand naively as the occupations of $2p_{3/2}$- and $2p_{1/2}$-orbits for $N = 32$ and $N = 34$, respectively, above the closed core of $1f_{7/2}$-orbit for ^{48}Ca. Experimentally these sub-shell structures are further confirmed by the mass measurements and the small $B(E2)$ strengths between the ground states and the first 2_1^+ states. Figure 6.34 also shows the systematics of 2_1^+ excitation energies in Si isotopes, to be compared to energies for Ca isotopes. While at $N = 20$ both ^{34}Si and ^{40}Ca show high 2_1^+ excitation energies, their different behaviour is striking for $N = 28$: while ^{48}C shows a large excitation energy consistent with the established shell closure at $N = 28$ for stable

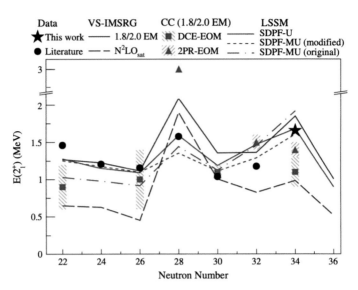

Fig. 6.35 Excitation energies of first 2^+ states in even-even Ar isotopes from ^{40}Ar to ^{52}Ar. Data are indicated with black markers and compared to shell-model predictions with phenomenological (red lines) and chiral-EFT-derived (blue lines) effective interactions, as well as to coupled-cluster predictions (purple symbols). Figure reprinted with permission from [38]. ©2021 by the American Physical Society

nuclei, the 2_1^+ excitation energy for ^{42}Si is low (742 keV) and shows a collapse of the $N = 28$ shell closure for neutron-rich nuclei. The origin of the disappearance of the $N = 28$ shell closure is understood as a coming from an onset of quadrupole collectivity in ^{42}Si together with a slight reduction of the $N = 28$ neutron shell gap, between the $1f_{7/2}$ and the $2p_{3/2}$ orbitals. The disappearance of $N = 28$ as a shell closure was first evidence from the in-beam γ spectroscopy of ^{42}Si at GANIL, France [37].

To illustrate the level of accuracy of state of the art theories for open-shell even-even nuclei, we show in Fig. 6.35 the systematics of first 2^+ excitation energies in Ar isotopes. Note the high excitation energy of the 2_1^+ state in ^{52}Ar, indicating the persistence of the $N = 34$ shell closure beyond ^{54}Ca. Predictions from the shell model with phenomenological interaction (SDPF-U and SDPF-MU interactions) as well as with chiral-EFT-derived interactions (N^2LO_{sat} and 1.8/2.0 EM) interactions are shown together with predictions from the coupled cluster (CC) formalism with the 1.8/2.0 EM ChEFT-derived interaction. In the case of CC, two methods to calculate the open shell nuclei are used. Depending on the formalism used and the interaction, the predictions vary within a typical value of ±500 keV. Effective interactions and methods to solve the nuclear many-body problem are discussed in Chap. 3.

6.3.2 Mean Field Approach for Shell Evolution

Nuclear shell structure was first interpreted in the mean field approximation whose main features are captured by a Woods-Saxon potential with spin-orbit potential. Figure 6.36 shows the neutron single-particle energies for very neutron-rich nuclei of $A = 10 - 90$ with the ratio of mass to proton numbers, $A/Z = 3$. It is known that the energy of the upper most occupied orbits (UMO) always around -8 MeV (indicated by the lower shaded area) for nuclei near the valley of stability line with $A > 10$. One can see clearly for the stable nuclei the closed shell structure at the magic numbers $N = 8$, 20, and 28 as shown also in Fig. 3.2 of Chap. 3. However, the closed shell structure is fragile in nuclei near the neutron drip line. In the upper shaded area with the binding energy below 1MeV in Fig. 6.36, the shell closure is changed from $N = 20$ of stable nuclei to $N = 16$ of weakly bound nuclei. This is due to the lowering of energy of the $2s_{1/2}$ orbit among $sd-$shell orbits. This is a characteristic feature of the single-particle states with low-ℓ values, $\ell \le 1$. Here we can see that the shell evolution is expected from an interchange between the $1d_{5/2}$ and $2s_{1/2}$ states for the case of loosely bound system, i.e., a new shell closure $N = 16$ was indeed predicted based on the mean field picture.

The shell evolution of $N = 20$ and $N = 28$ isotones was studied by deformed HFB model with a Gogny interaction G1S. The model assumes an axial symmetric shape for each nucleus in Fig. 6.37 and looks for the deformation minimum in the

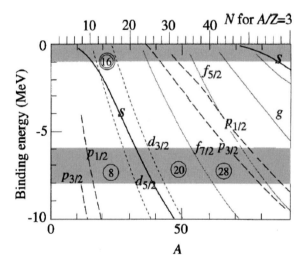

Fig. 6.36 Spectrum of single-neutron orbitals, obtained by the spherical Woods-Saxon potential for nuclei with the ratio of mass to proton numbers, $A/Z = 3$. The lines are labeled by the single-particle orbitals. The magic numbers are indicated by the numbers inside the circles. The shaded areas in the regions of a binding energy of about 6–8 MeV and below 1 MeV correspond to those for typical stable nuclei and for nuclei near the drip line, respectively. Figure reprinted with permission from [35]. ©2021 by the American Physical Society

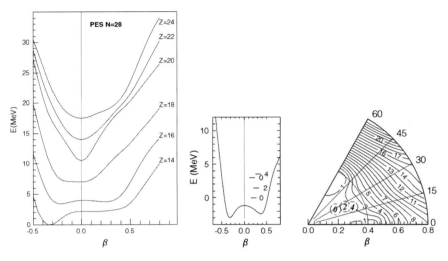

Fig. 6.37 (Left) Potential energy curves of the $N = 28$ isotones, as a function of the deformation parameter β. The minima of the curves are arbitrarily separated by 3.5 MeV. (Middle and right) Axial symmetric, and triaxial potential energy surfaces of ^{44}S. The energies of the quadrupole collective levels are displayed in the middle part. The circles in the β-γ plot indicate the mean deformation and triaxiality of the first 0^+, 2^+ and 4^+ levels. Equipotential lines are shown in 1 MeV intervals. All figures are reprinted with permission from [39]. ©2021 by Springer

potential energy surface (PES). Constrained HFB calculations with the deformation parameter β are shown for six $N = 28$ isotones in Fig. 6.37: ^{42}Si, ^{44}S, ^{46}Ar, ^{48}Ca, ^{50}Ti and ^{52}Cr. A nucleus ^{52}Cr is on the line of β-stability and will be considered as a reference nucleus. Three nuclei (^{48}Ca, ^{50}Ti and ^{52}Cr) display spherical ground states. ^{48}Ca being doubly magic is the most rigid of all six nuclei. For ^{46}Ar, the ground state corresponds to the very shallow oblate minimum near $\beta = -0.25$. For ^{44}S, the PES displays two minima in which an oblate minimum at $\beta = -0.2$ is 2 MeV deeper than a local prolate minimum at $\beta = 0.3$. The β-γ plane PES of this nucleus in the right panel shows that these two minima are genuinely axial ones, i.e., they are not connected by any triaxial valley and could be a candidate of shape coexistence. The absence of any spherical HFB minima in ^{46}Ar, ^{44}S, and ^{42}Si shows that the $N = 28$ spherical shell closure is broken.

6.3.3 Shell Model Approach: Monopole Interaction and Tensor Force

The shell evolution is also discussed in term of the monopole energy due to the tensor force by Otsuka et al.. The importance of the monopole interaction on the shell evolution was already pointed out by Talmi and Unna in the 1960th. The monopole interaction is defined as

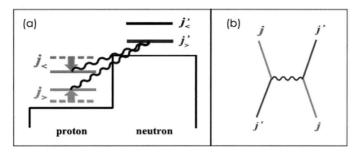

Fig. 6.38 a Schematic picture of the monopole interaction produced by the tensor force between a proton in $j_{>(<)} = l \pm 1/2$ and a neutron in $j'_{>(<)} = l' \pm 1/2$. **b** Feynman diagram of exchange process contributing to the monopole interaction by the tensor force. Figure reprinted with permission from [40]. ©2021 by the American Physical Society

$$V_{j,j'}^T = \frac{\sum_J (2J + 1)\langle (jj')JT|V|(jj')JT\rangle}{\sum_J (2J + 1)}, \qquad (6.29)$$

where $\langle (jj')JT|V|(jj')JT\rangle$ stands for the (diagonal) matrix element of a state. Two nucleons (jj') are coupled to an angular momentum J and an isospin T for the expectation value of a two-body interaction V. In the summation in Eq. (6.29), J takes values satisfying antisymmetrization. The monopole interaction is nothing but the interaction contributions to the HF energy added to the single-particle kinetic energy. If neutrons occupy j', the energy shift of the proton orbit $j (\neq j')$ is given by

$$\Delta \varepsilon_p(j) = \frac{1}{2}(V_{j,j'}^{T=0} + V_{j,j'}^{T=1})n_n(j'), \qquad (6.30)$$

where $n_n(j')$ is the occupation probability the of neutron j' orbit. The same formula can be obtained for $\Delta \varepsilon_n(j')$ replacing $n_n(j')$ by $n_p(j)$ in Eq. (6.30). The IS $T = 0$ channel is much stronger than IV $T = 1$ channel for the exchange interaction between j and j' in the right panel of Fig. 6.38. That is, the energy shift (6.30) for pn channel, having $T = 0$ and 1 contributions, is much larger than the one for nn and pp channels with $T = 1$ contribution only.

The monopole interaction is attractive (negative) between proton (neutron) $j_> (j'_>)$ and neutron (proton) $j'_< (j_<)$ orbitals, while repulsive (positive) between between proton $j_> (j_<)$ and neutron $j'_> (j'_<)$ orbitals due to the nature of tensor interaction. Therefore, if the neutron $j'_>$ orbital is gradually occupied, the proton $j_<$ orbital becomes more bound, while the proton $j_>$ orbital is less bound. That is, the energy splitting between proton $j_>$ and $j_<$ spin-orbit partners is narrower. Alternatively, if the neutron $j'_<$ orbital is gradually occupied, the splitting between proton $j_>$ and $j_<$ spin-orbit partners becomes wider.

Figure 6.39 shows N and Z number dependence of single-particle energies (SPEs). Figure 6.39a shows SPEs of the proton $1d_{5/2}$ and $2s_{1/2}$ orbitals relative to $1d_{3/2}$ as a function of N. As more neutrons occupy the $1f_{7/2}(j'_>)$ orbit, these proton orbits are

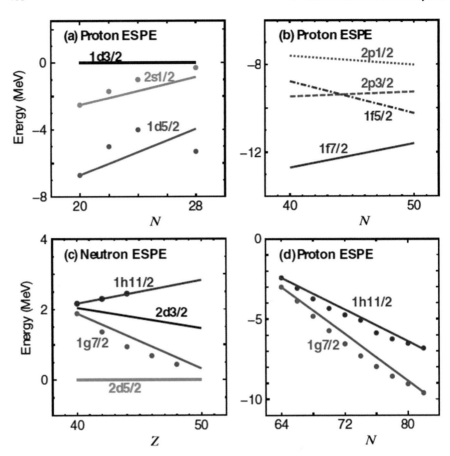

Fig. 6.39 Proton (neutron) SPE as a function of $N(Z)$. Lines in **a**, **b** and **c** show the change of SPE's calculated from the tensor force. Points represent the corresponding experimental data. **a** Proton ESPE's in Ca isotopes relative to $1d_{3/2}$ orbit. **b** Proton SPE's in Ni isotopes. **c** Neutron SPE's in $N = 51$ isotones relative to $2d_{5/2}$. **d** Proton ESPE's in Sb isotopes. Lines include a common shift of SPE as well as the tensor force effect. Figure reprinted with permission from [40]. ©2021 by the American Physical Society

shifted, i.e., the spin-orbit splitting between $1d_{5/2}$ and $1d_{3/2}$ is smaller. The empirical SPEs of Ca isotopes shown by dots follow this trend in a reasonable way. The same thing happens on pf-shell orbits in Fig. 6.39b. In this case, the $1g_{9/2}$ neutron orbit is gradually filled from $N = 40$ to $N = 50$ of Ni isotopes and $1f$ and $2p$ spin-orbit become narrower. It is seen that the shell gap in ^{68}Ni is much narrower in ^{78}Ni.

From $Z = 40$ to 50 in Fig. 6.39c, the $1g_{9/2}$ orbit is filled by protons. Through the tensor force, these protons lower the neutron $1g_{7/2}$, while they push up the $1h_{11/2}$. Such changes of SPEs are shown starting from experimental values at $Z = 40$. The empirical spacing change between neutron $1g_{7/2}$ and $1h_{11/2}$ SPEs is explained well by the calculations. Figure 6.39d shows proton $1h_{11/2}$ and $1g_{7/2}$ SPEs as a function of N. The changes due to the tensor force are indicated starting from experimental values for $N = 64$. The calculated results give a good account of experimental N dependence.

6.3.4 3N Force and Shell Evolution

The nucleus ^{18}O is the heaviest stable oxygen isotope: if one adds neutrons to produce heavier oxygen isotopes, they appear to be unstable and decay very quickly. The neutron drip-line is located at ^{24}O along the oxygen chain. This fact has been unexplained by most theories for a long time. Recently, a precise prediction of the neutron drip line for these oxygen nuclei has been given by an ab initio coupled cluster model approach with three-body force. The model adopted the interactions from chiral effective field theory and computed binding energies, excited states, and radii for isotopes of oxygen. The binding energies of oxygen isotopes are shown in Fig. 6.40 with and without the 3-body force. The calculation includes the effects of three-nucleon forces and of the particle continuum, both of which are claimed to be important for the description of neutron-rich isotopes in the vicinity of the nucleus ^{24}O.

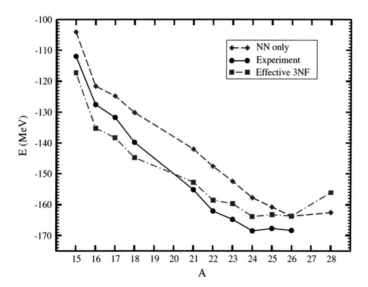

Fig. 6.40 Ground-state energy of the oxygen isotopes ^{A}O as a function of the mass number A. Black circles: experimental data; blue diamonds: results from nucleon-nucleon interactions; red squares: results including the effects of three-nucleon forces. The resonant ground state of ^{26}O at the energy of 18 keV above the two-neutron threshold was found at the RIBF in 2016. Figure reprinted with permission from [23]. ©2021 by the American Physical Society

The intrinsic three-body force originates from the internal structure of nucleons (see Sect. 2.6 of Chap. 2). In shell models, additional three-body force will be introduced by the renormalization procedure of realistic two-body force into effective interactions. The latter, the three-body force induced by the renormalization is model-dependent. The treatment and effects of three-body forces are presently at the heart of modern nuclear physics. Among related researches, an experimental program is carried at the RIBF of RIKEN with the objective to complete the systematics of the binding energy of oxygen isotopes from ^{14}O to ^{28}O. In 2020, these binding energies are known up to ^{26}O. Such a measurement for ^{28}O requires to populate ^{28}O from a nuclear reaction and measure all its decay particles ^{24}O and four neutrons to reconstruct its binding energy. The unique capabilities of the RIBF to produce radioactive beams in association with the most efficient detection devices gathered by an international collaboration made this measurement possible, today under analysis. It is important to quantify the role of three-body forces in nuclei as illustrated in Fig. 6.40. Indeed, the effect of three-body forces is magnified for very neutron-rich nuclei. The calculation with three-body forces predicts the correct location of the neutron drip-line at ^{24}O.

6.4 Island of Inversion

The shell structure can change very much towards the neutron drip line as mentioned already in Sect. 6.3. The shell evolution is interpreted as the result of competition between the shell gaps, and the monopole and also multipole correlations. The observation of irregularities in the binding energies of neutron-rich $A = 32$ nuclei and the suggestion of their deformations was first made by Catherine Thibault et al., in 1975. These authors measured the binding energies of Na isotopes and noted that ^{31}Na and ^{32}Na are considerably more bound than predicted theoretically. Xavier Campi et al. in 1975 performed constrained Hartree-Fock calculations on these Na isotopes and obtained results supporting this hypothesis. In particular they found large prolate deformations in ^{31,32}Na when the occupation of neutrons change from the $1d_{3/2}$ orbit to the $1f_{7/2}$ orbit. Later, it was found that both ^{31}Mg and ^{32}Mg are also much more bound than expected by the mass model evaluation. In addition, the first-excited state of ^{32}Mg was found to lie at the remarkably low energy of 885 keV compared with other $N = 20$ isotones, clearly indicating nuclear deformation because of Fig. 6.41. A large $B(E2)$ value was also observed by Coulomb excitation experiments by Tohru Motobayashi et al., $B(E2; 0_1^+ \rightarrow 2_1^+) = (454 \pm 78)$ e^{2}fm^4, and an experimental deformation parameter $\beta_2 = (0.522 \pm 0.041)$ is extracted. The low-energy and large $B(E2)$ value suggest an existence of large deformation and, at the same time, a disappearance of $N = 20$ magicity.

Shell model study of this anomaly in $A \sim 32$ region was performed in the $(2s, 1d)$ model space by Hobson Wildenthal and Wilton Chung in 1980. They found that the binding energies of $N = 20$ Na and Mg isotopes could not be understood on the basis of standard shell model interaction and named an "island of inversion", which is illustrated in Fig. 6.42. This abnormal phenomenon implies a melting of the $N = 20$ shell closure near the neutron drip line. In other words, the shell structure is broken by the deformation effect in nuclei near the island.

There are mainly two reasons which give rise to these anomalies in the ground states. One is the change of single-particle energies near the Fermi energy by the large neutron excess, which causes the loosely bound nature of single-particle states near the neutron threshold. Thus, it turns out that the diminishing or smaller energy gap between $1f_{7/2}$ and $1d_{3/2}$ neutron states than that of stable nuclei. Another factor is the strong $T = 0$ proton-neutron (p-n) interaction, which is a major trigger to cause the deformation in a wide region of mass table. This interaction is strong when the overlap between n-p wave functions is large. Certainly this is the case for the interaction between $\pi 1d_{5/2}$ and $\nu 1d_{3/2}$ states, and to a large extent, between $\pi 1d_{5/2}$ and $\nu 1f_{7/2}$ states.[4]

A systematic study of inversion of $2p$-$2h$ and $0p$-$0h$ states in $N = 20$ isotones was performed by shell model calculations with/without the correlations between $2p$-$2h$ and $0p$-$0h$ states. The results in Fig. 6.43 show a clear sign of the inversions at $Z = 10$, 11 and 12 isotones.

Two configurations for $^{32}_{12}\text{Mg}_{20}$ are displayed for 20 neutrons (Fig. 6.44, upper left); the closed-shell configuration (A), and two-particle-two-hole configuration (B) consisting of two neutrons excited from the $1d_{3/2}$ and $1s_{1/2}$ orbitals into the $1f_{7/2}$ and $2p_{3/2}$ orbitals across the $N = 20$ shell gap. Near the shell gaps, configurations of type (B) usually appear as excited states and are sometimes called pairing vibrational states. The energy of this state is lower than twice the shell gap due to the pairing correlations between the particles and between the holes.

In the island of inversion, configurations of the type (B) become the ground state rather than the excited state. This change is sudden, with (B) forming an excited state in $^{34}_{14}\text{Si}_{20}$ and then becoming the ground state for $^{32}_{12}\text{Mg}_{20}$ as is shown in Fig. 6.41.

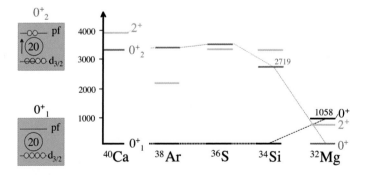

Fig. 6.41 Evolution of the 2_1^+ and 0_2^+ energies in the $N = 20$ isotones. A sudden inversion between the $0p - 0h$ and $2p - 2h$ configurations occurs between the ^{34}Si and ^{32}Mg nuclei

---

[4] The proton and neutron state are often denoted by "π" and "ν", respectively in the shell model configuration.

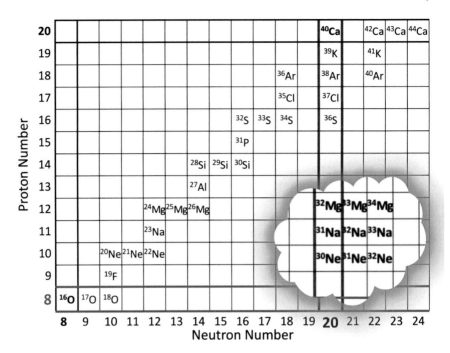

Fig. 6.42 Partial periodic table highlighting the "island of inversion" centered at ^{32}Na. The magic numbers $Z = 8$ and $N = 20$ are emphasized with double lines. Apart from the "island", only stable nuclei are shown. Figure reprinted with permission from [24]. ©2021 by the American Physical Society. Courtesy B.A. Brown, MSU

Fig. 6.43 Energy differences between $2p\text{-}2h$ and $0p\text{-}0h$ states in $N = 20$ isotones. The red circles are the results with the correlations between $2p\text{-}2h$ and $0p\text{-}0h$ states, while the black squares do not taken into account the correlations. Figure reprinted with permission from [58]. ©2021 by the American Physical Society

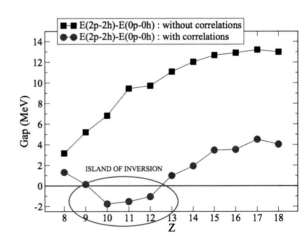

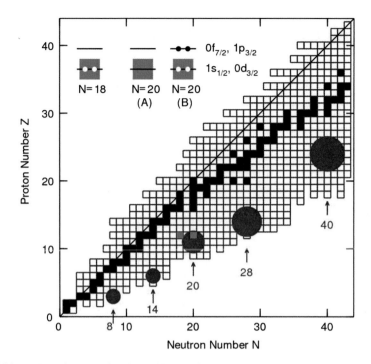

Fig. 6.44 Nuclear chart showing the stable nuclei (black) and those predicted to lie inside the proton and neutron drip lines (open blue). The archipelago of islands where the shell-model magic numbers are broken is shown in red. The shell model structure for neutrons for ^{30}Mg and ^{32}Mg involved in REX-ISOLDE experiments are shown in the upper left-hand corner. Courtesy B. A. Brown, MSU

As mentioned earlier, a gradual reduction in the spherical $N = 20$ shell gap as one approaches the neutron drip line, and the strong quadrupole correlations lead to a deformed configuration in $^{32}_{12}$Mg$_{20}$.

Many experiments have studied the states in the island of inversion with configurations similar to (B). The work at REX-ISOLDE at CERN provides the first confirmation that the predicted coexisting excited state (A) exists and shows some of its properties. It is the first experiment of its type where a radioactive beam of ^{30}Mg reacts with a target of triton, ^{30}Mg$(t, p)^{32}$Mg, using an inverse kinematics. When two neutrons are added to make a final state with spin and parity $J^\pi = 0^+$, there are two possibilities to make final states; either they can go into the $1d_{3/2}, 2s_{1/2}$ orbitals, making state (A), or into the $1f_{7/2}, 2p_{3/2}$ orbitals, making state (B). Kathrin Wimmer and collaborators observed two 0^+ states: the ground state and an excited state at 1.058 MeV. The energy of the excited state, which is presumed to correspond to configuration (A). The simple mind estimate based on the systematics for the energy of (B) from Fig. 6.41 is around 2 MeV above the ground state (A).

The full GCM, beyond-mean-field (BMF) calculation (see subsection 3.5.3) which incorporates configuration mixing using the finite-range density-dependent Gogny

Table 6.2 Results from GCM, beyond-mean-field calculations with Gogny force, for ^{30}Mg and ^{32}Mg (indicated as "Theo.") compared to experimental values ("Exp."). The monopole transition $\rho^2(E0)$ between $0_2^+ \to 0_1^+$ at 1788 keV is also shown together with $B(E2)$ transitions. Shell model results (SM) are also given in ^{32}Mg by [25]. The data are taken from [26] and [27]

		$E_x(2_1^+)$	$E_x(0_2^+)$	$B(E2 : 0_1^+ \to 2_1^+)$	$\rho^2(E0) \times 10^{-3}$	$B(E2 : 0_2^+ \to 2_1^+)$
		(MeV)	(MeV)	$(e^2\text{fm}^4)$		$(e^2\text{fm}^4)$
^{30}Mg	Theo. GCM	2.03	2.11	334.	6 46	181.5
	Exp.	1.482	1.788	241(31)	26.2(7.5)	53(6)
^{32}Mg	Theo. GCM	1.35	2.60	455.7	41	56.48
	Theo. SM	1.17	3.1	449.		
	Exp.	0.885	1.058	454(78)		<544

force with the D1S parametrization, reproduces reasonably this coexistence scenario as tabulated in Table 6.2. The BMF model predicts an excitation energy of 2.11 MeV for the 0_2^+ state in ^{30}Mg and only weak mixing between the two 0^+ states. In ^{31}Mg a recent measurement of the ground state spin $J^\pi = 1/2^+$ could only be explained by an intruder configuration in the ground state of ^{31}Mg as a result of strong prolate deformation $\beta_2 \sim 0.45$, thus placing ^{31}Mg exactly on the border of the island of inversion. The large $B(E2; 0^+ \to 2^+)$ for ^{32}Mg has clearly established its strongly deformed ground state in comparison with that for ^{30}Mg. The excited 0_2^+ observed at $E_x = 1.058$ MeV is nevertheless still much lower than the BMF prediction of 2.60 MeV. Modern shell model (SM) calculations that correctly describe the deformed ground state in ^{32}Mg predict the spherical 0_2^+ state at $E_x = (1.4 \sim 3.1)$ MeV. The energy of the excited state, which is presumed to correspond to configuration (A), is lower than any of the theoretical predictions mentioned in this section. Understanding the reason for this disagreement with theory will be crucial for improving the many-body models as they are used to predict the properties of nuclei in even more neutron-rich nuclei.

The island of inversion at $N \sim 20$ is now known to be part of an archipelago of "islands of shell breaking" associated with the magic neutron numbers 8, 14, 20, 28, and 40 as shown in Fig. 6.44. In order to accurately account for the effects of interactions and understand how the shells break down, more high-precision spectroscopic experiments are being pursued in these regions.

6.5 Di-neutron Correlations and Nuclear Superfluidity

6.5.1 Pairing Correlations

Pairing correlations act between pairs of spin-up and spin-down nucleons coupled with the total angular momentum $J = 0$. The importance of pair correlation has been widely recognized for nuclear many-body systems in various circumstances, in particular in open-shell nuclei and also in the inner crust of neutron stars (see Chap. 9). Typical phenomena of the superfluidity were already pointed out for the energy difference between the ground and the first excited states, the moment of inertia of deformed nuclei and the mass difference between even and odd nuclei (see Sect. 3.6.3 of Chap. 3 for details). Pairing correlation at low nucleon density is of special interest, since the theoretical predictions for low-density uniform matter suggest that the pairing gap[5] at around 1/10 of the normal nuclear density may take considerably larger values than that at the normal density.

Two types of pairing correlations have been known in condensed matter physics. One is called BCS (Bardeen-Cooper-Schrieffer) correlations in the weak pairing regime and another one is BEC (Bose-Einstein condensation) correlations in the strong pairing regime. BCS correlations take place in the momentum space so that the coherence length, which is related to the characteristic Cooper pair size, is very large in the coordinate space. This feature can be understood by an intuitive argument based on the Heisenberg's uncertainty principle,

$$\Delta r \, \Delta p \sim \hbar.$$

The correlation in the momentum space gives a very small momentum fluctuation Δp, which suggests a large fluctuation in the coordinate space Δr and the coherence length becomes very large (see Fig. 6.45).

On the one hand, the BEC correlation is a strong correlation in the coordinate space so that the coherence length is very small. The pairing correlations for stable nuclei discussed in Chap. 3 are the BCS-type correlations. On the other hand, the di-neutron in unstable nuclei, which will be presented hereafter, has a long tail of dilute density and show a similarity to BEC correlations since the pairing correlation is very strong in the region of dilute density.

We now discuss the pairing correlations in infinite matter to find out density dependence (or equivalently momentum dependence) of the correlations. In uniform nuclear matter and neutron matter, the gap equation and the quasi-particle energy are given in the momentum space as

---

[5] The pairing gap indicates the strength of pairing correlations between two nucleons in the pairing models. Experimentally, the paring gap is observed in the binding energy difference between even-even (N, Z) nuclei and neighbouring even-odd $(N, Z - 1)$ or odd-even $(N - 1, Z)$ nuclei.

BEC ←――――――――――――――――→ **BCS**

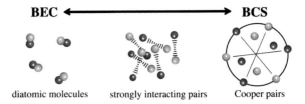

diatomic molecules   strongly interacting pairs   Cooper pairs

Fig. 6.45 The Bardeen-Cooper-Schrieffer (BCS)-Bose-Einstein condensation (BEC) crossover phenomenon. BCS superconducting phase is correlations in the momentum space, while BEC superconducting phase is those in the coordinate space. In the cold atoms, the cross-over is controlled changing the pairing strength using the novel Feshbach resonance technique. In nuclear physics, the density dependence of pairing correlations may take place a similar role to induce the crossover phenomena in halo nuclei, and also in the crust of neutron stars

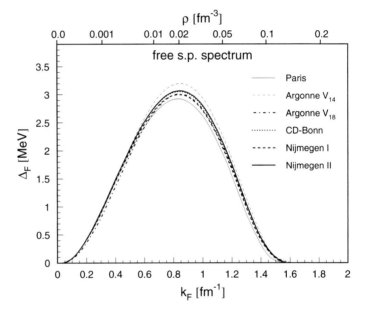

Fig. 6.46 Pairing gap Δ_F of $T = 1$ $^1S_0$ channel in the neutron matter as a function of the Fermi momentum k_F with several realistic NN interactions. The gap is calculated with BCS approximation with the effective mass m^*/m being unity (so called free single-particle spectrum). The equivalent neutron matter density ρ is also given in the upper axis. Figure reprinted with permission from [28]. Courtesy H. -J. Schultz, INFN Catania

$$\Delta(\mathbf{k}) = -\frac{1}{2} \sum_{k'} \langle \mathbf{k'} | \widetilde{V}(\mathbf{k} - \mathbf{k'}) | \mathbf{k} \rangle \frac{\Delta(\mathbf{k'})}{E(\mathbf{k'})}, \qquad (6.31)$$

$$E(\mathbf{k}) = \sqrt{[\epsilon(\mathbf{k}) - \mu]^2 + \Delta(\mathbf{k})^2}, \qquad (6.32)$$

where $\Delta(\mathbf{k})$ is the pairing gap at the single-particle momentum $\mathbf{k}$, $\epsilon(\mathbf{k})$ is the single-particle energy, μ is the chemical potential, and $E(\mathbf{k})$ is the quasiparticle energy.

The pairing gap is a critical quantity to indicate the phase transition from normal to superfluid nuclei. A realistic or phenomenological nucleon-nucleon interaction in the ^1S channel will be adopted as the two-body interaction $\tilde{V}(\mathbf{k} - \mathbf{k}')$. The gap equation is solved under the constraint of the number density

$$\rho\left(\equiv \frac{k_F^3}{3\pi^2}\right) = 2\sum_{\mathbf{k}} v(\mathbf{k})^2 = 2\frac{1}{2}\sum_{\mathbf{k}}\left[1 - \frac{\epsilon(\mathbf{k}) - \mu}{E(\mathbf{k})}\right], \tag{6.33}$$

where $v(\mathbf{k})^2$ is the occupation probability. Equation (6.33) provides the relation between ρ and the chemical potential μ. Notice that sum of Eq. (6.33) is proportional to the occupation probability $v(\mathbf{k})^2$ of BCS approximation.

Figure 6.46 shows the pairing gap Δ_F for neutron matter calculated with several realistic NN interactions. In general, the gaps of neutron matter show a similar overall density dependence for all realistic interactions. This is because the pairing gap is essentially determined by the ^1S phase shift function and all realistic interactions reproduce the experimental phase shift well. It is also confirmed that phenomenological interactions, such as the Gogny interaction, show the same density dependence of pairing gap.

The pairing gap becomes the maximum around $\rho/\rho_0 \sim 0.2$ in all the cases in Fig. 6.46. The gap decreases gradually with further decreases in the density. This feature of strong pairing correlations at the density $\rho \sim 0.02\,\text{fm}^{-3}$, i.e., $\rho/\rho_0 \sim 0.2$ is expected to have direct relevance to the properties of neutron stars, especially those associated with the inner crust. The strong pairing at low density may also be relevant to finite nuclei, if one considers neutron-rich nuclei near the drip line. This is because such nuclei often have a feature of low-density distribution of neutrons, so called the neutron skin or the neutron halo, surrounding the nuclear surface.

6.5.2 Three-Body Model and Borromian Nuclei

Arkady Beynusovich Migdal in 1973 suggested that the two-neutron system may be bound under the environment of dilute nuclear medium. This system is named "*di-neutron*" and the correlations which make two-neutrons bound are called the di-neutron correlation. Since the pairing correlations will be enhanced strongly in the dilute density region, two-neutrons may bound in the neutron-skin and/or halo nuclei far from the stability line. Especially the di-neutron correlations have been discussed in so-called *Borromean* nuclei such as ^{6}He and ^{11}Li, in which the core (^{4}He or ^{9}Li) plus one nucleon is unbound, but the core plus two nucleons is bound. This feature is identical to Borromean rings which consist of three topological circles linked together, while none of the pairs is linked. The name "Borromean rings" comes from their use in the emblem of the aristocratic Borromeo family in Northern Italy (Fig. 6.47).

Fig. 6.47 The term
"Borromean rings" is known
in the knot theory. The
Borromean rings consist of
three topological circles
which are linked but when
removing any one ring leaves
the other two unconnected,
but nonetheless all three are
linked. The name of
Borromean comes from the
emblem of Borromeo family,
the 13th century in Northern
Italy. ^{11}Li is analogous to the
Borromean rings since ^{10}Li
is unbound (two-boby
system), but ^{11}Li is bound as
a three-body system

2n Halo Nucleus

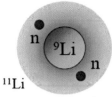

$S_{2n}=300keV$

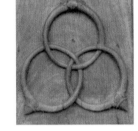

Knot theory   Emblem of Borromeo
Borromean Ring   Family, Italy

6.5.3   Di-neutron Correlations in ^{11}Li and Three-Body Model

On average nuclear radii increase as the one-third power of the nucleon number
A. It was therefore surprising that the measured interaction cross sections of Li
isotopes colliding with nuclear target ^{12}C displayed a very large increase for ^{11}Li.
This measurement was carried out by Tanihata et al., in 1985. This discovery was
soon followed in 1987 by the interpretation of the structure of this nucleus in terms of
a neutron halo, a dineutron weakly bound to the ^{9}Li core, which was first discussed
by Gregor Hansen and Björn Jonson. The large radius was explained by the weak
binding between ^{9}Li and the two neutrons with the two-neutron separation energy
$S_{2n} \simeq 300$ keV.

For the light neutron-rich nucleus ^{11}Li, a three-body model consisting of two
valence nucleons and an inert core was adopted to describe its ground state by
George Bertsch and Henning Esbensen. They handled a large single-particle model
space using a Green function technique with a contact effective interaction. Other
theoretical models such as the Faddeev three-body model, and the cluster orbital
shell model with the complex scaling method have also been used and are equally
successful to describe the exotic nuclei such as ^{11}Li and ^{6}He. These three-body
models are contrasted with the no-core shell model calculations, which do not require
an inert core, but have difficulty to deal with the large extension of the weakly bound
neutrons. Here we adopt the three-body model with Green's function approach in

which all the parameters in the Hamiltonian are constrained by the empirical single-particle energies of neighboring nuclei and the total binding energy. In this model, the Pauli blocking between the core and the valence particles is treated exactly, while it is evaluated approximately in the Faddeev model. Moreover we can describe not only the ground state, but also excited states in a consistent framework without any heavy computational burden.

In the three-body model for the A-body system, which consists of a structureless core nucleus with the number of core nucleons A_c and two valence neutrons, the model Hamiltonian reads

$$
H = \frac{p^2}{2m} + V_{nC}(\mathbf{r}_1) + \frac{p^2}{2m} + V_{nC}(\mathbf{r}_2) + V_{nn} + \frac{(\mathbf{p}_1 + \mathbf{p}_2)^2}{2A_c m}, \tag{6.34}
$$

where the last term is the recoil kinetic energy of the core obtained by the condition $\mathbf{P} = \mathbf{p}_1 + \mathbf{p}_2 + \mathbf{P}_c = \mathbf{0}$. Equation (6.34) is further rewritten to be

$$
H = h_{nC}(1) + h_{nC}(2) + V_{nn} + \frac{\mathbf{p}_1 \cdot \mathbf{p}_2}{A_c m}. \tag{6.35}
$$

Here h_{nC} is the single-particle Hamiltonian for a valence neutron interacting with the core,

$$
h_{nC} = \frac{p^2}{2\mu} + V_{nC}(\mathbf{r}), \tag{6.36}
$$

where $\mu = mA_c/(A_c + 1)$ is the reduced mass. The reduced mass μ, together with the last term in Eq. (6.35), originates from the recoil kinetic energy of the core. V_{nn} is the interaction between the valence neutrons taken as

$$
V_{nn}(\mathbf{r}_1, \mathbf{r}_2) = \delta(\mathbf{r}_1 - \mathbf{r}_2)v_0 \left(1 + \frac{v_\rho}{1 + \exp[(r_1 - R_\rho)/a_\rho]}\right), \tag{6.37}
$$

where the first term in the bracket is determined by the nucleon-nucleon scattering length and the second term mimics the medium effect in nuclei. The delta force (6.37) must be supplemented with an energy cutoff E_{cut} in the two-particle spectrum to get a reliable result. The typical cutoff energy is $E_{\text{cut}} = 30 \sim 40$ MeV to obtain the binding energies of $A + 2$ systems with a good convergence. In terms of the energy cutoff E_{cut} and the ^1S wave scattering length a_{nn} for nn scattering, the strength for the delta interaction v_0 is given by

$$
v_0 = \frac{2\pi^2 \hbar^2}{m} \frac{2a_{nn}}{\pi - 2k_c a_{nn}}, \tag{6.38}
$$

where k_c is the cutoff momentum determined by a relation $E_{\text{cut}} = \hbar^2 k_c^2/m$. The parameters for the density dependent part in Eq. (6.37), that is, v_ρ, R_ρ, and a_ρ, are adjusted in order to reproduce the known ground-state properties for each nucleus.

The single-particle basis is obtained from the eigenstates of the Hamiltonian h_{nC} which is a single-particle Hamiltonian consisting of a kinetic energy operator, a Woods-Saxon potential and a spin-orbit interaction,

$$V_{nC}(r) = V_0 \left(1 - 0.44 f_{so} r_0^2 (\mathbf{l} \cdot \mathbf{s}) \frac{1}{r} \frac{d}{dr} \right) \left[1 + \exp \left(\frac{r - R}{a} \right) \right]^{-1},$$

where V_0 is the depth of central potential, f_{so} is the spin-orbit coupling strength, r_0 is the radial parameter with $R = r_0 A_c^{1/3}$, and a is the surface diffuseness parameter. The model basis includes bound states as well as continuum states, but the bound states occupied by the core nucleons are explicitly excluded from the model space. In this way, the Pauli blocking between the core and the valence nucleons is treated exactly. In the Green's function method, the continuum is taken into account accurately with the proper asymptotic behavior of the wave functions. The coupling to the continuum is quite important to describe the spectra of excited states near the threshold of loosely-bound nucleus.

In order to find the solutions of the two-particle Hamiltonian (6.35), one can first define the unperturbed two-particle Green's function as

$$G_0^{(I^\pi)}(E) = \frac{1}{H_0 - E - i\eta} = \int \sum_{1,2} \frac{|(j_1 j_2)^{(I^\pi)}\rangle \langle (j_1 j_2)^{(I^\pi)}|}{\epsilon_1 + \epsilon_2 - E - i\eta}, \qquad (6.39)$$

where H_0 is the single-particle Hamiltonian $H_0 = h_{nC}(1) + h_{nC}(2)$, ϵ_i is the single-particle energy of the hamiltonian $h_{nC}(i)$, and η is an infinitesimal number. In Eq. (6.39), the completeness of two particle states

$$1 = \int \sum_{1,2} |(j_1 j_2)^{(I^\pi)}\rangle \langle (j_1 j_2)^{(I^\pi)}| \qquad (6.40)$$

is applied (see Section 3.5 for more discussions about the Green's function method). The sum includes all independent two-particle states coupled to the total angular momentum of I and the parity π including both the bound and continuum single-particle states, satisfying the condition $\epsilon_1 + \epsilon_2 \leq (A_c + 1) E_{cut} / A_c$. The correlated Green's function is defined as

$$G^{(I^\pi)}(E) = \frac{1}{H - E - i\eta} = \int \sum_{1,2} \frac{|\widetilde{(j_1 j_2)}^{(I^\pi)}\rangle \langle \widetilde{(j_1 j_2)}^{(I^\pi)}|}{\widetilde{E_{12}} - E - i\eta}, \qquad (6.41)$$

where $\widetilde{(12)}^{(I^\pi)}$ and $\widetilde{E_{12}}$ are the correlated two-particle state and its energy, respectively. Using the identity

$$1 + G_0 V_{nn} = G_0 (H_0 - E - i\eta + V_{nn}) = G_0 (H - E - i\eta), \qquad (6.42)$$

the correlated Green's function is expressed as

$$G^{(I^\pi)}(E) = (1 + G_0^{(I^\pi)}(E)V_{nn})^{-1}G_0^{(I^\pi)}(E). \tag{6.43}$$

The correlated Green's function (6.43) is an integrodifferential equation with two variables $\mathbf{r}_1$ and $\mathbf{r}_2$. To facilitate numerical calculations, the recoil term in Eq. (6.35) is discarded so that we have only one variable $\mathbf{r} = \mathbf{r}_1 - \mathbf{r}_2$ to solve the correlated Green's function (6.43) with the contact interaction (6.37).

Solving Eq. (6.43) for a given I^π, we can obtain both the ground-state and excited-state wave functions of neutron configurations in ^{11}Li. The proton configuration is frozen in the present three-body model calculations. The correlated ground state is specified by the total angular momentum of neutrons $I^\pi = 0^+$, while the states with $I^\pi = 1^-$ are excited by the electric dipole operator from the ground state. The correlated two particle wave function for the ground state is explicitly written as

$$\Phi_{gs}(\mathbf{r}_1, \mathbf{r}_2) \equiv \sum_{1,2} |\widetilde{(j_1 j_2)}^{(I^\pi = 0^+)}\rangle = \sum_{n' \le n, lj} \frac{\alpha_{nn'lj}}{\sqrt{2(1 + \delta_{n,n'})}} \mathcal{A}\{(\phi_{nlj}(\mathbf{r}_1)\phi_{n'lj}(\mathbf{r}_2))^{(I=0)}\}, \tag{6.44}$$

where $\mathcal{A}$ is the anti-symmetrization operator, $\phi_{nlj}(\mathbf{r})$ is the eigenstate of the Hamiltonian (6.36) with the radial quantum number n, and $\alpha_{nn'lj}$ is the expansion coefficient. This state is equivalent to a ket vector in the numerator of Eq. (6.41). The ground state two-body density is then defined by

$$\rho_{(2)}(\mathbf{r}_1, \mathbf{r}_2) = \langle \Phi_{gs}(\mathbf{r}_1, \mathbf{r}_2) | \Phi_{gs}(\mathbf{r}_1, \mathbf{r}_2) \rangle. \tag{6.45}$$

The two particle densities of ^{11}Li and ^{6}He are plotted as functions of distance from the core $r = r_2 = r_1$ and the angle between the two vectors $\theta_{12} = cos^{-1}(\mathbf{r}_1 \cdot \mathbf{r}_2)/r_1 r_2$ in Fig. 6.48 with a weight of $4\pi r^2 \cdot 2\pi r^2 \sin\theta_{12}$.

One can observe two maxima in the two-particle densities. The peaks at smaller and larger θ_{12} are referred to as "di-neutron" and "cigar-like" configurations, respectively. We see that the di-neutron part of the two-particle density has a long radial tail in ^{11}Li, and thus can be interpreted as a halo structure. In contrast, the cigar-like configuration has a rather compact radial shape with a large opening angle $\theta_{12} \ge 100°$. A mixing of different parity single-particle states plays an essential role to induce the localization of di-neutron wave function at the small angle of θ_{12}. In ^{11}Li, the $1p$ and $2s$ single-particle states outside of ^{9}Li core are close to each other in energy and are mixed strongly in the correlated ground state.

Problem

6.3 Calculate the mixing amplitudes a and b of the $1p_{1/2}$ and $2s_{1/2}$ two-particle states, respectively, outside of ^{9}Li core,

$$|\Phi\rangle = a|(1p)^2 : J = 0\rangle + b|(2s)^2 : J = 0\rangle \tag{6.46}$$

for the Hamiltonian

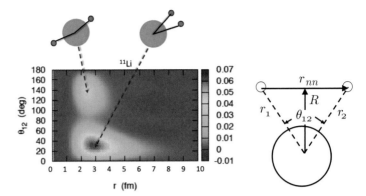

Fig. 6.48 The two-particle densities for ^{11}Li as a function of $r_1 = r_2 = r$ and the angle between the valence neutrons, θ_{12}. The two-particle density is weighted with a factor $4\pi r^2 \cdot 2\pi r^2 \sin\theta_{12}$. Geometry of di-neutron with respect to the core is shown also in the right panel. Two peaks in two-body densities are often called "di-neutron" and "cigarlike" configurations with small and large correlation angles, respectively. Figure reprinted with permission from [29]. ©2021 by the American Physical Society

$$H = \sum_i \varepsilon_i + \sum_{i,j} V_{ij}, \tag{6.47}$$

with the off-diagonal two-body matrix element $\langle (2s)^2 : J = 0, T = 1 | V | (1p)^2 : J = 0, T = 1 \rangle = -1$ MeV. Take three different cases of the single particle energies $\varepsilon_{2s} = 0.5$, 1 and 3 MeV with a fixed $\varepsilon_{1p} = 0.0$ MeV for the calculations.

6.5.4 Soft Dipole Excitation in ^{11}Li

The IV giant dipole resonance (GDR) is a mutual oscillation between protons and neutrons without changing their shapes and densities. In neutron-rich nuclei, it is known that the thick neutron skin or neutron halo is associated to the core. Then the neutron skin (halo) may oscillate against the core of nucleus and this peculiar oscillation is called "soft dipole excitation" (see Fig. 6.49). Since the oscillation of SDR has a lower frequency than that of GDR because of a weak coupling between excess neutrons and core, the excitation energy becomes much lower than that of the GDR. In the Coulomb breakup experiments of ^{11}Li, the strong peak of electric dipole (E1) excitation was observed just above the two neutron threshold energy with a very strong dipole strength. It is an unique and exciting phenomenon in the halo nucleus since the dipole strength of GDR is observed systematically at rather high energy $E_x \sim 80/A^{1/3}$ MeV, and the low-energy strength is in general very much hindered.

We study the dipole response by using the two-particle Green's function method taking into account the continuum effect. This model is essentially applied for all

Fig. 6.49 A schematic view of soft dipole and giant dipole resonances in ^{11}Li SDR is an oscillation between halo neutrons and core, while GDR is the one between neutrons and protons in the core. The excitation energy of SDR is seen in much lower energy than GDR. Figure reprinted with permission from [9]. ©2021 by Springer Nature

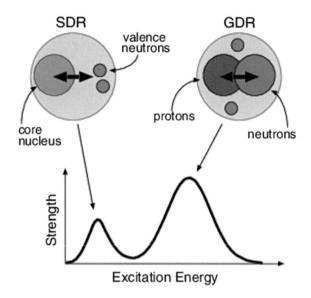

kind of multipole excitations at low energies (see next subsection for the application to 2^+ state in ^{26}O). Firstly, the matrix element for the dipole response is evaluated as

$$M^{(0)}(E1) = \sum_{j_1, j_2} \langle (j_1 j_2)^{(I^\pi = 1^-)} | D | \Phi_{g.s.} \rangle, \qquad (6.48)$$

where $\Phi_{g.s.}$ is the correlated ground state of the two valence particles given in Eq. (6.44) and D is the E1 transition operator. The E1 operator is defined with respect to the center of mass $\mathbf{R} = \sum_i^A \mathbf{r}_i / A$;

$$D = \sum_i^Z e(\mathbf{r}_i - \mathbf{R}) = \frac{N}{A} e \sum_i^Z \mathbf{r}_i - \frac{Z}{A} e \sum_i^N \mathbf{r}_i. \qquad (6.49)$$

For E1 excitation of halo neutrons, the first term in Eq. (6.49) drops out and the operator can be expressed as

$$D \rightarrow D_M = -\frac{Z}{A_c + 2} e \left(r_1 Y_{1M}(\hat{\mathbf{r}}_1) + r_2 Y_{1M}(\hat{\mathbf{r}}_2) \right), \qquad (6.50)$$

with the spherical harmonics $Y_{10}(\hat{\mathbf{r}}) = \sqrt{3/4\pi} z$. The dipole operator specified by $L = 1$ of the spherical harmonics remains in Eq. (6.50) because other terms are vanished in Eq. (6.49). The normalization factor $\sqrt{3/4\pi}$ is multiplied to Eq. (6.49) to obtain Eq. (6.50). The dipole operator (6.50) is thus entirely induced by the recoil effect of core protons to the valence neutrons.

The dipole strength distribution with the final state interaction can then be expressed by using the Green's functions (6.41) and (6.43) for $I^\pi = 1^-$ final states as,

$$
\begin{aligned}
\frac{dB(E1)}{dE} &= \sum_k |\langle \Phi_k^{(1^-)} | D_M | \Phi_{gs} \rangle|^2 \delta(E - E_k) \\
&= \frac{1}{\pi} \Im \sum_M \langle \Phi_{g.s.} | D_M^* G^{(I^\pi = 1^-)}(E) D_M | \Phi_{gs} \rangle
\end{aligned}
\tag{6.51}
$$

since the imaginary part of Eq. (6.41) gives

$$
\Im \left(G^{(I^\pi)}(E) \right) = \pi \sum_k |\Phi_k^{(1^-)}\rangle \langle \Phi_k^{(1^-)}| \delta(E - E_k).
\tag{6.52}
$$

Here $\Im$ denotes the imaginary part, and $|\Phi_k^{(1^-)}\rangle = |\widetilde{(j_1 j_2)}^{(1^-)}\rangle$ is the solution of the three-body Hamiltonian with the angular momentum $I^\pi = 1^-$ at the energy $E_k = \widetilde{E}_{12}$. If one takes only the unperturbed Green's function $G_0^{(I^\pi = 1^-)}$, that is, the first term of r.h.s. of Eq. (6.43), instead of $G^{(I^\pi = 1^-)}(E)$, Eq. (6.51) is equivalent to

$$
\frac{dB(E1)}{dE} = \sum_{j_1 j_2} |M^{(0)}(E1)|^2 \delta(E - \epsilon_1 - \epsilon_2),
\tag{6.53}
$$

where $M^{(0)}(E1)$ is the matrix element without the final state interaction given by Eq. (6.48). It has been shown that the effect of recoil kinetic energy, i.e., the last term in Eq. (6.35) is small on the dipole responses once it is taken into account in the ground-state wave function.

The dipole strength distributions for ^{11}Li calculated by Eq. (6.51) are shown by the solid bars in the left panel of Fig. 6.50. The height of the bars corresponds to the $B(E1)$ value. Also shown by the solid curves are the $B(E1)$ distributions smeared with the Lorenzian function with the width of $\Gamma = 0.2$ MeV, which mimics the continuum effect and also the experimental energy resolution. In the calculations, single particle continuum states are discretized by putting a nucleus in a large box. The wave functions for the ground state with $J^\pi = 0^+$ and for the excited $J^\pi = 1^-$ states are then obtained by diagonalizing the three-body Hamiltonian within a model space which is consistent with the cutoff energy of nn interaction (6.37).

The Coulomb breakup cross sections are also evaluated by paying particular attention to the recoil effect of the core nucleus. Based on the relativistic Coulomb excitation theory, the cross sections are obtained by multiplying the virtual photon number to the $B(E1)$ distribution;

$$
\frac{d\sigma(E1)}{dE} = \frac{16\pi^3}{9\hbar c} N_{E1}(E) \frac{dB(E1)}{dE}
\tag{6.54}
$$

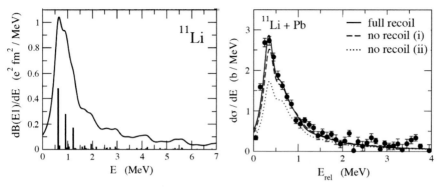

Fig. 6.50 (Left) $B(E1)$ distribution for ^{11}Li nuclei. The solid bars are obtained by the diagonalization of the Hamiltonian (6.35) with discrete basis. The solid curve is obtained by a smearing procedure with a Lorenzian function with the width parameter, $\Gamma = 0.2$ MeV. (Right) Coulomb breakup cross sections for the ^{11}Li +Pb at 70 MeV/nucleon. The solid line is the result of the full three-body calculations, while the dashed line denoted by "no recoil (i) " is obtained by neglecting the off-diagonal component of the recoil kinetic energy in the excited states. The dotted line denoted by "no recoil (ii)" is obtained by neglecting the off-diagonal recoil term both in the ground and the excited states. These results are smeared with an energy dependent width of $0.25 \cdot \sqrt{E}$ MeV. Figure reprinted with permission from [56]. ©2021 by the American Physical Society

where $N_{E1}(E)$ is the number of virtual photons at the photon energy E (see Chap. 6 for details on Coulomb excitation). The solid curve in the right panel of Fig. 6.50 shows the Coulomb breakup cross sections obtained for ^{11}Li+Pb reaction at 70 MeV/nucleon taking into account or not taking into account the recoil terms in Eq. (6.35). There are three lines in the right panel of Fig. 6.50. The dashed line denoted by "no recoil (i)" is obtained by neglecting the off-diagonal component of the recoil kinetic energy in the excited states, but included in the ground state. The dotted line denoted by "no recoil (ii)" is obtained by neglecting the off-diagonal recoil term both in the ground and the excited states. In order to facilitate the comparison with the experimental data, we smear the discretized cross sections with the Lorenzian function with the energy dependent width, $\Gamma = \alpha \cdot \sqrt{E}$. We take $\alpha = 0.25$ MeV$^{1/2}$ for ^{11}Li. One can see in the right panel of Fig. 6.50, the recoil terms has a substantial influence on the Coulomb breakup cross sections. However the recoil term can be neglected in the calculations of excited states if it is once included in the calculation of ground state. We see that the experimental breakup cross sections are reproduced remarkably well within the present three-body model for the ^{11}Li nucleus.

The strong dipole excitation near the threshold energy is a typical feature of the excitation of halo nuclei. Since the halo neutrons are extremely loosely bound to the core in ^{11}Li ($S_{2n} \sim 300$ keV), the coupling between the halo neutrons and the core is very weak so that the oscillation becomes very slow in an intuitive picture of the harmonic oscillation. The low-frequency mode provides a low-energy excited state, This is the reason of very low-excitation energy of the dipole excitation and called "*soft dipole excitation*". On top of that, the dipole excitation operator involves the coordinate operator $\mathbf{r}_i$ which induces an enhanced E1 strength for extended wave function such as halo neutrons. These two characteristics of halo are responsible for the strongly enhanced E1 strength near the threshold shown in Fig. 6.50.[6]

In the present three-body model, the core is frozen without any structure. The core excitations are also included in a more sophisticated coupled channel ^{9}Li$+n+n$ three-body model, which includes the coupling between the last neutron states and the various two-particle two-hole (2p-2h) configurations in ^{9}Li due to the tensor and pairing correlations. This coupled-channel model was applied to investigate the three-body Coulomb breakup of a two-neutron halo nucleus ^{11}Li [52].

6.5.5   Two Neutron Correlations in ^{26}O

The unbound nucleus ^{26}O has been investigated using invariant-mass spectroscopy following a one-proton removal reaction from a ^{27}F beam at the projectile energy 200 MeV/nucleon in 2016 at RIKEN. The ^{26}O ground-state resonance was found to lie only 18 keV above the two-neutron threshold. This resonant state will decay to the ground state of ^{24}O emitting two neutrons.

The angular distribution of the emitted neutrons can be calculated with the two-particle Green's function method. The amplitude for emitting two neutrons with spin components of s_1 and s_2 and momenta $\mathbf{k}_1$ and $\mathbf{k}_2$ reads

$$f_{s_1 s_2}(\mathbf{k}_1, \mathbf{k}_2) = \sum_{j,l} e^{-il\pi} e^{i(\delta_1 + \delta_2)} M_{j,l,k_1,k_2} \langle [\mathcal{Y}_{jl}(\hat{\mathbf{k}}_1) \mathcal{Y}_{jl}(\hat{\mathbf{k}}_2)]^{(00)} | \chi_{s_1} \chi_{s_2} \rangle, \quad (6.55)$$

where $\mathcal{Y}_{jlm}(\hat{\mathbf{k}})$ is the spin-spherical harmonics,

$$\mathcal{Y}_{jlm}(\hat{\mathbf{k}}) \equiv [Y_l(\hat{\mathbf{k}}) \times \chi_s]_{jlm} = \sum_{l_z, s_z} \langle l l_z \frac{1}{2} s_z | jm \rangle,$$

χ_s is the spin wave function, and δ is the nuclear phase shift. The symbol $[\mathcal{Y}_{jl}(\hat{\mathbf{k}}_1) \mathcal{Y}_{jl}(\hat{\mathbf{k}}_2)]^{(00)}$ denotes that the coupling angular momentum of two spin-spherical harmonics is zero. Here, M_{j,l,k_1,k_2} is a decay amplitude to a specific two-particle final state assigned by the quantum numbers j, l and the momenta k_1, k_2, and can be calculated by the correlated Green's function (6.43). The angular distribution is then obtained as

$$P(\theta_{12}) = 4\pi \sum_{s_1, s_2} \int dk_1 dk_2 \, |f_{s_1 s_2}(k_1, \hat{\mathbf{k}}_1 = 0, k_2, \hat{\mathbf{k}}_2 = \theta_{12})|^2, \quad (6.56)$$

---

[6] This threshold enhancement due to the loosely-bound wave function is typical for the soft dipole excitation. On the other hand, "Pigmy dipole mode" also appears below the giant dipole resonance (GDR), but not necessarily enhanced due to the extended wave function.

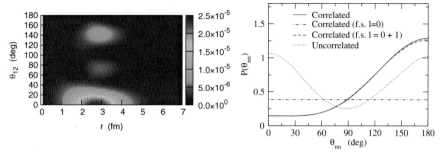

Fig. 6.51 (Left): The two-particle density (6.45) for the resonance state of ^{26}O. It is plotted as a function of $r_1 = r_2 = r$ and the angle between the valence neutrons, θ_{12} See the caption to Fig. 6.48 for details. (Right): The angular correlation for the two emitted neutrons from the ground state decay of ^{26}O. The probability distribution for the opening angle of the momentum vectors of the emitted neutrons is shown. The solid and the dotted lines denote the correlated and uncorrelated results with the full model space, respectively. The dot-dashed and the dashed lines denote the correlated results including the angular momenta of the final state up to $l = 0$ and $l = 1$, respectively. Figure reprinted with permission from [55]. ©2021 by the American Physical Society

where the z axis is set to be parallel to $\mathbf{k}_1$ and the angular distribution is evaluated as a function of the opening angle, θ_{12}, of the two emitted neutrons.

In the right panel of Fig. 6.51, the dotted line shows the angular distribution of two neutrons emitted from the resonant round state of ^{26}O obtained without including the final state nn interaction. Then the angular distribution is symmetric around $\theta_{12} = \pi/2$. This is because the two neutron configurations are coupled to $J^\pi = 0^+$ and each configuration is symmetric under the transformation $\theta_{12} = \pi - \theta_{12}$ in Eq. (6.56). In the presence of the final state interaction, many configurations with different angular momenta are mixed and the angular distribution turns to be highly asymmetric, in which the emission of two neutrons in the opposite direction (that is, $\theta_{12} = \pi$) is enhanced, as is shown by the solid line. The correlated angular distribution with $l = 0$ wave (dashed-dotted line) is flat since it is entirely governed by the Legendre polynomials $P_{l=0}(\theta) = 1$. The full correlated angular distribution (solid line) is almost identical to that of $l = 0 + 1$ (dashed line), which proves the $l = 0$ and $l = 1$ waves dominance in the angular distributions.

Let us discuss how the final state interaction changes the angular distributions by using a mixture of two configurations of different parity states. This behavior reflects properties of the resonance wave function of ^{26}O. That is, because of the continuum couplings, several configurations with opposite parity states mix coherently. Symbolically, let us write a two-particle wave function as

$$\Phi(\mathbf{r}, \mathbf{r}') = \alpha\, \Phi_{ee}(\mathbf{r}, \mathbf{r}') + \beta\, \Phi_{oo}(\mathbf{r}, \mathbf{r}'), \qquad (6.57)$$

where Φ_{ee} and Φ_{oo} are two-particle wave functions with l-even and l-odd angular momentum states in Eq. (6.55), respectively. The coefficients α and β are such that the interference term in the two-particle density, $\alpha^*\beta\Phi_{ee}^*\Phi_{oo} + c.c.$, is positive for $\mathbf{r}' = \mathbf{r}$

while it is negative for $\mathbf{r}' = -\mathbf{r}$. The negative sign is caused by the mixing of odd parity state $\Phi_{oo}(\mathbf{r}, \mathbf{r}')$ which shows a sign change $\Phi_{oo}(\mathbf{r}, -\mathbf{r}') = -\Phi_{oo}(\mathbf{r}, \mathbf{r}')$. Therefore, the two-particle density is enhanced for the nearside configuration with $\mathbf{r} \sim \mathbf{r}'$ as compared to the far side configuration with $\mathbf{r} \sim -\mathbf{r}'$. This correlation appears in the opposite way in the momentum space. In the Fourier transform of $\Phi(\mathbf{r}, \mathbf{r}')$,

$$\widetilde{\Phi}(\mathbf{k}, \mathbf{k}') = \int d\mathbf{r}d\mathbf{r}' \, e^{i\mathbf{k}\cdot\mathbf{r}} e^{i\mathbf{k}'\cdot\mathbf{r}'} \, \Phi(\mathbf{r}, \mathbf{r}'), \tag{6.58}$$

there is a factor i^l in the multipole decomposition of $e^{i\mathbf{k}\cdot\mathbf{r}}$ (see Eq. (2.28)). Since $\left(i^l\right)^2$ is +1 for even values of l and -1 for odd values of l, this leads to

$$\widetilde{\Phi}(\mathbf{k}, \mathbf{k}') = \alpha \, \widetilde{\Phi}_{ee}(\mathbf{k}, \mathbf{k}') - \beta \, \widetilde{\Phi}_{oo}(\mathbf{k}, \mathbf{k}'). \tag{6.59}$$

for the two particle wave function given by Eq. (6.57). If one constructs a two-particle density in the momentum space with this wave function, the interference term therefore acts in the opposite way to that in the coordinate space. That is, the two-particle density in the momentum space is hindered for $\mathbf{k} \sim \mathbf{k}'$, while it is enhanced for $\mathbf{k} \sim -\mathbf{k}'$ as is seen in the solid and dashed lines in Fig. 6.51.

We pointed out in Sect. 4.4.1 that the mixing of different parity states is the essential ingredient to induce the strong di-neutron correlations. From the argument above, we can conclude that, if an enhancement in the region of $\theta_{nn} \sim \pi$ of the angular distribution was observed experimentally, that would make a clear evidence for the di-neutron correlation in this nucleus, although such measurement would be experimentally challenging.

6.6 Isoscalar Spin-Triplet Pairing

In subsection 3.4.3, we discussed pairing correlations in nuclei, which is characterized by the isospin $T = 1$ and the spin $S = 0$. In nature, there is another type of pairing correlation occurring between a proton and a neutron specified by the isospin $T = 0$ and $S = 1$, which is even stronger than $T = 1$ pairing in nuclear matter calculations. An evidence of strong $T = 0$ correlations is provided by the bound deuteron system, which has $T = 0$ and $S = 1$ quantum numbers, while di-neutron or di-proton systems, having $T = 1$ and $S = 0$, are not bound. The momentum dependence of pairing correlations are shown in Fig. 6.52. It is clearly seen that the pairing gap of the isoscalar $T = 0$ channel is much stronger, about 3 times larger, than that of the isovector $T = 1$ channel. This strong $T = 0$ pairing correlations is mainly induced by the coupling between $^3S_1$ and $^3D_1$ channels by the tensor interactions.

There is an essential difference between the isoscalar and isovector pairing in the coupling scheme of two-particle configuration $(j_1 j_2)$. For the isovector pairing, the two particle LS coupling state has one-to-one transformation to jj coupling state,

$$|(L = 0, S = 0)J = 0, T = 1\rangle = |(j_1 = j_2)J = 0, T = 1\rangle, \quad (6.60)$$

where $\mathbf{L} = \mathbf{l}_1 + \mathbf{l}_2$ and $\mathbf{S} = \mathbf{s}_1 + \mathbf{s}_2$ in the case of LS coupling scheme and $\mathbf{j} = \mathbf{l} + \mathbf{s}$ for jj coupling wave function. On the other hand, for the isoscalar pairing, the mapping is not unique as

$$|(L = 0, S = 1)J = 1, T = 0\rangle = a|(l_1 = l_2, j_1 = j_2)J = 1, T = 0\rangle$$
$$+ b|(l_1 = l_2, j_1 = j_2 \pm 1)J = 1, T = 0\rangle, \quad (6.61)$$

since the spin-orbit partners $j_1 = j_2 \pm 1$ will contribute also to make the isoscalar spin-triplet pairing state. Since there is a strong spin-orbit interaction in nuclei, the isoscalar pairing is often hindered especially in heavy nuclei where large angular orbits, having large spin-orbit splittings, are very much involved in the nuclear structure calculations.

We discuss three examples of the isoscalar pairing correlations; the first one is the Wigner term in the mass formula, the second is the inversion of $J^\pi = 0^+$ and $J^\pi = 1^+$ states in $N = Z$ odd-odd nuclei and the third one is an enhancement of the pn pair transfer amplitude in $N = Z$ nucleus ^{40}Ca. Evidence of strong IS pairing is also found as strong low-energy GT transitions in $N = Z$ and $N = Z + 2$ nuclei by $(^3\text{He}, t)$ and (p, n) reactions.

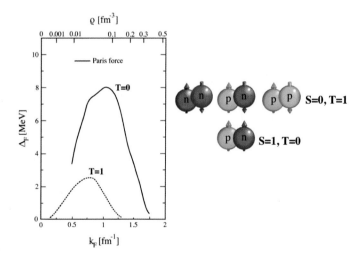

Fig. 6.52 Fermi momentum dependence of the pairing gaps in symmetric nuclear matter, for the $T = 0$ spin-triplet and the $T = 1$ spin-singlet channels calculated by using the Paris potential. The corresponding nuclear density is also given in the upper axis of the figure for the $T = 0$ case. Figure reprinted with permission from [50]. ©2021 by the American Physical Society

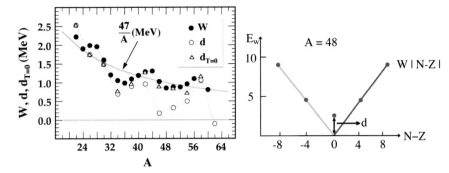

Fig. 6.53 (Left) Experimental values of W (filled circles) and d (open circles) in Eq. (6.62) in $N = Z$ nuclei extracted from measured binding energies. The triangles mark the values of d calculated using experimental binding energies of the lowest $T = 0$ states in odd-odd nuclei. The solid line represents the average value $W(\text{average}) = 47/A$ MeV. (Right) A sketch of Wigner energy E_W of A=48 nuclei. Figure reprinted with permission from [51]. ©2021 by Elsevier

6.6.1 Wigner Energy in the Mass Formula

Experimental masses of even-even and odd-odd nuclei exhibit a sharp slope discontinuity at $N = Z$. This cusp (Wigner energy) reflects an additional binding in nuclei in which neutrons and protons occupy the same shell model orbitals. It is usually attributed to neutron-proton pairing correlations.

The Wigner energy is decomposed into two parts:

$$E_W = W(A)|N - Z| + d(A)\pi_{np}\delta_{NZ}, \qquad (6.62)$$

where $\pi_{np} = 1$ for odd-odd nuclei and 0 otherwise. The $|N - Z|$-dependence in Eq. (6.62) was first introduced by Eugene Paul Wigner in his analysis of the SU(4) spin-isospin symmetry of nuclear forces. The Wigner energy refers to the apparent enhancement of the binding energy near $N = Z$ as sketched in the right panel of Fig. 6.53.

The empirical W and d terms in the Wigner energy are extracted from the binding energies of nuclei around $N = Z$ and shown in the left panel of Fig. 6.53. The two terms decrease smoothly as a function of A. For the sd-shell nuclei $A < 40$, the behavior of the d term follows rather closely that of W, but this trend is lost for the heavier systems. It is known experimentally that the nuclear ground states have the isospin $T_{gs} = |N - Z|/2$, except for heavy $N = Z$ odd-odd nuclei. Indeed, ^{34}Cl and nuclei heavier than ^{40}Ca (except for ^{58}Cu) have ground states of isospin $T_{gs} = 1$. To take into account the isospin $T = 0$ lowest state in the analysis consistently, the ground-state binding energies of these odd-odd $N = Z$ with $T_{gs} = 1$ are replaced by the binding energies of their lowest $T = 0$ states and modified d values with the $T = 0$ lowest states are obtained. These modified values are marked by triangles in Fig. 6.53. The resulting values $d_{T=0}$ follow very closely the values of

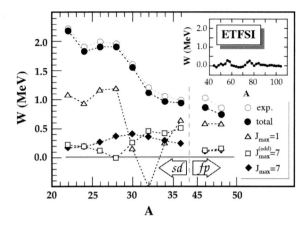

Fig. 6.54 The strength of the Wigner term, W, extracted using binding energies calculated with the $0\hbar\omega$ shell model. Full shell-model calculations (filled circles) agree very well with experimental data (open circles). The results of shell-model calculations with the ($J = 1, T = 0$) two-body matrix elements removed ($J_{max} = 1$ variant, triangles), with all $T = 0$ matrix elements removed ($J_{max} = 7$ variant, diamonds), and with ($J = 1, 3, 5, 7, T = 0$) two-body matrix elements removed ($J_{max}^{odd} = 7$ variant, squares) are also shown. In the fp-shell, only two points have been calculated due to practical limitations. The insert shows the values of W extracted from the extended Thomas-Fermi (ETFSI) mass formula with Skyrme EDF. They are practically zero for all nuclei considered. Figure reprinted with permission from [51]. ©2021 by Elsevier

W; i.e., $d_{T=0}(A)/W(A) \simeq 1$. This observation provides an important experimental conjecture that both terms constituting E_W might have the same microscopic origin. In particular, assuming that the structure of the lowest $T = 0$ states in odd-odd N=Z nuclei is strongly influenced by the effective interaction between $T = 0$ pairs, the close similarity between $d_{T=0}$ and $W(A)$ indicates that the $T = 0$ component of the nuclear interaction is responsible for the Wigner term.

In order to investigate the role played by the $T = 0$ component for the Wigner energy, and to study the relative importance of various np pairs, shell-model calculations were performed in both the $1f2p$ and $1d2s$ shell-model spaces with the KB3 effective interaction and the USD effective interaction, respectively. The results of the calculations are displayed in Fig. 6.54. These shell model results reproduce well experimental values of W both in the sd- as well as in the fp-shell (circles in Fig. 6.54). The results of calculations with the ($J = 1, T = 0$) two-body matrix elements removed, marked by triangles in Fig. 6.54, demonstrate that the impact of the deuteron-like correlations on the strength of the Wigner term is stronger in sd-nuclei than in fp-nuclei. Removing all $T = 0$ matrix elements with $J = 1, 3, 5, 7$ ($J_{max}^{odd} = 7$ variant), i.e., those which are predominantly due to the coupling between neutrons and protons in identical orbits, washes out W almost completely. Removing the remaining T=0 matrix elements with even values of J has the small impact on the heaviest sd-shell nuclei. These shell model studies confirm that the bulk part of Wigner energy comes from the $T = 0$ np pairs. It is also found that the contribution from deuteron-like ($J = 1, T = 0$) pairs is large but by no means dominant, i.e., the contribution from maximally aligned pairs is equally important for the Wigner energy. On top of these two large contributions, the $T = 0$ pairs with intermediate spins are also non-negligible, especially in the upper part of the sd-shell.

6.6.2   Isoscalar Pairing and Magnetic Dipole Transitions

A typical competition between the spin-singlet and spin-triplet pairing correlations in odd-odd nuclei is shown in Fig. 6.55. The calculations were done with a three-body model of neutron-proton np-pair outside of the core nuclei. Similar to the Hamiltonian (6.35), a three-body model Hamiltonian for odd-odd $N = Z$ nuclei, assuming the core + $p + n$ structure, is given by

$$H = \frac{\mathbf{p}_p^2}{2m} + \frac{\mathbf{p}_n^2}{2m} + V_{pC}(\mathbf{r}_p) + V_{nC}(\mathbf{r}_n) + V_{pn}(\mathbf{r}_p, \mathbf{r}_n), \tag{6.63}$$

where m is the nucleon mass, and V_{pC} and V_{nC} are the mean field potentials for the valence proton and neutron, respectively, which are represented by a Woods-Saxon potential. We use a contact pairing interaction between the valence neutron and proton, V_{np}, that is the sum of $T = 0$ and $T = 1$ pairing forces with inclusions of appropriate spin projectors,

$$V_{np}(\mathbf{r}_1, \mathbf{r}_2) = P_s v_s \delta(\mathbf{r}_1 - \mathbf{r}_2) \left(1 - x_s \left(\frac{\rho(r)}{\rho_0}\right)^\gamma\right)$$
$$+ P_t v_t \delta(\mathbf{r}_1 - \mathbf{r}_2) \left(1 - x_t \left(\frac{\rho(r)}{\rho_0}\right)^\gamma\right), \tag{6.64}$$

where P_s and P_t are the projectors onto the spin-singlet and spin-triplet channels, respectively, defined as $P_s = (1 - \sigma_n \cdot \sigma_p)/4$ and $P_t = (3 + \sigma_n \cdot \sigma_p)/4$. The density-dependent terms $(\rho(r)/\rho_0)^\gamma$ are introduced to take into account the effect of nuclear medium corrections. The strength parameters, v_s and v_t, are determined from the proton-neutron scattering length by using the T-matrix theory in the spin-singlet and spin-triplet channels, respectively. The three parameters x_s, x_t, and γ in the density dependent terms in Eq. (6.64) are determined to reproduce the energies of the ground state ($J = 1^+$), the first excited state ($J = 3^+$), and the second excited state ($J = 0^+$) in ^{18}F; $x_s = -1.24$, $x_t = -1.42$, and $\gamma = 1.23$.

The calculated spectra for ^{14}N, ^{18}F, ^{30}P, ^{34}Cl, ^{42}Sc, and ^{58}Cu nuclei are shown in Fig. 6.55 together with the experimental data. The spin-parity for the ground state of the nuclei in Fig. 6.55 are $J^\pi = 1^+$ except for ^{34}Cl and ^{42}Sc. This feature is entirely due to the interplay between the isoscalar spin-triplet and the isovector spin-singlet pairing interactions in these $N = Z$ nuclei. In the present calculations, the ratio

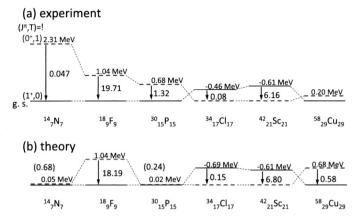

Fig. 6.55 The energies of the first 0_1^+ and the first 1_1^+ states in $N = Z$ nuclei. The upper panel a shows experimental data and the lower panel b corresponds to calculated results. The values with the arrows show the transition probabilities for the magnetic dipole transitions, $B(M1)$ (the calculated values are shown in brackets for ^{14}N and ^{30}P). This figure is taken from [31]

between the isoscalar and the isovector pairing interactions is $v_t/v_s = 1.9$. This ratio is somewhat larger than the value ≈ 1.6 extracted in the sd-shell and pf-shell model effective two-body interactions. It is remarkable that the energy differences $\Delta E = E(0_1^+) - E(1_1^+)$ are well reproduced in ^{34}Cl and ^{42}Sc both qualitatively (there is an inversion between the 0^+ state and the 1^+ state in the two nuclei), and quantitatively (the absolute value of the energy difference). The model description is somewhat poor in ^{14}N and ^{30}P because the cores of these two nuclei are deformed, nevertheless, the ordering of the two lowest levels are correctly reproduced.

The magnetic dipole transitions between 1^+ and 0^+ states may give us useful information on detailed structure of wave functions. The reduced magnetic dipole transition probability is given by (see also subsection 4.6.1)

$$B(M1 : J_i \to J_f) = \left(\frac{3}{4\pi}\right) \frac{1}{2J_i + 1} \left| \langle J_f || \sum_i (g_s(i)\mathbf{s}_i + g_l(i)\mathbf{l}_i) || J_i \rangle \right|^2 , (6.65)$$

where the double bar means a reduced matrix element in the spin space. We take the bare g factors $g_s(\pi) = 5.58$, $g_s(\nu) = -3.82$, $g_l(\pi) = 1$, and $g_l(\nu) = 0$ for the magnetic dipole transitions in unit of the nuclear magneton $\mu_N = e\hbar/2mc$. The $B(M1)$ transition from 0^+ to 1^+ in ^{18}F is the largest one so far observed in the entire region of the nuclear chart. We notice that the three-body calculations provide fine agreements not only for these strong transitions in ^{18}F and ^{42}Sc but also for the weak transitions in the other $N = Z$ nuclei such as in ^{14}N and ^{34}Cl. The shell model calculation in sd-shell model space [7] shows also a large enhancement for the $B(M1)$ transition in ^{18}F which is consistent with both the present study and the experiment.

In the case of ^{18}F, the 0^+ and 1^+ states are largely dominated by the $S = 0$ and $S = 1$ spin components, respectively, with orbital angular momentum $l = 2$ (d-orbit). Therefore, the two states can be considered as members of the SU(4) multiplet in the spin-isospin space. This is the main reason why the $B(M1)$ value is so large in this nucleus, since the isovector spin transition operator $g_s^{IV} s \tau_z$ is a generator of SU(4) group which give the maximum overlap between two states in the same SU(4) multiplet, that is, the transition is allowed, and the isovector g-factor is the dominant term in Eq. (6.65) with $g_s^{IV} = (g_s(\nu) - g_s(\pi))/2 = -4.70$. The configurations in ^{42}Sc are also similar to those in ^{18}F in terms of SU(4) multiplets, although they are dominated by $l = 3$ (f-orbit) wave functions. For ^{14}N and ^{34}Cl, the $B(M1)$ transitions do not acquire any enhancement, since the $S = 0$ component in the 0^+ state is suppressed due to the $j_< = l - 1/2$ coupling configuration: both the 0^+ and 1^+ states have very large $1p_{1/2}^2$ ($1d_{3/2}^2$) configuration in ^{14}N (^{34}C).

In the nuclei ^{30}P and ^{58}Cu, the 1^+ state is dominated by $1d_{3/2}2s_{1/2}$ and $2p_{3/2}1f_{5/2}$ configurations, respectively, while the 0^+ state is governed by the $2s_{1/2}^2$ and $2p_{3/2}^2$ configurations, respectively. Therefore the $B(M1)$ is much hindered in the two nuclei. The validity of SU(4) symmetry has been known already for a quite long time in p−shell nuclei. The two-body matrix element of standard shell models (for example, Cohen-Kurath two-body matrix elements) for the spin-triplet $(J, T) = (1, 0)$ interaction is certainly stronger than that for the spin-singlet $(J, T) = (0, 1)$ pairing interaction. Then, the structure of the two-body wave function of p-shell nuclei will be rather described by the LS coupling scheme than jj coupling scheme.

6.6.3 Pair Transfer Reactions and Pairing Correlations

As it is intuitive, two-particle transfer reactions are sensitive to the correlations between those particles, so that two-neutron transfer has been for long time used to pin down fingerprints of $T = 1$ pairing, and recently there is a strong interest to understand to which extent deuteron transfer can probe $T = 0$ pairing.

There are some basic problems, nonetheless, both at the conceptual and experimental level. Let us assume we can restrict to $L = 0$ states excited in transitions between even-even nuclei. Quasi-particle RPA (QRPA) (see subsections 3.4.3 and 3.4.4), or shell-model calculations (see subsection 3.6), can provide the wave functions of such excited states n; in particular, one can calculate the strength functions $S(E)$ associated either with the pair addition (ad) or pair removal (rm) operators. These strength functions read

$$S_{\text{ad}} = \sum_{n \in A+2} |\langle n | P^\dagger | 0 \rangle|^2 \, \delta(E - E_n), \qquad (6.66)$$

$$S_{\text{rm}} = \sum_{n \in A-2} |\langle n | P | 0 \rangle|^2 \, \delta(E - E_n), \qquad (6.67)$$

where the sums run over the appropriate set of final states in the $A + 2$ or $A - 2$ nuclei, and $|0\rangle$ is the ground state of the even-even nucleus with a mass A. The pair addition operator is defined by

$$P^\dagger_{T=1,S=0\,(T=0,S=1)} = \frac{1}{2} \int d\mathbf{r} \left[a^\dagger(\mathbf{r}\tau\sigma)a^\dagger(\mathbf{r}\tau'\sigma') \right]^{T=1,S=0(T=0,S=1)} , \quad (6.68)$$

where $a^\dagger(\mathbf{r}\tau\sigma) = a^\dagger(\mathbf{r})\chi_{t_z}\chi_{s_z}$ is the nucleon creation field operator in the second quatization with the isospin and spin wave functions. We have specifically projected out the T, S dependence in Eq. (6.68). One can consider the neutron-neutron case where only $J^\pi = 0^+$, $T = 1$, $S = 0$ is possible if $L = 0$. On the other hand, in the neutron-proton case either $J^\pi = 0^+$, $T = 1$, $S = 0$ or $J^\pi = 1^+$, $T = 0$, $S = 1$ are possible even for $L = 0$ case.

The proton-neutron $L = 0$ pair transfer strength in the $N = Z$ nuclei ^{40}Ca and ^{56}Ni has been studied by Kenichi Yoshida by using the proton-neutron RPA with a Skyrme EDF (see subsection 3.4). The pair addition strength of ^{40}Ca $\rightarrow$ ^{42}Sc associated with both the $J^\pi = 1^+$ and $J^\pi = 0^+$ states is shown in Fig. 6.56. It was found that the collectivity of the lowest $J^\pi = 1^+$ in ^{42}Sc is stronger than that of the lowest $J^\pi = 0^+$ state when the IS spin-triplet pairing ($T = 0$, $S = 1$) is taken to be equal or stronger than the IV spin-singlet pairing ($T = 1$, $S = 0$).

One sees that the excitation energy and the strength of the $J^\pi = 1^+$ states are strongly affected by the $T = 0$ pairing interaction. In the case of $f = 0$, without the $T = 0$ pairing interaction, the lowest 1^+ state in ^{42}Sc located at $Ex = 7.5$ MeV is a single-particle excitation $\pi f_{7/2}\nu f_{7/2}$. As the pairing interaction is switched on, and the pairing strength is increased, the 1^+ state is shifted downwards in energy with the enhancement of the transition strength. With increasing paring strength to $f \equiv v_s/v_t = 1.0$ or 1.3, the lowest 1^+ state is constructed by many particle-particle excitations involving $f_{5/2}$ and $p_{3/2}$ orbitals located above the Fermi levels as well as the $\pi f_{7/2}\nu f_{7/2}$ excitation. It is particularly worth noting that the hole-hole excitations from the $sd-$shell have an appreciable contribution to generate this $T = 0$ proton-neutron pair-addition vibrational mode, indicating ^{40}Ca core-breaking. The strong collectivity is associated by a coherent phase of these configurations of pp and hh excitations.

The comparison of the latest theoretical calculations with the experimental data have provided strong evidence that our current understanding of $T = 1$ pairing is confirmed by the analysis of the results of (p, t) transfer reactions on the stable Sn isotopes [49].

It is quite obvious also to analyze whether neutron-proton transfer reactions can play a similar role as discussed so far, to pin down better evidences for $T = 0$ pairing. One clear physics case would be to perform a reaction like $(^3$He, p) on an even-even $N = Z$ nucleus. $N = Z$ nuclei are stable only up to ^{40}Ca. As the pairing collectivity is expected to be more pronounced for heavier systems, experimental programs for $(^3$He, p) or $(^4$He, d) reactions with unstable beams such as ^{56}Ni in inverse kinematics are strongly desired and called for. In these experiments, one starts from a $T = 0$, $J^\pi = 0^+$ state and can probe the isospin invariance of $T = 1$ pairing by looking at

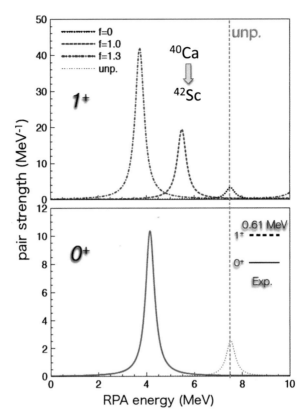

Fig. 6.56 RPA results of $L = 0$ neutron-proton pair addition strength (6.66) in the case of the transitions ^{40}Ca $\rightarrow$ ^{42}Sc. In the case of $J^\pi = 1^+$ states the operator $P_{T=0,S=1}$ is active, whereas in the case of $J^\pi = 0^+$ states drawn by the purple line in the lower panel, the strength is associated with $P_{T=1,S=0}$. The sharp peaks associated with the strength function (6.66) are smeared by means of Lorenzian functions with a width of 0.1 MeV. In the $(J, T) = (1, 0)$ channel the spin-triplet pairing strength is changed by scaling factors $f \equiv v_t/v_s = 0.0$, 1.0, and 1.3 [cf. Equation (6.64)] while the spin-singlet pairing is fixed. The unperturbed pair transfer strength is also shown by a dotted line. In the lower panel, the experimental level scheme is inserted. This figure is drawn based on the results in [32]. Courtesy K. Yoshida, Kyoto University

the $T = 1$, $J^\pi = 0^+$ states in the odd-odd system and investigate $T = 0$ pairing by looking at the $T = 0$, $J^\pi = 1^+$ states.

A full theory like the one developed for two-particle transfer reactions, that is, capable to predict absolute values of the cross section, still needs to be developed. A review of two-particle transfer reactions are given in Sect. 8.5.1.2 in Chap. 8. Meanwhile, if experimental data are available, relative cross sections $\sigma(T = 1, J = 0^+)/\sigma(T = 0, J = 1^+)$ can provide first valuable information of relative strength of IS and IV pairing correlations.

An experimental study of $(p, ^3$He$)$ reactions were performed with several even-even pf shell nuclei with $N = Z$. The cross sections to 0_1^+ and 1_1^+ states are extracted by second-order DWBA calculations with shell model wave functions. Experimental ratios $\sigma(0^+)/\sigma(1^+)$ are shown in Fig. 6.57. As a reference the ratios for the single-configuration case (two-particles in the $f_{7/2}$-orbit) are also reported. The trend of the experimental points is well reproduced by the DWBA with shell model two-nucleon wave functions calculated with effective interaction GXPF1. In the lower panel of Fig. 6.57, the cross-section of the T=1 channel is enhanced by a factor of 3 to 4 compared with the single-configuration limit when reaching the mid-shell. The enhancement is largely reduced at the shell closure in ^{56}Ni and reaches the single-configuration limit. On the other hand, the isospin-singlet channel shows no enhancement of the cross-section from mid-shell to closed shell and is always compatible with the single-configuration behavior. Given the good agreement of the theoretical ratios with the experimental results, one can conclude that no evidence for isospin-singlet pairing enhancement is observed neither for ^{52}Fe nor ^{56}Ni. One may speculate the reason of these features why the 1^+ strength is fragmented over the background quasi-particle states and/or much larger configuration space is needed for a large enhancement of the isoscalar 1^+ cross sections. Similar experiments involving the sdg-shell could be good candidates to search clear evidence of the enhancement of isoscalar 1^+ cross section, but one has to wait for new-generation radioactive-ion-beam facilities to this purpose.

6.7 Clusters in Neutron-Rich Nuclei

α clustering in nuclei is one of the important correlations to govern the structure of nuclear many-body systems. A variety of α cluster structures have been known in light stable nuclei. The most famous example is the Hoyle state: the excited 0_2^+ state of ^{12}C, whose energy is close to ^{4}He+^{8}Be threshold energy. This state has the dominant dilute gas-like 3α structure and was studied well both in experiments

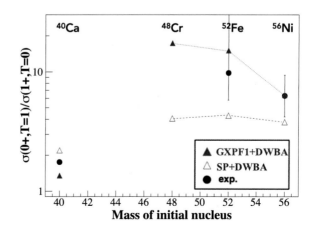

Fig. 6.57 Ratio $\sigma(0^+)/\sigma(1^+)$ obtained by (^{3}He, p) experiments (black dots), and for second order DWBA calculations with shell model wave functions (red triangle) and single particle picture (blue triangles). Figure and data are taken from [43]. Courtesy M. Assié, IJCLab

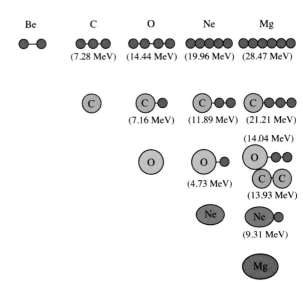

Fig. 6.58 Ikeda diagram. Chain configurations of cluster states and their threshold energies, where purple circles indicate the α clusters. The energy is given in unit of MeV. The chain structure is expected to appear near its decay threshold. Figure is based on an original figure from [41]

and theories (see Chap. 3 for details). The linear-chain configuration of 3α clusters in ^{12}C, in which α particles are linearly aligned, is another famous example of cluster structure which was firstly proposed by Haruhiko Morinaga in 1956. This idea of linear-chain configuration of α clusters was extended to other $4n$ nuclei from ^{8}Be (unbound) to ^{24}Mg by Kiyomi Ikeda 50 years ago. He predicted the existence of $n\alpha$ cluster structure, and the molecular-like structure of α clusters and residual nuclei near each threshold energy of the α cluster state shown in Fig. 6.58. The so-called Ikeda diagram indicates not only the linear-chain structure, but also bending chain structure, triangle shape in ^{12}C, and dilute cluster gas states near the threshold energies.

A qualitative argument of these cluster states is traced back to the tight binding of α particles. For example, the ground state of ^{16}O is predicted to be spherical by the standard shell model and HF calculations. When 4 particles (2 neutrons and 2 protons) are excited from the ground state, they can make a α cluster-like state with a certain probability. When the α cluster is created by the four particle excitation from the core, the residual nucleus will change its structure drastically and make a molecular-like structure of $\alpha+^{12}$C because of the tight binding of α cluster. This mechanism may cost some energy and the state will appear most probably near the α particle threshold. One can extend this argument on $4n$ particle excitations with ($n = 2, 3, \cdots$), continuously to make molecular-like and chain-like α cluster structure. This idea of the Ikeda diagram has been extended to $N = Z$ or $N = Z + 2$ nuclei with mass $A \sim 40$ and has been studied experimentally and theoretically with great efforts.

The instability of linear-chain structure against to the bending motion was pointed out recently by Anti-symmetrized Molecular Dynamics (AMD) and Fermion Molecular Dynamics (FMD) calculations (these models were discussed in details in

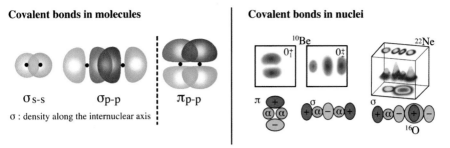

Covalent bonds in molecules

σ s-s   σp-p   πp-p

σ : density along the internuclear axis

Covalent bonds in nuclei

Fig. 6.59 Schematic representation of covalent bonds in molecules and nuclei. (Left) σ orbit between atoms with s- and p-states. π orbit between atoms with p-state. (Right) Covalent-bond neutron orbits in ^{10}Be and ^{22}Ne calculated by the AMD model [57]

Chap. 3) and thus the formation of linear chain structure is not so promising idea for the nuclei with mass $A = 4n$. The interest of the linear-chain structure configuration is revived by the recent progress of radioactive-ion-beam facilities because the valence neutrons of unstable nuclei may play a glue-like role to stabilize the linear-chain structure. These excess neutrons may exist in different types of geometrical orbits than the ordinary shell-model orbits classified by quantum numbers nlj. Contrarily they have a similarity to the orbits of molecular covalent bindings in chemistry.

6.7.1 Molecular Orbit and Linear-Chain Configuration

Two kinds of molecular orbits are discussed for neutrons in the multi-alpha chain configurations, so-called $\pi-$ and $\sigma-$orbits shown in Fig. 6.59a, b, which play glue-like roles in the chain-cluster formation. In analogy of molecular orbits in chemistry, σ and π orbits are defined with respect to the axis connecting cluster chain. In molecules, σ bonds are the strongest type of covalent chemical bond, also called molecular bond, which is a chemical bond that involves the sharing of electron pairs between atoms seen in Fig. 6.59b for an illustration. They are formed by a head-on overlapping between atomic orbitals. The σ bond is most simply defined for diatomic molecules using the language and tools of symmetry groups. In this formal approach, a σ-bond is symmetrical with respect to rotation about the bond axis. Another type of bond is called $\pi-$bond, shown in Fig. 6.59a, which changes phase with respect to rotation about the bond axis. Each of the atomic orbitals of $\pi-$bond has zero electron density at a plane, passing through the two-bonded nuclei.

A typical example of covalent bonding of α clusters in nuclei can be found in Be isotopes. A nucleus ^{8}Be has an extreme 2α cluster structure as shown in Fig. 6.58, but it is unstable against α decay. If one or two neutrons are added to ^{8}Be, the cluster structure is stabilized and the nuclei ^{9}Be and ^{10}Be are both bound. If more

excess neutrons are added, they enhance the clustering. ^{11}Be and ^{12}Be are known to
have pronounced 2α cluster core surrounded by valence neutrons and neutron magic
number $N = 8$ is broken in those isotopes. It was found that because of the two
center nature of the 2α cluster core, a special class of neutron single-particle orbits
different from the ordinary spherical shell is formed in Be isotopes. They are thus
called "molecular orbits" as an analogy to electron orbits for the covalent bonding
of atomic molecules.

Observed rotational bands in ^{8}Be, ^{9}Be and ^{10}Be are shown in Fig. 6.60. These
states have been populated by breakup reactions or resonant elastic scattering such
as ^{7}Li(^{3}He, p)^{9}Be*, ^{4}Be+^{4}Be→^{8}Be*, and ^{6}Be+^{4}Be → ^{10}Be*. In ^{9}Be, the ground state
is assigned by spin-parity $J^\pi = 3/2^-$, which is consistent with a naive shell model
picture; the last unpaired neutron occupies $1p_{3/2}$ orbit. In the Nilsson scheme (see
section 7.3.3), the last neutron is labeled by the Nilsson quantum number $K^\pi = 3/2^-$
The energy spectra is described by the rotational model with moment of inertia Θ,

$$Ex(J) = \frac{\hbar^2}{2\Theta}\left[J(J+1) + (-)^{J+1/2}a(J+\frac{1}{2})\delta_{K,1/2}\right] \quad (6.69)$$

where the second term is called Coriolis decoupling term, which has contributions
only for $K = 1/2$ band. In Fig. 6.60, the ground band of ^{9}Be has a sequence of states
$(3/2^-, 5/2^-, 7/2^-)$, which consist of a rotational band with the moment of inertia
$\hbar^2/2\Theta = 0.525$ MeV. $K^\pi = 1/2^+$ excited state at $E = 1.68$ MeV has a sequence of
positive parity states $(1/2^+, 3/2^+, 5/2^+, 7/2^+, 9/2^+)$, which is interpreted as a σ-
type molecular structure shown in Fig. 6.59b. As shown in the left panel of Fig. 6.60,
these two bands have similar average slopes although the $K^\pi = 1/2^+$ band has a
large staggering due to the Coriolis term. The moments of inertia extracted from the
slopes of two bands are close to that of ^{8}Be. This indicates that the $\alpha - \alpha$ molecular
structure of the core nucleus ^{8}Be is largely preserved in ^{9}Be. The right-hand panel
of Fig. 6.60 shows a systematics of negative parity band in ^{10}Be. This negative parity
configuration is also an indication of the molecular orbit associated with the $\alpha - \alpha$
molecular structure of core ^{8}Be. The rotational band in ^{10}Be shows a slower gradient
than those of ^{8}Be and ^{9}Be, which suggests a lager moment of inertia associated with
a more elongated shape.

An AMD wave function is expressed in general by a single Slater determinant.
However, the parity and angular momentum projected state for the ground state wave
function contains higher order correlations beyond the mean field approximation,
which is efficiently described by the variation after the angular-momentum and parity
projections (VAP) calculation. Indeed, cluster structures are remarkable in the present
VAP result in ^{9}Be and ^{10}Be, but they are relatively suppressed in calculations without
the projections. The spin-parity projections remove mixings of spurious components
in the wave functions and makes a clear localization of the intrinsic wave function.
For comparison with the present result obtained by the VAP, the result obtained by
the variation without the angular momentum and parity projections and that after the
parity projection without the angular-momentum projection are also demonstrated in
Fig. 6.61b and 6.61c, respectively. It is clearly seen that the result of ^{10}Be obtained by

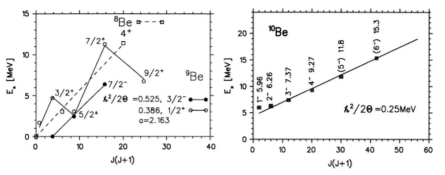

Fig. 6.60 (Left) Rotational states in ^{8}Be and ^{9}Be. (Right) Rotational states in ^{10}Be. The excitation energies are plotted as a function of $J(J+1)$. The moment of inertia $\hbar^2/2\Theta$ and Coriolis coupling constant are also given in the figures. Figure reprinted with permission from [45]. ©2021 by the American Physical Society

Fig. 6.61 Density distribution of the intrinsic wave functions of the ground states of ^{8}Be, ^{9}Be, and ^{10}Be obtained by **a** the AMD+VAP (variation after the angular-momentum and parity projections), **b** the variation after the parity projection without the angular-momentum projection, and **c** the variation without the angular-momentum and parity projections. The density integrated over the x-axis is shown on the z-y plane in the $|z| \leq 5$ fm and $|y| \leq 5$ fm region. Figure reprinted with permission from [21]. ©2021 by the American Physical Society

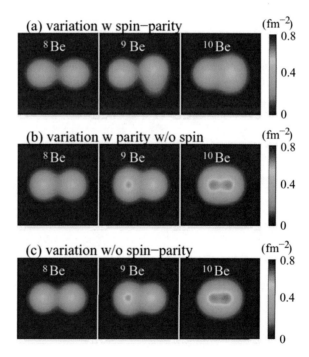

the variation without the projections shows weak clustering with a parity-symmetric intrinsic structure (see the right-hand panels of Fig. 6.61b, c). This indicates that the angular-momentum and parity projections in the energy variation are essential to obtain the parity-asymmetric structure with the ^{6}He+α correlation in ^{10}Be.

The molecular orbits play also an important role for the enhancement of the clustering and the breaking of the neutron magic number in C isotopes. This finding motivates the search for the very exotic cluster linear chain state in neutron-rich carbon isotopes which is composed of linearly aligned three α clusters with surrounding neutrons. AMD+GCM calculations were performed to study linear-chain configurations of C isotopes.

The density distributions of positive parity and negative parity states in ^{14}C are shown in Fig. 6.62. The intrinsic wave function of the ground state in Fig. 6.62a is located at the energy minimum at a normal deformation $\beta = 0.36$. Figure 6.62b–d show the density distributions of excited local minima having extremely large prolate deformations $\beta = 0.6$, 1.00 and 1.27, respectively. For the ground state in Fig. 6.62a, one can not see any clear localization of the wave function. On the other hand, the proton density distribution in Fig. 6.62b has a triangular shape showing a possible formation of 3α cluster core with a triangle configuration. At a larger deformation shown in Fig. 6.62c, there is another local energy minimum which corresponds to the axially symmetric deformed potential with frequency ratio equal to $(\omega_x = \omega_y = \omega_\perp : \omega_z) = (3 : 1)$. The density shows a linear-chain structure with higher energy than the minimum of Fig. 6.62(b). This intrinsic wave function has the most dominant component of the π-bond linear-chain configuration. The valence neutron wave functions indicated by a dark shadow show the character of π-orbit in the molecular picture illustrated schematically in Fig. 6.59. With a further increase of the deformation, a more elongated linear-chain configuration appears in Fig. 6.62d. From the density distribution in Fig. 6.62d, it is clear that this configuration has another valence configuration corresponding to the σ-orbit, analogous to Fig. 6.59.

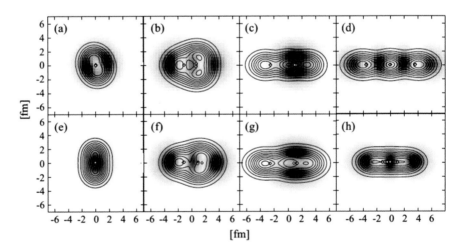

Fig. 6.62 The density distribution of **a–d** the positive parity states and **e–h** negative parity states in ^{14}C. The contour lines show the proton density distributions. The color plots show the single-particle orbits occupied by the most weakly bound neutron. Open small diamonds show the centroids of the Gaussian wave packets describing protons. Figure reprinted with permission from [46]. ©2021 by the American Physical Society

Figure 6.62e–h are the negative parity states corresponding to the positive parity states (a)–(d), respectively.

The calculated and experimental positive parity states are shown in the right panel of Fig. 6.63. In the three columns from the left, all observed positive-parity states are shown. In the fourth column from the left, the observed states by ^{8}Be+α breakup reactions are shown. The lower three states are strongly populated by the ^{8}Be(0_1^+)+α channel, while the upper 3 states are populated by ^{8}Be(2_1^+)+α channel. The calculated states are also classified by the components of wave functions and the connecting $B(E2)$ values. The two calculated bands from the left corresponds to the ground band and the $K^\pi = 0_2^+$ band, respectively. The experimental candidate of the $K^\pi = 0_2^+$ band is found in the second column from the left. The third band from the right corresponds to the π-bond linear chain cluster state (LCCS). Promising experimental

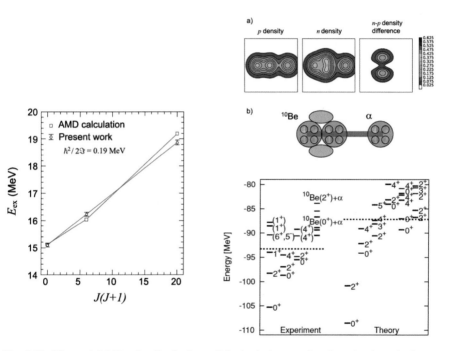

Fig. 6.63 (Upper right) Density distributions of the intrinsic wave functions dominant in the π-bond linear chain cluster state (LCCS) in ^{14}C calculated by the AMD method. **a** Proton density ρ_p, neutron density ρ_n and the difference between them, respectively. The box size is 10×10 fm^2 for all. **b** An intuitive picture of the above densities. (Bottom right) Energy levels of the positive-parity states in ^{14}C. The four columns on the left are the experimental data and the six columns on the right are the calculated results. The dotted lines on the left and right show the experimental and theoretical ^{10}Be + α threshold energies, respectively. The third band from the right corresponds to LCCS band. Figure reprinted with permission from [20]. ©2021 by the American Physical Society. (Left) The calculated and observed linear-chain candidates with positive parity in ^{14}C. The $J(J+1)$-dependence of E$_{ex}$ for the band identified by ^{10}Be+α inelastic scatterings, and the LCCS band predicted by AMD calculations. Figure reprinted with permission from [48]. ©2021 by Elsevier. AMD calculated results are taken from [20]

data are reported recently on the molecular orbit of C isotopes by ^{10}Be+α resonant scatterings with radioactive beams. The band head, the lowest energy state of the rotational band, is found at $E_x = 15.1$ MeV and the observed moment of inertia is very close to the calculated one shown in the left panel of Fig. 6.63. The σ–bond linear chain band and the negative parity LCCS are predicted at higher energy than the π–bond LCCS, but so far we do not have any clear experimental candidates. Further experimental efforts are still needed to establish these linear chain configurations assisted by the molecular orbits.

6.7.2   Oscillation of Cluster States

The dipole excitation in Be isotopes is studied in relation with the vibration between α clusters. The α-cluster configuration is taken care by a GCM with the distance between clusters as a generator coordinate (αGCM). The vibration is included by a shift of Gaussian centroid of each single-particle wave function (named sAMD). This method simulates effectively the RPA like $1p$-$1h$ excitations as the vibration between clusters. The AMD wave function was introduced in Sect. 3.8 of Chap. 3. Since the projected states are superposed in the sAMD, the couplings of the $1p$-$1h$ excitations with the rotation and parity transformation are taken into account. Therefore, sAMD contains, in principle, higher correlations beyond the RPA in the mean-field approximation.

In the ground-state wave functions obtained by the AMD+VAP for ^{8}Be, ^{9}Be, and ^{10}Be, α clusters are formed as shown in Fig. 6.61, even though any clusters are not a priori assumed in the framework. Consequently, Gaussian centroids Z_i in Eq. (3.178) for two protons and two neutrons are located at almost the same position. The intercluster motion of ^{5}He + α and ^{6}He + α structures in ^{9}Be and ^{10}Be can be excited by the dipole operators.

The sAMD + αGCM calculations are performed for E1 strength distributions of ^{9}Be in Fig. 6.64. The GDR in ^{9}Be shows a two-peak structure in the energy region $E = 20 \sim 40$ MeV. In addition to the GDR, low-lying $E1$ strength, so-called pigmy resonance, appears in $E = 10 \sim 20$ MeV. The origin of the two-peak structure of the GDR in Be isotopes is the prolate deformation of the 2α core. The longitudinal motion is responsible for the lower GDR peak at $E \sim 25$ MeV, while the transverse motion dominates the higher energy GDR at $E \sim 35$ MeV. The valence neutron modes couple with the transverse GDR mode of the 2α core broaden the higher peak of the GDR. The strength at $E \sim 10$ MeV in ^{9}Be is mainly contributed by the longitudinal motion of valence neutrons against the 2α core. Qualitatively, the structure of experimental data can be interpreted by the sAMD + αGCM calculations. However, quantitatively, the calculations predict almost twice more strength than the experimental ones in the GDR region. The peak position of pigmy resonance below GDR is also 10 MeV higher than the experiment. Because of these disagreements, it is not very certain whether the experimental data prove the vibration of molecular states. One needs

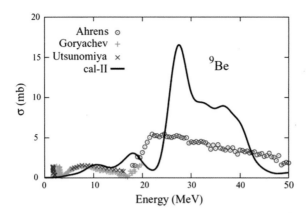

Fig. 6.64 Comparison of the calculated E1 cross section of ^{9}Be with the experimental photonuclear cross sections. The calculated values are those obtained with the sAMD + αGCM (denoted cal-II), smeared by a Lorenzian function with the width $\gamma = 2$ MeV. The experimental data are taken from the photonuclear cross sections by Ahrens et al., the bremsstrahlung data by Goryachev et al., and the photodisintegration cross sections by Utsunomiya et al. Figure reprinted with permission from [21]. ©2021 by the American Physical Society

further theoretical works to predict more quantitatively the energy spectra of both GDR and Pigmy states.

Summary

Radioactive-ion beams can be produced either by the in-flight technique, where the reaction residues are produced at the beam velocity in a thin target, or by the isotopic-separation on-line (ISOL) technique where the reaction residues are stopped in the production target and exit the target chamber after diffusion and effusion. ISOL residues need to be ionized before acceleration towards the experiment. The ISOL method is limited to radioactive nuclei with a half life greater than several tens of ms.

Both RIB production methods require an accelerated primary beam of stable isotopes impinging on a target. The acceleration of the primary beams, together with the building for the appratus, is the most expensive part of RIB productions. There are three types of proton and heay-ion accelerators commonly used for that purpose: (i) linear accelerators composed of a series of accelerating cavities as in the case of FRIB in the USA, (ii) cyclotrons in which the beam is bent in a constant magnetic field with increasing bending radii in between accelerating sections as in the case of the RIBF in Japan, (iii) synchrotrons where ions are accelerated in at a fixed bending radius with a magnetic field increasing proportionally to the acquired velocity. Synchrotrons are the basis of the acceleration of heavy-ions at GSI/FAIR.

A large enhancement of the reaction cross section is found in ^{11}Li and ^{6}He compared to neighbouring isotopes and confirmed as certain evidence of neutron halo. The very small neutron separation energy is a trigger to develop the halo structure in these nuclei.

The loose-bound single-particle wave function with the low angular momentum $l \le 1$ dominates the halo wave function, having a divergence feature in the mean square radius at the limit of single particle energy $\varepsilon \to 0$.

The shell structure is not universal across the nuclear chart but varies with the number of protons and neutrons. Shell closures established for stable nuclei may disappear away from stability, as it is the case for N=8, 20, 28 for neutron rich nuclei, while new shell closures may develop as evidenced for N=16, 32 and 34. Shell evolution is interpreted within the shell model as steaming from a competition between the monopole drift of single-particle orbitals (monopole part of the effective Hamiltonian) and correlations (multipole part of the Hamiltonian). The structure of neutron rich N=50 isotones, below ^{78}Ni, and around ^{60}Ca are today still an open question and should be addressed in the coming decades. Three-body interactions play an important role as well in the evolution of the structure of nuclei as shown, for example, in the binding energy of oxygen isotopes.

In very neutron-rich nuclei with $A \sim 32$, the shell structure is largely changed, and the ground-state configurations are dominated by $2p$-$2h$ configurations in the standard shell model picture. This region of mass chart is called "island of inversion".

The three-body model with the Hamiltonian

$$H = h_{nC}(1) + h_{nC}(2) + V_{nn} + \frac{\mathbf{p}_1 \cdot \mathbf{p}_2}{A_c m}$$

was introduced to study di-neutron and di-proton correlations in exotic nuclei near the drip lines. A large enhancement of Coulomb break-up cross sections is found experimentally just above the neutron threshold in the halo nuclei. The three-body model is applied successfully to describe the enhancement of Coulomb break-up cross sections in nuclei such as ^{11}Li and ^{6}He.

The resonant ground state of ^{26}O was observed just above the threshold by the inverse kinematics of $2n$ decay of ^{26}O, which was created by the proton pickup reaction on ^{27}F projectile. The fingerprint of strong di-neutron correlations is pointed out in $2n$ decay of ^{26}O.

The role of isoscalar (IS) spin-triplet pairing interaction is pointed out to enhance the magnetic dipole transitions in $N = Z$ odd-odd nuclei.

The pair transfer reactions $(p, ^3\text{He})$ or $(^3\text{He}, p)$ are proposed to study the strong pair correlations of IS pairing in the ground state of $N = Z$ nuclei.

Molecular type orbits play an important role to bind cluster chain-like states in neutron-rich nuclei, such as ^{16}C and ^{10}Be. sAMD+αGCM calculations were applied to study the molecular structure of these neutron-rich nuclei in the dipole excitation spectra.

Solutions of Problems

6.1 In the projectile rest frame the net momentum of the nucleons is zero $\mathbf{p}_A = 0$, i.e., $\sum_{n=1}^{A} \mathbf{p}_n = 0$. Using this requirement

$$\left[\sum_{n=1}^{A} \mathbf{p}_n\right]^2 = 0 = \sum_{n=1}^{A} \mathbf{p}_n \sum_{m=1}^{A} \mathbf{p}_m = \sum_{n=1}^{A} \mathbf{p}_n^2 + \sum_{n \neq m}^{A} \mathbf{p}_n \mathbf{p}_m. \tag{6.70}$$

We can re-write the equation in terms of averages. The average value $\langle x \rangle$ of a quantity x over a population of N occurrences is given by

$$\sum_{i=1}^{N} x_i = N \langle x \rangle, \tag{6.71}$$

which one can substitute into Eq. (6.70):

$$0 = \sum_{n=1}^{A} \mathbf{p}_n^2 + \sum_{n \neq m}^{A} \mathbf{p}_n \mathbf{p}_m = A \langle \mathbf{p}_n^2 \rangle + A(A-1) \langle\langle \mathbf{p}_n \mathbf{p}_m \rangle\rangle, \tag{6.72}$$

where the double bracket indicates an average over all $n \neq m$. One then gets

$$\langle\langle \mathbf{p}_n \mathbf{p}_m \rangle\rangle = -\frac{\langle \mathbf{p}_n^2 \rangle}{A-1}. \tag{6.73}$$

Next we consider the fragment momentum. In the projectile frame, one has

$$\mathbf{p}_f = -\sum_{n=1}^{\Delta A} \mathbf{p}_n. \tag{6.74}$$

The fragment momentum is Gaussian distributed with σ. By the variance definition $\sigma^2 = E[x^2] - E[x]^2$. In our case, taking $\mu = 0 \;\rightarrow\; \sigma^2 = \langle x^2 \rangle$. Writing explicitly

$$\sigma^2 = \langle \mathbf{p}_f^2 \rangle = \langle (\sum_{n=1}^{\Delta A} \mathbf{p}_n)^2 \rangle = \Delta A \langle \mathbf{p}_n^2 \rangle + \Delta A(\Delta A - 1) \langle\langle \mathbf{p}_n \mathbf{p}_m \rangle\rangle. \tag{6.75}$$

One can replace $\langle\langle \mathbf{p}_n \mathbf{p}_m \rangle\rangle$ with Eq. (6.73)

$$\sigma^2 = \Delta A \langle \mathbf{p}_n^2 \rangle - \frac{\Delta A (\Delta A - 1)}{A - 1} \langle \mathbf{p}_n^2 \rangle$$

$$= \langle \mathbf{p}_n^2 \rangle \frac{\Delta A}{A - 1} (A - 1 - \Delta A + 1)$$

$$= \langle \mathbf{p}_n^2 \rangle \frac{\Delta A}{A - 1} A_f, \tag{6.76}$$

where the number of nucleons in the fragment equals the difference between the A nucleus in the projectile and the number of removed nucleons, assuming no evaporation of nucleons.

Usually, one measures one component of the momentum distribution, for example the component along the beam axis. In the nucleus, all components of the intrinsic momentum are equal ($p_x = p_y = p_z$). Since $p^2 = p_x^2 + p_y^2 + p_z^2$, one needs to consider $\langle \mathbf{p}_n^2 \rangle / 3$ for a single Cartesian component of the momentum. In this case,

$$\sigma_z^2 = \frac{1}{3} \langle \mathbf{p}_n^2 \rangle \frac{\Delta A}{A - 1} A_f, \tag{6.77}$$

where we arbitrarily chose the z component.

The average mean square momentum of the nucleons is given by

$$\langle \mathbf{p}_n^2 \rangle = \frac{\int_0^k p^2 \rho(\mathbf{p}) d\mathbf{p}}{\int_0^k \rho(\mathbf{p}) d\mathbf{p}}. \tag{6.78}$$

We consider a constant density $\rho(\mathbf{p}) = C$, and take k the upper limit of the integral as the Fermi momentum: if we consider the nucleus as a Fermi gas at zero temperature, the nucleons fill the energy levels up to the Fermi energy, with p_F being its corresponds momentum.

$$\langle \mathbf{p}_n^2 \rangle = \frac{\int_0^{p_F} p^4 dp}{\int_0^{p_F} p^2 dp} = \frac{3}{5} p_F^2. \tag{6.79}$$

Substituting into Eq. (6.77) gives Eq. (6.20), i.e.,

$$\sigma_z^2 = \frac{1}{5} p_F^2 \frac{\Delta A}{A - 1} A_f. \tag{6.80}$$

The production of ^{10}Be from fragmentation of ^{12}C corresponds to $\Delta A = 2$. Taking $p_F = 200$ MeV/c and substituting into Eq. (6.20) gives

$$\sigma_z^2 = \frac{1}{5} (200)^2 \frac{2 \cdot 10}{11} \Rightarrow \sigma = 120 \text{ MeV/c}, \tag{6.81}$$

very close to the experimental value of the parallel momentum width of ^{10}Be after two nucleon removal from ^{12}C at 2.1 GeV/nucleon shown in Fig. 6.18.

6.2 A loosely-bond nucleon is bound in the one-dimensional potential in Fig. 6.65;

$$V(x) = \begin{cases} \infty & x < 0, \quad (\text{region I}) \\ V_0(< 0) & 0 \le x \le L, \quad (\text{region II}) \\ 0 & L < x, \quad (\text{region III}) \end{cases} \qquad (6.82)$$

This one-dimensional potential problem can be solved in a similar way to be described in Sect. 1.2. We consider only the wave function in the region III with $E < 0$. The wave function is given by

$$\phi_{\text{III}}(x) = Ae^{\kappa x} + Be^{-\kappa x} \qquad (6.83)$$

where

$$\kappa = \sqrt{\frac{2m|E|}{\hbar^2}}. \qquad (6.84)$$

Since the penetration probability is finite, it is enough to consider the second term in Eq. (6.83), but to discard the first term which gives a divergence. The mean square radius of the wave function (6.83) is evaluated as

$$< x^2 > = \frac{\int_L^\infty x^2 |B|^2 e^{-2\kappa x} dx}{\int_L^\infty |B|^2 e^{-2\kappa x} dx} \sim \frac{\int_0^\infty x^2 |B|^2 e^{-2\kappa x} dx}{\int_0^\infty |B|^2 e^{-2\kappa x} dx}$$

$$= \frac{1}{2\kappa^2} = \frac{\hbar^2}{4m|E|}, \qquad (6.85)$$

where the lower boundary $x = L$ of the integration is replaced by $x = 0$ since we consider the loosely bound wave function having a large extension of wave function beyond L. The radius will diverge in the limit of $|E| \to 0$. If one inserts $E = -400$ keV in Eq. (6.85) with the nucleon mass $mc^2 = 1000$ MeV and $\hbar c = 200$ MeV fm, the rms radius will be $\sqrt{< x^2 >} = 5, 0$ fm which is ~ 2 times larger than the radius of stable nuclei with mass $A = 10-20$.

Fig. 6.65 The one-dimensional square well potential for a loosely-bound nucleon. E is the binding energy of a particle, while V_0 and L are the potential depth and width, respectively

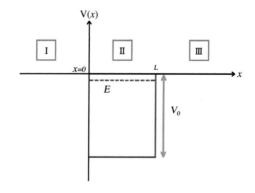

Table 6.3 Amplitudes a and b and eigen-energies for three different values of ε_{2s} state	ε_{2s}	$E(0^+)$	a	b
	0.5	-0.618	0.851	0.526
		1.618	-0.526	0.851
	1.0	-0.414	0.924	0.383
		2.414	-0.383	0.924
	3.0	-0.162	0.987	0.160
		6.162	-0.160	0.987

6.3 To obtain the eigenvalue of the Hamiltonian (6.47), we introduce the matrix form of the wave function

$$|\Phi\rangle = \begin{pmatrix} a \\ b \end{pmatrix}. \tag{6.86}$$

For this wave function, the secular equation for the eigenvalues becomes

$$\begin{pmatrix} \langle a|H|a\rangle & \langle a|H|b\rangle \\ \langle b|H|a\rangle & \langle b|H|b\rangle \end{pmatrix} \begin{pmatrix} a \\ b \end{pmatrix} = \begin{pmatrix} 2\varepsilon_{1p1/2} & -1.0 \\ -1.0 & 2\varepsilon_{2s1/2} \end{pmatrix} \begin{pmatrix} a \\ b \end{pmatrix} = E \begin{pmatrix} a \\ b \end{pmatrix} \tag{6.87}$$

The amplitudes a and b of ^{11}Li wave function (6.46) can be obtained by diagonalizing the above 2×2 matrix. The amplitudes and eigen-energies are tabulated in Table 6.3 for three different single-particle energies of $2s$-orbit.

Books for Further Readings

Chapter 9 of "100 years of subatomic physics" (edited by E. M. Henly and S. D. Ellis, World Scientific, 2013) by K. Hagino, I. Tanihata and H. Sagawa, reports on both theoretical and experimental progresses in the study of exotic nuclei far from the stability line.

References

1. M. Thoenessen, *The Discovery of Isotopes* (Springer, 2016)
2. J. Erler et al., Nature **486**, 509 (2012)
3. N. Sakudo et al., Rev. Sci. Inst. **49**, 940 (1978)
4. M. Leitner et al., in *Proceedings of the 16th International conference on RF superconductivity (SRF2013)*, vol. 1 (2013)
5. J.R.J. Bennett, Nucl. Inst. Meth. B **204**, 215 (2003)
6. P. Van Duppen, In *Lecture Notes in Physics* (Springer, 2006)
7. M.C. Etchegoyen, A. Etchegoyen, B.H. Wildenthal, B.A. Brown, J. Keinonen, Phys. Rev. C **38**, 1382 (1988)
8. A.C.C. Villari et al., NPA **787**, 126 (2007)

9. O. Bajeat et al., NIMB **317**, 411 (2013)
10. A.S. Goldhaber, Phys. Lett. B **53**, 306 (1974)
11. H. Geissel, M. Huyse, G. Münzenberg, P. Van Duppen, Exotic Nuclear Beam Facilities, Encyclopedia of Applied Physics (2014)
12. H. Takeda et al., Nucl. Instr. Meth. Res. B **463**, 515 (2020)
13. O. Tarasov, D. Bazin, NIM B **266**, 4647 (2008)
14. T. Kobayashi et al., Phys. Rev. Lett. **60**, 2599 (1988)
15. O. Tarasov et al., Phys. Rev. C **75**, 064613 (2007)
16. I. Tanihata et al., Phys. Rev. Lett. **55**, 2676 (1985)
17. O. Kiselev et al., Nucl. Inst. Res. Meth. A **641**, 72 (2011)
18. P. Egelhof et al., Eur. Phys. Jour. A **15**, 27 (2002)
19. A.V. Dobrovolsky et al., Nucl. Phys. A **766**, 1 (2006)
20. T. Suhara, Y. Kanada-En'yo, Phys. Rev. C **82**, 044301 (2010)
21. Y. Kanada-En'yo, Phys. Rev. C **93**, 024322 (2016)
22. M. Fukuda et al., Phys. Lett. B **268**, 339 (1991)
23. G. Hagen et al., Phys. Rev. Lett. **108**, 242501 (2012)
24. E.K. Warburton et al., Phys. Rev. C **41**, 1147 (1990)
25. N. Fukunishi et al., Phys. Lett. B **296**, 279 (1992)
26. W. Schwerdtfeger et al., Phys. Rev. Lett. **103**, 012501 (2009)
27. K. Wimmer et al., Phys. Rev. Lett. **105**, 252501 (2010)
28. U. Lombardo, H-J. Schulze, "Physics of Neutron Star Interiors", Lecture Notes in Physics, vol. 578, p.30 (Edited by D. Blaschke, N.K. Glendenning and A. Sedrakian)
29. K. Hagino, H. Sagawa, Phys. Rev. C **72**, 044321 (2005)
30. Y. Kondo et al., Phys. Rev. Lett. **116**, 102503 (2016)
31. Y. Tanimura et al., Prog. Theor. Exp. Phys. 053D02 (2014)
32. K. Yoshida, Phys. Rev. C **90**, 031303(R) (2014)
33. P. Sarriguren et al., Phys. Rev. C **76**, 044322 (2007)
34. I. Talmi, I. Unna, Phys. Rev. Lett. **4**, 469 (1960)
35. A. Ozawa et al., Phys. Rev. Lett. **84**, 5493 (2000)
36. K. Tshoo et al., Phys. Rev. Lett. **109**, 022501 (2012)
37. B. Bastin et al., Phys. Rev. Lett. **99**, 022503 (2007)
38. H. Liu et al., Phys. Rev. Lett. **122**, 072502 (2019)
39. S. Péru et al., Eur. Phys. J. A **9**, 35 (2000)
40. T. Otsuka et al., Phys. Phys. Lett. **95**, 232502 (2005)
41. K. Ikeda et al., Supplement of Prog. Theor. Phys. Extra Number, p. 464 (1968)
42. W. von Oertzen, M. Freer, Y. Kanada-En'yo, Phys. Rep. **432**, 43 (2006)
43. B. Le Crom et al., preprint arXiv: 2104.10708 (2021)
44. N. Itagaki et al., Phys. Rev. C **64**, 014301 (2001)
45. M. Freer, H. Horiochi, Y. Kanada-En'yo, D. Lee and Ulf.-G. Meißner, Rev. Mod. Phys. **90**, 035004 (2018)
46. T. Baba, M. Kimura, Phys. Rev. C **94**, 044303 (2016)
47. T. Baba, M. Kimura, Phys. Rev. C **95**, 064318 (2017)
48. H. Yamaguchi et al., Phys. Lett. B **766**, 11 (2017)
49. C. Potel et al., Phys. Rev. Lett. **107**, 092501 (2011)
50. E. Garrido et al., Phys. Rev. C **63**, 037304 (2001)
51. W. Satula, D.J. Dean, J. Gary, S. Mizutori, W. Nazarewicz, Phys. Lett. B **407**, 103 (1997)
52. Y. Kikuchi et al., Phys. Rev. C **87**, 034606 (2013)
53. K. Ikeda, T. Myo, K. Kato, H. Toki, Clusters in Nuclei. Lecture notes in Physics 818 (2010)
54. W. Schwerdtfeger et al., Phys. Rev. Lett. **103**, 012501 (2009)
55. K. Hagino, H. Sagawa, Phys. Rev. C **89**, 014331 (2014)
56. K. Hagino, H. Sagawa, Phys. Rev. C **76**, 047302 (2007)
57. Y. Kanada-En'yo, M. Kimura, and A. Ono, Prog. Theo. Exp. Phys. 01A202 (2012)
58. E. Caurier, F. Nowacki, A. Poves, Phys. Rev. C **90**, 014302 (2014)

Chapter 7
Deformation and Rotation

Abstract In this chapter, the microscopic origin of nuclear deformation is studied in terms of mean field and shell model. The observation of rotational spectra and electromagnetic moments are shown as evidence of nuclear deformation. Other phenomena related to nuclear shapes are also discussed such as backbendings in moments of inertia, shape coexistence and octupole deformation.

Keywords Nuclear deformation and spontaneous symmetry breaking ·
Backbending of moment of inertia · Superdeformation and hyperdeformation ·
Shape coexistence · Octupole deformation

7.1 Deformation of Molecules and Nuclei

A common feature of quantum systems which have rotational spectra is the existence of deformation, implying an anisotropy of their mass distribution that makes possible to specify the orientation of system as a whole. In the nucleus, the rotational degree of freedom is associated with the deformation in the nuclear equilibrium shape. In a molecule, as a solid body, the deformation reflects the highly anisotropic mass distribution in the intrinsic coordinate frame defined by the equilibrium positions of atoms. A similar rotational structure is also observed in the hadron spectra (referred to as Regge trajectories) in the isospin space and also in the particle number space.

7.1.1 Diatomic Molecule and Deuteron

A diatomic molecule, for example, H_2, has a dumbbell shape, while a molecule of three atoms, such as H_2O, has a triangular shape with fixed inter-ion distances. It seems very natural that molecules are deformed. What about nuclei? A nucleus is a system composed of a few to hundreds of nucleons, just like molecules are systems combined of atoms. At first sight, one may infer that nuclei are deformed as well. However, this naive argument is to be taken with caution.

© Springer Nature Singapore Pte Ltd. 2021
A. Obertelli and H. Sagawa, *Modern Nuclear Physics*, UNITEXT for Physics,
https://doi.org/10.1007/978-981-16-2289-2_7

Let us compare the scale of the nuclear force between nucleons to the chemical force between atoms. This comparison was first introduced by Bohr and Mottelson in [1]. All quantum mechanical systems undergo fluctuations even in their ground states and have an associated energy as a consequence of their wave-like nature. This fluctuation energy ΔE is called the zero-point energy and can be taken as the energy scale of a many-body quantum system. It can be estimated from the uncertainty principle $\Delta p \Delta x \sim \hbar$ (see Chap. 1) for the momentum and position fluctuations. For any strongly bound system, the zero-point fluctuation energy is small in comparison with its binding energy, while a weakly-bound system turns out to have a relatively large energy fluctuation. The zero-point fluctuation energy reads

$$\Delta E = \frac{(\Delta p)^2}{2m} \sim \frac{\hbar^2}{m(\Delta x)^2} \sim \frac{\hbar^2}{mr_c^2}, \tag{7.1}$$

where r_c is the size of core potential and m stands for the mass of the considered atom or nucleon. The mass of a hydrogen atom is almost equivalent to that of a nucleon. The chemical force between atoms is determined by the electronic structure of molecules. For a molecule whose inter-atom distance a is similar to the Bohr radius $\hbar^2/m_e e^2 \simeq 0.5\text{Å}$ ($1 \text{ Å} = 10^{-10}$ m), the interaction, which is governed by the Coulomb interaction between two bound electrons, is characterized as

$$V_{\text{mol}} \sim \frac{e^2}{a} \sim \frac{\hbar^2}{m_e a^2}, \tag{7.2}$$

where m_e is the electron mass. Compared to the zero-point energy (7.1) of atoms in the molecule, the strength of molecular interaction (7.2) is a factor $V_{\text{mol}}/\Delta E \sim m_p/m_e \sim 2000$ times stronger than the size of the zero-point fluctuation of an atom (taking $r_c \sim a$). For such a strong potential, the so-called Morse potential, the ground-state wave function is strongly peaked at the minimum of the potential as can be seen in the left panel of Fig. 7.1. In this kind of system, all atomic wave functions are well localized in space and behave like classical objects. This many-body wave function shows a vibrational-rotational spectrum at the minimum of the attractive potential since $\Delta E \ll V_{\text{mol}}$. What about the strength of the potential in a di-nucleon system? Following Eq. (7.1) and taking a repulsive core of nuclear potential $r_c \simeq 0.5$ fm, the energy scale will be $\Delta E = 156$ MeV in nuclei. On the other hand, the central nuclear force is relatively weak and not enough to make a bound state of deuteron without the additional tensor interaction (see Chap. 3). The binding energy of deuteron is $B_n(d) = 2.225$ MeV, which gives a ratio $B_n(d)/\Delta E \sim 10^{-2}$, two orders of magnitude smaller than the typical energy scale of nucleus as is seen in the right panel of Fig. 7.1. Even considering the average nuclear binding energy of $B_n \simeq 8$ MeV/nucleon in medium and heavy nuclei, the ratio $B_n/\Delta E \sim 1/20$ is much smaller than the energy scale of molecule ~ 2000, i.e., nuclear systems make barely tightly bound many-body systems and an independent-particle motion is a good approximation. Thus, deformation and rotational spectra should be conceptually very different from those of molecules.

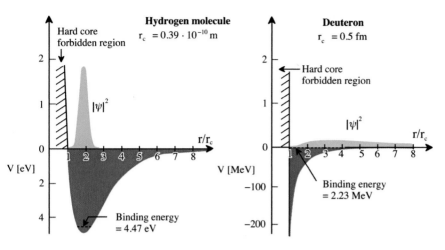

Fig. 7.1 Molecular and nuclear potentials and corresponding wave functions of diatomic molecule and deuteron, respectively. (Left) A Morse potential for a diatomic molecule which has a binding energy of 4.47 eV. An appropriate unit of energy for the molecule is $\hbar^2/m_p r_c^2 = 2.61 \times 10^{-2}$ eV where m_p is the mass of an atom (proton) and r_c is the extension of repulsive short-range potential. For the molecule, the value of r_c is taken as $r_c = 0.4\text{Å} = 0.4 \times 10^{-10}$ m, which is the order of the atomic radius $a \sim \hbar^2/e^2 m_e \sim 0.5$ Å. The binding energy corresponds to about 2000 times the energy unit of a molecule. (Right) A nuclear interaction represents a typical triplet-even central potential with a hard core at $r/r_c = 1$ where r_c is a typical size of the hard core $r_c = 0.5$ fm. The appropriate energy unit for nucleus is $\hbar^2/m r_c^2 = 156$ MeV where the nucleon mass is $mc^2 = 939$ MeV. The binding energy of deuteron 2.225 MeV corresponds to 1.4×10^{-2} in terms of the energy unit of nucleus. The probabilities of wave functions $|\phi(r/r_c)|^2$ are shown in both sides of figure

A schematic view of the change of the nuclear mean-field potential with the number of nucleons is illustrated in Fig. 7.2 as a function of axial quadrupole deformation, quoted β_2[1]. The nuclear potential of a closed-shell configuration shows a spherical minimum at the quadrupole deformation $\beta_2 = 0$. The potential becomes shallower when more particles are added to the closed configuration. Eventually, the potential has a deformed minimum at $\beta_2 \neq 0$ when the number of particles outside of the closed configuration reaches to the critical value to induce the "spontaneous symmetry breaking (SSB)" in the nuclear potential. This figure illustrates clearly the concept of SSB in nuclear physics when more and more particles are added to a closed shell core. The change from spherical to deformed nuclei as a function of mass number can be seen in the experimental spectra in Fig. 7.10. We will discuss a relation between the symmetry breaking mechanism and a Hamiltonian induced by the quadrupole vibration in the next subsection.

---

[1] β_2 is often used as an alternative to specify the quadrupole deformation, β. The definition of β_2 is the same as that of β. Hereafter, we adopt β_2 to specify the quadrupole deformation. The symbol δ specifies also the quadrupole deformation as defined in this section.

Potential energy V

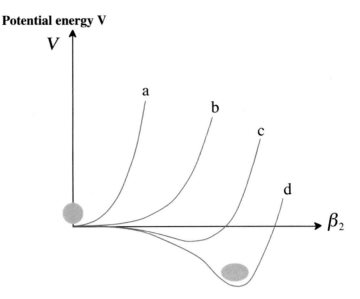

Fig. 7.2 Sketch of the nuclear potential as a function of the deformation parameter β_2. The potential energy surface a corresponds to the potential of a closed-shell spherical configuration. Adding more and more particles to the closed shell, the potential will change to the potentials b, c, d. Eventually the potential shows a deformed minimum at a finite β_2 value. This is an illustrative example of spontaneous symmetry breaking (SSB) in nuclear many-body systems

Problem

7.1 Evaluate typical energy scales of two atoms and two nucleons by using Eq. (7.1).

7.1.2 Harmonic Vibration and Particle-Vibration Coupling Hamiltonian

The Jahn–Teller effect (JT effect) is an important mechanism of spontaneous symmetry breaking (SSB) in molecular and solid-state systems. The effect is caused by an interplay between degenerate electron states and a molecular vibration, giving rise to a geometrical distortion of the molecule. In nuclear physics, an analogous situation is encountered in non-magic nuclei with nearly degenerate single-particle states, leading to the possible occurrence of an instability of the spherical symmetry and of a stable deformation as illustrated in Fig. 7.2. That is, the quadrupole collective motion will induce an instability of spherical shape and implement eventually a stable deformation as a nuclear Jahn–Teller effect. The key concept is the interaction between degenerate single-particle states and collective vibration as discussed in the following.

We study the nuclear Jahn–Teller effect through the change of the mean-field potential by the collective vibration, based on the idea of self-consistency between the vibrating potential and density by Bohr and Mottelson [1]. We start from the harmonic vibration model and introduce the particle-vibration Hamiltonian as a dynamical effect to induce the instability of the spherical shape. The Hamiltonian of the harmonic vibration is expressed by the vibrational amplitude α as

$$H = T + V = \frac{1}{2} B \dot{\alpha}^2 + \frac{1}{2} C \alpha^2, \tag{7.3}$$

where B is the mass parameter and C is the restoring force parameter. The momentum variable is obtained by using the Lagrangian $\mathcal{L}$,

$$\pi = \frac{\partial}{\partial \alpha} (\mathcal{L}) = \frac{\partial}{\partial \alpha} (T - V) = B \dot{\alpha}. \tag{7.4}$$

In the Hamiltonian, the vibrational amplitude α and its conjugate momentum π are considered as the operators, and satisfy the canonical commutation relation

$$[\pi, \alpha] = \frac{\hbar}{i}. \tag{7.5}$$

The variables π and α are quantized introducing the phonon creation and annihilation operators $O^\dagger$ and O as

$$\alpha = \left(\frac{\hbar}{2B\omega} \right)^{1/2} (O^\dagger + O), \tag{7.6}$$

$$\pi = i \left(\frac{\hbar B \omega}{2} \right)^{1/2} (O^\dagger - O), \tag{7.7}$$

where the frequency is defined by $\omega = (C/B)^{1/2}$, and the phonon operators obey the commutation relations,

$$[O, O^\dagger] = 1, \quad \text{and} \quad [O, O] = 0. \tag{7.8}$$

The transformation from the variables π and α to the phonon operators $O^\dagger$ and O in Eq. (7.6) is exactly the same as for the variables of the particle motion in the harmonic oscillator potential

$$x = \sqrt{\frac{\hbar}{2m\omega}} (c^\dagger + c), \tag{7.9}$$

$$p = i \sqrt{\frac{\hbar m \omega}{2}} (c^\dagger - c), \tag{7.10}$$

where the operators $c^\dagger$ and c are the creation and the annihilation operator of harmonic oscillator quanta. The Hamiltonian (7.3) is now expressed as a quantized form

$$H = \hbar\omega \left(O^\dagger O + \frac{1}{2} \right). \tag{7.11}$$

The Hamiltonian is diagonal to the n phonon state $|n\rangle = (O^\dagger)^n |0\rangle / \sqrt{n!}$ with the eigenenergy,

$$E_n = \hbar\omega \left(n + \frac{1}{2} \right) \quad n = 1, 2, 3, \ldots . \tag{7.12}$$

For the quadrupole vibration, the one phonon state, $n = 1$, has the angular momentum $J^\pi = 2^+$, and the two phonon state, $n = 2$, has $J^\pi = 0^+, 2^+$ and 4^+.

Problem

7.2 Obtain the quantized Hamiltonian (7.11) from Eq. (7.3).

7.1.3 Vibrational One-Body Potential and Separable Two-Body Interaction

The collective motions in a system governed by single-particle states can be understood in terms of the variation in the average one-body potential by an oscillation of nuclear density. Such variation in the one-body potential is responsible to excitations of the nucleonic motion, and a collective motion is described by using the induced density variation and the oscillating potential, which generates the density variation.

We consider the density variation by a phonon excitation $\alpha F(x)$, where $F(x) \equiv \sum_{i=1}^A F(\mathbf{r}_i, \sigma_i, \tau_i)$ is a one-body operator to excite the phonon and may depend on the space, the spin and the isospin degrees of freedom. The phonon leads to a vibration on the surface of nucleus so that we can take the deformed surface $r(1 - \alpha F)$ to include the variation of the static density $\rho(r)$.

We replace the radius r of the static density $\rho(r)$ by $r(1 - \alpha F)$, where αF induces a dynamical vibrational effect on the nuclear density, and changes the shape of the nuclear surface. Then, the density variation by the vibration is given by,

$$\rho_0(r(1 - \alpha F)) = \rho_0(r) + \delta\rho(r) \simeq \rho_0(r) - \alpha F r \frac{d\rho_0(r)}{dr}, \tag{7.13}$$

in the small amplitude limit of αF. The variation of the one-body potential is given in the same way as

$$V(r(1 - \alpha F)) = V(r) + \delta V(r) \simeq V(r) - \alpha F r \frac{dV(r)}{dr}. \tag{7.14}$$

The potential $\delta V(r)$, which is induced by the phonon excitation, is written in a form of the coupling between the vibrational amplitude α and the operator F,

$$\delta V(r) = \kappa \alpha F \quad \text{with} \quad \kappa = -r \frac{dV(r)}{dr}. \tag{7.15}$$

In this approximation, the one-body operator F excites many $1p - 1h$ states ($|ph\rangle$) from the ground state (a Slater determinant Ψ_0) and creates the phonon state with the same vibrational amplitude of the phonon operator α in Eq. (7.6),

$$\alpha_0^2 \equiv |\langle n = 1|\alpha|n = 0\rangle|^2 = \sum_{p,h} |\langle ph|F|\Psi_0\rangle|^2, \tag{7.16}$$

where α_0 represents the magnitude of vibrational motion. The two operators α and F can be treated as the same one to excite the phonon state. Then, the field coupling δV (7.15) gives a contribution to the vibrational Hamiltonian (7.3),

$$H' = \frac{1}{2}\kappa F^2 = \frac{1}{2}\kappa \alpha^2. \tag{7.17}$$

Here, the factor $1/2$ in Eq. (7.17) is needed to avoid a double counting of the sum $(i = 1 \ldots A)$ in the operator F. The two-body interaction H' is nothing but the quadrupole-quadrupole $(Q^\dagger \cdot Q)$ interaction for $F(r) = \sum_i r_i^2 Y_{2\mu}(\hat{r}_i)$, which is often adopted in the microscopic study of vibrational and rotational spectra. The Elliott SU(3) model also adopts a $(Q^\dagger \cdot Q)$ interaction, which depends on both the space and the momentum quadrupole operators, for the study of rotational spectra (see Sect. 7.3.1).

We discuss now specifically the quadrupole vibration. The idea can be extended not only to other multipole excitations (octupole, hexadecapole and higher), but also vibrations with spin and isospin degree of freedoms. The harmonic potential V in Eq. (7.3) has now the extra term (7.17)

$$V(r, \alpha) = \frac{1}{2}C\alpha^2 + \frac{1}{2}\kappa \alpha^2 = \frac{1}{2}(C + \kappa)\alpha^2. \tag{7.18}$$

Without the term δV, the potential (7.18) corresponds to the case (a) of Fig. 7.2. If κ is positive, the restoring parameter is larger and the oscillation frequency becomes higher. On the other hand, if κ is negative, the oscillation frequency is smaller. At a certain critical point $\kappa + C = 0$ near the case (b), the system has no restoring force to sustain and becomes unstable. The symmetry breaking effect is certainly realized in the case of $\kappa + C < 0$ as is schematically shown by the potential energy surface (d) in Fig. 7.2.

As a remark, the quadrupole-quadrupole interaction in Eq. (7.17) can be generalized to include the isospin degree of freedom

$$H = H_0 + \frac{1}{2}\kappa_0(Q_{IS}^\dagger \cdot Q_{IS}) + \frac{1}{2}\kappa_1(Q_{IV}^\dagger \cdot Q_{IV}), \qquad (7.19)$$

where H_0 is the one-body part, $\kappa_{0(1)}$ is the IS(IV) coupling constant, and $Q_{IS(IV)} = Q_n \pm Q_p$. The Hamiltonian (7.19) is rewritten as

$$H = H_0 + \frac{1}{2}(\kappa_0 + \kappa_1)(Q_n^\dagger \cdot Q_n + Q_p^\dagger \cdot Q_p) + \frac{1}{2}(\kappa_0 - \kappa_1)(Q_n^\dagger \cdot Q_p + Q_n^\dagger \cdot Q_p),$$
$$(7.20)$$

where $Q_{p(n)}$ is the proton (neutron) quadrupole operator. The sign and magnitude of the coupling constants κ_0 and κ_1 are determined by the self-consistent condition between vibration density and potential in [1]. The derived coupling constant is attractive for the IS channel κ_0, and repulsive for the IV one κ_1. Consequently, the neutron-proton interaction in Eq. (7.20) is much stronger than the neutron-neutron and proton-proton ones. These separable interactions have been applied successfully to describe both the low-lying collective and giant resonance excitations.

7.2 Nuclear Deformation

7.2.1 Parameterizations for Deformation/Nomenclatura

Any closed three-dimensional surface can be parameterized by using spherical harmonics. Describing the nucleus as a homogeneous density object, its surface can be described by the deformed radius,

$$R(\theta, \phi) = R_0 \left(1 + \sum_\lambda \sum_{\mu=-\lambda}^{\lambda} \alpha_{\lambda\mu} Y_{\lambda\mu}(\theta, \phi) \right), \qquad (7.21)$$

where $\alpha_{\lambda\mu}$ are the deformation amplitudes and $Y_{\lambda\mu}(\theta, \phi)$ are the spherical harmonics with the angular momentum λ and its z-component μ, respectively (see Appendix 7.10.2.1). The expression (7.21) is based on the same idea as the definition of F in Eq. (7.13). The multipoles $\lambda = 2, 3$ and 4 are called quadrupole, octupole, and hexadecapole deformations, respectively. For quadrupole deformation, the intrinsic coordinate system is commonly introduced such that its axis orientation is taken to be the same as the principal axis of the mass distribution of the system. The orientation angles $\omega = (\phi, \theta, \psi)$ relate the deformation amplitude $\alpha_{\lambda=2\mu}$ in the laboratory frame and the intrinsic frame amplitude $a_{\lambda=2\mu}$ as

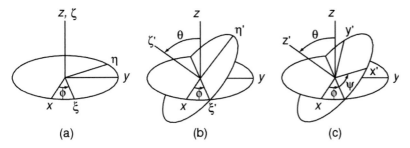

Fig. 7.3 Any orientation in 3D space can be specified by three angles (ϕ, θ, ψ) by the Euler theory of rotation. The rotation from the coordinate $\mathcal{H}(x, y, z)$ to the coordinate $\mathcal{H}'(x', y', z')$ is separated in three stages: **a** A rotation around z-axis by an angle ϕ. **b** A rotation around ξ (new x-axis) by an angle θ. **c** A rotation around ζ' (new z-axis) by an angle ψ. Euler angles are often denoted by a symbol $\omega \equiv (\phi, \theta, \psi)$. Courtesy A. L. Schwab, Delft University of Technology

$$\alpha_{\lambda=2\mu} = \sum_{\nu} \mathcal{D}_{\mu,\nu}^{\lambda=2}(\omega) a_{\lambda=2\nu}, \tag{7.22}$$

where $\mathcal{D}_{\mu,\nu}^{\lambda=2}(\omega)$ is the rotational matrix with three Euler angles $\omega = (\phi, \theta, \psi)$ (see Fig. 7.3). The deformed surface is then specified in the intrinsic frame as

$$R(\theta', \phi') = R_0 \left(1 + \sum_{\mu=-\lambda}^{\lambda} a_{\lambda=2\mu} Y_{\lambda=2\mu}(\theta', \phi')\right), \tag{7.23}$$

where we use a relation $Y_{\lambda=2\mu}(\theta', \phi') = \sum_{\nu} \mathcal{D}_{\mu,\nu}^{\lambda=2}(\omega) Y_{\lambda\nu}(\theta, \phi)$. The quadrupole deformation parameters are thus taken to be

$$a_{20} \neq 0, \quad a_{21} = a_{2-1} = 0, \quad a_{22} = a_{2-2} \neq 0. \tag{7.24}$$

Alternatively, the quadrupole amplitudes are also characterized by two parameters (β, γ) defined as

$$\begin{aligned} a_{20} &= \beta \cos\gamma, \\ a_{22} &= a_{2-2} = \frac{1}{\sqrt{2}}\beta \sin\gamma, \end{aligned} \tag{7.25}$$

where the quadrupole deformation parameter β is introduced as

$$\beta^2 \equiv \sum_{\mu} |a_{\lambda=2\mu}|^2 = |a_{\lambda=20}|^2 + 2|a_{\lambda=22}|^2. \tag{7.26}$$

Physical meaning of the parameters (β, γ) may be clarified to calculate the increment of the three principal axes from Eq. (7.23);

$$R_{x'} = R(\theta' = \frac{\pi}{2}, \phi' = 0) = R_0 \left(1 - \sqrt{\frac{5}{4\pi}} \left[-\frac{1}{2}\beta\cos\gamma + \frac{\sqrt{3}}{2}\beta\sin\gamma \right] \right)$$

$$R_{y'} = R(\theta' = \frac{\pi}{2}, \phi' = \frac{\pi}{2}) = R_0 \left(1 - \sqrt{\frac{5}{4\pi}} \left[-\frac{1}{2}\beta\cos\gamma - \frac{\sqrt{3}}{2}\beta\sin\gamma \right] \right)$$

$$R_{z'} = R(\theta' = 0, \phi' = 0) = R_0 \left(1 + \sqrt{\frac{5}{4\pi}}\beta\cos\gamma \right), \tag{7.27}$$

where $Y_{22}(\theta', \phi') = \frac{1}{4}\sqrt{\frac{15}{2\pi}}\sin\theta'^2 e^{2i\phi'}$ and $Y_{20}(\theta', \phi') = \sqrt{\frac{5}{4\pi}}\frac{1}{2}(3\cos\theta'^2 - 1)$ are used. The increment of the three principal axes, defined by $\delta R_\kappa \equiv R_\kappa - R_0$, can be written in a compact form

$$\delta R_\kappa = \sqrt{\frac{5}{4\pi}}\beta R_0 \cos\left(\gamma - \kappa\frac{2\pi}{3} \right), \tag{7.28}$$

where $\kappa = 1, 2, 3$ correspond to x', y', z', respectively.

For $\gamma = 0$ and $\beta > (<)0$, we have $R_{x'} = R_{y'}$ and $R_{z'} > (<)R_{x'}$, which corresponds to an axial symmetric prolate (oblate) deformation. Geometrically, prolate is elongated along the symmetry axis, while oblate is compressed along the symmetry axis (see Fig. 7.4). In the case of $\gamma \neq 0$, there is no axial symmetry and the corresponding quadrupole shape is called "triaxial". Notice that, for $\gamma = 60°$, we have an identical relation $R_{x'} = R_{z'}$ and $R_{y'} > (<)R_{x'}$ as that of $\gamma = 0$ and $\beta > (<)0$, interchanging the role between z' and y' axes. Because of the symmetry due to the relation (7.27), the (β, γ) plane is divided to six region and each region shows the same shape of nucleus. In practical study of the deformation, it is enough to study in the (β, γ) plane with γ ranging from $0°$ to $60°$. For octupole deformation, β_3 is assigned for an axial symmetric case,

$$a_{30} = \beta_3, \tag{7.29}$$

assuming $a_{33} = a_{32} = a_{31} = a_{3-1} = a_{3-2} = a_{3-3} = 0$. The axial symmetric quadrupole and octupole deformations are illustrated in Fig. 7.4. The octupole deformation occurs together with quadrupole deformation in general.

The deformation parameter of axial symmetric quadrupole deformation is defined by several different ways than Eq. (7.25). The collective deformation of the nucleus as a whole is conventionally characterized by a deformation parameter δ which is defined through the intrinsic electric quadrupole (E2) moment,

$$Q_0 = \sum_{i=1}^{Z}\langle 2z_i'^2 - x_i'^2 - y_i'^2 \rangle \equiv \frac{4}{3}\left\langle \sum_{i=1}^{Z} r_i'^2 \right\rangle \delta, \tag{7.30}$$

which quantifies the deviation from the spherical shape. For axial symmetric uniformly charged nucleus with z-axis as the symmetry axis, the parameter δ is expressed

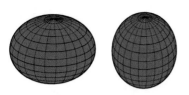
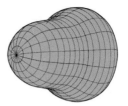

Axially symmetric **quadrupole**

oblate($\beta < 0$)   prolate($\beta > 0$)

$\lambda = 2$

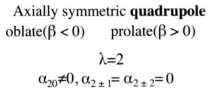

Axially symmetric **octupole**

$\lambda = 3$

$\alpha_{30} \neq 0$, $\alpha_{3 \pm 1,2,3} = 0$

$\alpha_{20} \neq 0$, $\alpha_{2 \pm 1,2} = 0$

Fig. 7.4 Sketch of axial symmetric quadrupole deformations with $\beta_2 \neq 0$ and octupole deformation with $\beta_3 \neq 0$. Octupole deformation is associated with quadrupole deformation in general. The quadrupole deformation ($\beta_2 > 0$) is named "prolate" shape, while the quadrupole deformation ($\beta_2 < 0$) is called "oblate" shape. The axial symmetric octupole deformation is often called "pear-shape" deformation

as

$$\delta = \frac{3}{2} \frac{R_{z'}^2 - (R_\perp)^2}{R_{z'}^2 + 2R_\perp^2} \approx \frac{\Delta R}{R} + \frac{1}{6}\left(\frac{\Delta R}{R}\right)^2, \qquad (7.31)$$

where $\Delta R = R_{z'} - R_\perp$, $R_\perp = R_{x'} = R_{y'}$, and the mean radius is defined by $R = (5\langle r^2\rangle/3)^{1/2} \approx (R_{z'} + 2R_\perp)/3$. In the leading order, the above two parameters are very close to each other with $\beta \approx 1.06\ \delta$.

Problem

7.3 Derive the r.h.s term of Eq. (7.31) from the definition $\Delta R = R_{z'} - R_\perp$.

7.2.2 Rotational Spectrum

The nuclear many-body Hamiltonian consists of a kinetic energy term and interactions among nucleons. Both terms do not depend on the orientation of the coordinate axes, and therefore the Hamiltonian is invariant under rotation of coordinate axis. However, this rotational symmetry in the laboratory frame does not exclude the possibility of non-spherical potential and density distribution in an intrinsic frame attached to the nucleus. Solutions of the Hamiltonian breaking the spherical symmetry are called deformed intrinsic states. The experimental observation of rotational spectra as shown in Fig. 7.7 supports the validity, as a first approximation, of a separation of the collective rotation as a whole from the intrinsic motion in the body-fixed frame as described by a Hamiltonian

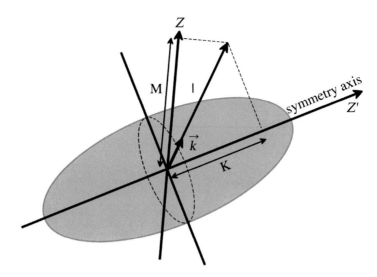

Fig. 7.5 An axial symmetric rotator. The symmetry axis is assigned to be the intrinsic z' axis. The projection of the angular momentum I on this axis is denoted K, and the projection on the z axis in the laboratory frame is M

$$H = H_{\text{int}} + H_{\text{rot}}, \tag{7.32}$$

where the intrinsic Hamiltonian H_{int} is deformed, i.e., the spherical symmetry is broken and x', y' and z' coordinates in the intrinsic frame are not equivalent.

The eigenfunction of the total Hamiltonian (7.32) is given by a product of a collective wave function $\Phi_{\text{rot}}(K)$ and an intrinsic wave function $\phi_{\text{int}}(q)$ as

$$\Psi = \phi_{\text{int}}(q)\Phi_{\text{rot}}(K), \tag{7.33}$$

where q describes the motion of individual particles with respect to the intrinsic coordinate system, while K is the projection of the total angular momentum I onto the symmetry axis. The K quantum number is conserved as a good quantum number for the axial symmetric deformed nuclei since the K value is constant for the rotation around the symmetry axis, z'-axis in Fig. 7.5. The eigenenergy of the Hamiltonian (7.32) has the form

$$E = E_{\text{rot}} + E_{\text{int}}. \tag{7.34}$$

For specific values of the three quantum numbers K, I, and M, the rotational wave function $\Phi_{\text{rot}}(K)$ is given by

$$\Phi_{KIM}(\omega) = \left(\frac{2I+1}{8\pi^2}\right)^{1/2} \mathcal{D}^I_{MK}(\omega), \tag{7.35}$$

where $\mathcal{D}^I_{MK}(\omega)$ is the so-called $\mathcal{D}$ function for the rotations with three Euler angles $\omega = (\phi, \theta, \psi)$. The wave function is a result obtained by a transformation from the fixed coordinate system (x, y, z) to the rotating intrinsic system (x', y', z'). In the particular case where the symmetry axis z' is perpendicular to the angular momentum I, i.e., $K = 0$, the $\mathcal{D}$ function becomes the spherical harmonics

$$\Phi_{K=0, IM}(\omega) = \left(\frac{1}{2\pi}\right)^{1/2} Y_{IM}(\theta, \phi). \qquad (7.36)$$

In the intrinsic system the deformed body is at rest and the symmetry can be either prolate, oblate or spherical. Nevertheless, in the laboratory system the symmetry is spherical because the deformed body rotates in all directions and there is no way to distinguish one of these directions from another. The angular momentum I is conserved since it commutes with the spherical Hamiltonian in the laboratory frame, so that I^2 and $I_z = M$ are good quantum numbers. Furthermore, in classical mechanics, one can distinguish whether a spherically symmetric body rotates or not, but in quantum mechanics all directions are equivalent and the spherically symmetry appears at the rest frame. Therefore, there is no rotational energy associated to such a system or, in general, to degrees of freedom corresponding to a spherical symmetry.

In classical mechanics, the moment of inertia of a rigid body is a quantity that determines the torque needed for an angular acceleration about a rotational axis; similar to how mass determines the force needed for an acceleration. The moment of inertia $\Im$ is defined as the ratio of the orbital angular momentum L of the system to its angular velocity ω_{rot} around the principal axis (rotational axis),

$$\Im \equiv \frac{L}{\omega_{\mathrm{rot}}}. \qquad (7.37)$$

The kinetic energy of a rotating body is expressed as

$$E = \frac{1}{2}L\omega^2_{\mathrm{rot}} = \frac{L^2}{2\Im}. \qquad (7.38)$$

Let us estimate the moment of inertia of a simple system: a point-like particle with mass m is rotating with the angular velocity ω_{rot} in a circle with radius r, as shown in the left panel of Fig. 7.6. The angular momentum is defined as

$$\mathbf{L} = \mathbf{r} \times \mathbf{p} = m\mathbf{r} \times \mathbf{v} = mr^2\omega_{\mathrm{rot}}\mathbf{k}, \qquad (7.39)$$

where $\mathbf{k}$ is a unit vector in the direction of rotational axis. Comparing between Eqs. (7.37) and (7.39), the classical moment of inertia is given by

$$\Im_{\mathrm{CM}} = mr^2, \qquad (7.40)$$

Point-like particle   Extended mass distribution

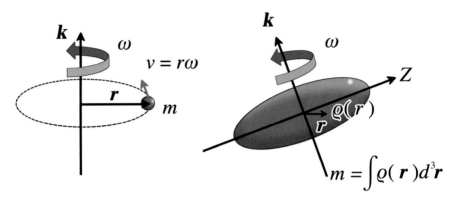

Fig. 7.6 Two rotating systems. (Left) A particle rotating in a circle of radius r around the axis $\mathbf{k}$. (Right) A system with an extended mass distribution $\rho(\mathbf{r})$ is rotating around the axis $\mathbf{k}$

which is proportional to the mass and the square distance between the object and the rotational axis. For a rotating system with the density $\rho(\mathbf{r})$, the formula (7.40) is generalized to be

$$\Im_{\text{CM}} = \int d\mathbf{r}\rho(\mathbf{r})r_\perp^2, \tag{7.41}$$

where $\rho(\mathbf{r})$ is the mass distribution and $r_\perp = |\mathbf{r} \times \mathbf{k}|$ is the distance perpendicular to the rotational axis (see the right panel of Fig. 7.6).

The rotational spectra of a deformed nuclear system is characterized by the moment of inertia. For an axially symmetric rigid body of mass M rotating around the x axis, the moment of inertia reads

$$\Im_{rid} = M\langle y^2 + z^2 \rangle = Am\frac{1}{5}(R_\perp^2 + R_z^2) = Am\frac{2}{5}R^2\left(1 + \frac{\delta}{3}\right), \tag{7.42}$$

where $\delta = (R_z - R_\perp)/R$ and $R \approx (R_z + 2R_\perp)/3$ as expressed in Eq. (7.31), A is the mass number, and m is the nucleon mass, approximating the mass $M \sim Am$. The moment of inertia generates an extra kinetic energy for the motion of nucleons in a uniform rotating potential. The one-body motion in the rotating potential is described by a Hamiltonian

$$H' = H_{int} - \hbar\omega_{\text{rot}}J_x \tag{7.43}$$

where H_{int} is the intrinsic hamiltonian without rotation, ω_{rot} is the rotational frequency around the x axis and $J_x = \sum_{k=1}^{A} j_x(k)$ is the sum of the x components of all single-particle angular momenta. The second term of the r.h.s. of Eq. (7.43) is called the Coriolis term.

In the general case without the axial symmetry, Eq. (7.43) may be extended to have a sum over all axes x, y, z with corresponding rotational frequencies. The energy increase due to the rotation can be evaluated by the second-order perturbation theory as

$$H_{\text{rot}} = \frac{1}{2}\Im \omega_{\text{rot}}^2, \tag{7.44}$$

$$\Im = 2\hbar^2 \sum_i \frac{|\langle i|J_x|0\rangle|^2}{E_i - E_0}, \tag{7.45}$$

where i labels the many-body excited state and E_i is the corresponding excitation energy. In analogy to classical mechanics, the moment of inertia in quantum mechanics is defined by the ratio between the angular momentum I and the rotational frequency ω_{rot},

$$\Im = \frac{\hbar I}{\omega_{\text{rot}}}, \tag{7.46}$$

where $\hbar I$ is a quantized angular momentum. Equation (7.46) has been often used to describe the rotational spectra of deformed nuclei. With Eq. (7.46), the energy formula (7.44) gives the energy spectrum of rotation $E_{\text{rot}} = \hbar^2 I^2/2\Im$.

Problem

7.4 Derive Eq. (7.45) by using the second order perturbation theory with the Hamiltonian (7.43).

The three-dimensional (3D) rotational Hamiltonian in quantum mechanics is also expressed in analogous to the classical kinetic energy (7.38). For an axial symmetric rotor, the Hamiltonian is given by

$$H_{\text{rot}} = \frac{\hbar^2 I_x'^2}{2\Im_x} + \frac{\hbar^2 I_y'^2}{2\Im_y} + \frac{\hbar^2 I_z'^2}{2\Im_z} = \frac{\hbar^2 I_x'^2 + \hbar^2 I_y'^2}{2\Im} = \frac{\hbar^2 I'^2 - \hbar^2 I_z'^2}{2\Im}, \tag{7.47}$$

where $I_i'(i = x, y, z)$ is the angular momentum in the intrinsic frame and $\Im_i$ is its associated moment of inertia, having $\Im_x = \Im_y = \Im$. Note that quantum mechanical systems can rotate only around the axis perpendicular to the symmetry axis, but cannot rotate around the symmetry axis.

The Hamiltonian (7.47) has its eigenvalue

$$E_{\text{rot}} = \frac{\hbar^2}{2\Im}(I(I+1) - K^2) \tag{7.48}$$

for the eigenfunction $|IMK\rangle$, which has all the symmetries of the system, including the axial symmetry, the reflection symmetry with respect to the (x, y) plane and the parity symmetry for the operation $\mathbf{r} \to -\mathbf{r}$.

If the intrinsic Hamiltonian is invariant with respect to the rotation of $180°$ around the axis perpendicular to the symmetry axis, a further reduction of the rotational degree of freedom follows, called $\mathcal{R}$ invariance. We take the intrinsic y' axis as a reference and denote the rotation of $180°$ around y' axis as $\mathcal{R} = \mathcal{R}_{y'}(\pi)$. The intrinsic state with $K = 0$ specifies the rotation perpendicular to the symmetry axis and becomes the most favorable rotational band in energy according to Eq. (7.48) since the lowest-energy level of the band has $I = K$, and therefore $E_{\text{rot}} = \hbar^2 K/2\mathfrak{I}$. The $K = 0$ intrinsic state in Eq. (7.36) is labeled by the eigenvalue of $\mathcal{R}$

$$\mathcal{R}\Psi_{r,K=0} = r\Psi_{r,K=0} \tag{7.49}$$

with the eigenvalue

$$r = \pm 1 \tag{7.50}$$

because $\mathcal{R}^2 = \mathcal{R}_{y'}(2\pi) = 1$, for a system of even-even or odd-odd proton-neutron numbers with integer angular momentum. The operation $\mathcal{R}$ acting on the wave function (7.36) inverts the direction of the symmetry axis ($\theta \rightarrow \pi - \theta, \phi \rightarrow \phi + \pi$), which gives rise to a phase,

$$\mathcal{R}Y_{IM} = (-1)^I Y_{IM}. \tag{7.51}$$

The $\mathcal{R}$ invariance, therefore, implies

$$r = (-1)^I \tag{7.52}$$

and the rotational spectrum contains states with either only even values or only odd values of I

$$
\begin{aligned}
I &= 0, 2, 4, 6, \ldots & \text{for } r = +1 \\
I &= 1, 3, 5, 7, \ldots & \text{for } r = -1.
\end{aligned}
\tag{7.53}
$$

In an even-even nucleus, the rotational band associated to $K = 0$, i.e., built on the lowest-energy 0^+ state, is only composed of even values of I.

The rotational energy spectra of deformed nuclei can be described by Eq. (7.48). The results of rotational bands in ^{166}Er, a well-deformed nucleus, are shown in the left panel of Fig. 7.7. Because of the $\mathcal{R}$ invariance for the $K = 0$ band, only even values of I with $r = +1$ are allowed as shown in Eq. (7.53). However for the $K = 2$ band, the $\mathcal{R}$ invariance does not hold and all integer values are allowed for I. Extracted moments of inertia are also shown in the right panel. As one can see in this figure, the moment of inertia is changing gradually for high spin states, i.e., the formula (7.48) is not a very accurate approximation in general and higher-order effects are important. This is because the rotational motion in nuclei is not completely independent of the intrinsic wave function, as it was first approximated with Eqs. (7.32) and (7.33). In molecules, atoms are well localized as shown in Fig. 7.1 and the rotational motion can be considered to be independent to the intrinsic wave functions. On the other hand,

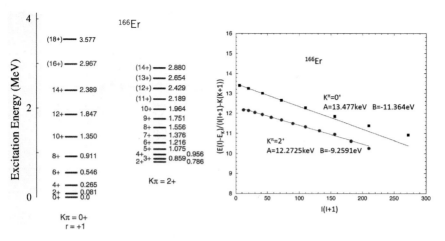

Fig. 7.7 (Left) Rotational spectra of ^{166}Er. Excitation energies are given in unit of MeV, while the spin-parity I^{π} are displayed on the l.h.s. The levels which belong to a band are characterized by approximately $I(I+1)$ energy spectra and connected by strongly-enhanced electric quadrupole (E2) transitions. Linked E2 transitions between two bands are also in the lower part of band members. The bands are classified by the quantum number K^{π} which is the projection of the total angular momentum on the symmetry axis. The data are taken from [5]. (Right) Rotational spectra are analyzed by an expansion formula (7.54). The expansion coefficients A and B are obtained by the energies of the lowest members of each band

nucleons in nucleus are strongly interacting each other and the collective motion and the intrinsic single-particle motions may influence each other.

Phenomenologically the rotational spectra are analyzed by an expansion formula

$$E(I, K) - E(I = K, K) = AI(I + 1) + BI^2(I + 1)^2 + CI^3(I + 1)^3 + \cdots,$$
$$(7.54)$$

where $E(I = K, K)$ is the lowest energy of each rotational band and the expansion coefficients $(A, B, C, \ldots)$ are determined from observed energies of rotational spectra. The term A is inversely proportional to the intrinsic moment of inertia and the other coefficients are higher-order effects of rotation. The results of rotational bands in ^{166}Er are shown in the right panel of Fig. 7.7, where the coefficients A and B are determined by the lowest members of each band. In this plot, a simple $I(I+1)$ dependence gives a horizontal straight line, and the inclusion of $I^2(I+1)^2$ term gives a slope for the straight line. These two terms describe the observed spectra with an accuracy of orders of 4% up to the spin $I = 16$ for $K^{\pi} = 0^+$ band and less than 1% up to $I = 12$ for $K^{\pi} = 2^+$ band, respectively. The deviation from the $I(I+1)$ law reflects the change of intrinsic structure of deformed nuclei by the rotation, which gives rise to the higher-order terms $I^2(I+1)^2, I^3(I+1)^3, \ldots$ in the rotational spectra.

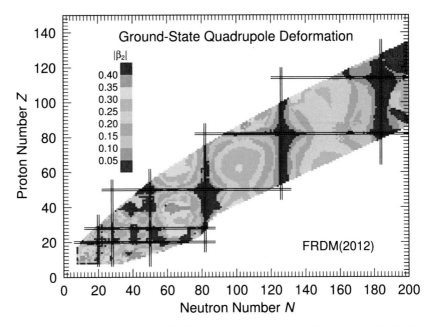

Fig. 7.8 Regions of deformed nuclei with mass $A > 16$ calculated by using the Finite-Range Droplet Model (FRDM). The straight lines parallel to the x and y axes show the magic numbers of protons and neutrons. The red color indicates a large deformation $\beta_2 > 0.4$, while the blue color shows a region with a small deformation or without any deformation. Reprinted with permission from [4]. ©2021 by Elsevier. Courtesy P. Möller

7.2.3 Quadrupole Deformation Across the Nuclear Chart

Regions of quadrupole deformation with mass $A > 16$ calculated within the Finite-Range Droplet Model (FRDM), a hybrid macro-microscopic model for the mass formula, are shown in Fig. 7.8. The FRDM is based on the macroscopic liquid drop model and takes into account the microscopic shell corrections. The energy is minimized in the multiple deformed space with the quadrupole, octupole and the hexadecupole deformations including also the triaxiality.

Several strongly deformed regions with $0.25 \leq \beta_2 \leq 0.35$ are predicted: rare-earth nuclei with $Z = 55 \sim 70$ such as Dy, Er, Yb isotopes, and heavy nuclei, U isotopes. Such large ground-state deformations are also encountered in light nuclei, ^{20}Ne and ^{24}Mg. A large deformation is also observed in the mass region $A < 16$ such as ^{7}Li, ^{8}Be and ^{10}Be. For well-deformed nuclei, the energy spectra exhibit a clear $I(I + 1)$ pattern given by Eq. (7.48). A straightforward evidence of quadrupole deformation is thus obtained by a ratio of energy between $E(4_1^+)$ and $E(2_1^+)$, which is

$$\frac{E(4_1^+)}{E(2_1^+)} = \begin{cases} 3.33 & \text{for rotational limit} \\ 2.0 & \text{for vibrational limit} \end{cases}. \tag{7.55}$$

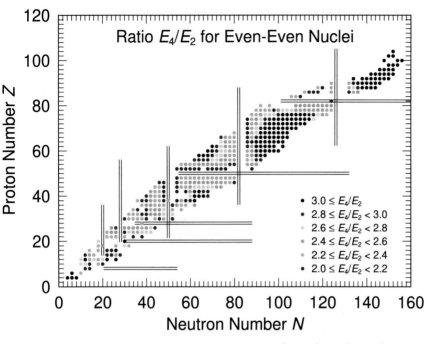

Fig. 7.9 Experimental ratio of the excitation energies between 4_1^+ and 2_1^+, $E(4_1^+)/E(2_1^+)$ in even-even nuclei. The circles with different colors represent nuclei with different values of the ratio $E(4_1^+)/E(2_1^+)$. The ideal rotor gives $E(4_1^+)/E(2_1^+) = 20/6 = 3.33$, while the ideal phonon spectra leads to $E(4_1^+)/E(2_1^+) = 2.0$. The thin straight lines parallel to the $x(y)$ axis show the magic numbers of protons (neutrons). Courtesy P. Möller and B. Shu

The rotational limit 3.33 is obtained by a ratio $E(4^+)/E(2^+) = 20/6$. On the other hand, the vibrational limit 2.0 is the outcome of the energy difference between 2 phonon state for 4^+ and one phonon state for 2^+, $E(4^+)/E(2^+) = 2/1$. In Fig. 7.9, nuclei with $E(4_1^+)/E(2_1^+) > 2.7$ are shown. The regions marked in Fig. 7.9 are consistent with the FRDM predictions in Fig. 7.8. As we can see in Fig. 7.8, most of deformed nuclei are mid-shell nuclei between the closed shells. The microscopic mechanism of deformation is discussed in Sect. 7.3.

The isotope dependence of energy spectra of Dy isotopes is shown in Fig. 7.10. These spectra show clearly the shape change from spherical to deformed when more and more neutrons are added to the closed neutron core $N = 82$. The nucleus ^{148}Dy is the neutron closed shell with $N = 82$ and shows a large energy gap between 0^+ and 2^+ states. The spectrum of ^{152}Dy is a typical vibrational spectrum with the ratio $E(4_1^+)/E(2_1^+) = 2.06$, while the ^{158}Dy spectrum gives a rotational band with $E(4_1^+)/E(2_1^+) = 3.2$. Other nuclei ^{154}Dy and ^{156}Dy stay in a transitional region between vibrational and rotational limits.

The electric quadrupole moment (E2) is an important observable to study static and dynamical properties of deformed nuclei. Detailed discussions of electromagnetic

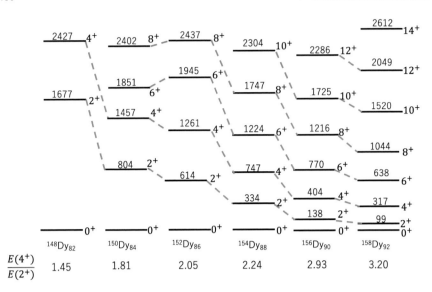

Fig. 7.10 Spectra of even-parity states in Dy isotopes. Energies are given in units of keV. The ratio $E(4^+_1)/E(2^+_1)$ is given at the bottom of the spectra of each isotope. The neutron number $N = 82$ corresponds to a magic number, which gives a large energy gap between the ground and the first excited states in ^{148}Dy. The ratio of 4^+_1 and 2^+_1 energies close 2.0 reflects the vibrational excitations of two states, while the ration close to 3.33 corresponds to that of the rotational band members of well deformed nuclei

transitions are found in Chap. 4. The E2 moment $M^{lab}(E2, \mu)$ in the laboratory frame is obtained by the rotational transformation of the tensor operator relative to the body-fixed intrinsic frame $\mathcal{M}^{intr}(E2, \nu)$,

$$M^{lab}(E2, \mu) = \sum_{\nu} \mathcal{D}^{\lambda=2}_{\mu\nu}(\omega)\mathcal{M}^{intr}(E2, \nu), \qquad (7.56)$$

with

$$\mathcal{M}^{intr}(E2, \nu) = e\sum_{i=1}^{Z} r_i'^2 Y_{2\nu}(\theta', \phi'), \qquad (7.57)$$

where the prime coordinates refer to the intrinsic body-fixed system. An axial symmetric quadrupole deformation is characterized by the intrinsic electric quadrupole moment $Y_{2\nu=0}(\theta', \phi')$ as

$$eQ_0 = e\langle K|\sum_{i=1}^{Z} r_i'^2(3\cos^2\theta_i' - 1)|K\rangle = \langle K|\int \rho_e(\mathbf{r}')r'^2(3\cos^2\theta' - 1)d\mathbf{r}'|K\rangle$$

$$= \left(\frac{16\pi}{5}\right)^{1/2} \langle K|\mathcal{M}^{intr}(E2, \nu = 0)|K\rangle, \qquad (7.58)$$

where the charge density operator is defined as $\rho_e(\mathbf{r}') = e \sum_{i=1}^{Z} \delta(\mathbf{r}_i - \mathbf{r}')$. The E2 matrix moment referring to the laboratory system is obtained from Eqs. (7.56) and (7.58) as

$$\langle K I_2 M_2 | M^{lab}(E2, \mu) | K I_1 M_1 \rangle = e Q_0 \left(\frac{5}{16\pi} \right)^{1/2} \sqrt{\frac{2I_1 + 1}{2I_2 + 1}}$$
$$\times \langle I_1 K 20 | I_2 K \rangle \langle I_1 M_1 2\mu | I_2 M_2 \rangle. \quad (7.59)$$

The E2 moment belonging to the same rotational band, referred to "collective" E2 moment, is given by the intrinsic moment as

$$B(E2; K I_1 \to K I_2) = \frac{1}{2I_1 + 1} \sum_{M_2, \mu, M1} |\langle K I_2 M_2 | M^{lab}(E2, \mu) | K I_1, M_1 \rangle|^2$$
$$= \frac{1}{2I_1 + 1} |\langle K I_2 || M^{lab}(E2) || K I_1 \rangle|^2$$
$$= \frac{5}{16\pi} e^2 Q_0^2 \langle I_1 K 20 | I_2 K \rangle^2, \quad (7.60)$$

where the double-bar symbol denotes the reduced matrix element for the angular momentum, and $\langle I_1 K 20 | I_2 K \rangle$ is the Clebsch–Gordan coefficient for the angular momentum (see Appendix 7.10.1 about the details of the reduces matrix element and the Clebsch–Gordan, especially, Eq. (7.150)).

The spectroscopic quadrupole moment is evaluated from Eq. (7.59) as

$$eQ = \left(\frac{16\pi}{5} \right)^{1/2} \langle K I I | M^{lab}(E2, \mu = 0) | K I I \rangle$$
$$= e Q_0 \langle I K 20 | I K \rangle \langle I I 20 | I I \rangle$$
$$= e Q_0 \frac{3K^2 - I(I + 1)}{(2I + 3)(I + 1)}, \quad (7.61)$$

where the Clebsh–Gordan coefficient is evaluted by using a formula $\langle I M 20 | I M \rangle = \frac{3M^2 - I(I+1)}{\sqrt{(2I-1)I(I+1)(2I+3)}}$ with $M = I$ and $M = K$. The ratio of Q to Q_0 is the expectation value of $\mathcal{D}_{\mu=0\nu=0}^{\lambda=2}(\omega) = P_2(\cos\theta)$ in Eq. (7.56). This means that the Q provides the average of intrinsic electric distortion Q_0 over the rotational motion being always $|Q| < |Q_0|$, but not the intrinsic moment itself. For the band head of the ground band $I = K = 0$, the spectroscopic quadrupole moment vanishes and becomes finite only for $I \neq 0$ band members. In the case of the $K \neq 0$ band, the spectroscopic quadrupole

Table 7.1 Observed Q moments and charge radii of $_{64}$Gd isotopes. The data of Q moments are taken from [6], and those of charge radii are from [7]

A	I^π	K	Q (b = 100 fm^2)	$\sqrt{r_c^2}$ (fm)
154	2$^+$	0	−1.82	5.1223
156	2$^+$	0	−1.93	5.1420
157	3/2$^-$	3/2	+1.36	5.1449
158	2$^+$	0	−2.01	5.1569
160	2$^+$	0	−2.08	5.1734

moment is finite even for the band head with $I = K$. The spectroscopic quadrupole moment is an important observable to indicate the sign of deformation. It should be noticed that the spectroscopic moment Q has the opposite sign to the intrinsic moment Q_0 for the $K = 0$ band member, i.e., $Q < 0$ for the prolate shape $Q_0 > 0$ and $Q > 0$ for the oblate shape $Q_0 < 0$.

Problem

7.5 Obtain the transition strength $B(E2; I_1 K \rightarrow I_2 K)$ in Eq. (7.60) by using an integration formula of the $\mathcal{D}$ function,

$$\int \mathcal{D}_{M_2 K}^{I_2 *}(\omega) \mathcal{D}_{\mu\nu=0}^{I}(\omega) \mathcal{D}_{M_1 K}^{I_1}(\omega) d\omega = \frac{8\pi^2}{2I_2 + 1} \langle I_1 M_1 I \mu | I_2 M_2 \rangle \langle I_1 K I 0 | I_2 K \rangle,$$
(7.62)

and the definition of the reduced matrix element (7.149). Notice also a relation between the spherical harmonics $Y_{2\mu}$ and the rotational $\mathcal{D}$ function, $\mathcal{D}_{\mu\nu=0}^{\lambda=2}$,

$$Y_{2\mu}(\theta\phi) = \left(\frac{5}{4\pi}\right)^{1/2} \mathcal{D}_{\mu\nu=0}^{\lambda=2}.$$
(7.63)

Problem

7.6 The measured Q moments and the charge radii of some Gd isotopes are listed in Table 7.1. Determine the deformation parameter $\delta \approx \Delta R/R$ of these nuclei from Eqs. (7.30) and (7.61).

7.3 Deformed Single-Particle States

In this section, we study the deformation and the rotational spectra taking the particle motion in a deformed harmonic oscillation potential. We first introduce the Elliott's SU(3) model and proceed to the Nilsson model.

7.3.1 Elliott SU(3) Model

One of the oldest and most elegant models of nuclear deformation is the Elliott SU(3) model [8]. This model has a unique and essential importance to understand of the relation between the collective rotation and the interacting particles in one-major shell of a harmonic oscillator potential. The Elliott SU(3) Hamiltonian contains one-body and two-body interactions. The one-body part is taken as harmonic oscillator Hamiltonian and the two-body part is the quadrupole-quadrupole interaction,

$$H = H_0 - \frac{1}{2}\kappa \sum_q Q_{2q}^\dagger Q_{2q}, \tag{7.64}$$

where H_0 is the harmonic oscillator Hamiltonian

$$H_0 = \frac{p^2}{2m} + \frac{1}{2}m\omega_0^2 r^2, \tag{7.65}$$

where m is the nucleon mass and ω_0 is the oscillator frequency. In Eq. (7.64), κ is the coupling strength and Q_{2q} is the quadrupole operator expressed as

$$Q_{2q} = \sqrt{\frac{4\pi}{5}} \frac{1}{b^2}(r^2 Y_{2q}(\Omega_r) + b^4 p^2 Y_{2q}(\Omega_q)), \tag{7.66}$$

where $Y_{2q}(\Omega)$ is the spherical harmonics and b^2 is the oscillation parameter $b^2 = \hbar/(m\omega_0)$. The $Q \cdot Q$ interaction of the Elliott model has both radial and momentum dependence. The presence of the momentum-dependent term in Q_{2q} ensures that there is no mixing of different oscillator shell configurations. Thus, the wave functions will be constructed within one-major shell configuration space. The SU(3) group has 8 generators: 3 angular momentum operator L_i $(i = x, y, z)$ and 5 quadrupole operators Q_q $(q = 2, 1, 0, -1, -2)$. The Casimir operator of the SU(3) group is expressed in a quadratic form,

$$C_{SU(3)} = \frac{3}{4}\mathbf{L}^2 + \frac{1}{4} \sum_{q=-2}^{+2} Q_{2q}^\dagger Q_{2q}, \tag{7.67}$$

which commutes with all generators and is diagonal for the SU(3) eigenstate. That is, the eigenstate is expressed by the irreducible representation in terms of the Lie group labelled by the quantum numbers (λ, μ). The eigenvalue of the Casimir operator is evaluated for the eigenstate $\phi(\lambda\mu)$ as

$$\langle\phi(\lambda\mu)|C_{SU(3)}|\phi(\lambda\mu)\rangle = \lambda^2 + \lambda\mu + \mu^2 + 3(\lambda + \mu). \tag{7.68}$$

The values λ and μ are the differences in the oscillator quanta between z and x directions, $\lambda = n_z - n_x$, and between x and y directions, $\mu = n_x - n_y$, respectively, and specify the intrinsic structure of the eigenstate. With the Casimir operator, the Hamiltonian (7.64) is rewritten to be

$$H = H_0 - \frac{1}{2}\kappa(4C_{SU(3)} - 3\mathbf{L}^2). \tag{7.69}$$

The eigenstates of the Hamiltonian (7.64) are labeled by the quantum numbers λ and μ together with the angular momentum L, and the eigenvalues are expressed as

$$E = E_0 - 2\kappa(\lambda^2 + \lambda\mu + \mu^2 + 3(\lambda + \mu)) + \frac{2}{3}\kappa L(L + 1), \tag{7.70}$$

where E_0 is the harmonic oscillator energy. For an attractive quadrupole-quadrupole interaction with $\kappa > 0$, the ground state belongs to the representation which maximizes the values of the Casimir operator, corresponding to maximizing λ or μ. For each combination of (λ, μ), different values of L are allowed and their energies satisfy the $L(L + 1)$ rule, a spectrum of rigid rotor. The wave function for a given quantum numbers (λ, μ) can be interpreted as a set of many intrinsic states. The SU(3) multiplet corresponding to rotational band members is created by the projection on good angular momentum state from the intrinsic states. We should notice that a $Q \cdot Q$ interaction is an outcome of the particle-vibration coupling model discussed in Sect. 7.1.2 and the coupling strength κ is determined by the self-consistent condition between vibrating density and potential.

The Elliot SU(3) model Hamiltonian is constructed in the laboratory frame, i.e., it is rotationally invariant. However, the Hamiltonian (7.64) gives a rotational spectra. The Elliott wave function contains only one-major shell configurations and can be expressed as $\phi(\lambda, \mu) = \phi(n_x, n_y, n_z)$. The quadrupole moments are diagonal for the wave function ϕ and evaluated as

$$\langle\phi|Q_{20}|\phi\rangle = \langle\phi|\frac{1}{2b^2}(2z^2 - x^2 - y^2) + \frac{b^2}{2}(2p_z^2 - p_x^2 - p_y^2)|\phi\rangle$$
$$= (2n_z - n_x - n_y) = (2\lambda + \mu), \tag{7.71}$$

$$\langle\phi|Q_{22}|\phi\rangle = \sqrt{\frac{3}{2}}\langle\phi|\frac{1}{2b^2}(x + iy)^2 + \frac{b^2}{2}(p_x + ip_y)^2)|\phi\rangle$$
$$= \sqrt{\frac{3}{2}}(n_x - n_y) = \sqrt{\frac{3}{2}}\mu. \tag{7.72}$$

These Q moments show explicitly that the eigenstate with $\mu = 0$ corresponds to axial symmetric state, while that with $\mu \neq 0$ is a triaxial deformed state. Assuming that the electric quadrupole (E2) transition operator is proportional to the group generator Q_{2q}, the E2 transition matrix has a non-vanishing matrix element only between the band members specified by the representation (λ, μ). Thus, the Elliott model provides successfully not only the rotational spectra but also strong transitions between the band members. The wave functions can be considered as an approximation of the wave functions in a deformed harmonic oscillator potential (see next section) without the mixing of different oscillator major shells N.

Let us consider a nucleus ^{8}Be as an example of the SU(3) model. Four particles of ^{8}Be occupy p-shell orbits, which raises the sum of the major quantum number of the harmonic oscillator wave function, $\sum N = 4$, since the major quantum number of p-shell orbit is $N = n_x + n_y + n_z = 1$. The SU(3) state with $(\lambda, \mu) = (4, 0)$ has the lowest energy in Eq. (7.70) and the largest Q_{20} value in Eq. (7.71). The L values are allowed to be $L = 0, 2, 4$, showing a rotational spectrum terminated by the maximum angular momentum $L_{\max} = 4$. The maximum L value is limited by the model space (one major shell) of the Elliott SU(3) model. Other intrinsic state having different combination of (λ, μ) will show β band and γ band structures. Thus, the SU(3) model provides an exact microscopic model, in which the interacting particles in one major shell with a quadrupole-quadrupole interaction, leads to the ground state with a large quadrupole moment and the corresponding rotational spectrum.

7.3.2 Deformed Harmonic Oscillator Model

We describe the shell structure of strongly deformed nuclei in terms of deformed single-particle states. The key concepts of shell structure effects in one-particle spectra can be illustrated in the case of spherical symmetry. For a spherical potential, the motion is separated into radial and angular momentum components, and the one-particle energies depend on the two quantum numbers n and l, where n is the radial quantum number and l is the orbital angular momentum. For a given angular momentum l, the radial quantum number orders the single-particle energy levels. Since the energy is independent of the magnetic quantum number m, each (n, l) has a degeneracy of $2l + 1$ as a consequence of the rotational invariance. The spin degree of freedom gives a further degeneracy of 2 for each (n, l) orbital.

In the case of axial-symmetric deformation, the single-particle spectrum can be described by an axial-symmetric harmonic-oscillator potential,

$$V(x', y', z') = \frac{m}{2}(\omega_\perp^2(x'^2 + y'^2) + \omega_z^2 z'^2) \tag{7.73}$$

where $\omega_x = \omega_y = \omega_\perp$. The energy eigenvalue of the deformed harmonic-oscillator hamiltonian $H = T + V(x', y', z')$ reads

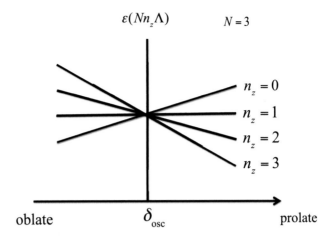

Fig. 7.11 Single-particle spectrum as a function of δ_{osc} for a Hamiltonian with an axial symmetric harmonic oscillator potential (7.73). The eigenvalue is given in Eq. (7.75). The case of major quantum number $N = 3$ is shown

$$\epsilon(N, n_z) = \hbar\omega_\perp(n_x + 1/2) + \hbar\omega_\perp(n_y + 1/2) + \hbar\omega_z(n_z + 1/2) \quad (7.74)$$

$$= \hbar\overline{\omega}\left(N + \frac{3}{2} - \frac{1}{3}\delta_{\mathrm{osc}}(3n_z - N)\right), \quad (7.75)$$

where the principal quantum number $N = n_x + n_y + n_z$ and $\overline{\omega} = (2\omega_\perp + \omega_z)/3$. The deformation parameter δ_{osc} is defined by

$$\delta_{\mathrm{osc}} = 3\frac{\omega_\perp - \omega_z}{2\omega_\perp + \omega_z} = \frac{\omega_\perp - \omega_z}{\overline{\omega}}$$

$$= \frac{R_z - R_\perp}{(2R_z + R_\perp)/3} = \frac{\Delta R}{R} + O(\Delta R^2), \quad (7.76)$$

where the last equation is derived to use a relation $\omega_i \approx 1/R_i$ of the harmonic oscillator potential. The single-particle spectrum is drawn in Fig. 7.11.

Problem

7.7 Derive Eq. (7.75) from Eq. (7.74) using the definitions of $\overline{\omega} = (2\omega_\perp + \omega_z)/3$ and δ_{osc} given by (7.76).

For prolate (oblate) deformation, $\omega_z < \omega_\perp$, equivalently $\delta_{\mathrm{osc}} > 0, (\omega_z > \omega_\perp, \delta_{\mathrm{osc}} < 0)$, the levels with larger (smaller) n_z become lower in energy as can be seen in Eq. (7.75). The eigenenergies have a linear dependence on δ_{osc} and the slope of the lowest level for a given N for a prolate shape ($n_z = N$) is twice steeper than for an oblate shape ($n_z = 0$) with an opposite sign. The configurations with an unfilled shell in the harmonic-oscillator potential have a deformed shape with high probability.

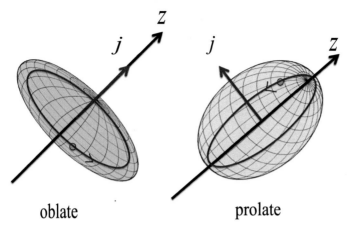

oblate   prolate

Fig. 7.12 Overlap between the single-particle wave function and the deformed core. The red arrow indicates the direction of the single-particle total angular momentum

The energy splitting of deformed single-particle states can also be interpreted by an intuitive picture shown in Fig. 7.12. A good overlap between the wave function and the core will produce more attractive energy for the single-particle state. For prolate deformation, the maximum overlap is obtained when the single-particle state is rotating around the axis perpendicular to the symmetry axis. That is the projection Λ of angular momentum on the symmetry axis is zero which means n_z is the maximum value. On the other hand, for the oblate deformation, the maximum overlap is obtained by the rotation of single-particle state around the symmetry axis, which gives the the projection Λ is the maximum $\Lambda = l$ and $n_z = 0$. The high degeneracy in the spherical limit is released at finite deformation. At the same time, the spherical symmetry is broken and a static deformation minimum is created in the potential energy surface.

In such a simple consideration of the shell filling of particles in the deformed harmonic-oscillator single-particle diagram, we expect equal number of prolate and oblate deformed nuclei in one major shell by the particle-hole symmetry. Experimentally, however, almost all deformed nuclei are identified as prolate axial-symmetric quadrupole deformed nuclei. In the known even-even nuclei in Fig. 7.9, the only exception is ^{12}C, which is identified as an oblate deformed nucleus by the sign of quadrupole moment of the first excited 2_1^+ state. There are several conjectures for the reason of prolate dominance in nuclear deformation such as the nuclear surface potential and the spin-orbit potential in nuclei. So far, the prolate dominance is still a challenging open question at a more microscopic level.

7.3.3 Nilsson Model for Deformed Single-Particle States

A realistic description of the deformed nuclear potential can be approximated by a modification of the harmonic oscillator potential (7.73) including the spin-orbit potential. The Hamiltonian reads

$$H = \frac{\mathbf{p}^2}{2m} + \frac{1}{2}m(\omega_\perp^2(x'^2 + y'^2) + \omega_z^2 z'^2) + v_{ll}\hbar\omega_0(\boldsymbol{\ell}^2 - \langle\boldsymbol{\ell}^2\rangle)$$
$$+ v_{ls}\hbar\omega_0(\boldsymbol{\ell}\cdot\mathbf{s}),\tag{7.77}$$

where the $\boldsymbol{\ell}^2$ term is also added with its expectation value $\langle\boldsymbol{\ell}^2\rangle = N(N+3)/2$. The $\boldsymbol{\ell}^2$ term mimics a sharp surface of the potential as described by a Woods–Saxon potential with low diffuseness. This term lifts the degeneracy within each major shell to favor the states with large l. The term $\langle\boldsymbol{\ell}^2\rangle$ is added to keep the average energy gap between major shells unchanged by the $\boldsymbol{\ell}^2$ term. The potential (7.77) is called Nilsson potential[2] and the energy levels calculated with the Hamiltonian (7.77) are shown as a function of the deformation parameter δ_{osc} (7.76) in Fig. 7.13, which is often called Nilsson diagram.

Single-particle wave functions, in general, have a large intrinsic anisotropy and will move up or move down in energy as a function of deformation in the single-particle deformed potential such as the Nilsson potential. The filling of the lowest orbitals above the Fermi surface may provide an energy gain, which stabilizes configurations under deformation. Most of deformations, both observed and predicted, have axial symmetry, while there are also evidence of non-axial deformation; triaxial deformation and octupole deformation at the mean field level.

The parity invariance and axial symmetry of the nuclear potential imply that the parity π and the projection Ω of the total angular momentum on the symmetry axis are good quantum numbers of the system. The single-particle wave function of axial symmetric harmonic oscillator Hamiltonian (7.73) is specified by the quantum numbers $[Nn_z n_\perp]$, where $n_\perp$ is the oscillator quantum number perpendicular to the symmetry axis. The degenerate states with the same values of $n_\perp$ can be specified by the component Λ of the orbital angular momentum along the symmetry axis,

$$\Lambda = \pm n_\perp, \pm(n_\perp - 2), \ldots, \pm 1, \quad \text{or} \quad 0.\tag{7.78}$$

In the absence of spin-orbit coupling, the single-particle states in the axial symmetric potential have a four-fold degeneracy ($\Lambda \neq 0$), corresponding to the sign of Λ and the component $\Sigma(= \pm 1/2)$ of the spin along the symmetry axis. The spin-orbit coupling in Eq. (7.77) produces a splitting of the states with different values of the component of total angular momentum

$$\Omega = \Lambda + \Sigma.\tag{7.79}$$

In the limit of large deformation, the spin-orbit coupling gives an energy splitting as

$$\Delta\varepsilon_{ls} \propto v_{ls}\hbar\omega_0\Lambda\Sigma,\tag{7.80}$$

where the expectation value of $\mathbf{l}\cdot\mathbf{s}$ is given by $\langle Nn_z\Lambda\Sigma|l_z s_z|Nn_z\Lambda\Sigma\rangle$. The parallel configuration of spin and angular momenta is favored in energy since the coefficient

---

[2] Named after Sven Gusta Nilsson, Swedish physicist who invented this potential.

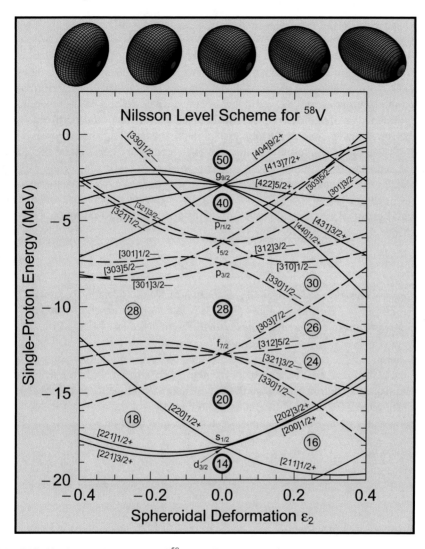

Fig. 7.13 Single-particle spectrum of ^{58}V as a function of deformation parameter ε_2 in an axial symmetric folded Yukawa potential [3, 4]. The orbitals are labeled by the asymptotic quantum number $[N, n_z, \Lambda]\Omega$ referring to large prolate deformations. The quantum number Λ is the projected component of angular momentum along the symmetry axis, while Ω is the component of total angular momentum $\Omega = \Lambda + \Sigma$ (Σ is the component of spin $\Sigma = \pm 1/2$). Levels with even (odd) parity are drawn with solid (dashed) lines. The closed shells at spherical shape are marked by numbers with red circles. The figure also shows large shell gaps at prolate and oblate deformations by numbers with circles. Courtesy P. Möller and T. Ichikawa

v_{ls} is negative. In the large deformation, the deformed single-particle state is thus specified by the quantum numbers $[Nn_z\Lambda\Omega]$ with $\Omega = \Lambda + \Sigma$.

The spectrum for $N < 40$ and $Z < 40$ is shown in Fig. 7.13. In addition to the spherical shell gaps corresponding to nucleon numbers 14, 20 and 28, new shell gaps at 16, 24, 26 and 30 arise at prolate deformation and 18 and 28 at oblate deformation. These shell gaps suggest $^{40}_{16}S_{24}$ could favour large prolate deformations, while $^{36}_{18}Ar_{18}$ is suggested to be oblate. These theoretical conjectures are proved experimentally by the measurement of quadrupole moments of the first excited 2^+_1 state (see Sect. 7.5).

The splitting of single-particle energies was studied by constrained HF calculations with the quadrupole deformation parameter. As an example, the HF and HF+BCS calculations are performed for ^{42}Ca, in which two neutrons are added in the $f_{7/2}$ orbit on top of the ^{40}Ca core. These two neutrons may induce a change of potential energy from the spherical minimum to a deformed minimum. The constrained Hamiltonian is given by

$$H' = H - \lambda Q, \tag{7.81}$$

where λ is a Lagrange multiplier and Q is the quadrupole deformation parameter $Q = (4\pi/3AR^2)r^2 Y_{20}$. Results of constrained HF calculations are shown in Fig. 7.14, where two neutrons occupy the states with different j_z states of $1f_{7/2}$ orbit denoted by $m = \pm 1/2, \pm 3/2, \pm 5/2$, or $\pm 7/2$. To compare with the deformed harmonic potential in Fig. 7.11, the values m is equivalent to the quantum number n_z as $n_z = 7/2 - |m|$. In Fig. 7.14a, at spherical symmetry $q = 0$, these states are degenerate. If the quadrupole deformation is applied, these states develop differently. Each solid curve for a fixed m value represents a potential energy for the quadrupole field. The potential energy curve may be approximated by a quadratic function,

$$V(q) = \frac{1}{2}C(q - q_0)^2,$$

where C is a stiffness parameter and q_0 is a minimum of each curve. The curve with $m = \pm 1/2$ shows a prolate minimum with $q_0 > 0$, while that with $m = \pm 7/2$ gives an oblate minimum with $q_0 < 0$. These curves can be considered as an indication of SSB effect shown by the case (c) in Fig. 7.2. We can mention from Fig. 7.14 that the spherical shell model states are unstable for the quadrupole deformation with the addition of two neutrons on top of the ^{40}Ca core and may become stable at a finite deformation when one adds more and more neutrons as seen in the case of Dy isotopes in Fig. 7.10.

7.4 Strutinsky's Shell Correction Method to Liquid Drop Model

The variation of the single-particle level density in the vicinity of the Fermi energy has an important impact on the stability of shapes in nuclei. In quantal systems, in

Fig. 7.14 a Potential energy curves for different configurations of $1f_{7/2}$ orbit in ^{42}Ca as a function of quadrupole moment $q = (4\pi/3AR^2)r^2Y_{20}$. The constrained HF (CHF) curves (solid lines) are obtained by using Skyrme SIII interaction. The BCS ground state is also shown by a dashed line. The shaded lines indicate the energies of giant resonance and the ground state. The CHF+BCS calculations are performed taking into account the pairing correlations within $1f_{7/2}$ shell orbits. **b** The ground state energy of the BCS state (dashed line) and quasi-particle excitation energies (dotted lines). Reprinted with permission from [9]. ©2021 by Elsevier

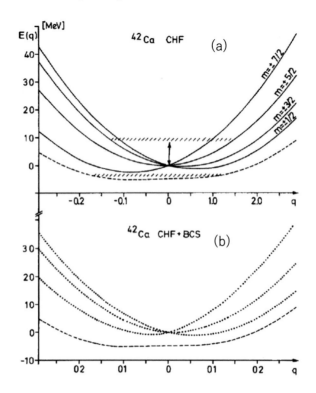

general, the degeneracy leads to reduced stability. This is because a small perturbation of a degenerate system produces a finite response in the system due to rearrangement of many close single-particle states and will cause an instability to a lower degeneracy of the system. Thus, the level density around the Fermi energy is intimately related to the symmetry breaking mechanism. In this respect, the situation of nuclear physics is analogous to the one met in molecular and solid state physics, i.e., the Jahn–Teller effect in the theory of molecules (see Sect. 7.1.2). The nucleus is expected to be more bound if the level density near the Fermi energy is low. In particular, the nuclear ground state, as well as any other excited state, might be stable or meta-stable, having the lowest possible degeneracy.[3]

The stability of nucleus with low level density around the Fermi energy is discussed by the shell correction method by Strutinsky. As is formulated in Eq. (4.3) in Chap. 4, the nuclear binding energy E is a smooth function of A and represented by

---

[3] The concept of "deformed magic number" is encountered in the literature. The accepted and historical definition of a magic number restricts to spherical shell closures which present an energy gap across the Fermi level that prevents correlations to develop, and the nucleus is well described by the Hartree–Fock model. The so-called deformed magic numbers are an extension to this first definition to deformed nuclei with a large energy shell gap in the deformed single-particle level scheme at the deformation that minimized the total energy of the nucleus. The structure of the nucleus is well described by a deformed mean-field model.

the liquid-drop mass formula E_{LDM}. Vilen Strutinsky introduced the shell effect on the binding energy by a formula

$$\delta E_{shell} = E_{shell} - \tilde{E}_{shell}, \tag{7.82}$$

where

$$E_{shell} = \int_{-\infty}^{\lambda_F} \varepsilon g_{sh}(\varepsilon), \tag{7.83}$$

with λ_F being the Fermi energy. The shell model level density is given by

$$g_{sh}(\varepsilon) = \begin{cases} \sum_i \delta(\varepsilon - \varepsilon_i) & \text{for } \varepsilon < \lambda_F \\ 0 & \text{for } \varepsilon > \lambda_F. \end{cases} \tag{7.84}$$

The particle number is calculated as

$$A = \int_{-\infty}^{\lambda_F} g_{sh}(\varepsilon) d\varepsilon. \tag{7.85}$$

The average shell model energy is defined as

$$\tilde{E}_{shell} = \int_{-\infty}^{\tilde{\lambda}} \varepsilon \tilde{g}_{sh}(\varepsilon) d\varepsilon, \tag{7.86}$$

where $\tilde{g}_{sh}$ is an average level density, which is a smooth function of ε. The $\tilde{\lambda}$ is defined by

$$A = \int_{-\infty}^{\tilde{\lambda}} \tilde{g}_{sh}(\varepsilon) d\varepsilon, \tag{7.87}$$

and different from λ_F. In the shell correction method by Strutinsky, the average part $\tilde{E}_{shell}$ is replaced by the Liquid drop energy E_{LDM} as

$$E_{shell} = E_{LDM} + \delta E_{shell}, \tag{7.88}$$

where

$$\delta E_{shell} = \int_{-\infty}^{\lambda_F} \varepsilon g_{sh}(\varepsilon) d\varepsilon - \int_{-\infty}^{\tilde{\lambda}} \varepsilon \tilde{g}_{sh}(\varepsilon) d\varepsilon. \tag{7.89}$$

The shell correction term is approximated by using the shell model occupation (7.84) as

$$\delta E_{shell} = \sum_{1=1}^{A} \varepsilon_i - \sum_{i=1}^{\infty} n_i \varepsilon_i, \tag{7.90}$$

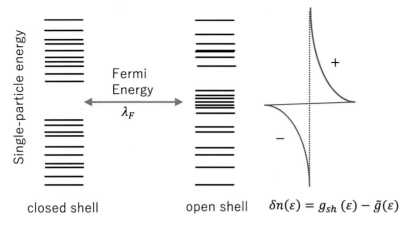

$$\delta n(\varepsilon) = g_{sh}(\varepsilon) - \tilde{g}(\varepsilon)$$

Fig. 7.15 Qualitative illustration of the connection between the level density at the Fermi energy and the binding energy of the nucleus: (Left) Closed shell in which there is a large shell gap at the Fermi energy. (Middle) Open shell with high degeneracy of single-particle states at the Fermi energy. (Right) The occupation probability $\delta n(\varepsilon) = g_{sh}(\varepsilon) - \tilde{g}_{sh}(\varepsilon)$ in the shell correction energy (7.89). $g_{sh}(\varepsilon) = 1$ (0) for $\varepsilon < \lambda_F$ ($\varepsilon > \lambda_F$) and $\tilde{g}$ is a positive smooth function changing from 1 at the bottom of potential to about 1/2 at the Fermi energy, and eventually to 0 at the very high energy above the Fermi energy

where n_i is a smooth single-particle occupation number at the energy ε_i varying from 1 to 0 in the energy interval of typically one or two major shells below and above the Fermi energy.

The E_{LDM} is the macroscopic energy and depends on the mass number A smoothly. In E_{LDM}, the surface energy dominates and tends to favor spherical shapes. The shell correction δE_{shell} is crucial to study the stability of shapes in terms of level density (see Fig. 7.15). If the level density is large (large degeneracy) around the Fermi energy, $\tilde{E}_{shell}$ is large and positive and will induce an instability of the shape. The opposite trend is expected in the case of low level density (small degeneracy): for magic nuclei, there is a large shell gap at the Fermi energy and the E_{shell} is large and negative, and $\tilde{E}_{shell}$ is expected to be small since only a few levels exist around the Fermi energy. That is, the stability of shape is implied in the case of low level density.

The shell correction is generalized in FRDM 2012 [4] to include the pairing correction and the zero-point energy correction E_{zp},

$$E_{mic} = \delta E_{shell} + \delta E_{pair} + E_{zp}. \tag{7.91}$$

The δE_{pair} is defined as

$$\delta E_{pair} = E_{pair} - \tilde{E}_{pair}, \tag{7.92}$$

where E_{pair} is calculated by a BCS model with the same single-particle energies ε_i in Eq. (7.90) and $\tilde{E}_{pair}$ is calculated by using an average smooth single-particle energy

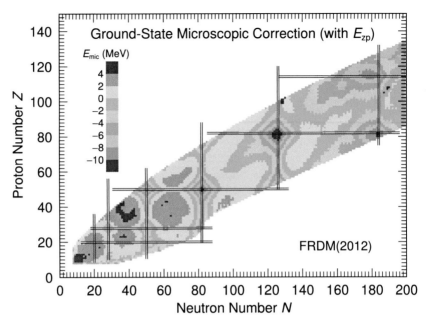

Fig. 7.16 Microscopic shell correction energy E_{mic} by FRDM 2012. The microscopic corrections include the δE_{shell} and also the pairing corrections δE_{pair} together with the zero-point energy E_{zp} of quadrupole vibration at the ground state shape. Reprinted with permission from [4]. ©2021 by Elsevier. Courtesy P. Möller

distribution. The zero-point energy E_{zp} takes into account the effect of zero-point motion of the quadrupole field at the ground-state shape.

In Fig. 7.16, the microscopic correction E_{mic} shows clear stabilities at ($Z = 50, N = 82$) and ($Z = 82, N = 126$) with large negative microscopic corrections, which are known as doubly-closed shells. There is also another stable region with ($Z = 116, N = 182$), corresponding to super-heavy elements. We can see also a stable region around ($Z = 82, N = 182$), which is close to or beyond the neutron-drip line. Near the proton drip line, one can find also a stable region near ($Z = 50, N = 50$).

Between the shell closures, the microscopic correction is positive and nuclei will be deformed to make more stable shapes with a lower degeneracy of level density.

The Strutinsky method for the shell correction, a combination of the shell model with the liquid drop model, has been applied successfully to the calculation of nuclear masses [4]. The method was also applied to calculate the energy surfaces as the function of deformations, not only quadrupole, but also octupole, hexadecupole and higher multipoles. The model predicts the deformation minima of the rare-earth nuclei as well as of the actinide nuclei. In some nuclei, the second minima are also found for very large deformations, referred to as fission isomers.

7.5 Measuring Shapes

The shape of a nucleus can be inferred from several indirect measurements, one example being the charge radius of a nucleus. It does not tell if the nucleus is deformed but its comparison to theoretical predictions can be used to validate the theory and provide support to the predicted deformation, although not a proof *strict senso*. Large variations of the charge radius from one isotope to another give hints for shape changes. An illustration is given in Fig. 5.3 of Chap. 5, where the change of the mean square charge radii $\delta\langle r^2\rangle$ between neighbouring nuclei with $(N + 1, Z)$ and (N, Z) measured from the isotopic shift of the atomic energy levels are shown for nuclei in the region of lead, well known to present coexistence and transitions from spherical, prolate, and oblate deformations. As an illustration, the strong increase of the isotopic shift in neutron-deficient gold isotopes [185,186]Au was interpreted as a steep onset of deformation, indicating a transition from spherical to prolate ground-state deformation in those isotopes.

Other indirect measurements such as the excitation energy of the first 2^+ state, the ratio of the first 4^+ and first 2^+ excitation energies, or the transition probability between the first 2^+ and the ground state $B(E2; 2^+ \to 0^+)$ for even-even nuclei are used to extract the nuclear deformation in a model dependent way. The direct determination of the shape of a nucleus requires the measurement of the multipole moments of its nucleon distribution. In most experiments, only the charge quadrupole moments can be measured. The strength of the nuclear interaction implies that protons and neutrons have the same deformation with the possible exception of semi-magic nuclei, in which the closed-shell species of nucleons may be stabilized into spherical shape, while the other species could present a different deformation.

7.5.1 Electric and Magnetic Moments

Electric and magnetic static moments of nuclei are measured via the interaction of the nuclear charge distribution and magnetism with external electromagnetic fields. These external fields can be of various kinds: either induced by the atomic electrons bound to the nucleus, by bulk electrons in a crystal in which the nucleus under study has been implanted, or by the electromagnetic field induced by the moving charge of a colliding object (nucleus, electron, . . .). Among these measurable moments, the electric multipole moments give a direct measure of the (charge) deformation. In the particular case of quadrupole deformation, the spectroscopic quadrupole moment can be accessed experimentally. Techniques can be divided into three families:

- laser spectroscopy to extract quadrupole moments from the atomic hyperfine structure (see Sect. 4.4.3 of Chap. 4),
- angular distributions of radiation decay in a strong electric and/or magnetic field after implantation of an oriented nucleus (see Sect. 4.4.5 of Chap. 4),

• low-energy Coulomb excitation to determine the quadrupole moment of states from their population probability.

7.5.2 Coulomb Excitation

The collision of two nuclei can be restricted to electromagnetic excitation if the center-of-mass energy is below the Coulomb barrier between the projectile and target, at the exception of low-probability quantum tunneling effect below the barrier. If so, the distance of least approach between the two colliding nuclei is long enough to avoid interferences with the short-range nuclear interaction. Following the historical development of Kurt Alder [2], the Coulomb excitation of a projectile in the electric field of a target can be obtained by solving the Schrödinger equation

$$i\hbar\frac{\Psi(t)}{dt} = [H_P + H_T + V(\mathbf{r}(t))]\Psi(t), \tag{7.93}$$

where $H_{P,T}$ are the intrinsic Hamiltonians of the projectile (P) and target (T) and $V(\mathbf{r}(t))$ is the electric potential that depends on the relative distance between the projectile and target during the process. For simplicity, we assume in the following that only the projectile can be excited in the reaction and the eigenstate and eigenenergy are given by

$$H_P\phi_n = E_n\phi_n. \tag{7.94}$$

The wave function Ψ can then be expressed by the projectile wave function and decomposed over the intrinsic states ϕ_n of the projectile,

$$\Psi(t) = \sum_n a_n(t)\phi_n e^{-iE_nt/\hbar}, \tag{7.95}$$

where $a_n(t)$ are time-dependent excitation amplitudes. Inserting Eq. (7.95) into Eq. (7.93) with the aid of Eq. (7.94), one obtains a set of coupled-channel equations,

$$i\hbar\frac{da_n(t)}{dt} = \sum_m \langle\phi_n|V(t)|\phi_m\rangle e^{\frac{i}{\hbar}(E_n-E_m)t}a_m(t). \tag{7.96}$$

These equations can be solved by a multipole expansion of the electromagnetic interaction as follows

$$V(\mathbf{r}) = \frac{Z_P Z_T e^2}{r} + \sum_{\lambda\mu} V_P(E\lambda, \mu) + \sum_{\lambda\mu} V_P(M\lambda, \mu), \tag{7.97}$$

where the first term is the Rutherford term (monopole) and the others are projectile electric (E) and magnetic (M) multipole excitations. The multipole excitations of the target have been omitted in the above formula. The electric multipole terms can be expressed as a function of the electric multipole moments $M(E\lambda, \mu)$,

$$V_P(E\lambda, \mu) = (-1)^\mu Z_T \frac{4\pi e}{2\lambda + 1} r^{\lambda+1} Y_{\lambda\mu}(\theta, \phi) M(E\lambda, \mu) \tag{7.98}$$

with

$$M(E\lambda, \mu) = \int_0^\infty \rho(\mathbf{r}') r'^\lambda Y_{\lambda\mu}(\hat{r}') d\mathbf{r}'. \tag{7.99}$$

It appears clearly that the Coulomb excitation cross section is sensitive to the electric multipole moments (at all orders). The transition amplitude b_{nm} from a state m to a state n is calculated from an action integral

$$b_{nm} = \frac{i}{\hbar} \int_0^\infty \langle a_n \phi_n | V(t) | a_m \phi_m \rangle e^{\frac{i}{\hbar}(E_n - E_m)t} dt. \tag{7.100}$$

The excitation probability to populate the state m from the initial state n is given by

$$P(I_n \rightarrow I_m) = \frac{|b_{nm}|^2}{2I_n + 1}. \tag{7.101}$$

In practice, the perturbation theory is used to calculate the excitation probabilities. In the case of only one state populated with small transition probability, the first-order perturbation can be used,

$$P^{(1)} = (I_n \rightarrow I_m) = \frac{|b_{nm}^{(1)}|^2}{2I_n + 1}, \tag{7.102}$$

with

$$b_{nm}^{(1)} = \langle I_m || M(E\lambda) || I_n \rangle, \tag{7.103}$$

where the double-bar matrix element implies the reduced matrix element. $P^{(1)}$ in Eq. (7.102) is nothing but the transition probability $B(E\lambda : I_n \rightarrow I_m)$ defined in Chap. 4. If several states can be populated from the ground state or when multiple-step excitations are possible (large excitation probabilities), an expansion to the second-order is mandatory. In the later case, the excitation probability follows

$$P^{(2)}(I_n \rightarrow I_m) = \frac{|b_{nm}^{(2)}|^2}{2I_n + 1}, \tag{7.104}$$

where the second-order amplitude $b_{nm}^{(2)}$ is the sum of the one-step excitation (first-order) and two-step processes via an intermediate state k given by

$$b_{nm}^{(2)} = b_{nm}^{(1)} + \sum_k b_{nk}^{(1)} b_{km}^{(1)} = b_{nm}^{(1)} + \sum_k b_{nkm}^{(2)}, \tag{7.105}$$

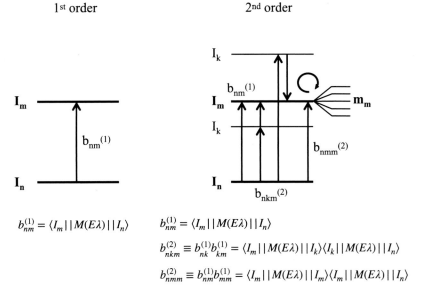

1st order   2nd order

$b^{(1)}_{nm} = \langle I_m || M(E\lambda) || I_n \rangle$ $b^{(1)}_{nm} = \langle I_m || M(E\lambda) || I_n \rangle$

$b^{(2)}_{nkm} \equiv b^{(1)}_{nk} b^{(1)}_{km} = \langle I_m || M(E\lambda) || I_k \rangle \langle I_k || M(E\lambda) || I_n \rangle$

$b^{(2)}_{nmm} \equiv b^{(1)}_{nm} b^{(1)}_{mm} = \langle I_m || M(E\lambda) || I_m \rangle \langle I_m || M(E\lambda) || I_n \rangle$

Fig. 7.17 Schematic view of Coulomb excitations at first- and second-order perturbation theory

where
$$b^{(2)}_{nkm} \equiv b^{(1)}_{nk} b^{(1)}_{km} = \langle I_m || M(E\lambda) || I_k \rangle \langle I_k || M(E\lambda) || I_n \rangle. \tag{7.106}$$

Among the the second-order processes, one encounters transitions involving a diagonal matrix element,

$$b^{(2)}_{nmm} \equiv b^{(1)}_{nm} b^{(1)}_{mm} = \langle I_m || M(E\lambda) || I_m \rangle \langle I_m || M(E\lambda) || I_n \rangle \tag{7.107}$$

for states $|I_m| \geq \lambda/2$. These matrix elements are the origin of the so-called "orientation effect".

From the above Eqs. (7.98), (7.100), (7.104), (7.105), one can see that at the second order, a component of the excitation probability depends on the diagonal matrix element and its sign

$$P^{(2)}_{\text{reorientation}} \propto \langle I_m || M(E\lambda) || I_m \rangle |\langle I_m || M(E\lambda) || I_n \rangle|^2, \tag{7.108}$$

as illustrated in Fig. 7.17. The reorientation effect gives the Coulomb excitation cross section a unique sensitivity to the nuclear shape. The second-order part of the excitation probability involving a change in the projection of the angular momentum is interpreted as a spatial *reorientation* of the projectile in the electrical field of the target, in the laboratory frame, that is, the orientation effect involves the diagonal matrix element $\langle I_m | M(E\lambda) | I_m \rangle$, which provides not only the information of magnitude but also the sign of deformation, i.e., prolate or oblate.

Problem

7.8 Derive relation (7.108) from Eqs. (7.104) and (7.105).

As an example, we consider the Coulomb excitation of ^{76}Kr from a target ^{208}Pb performed at the SPIRAL facility of GANIL at sub-Coulomb barrier energy of 4.4 MeV/nucleon [10]. The gamma spectrum of ^{76}Kr (top of Fig. 7.18) shows the levels populated during the process: the larger number of populated states and the populated spins (up to 8^+) evidence multiple-step excitations. The sensitivity of the differential cross section to the quadrupole moment of the first 2^+ state is illustrated in the bottom panel of the figure, in which the prolate shape ($Q > 0$) gives an enhancement of the cross section in comparison with the oblate case ($Q < 0$) case. This second-order reorientation effect requires sufficient statistics to allow for an extraction of the quadrupole moment which are usually made possible within few days at beam intensities at 10^5 particles per second or above.

7.6 Backbending Phenomenon in Moment of Inertia

In Eq. (7.46), the rotational frequency ω_{rot} is determined by the following relation between the moment of inertia and the angular momentum I,

$$\omega_{\text{rot}}\Im = \hbar I. \tag{7.109}$$

The rotational frequency $\omega_{\text{rot}} = d\theta/dt$ is not an observable. However, since the angle θ and the angular momentum are the canonical variables in classical mechanics, we can define ω_{rot} in a semiclassical way using a canonical equation between the angle θ and the angular momentum $\hbar I$,

$$\frac{d\theta}{dt} = \frac{\partial H(\theta, I)}{\hbar \partial I}, \tag{7.110}$$

where $H(\theta, I)$ is the rotational Hamiltonian. Then, the canonical equation (7.110) can be rewritten as

$$\omega_{\text{rot}} = \frac{1}{\hbar}\frac{dE_{\text{rot}}}{dI}, \tag{7.111}$$

for the rotational energy $E_{\text{rot}}(I)$. Once the rotational energy E_{rot} is measured experimentally, the rotational frequency can be deduced from E_{rot}. Consequently, the moment of inertia is also determined by Eq. (7.46). The angular momenta are integers or half-integers so that the empirical angular frequency is defined from experimental energies as

$$\omega_{\text{rot}}(\exp) \approx \frac{\Delta E}{\hbar\Delta\sqrt{I(I+1)}} \approx \frac{E(I) - E(I-2)}{\hbar(\sqrt{I(I+1)} - \sqrt{(I-2)(I-1)})}. \tag{7.112}$$

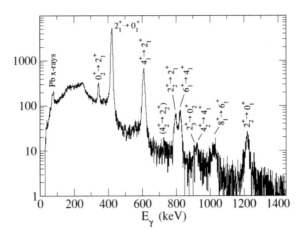

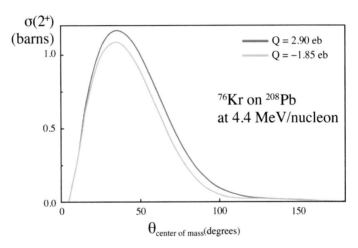

Fig. 7.18 (Top) Gamma spectrum obtained from the Coulomb excitation of ^{76}Kr from a ^{208}Pb target at 4.4 MeV/nucleon. The experiment was performed at the SPIRAL facility of GANIL, France. Reprinted with permission from [10]. ©2021 by the American Physical Society. (Bottom) Sensitivity of the differential Coulomb excitation cross section to the quadrupole moment. The calculation considers the excitation of the first 2^+ state of ^{76}Kr on ^{208}Pb at 4.4 MeV/nucleon

The moment of inertia is also obtained by

$$\Im = \frac{\hbar I}{\omega_{\text{rot}}} = I \left(\frac{dE}{dI} \right)^{-1} = \frac{1}{2} \left(\frac{dE}{d(I^2)} \right)^{-1} \approx \frac{2I - 1}{E(I) - E(I - 2)}. \qquad (7.113)$$

In realistic physical conditions, the moment of inertial varies as a function of rotational frequency because of the pairing correlations and the centrifugal stretching due to rotation. To describe the evolution of the rotational band in such circumstances, two moments of inertia have been introduced; they are the kinematic moment of inertia $\Im^{(1)}$ and the dynamic moment of inertia $\Im^{(2)}$ defined by

$$\Im^{(1)} = I \left(\frac{dE}{dI} \right)^{-1} \tag{7.114}$$

and

$$\Im^{(2)} = \left(\frac{d^2 E}{dI^2} \right)^{-1}, \tag{7.115}$$

respectively. The kinematic moment of inertia is related to the overall motion of the nucleus, and the dynamic moment of inertia describes its response to a torque. The kinematic moment of inertia (7.114) is identical to the definition (7.113). For a rigid rotor, the shape of object is not changed by the rotation and thus $\Im^{(1)} = \Im^{(2)}$. However, when there are changes in the internal structure of the nucleus, such as the alignment of the single-particle angular momentum to the rotational axis, $\Im^{(1)}$ will be different from $\Im^{(2)}$ reflecting the fact that the nuclear rotation is not the simple rotation of a rigid body.

For the angular momenta between $10\hbar$ and $20\hbar$, an anomaly is observed in the yrast states[4] of many deformed nuclei. It can be illustrated when the moment of inertia is plotted as a function of rotational frequency ω^2. As an example, the rotational states of ^{164}Er are shown in Fig. 7.19. A nucleus ^{164}Er is a typical prolate deformed nucleus. Energy spectra of the ground state and γ bands show clear $I(I + 1)$ band structures. However, the yrast levels change from the ground band to another rotational band with the band head of the excited state so-called "S-band"[5] at the angular momentum between $I = 14$ and $I = 16$. The moments of inertia of ground band and S-band are shown in Fig. 7.20. For a rigid rotor as is expected, the value of $\Im$ is constant. However, the moments of inertia in Fig. 7.20 gradually change for an increasing ω. The curve shows a steep increase, and at $I = 16$ the curve is even bending backwards, i.e., this is called "backbending" phenomenon. Such a phenomenon is understood as the crossing of two bands with different moments of inertia, i.e., the yrast states belong to the ground band for low-spin states, but change to the S-band after the band crossing. In general, the moment of inertia of the S-band is larger close to the rigid rotor one $\Im_{\text{rigid}}$, than that of ground band, which is typically less than half of $\Im_{\text{rigid}}$ because of the superfluidity phase of the ground state. Then, the two bands cross at certain angular momentum as shown in the right panel of Fig. 7.19. The S-band is understood as a two-quasi-particle band in which two particles get aligned along the axis of rotation. This alignment is induced by a Coriolis coupling term in the intrinsic Hamiltonian as will be seen in Eq. (7.120). This aligned angular

---

[4] The word "*yrast*" is the superlative of "yr", a Swedish adjective sharing the same root as the English whirl. Yrast can be translated whirlingest, although it literally means "dizziest" or "most bewildered". The word "*yrare*" is the comparative of "yr" and means the "whirlinger" or "dizzier". These Swedish words are often used to assign the lowest and the second lowest states of rotational bands for a given angular momentum.

[5] It has been said that the term "S-band" comes from an initial of "Stockholm" since this band was observed for the first time in Stockholm.

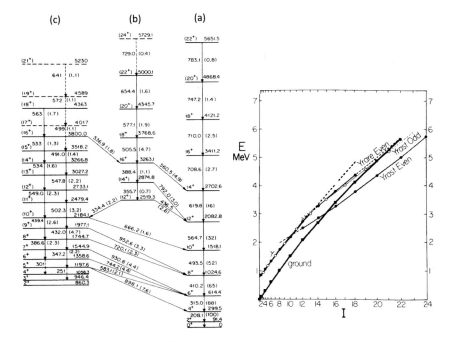

Fig. 7.19 (Left) Level scheme of ^{164}Er; **a** the ground band, **b** the S-band, and **c** the γ-vibrational band from the right to the left, populated in the reaction ^{150}Nd(^{16}O, 4n)^{164}Er. Each transition is shown with associated γ-ray energy and, in parentheses, the relative intensity. (Right) Experimental rotational bands of ^{164}Er as a function of I. The solid circles correspond to even-spin states and the open circles to odd-spin states. Reprinted with permission from [12]. ©2021 by the American Physical Society

momentum of two quasi-particles is the reason why the band head of S-band has a large angular momentum. The transition between two bands is smooth because the two bands are mixed by the coupling interaction between them.

Let us estimate approximately the crossing angular momentum between the ground band and the S-band. Firstly, the moment of inertia of the S-band is taken to be the rigid rotor value $\mathcal{I}_{S-band} = \mathcal{I}_{rigid}$, while that of the ground band is empirically almost a half of rigid rotor one $\mathcal{I}_{gb} \simeq \frac{1}{2}\mathcal{I}_{rigid}$ because of the superfluidity phase in the ground state. Secondly, the band head of S-band is considered as 2 quasi-particle angular momentum aligned state with the excitation energy E_{2qp}. Then the crossing angular momentum I_c is determined by an equation $E_{gb} = E_{S-band}$;

$$\frac{\hbar^2 I_c(I_c + 1)}{2\mathcal{I}_{gb}} = E_{2qp} + \frac{\hbar^2 I_c(I_c + 1)}{2\mathcal{I}_{S-band}}, \qquad (7.116)$$

which turns out to be

$$I_c(I_c + 1) = E_{2qp}\frac{2\mathcal{I}_{rigid}}{\hbar^2}. \qquad (7.117)$$

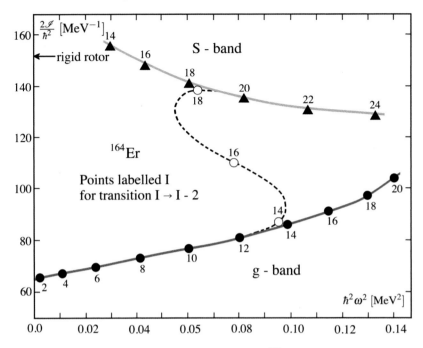

Fig. 7.20 Moment of inertia of ground band and S-band of ^{164}Er as a function of rotational frequency ω. The yrast state is changed from the ground band to the S-band at $I = 16$. The transition between two bands is occurred because the two bands are mixed by the coupling interaction between them. The data is taken from [12]

Taking a rigid rotor moment of inertia of $\frac{2\mathcal{I}_{\text{rigid}}}{\hbar^2} \approx 140\,\text{MeV}^{-1}$ for $A \approx 160$ and $\delta = 0.3$ in Eq. (7.42), and $E_{2qp} = 2\,\text{MeV}$, we can determine the crossing angular momentum as

$$I_c \approx 16, \qquad (7.118)$$

which is consistent with the observed band-crossing angular momentum shown in Fig. 7.20.

In the rotating frame, the individual particles are subject to the Coriolis force,

$$H' = H_0 - \hbar\omega_{\text{rot}} \cdot \mathbf{j}, \qquad (7.119)$$

where H_0 is the Hamiltonian without rotation and $\mathbf{j}$ is the angular momentum of single-particle state under consideration. Here, the rotation takes any direction of intrinsic cartesian coordinate, while the rotation is held around the x-axis in Eq. (7.43). To describe the single-particle motion in the rotation frame, the so-called "particle-rotor" microscopic model is introduced. In this model, the Hamiltonian for the Coriolis term is expressed as

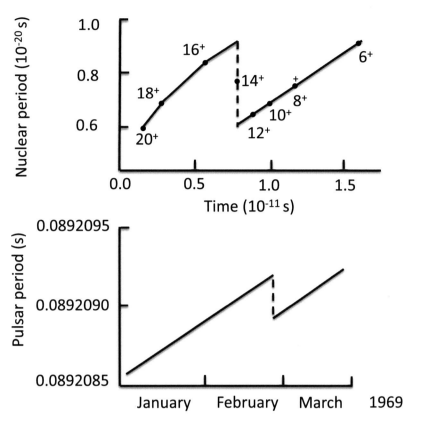

Fig. 7.21 An analogy of change in rotational period between the backbending phenomenon in ^{164}Er and the glitch in the Vela pulsar. See text for details. Data are taken from [13]

$$H_c = -\hbar\omega_{\text{rot}} \cdot \mathbf{j} = -\hbar^2 \frac{(I_+ j_- + I_- j_+)}{2\Im}, \qquad (7.120)$$

combining Eq. (7.46) with Eq. (7.120). The particle-rotor model is applied to describe the rotational spectra of ^{164}Er and gives a good description of the backbending phenomenon.

The backbending phenomenon is analogous to the glitch phenomenon in the rotational frequency of pulsar in the astrophysical observations,[6] in which the pulsar loses suddenly the rotational frequency during a short time period. The time period of rotation of the Vela pulsar is shown in the lower panel of Fig. 7.21. The rotational period of the pulsar is taken from the observation of a glitch in the Vela pulsar in between January and March 1969. One can see a clear drop of the rotational period at the end of February. Let us evaluate the analogous change in the rotational period of

---

[6] A pulsar is a highly magnetized rotating compact star, mostly neutron stars but also white dwarfs, that emits beams of electromagnetic radiation out of its magnetic poles.

rotation band in ^{164}Er. The time period of rotation for the nucleus ^{164}Er is calculated by

$$T = \frac{2\pi}{\omega_{\text{rot}}}. \qquad (7.121)$$

In the upper panel of Fig. 7.21, the time period (7.121) is given in the horizontal axis, which is obtained by the decay time from the yrast state $I^\pi = 20^+$ to $I^\pi = 6^+$ state by sequential decays with $E2$ transitions. A sudden analogous decrease in the moment of inertia in backbending phenomena of nuclear rotation appears in the top panel of Fig. 7.21. The glitch of the pulsar is considered due to the appearance of the pinned vortex in the crust of the neutron star. The rotation of neutron stars creates such vortices in the superfluid phase of the crust.

7.7 Shape Coexistence

The direct fingerprints for nuclear deformation are the diagonal E2 matrix elements (7.30). They therefore also provide the best way to evidence shape coexistence when states with different quadrupole deformations coexist at similar excitation energy. The observation is further evidenced when rotational bands built on different quadrupole moment configurations coexist. In the deformed mean field or shell model frameworks, the existence of low-lying excited 0^+ states in even-even nuclei is an indirect hint for shape coexistence since they can be interpreted as band heads of different shape configurations. The shape coexistence phenomenon often leads to a transition from one configuration to another as the favored shape for the ground-state configuration. The shape coexistence appears rather frequent across the nuclear landscape. Figure 7.22 shows regions where shape coexistence has been evidenced experimentally. The historical discovery of a triplet of differently shaped low-lying 0^+ states in the neutron-deficient ^{186}Pb is one of the most striking examples of shape coexistence [15].

In the following, as an illustration, we present some of the experimental evidence of shape transition and coexistence between prolate and oblate configurations in light Kr isotopes in Fig. 7.23, where experimental systematics of first excited 0^+, 2^+, 4^+ states in ^{72,74,76,78}Kr are shown. The ground state and the first excited 0^+ state are interpreted as the band heads of an oblate and a prolate shape intrinsic configuration in $^{72}_{36}$Kr, while the shape is interchanged in ^{74,76,78}Kr and the ground state and the first excited 0^+ are interpreted as a prolate and an oblate shape intrinsic configuration. Measured quadrupole moments from low-energy Coulomb excitation for ^{74,76,78}Kr give a coherent description of prolate-oblate shape coexistence at low excitation energy for these isotopes, i.e., a shape transition from the prolate ground states in ^{74,76,78}Kr to an expected oblate ground state in $^{72}_{36}$Kr$_{36}$. A precise measurement of the quadrupole moment of the first excited states of ^{72}Kr is still missing.

Theoretically, mean field Hartree–Fock–Bogolyubov (HFB) models indicate two or even three energy minima at different deformations in the $(\beta - \gamma)$ plane for some

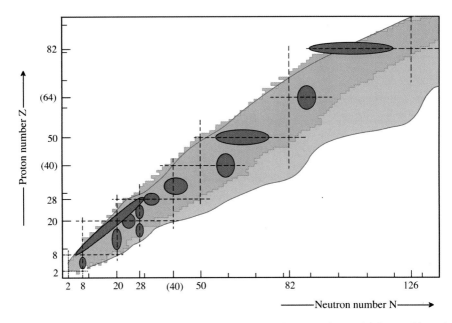

Fig. 7.22 Regions (in pink) of the nuclear landscape where shape coexistence has been evidenced experimentally. Nuclei discovered so far are shown in yellow. Reprinted with permission from [14]. ©2021 by the American Physical Society

medium-heavy and heavy nuclei. To study rotational spectra based on these deformed intrinsic states, a model beyond the mean field is necessary to project out good angular momentum states from intrinsic states with mixed angular momenta. One plausible model is the HFB+GCM model with triaxial degree of freedom for deformation. Here, we remind the basis of the HFB+GCM model, while details are given in Chap. 3. Trial quasi-particle vacuum wave functions are obtained from a variation principle for the constrained Hamiltonian

$$\delta\langle\phi_q|H - \sum_i \lambda_i Q_i - \lambda_Z \hat{Z} - \lambda_N \hat{N}|\phi_q\rangle = 0, \qquad (7.122)$$

where H is the mean field Hamiltonian, Q_i is a set of quadrupole operators including triaxial degree of freedom, and $\lambda_i, \lambda_Z, \lambda_N$ are Lagrange multipliers which are determined by the constraints

$$\langle\phi_q|Q_i|\phi_q\rangle = q_i, \quad \langle\phi_q|\hat{Z}|\phi_q\rangle = Z, \quad \langle\phi_q|\hat{N}|\phi_q\rangle = N, \qquad (7.123)$$

where q_i is the quadrupole deformation parameter. For the shape coexistence problem of quadrupole deformations, the GCM state is a superposition of trial functions ϕ_q,

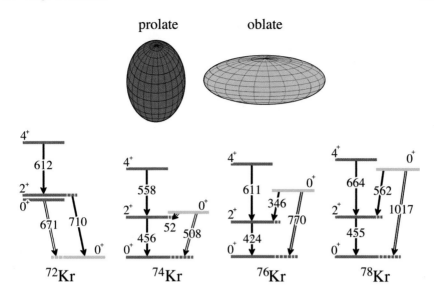

Fig. 7.23 Experimental systematics of first excited 0^+, 2^+, 4^+ states in ^{72,74,76,78}Kr. The ground state and the first excited 0^+ state are interpreted as band heads of an oblate and a prolate (or a prolate and an oblate) shape intrinsic configuration [16]. Measured quadrupole moments from low-energy Coulomb excitation for ^{72,74,76,78}Kr give a coherent description of prolate-oblate shape coexistence at low excitation energy for these isotopes. A shape transition is observed from the prolate ground states in ^{74,76,78}Kr to the oblate ground state in ^{72}Kr

$$|\Psi_k\rangle = \int f_k(q)|\phi_q\rangle dq, \tag{7.124}$$

where the generator coordinates q stand for five-hold collective coordinates; the axial symmetry shape parameter $Q_0 = r^2 Y_{20}$, triaxiality $Q_2 = r^2(Y_{22} + Y_{2-2})$, and three Euler angles. The amplitude $f_k(q)$ is obtained by solving the Hill–Wheeler–Griffin equation (Eq. (3.140) of Chap. 3).

The self-consistent HFB calculations have been performed to obtain deformed single-particle states in Fig. 7.24 and then the GCM approach is applied with Gaussian overlap approximation. There are two GCM calculations in Fig. 7.25 with Skyrme and Gogny interactions. Bender et al. used Skyrme SLy6 interaction and a density-dependent pairing interaction for the HFB calculations [17]. Figure 7.24a shows the deformation dependence of neutron single-particle energies with positive (solid lines) and negative (dotted lines) parities as a function of the intrinsic quadrupole deformation β_2 for $^{74}_{36}$Kr$_{38}$.

We can see large shell gaps with $N = 38$ both for the prolate side $\beta_2 > 0$ and the oblate side $\beta_2 < 0$. Figure 7.24b shows the mean field and $J = 0$ projected energy curves for ^{74,76}Kr. The mean-field energy curves (dotted lines in Fig. 7.24b) show that the energies of the two nuclei vary quite slowly with deformation. The calculations predict the two nuclei with coexisting prolate and oblate minima. The two minima

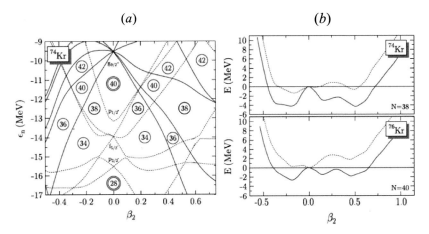

Fig. 7.24 a: Deformation dependence of neutron single-particle energies with positive (solid lines) and negative (dotted lines) parity as a function of the intrinsic quadrupole deformation β_2 for $^{74}_{36}\text{Kr}_{38}$. The deformed HFB calculations are performed with a Skyrme interaction SLy6 with a density dependent pairing interaction. **b:** Mean-field (dotted) and the projected deformation energy curves on $J = 0$ state (solid) for ^{74,76}K. Reprinted with permission from [17]. ©2021 by the American Physical Society. See also [18] for the HFB calculations with Gogny interaction D1S

in the mean-field energy curve (dotted line) of $^{74}_{36}\text{Kr}_{38}$ reflect the $N = 38$ gaps in the Nilsson diagram in Fig. 7.24a at small oblate and large prolate deformations. The mean field wave functions are projected on the good angular momentum and particle number state as

$$|\Psi^J\rangle = P^J P^A |\Psi\rangle, \qquad (7.125)$$

where P^J and P^A are the projection operators of angular momentum and particle number, respectively. As can be seen from the solid lines in Fig. 7.24b, the projection of a deformed mean-field state on the $J = 0$ ground state always leads to an energy gain and many local minima appear in both the prolate and oblate deformation side.

In Fig. 7.25, the GCM energy levels are shown in comparison with experimental data. The calculations performed with the Skyrme interaction are restricted to axially symmetric shapes, while those with the Gogny force performed with a five-dimensional collective Bohr Hamiltonian accounting for the axial and triaxial deformation (β, γ) and the three rotational degrees of freedom $(\mathcal{I}_1, \mathcal{I}_2, \mathcal{I}_3)$. It is observed in Fig. 7.25 that the calculation based on the Skyrme force finds a predominantly oblate ground state and an excited prolate configuration for both $^{74}_{36}\text{Kr}_{38}$ and $^{76}_{36}\text{Kr}_{40}$ in contrast to the experimental results, which give firm evidence for the prolate character of the ground-state band. This might be due to the fact that the relative position of prolate and oblate minima in the HFB is not in right order in Fig. 7.24b, which could be caused by the restriction of the Skyrme calculation to axial symmetric shapes. The five-dimensional calculation based on the Gogny force reproduces satisfactorily the excitation energies of the prolate and oblate rotational bands, but finds also a

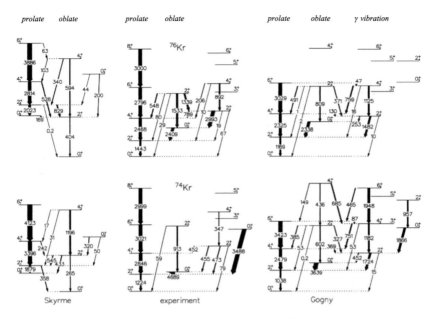

Fig. 7.25 Comparison between the theoretical and experimental level schemes for the oblate and prolate bands in $^{76}_{36}Kr_{40}$ (top) and $^{74}_{36}Kr_{38}$ (bottom). The excitation energies of the states are drawn to scale and the widths and labels of the arrows represent the calculated and measured $B(E2)$ values, respectively. See the text for details. Reprinted with permission from [10]. ©2021 by the American Physical Society

2^+, 4^+, and 6^+ state with predominant $K = 2$ character in ^{76}Kr very close to the experimentally observed states of the presumed γ-vibrational band. The transition strengths within the prolate band and in particular the decrease toward the ground state are well reproduced by both calculations.

The shape transition from the prolate ground states in $^{74,76}Kr$ to the oblate state in ^{72}Kr has been interpreted from the systematics of low energy 0^+_2 states and conversion electron decay probabilities $\rho(E0)$ to the ground state shown in Fig. 7.23. The coexistence of prolate and oblate states at low excitation energy was also suggested from low-energy Coulomb excitation at SPIRAL/GANIL, France, in ^{74}Kr and ^{76}Kr. A detailed comparison of the spectroscopic quadrupole moments (7.61) may give a direct evidence of shape changes in the ground and excited bands. The spectroscopic quadrupole moments for the ground and excited bands are shown in Fig. 7.26 together with $B(E2)$ values of the ground-state band. The experimental data show a clear sign of prolate deformation for the ground-state band in both ^{74}Kr and ^{76}Kr, while the deformation of excited 2^+_2 in ^{74}Kr and 2^+_3 in ^{76}Kr are identified to be oblate from the measured Q moment. However, the experimental uncertainty is still large. While the oblate configuration is predicted to become favored for the ground state of ^{72}Kr, current beam intensities do now allow to perform a similar Coulomb exci-

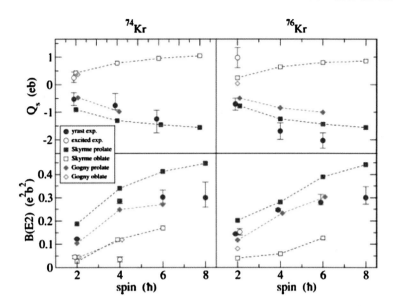

Fig. 7.26 Spectroscopic quadrupole moments (top) and $B(E2)$ values (bottom) for ^{74}Kr (left) and ^{76}Kr (right). The experimental values are compared to microscopic calculations with Skyrme and Gogny interactions. See the text for more details. Reprinted with permission from [10]. ©2021 by the American Physical Society

tation of ^{72}Kr. Further experimental information is needed to confirm the possible prolate-oblate deformation including the predicted gamma instability of shapes.

7.8 Superdeformation and Hyperdeformation

The observed and predicted deformed nuclei have typical values of the quadrupole deformation $\beta_2 = (0.25 - 0.30)$ in rare-earth nuclei and $\beta_2 = (0.20 - 0.25)$ in actinide nuclei. While light nuclei such as Be isotopes show a very large quadrupole deformation $\beta_2 > 0.6$ as a cluster-like structure, large deformations with $\beta_2 \sim 0.6$ have not been observed in the ground state in nuclei with $A \geq 16$, but observed in fission isomers in actinide nuclei and also in superdeformed bands at high-spin states of several medium and medium-heavy nuclei. The occurrence of these large deformations is not understood by the idea of spontaneous symmetry breaking due to the nuclear Jahn–Teller effect in the mean field, but can be understood in terms of the shell structure and the associated energy gaps in the deformed single-particle potential.

7.8.1 Shell Structure of Deformed Single-Particle Potential and Super- and Hyperdeformation

In Fig. 7.27, the calculated single-particle spectrum of the axially deformed harmonic oscillator potential (7.73) is shown as a function of the deformation parameter δ_{osc}. The strong bunching of levels occurs not only for spherical shapes but also for large deformations at specific ratios of $\omega_\perp : \omega_z = 2 : 1$ (prolate, $\delta_{osc} = 0.6$) or $1 : 2$ (oblate $\delta_{osc} = -0.75$). A shell structure also appears at $\omega_\perp : \omega_z = 3 : 1$ (prolate, $\delta_{osc} = 0.86$) or $1 : 3$ (oblate $\delta_{osc} = -1.2$) in the harmonic oscillator potential. The energy gaps at large deformation occur at nucleon numbers different from the magic numbers associated to spherical shape. We will discuss the reason why these new shell structures appear for large deformations. and relate with the geometry of nuclear shapes.

Let us discuss the relation between the shell structure and the geometry of nuclei using a simple one-body potential (7.73). The eigenenergy of this deformed potential is given in Eq. (7.75) as

$$\epsilon(N, n_z) = \hbar\overline{\omega}\left(N + \frac{3}{2} - \frac{1}{3}\delta_{osc}(3n_z - N)\right). \tag{7.126}$$

In the cases of integer ratios of $\omega_\perp$ to ω_z, the eigenenergies are proportional to simple combinations of N and n_z as given in the fourth column of Table 7.2. In a spherical case $\delta_{osc} = 0$, the energies have Nth fold degeneracies as $n_z = N, N - 1, N - 2, \ldots, 0$ for a given major quantum number N. For the ratio $\omega_\perp : \omega_z = 2 : 1$, the deformation parameter is $\delta_{osc} = 0.6$ and the single-particle levels show double-fold, triple-fold and more degeneracies for a specific combination of $2N - n_z$ as listed in the table. This deformation corresponds to a prolate superdeformation. These degeneracies occur also for $\omega_\perp : \omega_z = 3 : 1, 1 : 2, 1 : 3$ cases as shown in the fifth column of Table 7.2.

The characteristics of the shell structure for spherical potentials can be extended to any potential which permits a separation of the motion in the three dimensions. The separability implies that the eigenstates for single-particle motion can be characterized by three quantum numbers (n_x, n_y, n_z). Familiar examples of such potentials are the anisotropic harmonic oscillator as in Eq. (7.73) and also the rectangular box with infinite walls. The eigenenergy of the anisotropic harmonic oscillator potential is given in Eq. (7.74);

$$\epsilon(N, n_z) = \hbar\omega_x(n_x + 1/2) + \hbar\omega_y(n_y + 1/2) + \hbar\omega_z(n_z + 1/2),$$

where $\omega_x = \omega_y \equiv \omega_\perp$ in the case of axial symmetric potential. An expansion of the energy around a point $(n_x, n_y, n_z)_0$ exhibits the occurrence of shells of maximum degeneracy, when the first derivatives of the energy with respect to the three quantum numbers are in ratios of rational numbers

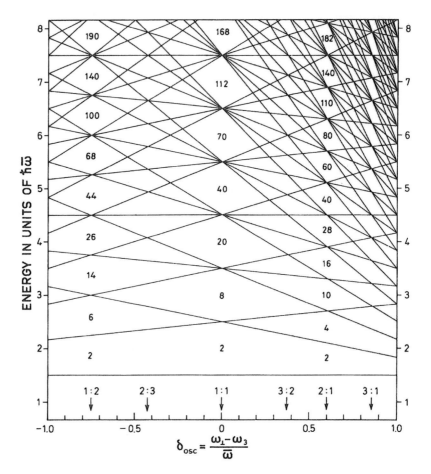

Fig. 7.27 Single-particle spectrum for a Hamiltonian with an axial symmetric harmonic oscillator potential (7.73). The eigenvalue is given in Eq. (7.75). Each level has a degeneracy of $2(n_\perp + 1)$ where a factor 2 is due to the spin degree of freedom, while $(n_\perp + 1)$ from that of degeneracy due to the axial symmetry, i.e., $n_\perp = n_x + n_y (n_x = 0, 1, 2, \ldots, n_\perp)$. The total particle number for each closed shell is given in the figure for the potentials with $\omega_\perp : \omega_z = 1 : 1$ (spherical), $2 : 1$ (prolate) and $1 : 2$ (oblate). The arrows indicate the deformation corresponding to the rational ratios of oscillator frequencies $\omega_\perp : \omega_z$. Figure reprinted with permission from [11]. ©2021 by Elsevier

$$\left(\frac{\partial \varepsilon}{\partial n_x}\right)_0 : \left(\frac{\partial \varepsilon}{\partial n_y}\right)_0 : \left(\frac{\partial \varepsilon}{\partial n_z}\right)_0 = \alpha : \beta : \gamma, \tag{7.127}$$

where α, β, and γ are integers. When the condition (7.127) is fulfilled, the bunching of the single particle levels appears in the deformed single-particle spectrum. This high degeneracy of the single-particle orbitals is deeply related to the geometry of the mean field potential. In the anisotropic potential, the condition (7.127) can be interpreted as for the oscillator frequency $\omega_x : \omega_y : \omega_z = \alpha : \beta : \gamma$. The highest

Table 7.2 Degenerate single-particle energies (7.75) in the deformed harmonic oscillator potential (7.73). The deformation parameter is defined as $\delta_{osc} = (\omega_\perp - \omega_z)/\bar{\omega}$ with the average angular frequency $\bar{\omega} = (2\omega_\perp + \omega_z)/3$. Two examples of the degenerate states are listed for each deformation

$\omega_\perp$	ω_z	δ_{osc}	$\epsilon(N, n_z)/\hbar\bar{\omega} - 3/2$	Degenerate energies $\epsilon(N, n_z)$
1	1	+0.0	N	$\epsilon(1,1) = \epsilon(1,0)$, $\epsilon(2,2) = \epsilon(2,1) = \epsilon(2,0)$
2	1	+0.6	$\frac{3}{5}(2N - n_z)$	$\epsilon(1,0) = \epsilon(2,2)$, $\epsilon(2,0) = \epsilon(3,2) = \epsilon(4,4)$
3	1	+0.86	$\frac{3}{7}(3N - 2n_z)$	$\epsilon(1,0) = \epsilon(3,3)$, $\epsilon(2,1) = \epsilon(4,4)$
1	2	−0.75	$\frac{3}{4}(N + n_z)$	$\epsilon(1,1) = \epsilon(2,0)$, $\epsilon(2,2) = \epsilon(3,1) = \epsilon(4,0)$
1	3	−1.2	$\frac{3}{5}(N + 2n_z)$	$\epsilon(1,1) = \epsilon(3,0)$, $\epsilon(2,1) = \epsilon(4,0)$

degeneracy may occur at a ratio corresponding to the smallest number $\alpha : \beta : \gamma = 1 : 1 : 1$, which is nothing but the spherical shape. The second smallest ratios for the axial symmetric cases are $\alpha : \beta : \gamma = 2 : 2 : 1$ or $\alpha : \beta : \gamma = 1 : 1 : 2$. The former (the latter) corresponds to a prolate (an oblate) shape with the ratio of axis $R_\perp : R_z = 1 : 2$ $(2 : 1)$. One can clearly see in Fig. 7.27 that the high degeneracy of single-particle orbitals occurs at both $\omega_\perp : \omega_z = 2 : 1$ ($\delta_{osc} = 0.6$) and $1 : 2$ ($\delta_{osc} = -0.75$) for prolate and oblate shapes, respectively. These geometrical arguments provide the reason why the superdeformation occurs at large deformation $\beta_2 \sim 0.6$. The next high degeneracy of single-particle levels might occur $\alpha : \beta : \gamma = 3 : 3 : 1$ or $1 : 1 : 3$ and the former is marked $\omega_\perp : \omega_z = 3 : 1$ in Fig. 7.27 and called "hyperdeformation". There have been no clear experimental evidence of oblate superdeformation and of prolate or oblate hyperdeformation so far, but it is extremely desired to explore these exotic deformations with future experiments. These exotic shapes are illustrated in Fig. 7.28.

7.8.2 *Superdeformed Band in* ^{152}*Dy*

The first observation of decay intra-band γ transitions from a superdeformed band was carried out in 1986 at Daresbury Laboratory, United Kingdom. The highest spin reached in the superdeformed band of ^{152}Dy was 60 $\hbar$. The gamma-ray energy spectrum between the states $(I \rightarrow I - 2)$ of the superdeformed band is shown in Fig. 7.29. The intrinsic quadrupole moment Q_0 of the band was found to be $Q_0 = (19 \pm 3)$ eb and the resulting electric quadrupole transition strength $B(E2)$ was enhanced 2000 times more than the single-particle unit (Weisskopf unit, W.u., see Sect. 4.6.4 for definition). These values are compared to a typical value of $Q_0 =$

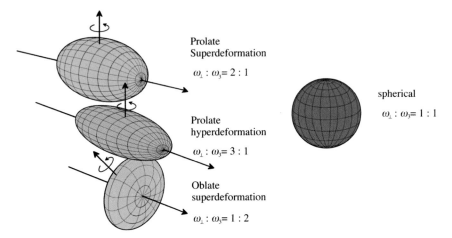

Fig. 7.28 Spherical, superdeformed and hyperdeformed nuclei

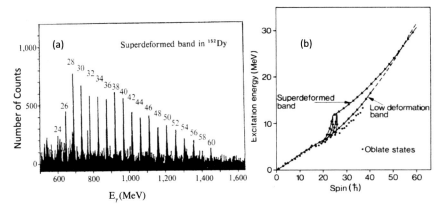

Fig. 7.29 **a** Gamma-ray spectra in ^{152}Dy of a superdeformed band. The numbers indicate the angular momenta of the decay states. **b** The decay paths of the superdeformed and normal deformed bands in ^{152}Dy. The superdeformed band indicated by the line with arrows terminates at spin $I = 24$ and decays to the oblate states shown by black points below the low deformation band. It is seen from the decay slopes of the two bands that the moment of inertia of the superdeformed band is larger than that of the normal deformed band. Reprinted with permission from [19]. ©2021 by the American Physical Society

(5−7) eb and the enhancement factor of (180–350) W.u. of $B(E2)$ within the normal deformed band in rare-earth nuclei with $\beta_2 \sim (0.25-0.3)$.

In Fig. 7.29b, it is seen from the decay slopes of the two bands that the moment of inertia of the superdeformed band is much larger than that of the normal deformed band. There are mainly two reasons for the larger moment of inertia of superdeformed band. One is a larger deformation that increases the moment of inertia, as is given in Eq. (7.42). The second is a weaker pairing correlation in the superdeformed band,

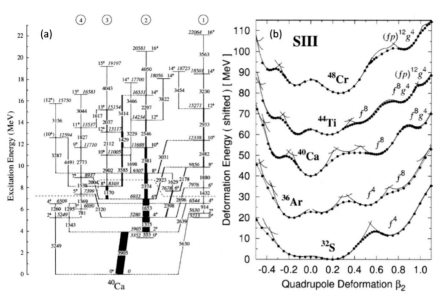

Fig. 7.30 **a** Level scheme of ^{40}Ca. The energy labels are given in keV and the widths of the arrows are proportional to the relative intensities of γ rays. The band (1) is identified as SD band with a quadrupole deformation $\beta_2 = 0.59^{+0.11}_{-0.07}$, while the band (2) is a normal deformed band with $\beta_2 \approx 0.27$. **b** Deformation energy curves as functions of the quadrupole deformation β_2 calculated by a constrained Hartree–Fock method with a Skyrme interaction SIII. This figure is obtained by deformed Hartree–Fock calculations including the spin-orbit interaction and the surface effect. The curves for different nuclei are shifted by 20MeV to accommodate them in a single figure. The notations $f^n g^m$ and $(fp)^n g^m$ indicate the configurations $f_{7/2}$ shell (or fp shell) and $g_{9/2}$ shell are occupied by n and m particles, respectively. The figure **a** is reprinted with permission from [20]. ©2021 by the American Physical Society. **b** is reprinted with permission from [21]. ©2021 by Elsevier

which will also give rise to a larger moment of inertia. This is due to the fact that the superfluidity decreases the moment of inertia and the loss of superfluidity will increase the moment of inertial in turn. Inserting the values of E_γ in Fig. 7.29a into Eq. (7.112), the rotational frequency can be deduced as $\hbar\omega_{\mathrm{rot}} = (0.3 - 0.7)$ MeV. The moment of inertia of the superdeformed band is then extracted from Eq. (7.113) to be $\Im \approx 80\hbar^2$ MeV^{-1}. For comparison, we will evaluate the moment of inertia of the rigid rotor from Eq. (7.42) for $A = 152$ with $\delta = 0.6$. It can be read as $\Im_{\mathrm{rigid}} = 60(1 + \delta/3) = 72\hbar^2$ MeV^{-1}, which is close to the experimental value of the superdeformed band in ^{152}Dy. We should notice that the empirical moment of inertia of the normal deformed band in rare-earth nuclei is almost a half of the rigid rotor value because of the strong pairing correlations, which implement the superfluidity phase in nuclei. Thus, the value of the moment of inertia for the superdeformed band suggests a disappearance of the superfluid phase.

7.8.3 Superdeformed Band in ^{40}Ca and Theoretical Models

In the mass region of $A \leq 40$, the superdeformed bands have been also found in several nuclei such as ^{36}Ar and ^{40}Ca. A doubly-closed shell nucleus ^{40}Ca is spherical in the ground state and the first-excited 2^+ state appears at a rather high excitation energy of $E_x = 3.905$ MeV as a typical feature of the closed shell nucleus. In Fig. 7.30a, the band head of a superdeformed band is found at $E_x = 5.213$ MeV together with that of a normal deformed band at $E_x = 3.352$ MeV. These two deformed bands marked (1) and (2), respectively, have the deformation $\beta_2 = 0.59$ and $\beta_2 = 0.27$. The intrinsic quadrupole moment was extracted from the transition strength between the members of superdeformed band to be $Q_0 = 1.80$ eb.

In Fig. 7.30b, the calculated deformed energy surfaces are shown as a function of the quadrupole deformation β_2 for various $N = Z$ even-even nuclei by using a deformed Hartree–Fock model. In all nuclei shown, the superdeformed band was predicted as a many-particle many-hole excited state from the ground-state configuration. In ^{40}Ca, the superdeformed state is predicted as a 8-particle 8-hole state excited from the sd shell to $f_{7/2}$ shell orbitals at $\beta_2 \simeq 0.65$. The superdeformed state in ^{36}Ar is also predicted as a 4-particle 8-hole state at $\beta_2 \simeq 0.55$.

So far the nuclear deformation has been discussed in the intrinsic frame Hamiltonian, i.e., Nilsson model or deformed HF model + the beyond mean field approach such as GCM with angular momentum projection and configuration mixing. The interacting large-scale shell-model calculations is also an approach to describe the deformation and rotational spectra in the laboratory framework. Large-scale shell-model calculations, with dimensions reaching 10^9, are carried out to describe the observed normal deformed (ND) and superdeformed (SD) bands on top of the first and second excited 0^+ states of ^{40}Ca at 3.35 and 5.21 MeV, respectively. A valence space is comprised by two major oscillator shells, ($s_{1/2}d_{3/2}$)- and full (pf)-shells and the configuration space includes up to $12p$-$12h$ states from the ^{40}Ca core. In the left panel of Fig. 7.31, the decay γ-ray energies between the superdeformed band members,

$$E_\gamma = E(J) - E(J - 2) = \frac{2J - 1}{\mathcal{I}} \qquad (7.128)$$

are shown against to J. The shell model calculations were done in a limited model space taking only $8\hbar\omega$ excitations. The calculated energies reproduce well the experimental data up to $J = 16$. In the right panel of Fig. 7.31, the calculated and experimental results of the intrinsic Q_0 moments are plotted. The intrinsic Q_0 moments of the SD band are extracted from the observed $B(E2)$ values by using Eq. (7.60). The deformation parameter is approximately determined to be $\beta_2 \simeq 0.6$ from the empirical Q_0 moment. The calculated results agree with the experimental data within the large experimental error bars. It is noticed that as J grows, the shell model results lose some collectivity, whereas the experimental data show a constant transition quadrupole moment keeping a solid collectivity up to the high angular momentum. The ND band is dominated by the configurations with four particles promoted to the pf-shell ($4p - 4h$ configurations), while the major configurations for the SD

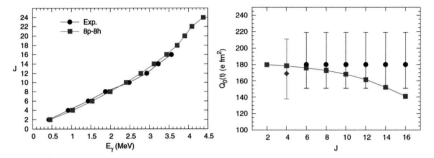

Fig. 7.31 (Left) Decay energy $E_\gamma(J \rightarrow J-2)$ against J of the SD band of ^{40}Ca calculated by an interacting shell model (ISM). (Right) The intrinsic Q_0 moment extracted from $B(E2)$ transitions in the SD band of ^{40}Ca. Reprinted with permission from [22]. ©2021 by the American Physical Society

band are $8p-8h$ ones. It is shown in [22] that the ground state of ^{40}Ca is strongly correlated, but the closed-shell configuration is still the major configuration having the 65% probability.

7.9 Octupole Deformation

Octupole deformation (sometimes referred as "pear shape") is attributed to strong shell effects present in the mean field potential of finite Fermion systems, like atomic nuclei or metallic clusters. The octupole deformation parameter $\alpha_{30} = \beta_3$ is defined in Eq. (7.29). Nuclear pear shapes in the ground state occur mainly in two mass regions: the Ra-region ($Z \approx 88$) and the Ba-region ($Z \approx 56$). For example, in the $Z \approx 56$ region, proton $2d_{5/2}$ and $1h_{11/2}$ orbitals becomes close in energy in the single-particle potential (see Fig. 3.2 in Chap. 2) and would be a driving force of collective octupole vibration. These single-particle states will induce the octupole instability, and eventually the intrinsic octupole deformation by possible strong coupling with $\Delta l = 3$. An associated neutron configuration of $2f_{7/2}$ and $1i_{13/2}$ orbitals will favor also the octupole instability around $N \approx 88$. This is an octupole Jahn–Teller effect analogous to the case of quadrupole deformation discussed in Sect. 7.3. In a similar way, in the Ra mass region with $Z \approx 88$, proton $2f_{7/2}$ and $1i_{13/2}$ orbitals play the role of octupole instability, together with $N \approx 134$ which region has neighboring neutron $2g_{9/2}$ and $1j_{15/2}$ orbitals. The systematic calculations of the FRDM model is shown in Fig. 7.32.

A strong octupole correlation, leading to a pear shape, can arise for certain proton and neutron numbers Z and N. When both proton and neutron numbers have these values, such as for $Z \approx 56$, $N \approx 88$ and $Z \approx 88$, $N \approx 134$, we expect a pear shape leading to the breaking of the reflection symmetry of the shape. If nuclei are reflection-symmetric, the intrinsic system is invariant under both the parity ($\mathcal{P}$) operation

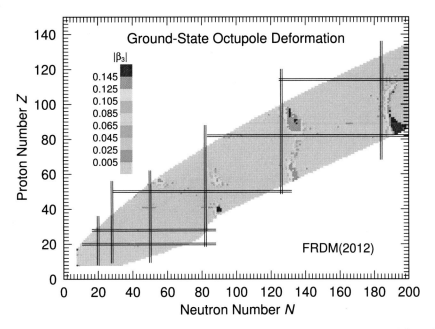

Fig. 7.32 Octupole deformation calculated by the FRDM model. A strong octupole deformed nucleus can arise for certain proton and neutron numbers Z and N such as for $Z \approx 56$, $N \approx 88$ and $Z \approx 88$, $N \approx 134$ as are expected by the shell structure of single-particle orbitals. Reprinted with permission from [4]. ©2021 by Elsevier. Courtesy P. Möller

and rotation of the system ($\mathcal{R}$) through $180°$ about an axis perpendicular to the symmetry axis. The symmetry under the $\mathcal{R}$ invariance implies the constrain $(-)^I = r$ as is shown in Eq. (7.52), so that the rotational spectrum contains states with only even values or only odd values of I. Thus, the nuclei with reflection-symmetric shape gives rise to laboratory projections of the angular momentum and parity $I^\pi = 0^\pm, 2^\pm, 4^\pm, \ldots$ or $I^\pi = 1^\pm, 3^\pm, 5^\pm, \ldots$. Notice that $\mathcal{R}$ invariance does not give any restriction on the parity.

For nuclei asymmetric under reflection to the plane orthogonal to the symmetry axis (for example shape deformations with odd-multipole order), neither the parity invariance ($\mathcal{P}$) nor the reflection invariance ($\mathcal{R}$) hold, but the symmetry defined by the combined operation ($\mathcal{I} = \mathcal{P}\mathcal{R}^{-1}$) is conserved. This is because the geometrical meaning of the combined operation $\mathcal{I}$ represents a reflection in a plane containing the symmetry axis;

$$(x', y', z') \xrightarrow{\mathcal{R}} (-x', y', -z') \xrightarrow{\mathcal{P}} (x', -y', z'),$$

where z'-axis is the symmetry axis (see also Fig. 7.5). Diatomic molecules with nuclei of different charges such as HCl have a deformation of this type as well as the axial symmetric octupole nuclear deformation with $\alpha_{30} = \beta_3$. The quantum number associated with $\mathcal{I}$ is specified by the parity π and the angular momentum I,

$$\pi(-1)^I = s \tag{7.129}$$

which is called the "simplex" quantum number. The rotational spectra of the axial symmetric deformation containing of odd multipole order are thus classified as

$$I^\pi = 0^+, 1^-, 2^+, \cdots \quad s = +1,$$
$$I^\pi = 0^-, 1^+, 2^-, \cdots \quad s = -1, \tag{7.130}$$

in even-even nuclei. In Eq. (7.130), the $\mathcal{I}$ invariance restricts a combination of angular momentum and parity in contrast to the $\mathcal{R}$ invariance, in which only the angular momentum is constrained. For odd-even nuclei, the rotational spectra are specified as

$$I^\pi = 1/2^+, 3/2^-.5/2^+, 7/2^- \cdots \quad s = +i,$$
$$I^\pi = 1/2^-, 3/2^+.5/2^-, 7/2^+ \cdots \quad s = -i. \tag{7.131}$$

In Fig. 7.33, the rotational spectra of positive and negative bands in ^{226}Ra are shown together with those of a molecule HCl. Both bands belong to the simplex quantum number $s = +1$.

For the octupole deformation which violates P invariance, the positive and the negative deformation for β_3 are equivalent as is seen at the top part of left panel of Fig. 7.33. To recover the parity conservation ensured by the Hamiltonian, the following combinations of the two deformations are adopted to provide the spectra of parity doublets. Namely, the positive parity and negative parity states are constructed by a coherent and decoherent sum of positive and negative octupole deformed states,

$$|\Psi_\pm\rangle = \frac{1}{\sqrt{2}}(|\Psi\rangle \pm P|\Psi\rangle), \tag{7.132}$$

where P is the parity operator and $P|\Psi\rangle$ corresponds to the state with opposite sign of octupole deformation illustrated in the right panel of Fig. 7.33. Consequently, $P|\Psi\rangle \neq |\Psi\rangle$ for the octupole deformed nucleus. The two configurations with opposite signs are separated by a potential barrier in the mean field potential. In a quantal system, two parity doublets might couple by the tunneling effect through the barrier. If the barrier height between positive and negative deformation is very high, the parity doublets degenerate without the tunneling effect.

In Fig. 7.34, the level scheme of ^{223}Th is shown. In this figure, the parity doublets of $s = +i$ are shown in Fig. 7.342a and 2b and the parity doublets of $s = -i$ are shown in Fig. 7.341a and 1b. The ground state is assigned as $I = 5/2^+$. Non-yrast bands are also shown in Fig. 7.343a and 3b with the $s = -i$.

Atoms with octupole-deformed nuclei will bring very important information in the search for permanent atomic electric dipole moments (EDMs). The observation

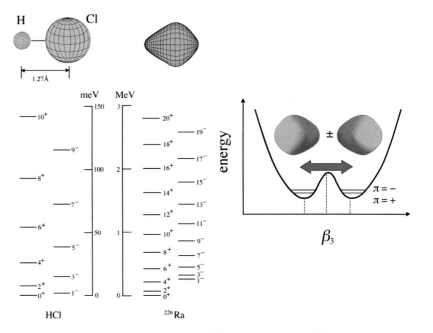

Fig. 7.33 Parity doublet of HCl molecule and ^{226}Rn. In the spectra of ^{226}Rn, only the members of simplex quantum number $s = +1$ have been observed so far. The right panel is a schematic representation of positive and negative parity states as the coherent and de-coherent sum of positive and negative octupole deformed states

of a non-zero EDM would indicate time-reversal (T) or equivalently charge-parity (CP) violation due to physics beyond the standard model. This research topic will be discussed in Sect. 10.5 of Chap. 10.

7.10 Appendix

7.10.1 Clebsch–Gordan Coefficient and Wigner–Eckart Theorem

The components of an angular momentum operator $\mathbf{j}$ (which can represent the orbital, spin or total angular momentum of a particle) obey the commutation relations (in natural units of $\hbar = 1$),

$$[j_x, , j_y] = i j_z \text{ and cyclic permutations.} \tag{7.133}$$

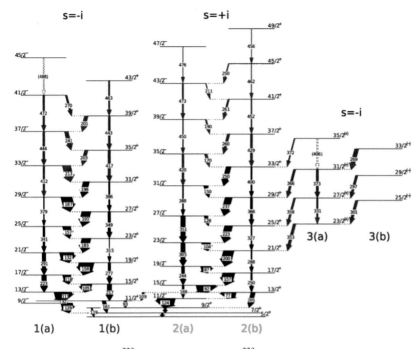

Fig. 7.34 Parity doublets in ^{223}Th. The level scheme of ^{223}Th is determined by γ-ray spectroscopy via the ^{208}Pb(^{18}O,$3n$)^{223}Th reaction. The widths of the arrows represents the γ-ray intensity of the transitions. The newly observed transitions are colored in red. The two main bands of alternating parity are classified in four sequences labeled 1**a**, 1**b** for the band of simplex $s = -i$ and 2**a**, 2**b** for the band of simplex $s = +i$ (each sequence has a fixed parity). A similar notation is used for the non-yrast structure, formed of two sequences labeled 3(**a**) and 3(**b**). Spin-parity assignment for the main structure is based on the assumption of a $5/2^+$ ground state. Each of the two main bands presents intense stretched E1 inter-sequence transitions as well as stretched E2 intra-sequence transitions. The parities of the non-yrast structure levels are tentative. Reprinted with permission from [23]. ©2021 by the American Physical Society

The Hamiltonian is in general invariant under rotation. That is, the eigenstates of the Hamiltonian are labelled by the angular momentum $\mathbf{j}^2$;

$$\mathbf{j}^2|jm\rangle = j(j+1)|jm\rangle, \quad j = 0, 1/2, 1, 3/2, \ldots, \tag{7.134}$$

$$j_z|jm\rangle = m|jm\rangle, \quad m = j, j-1, \ldots, -j+1, -j, \tag{7.135}$$

where the eigenstate is diagonal for j_z. The quantum numbers j, m characterize the transformation of the state under rotation of the coordinate system. Different m states will be rotated into each other by using the raising and lowering operators

$$j_\pm|jm\rangle = (j_x \pm ij_y)|jm\rangle = \sqrt{(j \mp m)(j \pm m + 1)}|jm \pm 1\rangle. \tag{7.136}$$

If two particles are specified by their angular momenta j_1 and j_2 and their z-components m_1 and m_2, the coupling of these two particles produces states with angular momenta

$$|j_1 - j_2| \le J \le j_1 + j_2 \quad \text{or} \quad J = |j_1 - j_2|, |j_1 - j_2| + 1, \ldots, j_1 + j_2. \quad (7.137)$$

The coupled states can be written as the eigenstates of total angular momentum and its z-component $J M$ as

$$|(j_1 j_2) J M\rangle = \sum_{m_1 m_2} \langle j_1 m_1 j_2 m_2 | J M\rangle | j_1 m_1, j_2 m_2 \rangle, \quad (7.138)$$

where the coupling coefficients are called Clebsch–Gordan coefficients or Vector-addition coefficients. The orthogonality conditions of basis states $|J M\rangle$ and $|j_1 m_1, j_2 m_2\rangle$ lead to the orthogonal relations of for the Clebsch–Gordan coefficients,

$$\sum_{m_1 m_2} \langle j_1 m_1 j_2 m_2 | J M\rangle \langle j_1 m_1 j_2 m_2 | J' M'\rangle = \delta(J, J')\delta(M, M'), \quad (7.139)$$

$$\sum_{J M} \langle j_1 m_1 j_2 m_2 | J M\rangle \langle j_1 m_1' j_2 m_2' | J M\rangle = \delta(m_1, m_1')\delta(m_2, m_2'). \quad (7.140)$$

The coupling coefficient in Eq. (7.138) is also expressed by a $3j$ symbol which is defined by

$$\begin{pmatrix} j_1 & j_2 & j_3 \\ m_1 & m_2 & m_3 \end{pmatrix} = (-1)^{j_1 - j_2 - m_3}(2j_3 + 1)^{-1/2} \langle j_1 m_1 j_2 m_2 | j_3 - m_3 \rangle. \quad (7.141)$$

The $3j$ symbol is also referred as the Wigner symbol.

There are several symmetry properties between the Clebsch–Gordan coefficients;

$$\langle j_1 m_1 j_2 m_2 | j_3 m_3 \rangle = (-1)^{j_1 + j_2 - j_3} \langle j_1 - m_1 j_2 - m_2 | j_3 - m_3 \rangle \quad (7.142)$$

and

$$\begin{aligned} \langle j_1 m_1 j_2 m_2 | j_3 m_3 \rangle &= (-1)^{j_1 + j_2 - j_3} \langle j_2 m_2 j_1 m_1 | j_3 m_3 \rangle \\ &= (-1)^{j_1 - m_1} \left(\frac{2j_3 + 1}{2j_2 + 1} \right)^{1/2} \langle j_1 m_1 j_3 - m_3 | j_2 - m_2 \rangle \\ &= (-1)^{j_2 + m_2} \left(\frac{2j_3 + 1}{2j_1 + 1} \right)^{1/2} \langle j_3 - m_3 j_2 m_2 | j_1 - m_1 \rangle \quad (7.143) \end{aligned}$$

for the permutations of the angular momenta. A useful formula for the $j_3 = 0$ case can be found from these symmetry relations;

$$\langle j_1 m_1 j_2 m_2 | 00 \rangle = (-1)^{j_1 - m_1} \left(\frac{1}{2j_2 + 1} \right)^{1/2} \langle j_1 m_1 00 | j_2 - m_2 \rangle$$

$$= (-1)^{j_1 - m_1} \frac{1}{\sqrt{2j_1 + 1}} \delta(j_1, j_2) \delta(m_1, -m_2). \qquad (7.144)$$

The most important single-particle operators in nuclear physics are expressed by spherical tensors. These are operators which transform between the initial and final states having definite angular momenta. Examples are the beta-decay operators and multipole operators of the electromagnetic transitions. The operator is expressed by the so-called spherical tensor of rank λ which is a set of operators $T_{\lambda,\mu}$ with an angular momentum λ and its z-components $\mu = \lambda, \lambda - 1, \ldots, -\lambda + 1, -\lambda$. The basic matrix element of any tensor operator has the form

$$\mathcal{M} = \langle j'm' | T_{\lambda,\mu} | jm \rangle. \qquad (7.145)$$

Static moments, such as the magnetic moment, are given by the matrix elements with $j = j' = m = m'$ and $\mu = 0$. Transition matrix elements generally require sums of m, m' of the matrix elements squired.

Matrix elements of the type (7.145) can be expressed conveniently in a form independent of m. To this end, we decompose a tensor product $T_{\lambda,\mu} | jm \rangle$ into angular momentum eigenstates,

$$T_{\lambda,\mu} | jm \rangle = \sum_J \langle \lambda \mu jm | J M = m + \mu \rangle | [T_\lambda j]^{JM} \rangle, \qquad (7.146)$$

where

$$| [T_\lambda j]^{JM} \rangle \equiv \sum_{\mu,m} \langle \lambda \mu jm | J M = m + \mu \rangle T_{\lambda,\mu} | jm \rangle. \qquad (7.147)$$

According to the angular momentum conservation, only the term with $J = j'$ will contribute to the matrix element (7.145) and it can be written without m subscription,

$$\mathcal{M} = \langle j'm' | T_{\lambda,\mu} | jm \rangle = \langle \lambda \mu jm | j'm' \rangle \langle j'm' | [T_\lambda j]^{j'm'} \rangle, \qquad (7.148)$$

which is nothing but the Wigner–Eckart theorem. This form of matrix element is easy to use for a practical application of β-decay and electromagnetic transitions. The matrix element $\langle j'm' | [T_\lambda j]^{j'm'} \rangle$ is just the transition matrix element from an initial state j to a final state j', including statistical factors and the sum over final m' quantum number.

A conventional way to define the reduced matrix element was proposed by Racah,

$$\langle j' | [T_\lambda j]^{j'} \rangle = (-1)^{\lambda - j + j'} \frac{\langle j' || T_\lambda || j \rangle}{\sqrt{2j' + 1}},$$

The matrix element is eventually expressed as,

$$\mathcal{M} = \frac{\langle jm\lambda\mu | j'm'\rangle}{\sqrt{2j'+1}} \langle j' || T_\lambda || j\rangle. \tag{7.149}$$

The essential point of this formula is that the dependence of magnetic quantum numbers m, m', and μ is expressed by the Clebsch–Gordan coefficients. The total transition probability from $j \to j'$, summed over μ and m', becomes independent of magnetic quantum numbers and is given by the reduced transition probability,

$$
\begin{aligned}
B(T_\lambda; j \to j') &\equiv \sum_{\mu m'} |\langle j'm' | T_{\lambda\mu} | jm\rangle|^2 = \sum_{\mu m'} \frac{\langle jm\lambda\mu | j'm'\rangle^2}{2j'+1} |\langle j' || T_\lambda || j\rangle|^2 \\
&= \sum_{\mu m'} \frac{\langle j'-m'\lambda\mu | j-m\rangle^2}{2j+1} |\langle j' || T_\lambda || j\rangle|^2 = \frac{|\langle j' || T_\lambda || j\rangle|^2}{2j+1}, \quad (7.150)
\end{aligned}
$$

where the symmetry relation of Clebsch–Gordan coefficients (7.143) and the orthogonality condition (7.139) are used.

7.10.2 Spherical Harmonics and Spherical Bessel Functions

Here we summarize useful formulas of spherical harmonics and spherical Bessel functions.

7.10.2.1 Spherical Harmonics

The spherical harmonics $Y_{lm}(\theta, \phi)$ are the eigenfunctions of angular momentum l and its projection on z-axis l_z,

$$l^2 Y_{lm}(\theta, \phi) = l(l+1) Y_{lm}(\theta, \phi), \tag{7.151}$$
$$l_z Y_{lm}(\theta, \phi) = m Y_{lm}(\theta, \phi). \tag{7.152}$$

The angular momentum raising (lowering) operator $l_\pm = l_x \pm i l_y$ acts

$$l_\pm Y_{lm}(\theta, \phi) = \sqrt{(l \mp m)(l \pm m + 1)} Y_{lm\pm1}(\theta, \phi). \tag{7.153}$$

The spherical harmonics satisfy the orthonormalization condition,

$$\int Y_{lm}^*(\Omega) Y_{l'm'}(\Omega) d\Omega = \int_0^\pi \sin\theta d\theta \int_0^{2\pi} d\phi Y_{lm}^*(\theta, \phi) Y_{l'm'}(\theta, \phi) = \delta_{l,l'}\delta_{m,m'}, \tag{7.154}$$

where Ω stands for two Euler angles θ and ϕ, $\Omega = (\theta, \phi)$. The spherical harmonics have also a parity,

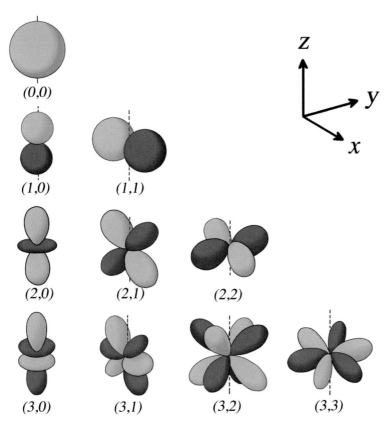

Fig. 7.35 Illustrations of the real part of spherical harmonics $\Re\{Y_{lm}(\theta, \phi)\}$ in 3D space. The (l, m) value is indicated next to each shape. Green color implies positive values, while purple color corresponds to negative values of spherical harmonics

$$Y_{lm}(\pi - \theta, \pi + \phi) = (-1)^l Y_{lm}(\theta, \phi), \qquad (7.155)$$

and the complex conjugate relation,

$$Y_{l-m}(\theta, \phi) = (-1)^m Y_{lm}^*(\theta, \phi). \qquad (7.156)$$

Some typical values of spherical harmonics are listed in Table 7.3 and shown in Fig. 7.35

Legendre polynomials are the generating functions for spherical harmonics;

$$P_l(\cos \theta_{12}) = \sum_m \frac{4\pi}{2l + 1} Y_{lm}(\Omega_1) Y_{lm}^*(\Omega_2), \qquad (7.157)$$

Table 7.3 Spherical harmonics Y_{lm} and solid spherical harmonics $r^l Y_{lm}$

(lm)	Y_{lm}	$r^l Y_{lm}$
(00)	$\sqrt{\frac{1}{4\pi}}$	$\sqrt{\frac{1}{4\pi}}$
(10)	$\sqrt{\frac{3}{4\pi}}\cos\theta$	$\sqrt{\frac{3}{4\pi}}z$
(11)	$-\sqrt{\frac{3}{8\pi}}\sin\theta e^{i\phi}$	$-\sqrt{\frac{3}{8\pi}}(x+iy)$
(20)	$\sqrt{\frac{5}{4\pi}}\frac{1}{2}(3\cos\theta^2-1)$	$\sqrt{\frac{5}{4\pi}}\frac{1}{2}(3z^2-r^2)$
(21)	$-\sqrt{\frac{15}{8\pi}}\sin\theta\cos\theta e^{i\phi}$	$-\sqrt{\frac{15}{8\pi}}z(x+iy)$
(22)	$\frac{1}{4}\sqrt{\frac{15}{2\pi}}\sin\theta^2 e^{2i\phi}$	$\frac{1}{4}\sqrt{\frac{15}{2\pi}}(x+iy)^2$

where θ_{12} is an angle between two vectors $\mathbf{n}_1$ and $\mathbf{n}_2$ specified by Euler angles Ω_1 and Ω_2 in 3D space, respectively. Typical values of Legendre polynomials are

$$P_0(\cos\theta) = 1, \quad P_1(\cos\theta) = \cos\theta, \quad P_2(\cos\theta) = \frac{1}{2}(3\cos^2\theta - 1). \tag{7.158}$$

Legendre polynomials satisfy the orthogonal condition,

$$\int_0^\pi P_l(\cos\theta)P_l'(\cos\theta)\sin\theta d\theta = \frac{2}{2l+1}\delta_{l,l'}. \tag{7.159}$$

7.10.2.2 Spherical Bessel Function

The plane wave can be expanded by Legendre polynomials and spherical Bessel functions,

$$e^{i\mathbf{q}\cdot\mathbf{r}} = e^{iqr\cos\theta} = \sum_l i^l(2l+1)j_l(qr)P_l(\cos\theta). \tag{7.160}$$

Equation (7.160) is further expressed by spherical harmonics as

$$e^{i\mathbf{q}\cdot\mathbf{r}} = 4\pi\sum_{lm} i^l j_l(qr)Y_{lm}^*(\Omega_q)Y_{lm}(\Omega_r). \tag{7.161}$$

The spherical Bessel function with rank l is generated by a formula from $j_0(z)$,

$$j_l(z) = (-z)^l\left(\frac{d}{zdz}\right)^l j_0(z), \tag{7.162}$$

where $j_0(z) = \sin z/z$. Some low rank spherical Bessel function are given as

$$j_0(z) = \frac{\sin z}{z}, \quad j_1(z) = \frac{\sin z - z \cos z}{z^2}, \quad j_2(z) = \frac{3(\sin z - z \cos z) - z^2 \sin z}{z^3}.$$

$$(7.163)$$

The spherical Bessel function is expressed by the ordinary Bessel function as

$$j_l(z) = \sqrt{\frac{\pi}{2z}} J_{l+1/2}(z). \tag{7.164}$$

In the limits $z \to 0$ and $z \to \infty$, $j_l(z)$ becomes

$$j_l(z) \sim \frac{z^l}{(2l + 1)!!}, \quad \text{for } z \to 0, \tag{7.165}$$

$$j_l(z) \sim \frac{1}{z} \cos\left(z - \frac{(n + 1)\pi}{2}\right) \quad \text{for } z \to \infty, \tag{7.166}$$

where $(2n + 1)!! = (2n + 1)(2n - 1) \cdots 5 \cdot 3 \cdot 1$.

Summary

Nuclear deformation is a phenomenon induced by spontaneous symmetry breaking (SSB) in the mean field potential. Most nuclei are deformed, with the exception of nuclei near a closed shell, and the quadrupole deformation is the most encountered one. The axial symmetric quadrupole deformation is characterized by a prolate shape (elongated along the symmetry axis) and an oblate shape (compressed along the symmetry axis). The prolate shape is dominant across the nuclear landscape.

It is shown in the harmonic vibration model that the particle-vibration coupling Hamiltonian can be interpreted as the quadrupole-quadrupole interaction by the self-consistent condition between the vibrating density and potential. In a strong coupling limit, the Hamiltonian will induce an instability of spherical shape and also further the symmetry breaking from spherical to deformed shape.

Elliott's SU(3) model has a unique and essential importance to understand of the relation between the collective rotation and the interacting particles with the quadrupole-quadrupole interaction in one-major shell of a harmonic oscillator potential. The SU(3) Hamiltonian has the quadrupole-quadrupole interaction in the laboratory frame and gives rise to the rotational spectrum in the degenerate single-particle limit. The SU(3) wave function of the maximum aligned configuration of angular momentum shows a large Q moment with an axial symmetric shape.

The variation of the single-particle level density near the Fermi energy is intimately related with the symmetry breaking mechanism (SSB). A large degeneracy of levels leads to a reduction of the stability and will induce the deformation change by the particle-vibration coupling Hamiltonian. This is named "nuclear Jahn–Teller effect" in analogy to the Jahn–Teller effect in the molecules and solid-state physics.

The deformation in the mean field is included by the Nilsson model and the deformed Woods-Saxon model, and these models successfully describe many spectroscopic properties of deformed nuclei.

The Strutinsky shell correction method is a combination of the shell model and the macroscopic liquid drop model. The total energy is written by $E_{shell} = E_{LDM} + \delta E_{shell}$, where δE_{shell} is obtained by subtracting a smooth single-particle energy from a shell model single-particle energy in Eq. (7.82). The shell correction is characterized by large negative values for lower level density cases and makes systems more stable, while it is positive for highly degenerate single-particle states and induces instabilities of shapes.

In analogy to the classical mechanics, the moment of inertia in quantum mechanics is defined by the ratio between the angular momentum I and the rotational frequency ω_{rot},

$$\Im = \frac{\hbar I}{\omega_{rot}},$$

where $\hbar I$ is a quantized angular momentum.

For an axial symmetric rotor, the excitation energy of a rotational state $|IMK\rangle$ is given by

$$E_{rot} = \frac{\hbar^2}{2\Im}(I(I+1) - K^2).$$

In an even-even nucleus, the rotational band associated to $K = 0$, i.e., built on the lowest-energy 0^+ state, is only composed of even values of I.

Backbending phenomena of the moment of inertia were found in many rare-earth nuclei with prolate deformation. These phenomena are interpreted as a results of band crossing between yrast and yrare bands having different moments of inertia.

Shape coexistence is encountered when different shape configurations compete in the energy spectrum near the ground state. This quantum effect is described as an interplay of single-particle and collective properties of the nucleus. It occurs in several regions of the nuclear landscape.

Quadrupole moments can be directly determined experimentally from the atomic hyperfine structure, splitting of nuclear states under strong external magnetic and electric fields, and low-energy Coulomb excitation. The first two techniques are restricted to ground state or long-lived states while Coulomb excitation can access the quadrupole moment of short lived excited states as well as its sign through the so-called *re-orientation* effect. When the above techniques are not applicable, nuclear shapes can also be inferred from indirect measurements.

Large scale shell model calculations are feasible to describe the normal- and super-deformed states. It was shown that the normal-deformed states are dominated by $4\hbar\omega$ excitations from the ^{40}Ca case, while the super-deformed ones have a character of $8\hbar\omega$ excitations.

Superdeformation and hyperdeformation are deeply related with the symmetry of shapes with the ratio of elongated and short axes 2:1 and 3:1, respectively. The superdeformed bands are experimentally found in nuclei in Ca and also in rare-earth regions. Hyperdeformation bands are still to be evidenced experimentally.

The existence of octupole deformed nuclei was predicted in the mass region of Ba-isotopes with $Z \approx 56$ and $N \approx 88$ and Ra-isotopes with $Z \approx 88$ and $N \approx 134$. Rotational bands of parity doublets are predicted for the octupole deformed nuclei and confirmed experimentally.

Solutions of Problems

7.1 We consider the zero-point fluctuation around the energy minimum. The kinetic energy of zero-point motion can be considered as a typical energy scale of a bound many-body system. The kinetic energy of an atom is given by

$$E = \frac{p^2}{2M} \tag{7.167}$$

where M is the mass of the atom, equivalent to the mass of a proton ($M \simeq m_p$) for the hydrogen atom. The uncertainty principle provides a scale of zero-point fluctuation $\Delta p \Delta x \sim \hbar$, so that the energy (7.167) can be evaluated using Eq. (7.1). For a diatomic molecule, the size of the repulsive core is $r_c \sim 0.5$Å $= 0.05$ nm so that the energy scale is

$$\Delta E \sim \frac{\hbar^2}{m_p r_c^2} = \frac{\hbar^2 c^2}{m_p c^2 r_c^2} \sim \frac{(200 \, \text{nm} \cdot \text{eV})^2}{1000 \times 10^6 \, \text{eV}(0.05 \, \text{nm})^2} \sim 0.016 \, \text{eV}. \tag{7.168}$$

For a deuteron, the size of the repulsive core of the nuclear potential is $r_c \sim 0.5$ fm,

$$\Delta E \sim \frac{\hbar^2 c^2}{m_p c^2 r_c^2} \sim \frac{(200 \, \text{MeV} \cdot \text{fm})^2}{1000 \, \text{MeV}(0.5 \, \text{fm})^2} \sim 160 \, \text{MeV}. \tag{7.169}$$

7.2 The Hamiltonian (7.3) is rewritten as

$$H = \frac{\pi^2}{2B} + \frac{1}{2} C \alpha^2 \tag{7.170}$$

with Eq. (7.4). Inserting (7.6) and (7.7) into Eq. (7.170), the Hamiltonian is expressed by the operators $O^\dagger$ and O as

$$\begin{aligned}
H &= \frac{1}{2B}(-)\frac{\hbar B \omega}{2}(O^\dagger O^\dagger - O^\dagger O - OO^\dagger + OO) + \frac{C}{2}\frac{\hbar}{2B\omega}(O^\dagger O^\dagger + O^\dagger O + OO^\dagger + OO) \\
&= \frac{1}{4}(O^\dagger O + OO^\dagger) + \frac{\hbar}{4\omega}\frac{C}{B}(O^\dagger O + OO^\dagger) \\
&= \hbar\omega(O^\dagger O + \frac{1}{2}),
\end{aligned} \tag{7.171}$$

where $C/B = \omega^2$ and the commutation relation (7.8) are used.

7.3 From $\Delta R = R_z - R_\perp$ and the mean radius $R = (5\langle r^2 \rangle/3)^{1/2} \approx (R_z + 2R_\perp)/3$, we have

$$R_z = R + \frac{2}{3}\Delta R, \quad \text{and} \quad R_\perp = R - \frac{1}{3}\Delta R. \tag{7.172}$$

Inserting the above equations into Eq. (7.31), we obtain

$$\delta = \frac{3}{2}\frac{2R\Delta R + \Delta R^2/3}{3R^2 + 2\Delta R^2/3} = \frac{\Delta R}{R} + \frac{1}{6}\left(\frac{\Delta R}{R}\right)^2 + O((\Delta R/R)^3), \tag{7.173}$$

where the higher order terms $O((\Delta R/R)^3)$ can be neglected.

7.4 We denote the ground state and the excited states of the intrinsic deformed Hamiltonian h_0 as $|0\rangle$ and $|i\rangle$ with the energy E_0 and E_i, respectively. The perturbed wave function by the cranking term $\hbar\omega_{\text{rot}}J_x$ up to the first order of rotational frequency ω_{rot} will be

$$|\Phi\rangle = |0\rangle + \sum_i \frac{\hbar\omega_{\text{rot}}\langle i|J_x|0\rangle}{E_i - E_0}|i\rangle. \tag{7.174}$$

The expectation value of J_x for the state Φ is expressed as

$$J \equiv \langle\Phi|J_x|\Phi\rangle = 2\sum_i \frac{\hbar\omega_{\text{rot}}|\langle i|J_x|0\rangle|^2}{E_i - E_0}. \tag{7.175}$$

With the relation (7.46), the rotational frequency and the angular momentum is expressed as, $\hbar J = \omega_{\text{rot}}\Im = \hbar\sqrt{I(I+1)}$, and the moment of inertia (the so-called Inglis cranking moment of inertia) is given by

$$\Im = 2\hbar^2 \sum_i \frac{|\langle i|J_x|0\rangle|^2}{E_i - E_0}. \tag{7.176}$$

The energy increase due to the cranking is evaluated by taking the expectation value of the h_0 with the wave function $|\Phi\rangle$ in the second order perturbation,

$$E = \langle\Phi|h_0|\Phi\rangle = \langle\Phi|h' + \hbar\omega_{\text{rot}}J_x|\Phi\rangle = E(\omega_{\text{rot}} = 0) + \hbar^2\omega_{\text{rot}}^2 \sum_i \frac{|\langle i|J_x|0\rangle|^2}{E_i - E_0}$$

$$= E(\omega_{\text{rot}} = 0) + \frac{1}{2}\Im\omega_{\text{rot}}^2 = E(\omega_{\text{rot}} = 0) + \frac{I(I+1)}{2\Im}. \tag{7.177}$$

7.5 The electric quadrupole matrix element is expressed by using Eqs. (7.35) and (7.62) as,

Table 7.4 Observed intrinsic Q_0 moments and deformation parameters δ of $_{64}$Gd isotopes

A	I^π	K	Q(b)	$\sqrt{r_c^2}$ (fm)	Q_0(fm^2)	δ
154	2^+	0	−1.82	5.1223	+637.	+0.285
156	2^+	0	−1.93	5.1420	+676.	+0.300
157	$3/2^-$	3/2	+1.36	5.1449	+680.	+0.301
158	2^+	0	−2.01	5.1569	+704.	+0.310
160	2^+	0	−2.08	5.1734	+728	+0.319

$$\langle K I_2 M_2 | M(E2,\mu) | K I_1 M_1 \rangle$$
$$= \left(\frac{2I_1+1}{8\pi^2}\right)^{1/2} \left(\frac{2I_2+1}{8\pi^2}\right)^{1/2} \left(\frac{5}{16\pi}\right)^{1/2} e Q_0 \int \mathcal{D}^{I_2 *}_{M_2 K}(\omega) \mathcal{D}^2_{\mu 0}(\omega) \mathcal{D}^{I_1}_{M_1 K}(\omega) d\omega$$
$$= \left(\frac{2I_1+1}{2I_2+1}\right)^{1/2} \left(\frac{5}{16\pi}\right)^{1/2} e Q_0 \langle I_1 M_1 2\mu | I_2 M_2 \rangle \langle I_1 K 20 | I_2 K \rangle. \tag{7.178}$$

The $B(E2)$ reads

$$B(E2; I_1 K_1 \to I_2 K_2) = \frac{1}{2I_1+1} \sum_{M_2,\mu,M1} |\langle K I_2 M_2 | M(E2,\mu) | K I_1, M_1 \rangle|^2$$
$$= \frac{1}{2I_2+1} \frac{5}{16\pi} \sum_{M_2,\mu,M1} \langle I_1 M_1 2\mu | I_2 M_2 \rangle^2 \langle I_1 K 20 | I_2 K \rangle^2 e^2 Q_0^2$$
$$= \frac{5}{16\pi} \langle I_1 K 20 | I_2 K \rangle^2 e^2 Q_0^2. \tag{7.179}$$

7.6 The intrinsic moments Q_0 are obtained by using Eq. (7.61),

$$Q = -\frac{2}{7} Q_0, \quad \text{for } K = 0,$$

$$Q = \frac{1}{5} Q_0, \quad \text{for } K = 3/2.$$

The empirical Q_0 moments and the deformation parameter δ (7.30) are listed in Table 7.4.

7.7 From $\overline{\omega} = (2\omega_\perp + \omega_z)/3$ and δ_{osc} (7.76), we get

$$\omega_\perp = \overline{\omega}\left(1 + \frac{1}{3}\delta_{\text{osc}}\right), \quad \text{and} \quad \omega_z = \overline{\omega}\left(1 - \frac{2}{3}\delta_{\text{osc}}\right). \tag{7.180}$$

Inserting Eq. (7.180) into Eq. (7.74), we can show

$$\epsilon(N, n_z) = \hbar\omega_\perp(n_x + 1/2) + \hbar\omega_\perp(n_y + 1/2) + \hbar\omega_z(n_z + 1/2)$$

$$= \hbar\overline{\omega}\left(1 + \frac{1}{3}\delta_{\text{osc}}\right)(n_x + n_y + 1) + \hbar\overline{\omega}\left(1 - \frac{2}{3}\delta_{\text{osc}}\right)(n_z + 1/2)$$

$$= \hbar\overline{\omega}\left(N + \frac{3}{2} - \frac{1}{3}\delta_{\text{osc}}(3n_z - N)\right),$$

where $N = n_x + n_y + n_z$.

7.8 The transition probability $P^{(2)}(I_n \to I_m)$ calculated with the second order of the transition amplitude is obtained by inserting Eq. (7.105) in Eq. (7.104):

$$P^{(2)}(I_n \to I_m) = \frac{1}{2I_n + 1}|b_{nm}^{(1)} + \sum b_{nk}^{(1)}b_{km}^{(1)}|^2, \qquad (7.181)$$

leading to

$$P^{(2)}(I_n \to I_m) = \frac{1}{2I_n + 1}(\underbrace{|b_{nm}^{(1)}|^2}_{P^{(1)}(I_n \to I_m)} + \underbrace{2b_{nm}^{(1)}\sum b_{nk}^{(1)}b_{km}^{(1)}}_{\text{interference term}} + \underbrace{|\sum b_{nk}^{(1)}b_{km}^{(1)}|^2}_{\text{higher order}}).$$

$$(7.182)$$

The second term of the r.h.s of the above relation is an interference between the first order amplitude and the second order amplitude. The intermediate states k run over all states and their magnetic substates. In particular, the sum contains $b_{nm}^{(1)}b_{mm}^{(1)}$, for $k = m$, which corresponds to the case where the intermediate state is a magnetic substate of the state I_m. This particular term corresponds to the re-orientation effect:

$$P_{reorientation}^{(2)}(I_n \to I_m) \propto b_{nm}^{(1)} \times b_{nm}^{(1)}b_{mm}^{(1)} = b_{mm}^{(1)}|b_{nm}^{(1)}|^2, \qquad (7.183)$$

which is Eq. (7.108), where $b_{mm}^{(1)} = \langle I_m||M(E\lambda)||I_m\rangle$, the quadrupole moment (including its sign) of the state I_m.

Books for Further Readings

"Shapes, Shells, Nuclear Structure" by I. Ragnarsson (Cambridge University Press, 2008) gives an overview of models of nuclear structure and dynamics, concentrating in particular on a description of deformed and rotating nuclei.

References

1. A. Bohr, B.R. Mottelson, *Nuclear Structure*, vol. II, chap. 6 (World Scientific, Singapore, 1998)
2. K. Alder et al., Rev. Mod. Phys. **28**, 432 (1956)
3. M. Bolsterli, E.O. Fiset, J.R. Nix, J.L. Norton, Phys. Rev. C **5**, 1050 (1972)

4. P. Moller, A.I. Sierk, T. Ichikawa, H. Sagawa, Nuclear ground-state masses and deformations: FRDM(2012). Atomic Data Nucl. Data Tables **109–110**, 1–204 (2016)
5. The data are taken from http://www.nndc.bnl.gov/nudat2/
6. N.J. Stone, Atomic Data Nucl. Data Tables **90**, 75 (2005)
7. K. Marinova, I. Angeli, Atomic Data Nucl. Data Tables **99**, 69 (2013)
8. J. P. Elliott, Proc. Roy. Soc. Ser. A **245**, 128 (1958); ibid., A **245**, 562 (1958)
9. P.-G. Reinhard, E.W. Otten, Nucl. Phys. A **420**, 173 (1984)
10. E. Clément et al., Phys. Rev. C **75**, 054313 (2007)
11. I. Ragnarsson, S.G. Nilsson, R.K. Sheline, Phys. Rep. **45**, 1 (1978), "Shell structure in nuclei"
12. N.R. Johnson et al., Phys. Rev. Lett. **40**, 151 (1978)
13. F.S. Stephens, in *Frontiers in Nuclear Dynamics* (Plenum Press, New York, 1985). Data of glitches in Vela pulsar are taken from V. Radhakrishnan and R. N. Manchester. Nature **222**, 228 (1969)
14. K. Heyde, J.L. Woods, Rev. Mod. Phys. **83**, 1467 (2011)
15. A.N. Andreyev et al., Nature **405**, 430 (2000)
16. E. Bouchez et al., Phys. Rev. Lett. **90**, 082502 (2003)
17. M. Bender, P. Bonche, P.-H. Heenen, Phys. Rev. C **74**, 024312 (2006)
18. M. Girod, J.-P. Delaroche, A. Görgen, A. Obertelli, Phys. Lett. B **676**, 39 (2009)
19. P.J. Twin et al., Phys. Rev. Lett. **57**, 811 (1986)
20. E. Ideguchi et al., Phys. Rev. Lett. **87**, 222501 (2001)
21. T. Inakura et al., Nucl. Phys. A **710**, 261 (2002)
22. E. Caurier et al., Phys. Rev. C **75**, 054317 (2007)
23. G. Maquart et al., Phys. Rev. C **95**, 034304 (2017)

Chapter 8
Nuclear Reactions

Abstract Nuclear reactions are essential to populate the ground state and specific excited states in nuclei and to obtain nuclear structure information from experiments. There is a broad diversity of reactions ranging from elastic scattering, transfer reactions, knockout to nuclear fusion. In this chapter, several reactions are described and important concepts such as the distorted wave Born approximation (DWBA), optical potentials or the sudden approximation. We further extend the study to compound nucleus reactions. Nuclear fission and fusion processes are discussed in the context of energy production.

Keywords Direct reactions · Distorted wave Born approximation (DWBA) · Elastic scattering · Transfer reactions · Quasifree scattering · Knockout reactions · Compound nuclei · Fusion · Fission

8.1 Diversity of Reaction Mechanisms

Nuclear reactions have been studied since the discovery of the nucleus: the atomic nucleus was actually evidenced by use of elastic scattering, the simplest nuclear reaction. In 1911, Rutherford, Geiger and Marsden, in England, investigated the nature of gold atoms by bombarding them with alpha particles emitted from a collimated radioactive source. It was quite a surprise when they noticed that a fraction of these alpha particles, known at the time to be penetrating particles, were scattered at large angles and even backward. This observation of backward angle elastic scattering was the first use of nuclear reactions to evidence the existence of a heavy and tiny entity inside the atom, its nucleus [1].

The understanding of the structure of a nucleus usually requires several probes. Nuclear reactions play a central role in building our representation of nuclear physics, in several ways:

- as a probe to investigate the nuclear structure of ground states: binding energies, sizes and shapes,
- as a tool to produce radioactive ions from stable beams,

© Springer Nature Singapore Pte Ltd. 2021
A. Obertelli and H. Sagawa, *Modern Nuclear Physics*, UNITEXT for Physics,
https://doi.org/10.1007/978-981-16-2289-2_8

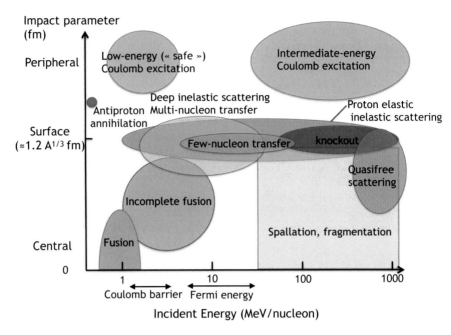

Fig. 8.1 Overview of nuclear reactions as a function of the incident energy and their centrality

- as a tool to excite nuclei, a necessary step to go beyond the study of ground-state properties,
- as a subject in itself to understand the dynamics of hadronic systems and the nuclear equation of state.

An overview of reactions separated in terms of impact parameter and incident energy is illustrated in Fig. 8.1. Coulomb excitation reactions are peripheral reactions due to the long range of the electromagnetic force. Reactions that are central will favor multiple scattering and strongly dissipative processes which will lead to high excitation energy in the residual nucleus, and statistical decay by nucleon emission. These dissipative reactions span from low energy (fusion, deep inelastic) to high energies (spallation, fragmentation) and are generally modeled as compound-nucleus reactions.

As a first attempt to classify reactions, one can define two kinds of reactions when two nuclei collide: *direct* reactions and *compound-nucleus* reactions. In the first case, direct reactions are short time reactions (about 10^{-22} s, the typical time for the projectile to fly a distance equivalent to the target size) which occur at the surface of the target nucleus. On the other hand, in the second case, the two colliding nuclei form an intermediate system which is most often highly excited and lives a sufficient time so that the excitation energy is shared by all nucleons. This compound nucleus does not keep memory of the initial state and the decay channels of the compound do not depend on the structure of the initial nuclei. In these reactions, the projectile

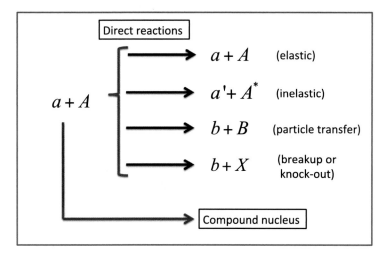

Fig. 8.2 Illustration of different reaction channels of direct reaction and the compound-nucleus. a and A denote projectile and target, respectively, and a', A^* means the excited states of the same nuclei. b and B are new projectile and target after the particle transfer reaction. The particle transfer reaction has two channels: particle pickup and particle stripping reactions, and both b and B are measured. In the "breakup" and "knockout" reactions only fragment b is measured, while X represents all other possible products of the reaction

may gain, lose or exchange few nucleons with the target. Due to the short time of the reaction and the few steps involved in the process, one can extend the definition of a direct reaction as a reaction that disturbs few degrees of freedom of the initial wave function, i.e., the populated final state keeps memory of the initial wave function. In other words, direct reaction cross sections to individual states can be used to probe the overlap between the initial and final states. These reaction processes are illustrated in Fig. 8.2.

Commonly studied direct reactions are "elastic", "inelastic", "particle transfer" and "break-up" channels and written symbolically as

$$a + A \rightarrow a + A \quad \text{or} \quad A(a, a)A \quad \text{(elastic scattering)},$$
$$a + A \rightarrow a' + A^* \quad \text{or} \quad A(a, a')A^* \quad \text{(inelastic scattering)},$$
$$a + A \rightarrow b + B \quad \text{or} \quad A(a, b)B \quad \text{(particle transfer reaction)},$$
$$a + A \rightarrow b + X \quad \text{or} \quad A(a, b)X \quad \text{(breakup/knockout reaction)}, \quad (8.1)$$

where a and A denote a projectile and a target, respectively, and a', A^* means the excited states of the same nuclei in inelastic scattering. The particle transfer reaction has two channels: particle pickup and particle stripping reactions. For particle transfer reactions, b and B are a new projectile and target different from a and A, respectively. More than two fragments exist in the final state of breakup and knock-out reactions; X represents two or more fragments after the reaction.

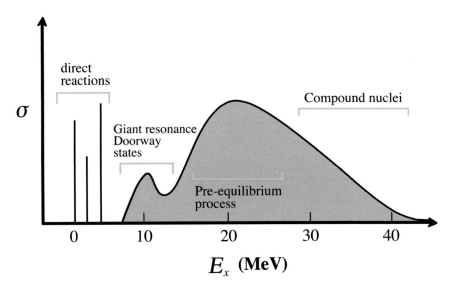

Fig. 8.3 Illustration of the different reaction mechanisms as a function of excitation energies. See the text for details

Today, due to their inherent complexity, we still do not know how to treat most of nuclear reactions in a self consistent and coherent way, with the exception of very recent *ab initio* treatment of very-low-energy elastic scatterings or transfer reactions involving few-body systems. When one deals with nuclear reactions, direct reactions as well as others, approximations have to be made. The validity of approximations are most of the time verified, or disqualified, by ad hoc comparisons with data. These comparisons can then be used to refine approximations. This trial-and-error approach is unavoidable in nuclear physics today. The complexity of treating dynamical nuclear systems (reactions) microscopically leads to several different models in alternative energy regions as shown in Fig. 8.3. The transition from direct to compound-nucleus reactions is progressive. At low excitation energies, few degrees of freedom, either single-particle or collective, are excited in direct reactions clearly. One particle transfer cross sections are also described by direct reactions. As we cross the particle-emission threshold, more degrees of freedom of the nucleus are excited, although for lower energies, the reactions can be understood as taking place via a few doorway states, such as giant resonances (GR). For larger energies, the process is more complex in nature, and it is usually described as a pre-equilibrium process that reaches thermodynamical equilibrium as a (maybe unbound) compound nucleus.

Figures 8.1 and 8.3 are simply illustrative pictures and the distinction between different reaction processes is not always strict. Sometimes there are overlaps between reaction processes. In such circumstances, one reaction process will be strongly affected by another one, e.g., fusion reactions are strongly affected by inelastic excitations and nucleon transfers.

8.2 Direct Reactions

Direct reactions have been used to study the nucleus and its structure [2]. These reactions can be divided into elastic scattering that leaves the nucleus in its initial quantum state after the reaction, inelastic scattering where the initial nucleus is excited to bound and unbound states of the same nucleus, and nucleon stripping (pickup) reactions where the final nucleus is obtained from the stripping (addition) of one or few nucleons from the initial nucleus. Typical direct reactions are

- elastic and inelastic reactions: (p, p), (n, n), (d, d), (t, t), $(^3\text{He}, {}^3\text{He})$ and (α, α),
- charge exchange reactions: (p, n), (n, p), $(^3\text{He}, t)$, $(t, {}^3\text{He})$,
- one-particle transfer reactions: (d, p), (d, n), (t, d), $(^3\text{He}, d)$, $(\alpha, {}^3\text{He})$, (α, t) and their inverse processes,
- two-particle transfer reactions: (t, p), $(^3\text{He}, p)$, $(^3\text{He}, n)$, (α, d) and their inverse processes,
- three-particle transfer reactions: (α, p) and (p, α),
- four-particle or α particle transfer reactions: $(^6\text{Li}, d)$, $(^{16}\text{O}, {}^{12}\text{C})$.

Common features of direct reactions are given below.

- The momentum and energy transfer are small between the initial and the final states. Therefore the reaction mechanism is relatively simple compared to the compound nuclei; single-step or at most double-step processes are dominant. Consequently, the final state of the projectile scatters dominantly to the forward direction.
- The cross sections depend very much on the structure of the projectile and the target since the reaction mechanism is relatively simple, i.e., the angular differential cross section depends clearly on the spin and parity of initial and final states and gives useful information on these quantum numbers.
- The direct reaction occurs dominantly at the nuclear surface and excites single-particle, vibrational or rotational collective states efficiently.

8.2.1 General Formalism

A two-body reaction between a target and a projectile nucleus is a dynamical many-body problem where the number of constituents is the sum of the nucleons in the target and in the projectile. The coordinates of all nucleons in the entrance channel can be decomposed into internal coordinates $\mathbf{r}_a$ and $\mathbf{r}_A$ of nucleons, which are the coordinates of nucleons in nuclei a and A respectively, and into the relative coordinates $\mathbf{r}_\alpha$ between the centers of mass of the two nuclei defined as

$$\mathbf{r}_\alpha = \frac{1}{A} \sum_{i=1}^{A} \mathbf{r}_i - \frac{1}{a} \sum_{j=A+1}^{A+a} \mathbf{r}_j \tag{8.2}$$

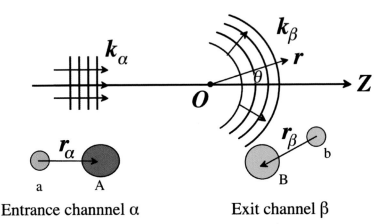

Entrance channnel α   Exit channel β

Fig. 8.4 Illustration of a scattering process starting from an entrance channel α composed of $(a + A)$ and leading to an exit channel β composed of $(b + B)$. The entrance channel is presented by a plane wave, while the exit channel is described by a spherical wave

as seen in Fig. 8.4. The internal structure of the nuclei a, A are driven by the intrinsic Hamiltonians h_A, h_a and the internal wave functions Φ_A, Φ_a are given by

$$h_a \Phi_a = E_a \Phi_a, \tag{8.3}$$

$$h_A \Phi_A = E_A \Phi_A, \tag{8.4}$$

where E_a and E_A are the total energy of projectile a and target A, respectively, in their intrinsic frame. The intrinsic state of the $A + a$ system is given by a product $\Phi_A \Phi_a$.

In a spatial representation, the Schrödinger equation to solve for the $a + A$ system can be expressed as:

$$H(\mathbf{r}_A, \mathbf{r}_a, \mathbf{r}_\alpha)\Psi_\alpha(\mathbf{r}_A, \mathbf{r}_a, \mathbf{r}_\alpha) = E\Psi_\alpha(\mathbf{r}_A, \mathbf{r}_a, \mathbf{r}_\alpha), \tag{8.5}$$

with $\Psi_\alpha(\mathbf{r}_A, \mathbf{r}_a, \mathbf{r}_\alpha)$ being the total wave function of the $A + a$ system, including the intrinsic and relative degrees of freedom. The full Hamiltonian $h(\mathbf{r}_A, \mathbf{r}_a, \mathbf{r}_\alpha)$ can be decomposed into a sum of the intrinsic Hamiltonians $h_{A,a}$ and a Hamiltonian driving the dynamical evolution of the projectile and target

$$H = h_A + h_a + H_\alpha = h_A + h_a + T + V, \tag{8.6}$$

where $H_\alpha = T + V$ is the Hamiltonian for the relative motion of A and a, decomposed into a kinetic term T and an interaction potential V between A and a. The main approximation used in most treatments of reactions is the *separation of degrees of freedom*, considering that V depends on a few degrees of freedom, for example, the valence or the core coordinates in transfer or breakup reactions. We will consider this approximation from now on. V is complex and contains a non-vanishing imaginary part when not all reaction channels are explicitly treated. The representation of Eq. (8.6) considers the partition in the entrance channel where nucleons are grouped together into the entrance-channel nuclei A and a. h can be written assuming a different partition of the nucleons such as the exit-channel partition. This approximation is exact at large distances beyond the nuclear interaction range for nuclear reactions by definition. However its validity of course depends on the considered reaction and energy, for example, for the Coulomb breakup reaction, the range of Coulomb interaction is infinite and a special care is needed.

We discuss hereafter the relative motion between projectile and target. The total wave function is written as

$$\Psi_\alpha(\mathbf{r}_A, \mathbf{r}_a, \mathbf{r}) = \psi(\mathbf{r})\Phi_\alpha(\mathbf{r}_A, \mathbf{r}_a), \tag{8.7}$$

where $\Phi_\alpha(\mathbf{r}_A, \mathbf{r}_a)$ is intrinsic wave function of the projectile a and the target A, and the $\psi(\mathbf{r})$ denotes the wave function of relative motion. The relative coordinate $\mathbf{r}_\alpha$ is simply denoted by $\mathbf{r}$. A wave function for the incoming particle is described by a plane wave in Fig. 8.4 and the outgoing wave has the nature of a spherical wave as an asymptotic form far from the scattering center O. The scattering wave function thus can be given by

$$\psi(\mathbf{r}) \xrightarrow{r\to\infty} \frac{1}{(2\pi)^{3/2}}\left[e^{ikz} + \frac{f(\theta)}{r}e^{ikr}\right], \tag{8.8}$$

where $f(\theta)$ is called the scattering amplitude. The first term of r.h.s. of Eq. (8.8) is the incoming plane wave to the z direction and the second term is the outgoing spherical wave from the scattering center O.

Elastic, inelastic and transfer scattering problems can be described by the scattering amplitude f from an initial state α to a final state β. In the case of elastic scattering, α and β both represent the same intrinsic state. In the general case, the final state wave function Ψ can be described as the following asymptotic behavior at $r \to \infty$:

$$\psi^{(+)} \to \begin{cases} \Phi_\alpha\left[e^{i\mathbf{k}\cdot\mathbf{r}} + f_{\alpha\alpha}(\theta)\frac{1}{r_\alpha}e^{ik_\alpha r_\alpha}\right] & \text{elastic} \\ \sum_{\alpha'\neq\alpha} \Phi_{\alpha'} f_{\alpha'\alpha}(\theta)\frac{1}{r_\alpha}e^{ik_{\alpha'} r_{\alpha'}} & \text{inelastic} \\ \sum_{\beta} \Phi_\beta f_{\beta\alpha}(\theta)\frac{1}{r_\beta}e^{ik_\beta r_\beta} & \text{transfer} \end{cases} \tag{8.9}$$

where α' and β designate inelastic and transfer channels, respectively. These formulas give concise expressions for the scattering amplitudes for different channels; equivalently, they are later formulated by the T matrix expressed in Eq. (8.32). For simplicity, we do not consider the Coulomb potential in Eqs. (8.7) and (8.9).

We first consider the two-body scattering problem in the center-of-mass system. The Schrödinger equation for the relative motion is given by

$$H\psi(\mathbf{r}) = E\psi(\mathbf{r}), \tag{8.10}$$

where $\mathbf{r}$ is the relative coordinate between the target A and the projectile a, and the Hamiltonian is given by

$$H = T + V(r) = -\frac{\hbar^2}{2m}\nabla^2 + V(r), \tag{8.11}$$

with the reduced mass $m = m_a m_A/(m_a + m_A)$. The Schrödinger equation (8.10) can be rewritten in the form of a Helmholtz equation,

$$(\nabla^2 + k^2)\psi(\mathbf{r}) = U(r)\psi(\mathbf{r}), \tag{8.12}$$

where $k^2 = 2mE/\hbar^2$ and $U = 2mV/\hbar^2$. Without the interaction term U, the solution of Eq. (8.12) is given by a plane wave

$$\phi_{\mathbf{k}}(\mathbf{r}) = \frac{1}{(2\pi)^{3/2}}e^{i\mathbf{k}\cdot\mathbf{r}}, \tag{8.13}$$

where the factor $(2\pi)^{-3/2}$ is for normalization of the plane wave. The scattering wave function $\phi_{\mathbf{k}}$ is normalized as

$$\langle\phi_{\mathbf{k}}|\phi_{\mathbf{k}'}\rangle = \int d\mathbf{r}\phi_{\mathbf{k}'}^*(\mathbf{r})\phi_{\mathbf{k}}(\mathbf{r}) = \delta(\mathbf{k} - \mathbf{k}'). \tag{8.14}$$

For the complete set of scattering wave functions, we have the identity

$$\int d\mathbf{k}|\phi_{\mathbf{k}}\rangle\langle\phi_{\mathbf{k}}| = \mathbf{1}. \tag{8.15}$$

The complete solution of Eq. (8.12) is given by a combination of a plane wave and a particular solution $\chi(\mathbf{r})$;

$$\psi(\mathbf{r}) = \phi(\mathbf{r}) + \chi(\mathbf{r}). \tag{8.16}$$

To obtain the particular solution of Eq. (8.12), we introduce a Green's function G_0, which satisfies the equation,

$$(\nabla^2 + k^2)G_0(\mathbf{r} - \mathbf{r}') = \delta(\mathbf{r} - \mathbf{r}'). \tag{8.17}$$

Then a particular solution $\chi(\mathbf{r})$ of (8.12) is expressed as

$$\chi(\mathbf{r}) = \int d\mathbf{r}' G_0(\mathbf{r} - \mathbf{r}')U(\mathbf{r}')\psi(\mathbf{r}'). \tag{8.18}$$

It is straightforward to check that the wave function (8.18) satisfies the Helmholtz equation (8.12).

We introduce the Fourier transform $G_0(\mathbf{k})$ of the Green's function $G_0(\mathbf{r})$ defined in Eq. (8.17) to obtain an explicit functional form of the Green's function:

$$G_0(\mathbf{r}) = \int G_0(\mathbf{k}')e^{i\mathbf{k}'\cdot\mathbf{r}}d\mathbf{k}'. \tag{8.19}$$

Inserting Eq. (8.19) into Eq. (8.17), one can derive the identity,

$$(k^2 - k'^2)G_0(\mathbf{k}') = \frac{1}{2\pi^3}, \tag{8.20}$$

with the help of the definition of the δ function,

$$\delta(\mathbf{r}) = \frac{1}{(2\pi)^3}\int e^{i\mathbf{k}'\cdot\mathbf{r}}d\mathbf{k}'. \tag{8.21}$$

Inserting $G_0(\mathbf{k}')$ of Eq. (8.20) into Eq. (8.19), the coordinate representation of Greens's function is expressed as

$$G_0(\mathbf{r}) = \frac{1}{(2\pi)^3}\int d\mathbf{k}'\frac{e^{i\mathbf{k}'\cdot\mathbf{r}}}{k^2 - k'^2}. \tag{8.22}$$

Equation (8.22) is further rewritten as

$$G_0(\mathbf{r}) = \frac{1}{(2\pi)^3}\int_0^\infty k'^2 dk'\int_{-1}^1 d(\cos\theta)\int_0^{2\pi}d\phi\frac{e^{ik'r\cos\theta}}{k^2 - k'^2} = \frac{1}{4\pi^2 r}\int_{-\infty}^\infty\frac{k'\sin k'r}{k^2 - k'^2}dk'. \tag{8.23}$$

The integration of Eq. (8.23) can be performed with the aid of Cauchy's integral theorem and produces two solutions,

$$G_0^{(\pm)}(\mathbf{r}) = -\frac{1}{4\pi r}e^{\pm ikr}, \tag{8.24}$$

which correspond to the outgoing wave $(+)$ and incoming wave $(-)$, respectively. Formal solutions of the Helmholtz equation (8.12) are two-fold and expressed as

$$\psi_k^{(\pm)}(\mathbf{r}) = \phi(\mathbf{r}) + \chi^{(\pm)}(\mathbf{r}) = \frac{1}{(2\pi)^{3/2}}e^{ikz} - \frac{1}{4\pi}\int\frac{e^{\pm ik|\mathbf{r}-\mathbf{r}'|}}{|\mathbf{r} - \mathbf{r}'|}U(r')\psi_k^{(\pm)}(\mathbf{r}')d\mathbf{r}'. \tag{8.25}$$

It should be noticed that this solution is an integral equation of $\psi_k^{(\pm)}$ which appears on both sides of the equation. Although this is a basic equation of scattering theory, it is not practical to solve this equation to obtain the wave function $\psi_k^{(\pm)}(\mathbf{r}')$ in the present form. A solvable equation, in which the solution appears only one time in

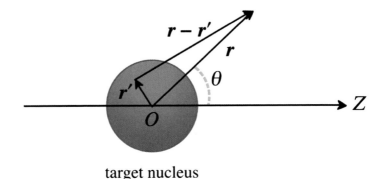

target nucleus

Fig. 8.5 Illustration of coordinates of scattering process. The coordinate **r** is the direction of scattered projectile with the angle θ from the incident direction z-axis. The coordinate **r**′ is restricted within a potential of the target nucleus

the equation, will be obtained by use of the Green's function formalism or T matrix theory.

Problem

8.1 Using Cauchy's integral theorem in the complex plane, derive Eq. (8.24) from Eq. (8.23),

$$G_0(\mathbf{r}) = \frac{1}{4\pi^2 r} \int_{-\infty}^{\infty} \frac{k' \sin k' r}{k^2 - k'^2} dk' = -\frac{1}{16\pi^2 i r}[I_+ - I_-], \qquad (8.26)$$

where

$$I_\pm = \int_{-\infty}^{\infty} \left[\frac{1}{k' + k} + \frac{1}{k' - k}\right] e^{\pm i k' r} dk'. \qquad (8.27)$$

The residue integral is summarized as

$$\lim_{\varepsilon \to 0} \frac{1}{2\pi^3} \int d\mathbf{k}' e^{i\mathbf{k}' \cdot (\mathbf{r} - \mathbf{r}')} \frac{1}{k^2 - k'^2 \pm i\varepsilon} = -\frac{1}{4\pi} \begin{cases} e^{ik|\mathbf{r} - \mathbf{r}'|}/|\mathbf{r} - \mathbf{r}'| & \text{if } + i\varepsilon \\ e^{-ik|\mathbf{r} - \mathbf{r}'|}/|\mathbf{r} - \mathbf{r}'| & \text{if } - i\varepsilon \end{cases}. \qquad (8.28)$$

Since the potential $U(r')$ is active within a range R, which is close to the radius of the nucleus, the asymptotic form can be obtained by the following approximations,

$$|\mathbf{r} - \mathbf{r}'| \approx [r^2 + r'^2 - 2(\mathbf{r}' \cdot \hat{\mathbf{r}})r]^{1/2} \approx r\left[1 - 2\frac{(\mathbf{r}' \cdot \hat{\mathbf{r}})}{r}\right]^{1/2} \approx r\left[1 - \frac{(\mathbf{r}' \cdot \hat{\mathbf{r}})}{r}\right],$$

$$\frac{1}{|\mathbf{r} - \mathbf{r}'|} \approx \frac{1}{r}\left[1 + \frac{(\mathbf{r}' \cdot \hat{\mathbf{r}})}{r}\right] \approx \frac{1}{r},$$

in the limit $r \gg r'$. Here $\hat{\mathbf{r}}$ is the unit vector in the direction of $\mathbf{r}$ (see Fig. 8.5). Using the above relations, the outgoing wave function is expressed as

$$\psi_k^{(+)} = \frac{1}{(2\pi)^{3/2}} \left(e^{ikz} + \frac{e^{ikr}}{r} \left[-\frac{(2\pi)^{3/2}}{4\pi} \int e^{-i\mathbf{k}'\cdot\mathbf{r}'} U(r') \psi_k^{(+)}(\mathbf{r}') d\mathbf{r}' \right] \right), \quad (8.29)$$

where $\mathbf{k}' \equiv k\hat{\mathbf{r}}$. Comparing Eq. (8.29) with Eq. (8.8), the scattering amplitude is given by

$$\begin{aligned} f^{(+)}(\mathbf{k}', \mathbf{k}) &= -\frac{(2\pi)^{3/2}}{4\pi} \int e^{-i\mathbf{k}'\cdot\mathbf{r}'} U(r') \psi_k^{(+)}(\mathbf{r}') d\mathbf{r}' \\ &= -2\pi^2 \frac{2m}{\hbar^2} \langle \phi_{k'} | V | \psi_k^{(+)} \rangle, \end{aligned} \quad (8.30)$$

where $\phi_{k'}$ is the normalized plane wave $\phi_{k'}(\mathbf{r}') = e^{i\mathbf{k}'\cdot\mathbf{r}'}/(2\pi)^{3/2}$. We remind that $U(r) = \frac{2mV(r)}{\hbar^2}$. The scattering amplitude of the incoming wave is also obtained in the same manner replacing $(+)$ by $(-)$.

8.2.2 T Matrix and Plane Wave Born Approximation (PWBA)

Hereafter, we introduce several important concepts of reaction theory: plane wave Born approximation (PWBA), distorted wave Born approximation (DWBA), T matrix, S matrix and associated formalisms, Lippman-Schwinger equation and Green's function as well.

First, we introduce the T matrix and the PWBA. In the case of a weak potential $V(r)$ (for example, the Coulomb potential) compared with the mean field potential, the wave function $\psi_k^{(+)}$ can be replaced by a plane wave $\phi_k^{(+)}$. The scattering amplitude is then simplified to

$$\begin{aligned} f^{(1)}(\mathbf{k}', \mathbf{k}) &= -2\pi^2 \frac{2m}{\hbar^2} \langle \phi_{k'} | V | \phi_k^{(+)} \rangle \\ &= -\frac{m}{2\pi\hbar^2} \int d\mathbf{r} e^{-i\mathbf{q}\cdot\mathbf{r}} V(r), \end{aligned} \quad (8.31)$$

where $\mathbf{q} = \mathbf{k}' - \mathbf{k}$. Equation (8.31) is called the first-order "Born" approximation or the plane wave Born approximation (PWBA). In other words, the first-order scattering amplitude is proportional to the Fourier transform of the potential V with respect to the momentum transfer $\mathbf{q}$. This approximation has very limited applicability for hadron scattering since the nucleon-nucleus interaction is rather strong. However, it has been applied for the electron scattering experiments since the Coulomb interaction is weak and can be treated perturbatively.

Equation (8.30) provides an explicit relation between the scattering matrix and the matrix element of potential V, which is called the T matrix, defined by

$$T(\mathbf{k}', \mathbf{k}) \equiv \langle \phi_{k'} | V | \psi_k^{(+)} \rangle. \tag{8.32}$$

The above equation is exact and contains the full solution $\psi_k^{(+)}$ of the scattering problem. The T matrix gives access to scattering quantities such as the cross sections, as derived in the following and expressed in the forthcoming Eq. (8.64). In practice, some approximations are necessary to calculate the T matrix.

The scattering amplitude is related to the T matrix as

$$f^{(+)}(\mathbf{k}', \mathbf{k}) = -2\pi^2 \frac{2m}{\hbar^2} T(\mathbf{k}', \mathbf{k}). \tag{8.33}$$

The formal solutions of the Helmholtz equation are given in Eq. (8.25) in the integral form. We will derive a more convenient form for general scattering problems by using the operator equation. We introduce the free-particle Green's function,

$$G_0^{(\pm)}(E) = \frac{1}{E - H_0 \pm i\varepsilon}, \tag{8.34}$$

where the inverse operator form is essential to develop a soluble equation of the scattering wave function, and equivalently the scattering amplitude in the T-matrix formalism. The infinitesimal imaginal part $\pm i\varepsilon$ in the denominator is necessary to obtain the incoming and outgoing scattering wave functions at the pole of the Green's function. The fully correlated Green's function is expressed as

$$G^{(\pm)}(E) = \frac{1}{E - H \pm i\varepsilon}, \tag{8.35}$$

where $H = H_0 + V$. By construction, we have a relation between the two Green's functions,

$$G(E) = G_0(E) + G_0(E)VG(E) \tag{8.36}$$

and

$$G(E) = G_0(E) + G(E)VG_0(E). \tag{8.37}$$

In the spectral representation, Eq. (8.34) is expressed as

$$G_0^{(\pm)}(E_k) = \int d\mathbf{k}' \frac{|\phi_{\mathbf{k}'}\rangle \langle \phi_{\mathbf{k}'}|}{E_k - E_k' \pm i\varepsilon}, \tag{8.38}$$

with the aid of the identity (8.15). The coordinate space representation is given by

$$G_0^{(\pm)}(E_k; \mathbf{r}, \mathbf{r}') \equiv \langle \mathbf{r} | G_0^{(\pm)}(E_k) | \mathbf{r}' \rangle = \int d\mathbf{k}' \frac{\langle \mathbf{r} | \phi_{\mathbf{k}'} \rangle \langle \phi_{\mathbf{k}'} | \mathbf{r}' \rangle}{E_k - E_k' \pm i\varepsilon}$$

$$= \int d\mathbf{k}' \frac{\phi_{k'}(\mathbf{r}) \phi_{k'}^*(\mathbf{r}')}{E_k - E_k' \pm i\varepsilon}, \tag{8.39}$$

where $\langle \mathbf{r} | \phi_{\mathbf{k}} \rangle = \phi_k(\mathbf{r}) = \exp(i\mathbf{k} \cdot \mathbf{r})/(2\pi)^{3/2}$.

Problem

8.2 Prove Eqs. (8.36) and (8.37) by using an identity for two operators A and B

$$\frac{1}{A} = \frac{1}{B} + \frac{1}{B}(B - A)\frac{1}{A}, \tag{8.40}$$

inserting $A = E - H$ and $B = E - H_0$ or $A = E - H_0$ and $B = E - H$.

With the aid of Eq. (8.28), Eq. (8.25) is rewritten as

$$\psi_k^{(\pm)}(\mathbf{r}) = \frac{1}{(2\pi)^{3/2}} e^{ikz} + \frac{1}{(2\pi)^3} \int d\mathbf{r}' \int d\mathbf{k}' \frac{e^{i\mathbf{k}' \cdot \mathbf{r}} e^{-i\mathbf{k}' \cdot \mathbf{r}'}}{E_k - E_{k'} \pm i\varepsilon} V(r') \psi_k^{(\pm)}(\mathbf{r}'). \tag{8.41}$$

Equation (8.41) is further expressed by using Eq. (8.39) as

$$\psi_k^{(\pm)}(\mathbf{r}) = \frac{1}{(2\pi)^{3/2}} e^{ikz} + \int d\mathbf{r}' G_0^{(\pm)}(E_k, \mathbf{r}, \mathbf{r}') V(r') \psi_k^{(\pm)}(\mathbf{r}'). \tag{8.42}$$

This equation (8.42) is solvable in the framework of Green's function theory as it will be shown hereafter.

Equation (8.42) can be expressed by an operator form of Green's function as

$$|\psi_k^{(\pm)}\rangle = |\phi_k\rangle + G_0^{(\pm)}(E_k) V |\psi_k^{(\pm)}\rangle. \tag{8.43}$$

Equation (8.43) is called the "Lippman-Schwinger" equation for scattering theory. We will show how the Lippman-Schwinger equation is applied for the scattering problem in a symbolic approach with the transition operator T, which is equivalent to the T matrix in Eq. (8.32). The operator T is defined by the equation

$$V |\psi_k^{(+)}\rangle = T |\phi_k\rangle. \tag{8.44}$$

Multiplying Eq. (8.43) by V, we have

$$V |\psi_k^{(+)}\rangle = T |\phi_k\rangle = V |\phi_k\rangle + V G_0^{(+)}(E_k) T |\phi_k\rangle, \tag{8.45}$$

which gives

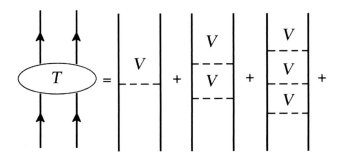

Fig. 8.6 Ladder diagram of T matrix. The horizontal lines indicate two particles interacting with T or V, while the vertical dashed lines denote mediating mesons to induce the two-body interaction V

$$T = V + V G_0^{(+)}(E_k) T. \tag{8.46}$$

The equation (8.46) leads to an iterative equation for T as

$$T = V + V G_0^{(+)}(E_k) V + V G_0^{(+)}(E_k) V G_0^{(+)}(E_k) V + \cdots. \tag{8.47}$$

The T matrix (8.47) is illustrated diagrammatically in Fig. 8.6. Correspondingly, we can expand the scattering amplitude (8.33) as

$$f(\mathbf{k}, \mathbf{k}') = \sum_{n=1}^{\infty} f^{(n)}(\mathbf{k}, \mathbf{k}'), \tag{8.48}$$

where n represent the number of times that the interaction V enters (see Fig. 8.6). For $n=1$ and 2, the scattering amplitudes are given by

$$f^{(1)}(\mathbf{k}, \mathbf{k}') = -2\pi^2 \frac{2m}{\hbar^2} \langle \mathbf{k}' | V | \mathbf{k} \rangle, \tag{8.49}$$

$$f^{(2)}(\mathbf{k}, \mathbf{k}') = -2\pi^2 \frac{2m}{\hbar^2} \langle \mathbf{k}' | V G_0^{(+)}(E_k) V | \mathbf{k} \rangle \tag{8.50}$$

The $f^{(1)}$ and $f^{(2)}$ amplitudes, as illustrated in Fig. 8.7, correspond to the first-order and second-order Born approximations, respectively, For the first-order Born approximation, a physical interpretation of $f^{(1)}$ is that the incident particle interacts at $\mathbf{r}'$ with $V(\mathbf{r}')$ and the particle is scattered into the direction $\mathbf{k}'$. The $f^{(2)}$ represents a two step process; the incident particle interacts at $\mathbf{r}''$ with $V(\mathbf{r}'')$, and propagates from $\mathbf{r}''$ to $\mathbf{r}'$ via Green's function G_0. Subsequently a second interaction occurs at $\mathbf{r}'$ with $V(\mathbf{r}')$, and finally the particle is scattered into the direction $\mathbf{k}'$. Likewise, a higher-order term $f^{(3)}$ is a three-step process, and so on.

The Lippman-Schwinger equation (8.43) has the same structure as Eq. (8.29) since $|\psi_k^{(\pm)}\rangle$ appears on both sides of the equation. However, it is possible to rearrange this

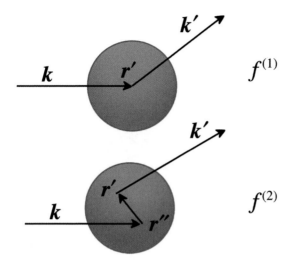

Fig. 8.7 Physical interpretation of the first-order and second-order Born approximations. $f^{(1)}$ and $f^{(2)}$ stand for the first-order and second-order processes, respectively

equation in a solvable form by using an operator equation, as show in the following. We have the identities (see problem 8.2),

$$G = G_0 + G_0 V G \quad \text{and} \quad G = G_0 + G V G_0, \tag{8.51}$$

for the free and correlated Green's functions. The Lippman-Schwinger equation (8.43) is rewritten by using the latter part of Eq. (8.51),

$$G_0 = G(1 - V G_0), \tag{8.52}$$

as

$$
\begin{aligned}
|\psi_k^{(\pm)}\rangle &= |\phi_k\rangle + G^{(\pm)}(E_k)(1 - V G_0^{(\pm)}(E_k)) V |\psi_k^{(\pm)}\rangle \\
&= |\phi_k\rangle + G^{(\pm)}(E_k) V (1 - G_0^{(\pm)}(E_k) V) |\psi_k^{(\pm)}\rangle \\
&= (1 + G^{(\pm)}(E_k) V) |\phi_k\rangle.
\end{aligned} \tag{8.53}
$$

To derive the final equation (8.53) from the second line, we use the equation $(1 - G_0^{(\pm)}(E_k) V)|\psi_k^{(\pm)}\rangle = |\phi_k\rangle$ obtained from Eq. (8.43). It should be noticed that only the initial plane wave $|\phi_k\rangle$ appears on the right-hand side. This is quite important for practical applications of the Lippman-Schwinger equation to obtain the scattering wave function of the direct reaction, i.e., the distortion effect on the scattering wave can be accommodated to the incoming wave ϕ_k solving Eq. (8.53).

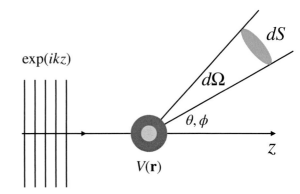

Fig. 8.8 Differential cross section and solid angle. An incident plane wave arrives along the z direction, from the left, and is scattered by a potential $V(\mathbf{r})$. The differential cross section $d\sigma/d\Omega(\theta)$ as function of the scattering angle θ is equal to the square of the scattering amplitude at the angle θ (see Eq. (8.58))

8.2.3 Scattering Amplitude and Cross Sections

Both classical and quantum mechanical scattering phenomena are characterized by the scattering cross section σ. The differential cross section scattered to the angles θ and ϕ is defined as the flux of scattered particles in the area dS in the direction of (θ, ϕ), per unit incident flux. Imagine a collision experiment in which a detector measures the number of particles per unit time scattered into a surface element dS subtending by a solid angle $d\Omega = dS/r^2$ in direction (θ, ϕ) (see Fig. 8.8). We consider first the number of incoming particles crossing a plane perpendicular to the incident direction per unit area and per time. This is expressed by the current density $\mathbf{j}$ defined by

$$\mathbf{j}(\mathbf{r}) = Re\left[\phi_k^* \mathbf{v} \phi_k\right] = Re\left[\phi_k^* \frac{\hbar}{im} \nabla \phi_k\right] = \frac{\hbar}{2im}\left[\phi_k^* \nabla \phi_k - \phi_k \nabla \phi_k^*\right], \quad (8.54)$$

where ϕ_k is the incoming wave, and $\mathbf{v}$ is the relative velocity of incoming projectile to the target. Then the number of scattered particles, going into a small area dS subtending a differential solid angle element $d\Omega$, is also given by the current density multiplied by dS. The differential cross section is thus given by

$$d\sigma \equiv \frac{|\mathbf{j}_{out}|dS}{|\mathbf{j}_{in}|} = \frac{r^2|\mathbf{j}_{out}|}{|\mathbf{j}_{in}|}d\Omega. \quad (8.55)$$

For the plane wave e^{ikz}, $\mathbf{j}_{in}$ is calculated as

$$|\mathbf{j}_{in}| = \frac{1}{(2\pi)^3}\frac{\hbar k}{m} = \frac{v}{(2\pi)^3}. \quad (8.56)$$

The current density for scattered particles is also calculated from the outgoing wave function in Eq. (8.8), $\chi(r) = (2\pi)^{-3/2} f(\theta)e^{ikr}/r$,

$$|\mathbf{j}(r)| = \frac{\hbar}{2im}\left[\chi^* \frac{\partial \chi}{\partial r} - \chi \frac{\partial \chi^*}{\partial r}\right] = \frac{v}{(2\pi)^3 r^2}|f(\theta)|^2. \quad (8.57)$$

Then the differential cross section $d\sigma(\theta)$ at $d\Omega$ is expressed by the scattering amplitude as,

$$\frac{d\sigma(\theta)}{d\Omega} = |f(\theta)|^2. \tag{8.58}$$

From the differential cross section, we can obtain the total cross section by integrating over all solid angles

$$\sigma = \int d\Omega \frac{d\sigma(\theta)}{d\Omega} = \int d\phi \int d\theta \sin\theta \frac{d\sigma(\theta)}{d\Omega}. \tag{8.59}$$

The cross section, which depends strongly on the energy of incoming particles, has the dimension of an area and can be measured for many channels, for example, elastic scattering, inelastic scattering, particle transfer, break-up, and absorption cross sections.

The generalization to an inelastic scattering or transfer channel

$$a + A \rightarrow b + B \tag{8.60}$$

can be formulated as follows. The final state of a transfer channel is now described by the wave function,

$$\Psi_\beta = \psi_\beta(\mathbf{r}_\beta)\Phi_\beta \tag{8.61}$$

where Φ_β is the intrinsic wave function of final state $(b + B)$ and $\psi_\beta(\mathbf{r}_\beta)$ is the wave function of relative motion. The asymptotic form of the wave function $\psi_\beta(\mathbf{r}_\beta)$ at $(r \rightarrow \infty)$ is given by

$$\psi_\beta^{(+)}(\mathbf{r}_\beta) \sim f_{\beta\alpha}(\theta_\beta)\frac{1}{r_\beta}e^{ik_\beta r_\beta}, \tag{8.62}$$

where the scattering angle θ_β is the one between $\mathbf{k}_\beta$ and $\mathbf{k}_\alpha$.

$$f_{\beta\alpha}(\theta_\beta) = -\frac{(2\pi)^{3/2}}{4\pi} \int e^{-i\mathbf{k}_\beta \cdot \mathbf{r}'} U(r')\psi_{k_\alpha}(\mathbf{r}')d\mathbf{r}' \tag{8.63}$$

The cross section for the channel $\alpha \rightarrow \beta$ is also given in a similar manner to the elastic channel (8.58) as

$$\frac{d\sigma_{\beta\alpha}(\theta_\beta)}{d\Omega_\beta} = \frac{v_\beta}{v_\alpha}|f_{\beta\alpha}(\theta_\beta)|^2 = \frac{m_\alpha m_\beta}{(2\pi\hbar^2)^2}\frac{k_\beta}{k_\alpha}|T_{\beta\alpha}|^2, \tag{8.64}$$

where v_α and v_β are the relative velocities of the initial and final channels, k_α and k_β are their wave numbers, and m_α and m_β are their reduced masses, respectively. Equation (8.64) is the fundamental equation for all reactions in general. Approximations are necessary to solve the $T_{\beta\alpha}$ matrix for each reaction channel.

8.2.4 S Matrix and Scattering Amplitude

In the previous section, we have reached the fundamental equation of reaction theory, leading to the scattering cross section. Here, we discuss S matrices and optical potentials, which are common quantities and ingredients used for practical applications of reaction theory. We introduce the S matrix operator, which has a relation with the T matrix operator in an operator form

$$S = 1 - 2\pi i T. \tag{8.65}$$

As is shown hereafter, the S matrix has several advantages to describe scattering problems:

- The S matrix operator is unitary ($S^{\dagger}S=1$) if the potential is real.
- Consequently, the S matrix is useful to describe the conservation law in the scattering process, and has a direct link to the optical theorem for the total cross section (see Eq. (8.86)).
- S matrix is useful to describe the scattering in terms of the phase shift, especially the resonance states.

The S matrix has a direct link with the time evolution of a scattering system, which is described by the time-dependent Schrödinger equation

$$i\hbar\frac{\partial \Psi_s(t)}{\partial t} = H\Psi_s(t) = (H_0 + V(r))\Psi_s(t), \tag{8.66}$$

where H_0 is the kinetic energy of the relative motion between projectile and target, and $V(r)$ is the interaction between them. The wave function $\Psi_s(t)$ is that for the Schrödinger representation, in which the wave function describes the time evolution of the system.[1] We introduce a wave function in a new representation

$$\Psi(t) = e^{iH_0t/\hbar}\Psi_s(t), \tag{8.67}$$

which is referred to "interaction representation" to be different from the ordinary Schrödinger representation of Eq. (8.66), but useful for the many-body scattering problem. The wave function (8.67) satisfies the equation

$$i\hbar\frac{\partial \Psi(t)}{\partial t} = V(t)\Psi(t), \tag{8.68}$$

where

$$V(t) \equiv e^{iH_0t/\hbar}V(r)e^{-iH_0t/\hbar}. \tag{8.69}$$

---

[1] Alternatively, the time evolution of the system is represented by the physical observables in the Heisenberg representation.

To describe the time evolution of the wave function in the interaction representation, we introduce a unitary operator U as

$$\Psi(t) = U(t, t_0)\Psi(t_0). \tag{8.70}$$

Inserting (8.70) in (8.68), we obtain the time evolution of the unitary operator U as

$$i\hbar\frac{\partial U(t, t_0)}{\partial t} = V(t)U(t, t_0). \tag{8.71}$$

Notice that the interaction $V(t)$ appears in Eq. (8.71), while both H_0 and $V(r)$ are involved in the Schrödinger equation (8.66).

The time evolution of the many-body system $\Psi(t)$ at t is expressed by using the S matrix operator as,

$$\Psi(t) = S\Psi_i(t_0), \tag{8.72}$$

where $\Psi_i(t_0)$ is the initial state at t_0. The scattering amplitude of the state Ψ_i is given by the operator U in Eq. (8.71) at the limit $t \to \infty$ (no effect of the interaction $V(t)$) starting from $t_0 \to -\infty$ (before the reaction) as

$$S \equiv \lim_{\substack{t \to +\infty \\ t_0 \to -\infty}} U(t, t_0) = U(+\infty, -\infty). \tag{8.73}$$

One can obtain the operator U to integrate Eq. (8.71) as

$$U(t, t_0) = 1 - \frac{i}{\hbar}\int_{t_0}^{t} V(t')U(t', t_0)dt', \tag{8.74}$$

where the initial condition is set to be $U(t_0, t_0)=1$. From Eqs. (8.73) and (8.74), the S matrix is expressed by a sequential expansion in terms of the interaction $V(t)$ as

$$S = 1 - \frac{i}{\hbar}\int_{-\infty}^{\infty} V(t)dt - \frac{1}{\hbar^2}\int_{-\infty}^{\infty} dt_1 \int_{-\infty}^{t_1} dt_2 V(t_1)V(t_2) + \cdots. \tag{8.75}$$

The first-order term of S matrix operator for a transition from $|m\rangle$ to $|n\rangle$ is given by

$$\langle n|S^{(1)}|m\rangle = -\frac{i}{\hbar}\int_{-\infty}^{\infty} dt\, e^{i(E_n - E_m)t/\hbar}\langle n|V(r)|m\rangle$$
$$= -2\pi i\delta(E_n - E_m)\langle n|V(r)|m\rangle, \tag{8.76}$$

where E_n and E_m are the eigenenergies of the Hamiltonian H_0 for the eigenstates $|m\rangle$ and $|n\rangle$.[2] Equation (8.76) guarantees the energy conservation between the initial and final states. The matrix element is equivalent to that of the first-order term of the

---

[2] In Eq. (8.76), we use an integral form of the δ function, $\delta(x) = \frac{1}{2\pi}\int_{-\infty}^{\infty} e^{ixt}dt$.

T matrix in Eq. (8.32) with a factor $-2\pi i$. Let us evaluate next the matrix element of the second-order term in Eq. (8.75) (proportional to V^2). The matrix element reads,

$$
\begin{aligned}
\langle n|S^{(2)}|m\rangle &= -\frac{1}{\hbar^2}\sum_l \int_{-\infty}^{\infty} dt_1 \int_{-\infty}^{t_1} dt_2 \langle n|V(t_1)|l\rangle\langle l|V(t_2)|m\rangle \\
&= -\frac{1}{\hbar^2}\sum_l \int_{-\infty}^{\infty} dt_1 \int_{-\infty}^{t_1} dt_2 e^{i(E_n-E_l)t_1/\hbar}\langle n|V|l\rangle e^{i(E_l-E_m-i\varepsilon)t_2/\hbar}\langle l|V|m\rangle \\
&= -\frac{1}{i\hbar}\sum_l \int_{-\infty}^{\infty} dt_1 e^{i(E_n-E_m)t_1/\hbar}\langle n|V|l\rangle \frac{1}{E_l-E_m-i\varepsilon}\langle l|V|m\rangle \\
&= -2\pi i\delta(E_n-E_m)\sum_l \langle n|V|l\rangle \frac{1}{E_m-E_l+i\varepsilon}\langle l|V|m\rangle. \quad (8.77)
\end{aligned}
$$

This matrix element is again equivalent to that of the second-order term in the T matrix (8.32) with a factor $-2\pi i$ since the free Green's function (8.34) has the denominator $(E_k - H_0 + i\varepsilon)$ which turns out to be $(E_m - E_l + i\varepsilon)$ in Eq. (8.77). The $i\varepsilon$ is introduced to avoid a divergence of the integral with respect to t_2. The equivalence between S and T matrices holds also for higher-order terms in the expansion in terms of V and, as the extension of (8.76), S can be expressed by the T matrix operator as

$$
\langle n|S|m\rangle = \delta_{nm} - 2\pi i\delta(E_n - E_m)\langle n|T|m\rangle. \quad (8.78)
$$

Equation (8.78) is equivalent to the operator equation

$$
S = 1 - 2\pi i T. \quad (8.79)
$$

Since the S matrix is defined by the unitary operator U in Eq. (8.73), S also satisfies the unitarity condition

$$
S^\dagger S = 1, \quad (8.80)
$$

and can be parameterized for the elastic scattering for the channel α to be

$$
S_\alpha \equiv \langle \alpha|S|\alpha\rangle = e^{2i\delta_\alpha}, \quad (8.81)
$$

where δ_α is the phase shift. The phase shift was discussed in more details in Sect. 2.5 in Chap. 2.

With the S matrix for a scattering wave of angular momentum ℓ, denoted by S_l, the scattering amplitude can be expressed by partial waves as

$$
f(\theta) = \frac{1}{2ik}\sum_l (S_l - 1)(2l + 1)P_l(\cos\theta), \quad (8.82)
$$

and the differential cross section is then given by

$$d\sigma(\theta) = |f(\theta)|^2 d\Omega = \frac{1}{4k^2} |\sum_l (2l+1)(S_l - 1) P_l(\cos\theta)|^2 d\Omega. \qquad (8.83)$$

The elastic total cross section is evaluated to be

$$\sigma_{el}(\theta) = \int d\sigma(\theta) = \frac{\pi}{k^2} \sum_l (2l+1) (|S_l - 1|)^2, \qquad (8.84)$$

where the orthogonal condition of Legendre polynomials

$$\int d\Omega \, P_l(\cos\theta) P_{l'}(\cos\theta) = \delta_{ll'} \frac{4\pi}{2l+1} \qquad (8.85)$$

is used. The total cross section is calculated from the scattering amplitude in the forward direction, since the attenuation of the incident wave, which is proportional to the total cross section, is the result of interferences between the incident wave and the scattered wave in the forward direction. The relation between the total cross section and the scattering amplitude $f(0)$ then reads

$$\sigma_t = \frac{4\pi}{k} \text{Im}\{f(0)\} = \frac{2\pi}{k^2} \sum_l (2l+1)(1 - \Re\{S_l\}), \qquad (8.86)$$

where $P_l(0) = 1$ is used. The formula (8.86) is called the "optical theorem". Eventually the total reaction cross section is given by

$$\sigma_R = \sigma_t - \sigma_{el} = \frac{\pi}{k^2} \sum_l (2l+1)(1 - |S_l|^2), \qquad (8.87)$$

where $|S_l| < 1$ if the potential is complex.

8.2.5 Scattering Length and Effective Range

We consider how to find the scattering amplitude in low energy S-wave scattering in this subsection. For the spherical symmetric system, the Schrödinger equation (8.12) depends only on the coordinate $r = |\mathbf{r}|$ and reads

$$\left(\frac{d^2}{dr^2} + k^2\right) u(r) = U(r)u(r), \qquad (8.88)$$

where $u(r) = r\psi(r)$, $k^2 = 2mE/\hbar^2$ and $U = 2mV/\hbar^2$. Here, we use a relation

$$\nabla^2 = \frac{1}{r}\frac{\partial^2}{\partial r^2} r + \frac{l(l+1)}{r^2}. \qquad (8.89)$$

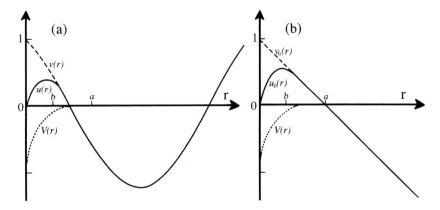

Fig. 8.9 **a** The radial wave function $u(r)$ at a positive energy E is equal to the asymptotic wave function $v(r)$ in Eq. (8.91) outside of the range b of the nuclear interaction $V(r)$. $v(r)$ is normalized to be unity at $r = 0$, while $u(r)$ is zero at $r = 0$ by definition. **b** The radial wave function at $E=0$. The wave function $u_0(r)$ is equal to the asymptotic one $v_0(r)$ in Eq. (8.92) outside of the range of b of the nuclear interaction. Inside the range $r < b$, $u_0(r)$ and $v_0(r)$ are close to $u(r)$ and $v(r)$, respectively

The solution $u(r)$ can be found with the boundary condition $u(r = 0) = 0$ and

$$u(r) \to \sin(kr + \delta), \quad (r \to \infty). \qquad (8.90)$$

In principle, the phase shift δ can be determined uniquely by solving the Schrödinger equation (8.88) with the boundary condition $u(r = 0) = 0$. Let us introduce a wave function $v(r)$ which has the same asymptotic behavior at $r = \infty$ as $u(r)$.

$$v(r) = \frac{\sin(kr + \delta)}{\sin \delta}, \qquad (8.91)$$

which is different from u(r) at $r = 0$ to be $v(r = 0) = 1$. The schematic behavior of $u(r)$ and $v(r)$ is given in Fig. 8.9a.

Our problem is to find the energy dependence of the S-wave phase shift. To this end, we study the wave function $u(r)$ at the zero energy limit $k = \sqrt{2mE/\hbar^2} \to 0$. We denote the wave function u_0 at $E = 0$ limit. Beyond the range of potential $U(r)$ at $E = 0$, the Schrödinger equation becomes $d^2u/dr^2 = 0$, so that the solution can be given by $u_0(r \to \infty) = a + br$, where a and b are constant. The important feature of $u_0(r)$ is the point at which the wave function intersects the r-axis. Thus we introduce a wave function $v_0(r)$ which has the same asymptotic behavior as $u_0(r)$ and normalized to be 1 at $r = 0$;

$$v_0(r) = 1 - \frac{r}{a}, \qquad (8.92)$$

where "a" is called "the scattering length" (see Fig. 8.9b). At the first-order expansion of kr, the wave function (8.91) can be expressed as $v \simeq kr \cot \delta + 1$, and we obtain a relation from the comparison with Eq. (8.92);

$$k \cot \delta = -\frac{1}{a} \quad \text{for } (k \to 0). \tag{8.93}$$

In the low energy limit, we can make an approximate value for the phase shift

$$\tan \delta \simeq \sin \delta \simeq \delta = -ak. \tag{8.94}$$

Inserting (8.94) into (8.83), we obtain the total scattering cross section at zero energy,

$$\sigma_0 = 4\pi a^2, \tag{8.95}$$

since the scattering amplitude for S wave becomes $f = \frac{1}{k} e^{i\delta} \sin \delta$. The cross section (8.95) is identical to the scattering cross section of a black-body sphere of radius a. Such a sphere requires the wave function to vanish at $r = a$ as was imposed on $v_0(r)$ of Eq. (8.92). Equation (8.95) suggests that the measurement of the cross section at zero energy determines the magnitude of the scattering length, but not its sign.

Next, we will obtain a formula for the phase shift, consequently the cross section, at an arbitrary energy E. The wave function $u(r)$ at an arbitrary energy can be obtained from Eq. (8.88) and compared with the wave function $u_0(r)$ at zero energy;

$$\frac{d^2}{dr^2} u_0(r) = U(r) u_0(r). \tag{8.96}$$

Multiplying (8.88) by $u_0(r)$ and also (8.96) by $u(r)$ and subtracting two equations, we have,

$$\frac{d}{dr} \left(u \frac{du_0}{dr} - u_0 \frac{du}{dr} \right) = k^2 u u_0. \tag{8.97}$$

A similar relation holds for the asymptotic wave function $v(r)$ in Eq. (8.91) and $v_0(r)$ in Eq. (8.92);

$$\frac{d}{dr} \left(v \frac{dv_0}{dr} - v_0 \frac{dv}{dr} \right) = k^2 v v_0, \tag{8.98}$$

since $v(r)$ and $v_0(r)$ fulfill the Schrödinger equations (8.88) and (8.96), respectively, for $V{=}0$. Subtracting (8.98) from (8.97) and integrating over r from zero to infinity, we get

$$\left[v \frac{dv_0}{dr} - v_0 \frac{dv}{dr} \right]_{r=0}^{r=\infty} - \left[u \frac{du_0}{dr} - u_0 \frac{du}{dr} \right]_{r=0}^{r=\infty} = k^2 \int_0^\infty (v v_0 - u u_0) dr. \tag{8.99}$$

The left-hand side of this equation can be evaluated by using $u(0) = 0$, $v(0) = 1$, $dv/dr|_{r=0} = k \cot \delta$ and $dv_0/dr|_{r=0} = -1/a$, together with $v(r) = u(r)$ and $v_0(r) = u_0(r)$ at $r \to \infty$. Then we can obtain a relation,

$$k \cot \delta = -\frac{1}{a} + k^2 \int_0^\infty (v v_0 - u u_0) dr. \tag{8.100}$$

Equation (8.100) is exact. We notice that large contributions to the integral come from the region where $u(r)$ and $u_0(r)$ differ from the asymptotic forms $v(r)$ and $v_0(r)$, respectively. It is clear from Fig. 8.9 that the contributions come from the inner part where the nuclear force is effective. In the low energy reaction, $E << |V|$, the behavior of wave functions is independent of E so that we can replace in a good approximation $u(r)$ and $v(r)$ by the zero energy values $u_0(r)$ and $v_0(r)$ in the integral;

$$k \cot \delta = -\frac{1}{a} + k^2 \int_0^\infty (v_0^2 - u_0^2) dr \equiv -\frac{1}{a} + \frac{1}{2} r_0 k^2, \tag{8.101}$$

where

$$r_0 = 2 \int_0^\infty (v_0^2 - u_0^2) dr \tag{8.102}$$

is called "the effective range". The effective range r_0 depends only on the potential, but not on the energy k^2. The integral (8.100) is zero outside of the range of the potential and is finite, of the order of unity, inside the range. In fact, the integral reflects the mean distance of interaction. This is the reason why r_0 is called the effective range.

The scattering length gives a critical index for deciding whether a two particle system is bound or not, i.e., a two-particle system with a specific channel assigned by the spin S and the isospin T will lead to a bound state if the scattering length is positive, while no bound state appears for the channel with negative scattering length. In proton-neutron scattering, the scattering length and the effective range are determined as

$$a(^1S_1) = -23.71 \pm 0.01 \text{fm} \quad \text{and} \quad r_0(^1S_1) = 2.70 \pm 0.09 \text{fm}, \tag{8.103}$$

$$a(^3S_1) = 5.432 \pm 0.005 \text{fm} \quad \text{and} \quad r_0(^3S_1) = 1.726 \pm 0.014 \text{fm}, \tag{8.104}$$

for the single S-state, $^{(2S+1)}L_J = {}^1S_1$, and the triplet S-state, $^3S_1$, respectively. These values imply that the deuteron is bound in $^3S_1$ channel with spin $S = 1$ and isospin $T = 0$ state, but becomes only a resonant state in $^1S_1$ channel with $S = 0$ and $T = 1$ state. The scattering length and the effective range are also determined empirically in nn and pp channels; for the spin-singlet S wave, $^1S_0$ channel, empirical scattering lengths and effective ranges are

$$a_0^{pp} = -17.3 \pm 0.4 \text{ fm}, \quad \text{and} \quad r_0(^1S_1) = 2.85 \pm 0.04 \text{fm} \tag{8.105}$$

$$a_0^{nn} = -18.7 \pm 0.7 \text{ fm}, \quad \text{and} \quad r_0(^1S_1) = 2.83 \pm 0.11 \text{fm} \tag{8.106}$$

The difference between a_0^{pp} and a_0^{nn} is an evidence of charge symmetry breaking (CSB) nuclear force, while the difference between a_0^{pn} and the average $(a_0^{pp} + a_0^{nn})/2$ originates from charge independence breaking (CIB) force. In the nn channel, the scattering length is determined to be slightly model-dependent because of the experimental difficulties to measure the nn scattering. Some other analysis report a 10% smaller value $a_0^{nn} = -16.6 \pm 1.2$ fm. The experimental determination of a precise value of the nn scattering length is still an open question. These negative scattering lengths of $^1S_0$ channel show that there are no bound states for the nn, pp and np two-nucleon systems in $J = 0$ channel.

8.2.6 Optical Model

The optical model plays an important role in the description of nuclear reactions. It is a model that provides an effective interaction which replaces the complicated many-body problems between two colliding nuclei by a much simpler problem of two objects interacting through a potential. The model provides also the wave function for the relative motion of the colliding pair. These wave functions are used as ingredients for more complicated reaction calculations. In this context, the optical model provides a useful basis in analogy to the bases for nuclear structure calculations by shell models or collective models.

It is almost impossible to obtain the exact solution for many-body scattering problems. We divide the full space into the active space P and the non-active space Q. The scattering problem is solved in the P space, while Q space is renormalized in the effective Hamiltonian for the P space (see Fig. 8.10). $P + Q$ covers the full space so that $P + Q = 1$. The model wave function is given by

$$\Psi_{\text{model}} = P\Psi, \tag{8.107}$$

where Ψ is the exact wave function. The truncated space $Q\Psi$ gives rise to the effective Hamiltonian,

$$H_{\text{eff}} = PHP + PHQ\frac{1}{E - QHQ}QHP, \tag{8.108}$$

where E is the solution of a Schrödinger equation of P space

$$(H_{\text{eff}} - E)P\Psi = 0. \tag{8.109}$$

This is called the "Feshbach" projection method for the effective model space. More detailed discussions on the Feshbach projection method is given in Sect. 3.6.1 in Chap. 3. The second term of Eq. (8.108) is a correction due to the truncation of the model space. In real calculations, although we treat explicitly only the transitions in the P space, the Hamiltonian can couple between P and Q spaces, that is, the structure of the second term of Eq. (8.108) reflects the coupling to the Q space. If the Q space has open channels such as continuum and resonance states, the coupling of P and Q space results in a loss of flux from the P space to the Q space. Then, the effective Hamiltonian h_{eff} is absorptive and becomes complex. Mathematically,

Fig. 8.10 Feshbach projection method and optical potential. The P space is the active space, while the Q space is the non-active space. The coupling between the two spaces is renormalized in the P space effectively

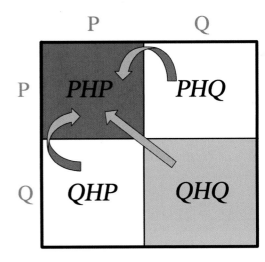

this property is expressed by the Cauchy theorem with an infinitesimal value $i\varepsilon$ in the denominator of $1/(E - QHQ)$ as

$$\lim_{\varepsilon \to 0} \frac{1}{E - QHQ + i\varepsilon} = \mathcal{P}\frac{1}{E - QHQ} - i\pi\delta(E - QHQ), \qquad (8.110)$$

where $\mathcal{P}$ denotes the Cauchy principal value. The infinitesmal value $i\varepsilon$ does not appear if all channels of Q are closed states, and consequently the effective Hamiltonian becomes real. We adopt the Hamiltonian $H = H_P + V$, where H_P acts on only the P space and V is a residual interaction acting both in P and Q spaces. Suppose the basis states in the P space is defined by the eigenstates of the Hamiltonian H_P, the effective Hamiltonian can be written as

$$H_{eff} = PH_PP + V_{eff}, \qquad (8.111)$$

where the effective interaction V_{eff} is given by

$$V_{\text{eff}} = PVP + PVQ\frac{1}{E - QHQ + i\varepsilon}QVP \equiv PVP + \Delta V(E). \qquad (8.112)$$

The optical potential V_{eff} takes into account the $\Delta V(E)$ part, the so-called dynamical polarization potential. The V_{eff} has the following characteristics;

- complex with an absorptive imaginary part describing the loss of flux for the model space into the open channels in the Q space,
- non-local (or momentum dependence) because of couplings to the eliminated channels,
- energy dependent because of couplings to the eliminated channels,
- dependent on the adopted P space.

8.2.7 Optical Potentials

We now introduce the so-called "optical potential", which is an effective potential including the effect of Q space and identical to V_{eff} in Eq. (8.112) in the above formal derivation. The optical potential can be derived in principle starting from realistic NN interactions. In practice, an optical potential is often determined in a phenomenological way to fit experimental cross sections. The terms, "optical potential" and "optical model" in nuclear physics, are due to the analogous use of a complex index of refraction in the analysis of the transmission of electromagnetic waves through matter.

The optical potential has a central and a spin-orbit potential given by

$$V_{\text{opt}} = V_C(r) + V_{ls}(r)\mathbf{l}\cdot\mathbf{s} \tag{8.113}$$

where $V_C(r)$ and $V_{ls}(r)$ are central and spin-orbit potentials, respectively. The optical potential V_{opt} is local and energy-independent. A standard parameterization of $V_C(r)$ is complex,

$$V_C(r) = U(r) + iW(r) \tag{8.114}$$

where $U(r)$ is the real part of central potential and has a radial dependence,

$$U(r) = U_0 f(r, R_U, a_U), \quad f(r, R, a) = \frac{1}{1 + exp((r - R)/a)}, \tag{8.115}$$

where U_0 is the potential depth ($U_0 < 0$), and R_U and a_U are the radial and surface diffuseness parameters, respectively. Due to the short range of nuclear forces, the optical potential is expected to follow the nuclear density; it is common to use the Woods-Saxon form for the real part of the optical potential $U(r)$. The interior part of $U(r)$ is flat and attractive, and rises quickly and monotonically to zero in the surface region. For the imaginary absorptive part of the potential, some absorptive processes are expected to follow the nuclear density, and are modeled via a volume term that is usually taken as a Woods Saxon form, while other absorptive effects, like collective excitations and nuclear breakup are expected to be concentrated in the nuclear surface, and are modeled with a surface term, usually the derivative of a Woods Saxon form. Thus, the imaginary part of the central potential $W(r)$ is taken in three different ways, to be "volume" or "surface" type, or a mixture of the two types. The functional form of volume type is the same as (8.115),

$$W(r) = W_V f(r, R_W, a_W), \tag{8.116}$$

where W_V, R_W and a_W are the potential depth, the radial and surface diffuseness parameters, respectively, and they are usually taken different values to those for the real part of potential, U_0, R_U and a_U. The surface type is mostly taken to be proportional to the derivative of Woods-Saxon form,

$$W_s(r) = 4a_W W_s \frac{df(r, R_W, a_W)}{dr}. \tag{8.117}$$

The spin-orbit potential is also introduced in the optical potential (8.113) as a surface type functional form,

$$V_{ls}(r) = V_{ls} r_{ls}^2 \frac{1}{r} \frac{df(r, R_{ls}, a_{ls})}{dr}, \tag{8.118}$$

where V_{ls}, R_{ls} and a_{ls} are the depth, the radius and the surface diffuseness parameters of the spin-orbit potential, respectively. The Coulomb potential should be added for the scattering with charged object, i.e., protons, deuterons and heavy-ions. Calculated elastic cross sections and reaction cross sections by the optical model are given in Fig. 8.14 and Table 8.2.

The damping effect of an imaginary potential is seen directly from the time-dependent Schrödinger equation,

$$i\hbar \frac{\partial \phi(\mathbf{r}, t)}{\partial t} = \{\frac{p^2}{2m} + U(\mathbf{r}) + iW(\mathbf{r})\}\phi(\mathbf{r}, t), \tag{8.119}$$

and its complex conjugate,

$$-i\hbar \frac{\partial \phi^*(\mathbf{r}, t)}{\partial t} = \phi^*(\mathbf{r}, t)\{\frac{p^2}{2m} + U(\mathbf{r}) - iW(\mathbf{r})\} \tag{8.120}$$

Multiplying ϕ^* on the left of (8.119), we have

$$i\hbar\phi^*(\mathbf{r}, t) \frac{\partial \phi(\mathbf{r}, t)}{\partial t} = -\phi^*(\mathbf{r}, t)\frac{\hbar^2}{2m}\nabla^2\phi(\mathbf{r}, t) + U(\mathbf{r})\rho(\mathbf{r}, t) + iW(\mathbf{r})\rho(\mathbf{r}, t), \tag{8.121}$$

where $\rho = \phi^*\phi$. Similarly, multiplying ϕ from the right of (8.120), we obtain

$$-i\hbar \frac{\partial \phi^*(\mathbf{r}, t)}{\partial t}\phi(\mathbf{r}, t) = -\frac{\hbar^2}{2m}\left(\nabla^2\phi^*(\mathbf{r}, t)\right)\phi(\mathbf{r}, t) + U(\mathbf{r})\rho(\mathbf{r}, t) - iW(\mathbf{r})\rho(\mathbf{r}, t). \tag{8.122}$$

Subtracting Eq. (8.122) from Eq. (8.121), the continuity equation becomes

$$\nabla \cdot \mathbf{j}(\mathbf{r}, t) + \frac{\partial \rho(\mathbf{r}, t)}{\partial t} = \frac{2}{\hbar}W(\mathbf{r})\rho(\mathbf{r}, t), \tag{8.123}$$

where $\mathbf{j}$ is the current density defined in Eq. (8.54) and its divergence is given by,

$$\nabla \cdot \mathbf{j} = \frac{\hbar}{2im}(\phi^*\nabla^2\phi - \phi\nabla^2\phi^*).$$

From Eq. (8.123), it is seen that the particle is absorbed by a rate per unit time

$$- \frac{2}{\hbar} W(\mathbf{r}) \quad (W(\mathbf{r}) < 0), \tag{8.124}$$

which is expressed by the velocity v and the mean free path λ of the particle as

$$\frac{v}{\lambda} = -\frac{2}{\hbar} W(\mathbf{r}). \tag{8.125}$$

Equation (8.125) gives the mean free path,

$$\lambda = -\frac{\hbar v}{2 W(\mathbf{r})}. \tag{8.126}$$

By definition, the mean free path is inversely proportional to the absorption rate.

Instead of using a Woods-Saxon potential, the radial dependence of the real part U of the optical potential can be obtained by a folding potential model for the optical potential. The folding potential is defined as

$$U(\mathbf{r}) = \int \int \rho_A(\mathbf{r}_1)\rho_a(\mathbf{r}_2) V(\mathbf{r}_{12}) d\mathbf{r}_1 d\mathbf{r}_2 + \text{exchange terms}, \tag{8.127}$$

where ρ_A and ρ_a are the mass densities of A and a nuclei, respectively, and the geometry of coordinates is $\mathbf{r} = \mathbf{r}_1 + \mathbf{r}_{12} - \mathbf{r}_2$ (see Fig. 8.11b). The $V(\mathbf{r}_{12})$ is the two-body nucleon-nucleon interaction. The potential (8.127) is called "the double-folding potential" since the integrations run over two densities. If we assume the projectile is a point-like object, $\rho_a = \delta(\mathbf{r})$, for example, in the case of a proton or α particle, Eq. (8.127) reduces to the form (Fig. 8.11a)

$$U(\mathbf{r}) = \int \rho_A(\mathbf{r}_1) V(\mathbf{r} - \mathbf{r}_1) d\mathbf{r}_1. \tag{8.128}$$

This is called "the single-folding potential". There is an attempt to obtain the global optical potential from Skyrme and RMF EDFs by using the single-folding potential approach of nucleon-nucleus scattering [3]. There is also an attempt to obtain the global optical potential for nucleus-nucleus scattering from the same EDFs by the double-folding model [4].

More challenging approaches based on ab initio models have been taken to obtain the optical potential from chiral EFT interactions. One is a Green's function approach combined with the coupled cluster model [6]. Another is the self-consistent Green's function (SCGF) formalism with a ChEFT interaction (The two models are introduced in Chap. 3). The calculated low-energy differential cross sections to neutron scattering data by the SCGF model are shown in Fig. 8.12 for ^{16}O at 3.286 MeV and ^{40}Ca at 3.2 MeV. Both the structure wave functions and the optical potential are obtained by using a chiral next-to-next-to-leading order $NN + 3N$ interactions, which was explicitly constructed to reproduce correct nuclear saturation properties of medium mass nuclei, including binding energies and radii (named $NNLO_{\text{sat}}$

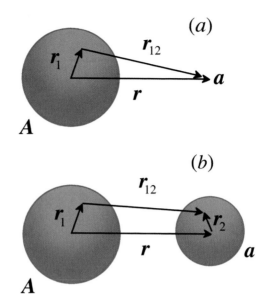

Fig. 8.11 Coordinates for **a** single-folding **b** double-folding potential

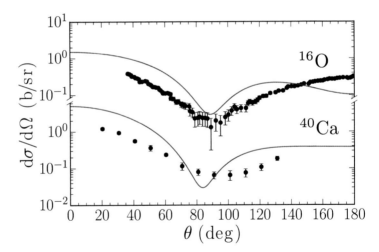

Fig. 8.12 Differential cross section for neutron elastic scattering of ^{16}O (^{40}Ca) at 3.286 (3.2) MeV of neutron energy. Calculated results are shown by red curves, while experimental data are shown by black dots with experimental uncertainties. A chiral $NN+3N$ interaction, called $NNLO_{sat}$, is used to calculate the structure wave functions and the optical potentials consistently. Figure is reprinted with permission from [5]. ©2021 from the American Physical Society

interaction). The diffraction minima are reproduced well for ^{16}O, and close to the experimental data for ^{40}Ca, confirming reasonable predictions of density distributions. However, the calculated results are somewhat overestimated, and suggest a lack of absorption that is usually faced by attempts at computing the ab initio optical potentials.

8.3 Born Approximation

8.3.1 Plane Wave Born Approximation (PWBA)

We discussed the first-order plane wave Born approximation (PWBA) for the elastic scattering in Sect. 8.2.1. In the following, we will solve the scattering amplitude (8.31) for a Yukawa potential and Coulomb potential and obtain the elastic cross section. For elastic scattering, the amplitude $f^{(1)}(\mathbf{k}', \mathbf{k})$ is a function of the momentum transfer q, which is given by

$$q = |\mathbf{k}' - \mathbf{k}| = 2k \sin \frac{\theta}{2}, \qquad (8.129)$$

where we use the energy conservation $k = k'$ (see Fig. 8.13). The amplitude (8.31) for the central potential $V(r)$ is expressed after the integrations over the angles as

$$f^{(1)}(\theta) = -\frac{m}{\hbar^2} \int_0^\infty \frac{r^2}{iqr} V(r)(e^{iqr} - e^{-iqr})dr$$
$$= -\frac{2m}{\hbar^2} \frac{1}{q} \int_0^\infty rV(r)\sin(qr)dr. \qquad (8.130)$$

As an example, we consider the scattering by a Yukawa potential,

$$V(r) = V_0 \frac{e^{-\mu r}}{\mu r}, \qquad (8.131)$$

where V_0 is the potential strength and μ is the range of the potential, typically $1/\mu \sim$ 1fm. It can be seen that V goes to zero very rapidly for $r >> 1/\mu$. For this potential, we can perform the radial integration analytically[3] and obtain

$$f^{(1)}(\theta) = -\frac{2mV_0}{\mu\hbar^2} \frac{1}{q^2 + \mu^2}. \qquad (8.133)$$

---

[3] The integration in Eq.(8.130) can be performed as

$$\int_0^\infty e^{-\mu r} \sin qr\, dr = \text{Im}[\int_0^\infty e^{-\mu r} e^{iqr} dr] = -\text{Im}[(1/(-\mu + iq)] = q/(q^2 + \mu^2). \qquad (8.132)$$

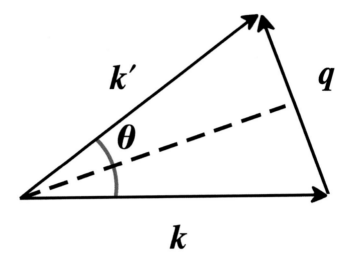

Fig. 8.13 Geometry of scattering. The momentum transfer is defined by $\mathbf{q} = \mathbf{k}' - \mathbf{k}$

Since $q^2 = 4k^2 \sin^2 \theta/2$, the differential cross section (8.58) for the scattering by a Yukawa potential is given by

$$\frac{d\sigma}{d\Omega} = \left(\frac{2m V_0}{\mu \hbar^2}\right)^2 \frac{1}{(4k^2 \sin^2 \theta/2 + \mu^2)^2}. \tag{8.134}$$

In the limit of $\mu \to 0$, the Yukawa potential can be converted to the Coulomb potential, provided the ratio V_0/μ is replaced by a factor $ZZ'e^2$, where Z and Z' stand for the proton numbers of projectile and target, respectively. In this limit, the cross section for a Yukawa potential (8.134) reads

$$\frac{d\sigma}{d\Omega} = \frac{(2m)^2 (ZZ'e^2)^2}{\hbar^4} \frac{1}{16k^4 \sin^4 \theta/2}. \tag{8.135}$$

The formula (8.135) is further rewritten to be

$$\frac{d\sigma}{d\Omega} = \frac{1}{16} \frac{(ZZ'e^2)^2}{E^2} \frac{1}{\sin^4 \theta/2}, \tag{8.136}$$

where E is the kinetic energy expressed as $E = p^2/2m = (\hbar k)^2/2m$. The formula (8.136) is nothing but "the Rutherford scattering cross section" for the Coulomb potential, which can be obtained also from classical scattering theory. The Rutherford formula has several characteristic features of Coulomb scattering;

- The angular distribution depends on the momentum transfer q only; $f(\theta)$ depends on the energy $E = \hbar k/2m$ and θ only through the combination of $q^2 = 4k^2 \sin^2 \theta/2$.
- The cross section is determined by the absolute value of the potential and does not depend on the sign, i.e., whether attractive or repulsive potentials give the same result.

- The cross section decreases proportional to E^2, i.e., the higher the incident energy, the lower the cross section.
- The differential cross section diverges in the limit $\theta \to 0$.

8.3.2 Distorted Wave Born Approximation (DWBA)

The formalism of Born approximation can be extended to inelastic channels as well as transfer channels,

$$(a + A) \to (a' + A^*) \text{ or } (b + B), \tag{8.137}$$

where a and A denote the initial projectile and target, and a' and A^* are excited states of the same nuclei in the inelastic scattering, while b and B are new nuclei after the transfer reaction. In these reactions, DWBA is adopted to obtain the cross sections. In most approaches, the corresponding Schrödinger equation is solved for the intrinsic and relative Hamiltonians as was given in Eq. (8.6) and the wave function is the product of the relative motion and the intrinsic motion (8.7),

$$\Psi_\alpha(\mathbf{r}_A, \mathbf{r}_a, \mathbf{r}_\alpha) = \chi(\mathbf{k}_\alpha, \mathbf{r}_\alpha)\Phi_\alpha(\mathbf{r}_A, \mathbf{r}_a),$$

for the entrance channel and,

$$\Psi_\beta(\mathbf{r}_B, \mathbf{r}_b, \mathbf{r}_\beta) = \chi(\mathbf{k}_\beta, \mathbf{r}_\beta)\Phi_\alpha(\mathbf{r}_B, \mathbf{r}_b),$$

for the exit channel. Here $\mathbf{k}_{\alpha(\beta)}$ is the relative momentum and $\mathbf{r}_{\alpha(\beta)}$ is the separation of the centers of mass of nuclei a and A (b and B) (see Fig. 8.4 for the reaction geometry). The interaction between the target and the projectile can be expressed as the sum of an optical potential $U(\mathbf{r}_\alpha)$ responsible for the relative motion, and the transition potential $\Delta U(\mathbf{r}_\alpha, \mathbf{r}_A, \mathbf{r}_a)$ responsible for the excitation of intrinsic states,

$$V = U + \Delta U, \tag{8.138}$$

where ΔU depends on both the relative motion degree of freedom and internal degrees of freedom. The DWBA is the most commonly used approximation for direct reactions. It takes into account the distortion of the incoming and outgoing waves caused by the nuclear optical potential U between the projectile and the target in the entrance and exit channels. The potential ΔU is considered to be weak compared with U and taken into account by the perturbation. The DWBA calculations are separated into two steps. First, the incoming and outgoing waves are calculated by solving the Schrödinger equation,

$$(\frac{p^2}{2\mu} + U)\chi(\mathbf{k}_{\alpha(\beta)}, \mathbf{r}) = E_{\alpha(\beta)}\chi(\mathbf{k}_{\alpha(\beta)}, \mathbf{r}), \tag{8.139}$$

where μ is the reduced mass. In the inelastic and transfer reactions, the transition potential ΔU induces the transition between the initial and final intrinsic states as

$$\langle \Phi_\beta | \Delta U | \Phi_\alpha \rangle. \tag{8.140}$$

If the reaction is assumed to occur in a single step, the transition amplitude (see Eq. (8.32)) can be written as

$$T_{\beta\alpha} = \int \chi^{(-)}(\mathbf{k}_{fi}, \mathbf{r}_\beta) \langle \Phi_\beta | \Delta U | \Phi_\alpha \rangle \chi^{(+)}(\mathbf{k}, \mathbf{r}_\alpha) d\mathbf{r}_\alpha d\mathbf{r}_a d\mathbf{r}_A, \tag{8.141}$$

where one can take equivalently another set of variables $d\mathbf{r}_\beta d\mathbf{r}_b d\mathbf{r}_B$ for the integrations. Since the interaction ΔU is of short range, one can take the approximation $\mathbf{r}_\alpha \sim \mathbf{r}_\beta$ if the target $A(B)$ is much heavier than the projectile $a(b)$. The wave function $\chi^{(+)}(\mathbf{k}, \mathbf{r})$ is the "distorted" wave composed of an incoming plane wave in the state α and outgoing scattered waves. Similarly, $\chi^{(-)}(\mathbf{k}_{fi}, \mathbf{r})$ is the outgoing distorted wave in the channel β. In the case of inelastic scattering, the form factor given by $\langle \Phi_\beta | \Delta U | \Phi_\alpha \rangle$ does not depend anymore on internal degrees of freedom since they have been integrated out, but they depend only on the projectile-target relative coordinates. The shape of this form factor is strongly related to the matter distribution of the involved nuclear densities which are intrinsic properties of the nuclei. The calculation of the potential can either be based on (i) microscopic nuclear densities, or on (ii) a semi-microscopic collective model such as a vibrational model or a rotational model. In the latter case, the amplitude of the excitation is governed by a parameter which is often given as a deformation length δ_ℓ for an angular momentum transfer ℓ during the inelastic excitation.

There are several ways to extend the first-order DWBA method. As was discussed in Eq. (8.50) and also shown in Fig. 8.7, the second-order Born approximation is an extended model beyond the first-order DWBA. The Coupled Channel Born Approximation (CCBA) is also an extension of the DWBA that includes higher-order processes, such as virtual excitations of the projectile or the target before or after the scattering. One has therefore to consider explicitly the coupling of excited states to the ground state through the projectile-target potential that modifies the distorted waves. These couplings are usually treated through coupled Schrödinger equations.

8.4   Applications of Direct Reaction Theory

8.4.1   Proton Elastic Scattering

An optical model analysis of proton elastic scattering cross sections on various nuclei at 30 MeV is shown in Fig. 8.14. The cross sections are plotted divided by Rutherford cross sections (8.136). The data are the differential cross sections for nuclei ranging from ^{40}Ca to ^{208}Pb. The Woods-Saxon forms (8.113)–(8.118) are adopted for the optical potentials. The optical model parameters are given in Table 8.1. The analyses include the Coulomb potential obtained by an uniformly charged sphere of the radius R_C. In fitting the experimental data, it was found that the results are not sensitive to the Coulomb radius R_C and set to be equal to $R_C = R_V$. The analyses were performed under the assumption of the same diffuseness parameter, $a = 0.7$ fm, for all nuclei and for all channels. The radius parameters, R, have a $A^{1/3}(R = rA^{1/3})$ dependence with different choices for r values; r_U=1.2 fm, r_W=1.25 fm, and r_{ls}=1.10 fm. With these geometrical parameters R and a, the potential depth parameters are adjusted to give an optimal fit to the observed cross sections. The resulting parameters are given in Table 8.1. We can see that the potential depth U_0 and the spin-orbit coupling potential V_{ls} are rather constant for all nuclei, i.e., $U_0 \sim -50$ MeV and $V_{l \cdot s} \sim 20$ MeV. The features justify the mean-field description of elastic scattering by optical potentials. This idea is extended further to inelastic and particle-transfer reactions.

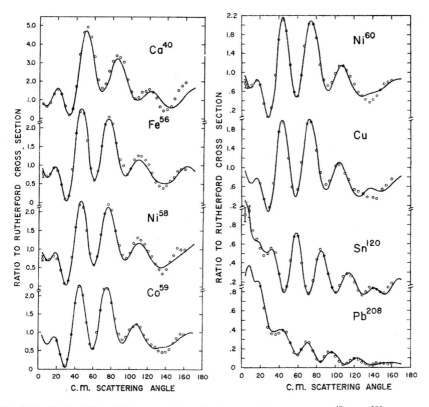

Fig. 8.14 Elastic differential cross sections of various nuclei ranging from ^{40}Ca to ^{208}Pb with 30 MeV proton projectile. The differential cross sections are plotted to the ratio of the Rutherford cross sections (8.136). The optical model parameters are listed in Table 8.1. Figure is reprinted with permission from [7]. ©2021 from the American Physical Society

Table 8.1 Optical model parameters (in MeV) for 30 MeV proton elastic scatterings. The same diffuseness parameter $a = 0.7$ fm is adopted for all nuclei and for all channels. The radius parameters, R, have a $A^{1/3}$ ($R = rA^{1/3}$) dependence with different choices for r values; r_U=1.2 fm, r_W=1.25 fm, and r_{ls}=1.10 fm and are common in all nuclei. These parameters are used to calculate the cross sections in Fig. 8.14, and the reaction cross sections in Table 8.2. Numerical values from [7]

	^{40}Ca	^{56}Fe	^{58}Ni	^{59}Co	^{60}Ni	Cu	^{120}Sn	^{208}Pb
$-U_0$	46.1	46.4	47.0	47.5	47.6	47.7	51.1	53.4
$-W_V$	0.4	2.7	3.4	2.8	2.8	1.8	1.2	4.0
W_S	5.96	5.2	4.4	5.7	5.5	6.1	8.7	7.6
V_{ls}	20.1	19.5	14.8	19.5	18.2	19.5	20.2	17.2

Table 8.2 Reaction cross sections (8.87) for 30 MeV protons. The cross sections are given in unit of mb (10^{-27} cm^2). Numerical values from [7]

	^{40}Ca	^{56}Fe	^{58}Ni	^{59}Co	^{60}Ni	Cu	^{120}Sn	^{208}Pb
σ_R (exp.)	915±38	1140±43	1038±32	1169±39	1053±51	1124±40	1638 ±68	1865± 98
σ_R (calc.)	941	1137	1117	1162	1174	1215	1604	1838

The oscillations in the angular distribution of the cross section are characteristic of a "Fraunhofer diffraction" pattern, similar to the scattering of light by an opaque disk. The Fraunhofer diffraction was discussed already in the case of elastic electron scattering in Sect. 4.3.1 in Chap. 4. In the nuclear case, one considers a black body with the diameter $D = 2R$ (R is the radius of nucleus). For the impact parameter, b, smaller than R, everything is absorbed, and for $b > R$ nothing happens. Elastic scattering occurs at the two edges of the black body at $b = \pm R$. The coherence of condition between the waves is given by

$$2R \sin \theta = m\lambda, \quad m = 1, 2, \cdots . \tag{8.142}$$

The angular distance $\Delta\theta$ between the diffraction maxima follows closely the expression,

$$\Delta\theta = \frac{\pi\hbar}{pR}, \tag{8.143}$$

where we use the de Broglie relation $\lambda = h/p$. The results in Fig. 8.14 follow qualitatively the prediction (8.143), i.e., the angle distance between the diffraction maxima is smaller for heavy nuclei and larger for lighter nuclei. The observed angle distances are $\Delta\theta \sim 35°$ for ^{48}Ca and $\sim 25°$ for ^{208}Pb, which are consistent with Eq. (8.143), although the proportionality is somewhat masked by the strong absorptive central potentials iW in Eq. (8.114). The diffraction angle distance is also smaller for higher energy reaction. In this analysis, both the volume and surface type imaginary potentials are introduced. The use of pure surface type ($W_V = 0$) gives calculated cross sections with refitted parameters almost as good as those shown in Fig. 8.14.

On the other hand, pure volume absorption ($W_S = 0$) leads to somewhat poorer fits. This suggests that the surface absorption might be more plausible at this energy to describe the open channel effects in the Q space than the volume-type one.[4]

8.4.2 Inelastic Scattering

Nuclear inelastic scattering has provided a wealth of information on nuclear structure and interaction potentials. A wide variety of inelastic scattering angular distributions have been observed. They are classified by spin transfer Δs and angular momentum transfer Δl. The angular distributions have characteristic features depending on the angular momentum ℓ and parity π, which gives the important information on the structure of the populated states. It was already suggested in 1950s that inelastic scattering for nuclei would excite collective vibrational and rotational states; especially, a significant correlation was found between electric multipole transitions $B(El)$ and the cross sections for inelastic scatterings. For charge exchange reactions such as (p, n) reactions, the Gamow-Teller (GT) giant resonances ($\Delta s = 1$, $\Delta l = 0$) were also measured at the forward angle of cross sections.

We consider the inelastic scattering of a structureless projectile (a) from a target (A) using the DW approximation. The transition amplitude can be formulated as

$$T_{\alpha\alpha'}^{DW}(\mathbf{k}_{\alpha'}, \mathbf{k}_\alpha) = \langle \psi_A' | P(\mathbf{k}_{\alpha'}, \mathbf{k}_\alpha; x_A) | \psi_A \rangle, \tag{8.144}$$

where P is the operator

$$P(\mathbf{k}_{\alpha'}, \mathbf{k}_\alpha; x_A) = \int d\mathbf{r}_\alpha \chi^{(-)*}(\mathbf{k}_{\alpha'}, \mathbf{r}_\alpha) \Delta U \chi^{(+)}(\mathbf{k}_\alpha, \mathbf{r}_\alpha), \tag{8.145}$$

where x_A is the intrinsic coordinate of a nucleus A and the interaction between A and a is considered to be local,

$$\Delta U = \sum_{i=1}^{A} V(\mathbf{r}_\alpha - \mathbf{r}_i). \tag{8.146}$$

We examine the operator P in the plane wave (PW) approximation $\chi(\mathbf{k}_\alpha, \mathbf{r}_\alpha) = \exp(i\mathbf{k} \cdot \mathbf{r}_\alpha)$ for an illustration. Within the PW assumption, the P operator is simply a Fourier transform of the interaction ΔU;

$$P^{PW}(\mathbf{q}; x_A) = \int e^{i\mathbf{q}\cdot\mathbf{r}_\alpha} \Delta U d\mathbf{r}_\alpha \quad \mathbf{q} = \mathbf{k}_\alpha - \mathbf{k}_{\alpha'}. \tag{8.147}$$

---

[4] There are attempts to obtain the global optical potential of nucleon-nucleus scattering, which can be applied for different nuclei with different projectile energies (see for example [8]). The parameters of interaction strength U_0, W_V, W_S and V_{ls} are the energy dependent and the potential parameters r and a depend on the nuclear mass A for the global optical potential.

For a contact interaction (CI), $\Delta U = V_0 \sum_{i=1}^{A} \delta(\mathbf{r}_i - \mathbf{r}_\alpha)$, Eq. (8.147) becomes

$$P_{CI}^{PW}(\mathbf{q}; x_A) = V_0 \sum_{i=1}^{A} \int \delta(\mathbf{r}_i - \mathbf{r}_\alpha) e^{i\mathbf{q}\cdot\mathbf{r}_\alpha} d\mathbf{r}_\alpha$$

$$= V_0 \sum_{i=1}^{A} e^{i\mathbf{q}\mathbf{r}_i} = V_0 4\pi \sum_{i,(l,m)} i^l j_l(qr_i) Y_{lm}(\hat{r}_i) Y_{lm}^*(\hat{q}). \quad (8.148)$$

In Eq. (8.148), the spherical bessel function is expressed approximately as $j_l(qr) \rightarrow (qr)^l/(2l+1)$ in the long wave length limit $qr \ll 1$. Then, the operator P_{CI}^{PW} is proportional to the multipole operator for nuclear excitation as

$$P_{CI}^{PW} \propto \sum_i r_i^l Y_{lm}(\hat{r}_i), \quad (8.149)$$

where $\mathbf{r}_i$ is the internal coordinate of the target nucleus A. The operator form (8.149) is similar to the electric ℓth-pole transition operator (7.145) so that the cross section gives the information of enhancement or hindrance of transition strength to the excited state similar to the electric transition probability. Since the nuclear interaction is short-range, the contact interaction adopted in Eq. (8.148) is a reasonable approximation in many cases. The inelastic scattering will thus provide an important information on nuclear excited states through the operator P, concerning about the collectivity of vibration and rotation.

8.4.2.1 Bohr-Mottelson Model for Inelastic Transition Potential

The collective model proposed by Bohr and Mottelson is a powerful tool to describe vibrational and rotational motions in nuclei. To construct the transition densities and transition potentials, we start from the spherical density distribution $\rho(r)$ and introduce multipole deformation parameters $\alpha_{\lambda\mu}$. These parameters are considered dynamical variables $\alpha_{\lambda\mu}$, which excite vibrational and rotational states. The two limiting cases are considered, either harmonic oscillations around a spherical shape, or a static stable deformation in which the deformation parameters have non-vanishing values $\langle \alpha_{\lambda\mu} \rangle$ for the expectation value of an intrinsic deformed ground state.

Suppose that the density $\rho(r)$ has a distribution which is flat in the interior with a sharp cutoff at $r = R_0$ like a hard sphere. The deformation is introduced by making the surface at $r = R_0$ to depend on the direction (θ, ϕ) of the coordinate vector $\mathbf{r}$:

$$R(\theta, \phi) = R_0(1 + \sum_{\lambda\mu} \alpha_{\lambda\mu} Y_{\lambda\mu}^*(\theta, \phi)) \equiv R_0 + \delta R(\theta, \phi). \quad (8.150)$$

For the Woods-Saxon form of the potential, $f(r) = 1/(1 + \exp(r - R)/a)$, the parameter R is considered as a dynamical parameter and the density can be expanded in the first order

$$\rho(r, R(\theta, \phi)) = \rho(r, R_0) + R_0 \frac{\partial \rho}{\partial R_0} \sum_{\lambda\mu} \alpha_{\lambda\mu} Y^*_{\lambda\mu}(\theta, \phi) \equiv \rho(r, R_0) + \sum_{\lambda\mu} \delta\rho_{\lambda\mu} Y^*_{\lambda\mu}(\theta, \phi),$$
(8.151)

where the second term is called "the transition density" to induce the excitation to the λth-multipole state. Although higher-order terms, such as $|\alpha_{\lambda\mu}|^2$, are required to ensure volume conservation, we will keep only first-order terms to simplify the discussion. The potential is also expanded in a similar manner to the first order,

$$U(r, R(\theta, \phi)) = U(r, R_0) + R_0 \frac{\partial U}{\partial R_0} \sum_{\lambda\mu} \alpha_{\lambda\mu} Y^*_{\lambda\mu}(\theta, \phi),$$
(8.152)

where the first and second terms induce the elastic and inelastic scatterings, respectively. The second term corresponds to the transition potential ΔU for inelastic scattering in Eq. (8.138),

$$\Delta U = R_0 \frac{\partial U}{\partial R_0} \sum_{\lambda\mu} \alpha_{\lambda\mu} Y^*_{\lambda\mu}(\theta, \phi).$$
(8.153)

In the collective vibrational model, $\alpha_{\lambda\mu}$ represents the phonon excitation operator of the λth-multipole state. For deformed nuclei, the expectation value of $\alpha_{\lambda\mu=0}$ gives the deformation parameter β_λ;

$$\langle \alpha_{\lambda\mu=0} \rangle = \beta_\lambda,$$
(8.154)

for the axial symmetric deformation (see Chap. 7 for applications of the Bohr-Mottelson collective model).

8.4.2.2 Proton and Neutron Transition Matrix Elements M_p and M_n

Proton-induced inelastic scattering has the advantage that the proton can be considered as structureless in most energy domains at which direct reaction measurements are performed, since the first-excited state of the nucleon, the Δ resonance, lies at an excitation energy of 290 MeV. Proton-induced reactions can therefore be expected to lead to more reliable structure information compared to heavy-ion induced inelastic scattering measurements. Proton inelastic scattering, being a hadronic probe, can excite both protons and neutrons. It is therefore seen as a complementary probe to electromagnetic excitations of nuclei. The combination of both measurements brings unique information on the *isoscalar* (protons and neutrons play the same role) and also *isovector* (protons and neutrons play a different role) nature of excitations.

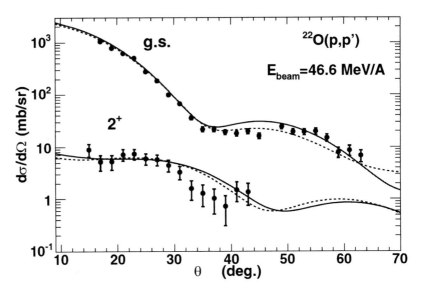

Fig. 8.15 Differential cross sections of proton elastic and inelastic scattering on ^{22}O with the beam energy at 46.6 MeV/nucleon measured at GANIL. DWBA using the phenomenological KD global optical potential (solid lines) and folding model (dashed lines) calculations are shown. Figure is reprinted with permission from [9]. ©2021 from the American Physical Society

The transition from one state to another can be decomposed into a proton excitation and a neutron excitation and is driven by the so-called transition matrix elements M_n and M_p. They can be defined by using Eq. (8.149) as

$$M_{n(p)} = \langle \Psi_f | O_{lm}^{n(p)} | \Psi_i \rangle = \sum_{i=1}^{N(Z)} \langle \Psi_f | r_i^l Y_{lm}(\hat{r}_i) | \Psi_i \rangle = \int \delta\rho_{n(p)}^{(fi)}(r) r^l Y_{lm}(\hat{r}) d\mathbf{r},$$
(8.155)

where

$$\delta\rho_{n(p)}^{(fi)} \equiv \langle \Psi_f | \sum_{i=1}^{N(Z)} \delta(\mathbf{r} - \mathbf{r}_i) | \Psi_i \rangle$$
(8.156)

is the neutron (proton) transition density between the state Ψ_i (usually the ground state) and the final excited state Ψ_f and ℓ is the multipolarity of the transition.[5] The transition matrix elements $M_{n(p)}$ contain only nuclear structure information and are independent of the probe used to excite the nucleus.

Elastic and 2_1^+ proton-induced inelastic scattering angular distributions for ^{22}O at 46.6 A MeV are shown in Fig. 8.15. DWBA optical potential analyses are performed and the results are given by solid lines with the phenomenological nucleon-nucleus global potential parameterization elaborated by Koning and Delaroche (KD) [8].

---

[5] The identity $\sum_{i=1} r_i^l Y_{lm}(\hat{r}_i) = \int d\mathbf{r} \sum_{i=1} \delta(\mathbf{r} - \mathbf{r}_i) r^l Y_{lm}(\hat{r})$ is used to derive Eq. (8.155).

In order to compare densities calculated by HFB and QRPA models to the proton scattering data, the single-folding model is also used to generate the optical potentials from these microscopic densities (dashed lines). The elastic angular distribution is well described, even at large angles by the two theoretical models.

To study microscopic model predictions on the wave functions of the excited state, the transition matrix elements M_n and M_p in Eq. (8.155) may give the important information. The neutron (proton) transition amplitude is proportional to the numbers of neutrons (protons) involved in the excitation process. In a very simple picture, they are the numbers of protons and neutrons in the nucleus. This conjecture gives rise to the ratio $(M_n/M_p)/(N/Z) = 1$.

The proton inelastic scattering gives the information on combination of M_n and M_p because the hadronic probe excites both protons and neutrons. To extract the empirical extraction of M_n/M_p ratio in 2_1^+, we need not only proton scattering data, but also other data with a heavy projectile with high Z, whose reaction will be dominated by Coulomb field and the proton transition matrix M_p. To this end, the inelastic scattering of ^{22}O from ^{197}Au at 50 MeV/nucleon, performed at MSU, was analyzed to extract the electric quadrupole transition strength $B(E2)$ value, and $B(E2)_{exp} = 21 \pm 8$ e^2fm^4 was obtained. Eventually, the experimental value of the M_n/M_p ratio for the 2_1^+ state is obtained from the combination of the heavy ion and the (p, p') measurements to be $M_n/M_p = 2.5 \pm 1.0$, or, $(M_n/M_p)/(N/Z) = 1.4 \pm 0.5$. The observed value is significantly different from unity and shows the importance of neutron contributions to the transition matrix element. Thus, the 2_1^+ state in ^{22}O is considered to be dominated by neutron excitations from the ground state. A similar enhancement of M_n is also observed in the inelastic cross sections to 2_1^+ state in ^{20}O.

8.5 Nucleon Transfer Reactions

8.5.1 One Nucleon Transfer Reactions and Spectroscopic Factors

The transfer of one nucleon *from* a nucleus (stripping) or *to* a nucleus (pickup) is a powerful tool to determine the nature of the populated states. More quantitatively, the final states after pickup and stripping from the same nucleus and the corresponding population cross sections give access, in principle, to single-particle energies. Effective single-particle energies are the backbone of many nuclear models. Two-nucleon transfer has been extensively used to study two-nucleon correlations inside the nucleus. Transfer cross sections are often analyzed through the DWBA and coupled-channel formalisms.

The independent particle model (IPM) of nuclear structure assumes nucleons lying on single-particle orbitals with no correlation among them. The shell model description of nuclei is based on single-particle configurations on top of which nucleon correlations are built. Mean field models also lead to the definition of single-particle orbitals from which beyond-the-mean-field long-range correlations can be built. The energies of these single-particle states are not observed directly by experiments since real nuclei are correlated systems by nature. Empirical information of these single-particle states nevertheless can be extracted from data if one relies entirely on the definition by M. Baranger [10]:

$$e_p = \frac{\sum_k S_k^{p+}(E_k - E_0) + S_k^{p-}(E_0 - E_k)}{\sum_k S_k^{p+} + S_k^{p-}}, \tag{8.157}$$

where S_k^{p+} (S_k^{p-}) are the spectroscopic factors (squares of spectroscopic amplitudes) for the population of a final state k following the creation (annihilation) of a nucleon with quantum numbers $p = \{n\ell j\}$, defined as

$$S_k^{p+} = |\langle \Psi_k^{A+1} | a_p^\dagger | \Psi_0^A \rangle|^2, \; S_k^{p-} = |\langle \Psi_k^{A-1} | a_p | \Psi_0^A \rangle|^2, \tag{8.158}$$

where $|\Psi_k^{A(\pm 1)}\rangle$ is the wave function of the nucleus with $A(\pm 1)$ nucleons in the state k. In the case of the IPM spectroscopic factors are either 1 or 0, and the single-particle energies coincide with single nucleon excitations (either particle or hole states) from nucleon stripping or pickup reactions. In the case of physical (i.e., correlated) nuclei, one-nucleon stripping or pickup from a given state leads to the population of several states in the residual nucleus: the spectroscopic strength of a given orbital is spread over several final states, as illustrated in Fig. 8.16.

If one observes experimentally all E_k and S_k^p, the single-particle energies would then be extracted from Eq. (8.157). Today, direct reactions are used to extract spectroscopic factors from cross sections to individual final states. In case the theoretical description of the reaction mechanism has been sufficiently benchmarked and is under control, one can aim at testing our understanding of the nuclear structure from such measurements.

8.5.1.1 PWBA and DWBA Descriptions of Transfer Reactions

In the traditional treatment of transfer reactions, the target or projectile are most often described as two-body systems composed of a core and the nucleon to be transferred as illustrated in Fig. 8.17. In a nucleon transfer reaction also, the potential V between the projectile and the target is decomposed into two parts as was given in Eq. (8.138) for inelastic scattering; the optical potential $U(\mathbf{R})$ and the transition potential $\Delta U(\mathbf{R}, \mathbf{r})$, that will be responsible for the nucleon transfer itself.

This separation form of the potential $V = U + \Delta U$ is exactly the same as for inelastic scattering as was shown in Eq. (8.138). In inelastic scattering, the potential ΔU induces the excitation of the intrinsic systems (A) and/or (a), while in the transfer reaction, the potential ΔU plays the role to transfer one nucleon from the initial state to the final state. In the DWBA approach, the transfer process is seen as a one step "perturbation" of the distorted trajectory. The homogeneous and inhomogeneous

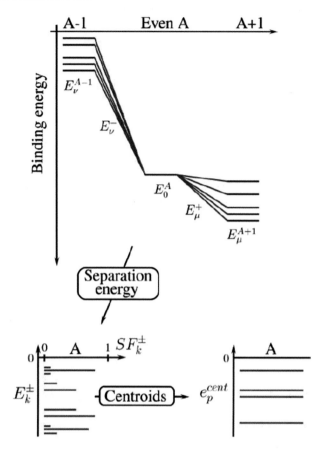

Fig. 8.16 (Top) Total binding energies and associated one-nucleon stripping and pickup energies from the ground state of the initial nucleus. (Bottom) Single-particle energies (right) are obtained from the centroid of the measured spectral function (left) following the relation (8.157). Reprinted with permission from [11]. ©2021 from the American Physical Society

equations are

$$(T + U - E)\chi = 0, \qquad (8.159)$$

$$(T + U - E)\phi = \Delta U \phi. \qquad (8.160)$$

The transition matrix element for transfer is written as

$$T_{\beta\alpha} = \langle \phi_\beta \Phi_\beta | \Delta U | \phi_\alpha \Phi_\alpha \rangle, \qquad (8.161)$$

where, following the notation of Sect. 8.2.1, $\Phi_{\alpha,\beta}$ are the intrinsic wave functions of the entrance and exit channels α, β.

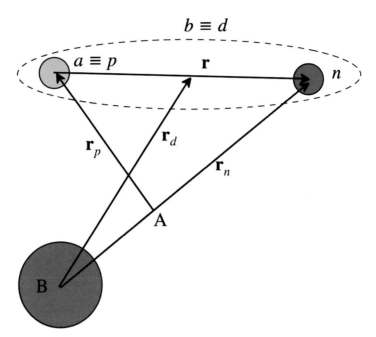

Fig. 8.17 Illustration of a transfer reaction during which one nucleon is transferred from the projectile A to the target a. The exit channel β is then composed of the target-like residue b and B, the core of the projectile A: $A + a \rightarrow B + b$. In particular, one neutron pick-up reaction, $A(p, d)B$, is illustrated in the figure. $\mathbf{r}_p$ is the relative coordinate between the proton and the center of mass of target A, $\mathbf{r}_d$ is the coordinate between the core B and the deuteron, and $\mathbf{r}$ and $\mathbf{r}_n$ are the distances between n and p, and between B and n, respectively

To be explicit, we now treat the specific case of the one neutron pickup reaction from a target nucleus A by an incident proton through the reaction (p, d) to populate a state i in the residual nucleus B. In the representation of Fig. 8.17, the target a is a proton and the residual nucleus b is a deuteron. If a target is heavy enough, the transfer potential ΔU can be approximated by the potential between the proton and the neutron to be picked up with $\mathbf{r} = \mathbf{r}_n - \mathbf{r}_p$ being the relative coordinates between the picked-up neutron and the proton.

In the plane wave Born approximation, when the incoming and outgoing waves are treated as unperturbed by the potential (i.e., assuming $U = 0$ in Eq. (8.160)), the initial and final wave functions are

$$|\Psi_\alpha\rangle = e^{i\mathbf{k}_p \cdot \mathbf{r}_p} \Phi_A, \tag{8.162}$$

$$|\Psi_\beta\rangle = e^{i\mathbf{k}_d \cdot \mathbf{r}_d} \Phi_B \Phi_d, \tag{8.163}$$

where $\mathbf{r}_p$ is the relative coordinate between the proton and the center of mass of target A, and $\mathbf{r}_d$ is the coordinate between the core B and the deuteron, respectively. In the case of a pure single particle neutron state with quantum numbers $n\ell j$, the wave function of the final state can be written as $\Phi_A = \Phi_B \phi_{n\ell j}(\mathbf{r}_n)$ and the transition matrix element of Eq. (8.141) is then given by

$$T_{\beta\alpha}^{n\ell j} = \int e^{-i\mathbf{k}_d \cdot \mathbf{r}_d} \Phi_d^*(\mathbf{r}) \Phi_B^* \Delta U(\mathbf{r}) e^{i\mathbf{k}_p \cdot \mathbf{r}_p} \phi_{n\ell j} \Phi_B d\mathbf{r}_B d\mathbf{r}_n d\mathbf{r}_p,$$

$$= \int e^{-i\mathbf{k}_d \cdot \mathbf{r}_d} \Phi_d^*(\mathbf{r}) \Delta U(\mathbf{r}) e^{i\mathbf{k}_p \cdot \mathbf{r}_p} \phi_{n\ell j} d\mathbf{r}_n d\mathbf{r}_p. \tag{8.164}$$

Considering the relations

$$\mathbf{r}_d = \frac{1}{2}\mathbf{r}_n - \mathbf{r} \quad \text{and} \quad \mathbf{r}_p = \frac{A-1}{A}\mathbf{r}_n - \mathbf{r}, \tag{8.165}$$

one gets

$$\mathbf{k}_p \cdot \mathbf{r}_p - \mathbf{k}_d \cdot \mathbf{r}_d = -(\mathbf{k}_d - \frac{A-1}{A}\mathbf{k}_p) \cdot \mathbf{r}_n - (\mathbf{k}_p - \mathbf{k}_d/2) \cdot \mathbf{r}$$

$$\equiv -\mathbf{q} \cdot \mathbf{r}_n - \mathbf{K} \cdot \mathbf{r}. \tag{8.166}$$

where $\mathbf{q} \equiv \mathbf{k}_d - \frac{A-1}{A}\mathbf{k}_p$ is the momentum carried by the picked-up neutron, and $\mathbf{K} = \mathbf{k}_p - \mathbf{k}_d/2$. The transition matrix element of Eq. (8.164) can then be formulated as

$$T_{\beta\alpha}^{n\ell j} = \int e^{-i\mathbf{K} \cdot \mathbf{r}} \Phi_d^*(\mathbf{r}) \Delta U(\mathbf{r}) d\mathbf{r} \times [\int_R^\infty e^{-i\mathbf{q} \cdot \mathbf{r}_n} \phi_{n\ell j} d\mathbf{r}_n]. \tag{8.167}$$

Equation (8.167) can be intuitively interpreted as the product of two Fourier transforms: the first one of the deuteron wave function and the potential, and the second one of the picked-up neutron wave function. Under the PWBA, it appears explicitly that the transition matrix element is a product of a reaction term and a structure term. Note that the integration limit of the second integral in Eq. (8.167) has been set *by hand* to the nuclear radius R since the PWBA does not take into account any absorption of the projectile during the reaction, i.e., the imaginary part of the optical potential W is set to 0. In the more realistic DWBA approximation, the absorption is taken care of by the imaginary part of the optical potential W, neglected in the previous derivation. By assuming the initial intrinsic wave function Φ_A as a single-particle state with quantum numbers $n\ell j$ weighted by a certain spectroscopic amplitude A_{nlj}

$$|\Phi_A\rangle = \sum_\beta \sum_{n\ell j} A_{nlj}^\beta |\phi_{n\ell j} \Phi_B^\beta\rangle, \tag{8.168}$$

with

$$\sum_\beta |A_{nlj}^\beta|^2 = S_k^{nlj+}, \tag{8.169}$$

following the definitions of Eq. (8.158). The transition matrix element for the one neutron pick-up is then given by

Table 8.3 Occupation probabilities of shell model orbits in ^{39}Ca. The values extracted from (p, d) cross sections are analysed under the DWBA and the zero-range approximation. The occupancies obtained from the data can be considered in such way that the single-particle strength for a given orbit is normalized to 1

orbit	(p, d)	BCS	Second RPA	mean field theory
$2p_{3/2}$	0.01 ± 0.001	0.02	0.03	0.08
$1f_{7/2}$	0.03 ± 0.003	0.04	0.06	0.12
$1d_{3/2}$	0.93 ± 0.05	0.86	0.89	0.88
$2s_{1/2}$	0.89 ± 0.08	0.94	0.92	0.89
$1d_{5/2}$	> 0.79	0.99	0.93	0.90

$$T^{n\ell j}_{\beta\alpha} = A^{\beta}_{n l j} \int \chi^{(-)*}_d (\mathbf{k}_d, \mathbf{r}_d) \Phi^*_d(\mathbf{r}) \langle \Phi^{\beta}_B | \Delta U(\mathbf{r}) | \Phi^{\beta}_B \rangle \phi_{n\ell j} \chi^{(+)}(\mathbf{k}_p, \mathbf{r}_p) d\mathbf{r}_p d\mathbf{r}_d,$$

(8.170)

where the implicit integrals run over the integral degrees of freedom $\mathbf{r}_B$ of nucleus B. In this equation, we note that structure and reaction information are factorized in the DWBA approach.

The population of the final state β from α may occur via the pickup from several neutron orbitals. The pickup cross section will then be a sum over all neutron orbitals $i = (n\ell j)$, following

$$\sigma_{\beta\alpha} = \sum_i S^{n\ell j} \sigma^{n\ell j}_{\beta\alpha},$$

(8.171)

where $\sigma^{n\ell j}_{\beta\alpha}$ is the single-particle cross section calculated with the transition matrix element of Eq. (8.170), assuming a single-particle neutron state (with $S^{n\ell j} = 1$). The cross section from α to β is then decomposed in structure information ($S^{n\ell j}$) and terms that are calculated in the reaction formalism ($\sigma^{n\ell j}$).

Figure 8.18 shows the differential cross sections of protons from the one-particle pick-up reactions ^{40}Ca$(p, d)^{39}$Ca with the proton energy E_p=65 MeV. The angular distributions are calculated by DWBA with the zero-range optical potentials. The absolute normalization of single-particle cross sections is still today under discussion ([13]). In the case of zero-range potentials, the T matrix may not give any absolute value and the resulting cross sections should be considered relative. Often, the normalization factor is taken to lead to a unity spectroscopic factor for the pickup from a (considered) closed shell nucleus. On the contrarily, one would expect the absolute spectroscopic factor from the $(e, e' p)$ experiment, which is reported always smaller than the unity by many-body correlations and/or the short-range correlations. The DWBA results show reasonable agreements with experimental data for final states with different angular momenta. The populations of final states show characteristic angular momentum distributions depending on the transferred angular momentum ℓ; the lower ℓ of $s_{1/2}$ state shows a strong maximum at zero degree in Fig. 8.18(c) and the large ℓ of $f_{7/2}$ orbit shows rather flat angular distributions in Fig. 8.18(e).

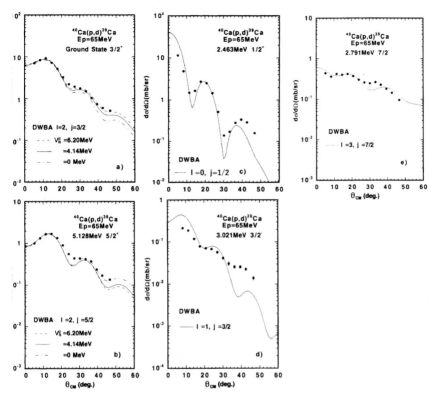

Fig. 8.18   Differential cross sections of deuterons from the ^{40}Ca$(p, d)^{39}$Ca reactions with proton energy $E_p = 65$ MeV: **a** the ground $3/2^+(l = 2, j = 3/2)$ state, **b** $5/2^+(l = 2, j = 5/2)$ state at 5.128 MeV, **c** $1/2^+(l = 0, j = 1/2)$ state at 2.463 MeV, **d** $3/2^-(l = 1, j = 3/2)$ state at 3.021 MeV, and **e** $7/2^-(l = 3, j = 7/2)$ state at 2.791 MeV in ^{39}Ca. The curves show predictions of the DWBA calculations with the zero-range optical potentials. The spin-orbit strength V_{ls} is taken in three cases 12.8, 8.28 and 0.0 MeV in unit of $(\hbar/m_\pi c)^2$ (instead of r_{ls}^2 in Eq. (8.118)) to study the effect of spin-orbit potential in **a** and **b**. Reprinted with permission from [12]. ©2021 from the American Physical Society

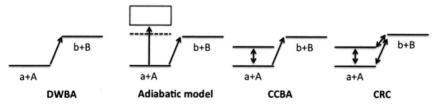

Fig. 8.19   Illustration of the different approaches to treat a transfer reaction from an entrance channel $\alpha = a + A$ to an exit channel $\beta = b + B$. The DWBA, the CCBA and the CRC are abbreviations of Distorted Wave Born Approximation, Coupled Channel Born Approximation and Coupled Reaction Channel, respectively. The adiabatic model here assumes that a is a deuteron (that breaks up easily)

The spectroscopic factors of shell orbits were also obtained by the comparison between the experimental and DWBA cross sections. The occupation probability of each orbit is the sum of spectroscopic factors in the energy range $E_x = (0 - 10)$ MeV. The empirical values are compared with three theoretical models, BCS, second RPA and mean field theory in Table 8.3. The errors are estimated from those of the empirical spectroscopic factors. Errors from DWBA analysis are not included. The fragmentation of spectroscopic factors are very small for $1d_{3/2}$ and $2s_{1/2}$ orbits and the two states shown in Fig. 8.18 (a) and (c) exhaust large portions of the spectroscopic factors.

As seen previously for the elastic scattering reactions, several approximations can be used to describe transfer reactions. Some of them are introduced below and illustrated in Fig. 8.19:

- The *Distorted Wave Born Approximation* (DWBA) described above is the simplest (but useful) method. It assumes a direct one-step process that is weak compared to the elastic channel and may be treated by perturbation theory.

- The *adiabatic model* is a modification of the DWBA formalism for (d, p) and (p, d) that takes deuteron breakup effects into account in an approximate way.

- The *Coupled Channel Born Approximation* (CCBA) is used when the one-step approximation breaks down. Strong inelastic excitations are treated in coupled channel method, while transfer is still treated with DWBA.

- The *Coupled Reaction Channel* (CRC) does not assume one-step or weak transfer process. All processes are taken into account on equal footing. It is the most complete treatment of transfer reactions.

8.5.1.2   Two-Particle Transfer Reaction

The pair addition or pair removal strength is measured by the two-particle transfer cross section, which is not simply a factorized product of such matrix element times kinematical factors since the reaction process is quite involved. Sophisticated calculations of absolute cross sections have been published starting from the 1960s until recent state-of-the-art schemes.

In a DWBA picture, the reaction cross section associated with $A + a \rightarrow B + b$ will be proportional to the square of the transition amplitude. At variance with the inelastic case, the nucleons are in different partitions in the initial Aa and final Bb channels. One can define a coordinate system in each of these two initial and final channels depicted in Fig. 8.20. The projectile-target interaction can be defined either in the initial or final channel. In the latter case, the terms ejectile-residue are more appropriate. One refers to the two choices as prior or post representations, respectively, but two representations are equivalent for the final results. The transition amplitude can be expressed as

$$T_{A+a \to B+b} \approx \int d\mathbf{r}_i d\mathbf{r}_f \chi_{Bb}^\dagger(\mathbf{r}_f, \mathbf{k}_f) F(\mathbf{r}_i \mathbf{r}_f) \chi_{Aa}(\mathbf{r}_i, \mathbf{k}_i), \tag{8.172}$$

where the χ are distorted wave functions that carry appropriate momentum labels of the initial and final states, while F is the reaction form factor. Taking a pair transfer reaction (p, t), or $(p, {}^3\text{He})$ (the same geometry for a pair pickup reaction such as (t, p) or $({}^3\text{He}, p)$), the reaction form factor can be given by

$$F(\mathbf{r}_i \mathbf{r}_f) = \langle \Phi_B \Phi_b | V_{b,1} + V_{b,2} | \Phi_A \Phi_a \rangle, \tag{8.173}$$

where $V_{b,i}$ is the interaction between b and the particle $i = 1, 2$ of the transferred pair. The differential cross section is then given by

$$\left(\frac{d\sigma}{d\Omega} \right) = \frac{m_a m_b}{(2\pi\hbar^2)^2} \frac{k_b}{k_a} |T_{A+a \to B+b}|^2. \tag{8.174}$$

In the form factor (8.173), the effective interaction V is active only in the range allowed by the reaction mechanism and acts as a kind of filter that makes the connection between pairing correlations and reaction cross sections quite indirect. The residue B can be described by

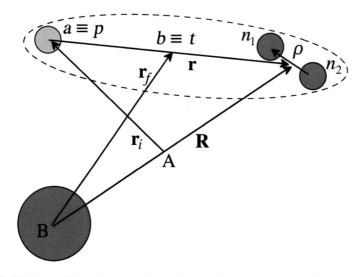

Fig. 8.20 Relative positions of two-particle transfer reaction $a + A \to b + B; b = a + x$ and $A = B + x$ where x will be $2n$ or $2p$ or np pair. In this figure, a $2n$ transfer reaction, $A(p, t)B$, is illustrated. $\mathbf{r}_i$ and $\mathbf{r}_f$ are the relative coordinates A and a (proton), B and b (triton), respectively, while $\mathbf{R}$ and $\mathbf{r}$ are the relative coordinates between B and $x(2n)$, and p and $x(2n)$, respectively. The vector ρ corresponds to the relative coordinate between two neutrons

$$\Phi_B(A+2) = \sum_{j_1 j_2 J'_A T'_A} C_{j_1 j_2 JT}(B, A') \mathcal{A}[\Phi_{J'_A T'_A} \Phi_{(j_1 j_2)JT}(1,2)]_{J_B T_B}, \qquad (8.175)$$

where $\mathcal{A}$ means the anti-symmetrization in the case of identical transferred particles, $C_{j_1 j_2 JT}(B, A')$ is a parentage coefficient, and $\Phi_{(j_1 j_2)JT}$ is a two-particle wave function calculated by either shell model or RPA. The symbol A' denotes the target state which will be either the ground state or the excited state. Consider the target A to be always the ground state, the form factor is given by

$$F(\mathbf{R}, \mathbf{r}) \propto \int d\rho \int d\mathbf{b} \langle \Phi_{(j_1 j_2)JT} \Phi_b | V | \Phi_a \rangle, \qquad (8.176)$$

where the second integral is the coordinate of projectile residue b. The relations $\mathbf{r}_i = \frac{A-2}{A} \mathbf{R} - \mathbf{r}$ and $\mathbf{r}_f = \mathbf{R} - \mathbf{r}/3$ are used to rewrite Eq. (8.173) to Eq. (8.176). For light ions, $d, t, {}^3$He or ^{4}He, the transferred two nucleons are dominantly in the lowest $l = 0$ state ($1s$ state) of relative motion, which can be represented by a harmonic oscillator wave function $\Phi_a \propto \phi_{1s}(\alpha\rho)\Phi_b$, where ρ is the relative distance of two nucleons and α is a scaling factor.

We assume a structureless projectile a such as triton and the contact interaction V between b and n_1, and between b and n_2,

$$V = V_0 \delta(\mathbf{r}_{bi}), \quad (i = 1, 2) \qquad (8.177)$$

where V_0 is an interaction strength. These assumptions make the geometry simpler in this limit $\mathbf{r} \to 0$;

$$\mathbf{r}_i = \frac{A-2}{A} \mathbf{R}, \quad \mathbf{r}_f = \mathbf{R}, \quad \text{and} \quad \rho = 0. \qquad (8.178)$$

Without any geometrical, spin and isospin projection factors, the form factor is given by

$$F(\mathbf{R}, \mathbf{r}) \propto \int d\rho \Phi_{(j_1 j_2)JT}(\rho, R) \phi_{1s}(\alpha\rho), \qquad (8.179)$$

where $\Phi_{(j_1 j_2)JT}(\rho, R)$ is expressed by the coordinates $\mathbf{r}_1 = \mathbf{R} + \rho/2$ and $\mathbf{r}_2 = \mathbf{R} - \rho/2$. Even the correlated wave function $\Phi_{(j_1 j_2)JT}$ gives a large enhancement for the pair transfer matrix element (6.66) discussed in Chap. 6, it is not straightforward that the cross section shows also the same amount of enhancement because the overlap integral in Eq. (8.179) might be quenched in the case of light ions. To have a large overlap integral in Eq. (8.179), the transfer reactions with heavy projectiles might be better because of similar large orbital angular momenta of transferred particles to those of correlated two-particles in the target. On the other hand, the reaction mechanism of heavy projectiles will be more complicated that that of light ions.

8.5.2 Quasi-free Scattering

During a quasifree scattering reaction, a nucleon from the nucleus (target nucleus in direct kinematics or projectile in inverse kinematics) is knocked out with a high momentum transfer compared to its intrinsic momentum from the interaction with a particle. The emblematic quasifree scattering is the $(e, e'p)$ reaction where a proton is knocked out from a nucleus by an incoming electron. $(e, e'p)$ is restricted to proton knockout and can only be performed for stable nuclei so far. For these reasons, proton-induced quasi-free scattering (p, pN) (with $N = p, n$) at energies beyond 300 MeV/nucleon is considered the cleanest stripping probe to investigate the structure for both stable and unstable nuclei. A formalism for the proton-induced quasifree scattering was first proposed by Jacob and Maris in 1966 [14]. The best energy for quasi-free scattering minimizes initial-state and final-state interactions, i.e., the distortions of the incoming and outgoing protons in the nuclear potential of the target nucleus and the residue, respectively. This is achieved when both the incident energy and the kinetic energy of the scattered protons are close to the minimum of the nucleon-nucleon reaction cross section. Due to the equal mass of the recoil proton and ejected nucleon, when both are detected at about 45° in the laboratory frame, they approximately share equally the energy of the incoming proton. For example, a quasi-free scattering performed at 400 MeV/nucleon will lead to two nucleons scattered at about 200 MeV when both detected at 45°, due to momentum conservation. Figure 8.21 shows the value of nucleon-nucleon reaction cross sections as a function of the energy (laboratory frame). The minimum for the pn channel is located around 400 MeV/nucleon, while the reaction cross section takes the maximum at 900 MeV for pp channel. According to this plot, the best energies for quasi-free scattering are between 500 to 1000 MeV/nucleon (for both direct or inverse kinematics). The measured momenta of the two scattered protons in a quasi-free scattering experiment give access to the intrinsic momentum $\mathbf{q} = (q_\parallel, q_\perp)$ of the removed proton inside the projectile as illustrated in Fig. 8.22. By use of the missing-mass technique, quasi-free scattering allows to measure the separation energy E_s of the populated bound and unbound states of the residue:

$$\mathbf{q}_\perp = \mathbf{p}_{1\perp} + \mathbf{p}_{2\perp}, \tag{8.180}$$

$$\mathbf{q}_\parallel = \frac{(\mathbf{p}_{1\parallel} + \mathbf{p}_{2\parallel}) - \gamma\beta(M_A - M_{A-1})}{\gamma}, \tag{8.181}$$

$$E_s = T_0 - \gamma(T_1 + T_2) - 2(\gamma - 1)m_p + \beta\gamma(p_{1\parallel} + p_{2\parallel}) - \frac{q^2}{2M_{A-1}}, \tag{8.182}$$

where β, γ are the velocity in units of c and Lorentz factor of the projectile in the laboratory frame, T_0 and $T_{1,2}$ the kinetic energies of the projectile and the two protons, respectively. The separation energy of the final state in the residual nucleus is linked to the excitation energy E_x of the populated state by $E_s = E_x - S_p$, where S_p is the proton separation energy of the residual nucleus in its ground state. The example of

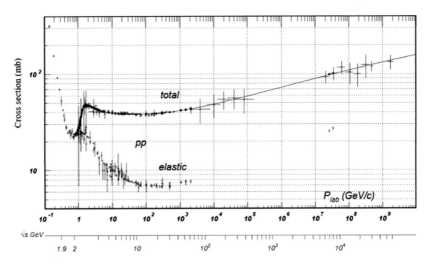

Fig. 8.21 Measured proton-proton reaction cross sections as a function of the laboratory momentum of the incident proton in GeV/c. Elastic and inelastic scattering cross sections are shown separately. The center of mass energy, defined as the square root of the Mandelstram variable $s = (p_1 + p_2)^2$, the invariant-mass norm of the quadrivector momentum of the two colliding protons 1 and 2, is given as a scale in red. Figure taken from [15]

$^{12}C(p, 2p)^{11}B$ at 392 MeV incident energy in direct kinematics measured at RCNP, Japan, is given in Fig. 8.23, which shows a $(p, 2p)$ proton kinematics. Indeed, in the non-relativistic limit and assuming that the removed proton has a negligible binding energy in the initial nucleus, the energy conservation of Eq. (8.182) simplifies to $E_{p1} + E_{p2} \sim T_0 - E_s$, i.e., for a given final state of separation energy E_s the initial kinetic energy T_0 subtracted by the cost E_s is shared among the two recoiling protons. This leads to anti-diagonal lines in the $E_{p1} - E_{p2}$ plot, one for each final state of the residue. Very recently, quasi-free scattering $^{12}C(p, 2p)$ was measured for the first time exclusively in complete and inverse kinematics at an energy of about 400 MeV/nucleon at GSI.

Problem

8.3 (a) Derive Eqs. (8.180), (8.181), and (8.182) from energy-momentum conservation.

(b) Calculate the non-relativistic limit of these equations by considering $T = p^2/2m$.

Quasi-free scattering is usually described theoretically based on one main assumption, the *impulse approximation* (IA), assuming that the knockout process occurs in one single step and involves only the two interacting nucleons: the incoming proton and the to-be-knocked-out nucleon. The quasi-free process is the transition from an initial state $|i\rangle$ consisting of a nucleus A and an incoming proton, to a final state $|f\rangle$ with a recoil nucleus $(A - 1)$ and two outgoing protons or a proton and a neutron, as is shown in Fig. 8.22. Assuming direct kinematics with (E_0, p_0c) for the incoming proton in the laboratory frame, the energy and momentum conservations require

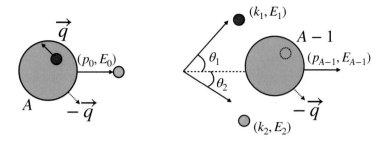

Initial state   **Final state**

Fig. 8.22 (Left) Scheme of a quasi-free scattering reaction in inverse kinematics. A nucleus composed of A nucleons (green) at momentum p_0 and total energy E_0 interacts with a proton (yellow) at rest in the laboratory. A nucleon in the projectile with intrinsic momentum $\mathbf{q}$ (in the mass frame of the nucleus A) is identified in red. Due to momentum conservation, in the center of mass frame of the nucleus A, the remaining system composed of $A - 1$ nucleons has a momentum $-\mathbf{q}$. (Right) after a quasielastic scattering, the knocked-out nucleon has momentum k_1 and total energy E_1 in the laboratory frame, while the recoil target nucleon has momentum and energy k_2 and E_2, respectively. Note that in the case of a proton knockout, the two protons are not distinguishable. The recoil nucleus $A - 1$ keeps its initial momentum $-\mathbf{q}$ in the center of mass frame of the projectile since it is considered as a spectator

$$E_0 + M_A c^2 = E_1 + E_2 + E_{A-1}, \tag{8.183}$$

and

$$\mathbf{p}_0 = \mathbf{p}_1 + \mathbf{p}_2 + \mathbf{p}_{A-1}. \tag{8.184}$$

The energy E_{A-1} of the heavy-ion residue of ground-state mass M_{A-1} is given by

$$E_{A-1} = M_{A-1} c^2 + E_x + T_{A-1}, \tag{8.185}$$

where E_x is the excitation energy of the residual nucleus. The general expression for the cross section is given by

$$\frac{d^9 \sigma}{d^3 p_1 d^3 p_2 d^3 p_{A-1}} = \frac{4\pi^2}{\hbar^{10}} \frac{E_0 E_A}{F} |t_{fi}|^2 \delta(\mathbf{p}_1 + \mathbf{p}_2 + \mathbf{p}_{A-1} - \mathbf{p}_0 - \mathbf{p}_A)$$
$$\times \delta(E_1 + E_2 + E_{A-1} - E_0 - E_A), \tag{8.186}$$

where

$$F = c \left[(E_o E_A - c^2 \mathbf{p}_0 \cdot \mathbf{p}_A)^2 - c^8 m^2 M_A^2 \right]^{1/2}, \tag{8.187}$$

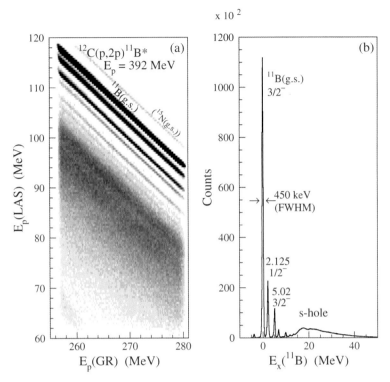

Fig. 8.23 (Left) Energy correlation of the two scattered protons from $^{12}C(p, 2p)^{11}B$ at 392 MeV incident energy in direct kinematics. One of the protons is measured by the Large Acceptance Spectrometer (LAS) and the other by the Gran Raiden spectrometer (GR). (Right) Corresponding missing mass spectrum following Eq. (8.182). Data are taken from [16]. Courtesy M. Yosoi, RCNP

where m is the mass of the target proton. The matrix element t_{fi} is associated with the T matrix in the first-order Born approximation, as

$$\langle f|T|i\rangle = \langle f|V|i\rangle = \frac{1}{2\pi}\delta(\mathbf{p}_1 + \mathbf{p}_2 + \mathbf{p}_{A-1} - \mathbf{p}_0 - \mathbf{p}_A)t_{fi}. \qquad (8.188)$$

The transition matrix elements for (p, pN) quasi-free scattering can be expressed in the impulse approximation as

$$T_{(p,pN)} = \langle \chi_{p_2}\chi_{p_1}|V_{pN}|\chi_{p_0}\Psi_{jlm}\rangle, \qquad (8.189)$$

where V_{pN} is the potential between the incoming proton and the nucleon N to be removed, $\chi_{p_{1,2}}$ are the distorted waves for the outgoing and scattered-off nucleons in the potential field of the residual nucleus $A - 1$, χ_{p_0} is the incoming proton wave function distorted by the presence of the target nucleus A, and Ψ_{jlm} is the bound state wave function of the knocked out nucleon.

In the extreme limit of the plane-wave approximation and a zero-range interaction, the above transition matrix element reduces to the Fourier transform of the single-particle wave function $\Phi_{n\ell j}$, showing that there is a very close connection between the quasi-free scattering cross section and the wave function of the removed nucleon, similar to $(e, e'p)$ reactions.

Recent developments have been made to predict (p, pN) cross sections under the eikonal [17] and PWIA approximations, DWIA [18] or within a so-called "ab initio" Fadeev multipole scattering reaction framework [20].

Two major facilities can today produce radioactive ion beams at adequate energy for quasi-free scattering: the RIBF of RIKEN in Japan with beams of about 250 MeV/nucleon and, at energies up to ~ 1 GeV/nucleon at GSI. In the coming decade, the FAIR facility should provide the best kinematical conditions to investigate nuclear structure from quasi-free scattering in inverse kinematics with the R3B setup. Indeed, the incident energies available at R3B, from 400 MeV/nucleon to about 2 GeV/nucleon, allow to realize the experimental conditions at which the sudden approximation is the best verified while the initial and final state interactions are minimized.

8.5.3 Heavy-Ion Induced Inclusive Nucleon Removal Reactions at Intermediate Energy

As a tool for the spectroscopy of unstable nuclei, heavy-ion induced nucleon-stripping reactions have been so far more popular than quasi-free scattering because they are easier to implement experimentally. Many important results and theoretical developments have been recently accomplished with this method. In the literature, these reactions are often referred to as "knockout" although the exact reaction process for the nucleon removal may be, in some situations, more complex than a single-step nucleon stripping resulting from a hard-core nucleon-nucleon collision. In most cases, the experimental technique uses in-beam gamma spectroscopy at the secondary target to tag the final state of the populated projectile-like residue. In these measurements, the final state of the target, most often ^{9}Be or ^{12}C, is not measured. Note that there can be ambiguity in the use of the term "inclusive" in such experiments. Regarding the detection of reaction products, they are always inclusive since the target-like recoil and its products are not measured. On the other hand, when one focuses on the final state of the projectile-like residue, if the cross sections to individual final bound states of the residue are measured (by gamma tagging), one speaks of exclusive cross sections. The cross section to all bound states of the residue are then referred as "inclusive". From here, the term "inclusive" will be used in the latter definition.

A sketch of an inverse-kinematics spectroscopy experiment based on the inclusive one-nucleon removal reaction mechanism is shown in Fig. 8.24. One asset of inverse-kinematics of nucleon-stripping reactions is given by the momentum distri-

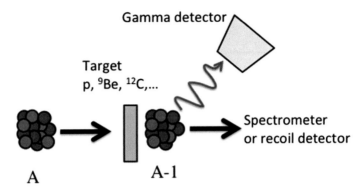

Fig. 8.24 Schematic representation of a typical setup for inclusive knockout reactions combined with in-beam gamma spectroscopy. A projectile composed of A nucleons is sent onto a light-ion target (proton, ^{9}Be, ^{12}C in most cases). Nuclear reactions take place inside the target. Some of the reactions lead to the projectile with one nucleon removed, leading to a nucleus composed of $A - 1$ nucleons (and the same number of protons Z if a neutron is removed, or $Z - 1$ protons in the case of a proton removal). The recoil nucleus can be identified by a recoil detector or a spectrometer. In the case that the $A - 1$ nucleus is populated in a bound excited state, it may decay via a prompt γ decay. The emitted photon can be detected by placing γ detectors around the target. By correcting the energy of the in-flight-emitted photons measured in the laboratory from the Doppler distortion, one can determine the energy of the emitted photon in the center of mass of the $A - 1$ nucleus. See text for details

bution of the residue. By measuring the total, parallel or perpendicular momentum distributions, one has direct access to the intrinsic momentum distribution of the removed nucleon provided the final state interaction is negligible. This quantity is highly connected to the angular moment ℓ of the removed nucleon and therefore it gives important information on the shell structure of the nucleus under study.

The technique has been pioneered and largely developed in inverse kinematics from the 90s at the NSCL at energies around or below 100 MeV/nucleon, at GANIL, and at GSI at higher incident energies, better suited to satisfy the eikonal and sudden approximations since the beam velocity is much higher than the intrinsic velocity of nucleons in the projectile.

An example of such data in coincidence with the detection of prompt de-excitation γ rays from the residue is illustrated in Fig. 8.25: the one neutron removal ^{12}Be(^{9}Be,X)^{11}Be performed at an incident energy of 78 MeV/nucleon. The γ coincidences allow the separation of the measured inclusive longitudinal momentum distribution into the constituent distributions for the $1/2^+$ and $1/2^-$ states as shown in the right panel of Fig. 8.25. Here, the momentum distribution for the $1/2^-$ state was obtained by gating on the 320 keV γ ray and subtracting this contribution from the background. The momentum distribution to the $1/2^+$ ground state is also shown in the same figure. The momentum distributions reflect essentially the Fourier transform of the single-particle wave function in the coordinate space. That is, the large spacial extension gives a narrow width in the momentum distribution, while narrow spacial distribution provides a wider width in the momentum space. The narrow momentum

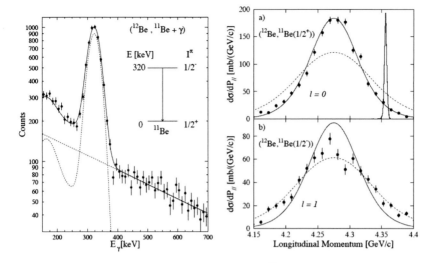

Fig. 8.25 (Left) Gamma spectrum from the de-excitation of ^{11}Be populated by one-neutron knock-out ^{12}Be(^{9}Be,X)^{11}Be at 78 MeV/nucleon. (Right) Laboratory frame longitudinal momentum distributions for ^{11}Be residues in the ground (**a**) and excited (**b**) states. The solid (dashed) curves are calculated for $l=0$ (1) neutron stripping momentum distributions with shell model wave functions. The narrow line in (a) is the line profile of the spectrograph. The intrinsic angular momentum of the removed nucleon can be easily disentangled in both cases. Figures reprinted with permission from [21]. ©2021 by the American Physical Society

distribution for the ground $1/2^+$ state in the right panel of Fig. 8.25 confirms the halo nature with a dominant $s_{1/2}$ configuration.

8.5.4 Eikonal Approximation and S Matrices

Intermediate-energy nucleon-stripping cross sections are often interpreted by comparison to shell-model spectroscopic factors and single-particle cross sections calculated under the eikonal approximation. The first nuclear knockout model in the eikonal approximation was presented in [22]. The S-matrix formalism is commonly used to derive these cross sections. The basics of the formalism, as most often used to calculate heavy-ion induced knockout cross sections, are given below. Other approaches have been developed and the most appropriate method to describe nuclear breakup is still under debate in some cases.

We treat the general case of an arbitrary central potential $U(\mathbf{r})$ in the eikonal approximation. The Schrödinger equation to solve can be expressed as

$$(-\frac{\hbar^2 \nabla^2}{2\mu} + U(\mathbf{r}) - E)\Psi(\mathbf{r}) = 0. \tag{8.190}$$

In the classical limit, valid for a fast projectile with $U/E \ll 1$, one can approximate the total energy to the kinetic energy: $E = \frac{\hbar^2 k^2}{2\mu}$. In the case of fast beams, we can make the approximation that the ejectile propagates along the beam direction. This straight line approximation is called the *eikonal approximation* in reference to the greek term used in optics to describe light waves that propagate along straight trajectories. Under the eikonal approximation, the wave function Ψ is simplified into

$$\Psi(\mathbf{r}) = e^{i\mathbf{k}\cdot\mathbf{r}}\phi(\mathbf{r}), \tag{8.191}$$

where all effects of the potential $U(r)$ are now contained in the modulation $\phi(\mathbf{r})$ which is a solution of the differential equation

$$2i\mathbf{k}\cdot\nabla\phi(\mathbf{r}) - \frac{2\mu}{\hbar^2}U(\mathbf{r})\phi(r) + \nabla^2\phi(\mathbf{r}) = 0. \tag{8.192}$$

This equation can be further simplified, following two approximations we already made:

1. the classical approximation $U/E \ll 1$ allows to assume that $\mathbf{k}\cdot\nabla\phi(\mathbf{r}) \gg \nabla^2\phi(\mathbf{r})$,
2. the eikonal approximation which restricts the propagation along the incident z axis.

The differential equation then becomes

$$\frac{d\phi(\mathbf{r})}{dz} = -i\frac{\mu}{\hbar^2 k}U(r)\phi(\mathbf{r}), \tag{8.193}$$

leading to the solution

$$\phi(\mathbf{r}) = exp\left[-i\frac{\mu}{\hbar^2 k}\int_{-\infty}^{z}U(\mathbf{b}, \mathbf{z'})\mathbf{dz'}\right], \tag{8.194}$$

where we adopt the cylindrical coordinates $\mathbf{r} = (\mathbf{b}, \mathbf{z})$. The wave function can then be expressed as

$$\Psi(\mathbf{r}) = exp\left[-i\frac{\mu}{\hbar^2 k}\int_{-\infty}^{z}U(\mathbf{b}, \mathbf{z'})\mathbf{dz'}\right] \times e^{i\mathbf{k}\cdot\mathbf{r}} = S(\mathbf{b})e^{ikr}. \tag{8.195}$$

We introduce here the S matrix as a function of the impact parameter $\mathbf{b}$, defined as $\mathbf{r} = \mathbf{b} + z\hat{e}_z$, as

$$S(\mathbf{b}) = exp\left[-i\frac{\mu}{\hbar^2 k}\int_{-\infty}^{z}U(\mathbf{b}, \mathbf{z'})\mathbf{dz'}\right], \tag{8.196}$$

which is actually identical to Eq. (8.194). Most reaction observables can be obtained from the S matrix. The behavior of $S(b)$ is intuitive: in the presence of an absorptive potential, the norm of the S matrix is equal to 1 when $b \to \infty$ since the particle does not feel any absorptive potential. Contrarily, it is equal to zero for small impact param-

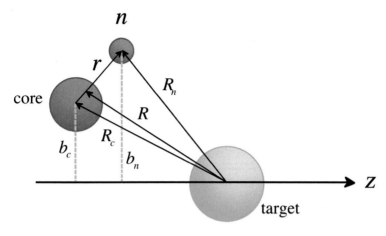

Fig. 8.26 Definition of variables used to describe the one neutron-stripping reaction

eters because the particle is captured completely by the strong absorptive potential. The S matrix is calculated microscopically.

The S matrix, in general, relates the initial and the final states of a many-body system in a scattering process. In Eqs. (8.72) $\sim$ (8.75), the S matrix is formulated through the time-evolution unitary operator $U(t, t_0)$ in the interaction representation. In this subsection, the S matrix is defined in a time-independent picture and is given by Eqs. (8.196). Nevertheless, the physical meaning of S matrix is the same in the two representations.

The one-nucleon stripping cross section is calculated using the eikonal formalism. The projectile wave function is defined as a core wave function $|\Phi_c\rangle$ complemented with the wave function of the removed nucleon $|\Phi_n\rangle$. Calculations are based on two main quantities: the S matrices for the core (S_c) and the removed nucleon (S_n), as described above. The final state wave function of projectile is thus given by

$$\Psi_f(\mathbf{r}, \mathbf{R}) = S_n(\mathbf{b}_n) S_c(\mathbf{b}_c) \phi_0(\mathbf{r}) e^{ikz}, \tag{8.197}$$

where $\mathbf{r}$ is the coordinate of the center of mass of the projectile nucleus, and $\mathbf{b_c}$ and $\mathbf{b_n}$ are the impact parameters of the core and the nucleon with respect to the target nucleus, i.e., $b_n = R_\perp + r_\perp A c/(Ac + 1)$ and $b_c = R_\perp - r_\perp/(A_c + 1)$, where A_c is the mass number of the core and the designation "$\perp$" refers to components transverse to either z-axis or to R_n and R_c (see Fig. 8.26 for the variables of one neutron-stripping reaction). Here, $\phi_0(\mathbf{r})$ is the nucleon wave function in the ground state of the projectile. The scattering wave function is the difference between Eq. (8.197) and the wave function of the undisturbed beam $\phi_0(\mathbf{r})$,

$$\Psi_{\text{scat}} = (S_n(\mathbf{b}_n) S_c(\mathbf{b}_c) - 1) \phi_0(\mathbf{r}) e^{ikz}. \tag{8.198}$$

Scattering cross sections are calculated by taking overlaps of Ψ_{scat} with different final states. The one-nucleon stripping cross section consists of mainly a stripping and a diffractive part

$$\sigma = \sigma_{\text{str}} + \sigma_{\text{diff}}. \tag{8.199}$$

The stripping part σ_{str} corresponds to reactions where the target is excited by the nucleon or the nucleon is absorbed by the target, while the diffractive part corresponds to breakup events during which both the target and the core remain in their ground states. The diffractive part of the nucleon stripping cross section can be seen as events where both the core and the nucleon are elastically scattered off the target and for which the overlap of the final core and nucleon with the incoming projectile ground state does not equal to 1 since the final core-nucleon system is unbound.

In an impact-parameter representation, the stripping part of the cross section is calculated as

$$\sigma_{\text{str}} = 2\pi \int b\,db \int d\mathbf{r} |\phi_0(\mathbf{r})|^2 |S_c(\mathbf{b}_c)|^2 (1 - |S_n(\mathbf{b}_n)|^2). \tag{8.200}$$

The S matrix for the core-target system S_c is defined from the target and core densities as well as from the in-medium NN cross section that depends on the incident energy. Equation (8.200) is intuitive: it is the sum over all possible impact parameters to preserve the core (probability $|S_c|^2$ by definition of the S matrix) and to strip off the valence nucleon (probability $1 - |S_n|^2$). There is another possibility for the stripping reaction which is the stripping process from the core. This process is also formulated from Eq. (8.200) interchanging the subscripts ($c \leftrightarrow n$).

The diffractive part corresponds to the cross section for the final state being in the continuum. The diffractive dissociation is one of the major processes of nucleon removal reactions. For simplicity, we consider here the special case where $|\Psi_0\rangle$ is the only bound state of the projectile. Then, the wave function in the continuum of the final state is given by

$$|\Psi_f^{\text{cont}}\rangle = |\Psi_f\rangle - |\Psi_0\rangle\langle\Psi_0|\Psi_f\rangle. \tag{8.201}$$

The diffractive dissociation probability is given by

$$P_{\text{diff}}(\mathbf{b}) = \langle\Psi_f|\Psi_f^{\text{cont}}\rangle = \langle\Psi_f|\Psi_f\rangle - |\langle\Psi_0|\Psi_f\rangle|^2. \tag{8.202}$$

The diffractive cross section then can be expressed as

$$\sigma_{\text{diff}} = 2\pi \int b\,db \left[\langle\Phi_0||S_C S_N|^2|\Phi_0\rangle - |\langle\Phi_0|S_C S_N|\Phi_0\rangle|^2 \right]. \tag{8.203}$$

The derivation of Eq. (8.203) can be extended to the case where several bound states exist below the particle threshold, but its form gets then less intuitive. A more detailed derivation of the cross sections can be found, for example, in [13]. The single-particle stripping cross section $\sigma = \sigma_{str} + \sigma_{diff}$ combined with spectroscopic factors calculated from a nuclear-structure formalism can be compared to the experimental cross section.

The distribution of spectroscopic strength in nuclei can be extracted from direct reaction cross section measurements, assuming a model of the reaction mechanism. A compilation of one-nucleon stripping at intermediate energies from sd-shell exotic nuclei showed that the measured cross sections for knocking out a valence neutron from a very neutron-rich nucleus or a proton from a very neutron-deficient nucleus (such as a neutron in ^{32}Ar, ^{28}Ar and ^{24}Si) are about four times smaller than predictions from state-of-the-art calculations [23]. The predictions of inclusive cross sections based on shell-model spectroscopic factors and the above-described reaction model based on the eikonal and sudden approximations show a disagreement with experimental data of heavy-ion induced cross sections measured mostly at incident energies around 100 MeV/nucleon. The ratio of experimental to predicted cross sections are illustrated in the left panel of Fig. 8.27 as a function of the asymmetry in proton-to-neutron separation energy

$$\Delta S = \epsilon(S_p - S_n), \tag{8.204}$$

where $\epsilon = +1$ in the case of a neutron removal and $\epsilon = -1$ in the case of a proton removal, so that the stripping of deeply bound nucleons is located at large positive ΔS values.

On the other hand, at low energy, a study of the (p, d) neutron transfer on the proton-rich ^{34}Ar and on the neutron-rich ^{46}Ar provides experimental spectroscopic factors in agreement with large-basis shell model calculations within 20%. These findings, together with a more recent study with stable and radioactive oxygen isotopes in agreement with systematics of transfer reactions, do not confirm the trend deduced from the analysis of nucleon stripping cross sections at about 100 MeV/nucleon. A systematic study over transfer reactions with stable nuclei did not evidence any dependence of the effect of short range or beyond model space correlations with the transferred angular momentum ℓ, mass of target nuclei or asymmetry ΔS. Therefore, it is suggested that these two probes, transfer and knockout, lead systematically to different spectroscopic factors when analyzed in the above mentioned frameworks, namely DWBA or CRC for transfer and under the sudden and eikonal approximations for knockout. The origin of this difference still has to be understood.

The corresponding transfer stripping reactions (d, t) and $(d, {}^3\text{He})$ from the same ^{14}O nucleus at 18 MeV/nucleon performed at GANIL and analyzed within the framework of coupled reaction channel formalism with a set of optical potentials. Matter radii and spectroscopic factors did not show any strong systematic reduction for deeply-bound nucleon stripping. This analysis tends to show that the mostly used reaction-mechanism models for transfer (DWBA, CRC) and heavy-ion induced knockout (sudden and eikonal approximation) do not lead to the same structure information in some cases. These discrepancies still need to be quantitatively understood. A more systematic study of deeply-bound nucleon stripping reactions from weakly

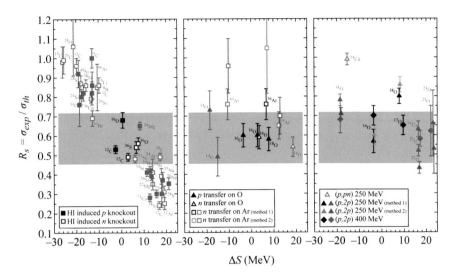

Fig. 8.27   Ratio of experimental to theoretical one-nucleon stripping cross sections with radioactive nuclei for heavy-ion-induced (mostly ^{9}Be target) nucleon removal around 100 MeV/nucleon (Left), transfer (Middle) and quasifree scattering (Right). The ratio is plotted as a function of the energy-separation asymmetry ΔS as defined in Eq. (8.204). The green horizontal band represents the ratio values obtained from the $(e, e'p)$ reference measurements and analysis as discussed in Chap. 5. The data and theoretical predictions can be found in [13, 24–29] and references therein. Courtesy B. Kay, ANL

bound nuclei will definitely help in understanding the limits of current direct reaction models.

8.6   Compound Nuclear Reactions

While the previous sections focus on direct reactions and their applications, being described as mainly one step process, many reactions are not direct and are characterized by a high density of energy levels. The nuclear level densities at high excitation energies are of the order of 10^6 times larger than those corresponding to single-particle motion (in a nucleus with mass $A \sim 100$). The high level densities imply, for example, that slow neutrons entering in a nucleus will share their energies with a large number of degrees of freedom of the target, thus establishing a highly complex state of motion leading to the so-called "compound nucleus". The compound nucleus corresponds to very large-scale configuration mixings in term of Fermi gas model or shell model. The idea of compound nuclei was proposed by Niels Bohr in 1936. The compound nuclear reaction model is a description of nuclear reactions as a two-stage process comprising the formation of a relatively long-lived intermediate nucleus and its subsequent decay. First, a bombarding particle loses all its energy to the target nucleus, which becomes a new excited unstable nucleus, called a compound nucleus. The formation stage takes a period of time approximately equal to the time interval for the bombarding particle to travel across the diameter of the target nucleus:

$$\Delta t \sim R/v \sim 10^{-21} s. \tag{8.205}$$

Second, after a relatively long period of time (typically from 10^{-18} to 10^{-16}s) and independent of the properties of the reactants, the compound nucleus disintegrates, usually into few ejected light particles and a product nucleus. For the calculation of properties for the decay of this system by particle evaporation, one may thus borrow from the techniques of statistical mechanics. Features of compound nuclei are summarized as follows;

- The compound nucleus is usually created if the projectile has low energy and the incident particle interacts inside of the nucleus.
- The time scale of compound-nucleus reactions is much longer than the time scale of the projectile passing through the nucleus.
- A compound nucleus is a relatively long-lived intermediate state of the particle-target composite system compared to the time scale of electromagnetic or beta decays.
- Compound-nucleus reactions involve a large number of nucleon-nucleon collisions, which lead to a thermal equilibrium inside the compound nucleus.
- Products of the compound nucleus reactions are emitted rather isotropically in angle since the nucleus loses memory of how it was created. It is called the Bohr's hypothesis of independence.
- The decay mode of the compound nucleus does not depend on the way in which the compound nucleus is formed. Resonances in the cross section are typical for the compound nucleus reaction.

8.6.1 Resonant Scattering of Neutrons

Let us consider the resonant scattering of S-wave neutrons. The resonant scattering is a common phenomenon of both atomic and nuclear collisions, and occurs when the projectile energy is the same as or very close to the excitation energy of a resonance in the target-projectile system. In atomic collisions, the Feshbach resonance is a typical phenomenon of resonant scattering. While the continuum wave function is usually varying slowly as a function of energy, the wave function changes suddenly at a particular energy and gives a sharp peak in the continuum above the threshold energy of particle emission. This is called a "resonant scattering" and gives an important contribution to the cross sections. The resonance appears as a rapid change of the phase shift. There are two kinds of resonant scatterings; elastic resonant and inelastic resonant scatterings.

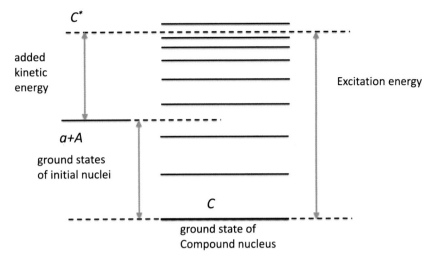

Fig. 8.28 Compound reaction $a + A \to C^* \to b + B$. The symbol C^* stands for a compound nucleus. The excitation energy is the sum of the kinetic energy of incident particle and the binding energy difference between $(a + A)$ system and the ground state of C

8.6.1.1 Elastic Compound Scattering

For the potential scattering, the S-wave neutron is a free particle above the neutron threshold energy and never made the resonance state since the neutron has neither Coulomb nor centrifugal barriers. However, it is possible in the compound nucleus since it involves a complex reaction mechanism (see Fig. 8.28). The incoming and outgoing wave functions are given by

$$\psi_0(r) = \frac{1}{2ikr}(S_0 e^{ikr} - e^{-ikr}) \equiv \frac{u_0(r)}{r}, \tag{8.206}$$

where S_0 stands for the S matrix for the S-wave ($l = 0$). The resonance appears as the pole of the S matrix or, equivalently, it occurs at the energy where the phase shift $\delta_{l=0}$, defined as $S_0 = e^{2i\delta_0}$, passes through $\pi/2$ upward from below. The phase shift is determined by a matching of the wave functions inside and outside of the potential; it is given by the continuity condition of wave function and its derivate across the surface (see Sect. 2.5). This condition is represented by the logarithmic derivative of u_0 at the surface $r = R$, where the radius can be defined approximately by $R = r_0 A^{1/3}$ fm with a radius parameter $r_0 = 1.2$ fm. Denoting the logarithmic derivate of u_0 by L_0, the value L_0 satisfies the condition at the resonance energy E_r,

$$L_0(E = E_r) = R\frac{1}{u_0(R)}\left(\frac{du_0}{dr}\right)_{r=R} = ikR\frac{S_0 + e^{-2ikR}}{S_0 - e^{-2ikR}} = 0. \tag{8.207}$$

The condition can be understood as the requirement to maximize the probability of the wave function inside the nucleus $r < R$ to induce the resonance. The S matrix is expressed from Eq. (8.207) as

$$S_0 = \left(1 + \frac{2ikR}{L_0 - ikR}\right) e^{-2ikR}. \tag{8.208}$$

Around the resonance energy E_r, the logarithmic derivative can be expanded by a power series,

$$L_0(E) = L_0(E = E_r) + \frac{dL_0}{dE}\bigg|_{E=E_r} (E - E_R) + \cdots$$

$$= -\frac{2kR}{\Gamma}(E - E_R) + \cdots, \tag{8.209}$$

where Γ is defined by

$$\Gamma \equiv -2kR / \left(\frac{dL_0}{dE}\bigg|_{E=E_r}\right), \tag{8.210}$$

which will be identified as the width of the resonance cross section later. Inserting Eq. (8.209) into Eq. (8.208), we obtain a formula

$$S_0 = \left(1 - \frac{i\Gamma}{(E - E_r) + i\Gamma/2}\right) e^{-2ikR}. \tag{8.211}$$

The cross section σ is obtained by the S matrix as (see Eq. (8.84))

$$\sigma = \frac{\pi}{k^2}|1 - S_0|^2 = \frac{4\pi}{k^2} \left| e^{ikR} \sin kR + \frac{\Gamma/2}{E - E_r + i\Gamma/2} \right|^2. \tag{8.212}$$

Near the resonance energy, the second term of Eq. (8.212) dominates and the cross section can be expressed by only the second term of the S matrix to be

$$\sigma_{ce} = \frac{\pi}{k^2} \frac{\Gamma^2}{(E - E_r)^2 + \Gamma^2/4}, \tag{8.213}$$

which is called the "Breit-Wigner" formula for the resonance. The cross section is maximum at the energy $E = E_r$ and reaches half-maximum for $|E - E_r| = \Gamma/2$, i.e., Γ is the width at half maximum. The cross section far from the resonance is obtained from the first term of Eq. (8.212) as

$$\sigma_{c-elastic} = \frac{4\pi}{k^2} sin^2 kR \sim \frac{4\pi}{k^2}(kR)^2 = 4\pi R^2, \tag{8.214}$$

where $kR \ll 1$ is adopted for S-wave scattering. The cross section (8.214) is nothing but the potential scattering cross section. This is called the compound elastic scattering cross section.

8.6.1.2 Compound Inelastic Scattering

The resonant formula can be extended to the case with open channels. Then the potential is given by a complex form as we discussed in the case of the inelastic direct reaction; $V = U + iW$. The logarithmic derivative is expanded then both at $E = E_r$ and $W = 0$,

$$L_0(E, W) = L_0(E = E_r, W = 0) + \left.\frac{\partial L_0}{\partial E}\right|_{E=E_r, W=0} (E - E_r) + W \left.\frac{\partial L_0}{\partial W}\right|_{E=E_r, W=0} + \cdots.$$
(8.215)

With the same condition as (8.207),

$$R \frac{1}{u_0(R)} \left(\frac{du_0}{dr}\right)\bigg|_{r=R} = 0 \ \text{ at } \ E = E_r; \ W = 0,$$

we define the widths for the entrance channel

$$\Gamma_\alpha \equiv -2kR / \left(\left.\frac{dL_0}{dE}\right|_{E=E_r; W=0}\right),$$
(8.216)

and also for the absorption channel,

$$W \left.\frac{\partial L_0}{\partial W}\right|_{E=E_r, W=0} = -iW \left.\frac{\partial L_0}{\partial E}\right|_{E=E_r, W=0} = -i \frac{kR}{\Gamma_\alpha} \Gamma_r,$$
(8.217)

where the width Γ_r is defined by

$$\Gamma_r \equiv 2|W|.$$
(8.218)

From the first to the second derivative in Eq. (8.217), the relation $K = \sqrt{2m(E - U - iW)}/\hbar$ is implicitly used. Then, the logarithmic derivative (8.215) is expressed as

$$L_0(E, W) = -\frac{2kR}{\Gamma_\alpha} (E - E_r) - i \frac{kR}{\Gamma_\alpha} \Gamma_r + \cdots.$$
(8.219)

The S matrix is now obtained as

$$S_0 = \left(1 - \frac{i \Gamma_\alpha}{E - E_r + i \Gamma_{tot}/2}\right) e^{-2ikR}$$
(8.220)

where the total width is defined by the sum of two widths,

$$\Gamma_{tot} = \Gamma_\alpha + \Gamma_r. \tag{8.221}$$

Then the compound-nucleus cross section with open channels is divided into two parts: the compound-elastic cross section and the absorption cross section.[6] The former is the resonance part of the elastic cross section, which is proportional to the factor $|1 - S_0|$ since the S matrix represents the scattering amplitude by the reaction

$$\sigma_{ce} = \frac{\pi}{k^2}|1 - S_0|^2 = \frac{4\pi}{k^2}\left|\frac{\Gamma_\alpha/2}{E - E_r + i\Gamma_{tot}/2} + e^{ikr}\sin kR\right|^2. \tag{8.222}$$

The resonance part of Eq. (8.222) is given by

$$\sigma_{ce} = \frac{\pi}{k^2}\frac{\Gamma_\alpha^2}{(E - E_r)^2 + \Gamma_{tot}^2/4}. \tag{8.223}$$

The absorption cross section is the difference between the total cross section (8.86) and the elastic cross section (8.222), which can be expressed as,

$$\sigma_r = \frac{\pi}{k^2}\left[2(1 - \Re\{S_0\})) - |1 - S_0|^2\right] = \frac{\pi}{k^2}\left[1 - |S_0|^2\right] = \frac{\pi}{k^2}\frac{\Gamma_\alpha\Gamma_r}{(E - E_r)^2 + \Gamma_{tot}^2/4}. \tag{8.224}$$

The cross section for the compound nucleus formation from the initial state α is obtained by adding the two channels,

$$\sigma_{CN} = \sigma_{ce} + \sigma_r = \frac{\pi}{k^2}\frac{\Gamma_\alpha\Gamma_{tot}}{(E - E_r)^2 + \Gamma_{tot}^2/4}. \tag{8.225}$$

The compound-nucleus reaction has two stages, the formation and decay processes, which allows to write the cross section as a product

$$\sigma(\alpha \to \beta) = \sigma_{CN}(\alpha)P(\beta), \tag{8.226}$$

where $\sigma_{CN}(\alpha)$ is the cross section of the formation of the compound nucleus from the initial channel α, and $P(\beta)$ is the decay probability of channel β. The decay probability $P(\beta)$ can be written as

$$P(\beta) = \frac{\Gamma(\beta)}{\Gamma_f}, \tag{8.227}$$

where Γ_f is the total decay rate of the compound nucleus,

---

[6] See Eq. (8.84) for the elastic cross section. The absorption cross section is equivalent to the reaction cross section in (8.87).

$$\Gamma_f = \sum_\beta \Gamma(\beta). \tag{8.228}$$

In many available channels for β, the neutron emissions are the most dominant channels since they have no Coulomb barrier and they are also faster than the electromagnetic decays. As an example, we refer the reader to heavy-ion fusion used to synthesise super-heavy elements, as detailed in Sect. 5.2 of Chap. 5.

We demonstrate how the cross section of compound systems behaves at very low-energy neutron energy where neutron capture occurs completely inside the nucleus. It is a crude approximation, but it illustrates some important features of low-energy neutron reactions. We will write the wave function (8.206) only for the incoming wave,

$$u_0 \sim e^{-iKr}, \tag{8.229}$$

where $K = \sqrt{2m(E - V_o)}/\hbar$ ($V_0 < 0$) and V_0 is the potential depth. This wave function is justified only in the case of strong absorption since the scattering wave in (8.206) is disappeared. Then the logarithmic derivative becomes purely imaginary

$$L_0 = -iKR. \tag{8.230}$$

For such a L_0, the S matrix (8.208) is expressed as

$$S_0 = \frac{K - k}{K + k} e^{-2ikR}, \tag{8.231}$$

where $k = \sqrt{2mE}/\hbar$. Since the L_0 is imaginary, the compound-nucleus formation cross section is only composed of the reaction cross section

$$\sigma_{CN} = \sigma_r = \frac{4\pi}{k^2} \frac{Kk}{(k + K)^2}. \tag{8.232}$$

For very small neutron energy $E << |V_0|$, the wave number is $K >> k$ so that the cross section is approximated to be

$$\sigma_r \sim \frac{4\pi}{kK} \propto \frac{1}{v} \tag{8.233}$$

where we use a relation $k \propto v$. Thus we can show the slow-neutron-induced cross section is inversely proportional to the neutron velocity. This is the well-known $1/v$ law that governs the capture cross section of low energy neutrons. Experimental data of low energy neutron capture is shown in Fig. 8.29. The square-well optical-potential model calculation of slow-neutron captures is also discussed in Sect. 9.7 of Chap. 9.

In Fig. 8.29, the total neutron capture cross section for the reaction $n+^{232}$Th is shown as a function of neutron energy $E_n = 20 - 4000$ eV. The average spacing of the resonances is about 17 eV, which is about 10^5 times smaller than the single

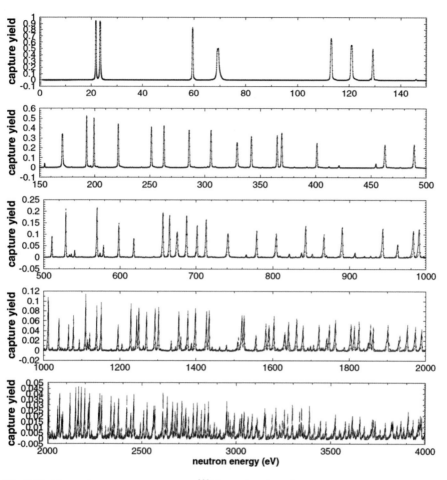

Fig. 8.29 The total neutron-capture rate on ^{232}Th as a function of the neutron energy E_n (eV). The experimental caption rate is shown by black dots together with the rate calculated from the evaluated data including Doppler and resolution broadening (red lines). Figure reprinted with permission from [30]. ©2021 by the American Physical Society

particle spacing, which is of the order of few MeV. This high density of resonances is clear evidence of many degrees of freedom involved in the formation of compound nuclei.

8.6.2 Statistical Distributions of Resonance Energies

The statistical distributions of the resonance energies and widths provide information on the structure of the compound states. In the Femi gas model, the compound nucleus corresponds to a system of large configuration mixing in the bound states and resonances. Such large configuration mixing implies that the single-particle motion in terms of shell model is distributed in the wave functions of a large number of different states. Then the distribution of the spacing D between adjacent levels with the same angular momentum and parity I^π is given by the Wigner distribution,

$$P_{\text{Wigner}}(D) = \frac{\pi D}{2\overline{D}^2} \exp(-\frac{\pi D^2}{4\overline{D}^2}), \tag{8.234}$$

where $\overline{D}$ is the average spacing. The Wigner distribution is implemented because the levels with the same I^π repel each other by the coupling matrices to cause the configuration mixings, so that the distribution $P_{\text{Wigner}}(D)$ is zero probability at $D = 0$ and has a peak at a certain value of D, depending on the shell structure of target, and decreasing to be zero at $D \to \infty$. The Wigner distributions are contrasted with the Poisson distribution

$$P(D) = \frac{1}{\overline{D}} \exp(-\frac{D}{\overline{D}}), \tag{8.235}$$

which is derived from the assumption that the eigenvalues E_i are randomly distributed, as would be expected in the Fermi gas model with no configuration mixing. In the case of no configuration mixing, the levels distribute at random, and its distribution has a peak at $D = 0$ and decreases exponentially as a function of D.

Figure 8.30 shows the histogram of observed spacings between nearest neighbor resonances in $n+^{232}$Th reaction. The Wigner distributions and the Poisson distribution are also shown. The data exhibit the absence of small spacings which implies extensive configuration mixing and is consistent with the Wigner distributions. On the other hand, the data is inconsistent with the Poisson distribution, which would apply to the circumstances where there are a large number of conserved quantities beside the angular momentum and the parity, and the configuration mixing hardly takes place.

The Wigner distribution of level spacing is considered as a manifestation of chaotic behavior of quantum many-body systems[7] since a small perturbation between adjacent levels distributes the single-particle motion into a large number of states, and the spacing is only predicted by a statistical function. Other chaotic behavior in nuclei has been discussed in nuclear masses [31], and also in quantum spectra in relation with the universality of level fluctuation laws of Sinai's billiard [32].

---

[7] The term "chaos" qualifies a system in which a small perturbation may change the many-body system drastically and the final result is not predictable in a deterministic way, but only in a statistical way.

8.7 Fusion, Energy from the Stars

So far we have discussed the direct and the compound reactions. In this section, we study fusion reactions, which was found the key reaction mechanism to create the energy of the sun. The central star of our planetary system, the sun (Fig. 8.31), survives against gravitational collapse and shines thanks to nuclear fusion reactions. Indeed, due to its mass, the pressure in the center of the sun amounts to 250 billion atmospheres which could only be sustained with a counterbalancing source of energy. Under this very high pressure, the innermost core of the sun reaches temperatures that permit nuclear fusion to take place. The core of the sun is mostly composed of hydrogen nuclei that fuse into helium nuclei following a series a reactions. Most of the solar energy is released by the following sequence of fusion reactions

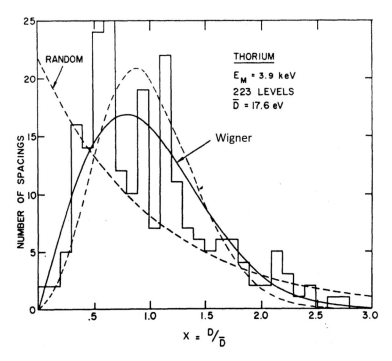

Fig. 8.30 Histogram of the observed distribution of nearest neighbour level spacings $x = D/(\overline{D})$ for ^{232}Th up to an energy $E_n = 3.9$ keV (223 resonances). The average level spacing is denoted by $\overline{D} = 17.6$ eV. The theoretical curves, long-dashed, and solid lines correspond to random (Poisson) distribution in Eq. (8.235), and Wigner distribution in Eq. (8.234), respectively. The short-dashed line shows another Wigner distribution, called the Gaussian Unitary Ensemble (GUE) distribution. Figure reprinted with permission from [33]. ©2021 by the American Physical Society

Fig. 8.31 Picture of the sun,
the central star of our
planetary system. The heat
source and subsequent light
emission of the sun comes
from nuclear fusion reactions
in its core

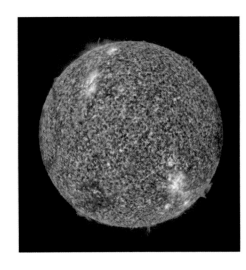

$$p + p \rightarrow d + e^+ + \nu,$$
$$d + p \rightarrow {}^3\text{He} + \gamma,$$
$${}^3\text{He} + {}^3\text{He} \rightarrow {}^4\text{He} + p + p. \qquad (8.236)$$

Most of the energy is released in the form of high-energy gamma rays of the order
of MeV. The photons escape then towards the outer shell of the sun by subsequent
scattering collisions.

By doing so, they lose energy and excite atoms in the bulk part of the sun. From
these interactions, X-rays, UV and visible light are created, giving the shining appearance of the sun. Several of these photons irradiate the surface of the earth and maintain
our planet at the temperature necessary to the creation of life as we know it.

The use of fusion at an industrial scale has been recently considered by scientists
to be a realistic solution to human energy needs for the next centuries. To utilise
this energy source on the earth as in stars, it is required to confine hydrogen ions
at temperatures reaching hundreds of thousands of degrees for long periods (few
seconds). The international community has deployed since 1995 the most important
scientific program for the coming decades: ITER. If the concept behind ITER is
demonstrated to be technologically viable, we may dream to access a clean and a
infinite source of energy in the coming decades.

Fusing two light nuclei, for example, two deuterium ions merging into an helium-
4, creates energy. The basic principle of energy production from fusion is to favor the
proper conditions for such exothermic reactions to occur in mass and develop a system
to convert the produced energy into heat, to be then transformed into electricity.
Fusion is much more difficult than fission: to make two deuterium ions fuse, they
needed to be at close distance to feel the nuclear attraction from their partner. This very
small distance is of few femtometers (10^{-15} meter). Ions are positively charged and
experience an electrical repulsion when at close distance. To penetrate the electrical

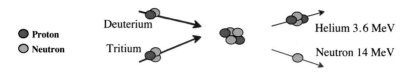

Fig. 8.32 Fusion of deuterium and tritium to produce energy. An α particle and a neutron are formed with mean kinetic energies of 3.6 MeV and 14 MeV, respectively. The reaction is very exothermic

repulsion, the ions need enough kinetic energy to overcome the Coulomb barrier originating in their positive charge. We can evaluate that fusion becomes possible when ions have a kinetic energy of 20 keV, corresponding to a velocity of about 1000 km per second. The temperature of a fluid (liquid, gas, plasma) is nothing else than a measure of the mean kinetic energy of the constituting particles. In a bucket of boiling water, the water molecules are strongly agitated (they have a rather high mean kinetic energy of meters per second). Their collisions with any probe deep into the liquid transfers energy, provide heat and give a measurement of the temperature. In a star or plasma of hydrogen ions of mean kinetic energy of 20 keV, the temperature is 100 million degrees! The relationship between the particle's kinetic energy and the temperature reads

$$E_{\text{kin}} = \frac{3}{2} k_B T, \tag{8.237}$$

where E_{kin} is the mean kinetic energy of the ions, T is the temperature of the plasma and k_B is the Boltzmann constant, $k_B = 8.617343 \times 10^{-5}$ eV/K.

Problem

8.4 Demonstrate the relation (8.239) from the ideal gas law $PV = Nk_B T$, where P is the pressure of the gas, V its volume, N its density, T its temperature and k_B the Boltzmann constant.

The reaction $d + d \rightarrow {}^4\text{He}$ is of course not the only fusion reaction that can create energy. It is actually not the one considered the most effective to produce energy. Indeed, its cross section, i.e., its probability to happen when two deuterium ions meet, is the maximum for a kinetic energy of about 1 MeV. Such conditions are extremely difficult and expensive to create. The reaction $d + {}^3\text{H}$ (t) $\rightarrow {}^4\text{He} + n + 17.6$ MeV, from the fusion of deuterium and tritium ions, presents on the other hand a very high cross section for kinetic energies close to 30 keV. In this fusion reaction, the α nucleus gains 3.5 MeV of kinetic energy and the produced neutron is emitted with a 14.1 MeV mean kinetic energy (Fig. 8.32).

The fusion reaction rate λ in the thermal equilibrium condition can be obtained by using the nuclear cross section σ and a normalized Maxell-Bolzmann distribution $\phi(v)$ as

$$\lambda = < \sigma v > = \int \sigma(E)v\phi(v)d\mathbf{v},$$

$$\phi(v)d\mathbf{v} = \left(\frac{\mu}{2\pi k_B T}\right)^{3/2} \exp\left(-\frac{\mu v^2}{2k_B T}\right) d\mathbf{v} = \frac{2}{\sqrt{\pi}}\beta^{3/2}\exp(-\beta E)\sqrt{E}dE, \quad (8.238)$$

where $\beta = 1/k_B T$ and $E = \mu v^2/2$ with the reduced mass μ of the projectile and the target. Since the thermal fusion reaction takes place at the energy much lower than the Coulomb barrier, "the astrophysical S-factor" is commonly introduced to take out a trivial exponent of the tunneling effect, the so called "Gamow factor", from the cross section as

$$\sigma(E) = S(E)\exp(-2\pi\eta)/E, \quad (8.239)$$

where

$$\eta = \frac{Z_a Z_b e^2}{\hbar v} = \frac{Z_a Z_b e^2 \sqrt{\mu}}{\sqrt{2}\hbar} \frac{1}{\sqrt{E}}. \quad (8.240)$$

In Eq. (8.240), Z_a and Z_b are the proton numbers of projectile and target, respectively, and η is called "Sommerfeld parameter". Then the reaction rate is expressed as

$$\lambda = \langle\sigma v\rangle = \sqrt{\frac{8}{\pi\mu(k_B T)^3}} \int_0^\infty e^{-b/E^{1/2}} e^{-E/k_B T} S(E)dE, \quad (8.241)$$

where

$$b = \frac{\pi Z_a Z_b e^2 \sqrt{2\mu}}{\hbar}. \quad (8.242)$$

The integrand contains the product of two exponentials. If $S(E)$ does not contain any resonance and varies as a smooth function of energy E, the product will result in a peak at the so-called Gamow energy,

$$E_G = \left(\frac{bk_B T}{2}\right)^{\frac{2}{3}}, \quad (8.243)$$

which is obtained by the method of steepest descent of two exponents,

$$\frac{d}{dE}\left[\frac{b}{E^{1/2}} + \frac{E}{k_B T}\right] = 0. \quad (8.244)$$

The ^{3}H$(d, n)^4$He reaction in Fig. 8.32 is a leading process in the primordial formation of the very light elements (mass number, $A < 7$), according to the predictions of big bang nucleosynthesis for light nucleus abundances. With its low activation energy and high yield, the ^{3}H$(d, n)^4$He is also the easiest reaction to achieve on the earth, and is pursued by research facilities directed toward developing fusion power by either magnetic (e.g., ITER) or inertial (e.g., NIF) confinement.

The *ab initio* many-body calculation of the fusion reaction ^{3}H$(d, n)^4$He is performed using NN and $3N$ forces derived in the framework of chiral effective field

Fig. 8.33 Unpolarized ^{3}H$(d, n)^4$He cross sections. **a** Astrophysical S-factor as a function of the energy in the center-of-mass (c.m.) frame, $E_{c.m.}$, compared to available experimental data (with error bars indicating the associated statistical uncertainties). **b** Angular differential cross section $\partial\sigma/\partial\Omega$ as a function of the deuterium incident energy, E_D, at the c.m. scattering angle of $\theta_{c.m.} = 0°$ compared to the evaluated data. In the figures "NCSMC" and "NCSMC-pheno" stand for the results of the no-core shell model with continuum calculations before and after a phenomenological correction of −5 keV to the position of the $3/2^+$ resonance. The figure is taken from [37]. ©2021 by Springer Nature

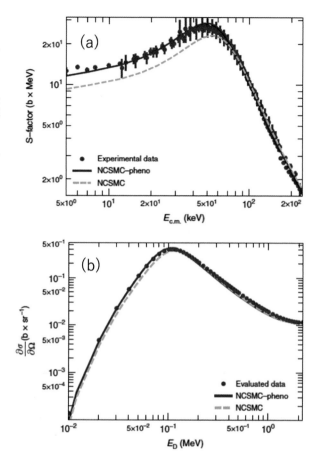

theory (EFT). The quantum-mechanical five-nucleon problem is solved using the no-core shell model with continuum (NCSMC), where the model space includes $d + t$ and $n+^4$He microscopic-cluster states, plus conventional static solutions for the aggregate ^{5}He system.

Figure 8.33a shows the astrophysical S-factor as a function of the energy in the center-of-mass (c.m.) frame, $E_{c.m.}$. The experimental peak at the c.m. energy of $E_{c.m.}$=49.7 keV corresponds to the enhancement from the $3/2^+$ resonance of ^{5}He. The calculations underpredict by 15% the experiment (green dashed line versus red circles).

To overcome this disagreement and arrive at an accurate evaluation of polarized reaction observables, a phenomenological correction of −5 keV is applied to the position of the resonance centroid, achieving a good agreement with the experimental S-factor over a wide range of energies (blue line). Figure 8.33 (b) presents the differential cross section in the center-of-mass frame at the scattering angle of $\theta_{c.m.} = 0°$ over a range of energies up to the deuterium breakup threshold. Calculated results (blue solid and green dashed lines) match the measured differential cross section well.

8.7.1 Producing Energy by Fusion

The $d+^3$H fusion reaction

$$d + t \rightarrow^4 \text{He } (3.52 \text{ MeV}) + n \text{ } (14.06 \text{ MeV}) \tag{8.245}$$

is considered the most cost effective for massive energy production. It is the main focus of fusion research today. A technological solution to implement this fusion process to produce energy efficiently at an industrial level is far from being realized today. On the other hand, the effort is worth since it might lead to a clean and almost infinite source of energy for humanity. Over the past decades, efforts and investments have increased towards such a goal.

8.7.1.1 Condition of Power Generation

Naively speaking, the problematics of energy production from fusion is quite simple: to confine a large number of deuterium and tritium at high temperature for long enough to produce sufficient energy to balance the energy cost of the apparatus. Let us estimate here the necessary condition to ignite fusion and maintain fusion. The reaction produces an energetic neutron of about 14 MeV. The collection of these energetic neutrons and the conversion of their kinetic energy into heat is at the basis of fusion reactors.

The cross section of the fusion reaction of Eq. (8.245) depends on the relative energy and is illustrated in Fig. 8.33. An efficient energy production from fusion therefore requires to reach at least several tens of keV in the deuterium-tritium center of mass. Such large kinetic energies can be reached by a mixed plasma of deuterium, tritium and electrons. The thermal energy of a plasma per unit volume is

$$\frac{3}{2}n(T_i + T_e), \tag{8.246}$$

where n is the charge density, i.e., the sum of the electron and ion densities, and T_i, T_e are the ion and electron temperatures, respectively. At equilibrium, $T_i = T_e \equiv T$. This thermal energy is lost by various processes inside the plasma: thermal conduction, convection and radiations such as electron bremsstrahlung. These energy losses per unit volume and unit time are quoted P_{loss}. The total energy confinement time τ is then defined as

$$\tau \equiv \frac{(3/2)n(T_i + T_e)}{P_{loss}} = \frac{3nT}{P_{loss}}. \tag{8.247}$$

To maintain the plasma and the energy production, it is therefore necessary that the input heating power P_{heat} equals P_{loss}. In the case of the two-body fusion reaction (8.245), the released kinetic energy per reaction is $Q = 3.52 + 14.06 = 17.58$ MeV. Assuming an equally-mixed plasma, i.e., where the density of deuterium and tritium are both equal to $n/2$, the number of fusions per unit time is given by $(n/2)(n/2)\langle\sigma v\rangle$, with $\langle\sigma v\rangle$ the reaction rate. The fusion output power per unit volume reads

$$P_{fus} = (n/2)(n/2)\langle\sigma v\rangle Q. \tag{8.248}$$

The fusion output power is converted to electrical power with a certain efficiency ϵ. The conditions of power generation are therefore given by

$$\frac{\epsilon P_{fus}}{P_{loss}} > 1, \tag{8.249}$$

which becomes, by use of Eqs. (8.247) and (8.248),

$$\epsilon(n/2)(n/2)\langle\sigma v\rangle Q > \frac{3nT}{\tau},$$

$$n\tau > \frac{12\,T}{\epsilon\langle\sigma v\rangle Q}. \tag{8.250}$$

A realistic overall energy-conversion efficiency is $\epsilon = 0.2$, while the fusion cross section requires $T \sim 10$ keV. With these values, the necessary condition to produce energy becomes $n\tau > 3.4 \times 10^{20}\text{m}^{-3}\text{s}$. There are two ways currently followed in the most forefront research to reach this condition for power generation: inertial confinement and magnetic confinement.

8.7.1.2 Inertial Confinement

Inertial confinement consists of confining deuterium and tritium at extremely high densities for very short times, of few nanoseconds. Confinement is obtained by sending very short impulses of laser waves at identical energy, exactly at the same time, on a millimeter wide solid ball made of deuterium and tritium at very low temperature. Hydrogen ice is extremely dense and many atoms of hydrogen are contained in the ball. The method is not intended to be used for large scale energy production but as a way to ignite in the laboratory fusion reactions in plasma conditions and study the fundamental aspects of the process.

The laser waves will compress the hydrogen ball by a factor of ten. The density will then increase by a factor of a thousand. The compression will create a shock wave inside the ball that will warm up the hydrogen up to a 10 keV equivalent temperature, leading to first fusion reactions. Residues from the first reactions will then warm the outer parts of the compressed hydrogen ball, then reaching proper conditions for fusion to take place. The phases of inertial fusion are illustrated in Fig. 8.34.

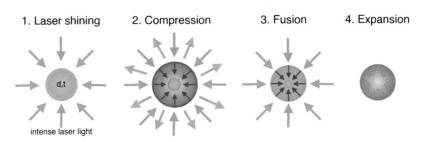

1. Laser shining   2. Compression   3. Fusion   4. Expansion

d,t

intense laser light

Fig. 8.34 Schematic description of the main steps of inertial fusion. Synchronized high-intensity laser pulses are sent on a cryogenic ball composed of deuterium and tritium ice. The laser pulses compress the target, increasing its inner density and temperature, leading to fusion. When the laser compression finishes, the plasma expands and the temperature and density conditions for fusion are quickly not fulfilled anymore

In theory, such a process could produce a hundred times more energy that it costs, but physicists and engineers are far from having demonstrated that currently available lasers can provide enough heat to the hydrogen for massive energy production. The most powerful facility dedicated to inertial confinement in the world has recently been finalized: the Laser Mega Joule facility of the french Commissariat à l'Energy Atomique et aux Energies Alternatives (CEA) was inaugurated in Bordeaux, France, in 2002 for first commissioning experiments (see Fig. 8.35). In 2019, 56 laser beams could be focused on a target. When completed, the 176 lasers of the LMJ (Laser Mega Joule) facility will deliver more than a million joule to a hydrogen *microball* in less than a nanosecond.

8.7.1.3 Magnetic Confinement

Nowadays, magnetic confinement is the most serious candidate for an economically-sustainable production of energy from fusion. The concept is to trap a plasma of hydrogen ions and electrons in a very strong magnetic field. The magnetic field is shaped to have a *torus* geometry where closed circular magnetic field lines have the shape of a donut. The most common apparatus for such magnetic field is called Tokamaks, invented in the 1950s by Russian scientists. Several heat sources are necessary to reach the ~ 10 keV temperature necessary to ignite fusion at a sufficient rate. The electrons in the plasma, by collisions with ions, will warm the plasma with a moderate heat transfer of about 3 keV. High-energy neutral atoms will be added into the plasma so that their kinetic energy can be converted into heat. Finally, oscillating electrical waves are produced to excite the plasma and provide additional heat. These different techniques have been validated in several small-size tokamaks in the last century.

The project ITER is the leader today in the domain. ITER is built over a 40 year-old scientific experience on plasma physics and tokamaks carried world-wide. The project gathers 34 countries. The overall objective of ITER is to reach a plasma of

Fig. 8.35 View of the inside of experimental reaction chamber of the the Laser Mega Joule (LMJ) facility. The design of the 140-ton aluminium reaction chamber allows the 176 laser beams to be pointed towards a cryogenic micro ball composed of deuterium and tritium at its center. In addition, diagnostics are inserted in the chamber to measure radiations from the laser-induced reactions. Credit: MS-BEVIEW-CEA

150 million degrees in a sufficient time for substantial fusion reactions and provide a definite proof that energy can be produced from nuclear fusion based on the magnetic confinement concept (see Fig. 8.36). The ITER facility is built in the south of France on a research site of CEA, a founding and main member of the collaboration, and should start operation from 2030. In addition to ITER, other facilities are built to develop the necessary research to overcome the technical difficulties related to the project. The behavior of material under very high flux of high-energy neutrons, as they will be encountered at ITER, is an unknown area. In order to characterize the materials foreseen to build the walls of the ITER tokamak and develop more robust materials, a dedicated high-intensity neutron-source facility, called IFMIF (International Fusion Materials Irradiation Facility), has been built in Japan. IFMIF will provide very high-intensity neutron beams to irradiate materials and study their properties during and after irradiation. IFMIF belongs to the ITER project.

While the goal of ITER is to produce 500 mega watts during 400 s, the next-generation system called DEMO aims at being four times more powerful (2 giga watts of thermal power) and produce energy in a continuous mode. DEMO, when built, will be the first fusion reactor. DEMO will open the way to the development of industrial fusion reactors. The technical difficulties are still to be overcome and DEMO is unfortunately not expected to be operational before 2060.

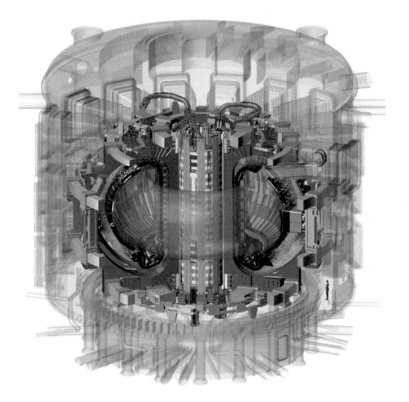

Fig. 8.36 Schematic view of the central solenoid of the ITER tokamak. The hot ion and electron plasma is colored in pink following the shape of the toroidal magnetic field. Credit: US-ITER

8.8   Fission

8.8.1   Brief History

In Roma, Italy, in 1934, Enrico Fermi and collaborators performed the first fission experiments while they were investigating neutron capture on Uranium nuclei. To their surprise, they observed that several nuclei were produced from the reactions but did not have any proper interpretation for the enigma. In these years, after the discovery of artifical radioactivity by Irène Joliot-Curie and her husband Frédéric Joliot, nuclear physicists were already predicting that soon researchers would be able to master energy production from nuclear reactions in chain. In 1938, the political situation in Europe was extremely tense. Germany and Italy were governed by totalitarian regimes, Spain was under civil war. That year, Enrico Fermi moved to the US with his family just after receiving the Nobel prize in physics. Several of his collaborators followed him. They later contributed very actively to the development of the first nuclear reactors. In late 1938, Otto Hahn, Lise Meitner and Fritz Strassmann

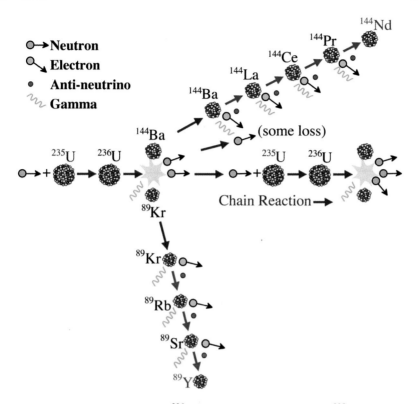

Neutron
Electron
Anti-neutrino
Gamma

^{144}Nd
^{144}Pr
^{144}Ce
^{144}La
^{144}Ba
^{144}Ba
(some loss)
^{235}U ^{236}U
^{235}U ^{236}U
Chain Reaction →
^{89}Kr
^{89}Kr
^{89}Rb
^{89}Sr
^{89}Y

Fig. 8.37 Example of fission decay of ^{236}U after a neutron is captured by ^{235}U. During fission, ^{236}U splits into two fragments, typically one heavier and one lighter nucleus, and several neutrons. The fission fragments are radioactive neutron-rich nuclei, ^{144}Ba and ^{89}Kr in this example

discovered that Barium is produced when a uranium sample is exposed to neutrons. They concluded that Uranium gave rise to a new phenomenon called "*nuclear fission*" in which heavy nuclei like Uranium are split into two fragments having almost equal mass, referred as "*fission fragments*". The fission reaction is written as

$$^{235}_{92}U + n \rightarrow {}^{140}_{56}Ba + {}^{93}_{36}Kr + 3n + Q_f, \tag{8.251}$$

where the last term of the r.h.s Q_f is the energy released by the fission reaction (see also Fig. 8.37). This is actually a huge amount of energy which we will estimate now. Several years before they found fission events, Lise Meitner attended to a lecture of special relativity by Albert Einstein in which he claimed the equivalence of mass and energy as an outcome of the concept of special relativity. Let us estimate the energy released by the fission process in Eq. (8.251) using the mass formula (4.3) in Chap. 4. The binding energy of a heavy nucleus ^{235}U is evaluated to be $B/A(A = 235, Z = 93) = 7.6$ MeV/A, while those of fission fragments are about 8.5 MeV/A. Then the mass difference between uranium and the fission fragments is

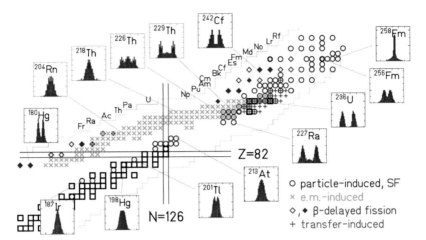

Fig. 8.38 An overview of fissioning systems investigated up to 2016 in low-energy fission with excitation energies up to 10 MeV above the fission barrier. In addition to the systems where fission-fragment mass distributions (FFMDs) were previously obtained in particle-induced and spontaneous fission (SF) (○), the nuclei for which the fission-fragment Z distributions after electromagnetic excitations were measured in an experiment in 1996 and in the recent SOFIA experiment in inverse kinematics at the FRS at GSI (×) and the fissioning daughter nuclei studied in β-delayed fission (◇) are also shown. Full diamonds mark systems produced after beta decay for which the FFMD were measured. Furthermore, 25 nuclei are marked (+) obtained from multi-nucleon-transfer-induced fission with the ^{18}O+^{232}Th and ^{18}O+^{238}U reactions. The primordial isotopes are indicated by squares. Figure modified from [34] and reprinted with permission from [35]. ©2021 by the American Physical Society

$$\Delta M (\equiv Q_f) = M(A = 235, Z = 92) - M(140, 56) - M(93, 36) - 2M_n$$
$$= B(A = 93, Z = 36) + B(140, 56) - B(235, 92)$$
$$= 8.5 \times 233 - 7.6 \times 235 = 195 \, \text{MeV}. \tag{8.252}$$

The ΔM is called the missing mass and is released from Uranium as the kinetic energies of fission fragments (80%). They lose their energy by heating the medium when slowed down. Light particles such as neutrons, and decay γ rays represent 15% of the released energy while neutrinos take away about 5% of the energy.

There are a full of variety in the way where a nucleus divides itself in a fission process. Nuclei tend not to split into equal fragments, but they split into two cluster zones. The present status of experimental fission research is exhibited in Fig. 8.38, and gives an overview of the observed fission-fragment mass and nuclear-charge distributions. These are among the most prominent signatures of nuclear structure in low-energy fission. The large variations of the fission-fragment distributions with the mass and nuclear charge of the fissioning system are shown clearly as a well-established evidence.

8.8.2 Models for Fission

To induce the fission process, there is a barrier to overcome. The barrier manifest through an increase in the potential energy for the increase of deformation. There are two possibilities to overcome the barrier: the first one is to give the nucleus an excitation energy greater than the barrier. This is called the *induced fission*. The second possibility is the *spontaneous fission*, where the barrier is surpassed by the tunneling effect starting from the ground state of the nucleus.

The most useful model for explaining fission phenomena intuitively is the liquid drop model (see subsection 4.1.2 of Chap. 4). This model permits to calculate the change in the potential energy when the nucleus suffers deformations from the spherical shape. Soon after the discovery of fission in 1938, the potential energy was described in terms of a liquid-drop model in which the potential energy is the sum of shape-dependent surface and Coulomb energy terms. In the first model of the fission barrier, the liquid-drop model, Bohr and Wheeler expanded the Coulomb and surface energies up to the fourth order in β_2. The change of the potential energy will be estimated by a simple constant density profile with the prolate deformation given by

$$r = R_0(1 + \beta_0 + \beta_2 Y_{20}(\theta, \phi)), \qquad (8.253)$$

where R_0 is the radius of the original spherical shape and β_2 is the quadrupole deformation parameter. The constant term β_0 is determined by the volume conservation of the liquid drop as $\beta_0 = -\beta_2^2/(4\pi)$ since the nuclear volume is invariant under this prolate deformation to be $V = (4\pi/3)R_0^3$. The surface and the Coulomb energy terms in the mass formula change with the deformation as

$$E_S(\beta_2) = E_S(\beta_2 = 0)(1 + \frac{2}{4\pi}\beta_2^2), \qquad (8.254)$$

$$E_C(\beta_2) = E_C(\beta_2 = 0)(1 - \frac{1}{4\pi}\beta_2^2), \qquad (8.255)$$

where the higher order terms more that $O(\beta_2^2)$ are discarded. Thus, the total energy gained by the deformation is

$$\Delta E = E_S(\beta_2 = 0)\frac{2}{4\pi}\beta_2^2 + E_C(\beta_2 = 0)(-\frac{1}{4\pi}\beta_2^2) = -a_S A^{2/3}\frac{2}{4\pi}\beta_2^2 + a_C \frac{Z^2}{A^{1/3}}\frac{1}{4\pi}\beta_2^2. \qquad (8.256)$$

If ΔE is positive, the spherical shape is stable, but unstable if ΔE is negative. The critical point is determined by the ratio,

$$\left(\frac{Z^2}{A}\right)_C = \frac{2a_s}{a_C} \approx 50. \qquad (8.257)$$

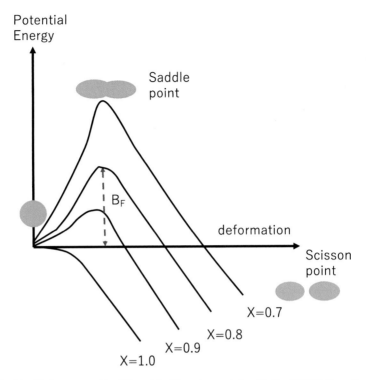

Fig. 8.39 Liquid-drop energy for different fissility values x defined in Eq. (8.258). Here, B_F denoted the fission barrier. The horizontal axis shows the magnitude of deformation

The quantity (8.257) has an important role to estimate the probability of the spontaneous fission and called the *fissionability parameter*. The *fissility* parameter x is further defined as,

$$x = \frac{Z^2/A}{(Z^2/A)_C},$$ (8.258)

which is useful to judge the fission probability of each nucleus. The formulas of Eqs. (8.254) and (8.255) can be extended to any higher multipole deformation. Figure 8.39 shows the change of the liquid drop energy for different values of fissility x.

Historical development of fission-barrier models are summarized in Fig. 8.40. In 1947, Stan Frankel and Nicolas Metropolis published an important paper in which they calculated the Coulomb and surface energies of more highly deformed nuclear shapes using numerical integration. This was one of the first basic physics calculations done on a digital computer. In 1955, Wladyslaw J. Swiatecki suggested that more realistic fission barriers could be obtained by adding a "correction energy" to the minimum in the liquid-drop-model barrier. The correction was calculated as the difference between the experimentally observed nuclear ground-state mass and the mass given by the liquid-drop model. In the middle of the 1960s, Vilen M. Strutinsky

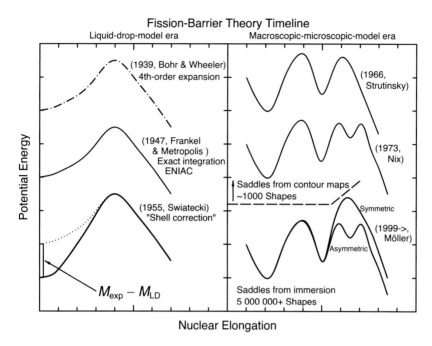

Fig. 8.40 Historical development of fission models. See the text for detailed explanations. Figure reprinted with permission from [36]. ©2021 by the American Physical Society. Coutesy P. Möller

presented a method to theoretically calculate these shell corrections as was studied in Chap. 7. His method led to the realization that actinide fission barriers are "double humped": beyond the ground-state minimum there are two saddles or maxima in the fission barrier, separated by a fission-isomeric second minimum. This is understood in the followings: the shell corrections give an extra energy in the region of high level density, while it is opposite in the case of a large shell gap. As a result, the potential energy shows double humped structure, i.e., one minimum corresponds to the ground state, and another one may give rise to the fission isomer. In the microscopic-macroscopic models, the fission potential energy was calculated for a few hundred different shapes in the 1970s and found three peaks in the barrier. In the modern calculations in [36], the potential energy for several million different shape points in a three dimensional deformation space ε_2 (elongation), ε_4 (neck coordinate), and γ (axial asymmetry) are considered for identifying relevant saddle points on the paths from the nuclear ground state to the separated fission fragments.

There are also many sophisticated microscopic theories to calculate the fission reactions. Microscopic models for fissions are developed on the density functional theories. Constrained and time-dependent Hartree-Fock calculations with Bardeen–Cooper–Schrieffer (BCS) pairing correlations were done recently in a three-dimensional Cartesian geometry including both the quadrupole and octupole deformations with a Skyrme energy density functional and a surface pairing interaction

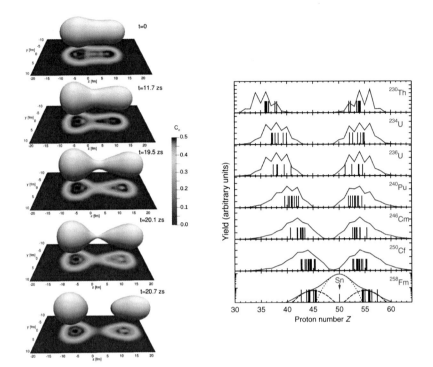

Fig. 8.41   (Left) Microscopic calculations of asymmetric fission of ^{240}Pu. In the figures, iso-density surfaces at 0.08 fm^{-3} (half the saturation density) are shown at different times. The localization function C_n of the neutrons is shown in the projections. The scale is given on the right side of the projections. In the time evolution, a scission occurs at t$\approx$20 zs (1 zs = 10^{-21} s). The quadrupole and octupole deformation parameters at scission are $\beta_2 \approx 0.16$ and $\beta_3 \approx 0.22$ for the heavy fragment (left) and $\beta_2 \approx 0.64$ and $\beta_3 \approx 0.4$ for the light fragment (right), respectively. (Right) Predictions of the expectation values $\langle Z \rangle$ of the number of protons in the fission fragments (vertical lines) are compared with fragment proton number distributions (solid lines) extracted from experimental results of thermal-neutron-induced fission (top six panels) and the spontaneous fission of ^{258}Fm (bottom panel). Dashed and dotted lines in the bottom panel are Gaussian fits of the asymmetric and symmetric components, respectively. Figure reprinted with permission from [38]. ©2021 by Springer Nature. Courtesy G. Scamps, Université Libre de Bruxelles

[38]. The left panel of Fig. 8.41 shows the dynamical evolution of isodensity surfaces of ^{240}Pu, starting from a configuration in the asymmetric-fission valley. The final state corresponds to a quantum superposition of different repartitions of the number of nucleons between the fragments, with the average $\langle Z \rangle \equiv 53.8$ protons and $\langle N \rangle \equiv 85.2$ neutrons in the heavy (left) fragment. In the figure, the scission occurs after about 20 zs (1 zs=10^{-21}s), but can reach up to 90 zs depending on the initial configuration. Despite large fluctuations in the time to the scission, almost all heavy fragments are formed with $Z = 52 - 56$ protons in the right panel of Fig. 8.41 in agreement with experimental data. The fact that the number of protons shows little

dependence on the initial elongation is due to the slow viscous motion of fissioning nuclei in the model.

8.8.3 Energy Production from Fission

At the very beginning of the second world war (WWII) in 1939, several European nuclear physicists who emigrated to the USA persuaded Albert Einstein, the most famous and recognized physicist of the time, to write to the US president Franklin Roosevelt and convince him that there was a great danger that Germany was mastering nuclear energy and developing a nuclear weapon that could be used against the USA and its coalition. Before his death, Einstein confessed that he regretted his letter which was the starting point of the Manhattan project and ended up in the nuclear bombs that killed thousands of Japanese citizens at the end of World War II in 1945. When the US decided to develop a nuclear weapon, no such technology existed. The so-called Manhattan project started from scratch in 1939. In 1945, more than 130000 people were working for it and reached an overall cost of 2 billion dollars of that time, equivalent to 30 billion dollars of today. More than 90% of the expenses were spent for the building of plants and the preparation of the necessary fissile materials. Less than 10% corresponded to the development of the weapons themselves. The project started in universities such as Columbia, Berkeley and Chicago. Several laboratories were then built such as Oak Ridge, Tennessee, and Los Alamos, New Mexico. The best nuclear physicists of the time worked on the project, including four recipients of physics Nobel prizes: Niels Bohr, James Chadwick, Enrico Fermi and Isidor Isaac Rabi. In July 1945, the first nuclear explosion of humankind's history was triggered in the Alamogordo desert of New Mexico (Fig. 8.42). The bomb was based on plutonium and released an incredible amount of energy equivalent to 19000 tons of Trinitrotoluene (TNT). On August 6th and 9th 1945, the most deadly acts of war were perpetrated with the bombing of Hiroshima and Nagasaki, respectively. They led to the official end of the second world war. The production of plutonium was one of the main objectives of the Manhattan project. To obtain significant quantities of Plutonium, the development of a nuclear reactor was necessary to irradiate uranium material with neutrons. ^{238}U was then transmitted into ^{239}U which rapidly decays into ^{239}Np and then into ^{239}Pu. Fermi was in charge of the plutonium production. In 1942, the first controlled nuclear reaction in chain took place at the University of Chicago, in a hidden laboratory below the terraces of an American-football field. The very first nuclear reactor was built.

Just after the war, in 1946, another small reactor was built for energy production. The first fission reactor for energy production was connected to the American electricity network in 1951. Three years later, the Soviet Union produced energy from its own nuclear reactors. Between the 1950s and 1960s, several other countries mastered nuclear science for energy production: France, Sweden, Japan, China, Germany and India. This was the start of the nuclear area for energy production.

Fig. 8.42 Explosion of the first nuclear bomb Trinity on July 16th, 1945, 25 ms after the nuclear explosion took place (left) and few seconds later (right) in the Alamogordo desert of New Mexico, USA. Courtesy Los Alamos National Laboratory

A reactor is composed of three main components: (1) the fuel, i.e., the fissile material which fissions after the absorption of a low-energy neutron, (2) the moderator used to slow down the neutrons produced during the fission reaction, (3) the coolant (heat transfer fluid) that moves the heat produced by the nuclear reactions to an external system dedicated to the conversion of the heat to electricity. Today, most reactors dedicated to energy production are built as *thermal neutron reactors*, meaning that the fission process is induced by neutrons slowed down to kinetic energies below 1 eV. In these systems, the moderator and the heat transfer fluid are the same fluid composed of water, either boiling or, more frequently, at a high pressure of 150 bars. A schematic view of a nuclear power plant is illustrated in Fig. 8.43. The fuel is mostly composed of enriched uranium (^{238}U with 3% of ^{235}U). A reactor requires an extreme care in the control of the reaction rate: the overall rate of neutrons produced by fission should exactly compensate the number of captured neutrons to trigger fission. Indeed, if too many neutrons are produced, the number of fissions will increase exponentially and lead to an explosion, while the chain reactions will stop if not enough neutrons are produced. These rates are governed by several factors that can be controlled and adjusted during the reactor operation.

Nuclear science has today about 80 years of developments. From the discovery of fission, incredible progresses have been made and nuclear science has been matured enough since the 1950s to become a workhorse of mass energy production on the planet. The principle of producing energy from nuclei is based on a simple concept: when nucleons are bound together to form a nucleus, the sum of their mass is less than the mass of the final nucleus. Mass is converted into binding energy. The binding energy of a nucleus finds its origin in the strong nucleon-nucleon interaction. The binding energy of a nucleus is detailed in Chap. 4.

The binding energy per nucleon involved is the maximum for ^{56}Fe: fusing two light nuclei will release energy, as seen in Sect. 8.7, whereas dividing a nucleus bigger than Fe will also produce energy. The latter is now a well established technology.

Any living system, vegetable or animal spends a large part of its time seeking and absorbing sources of energy to survive. Plants develop roots to seek miner-

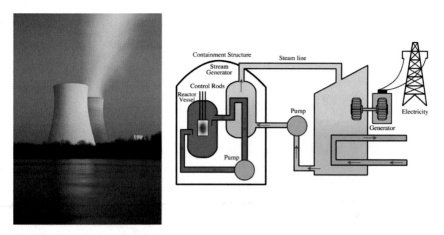

Fig. 8.43 (Left) Picture of the cooling chimneys of the Buguey nuclear power plant in France. Water vapor from the secondary fluid circuitry is released into the atmosphere. (Right) Scheme of a nuclear reactor. The combustion of uranium, the fuel, produces neutrons which, after moderation of their energy by water, will induce further fissions. The energy produced by nuclear reactions is conveyed by the coolant to an external circuit where turbins convert heat to electrical power

als and water in the ground, they grow leaves to capture light from the sun, while killer whales, for example, spend 80% of daylight hours foraging for fish and seals. Mankind is no exception. A sizable part of our days are spent eating and buying food, whereas a significant part of incomes is usually spent in electricity, gasoline and house heating. Modern societies have very high energy consumption. Let us define a few units to measure energy. The most common, at the level of our daily consumption, is the *calorie*. Historically, one calorie is the energy required to warm up one gram of water at an initial temperature of 14 Celsius degrees by one single degree. A human individual with normal activity should absorb about 2,000 kcal every day for a long-term survival. But the overall expenditure of energy in developed countries is about 100,000 kcal/person, well beyond primary needs.

The energy consumption shows very large variations from one country to another. While the mean energy consumption per individual is similar for Japan, France, Germany, it is almost twice in the USA and thirty times less in Bangladesh. This factor finds its origin in several sources: the industrial development of the country, its geography and cultural aspects. See Fig. 8.44 for an illustration of "light pollution" on the earth, where inequalities in population densities and energy consumption are visible.

Two main factors contribute to the increase of the world's energy demand in the coming decades: the population will increase and most countries are under significant development and energy needs. Realistic estimates give a 100% increase of energy demand in the next 30 years. If this is the case, there is no other solution today than the further development of nuclear energy. It is the responsibility of political leaders and scientists, especially those specialized in nuclear technology, to improve the current

Fig. 8.44 Light *pollution* on the earth, at night, viewed from space. The image has been built from different pictures

state-of-the-art to reach safer, cleaner and more efficient ways of producing energy, as well as discovering new ways to satisfy the population energy needs.

8.8.4 New Generation Reactors

Unfortunately, the power of nuclear technologies has not led to progresses only, but has played a major role in few of the darkest moments of humanity. The dreadful bombing of populations at Hiroshima and Nagasaki in Japan in 1945, as well as major nuclear accidents at Three Miles Islands in the US in 1957, Chernobyl in Ukraine in 1981 and in Fukushima in Japan in 2011 demonstrate that nuclear science development should be employed and developed with the utmost control and responsibility.

Although the safety has been part of nuclear power plants from the beginning, accidents occur. In avoiding such accidents the industry has been successful. In over 14,500 cumulative reactor-years of commercial operation in 32 countries, there have been only three major accidents to nuclear power plants as mentioned already. Review of accidents related to energy production from 1970 to 2000, reveals that nuclear energy has led to 50 deaths per produced TeraWatt/year (the amount of energy used by the world in about 5 months), while hydro-power accounts to 10,000 deaths (the largest accident resulted from the failure of the Chinese Banqiao dam in 1975 during which 26000 people died from the flooding), coal 800 deaths and natural gas 200 deaths. It of course does not mean that the nuclear energy community should be satisfied: a zero death level is still the ambition. Even safer and more robust power plants should be developed. There are also reasons for new schemes and more efficient reactors: world resources of uranium as they are known today can only fulfill our needs to produce energy for about 70 years with the current technology. Indeed, the fuel of current reactors contains 3% of ^{235}U which composes only 0.7% of natural uranium, most uranium being ^{238}U. Today's electrical consumption is expected to exhaust the known resources of ^{235}U in less than a century.

The international community has started the development of new generation nuclear reactors based on fission. The future reactors are named as 4th generation. Most of the foreseen new designs are not expected to be operational before 2030–2040. The list of 4th generation designs under study gathers six different systems: three are thermal neutron reactors, and three are based on the use of fast neutrons. These foreseen systems offer significant advances in sustainability, safety and reliability. They differ from their combination of fuel, moderator and coolant. As examples, two demonstrators for fast neutron reactors are foreseen in Europe: the ASTRID reactor, a sodium cooled reactor to be built in France from 2019, and the MYRRHA reactor, a liquid-lead cooled reactor under construction in Belgium. The latter has the original feature of requiring an external source of energy (a proton beam produced from an accelerator) to maintain the chain reactions. This aspect of the design makes the reactor extremely safe in regards of accidental divergence of the reactor. These fast neutron reactors have the strong advantage to use plutonium or ^{238}U as fuel. Their development would lead to more efficient reactors with less nuclear waste produced.

Summary

The direct reactions are commonly used to analyse the scattering problems in nuclear physics. The direct reactions are classified as "elastic", "inelastic", "particle transfer" and "break-up" channels and written symbolically as

$$a + A \rightarrow a + A \quad \text{or} \quad A(a, a)A \quad \text{(elastic scattering)}$$
$$a + A \rightarrow a' + A^* \quad \text{or} \quad A(a, a')A^* \quad \text{(inelastic scattering)}$$
$$a + A \rightarrow b + B \quad \text{or} \quad A(a, b)B \quad \text{(particle transfer reaction)}$$
$$a + A \rightarrow b + X \quad \text{or} \quad A(a, b)X \quad \text{(breakup/knock-out reaction)}$$

where a and A denote projectiles and targets, respectively, and a', A^* means the excited states. "b''" and "B''" are projectile and target after the particle transfer reaction. The particle transfer reaction has two channels: particle pickup and particle stripping reactions. More than two fragments exist in the final state of breakup and knock-out reactions. X corresponds to nuclei/nucleons after breakup (knock-out) reactions.

The scattering amplitude is expressed in Eqs. (8.33) and (8.32) by the T matrix as

$$f^{(+)}(\mathbf{k}', \mathbf{k}) = -2\pi^2 \frac{2m}{\hbar^2} T(\mathbf{k}', \mathbf{k}).$$

with

$$T(\mathbf{k}', \mathbf{k}) \equiv \langle \phi_{\mathbf{k}'} | V | \psi_k^{(+)} \rangle,$$

where $\psi_k^{(+)}$ is the full wavefunction with an incoming plane wave and an outgoing spherical waves and $\phi_{\mathbf{k}'}$ is the outgoing wave. The differential cross section $d\sigma(\theta)$ at the solid angle $d\Omega$ is expressed by the scattering amplitude (8.58) as,

$$\frac{d\sigma(\theta)}{d\Omega} = |f(\theta)|^2,$$

where θ is the angle between $\mathbf{k}$ and $\mathbf{k}'$. The cross section of the inelastic channel $\alpha \to \beta$ is also given by (8.64) in a similar manner to the elastic channel of Eq. (8.58) as,

$$\frac{d\sigma_{\beta\alpha}(\theta_\beta)}{d\Omega_\beta} = \frac{v_\beta}{v_\alpha} |f_{\beta\alpha}(\theta_\beta)|^2 = \frac{m_\alpha m_\beta}{(2\pi\hbar^2)^2} \frac{k_\beta}{k_\alpha} |T_{\beta\alpha}|^2,$$

where v_α and v_β are the relative velocities of the initial and final channels and k_α and k_β are their wave numbers, respectively. Equation (8.64) is the fundamental equation for all reactions in general. Approximations are necessary to solve the $T_{\beta\alpha}$ matrix for each reaction channel.

The scattering problem is essentially to obtain the wave function (8.25),

$$\psi_k^{(\pm)}(\mathbf{r}) = \phi(\mathbf{r}) + \chi^{(\pm)}(\mathbf{r}) = \frac{1}{(2\pi)^{3/2}} e^{ikz} - \frac{1}{4\pi} \int \frac{e^{\pm ik|\mathbf{r}-\mathbf{r}'|}}{|\mathbf{r}-\mathbf{r}'|} U(r') \psi_k^{(\pm)}(\mathbf{r}') d\mathbf{r}',$$

solving the Helmholtz equation (8.12),

$$(\nabla^2 + k^2)\psi(\mathbf{r}) = U(r)\psi(\mathbf{r}),$$

where $k^2 = 2mE/\hbar^2$ and $U = 2mV/\hbar^2$. Equation (8.25) is expressed in a convenient way as in (8.53),

$$|\psi_k^{(\pm)}\rangle = (1 + G^{(\pm)}(E_k)V)|\phi_k\rangle,$$

with the correlated Green's function (8.52),

$$G = \frac{G_0}{1 - VG_0}.$$

Equation (8.53) is referred to as the Lippman-Schwinger equation. The advantage of the Lippman–Schwinger equation is that the r.h.s. is expressed by using only the initial state $|\phi_k\rangle$ and is solvable when the correlated Green's function (8.52) is obtained.

In general, it is difficult to solve exactly (8.25) to obtain the scattering wave function $\psi^{(\pm)}$. There are several approximations to evaluate the scattered wave function and the cross section. A commonly used approximation is the Born approximation. In the plane wave first-order Born approximation (PWBA), the scattered wave function $\psi_k^{(+)}$ is replaced by the plane wave $\phi_k^{(+)}$ in the T matrix (8.32). In the first-order distorted wave Born approximation (DWBA), $\psi_k^{(+)}$ is replaced by the wave function $\chi(\mathbf{k}, \mathbf{r})$ in Eq. (8.139), which has a distortion effect by the static nuclear potential. As a characteristic feature of the Lippman-Schwinger equation, the higher-order terms of the Born approximation can be evaluated by the expansion of correlated Green's function as

$$GV = \frac{G_0}{1 - VG_0}V = G_0V + G_0VG_0V + \cdots,$$

where, in the r.h.s., the first, and second terms correspond to the first- and the second-order Born approximations, respectively.

The S matrix has a relation with the T matrix in the operator form (8.79),

$$S = 1 - 2\pi i T.$$

The S matrix has several advantages to describe scattering problems;

- The S matrix is unitary: $S^\dagger S = 1$ when the potential is real.
- Consequently, the S matrix is useful to describe the conservation laws in the scattering process, and has a direct link to the optical theorem.
- The S matrix is useful to describe the scattering in terms of phase shifts, especially the resonance state.

The scattering amplitude can be expressed by the partial wave expansion of the S matrix, S_l, as

$$f(\theta) = \frac{1}{2ik}\sum_l (S_l - 1)(2l + 1)P_l(\cos\theta), \tag{8.82}$$

where $P_l(\cos\theta)$ is the Legendre polynomial of ℓth order.

The optical model is a model that provides an effective interaction which replaces the complicated many-problem between two colliding nuclei by a simple problem of two objects interacting through a potential, the so-called optical potential. The optical potential is considered to take into account the effect of the Q space for the potential for the active P space expressed by the effective potential,

$$V_{\text{eff}} = PVP + PVQ\frac{1}{E - QHQ + i\varepsilon}QVP \equiv PVP + \Delta V(E). \tag{8.112}$$

The optical potential has the following characteristics;

- It is complex with an absorptive imaginary part describing the loss of flux from the model space P into the open channels in the Q space.
- The optical potential is non-local in space (or momentum dependence) because of couplings to the eliminated channels.
- It is also energy dependent because of couplings to the eliminated channels.
- It depends on the active model space P.

In the inelastic and transfer reactions, the interaction V is divided into two parts,

$$V = U + \Delta U,$$

where U is the optical potential for scattering states and ΔU is the transition potential for the excitation of the initial state or the transfer of particles.

The compound nucleus is characterized by the highly excited intermediate stage and does not keep any memory of the initial state so the decay channel does not depend on the structure of initial state. Main features of the reaction are

- The compound nucleus is usually created if the projectile has low energy and the incident particle interacts inside of the nucleus.
- The time scale of compound nucleus reactions is much longer than the time scale of the projectile passing through the target.
- A compound nucleus is a relatively long-lived intermediate state of the particle-target composite system.
- Compound nucleus reactions involve a large number of nucleon-nucleon collisions, which lead to a thermal equilibrium inside the compound nucleus.
- Products of the compound nucleus reactions are distributed nearly isotropically in angle since the nucleus loses memory of how it was created. This is called the Bohr's hypothesis of independence.
- The decay mode of a compound nucleus does not depend on the way in which the compound nucleus is formed. Resonances in the cross-section are typical for the compound nucleus reaction.

Macroscopic models based on the liquid drop model are applied for studying the nuclear fission processes. A microscopic TDHF model+BCS model is now feasible to describe the symmetric and asymmetric fissions of heavy nuclei $A \geq 230$.

Nuclear fission and fusion are discussed in terms of nuclear solar energy and nuclear power plants. Modern technology for nuclear energy production is also introduced, such as inertial and magnetic confinement techniques of nuclear fusion.

Solutions

8.1 Equation (8.23) can be expressed as

$$G_0(r) = \frac{1}{(2\pi)^3} \int_0^\infty k'^2 dk' \int_{-1}^1 d(\cos\theta) \int_0^{2\pi} d\phi \frac{e^{ik'r\cos\theta}}{k^2 - k'^2}$$

$$= \frac{1}{4\pi^2 r} \int_0^\infty \frac{k'}{k^2 - k'^2} \cdot 2 \cdot \left[\frac{e^{ik'r} - e^{-ik'r}}{2i} \right] dk' = \frac{1}{2\pi^2 r} \int_{-\infty}^\infty \frac{k'}{k^2 - k'^2} \sin(k'r) dk'.$$

$$(8.259)$$

Let us write explicitly the sum $[I_+ - I_-]$:

$$[I_+ - I_-] = \int_{-\infty}^\infty \left[\frac{1}{k' + k} + \frac{1}{k' - k} \right] \left(e^{ik'r} - e^{-ik'r} \right) dk'$$

$$= \int_{-\infty}^\infty \frac{k' - k + k' + k}{k'^2 - k^2} \sin(k'r) \cdot 2i dk' = -4i \int_{-\infty}^\infty \frac{k' \sin(k'r)}{k^2 - k'^2} dk'.$$

$$(8.260)$$

Equations (8.259) and (8.260) give

$$G_0(r) = \frac{1}{2\pi^2 r} \left(-\frac{1}{4i} \right) [I_+ - I_-] = -\frac{1}{8\pi^2 r i} [I_+ - I_-].$$

$$(8.261)$$

The Cauchy integral formula states

$$\oint_C \frac{f(z)dz}{z - z_0} = 2\pi i \begin{cases} f(z_0), & z_0 \text{ interior} \\ 0, & z_0 \text{ exterior} \end{cases},$$

$$(8.262)$$

with a pole at z_0, and $f(z)$ analytic on C and $f(z_0) \neq 0$.
In the case $f = e^{\pm ik'r}$, the poles exist at $\pm k$. Substituting (8.262) into (8.261), we prove (8.24);

$$G_0(r) = -\frac{1}{8\pi^2 r i} (2\pi i) e^{\pm ikr} = -\frac{1}{4\pi r} e^{\pm ikr}.$$

$$(8.263)$$

8.2 Let us start with the 1st case for $A = E - H$ and $B = E - H_0$ and substitute into Eq. (8.40):

$$\frac{1}{E - H} = \frac{1}{E - H_0} + \frac{1}{E - H_0} (E - H_0 - E + H) \frac{1}{E - H} = \frac{1}{E - H_0} + \frac{1}{E - H_0} V \frac{1}{E - H},$$

$$(8.264)$$

where $H = H_0 + V$. Using Eqs. (8.35) and (8.34), we obtain

$$G = G_0 + G_0 V G.$$

$$(8.265)$$

Similarly for the second case, $A = E - H_0$ and $B = E - H$, gives

$$\frac{1}{E - H_0} = \frac{1}{E - H} + \frac{1}{E - H} (E - H - E + H_0) \frac{1}{E - H_0}$$

$$= \frac{1}{E - H} - \frac{1}{E - H} V \frac{1}{E - H_0}, \tag{8.266}$$

and we can obtain

$$G_0 = G - GVG_0 \quad \Longrightarrow \quad G = G_0 + GVG_0. \tag{8.267}$$

8.3 a) In the initial state, the energy-momentum is expressed as

$$(E_A, \mathbf{p}_A) + (E_p, \mathbf{p}_p). \tag{8.268}$$

Using the relativistic relation $\mathbf{p} = \gamma\beta m$, and the fact that the proton is at rest in the initial state in the laboratory frame, Eq. (8.268) can be written as

$$(M_A + T_A, \gamma\beta M_A) + (m_p, \mathbf{0}) \tag{8.269}$$

(we use the natural unit $c \equiv 1$ and the fact that the proton is at rest hereafter). In the final state we have

$$(E_{A-1}, \mathbf{p}_{A-1}) + (E_1, \mathbf{p}_1) + (E_2, \mathbf{p}_2) \tag{8.270}$$

$$= (M_{A-1} + T_{A-1}, \mathbf{p}_{A-1}) + (m_p + T_1, \mathbf{p}_1) + (m_p + T_2, \mathbf{p}_2),$$

where M_{A-1} contains information about the excitation energy of the $(A - 1)$ system. From momentum conservation we get

$$\mathbf{p}_A = \gamma M_A \beta = \mathbf{p}_{A-1} + \mathbf{p}_1 + \mathbf{p}_2. \tag{8.271}$$

The $(A - 1)$ four-momentum in its system is

$$P_{A-1} = (E_{A-1}, \mathbf{p}_{A-1}) = (M_{A-1}, -\mathbf{q}). \tag{8.272}$$

The energy-momentum Lorentz transformation is given by:

$$E' = -\gamma\beta\mathbf{p}_{||} + \gamma E, \tag{8.273}$$

$$\mathbf{p}'_{||} = \gamma\mathbf{p}_{||} - \gamma\beta E, \tag{8.274}$$

$$\mathbf{p}'_{\perp} = \mathbf{p}_{\perp}. \tag{8.275}$$

We now move from the nucleus frame, at which $\mathbf{p}_{||} = -\mathbf{q}_{||}$ and $\mathbf{p}_{\perp} = -\mathbf{q}_{\perp}$, to the laboratory frame using Eqs. (8.273)–(8.275):

$$E_{A-1}|_{lab} = -\gamma\beta\mathbf{q}_{||} + \gamma M_{A-1} \qquad (8.276)$$

$$\mathbf{p}_{A-1||}|_{lab} = -\gamma\mathbf{q}_{||} + \gamma\beta M_{A-1} \qquad (8.277)$$

$$\mathbf{p}_{A-1\perp}|_{lab} = -\mathbf{q}_{\perp}. \qquad (8.278)$$

Substituting (8.278) into (8.271), and using $\beta = \beta_{||}$ gives

$$0 = -\mathbf{q}_{\perp} + \mathbf{p}_{1\perp} + \mathbf{p}_{2\perp} \quad \Rightarrow \quad \mathbf{q}_{\perp} = \mathbf{p}_{1\perp} + \mathbf{p}_{2\perp}. \qquad (8.279)$$

Substituting (8.277) into (8.271) gives

$$\gamma M_A \beta = -\gamma\mathbf{q}_{||} + \gamma\beta M_{A-1} + \mathbf{p}_{1||} + \mathbf{p}_{2||} \qquad (8.280)$$

$$\implies \quad \mathbf{q}_{||} = \frac{\mathbf{p}_{1||} + \mathbf{p}_{2||} - \gamma\beta(M_A - M_{A-1})}{\gamma}.$$

The separation energy of the removed proton is defined by

$$E_s = \left(m_p + M_{A-1}\right) - M_A. \qquad (8.281)$$

We can write the energy conservation from Eqs. (8.268) and (8.270) in the A frame nucleus as

$$M_A + E_p = (M_{A-1} + T_{A-1}) + E_1 + E_2. \qquad (8.282)$$

Following Eq. (8.273) the protons energies at this frame are

$$E_p|_A = \gamma m_p, \qquad E_{1(2)}|_A = -\gamma\beta\mathbf{p}_{1(2)||} + \gamma\left(T_{1(2)} + m_p\right). \qquad (8.283)$$

b) Substituting (8.283) into (8.282), and assuming that the kinetic energy of the $(A - 1)$ nucleus is, within the non-relativistic limit, i.e., $T \approx p^2/2m$, we get,

$$M_A + \gamma m_p = M_{A-1} + \frac{q^2}{2M_{A-1}} - \gamma\beta\left(\mathbf{p}_{1||} + \mathbf{p}_{2||}\right) + \gamma\left(T_1 + T_2\right) + 2\gamma m_p$$

$$(8.284)$$

$$\implies \quad M_{A-1} - M_A = \gamma\beta\left(\mathbf{p}_{1||} + \mathbf{p}_{2||}\right) - \gamma\left(T_1 + T_2\right) - \gamma m_p - \frac{q^2}{2M_{A-1}}.$$

Adding m_p to both sides and adding and subtracting $(\gamma - 1)m_p$ gives

$$E_s = (\gamma - 1)m_p + \gamma\beta\left(\mathbf{p}_{1||} + \mathbf{p}_{2||}\right) + \gamma\left(T_1 + T_2\right) - 2\left(\gamma - 1\right)m_p - \frac{q^2}{2M_{A-1}}.$$

$$(8.285)$$

To evaluate the $(\gamma - 1)m_p$ term let us calculate the invariant mass (which by its definition is equal in the laboratory and the A frames):

$$M^2 = \left(\sum_k E_k \right)^2 - \left\| \sum_k \mathbf{p}_k \right\|^2. \tag{8.286}$$

For two particles scattering we get

$$M^2 = m_1^2 + m_2^2 + 2 \left(E_1 E_2 - \mathbf{p}_1 \cdot \mathbf{p}_2 \right), \tag{8.287}$$

which reduces to

$$M^2 = m_A^2 + m_p^2 + 2 E_A E_p \tag{8.288}$$

in our case, since $\mathbf{p}_p |_{lab} = 0$ and $\mathbf{p}_A |_{lab} = 0$. Requiring $M^2 |_{lab} = M^2 |_A$:

$$(M_A + T_A) m_p = M_A \gamma m_p \quad \Longrightarrow \quad m_p (\gamma - 1) = \frac{T_A}{M_A} m_p = \frac{T_0 A}{A m_p} = T_A m_p. \tag{8.289}$$

Substituting (8.289) into (8.285) we get

$$E_s = T_0 + \gamma \beta (\mathbf{p}_1 + \mathbf{p}_2) - \gamma (T_1 + T_2) - 2 (\gamma - 1) m_p - \frac{q^2}{2 M_{A-1}}. \tag{8.290}$$

8.4 In an ideal gas, the particles are distributed according to the the Maxwell-Boltzmann distribution. In the velocity space, the distribution reads,

$$f(v)d\mathbf{v} = \left(\frac{m}{2\pi k_B T} \right)^{3/2} e^{-\frac{mv^2}{2k_B T}} d\mathbf{v}. \tag{8.291}$$

With (8.291), the mean velocity of the particles is calculated to be

$$\langle v^2 \rangle = \int v^2 f(v) d\mathbf{v} = \int_0^{2\pi} d\phi \int_{-1}^{1} d\cos\theta \int_0^\infty v^4 f(v) dv$$

$$= 4\pi \left(\frac{m}{2\pi k_B T} \right)^{3/2} \int_0^\infty v^4 e^{-\frac{mv^2}{2k_B T}} dv, \tag{8.292}$$

where the integration formula,

$$\int_0^\infty x^{2n} e^{-x^2/a^2} dx = (\pi)^{1/2} \frac{(2n)!}{n!} \left(\frac{a}{2} \right)^{2n+1} \tag{8.293}$$

is used. Substituting (8.293) with $n = 2$ and $a = (2k_B T/m)^{1/2}$ into (8.292), we obtain

$$\langle v^2 \rangle = 4\pi \left(\frac{m}{2\pi k_B T} \right)^{3/2} (\pi)^{1/2} \frac{4!}{2!} \left(\frac{k_B T}{2m} \right)^{5/2} = \frac{3k_B T}{m}. \tag{8.294}$$

Using the relation $E_{kin} = \frac{1}{2} m \langle v^2 \rangle$, we prove finally the relation,

$$E_{kin} = \frac{3}{2} k_B T. \tag{8.295}$$

Books for Further Readings

G. R. Satchler, "Direct nuclear reactions" (Clarendon Press, 1983), is a useful classics for direct reaction theory.

N. K. Glendening, "Nuclear direct reactions" (World Scientific, 2004), describes the theory of direct nuclear reactions emphasizing the microscopic aspects of the reactions induced by the motions of the individual nucleons.

C. Bertulani and P. Danielewicz, "Introduction to nuclear reactions" (CRC Press, 2004) gives a study on the basic scattering theory, followed by its applications in nuclear reaction processes.

References

1. E. Rutherford, Philos. Mag. **21**, 669 (1911)
2. H.B. Burrows, W.H. Gibson, J. Rotblat, Phys. Rev. **80**, 1095 (1950)
3. T. Furumoto, K. Tsubakihara, S. Ebata, W. Horiuchi, Phys. Rev. C **99**, 034605 (2019)
4. T. Furumoto, W. Horiuchi, M. Takashina, Y. Yamamoto, Y. Sakuragi, Phys. Rev. C **85**, 044607 (2012)
5. A. Idini, C. Barbieri, P. Navrátil, Phys. Rev. Lett. **123**, 092501 (2019)
6. J. Rotureau, P. Danielewicz, G. Hagen, G.R. Jansen, F.M. Nunes, Phys. Rev. C **98**, 044625 (2018)
7. G.W. Greenlees, G.J. Pyle, Phys. Rev. **149**, 836 (1966)
8. A.J. Koning, J.P. Delaroche, Nucl. Phys. A **713**, 231 (2003)
9. E. Becheva et al., Phys. Rev. Lett. **96**, 012501 (2006)
10. M. Baranger, Nucl. Phys. A **149**, 225 (1970)
11. T. Duguet, G. Hagen, Phys. Rev. C **85**, 034330 (2012)
12. M. Matoba et al., Phys. Rev. C **48**, 95 (1993)
13. T. Aumann et al., Prog. Part. Nucl. Phys. **118**, 103847 (2021)
14. G. Jacob, T.A.J. Maris, Rev. Mod. Phys. **38**, 1211 (1966)
15. K.A. Olive et al., Particle data group. Chin. Phys. C **38**, 090001 (2014)
16. M. Yosoi et al., Nucl. Phys. A **738**, 451 (2004)
17. T. Aumann, C. Bertulani, J. Ryckebusch, Phys. Rev. C. **88**, 064610 (2013)
18. T. Wakasa, K. Ogata, T. Noro, Prog. Part. Nucl. Phys. **96**, 32 (2017)
19. A.M. Moro, Phys. Rev. C **92**, 044605 (2015)
20. E. Cravo, R. Crespo, A. Deltuva, Phys. Rev. C **93**, 054612 (2016)
21. A. Navin et al., Phys. Rev. Lett. **85**, 266 (2000)
22. K. Hencken, G.F. Bertsch, H. Esbensen, Phys. Rev. C **54**, 3043 (1996)

23. A. Gade et al., Phys. Rev. C **77**, 044306 (2008)
24. J.A. Tostevin, A. Gade, Phys. Rev. C **90**, 057602 (2014)
25. F. Flavigny et al., Phys. Rev. Lett. **110**, 122503 (2013)
26. J. Lee et al., Phys. Rev. Lett. **104**, 112701 (2010)
27. F.M. Nunes, A. Deltuva, J. Hong, Phys. Rev. C **83**, 034610 (2011)
28. L. Atar et al., Phys. Rev. Lett. **120**, 052501 (2018)
29. S. Kawase et al., Prog. Theo. Exp. Phys. 021D01 (2018)
30. F. Gunsing et al., Phys. Rev. C **85**, 064601 (2012)
31. O. Bohigas, P. Leboeuf, Phys. Rev. Lett. **88**, 092502 (2002)
32. O. Bohigas, M.J. Giannoni, C. Schmit, Phys. Rev. Lett. **52**, 1 (1984)
33. J.B. Garg et al., Phys. Rev. **134**, B 985 (1964)
34. K.-H. Schmidt, B. Jurado, Phys. Proc. **31**, 147 (2012)
35. A.N. Andreyev, M. Huyse, P. Van Duppen, Rev. Mod. Phys. **85**, 1541 (2013)
36. P. Möller et al., Phys. Rev. C **79**, 064304 (2009)
37. G. Hupin, S.A. Quaglioni, P. Navrátil, Nat. Commun. **10**, 351 (2019)
38. G. Scamps, C. Simenel, Nature **564**, 382 (2018)

Chapter 9
Neutron Stars and Nucleosynthesis

Abstract In this chapter, we first focus on the equation of state (EoS) of nuclear matter and neutron matter. We discuss how one can disentangle the EoS of nuclear and neutron matter using terrestrial experiments. The current knowledge of the structure of neutron stars is presented. Empirical EoS is used to calculate predictions for the size and the weight of neutron stars by using a general relativistic model. In the latter part of this chapter, we discuss how, where and when nucleosynthesis has taken place in the universe starting from the Big Bang synthesis (BBS) up to the final stage of r- and s-process nucleosynthesis occuring in supernova explosions and neutron star mergers.

Keywords Equation of State (EoS) of nuclear matter · Incompressibility · Symmetry energy · Neutron stars · Nucleosynthesis

9.1 The Nuclear Equation of State (EoS)

Contemporary nuclear science aims at understanding the properties of strongly interacting bulk matter at the nuclear, hadronic and quark levels. In addition to their intrinsic interest in fundamental physics, such studies have enormous impact on astrophysics, from the evolution of the early universe to neutron star structure. For example, a precise knowledge of the equation of states (EoS) of neutron matter is essential to understand the physics of neutron stars and binary mergers, recently observed as strong sources of gravitational waves. A neutron star is created by the core collapse supernova explosion of a giant star with a mass from about 10 to 30 times the solar mass. This collapse is caused by a strong gravitational field of the massive star, and compresses the core as much as the density of a nucleus. Around 100 million neutron stars are expected to exist in the Milky Way. However, most of them are old and cold, and neutron stars can only be easily detected in certain circumstances, such as a pulsar neutron star or as a part of a binary system. Although the size difference between the nucleus and the neutron star is almost 10^{20} times,

© Springer Nature Singapore Pte Ltd. 2021
A. Obertelli and H. Sagawa, *Modern Nuclear Physics*, UNITEXT for Physics,
https://doi.org/10.1007/978-981-16-2289-2_9

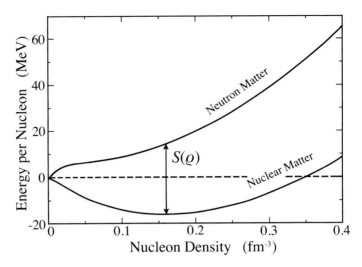

Fig. 9.1 Nuclear and neutron matter equation of states (EoS) near the saturation density. $S(\rho)$ is the symmetry energy in asymmetric nuclear matter. See the text for details

there are deep and intimate relations between the two objects because of their similar densities and constituents.

The EoS of symmetric nuclear matter consisting of equal amount of neutrons and protons has been studied near the saturation density using terrestrial experiments. The EoS of symmetric matter stems from attractive two-body forces and repulsive forces due to three-body effects and short-range correlations. The difference of the EoS for symmetric nuclear matter and pure neutron matter is shown in Fig. 9.1. The attractive and repulsive forces are balanced at certain density, i.e., the saturation density $\rho_0 = 0.16$ fm^{-3}, and lead to a stable nuclear matter which is characterized by a binding energy of 16 MeV per nucleon.

The difference between the neutron matter and the symmetric nuclear matter EoS is defined by the symmetry energy $S(\rho)$. The later is the energy necessary to separate protons from neutrons in nuclear matter. In other words, the nuclear symmetry energy characterizes the variation of the binding energy as the ratio of protons to neutrons in a nuclear system is varied. The properties of symmetric nuclear matter at saturation stem from the strongly attractive proton-neutron interaction. However, this interaction is not present in pure neutron matter so that $S(\rho)$ becomes always repulsive.

Thus, the symmetry energy reduces the nuclear binding energy not only in nuclear matter but also in nuclei, and is critical for understanding properties of nuclei including the existence of rare isotopes with extreme proton-to-neutron ratios. On top of that, in a neutron star, the strong gravitation force is balanced with the pressure of neutron matter. The hadronic contributions of pressure is derived by the slope of $S(\rho)$, as it will be shown in Eqs. (9.8) and (9.13). As for terrestrial experiments, the slope of $S(\rho)$ at saturation density is known to show a strong correlation with the

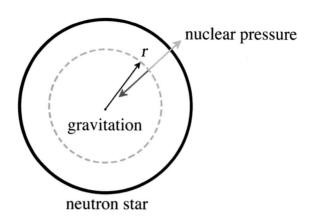

Fig. 9.2 A balance between two pressures in neutron stars. The gravitational force creates the pressure inwards, while the nucleon-nucleon force provides the pressure outwards. See text for details

neutron skin size of nuclei, which can be extracted, for examples, from polarized electron scattering or proton elastic scattering experiments (see Chap. 4).

Let us study hereafter the EoS more quantitatively. The total energy of a many-body system is defined by

$$E_T = \int E \, dV, \tag{9.1}$$

where E is the energy density in units of MeV/fm^3 and dV is the volume element. In nuclear matter, the energy per particle is defined as

$$E_T/A = EV/A = E/\rho, \tag{9.2}$$

where the density is defined by $\rho = A/V$ with the mass number A and its volume V.

The energy density E/ρ depends on the isoscalar density $\rho = \rho_n + \rho_p$, and also the asymmetric coefficient $\delta = (\rho_n - \rho_p)/\rho$, where ρ_n and ρ_p are neutron and proton densities, respectively. Taking ρ and δ as independent variables, the energy density of asymmetric nuclear matter is first expanded by δ as

$$\frac{E(\rho, \delta)}{\rho} = \frac{E(\rho, \delta = 0)}{\rho} + E_{sym}(\rho)\delta^2 + \cdots, \tag{9.3}$$

where

$$E_{sym}(\rho) = \frac{1}{2} \frac{\partial^2 (E/\rho)}{\partial(\delta)^2}\bigg|_{\delta=0}, \tag{9.4}$$

is the symmetry energy. In classical mechanics, the pressure is defined by the formula

$$P = F/S, \tag{9.5}$$

where F is the force acting orthogonal to the surface element with the area S. Equation (9.5) is rewritten as

$$P = \frac{F}{S} = \frac{F\,ds}{S\,ds} = -\frac{dE_T}{dV}, \tag{9.6}$$

where ds is the migration distance of the surface by the pressure and dE_T is the total energy due to this action. The minus sign of the r.h.s. of Eq. (9.6) is due to the fact that the force acts on the surface whose normal vector is opposite to the direction of force, i.e., $dE_T = \mathbf{F} \cdot \mathbf{ds} = -F\,ds$. If we consider that E_T is the energy density of nuclear matter and V is its volume, Eq. (9.6) is further rewritten as

$$P = \frac{\partial E_T}{\partial V} = \frac{\partial(E_T/A)}{\partial(V/A)}. \tag{9.7}$$

With the help of Eq. (9.2) and $V/A = 1/\rho$, we finally obtain

$$P = -\frac{\partial(E/\rho)}{\partial(1/\rho)} = \rho^2 \frac{\partial(E/\rho)}{\partial\rho}. \tag{9.8}$$

The pressure (9.8) corresponds to the value at zero temperature of nuclear matter.[1]

The symmetry energy E_{sym} in Eq. (9.4) is further expanded around the saturation density ρ_0 as

$$E_{sym}(\rho) \equiv S(\rho) = J + L\frac{(\rho - \rho_0)}{3\rho_0} + \frac{1}{2}K_{sym}\frac{(\rho - \rho_0)^2}{9\rho_0^2}, \tag{9.9}$$

where

$$J = S(\rho_0), \tag{9.10}$$

$$L = 3\rho_0 \frac{\partial S(\rho)}{\partial\rho}\bigg|_{\rho=\rho_0}, \tag{9.11}$$

$$K_{sym} = 9\rho_0^2 \frac{\partial^2 S(\rho)}{\partial\rho^2}\bigg|_{\rho=\rho_0}. \tag{9.12}$$

Since a neutron star contains a low fraction of protons, the inner crust as well as global neutron star properties are sensitive to the symmetry-energy parameters J and L. One can manifest the importance of the symmetry energy by evaluating the pressure of neutron matter ($\delta = 1$) at the saturation density,

$$P(\rho_0) = \rho^2 \frac{\partial(E/\rho)}{\partial\rho}\bigg|_{\rho=\rho_0} = \frac{\rho_0}{3}L. \tag{9.13}$$

The pressure acts at each radial position r of the neutron star and sustains it from the gravitational pressure as is shown schematically in Fig. 9.2.

---

[1] The temperature is defined by the total excitation energy of a many-body system. The state at zero temperature means the ground state.

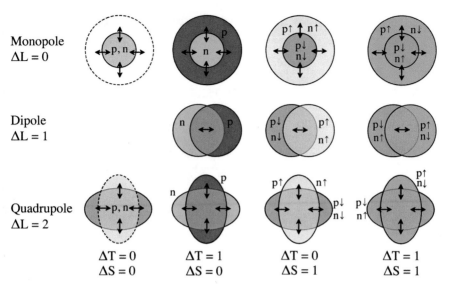

Fig. 9.3 Macroscopic representations of various giant resonances (GR) with the angular momentum ΔL, isospin ΔT and spin ΔS degrees of freedoms. For the monopole mode $\Delta L = 0$, there is no deformation of shapes for both neutrons and protons, and the oscillation preserves the spherical symmetry. The oscillation takes place in one particular direction without any deformation of proton and neutron densities in the case of dipole mode $\Delta L = 1$. For the quadrupole mode, the proton and neutron densities are deformed in the direction specified by the angular momentum operator. Giant resonances with higher angular momenta $\Delta L \geq 3$ have also deformed vibrational densities specified by each multipole. In the isoscalar mode with $\Delta T = 0$, protons and neutrons oscillate in phase, while they oscillate out of phase in the isovector mode $\Delta T = 1$. Since no deformation of oscillating densities in the $\Delta L = 1$, the IS GDR is a spurious center-of-mass motion in the lowest-order approximation. The spin mode with $\Delta S = 1$ involves the spin flip in the initial and final stages of the oscillation. The $\Delta L = 0$ and $\Delta L = 1$ charge exchange excitation with $\Delta T = 1$ and $\Delta S = 1$ are called Gamow–Teller and Spin-Dipole GRs, respectively

At the saturation point, $P(\rho_0) = 0$, the energy density in symmetric matter ($\delta = 0$), the first term of r.h.s. of Eq. (9.3), is expressed by the Taylor expansion around the saturation density ρ_0 as

$$\frac{E(\rho, \delta = 0)}{\rho} = \frac{E(\rho_0, \delta = 0)}{\rho_0} + \frac{1}{2}\left(\frac{\rho - \rho_0}{3\rho_0}\right)^2 K_\infty, \qquad (9.14)$$

where the incompressibility of nuclear matter is defined as the second derivative of the binding energy per particle with respect to the density at the saturation point,

$$K_\infty = 9\rho^2 \frac{\partial^2 (E/\rho)}{\partial \rho^2}\bigg|_{\rho=\rho_0}. \qquad (9.15)$$

9.2 Incompressibility and Giant Monopole Resonances

We discuss the incompressibility in relation with isoscalar giant monopole resonances (ISGMR) as terrestrial observables. The ISGMR is also called a breathing mode or a compression mode since it does not involve any deformation of shape and the protons and neutrons are oscillating in phase keeping the spherical symmetry. Because of these symmetry restrictions in the ISGMR, when the oscillation density increases at the surface, it simultaneously decreases at the interior, and vice versa. This is the reason why the ISGMR is called the breathing mode and the nuclear incompressibility plays a role to determine the oscillation frequency, equivalently the excitation energy. A macroscopic image of ISGMR is illustrated in Fig. 9.3 together with other types of giant resonances (GRs).

The incompressibility in a finite nuclei has an analytic relation with the excitation energy of ISGMR as

$$\hbar\omega = \sqrt{\frac{\hbar^2 K_A}{m\langle r^2 \rangle}}. \tag{9.16}$$

where K_A is the incompressibility of finite nucleus, m is the nucleon mass and $\langle r^2 \rangle$ is the mean square nuclear radius. This relation can be derived by using the energy-weighted sum rule (EWSR) technique without any approximation. The EWSR is classified by its weight on the excitation energy. The k-th moment of EWSR depends on $(E_n)^k$ and defined as

$$m_k(\lambda) = \sum_n (E_n)^k B(E\lambda; n), \tag{9.17}$$

where E_n is the excitation energy of the n-th state with multipolarity λ, and $B(E\lambda; n)$ is its transition strength. The formula (9.16) is derived by using two EWSRs, m_3 and m_1, for ISGMR as $\hbar\omega = \sqrt{m_3/m_1}$. Intuitively, this relation tells how the ISGMR can be affected by the rigidity of the nucleus; a hard system with a larger K_A has a higher excitation energy for the compression mode, while a soft system with a smaller K_A has a lower excitation energy for the compression mode. Since the relation is based on the assumption of all the sum rule strength in the numerator and the denominator, Eq. (9.16) provides a precise empirical information of incompressibility in finite nuclei when the ISGMR is a sharp single peak.

For the study of celestial observables such as supernovae or neutron stars, we need information of nuclear matter incompressibility. The incompressibility K_A in finite nuclei may have contributions from the surface, the symmetry energy, and the Coulomb energy on top of the infinite nuclear matter incompressibility as an analogy with the mass formula (4.3) in Chap. 4. The relation can be written as

$$K_A = K_\infty + K_{surf} A^{-1/3} + K_{sym}\delta^2 + K_{Coul}\frac{Z^2}{A^{4/3}}, \tag{9.18}$$

where $\delta = (N - Z)/A$.

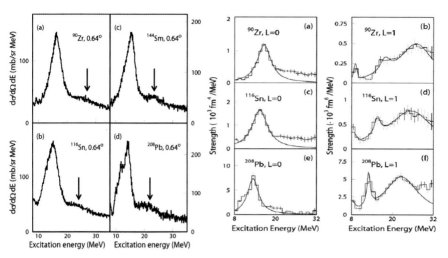

Fig. 9.4 (Left) The differential cross sections of (α, α') reactions for ^{90}Zr, ^{116}Sn, ^{144}Sm and ^{208}Pb at $\theta_{lab} = 0.64°$. The ISGMRs appear as sharp peaks at $E_x \sim 15$ MeV for all nuclei. The arrows indicate the locations of the high energy ISGDR. (Right) Experimentally obtained strength distributions of the ISGMR in ^{90}Zr, ^{116}Sn, and ^{208}Pb. The error bars are systematic errors due to the MDA. The figures are reprinted with permission from [1]. ©2021 by the American Physical Physics Society

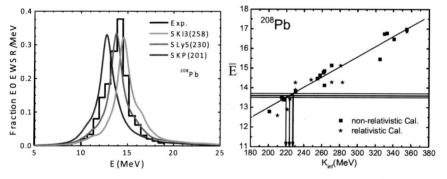

Fig. 9.5 (Left) The strength distributions of ISGMR in ^{208}Pb as a function of the excitation energy. The theoretical results are obtained by self-consistent HF+RPA calculations with SkP, SLy5 and SkI3 interactions which have $K_\infty = 201$, 230 and 258 MeV, respectively. The experimental data were taken at Texas A and M University, Cyclotron laboratory, see [2]. (Right) The average excitation energy $\bar{E}$ of ISGMR as a function of the incompressibility K_∞. The results are obtained by the self-consistent RPA calculations with 16 Skyrme EDFs and 10 RMF EDFs. The experimental data was adopted from recent compilation $\bar{E} = (13.6 \pm 0.1)$ MeV. Courtesy Ligang Cao

Experimental data of ISGMR have been obtained by inelastic scattering of isoscalar probes,[2] especially by (α, α') inelastic scattering. The experimental cross sections are analyzed by multipole decomposition analysis (MDA) to separate the monopole components from other multipoles with $L > 0$ (see more discussions in Chap. 4). The monopole component can be separated since the cross sections with $L = 0$ have the maximum at the forward angle $\theta \sim 0°$ and other multipoles have peaks at larger angles $\theta > 0°$. In Fig. 9.4, the differential cross sections of inelastic (α, α') reactions are shown at $\theta_{lab} = 0.64°$ for ^{90}Zr, ^{116}Sn, ^{144}Sm and ^{208}Pb. The α projectile is accelerated at a medium energy $E_{lab} = 386$ MeV. By the MDA technique, the strength distributions of ISGMRs are extracted from the differential cross sections at angles $(\theta_{lab} = 0.64° - 13.5°)$. The strength distributions are shown in the right panel of Fig. 9.4. The ISGMR is fitted by a Breit–Wigner function as indicated in the figure. The extracted peak strengths of ISGMR exhaust almost 100% of the energy-weighted sum rule value m_1 in Eq. (9.17) in the nuclei shown. The peak energies of ISGMR are determined to be

$$
\begin{aligned}
E_x(\text{ISGMR}) &= 16.6 \pm 0.1 \text{ MeV for } ^{90}\text{Zr}, \\
&= 15.4 \pm 0.1 \text{ MeV for } ^{116}\text{Sn}, \\
&= 13.4 \pm 0.2 \text{ MeV for } ^{208}\text{Pb}.
\end{aligned}
\tag{9.19}
$$

Existing experimental data sets of various nuclei are not enough to extract all the values of K_∞, K_{surf}, K_{sym} and K_{Coul} of the r.h.s of Eq. (9.18). Another plausible approach to extract the value K_∞ from the experimental data is the framework of self-consistent Hartree–Fock (HF) or Hartree+random phase approximation (RPA) model. In the self-consistent approach, a single Hamiltonian, which has good saturation properties, is adopted in all calculations of nuclear matter and finite nuclei simultaneously so that one can see a direct correlation between the incompressibility in nuclear matter and the excitation energy of ISGMR through the adopted Hamiltonian. This approach was quite successful to determine the incompressibility within microscopic Skyrme, Gogny and relativistic mean field (RMF) models.

In the left panel of Fig. 9.5, the experimental data of ISGMR is compared with the self-consistent HF+RPA calculations with three Skyrme interactions SkP, SLy5 and SkI3, which have the nuclear matter incompressibility K_∞=201, 230 and 258 MeV, respectively. The empirical strength distributions for ^{208}Pb are better reproduced by the SLy5 interaction than the other two interactions. The right panel of Fig. 9.5 the calculated ISGMR energies of ^{208}Pb with various energy density functionals (EDFs) and their nuclear matter incompressibilities K_∞. Both the excitation energy and K_∞ are calculated by using the same EDF. An optimal value of nuclear matter incompressibility is expected to be $K_\infty = 220 \pm 10$ MeV from this figure. However, there are some uncertainties on this value of K_∞ which, to some extent, come from the ambiguity of empirical determination of the energy and also from the theoretical models involved in the microscopic calculations. Another uncertainty comes from the fact

---

[2] Projectiles or targets having the isospin $T = 0$.

that the mass number dependence of the excitation energies is not perfectly regular. Thus, the proposed empirical incompressibility may depend on how to select the data set of excitation energies of ISGMR. A similar empirical value $K_\infty = 225$ MeV was also obtained from the data analysis of superfluid Sn and Cd isotopes. The current optimal value of nuclear incompressibility from ISGMR is

$$K_\infty = 225 \pm 20 \text{ MeV}, \tag{9.20}$$

including the statistical errors from the experiments and the systematic errors from the theoretical models.

9.3 Symmetry Energy and Terrestrial Experiments

The symmetry energy plays a decisive role to determine the EoS of neutron matter on top of the EoS of symmetric nuclear matter as we can see in Fig. 9.1. From the 1990s, tremendous experimental and theoretical efforts have been made to explore the symmetry energy at various nuclear matter densities $0 < \rho/\rho_0 < 3$. It was pointed out that the neutron skin thickness provides useful information to pin down the symmetry energy coefficients J and L. At a normal or lower density region, the parity violating electron scattering experiment has been performed to extract the neutron skin thickness of ^{208}Pb (see Sect. 4.3.5 in Chap. 4) as well as the isovector giant dipole resonances (GDR) and the proton elastic scattering. The multi-fragmentation process of heavy ion collisions (HIC) provides empirical information on the symmetry energy at higher density than the saturation density ρ_0. However, a large uncertainty still exists to extract reliable information of the EoS from very complicated multi-fragmentation data since the interpretation relies heavily on theoretical analysis using transport models. Because of this reason, we do not discuss any details of multi-fragmentation in this section. These experiments, where nuclear dynamics and statistics of macroscopic systems are combined together, remain a challenge for theory. We just remind that some HIC observables might be more sensitive to higher densities than others, for example, p/n flow in comparison with π^+/π^- production. Such experimental programs are being pursued.

In the following sections, we will first study the correlations between the electric dipole polarizability and the neutron skin thickness as well as the symmetry energy. Recently, the mass formula was discussed providing a useful information on symmetry energy around the saturation density. Next, we will study the mass formula constrains on the symmetry-energy coefficients.

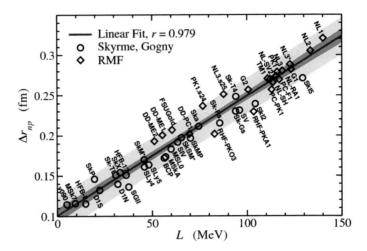

Fig. 9.6 Correlation between the neutron skin thickness Δr_{np} in ^{208}Pb and the slope parameter L of the symmetry energy at saturation density. The predictions of various nuclear energy density functionals are shown. The linear correlation coefficient of the results for Δr_{np} and L is $r = 0.979$. A linear fit gives $\Delta r_{np} = 0.101 + 0.00147L$, with Δr_{np} in fm and L in MeV. The thinner and thicker shadowed regions represent, respectively, the 95%-confidence band and the 95%-prediction band of the linear regression. Figure reprinted with permission from [3]. ©2021 by the American Physical Society

9.3.1 Electric Dipole Polarizability and Neutron Skin

The correlation between the neutron thickness for ^{208}Pb and the slope parameter L is shown in Fig. 9.6. The neutron skin thickness is defined by

$$\Delta r_{np} = \sqrt{< r^2 >_n} - \sqrt{< r^2 >_p}, \tag{9.21}$$

where $< r^2 >_{n,p}$ are the mean square radii of neutrons and protons, respectively. The various mean field models are adopted to calculate the correlation plot including Skyrme, Gogny and RMF EDFs. The values of the neutron skin thickness Δr_{np} in ^{208}Pb calculated in various models are plotted as a function of the slope parameter of the symmetry energy L at saturation density.

One can see a clear correlation between the neutron skin and L irrespective of the theoretical models. In general, the non-relativistic models give lower L, while RMF models predict higher values. The linear correlation coefficient[3] of the results for Δr_{np} and L is $r = 0.979$. A linear fit gives $\Delta r_{np} = 0.101 + 0.00147L$, with

---

[3] The linear correlation coefficient r is the measure of strength of linear relationships between two variables. The value r lies between $+1$ and -1. The value $r = +1$ means the perfect linear relationship between two variables, while $r = -1$ corresponds to the perfect anti-linear relationship. The value close to $r = 0$ suggests no linear correlations between two variables.

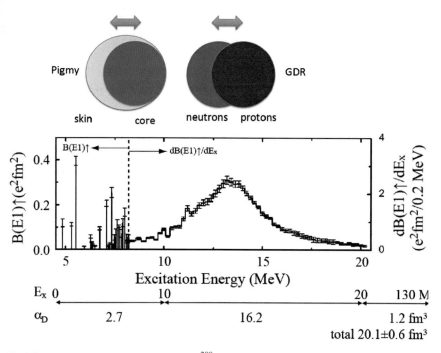

Fig. 9.7 Total $B(E1)$ strength distribution of ^{208}Pb deduced from (p, p') experiments at RCNP. The γ absorption cross section above $E_x > 20$ MeV measured at Mainz is added to the total dipole polarizability. The bump centered at $E_x \sim 13$ MeV corresponds to the giant dipole resonance (GDR), and the strength concentration at around (7–9) MeV is referred to the pygmy dipole resonance (PDR). The dashed vertical line in the lower panel indicates the neutron threshold energy. Below the threshold, the E1 value is shown by a dotted vertical line refered to the left scale, while, above the threshold, the E1 strength is refereed to the right scale. In the upper panel, the vibrations of PDR and GDR are illustrated. The data are taken from [4]. Courtesy A. Tamii, RCNP

Δr_{np} in fm and L in MeV. The thinner and thicker shadowed regions represent, respectively, the 95%-confidence band and the 95%-prediction band of the linear regression. Figure 9.6 implies that the neutron skin thickness increases for larger L value, i.e., a harder neutron matter EoS. This is understood by means of the simple argument that a harder EoS leads to higher neutron matter pressure, as shown by Eq. (9.13), so that the neutrons are pushed away from the core. This feature is implemented in the symmetry energy parameter L, which governs the slope of energy density functionals.

Let us discuss first the relation between GDR and the symmetry energy. The electric dipole (E1) response of nuclei is dominated by the isovector (IV) GDR, a highly excited collective mode above the particle emission threshold. In a macroscopic picture, IVGDR is a mutual oscillation between proton and neutron density as shown in Fig. 9.3 and top part of Fig. 9.7. In an analogy with dielectric substance, a nucleus will be polarized in the direction of external electric dipole (E1) field as shown in Fig. 9.7, where protons (neutrons) behave as holes (electrons). This phenomenon is called "dipole polarizability" in nuclei. Under the external field, protons oscillate against neutrons each other in the same direction as the external field. Since this is an IV-type oscillation, in a simple macroscopic picture, the restoring force is considered to be induced by the symmetry energy.

The properties of IVGDR have been discussed and are well understood by various theoretical models. Recent interest focuses on the soft mode in neutron-rich nuclei which appeared below the GDR in energy. It is called "Pygmy dipole resonance (PDR)" since its peak height is much smaller than the GDR. The physical mechanism of PDR can be understood as follows: because of the saturation of nuclear density, excess neutrons might form a skin whose oscillations against the core should give rise to a low-energy electric dipole (E1) mode since the restoring force is weaker than that of GDR as is illustrated in Fig. 9.7. Therefore, the PDR may shed light on the formation of neutron skins in nuclei.

Both for GDR and PDR modes, the symmetry energy acts as a restoring force to determine the oscillation frequencies. Therefore, the strength distributions of PDR and GDR may provide information on not well-known magnitude and density dependence of the symmetry energy, indispensable ingredients for the modeling of neutron stars.

The dipole polarizability is a measure of the distortion of nuclear density under an external electric dipole field. The induced dipole moment can be calculated by a perturbation theory with the perturbed Hamiltonian $H' = \alpha_{ext} F$ where α_{ext} is the external dipole field and F is the induced dipole moment field. The induced moment is proportional to the dipole polarizability α_D as $\langle F \rangle = \alpha_D$. The dipole polarizability is evaluated by the second-order perturbation theory as

$$\alpha_D = 2e^2 \sum_n \frac{|\langle n|O(E1)|0\rangle|^2}{\omega_n}, \tag{9.22}$$

where ω_n denotes the excitation energy of nth state and the E1 operator is defined by subtracting the center of mass correction as

$$O(E1) = \frac{N}{A} \sum_{i=1}^{Z} r_i Y_{1\mu}(\hat{r}_i) - \frac{Z}{A} \sum_{i=1}^{N} r_i Y_{1\mu}(\hat{r}_i), \tag{9.23}$$

where $r_i Y_{1\mu}(\hat{r}_i)$ are spherical harmonics. The dipole polarizability α_D is, thus, proportional to the inverse-energy-weighted sum rule m_{-1} in Eq. (9.17) so that it depends more the E1 strength at low energies than on the strength at high energies. Energy density functionals (EDFs) using Skyrme forces or a relativistic framework RMF suggest an intimate correlation between the nuclear dipole polarizability α_D and the neutron skin thickness as shown in Fig. 9.8.

The (p, p') experiments were performed with intermediate-energy polarized proton beams $E_p = 295$ MeV at RCNP to observe E1 strength in ^{208}Pb. The cross section measurements were done at angles close to and including $0°$, where the

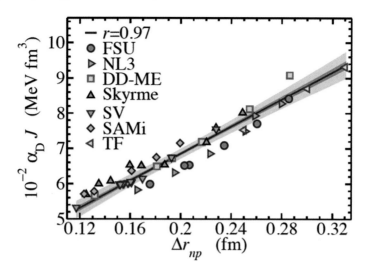

Fig. 9.8 Correlation between the dipole polarizability α_D times J and the neutron skin thickness Δr_{np} in ^{208}Pb for various mean field models; Skyrme and RMF EDF. The dipole polarizability α_D is calculated by the RPA model, while the neutron skin and J value are evaluated by HF for Skyrme and Hatree method for RMF models with the same EDF as that for RPA. The value r is the correlation coefficient and the two shadowed regions represent the 99.9% and 70% confidence bands. The dipole polarizability gives a band of the constraints for J and L which is shown in Fig. 9.10. Figure reprinted with permission from [5]. ©2021 by the American Physical Society

isovector spin-flip $M1$ transitions (the analog of the Gamow–Teller excitation) and the Coulomb excitation of non-spin-flip E1 transitions are dominating. A separation of these two contributions, necessary for an extraction of the E1 response, is achieved by two independent methods: a multipole decomposition analysis of the angular distributions (MDA) and the measurement of polarization transfer observables. The $B(E1)$ distribution determined by the (p, p') measurement is shown in Fig. 9.7. The bump centered at $E_x \sim 13$ MeV corresponds to the GDR and the strength concentration around $(7-9)$ MeV corresponds to the PDR. The $B(E1)$ sum rule of ^{208}Pb was determined by the strength between 5 and 20 MeV, which fully covers the PDR and GDR regions, as well as the region just above neutron separation energy. The observed $B(E1)$ strengths yield a dipole polarizability $\alpha_D = 18.9 \pm 1.3$ fm^3 up to 20 MeV. By including the gamma absorption data above 20 MeV, the dipole polarizability of ^{208}Pb up to 130 MeV is determined as $\alpha_D = 20.1 \pm 0.6$ fm^3.

The correlation between α_D times J and the neutron skin thickness Δr_{np} in ^{208}Pb is shown in Fig. 9.8. The dipole polarizability α_D is calculated by a RPA model, while the neutron skin and J value are evaluated by the HF method for Skyrme and Hartree for RMF with the same EDF as that for the RPA. Adopting $J = (31 \pm 2)$ MeV as a realistic range of values for the symmetry energy (see Sect. 9.4 more details of symmetry energy), the constraint on the neutron skin thickness of ^{208}Pb is extracted from $\alpha_D = 20.1 \pm 0.6$ fm^3 as

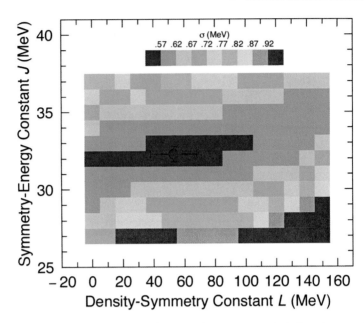

Fig. 9.9 Symmetry energy constants J and L versus the mean square deviation σ between the experimental mass and the FRDM mass formula predictions. Calculated mass model FRDM accuracy are shown by different colors for different values of J and L. The best accuracy region is indicated by a red dot with uncertainty bars. Reprinted with permission from [6]. ©2021 by Elsevier

$$\Delta r_{np} = 0.165 \pm (0.009)_{\text{expt}} \pm (0.013)_{\text{theor}} \pm (0.021)_{\text{est}} \ \text{fm}, \qquad (9.24)$$

where "expt" is the experimental error, while "theor", and "est" are theoretical uncertainties associated with the correlation plot, and J value in the mean field models, respectively. The value of the neutron skin thickness of ^{208}Pb is compatible with other hadronic observations, while the PREX experiment [7] gives a much larger neutron skin value with a large experimental uncertainty

$$\Delta r_{np}(\text{PREX}) = 0.283 \pm 0.071 \ \text{fm}.$$

The constraints on J and L imposed by the dipole polarizability (denoted as "DP") are shown by the bands in Fig. 9.10 together with the constraints from other experiments.

9.3.2 Symmetry Energy and Mass Formula

One of the decisive ingredients of the nuclear mass formula is the symmetry energy. The functional form of the mass formula is given in Eq. (4.3) in Chap. 4. The predicting power of the mass formula is sensitive to the symmetry energy coefficients. A recent study of symmetry energy in the mass formula was done by using the finite-range droplet mass model (FRDM). The FRDM is a hybrid model of macroscopic liquid drop energy and microscopic shell corrections. A finite-range liquid-drop model is adopted for the macroscopic energy and a folded-Yukawa single-particle potential is used for the microscopic shell corrections. FRDM is quite successful to predict not only masses of stable nuclei but also unstable nuclei. The mass parameters including the symmetry energy coefficients J and L are optimized by using all available experimental data of binding energies for several thousand nuclei. In Fig. 9.9, the smallest mean square deviation σ was obtained by the optimization process at the values,

$$J = 32.3 \pm 0.5 \text{ MeV},$$
$$L = 53.5 \pm 15 \text{ MeV}, \tag{9.25}$$

shown by a red dot with uncertainty bars in Fig. 9.9. The errors come from the systematic uncertainty in the optimization process of parameters in the FRDM. The values (9.25) are consistent with empirical values obtained from GDR, and also from the systematical analysis of excitation energies of isobaric analog states (IAS).

9.4 Summary of Constraints on Symmetry Energy

To summarize the current status of empirical symmetry energy coefficients, constraints on the slope L and magnitude J of the symmetry energy at saturation density from various experiments and theories are shown in Fig. 9.10. Each method is labeled next to the box with the estimated uncertainties. HIC is the constraint from the analysis of heavy-ion multi-fragmentation. The heavy-ion collisions could be useful to specify the EoS not only at the normal density but also at a higher density than the normal density as we can see in Fig. 9.10. However, the present experimental environment provides only the constraints with a large uncertainty around the normal density. The region denoted as IAS is the constrain from the analysis of isobaric analog states in broad mass region. The constrain denoted PDR comes from the Pygmy dipole states of ^{132}Sn and ^{68}Ni. The region of FRDM is due to the mass formula of Finite Range Droplet model as was discussed in Sect. 9.3.2. "n-star" denotes the constraint from the low mass binary neutron stars. The observed masses and radii of these neutron stars are calculated by using Tolman–Oppenheimer–Volkoff equation (see next section for details) and J and L are optimized to reproduce these data. χ_{EFT} is the constraint from ab initio chiral effective theory calculations. The region denoted

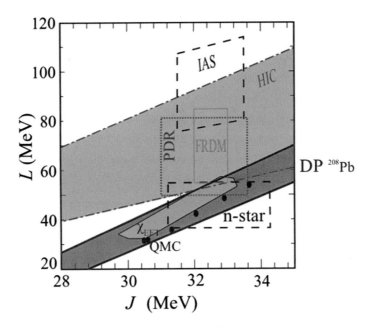

Fig. 9.10 Constraints on the slope L and magnitude J of the symmetry energy at saturation density from different experiments and theories. Each method is labeled next to the box with the estimated uncertainties. The symbols are results without the analysis of the errors. HIC is from the analysis of heavy ion multi-fragmentation. IAS is from the analysis of isobaric analog states. PDR is from the Pygmy dipole states of ^{132}Sn and ^{68}Ni. FRDM is the mass formula of Finite Range Droplet model (see Fig. 9.9). n-star is the constraint from quiescent low mass binary neutron stars. Two theoretical "ab initio" predictions are also shown as blue zone (χ_{EFT}) and black dots (QMC). χ_{EFT} is the constraint from chiral effective theory calculations. QMC is the constraint from quantum Monte-Carlo calculations with the Argonne AV8' interaction. DP is the constraint from dipole polarizability in ^{208}Pb taken from Fig. 9.8. Courtesy A. Tamii, RCNP

QMC corresponds to the constraint from Quantum Monte-Carlo calculations. The Argonne NN potential AV8', obtained by fitting nucleon-nucleon scattering data up to laboratory energies of 350 MeV, was used for QMC calculations. DP is the constraint from dipole polarizability in ^{208}Pb as was discussed in Sect. 9.3.1. In general, the J value is rather well constrained by several experiments and theoretical calculations to be $J = (31 - 33.5)$MeV. On the other hand, the L value has still a large uncertainty of $L = (20 - 110)$MeV.

9.5 Neutron Stars

There are millions of stars which are shinning in the sky. There are also stars which are not shinning and hide in the darkness. Among them, neutron stars are very curious objects of the universe. In general, the size of celestial objects exceed the scales we experience on our daily life by orders of magnitude. For example, the size of the sun is larger than that of the earth by more than hundreds of times. The distance from the earth to the sun is 1 astronomical unit (about 150,000,000 km denoted as "AU"). It takes 8 min 20 s for light to travel between them. The most distant planet of the solar system, Neptune, is 30 AU (4.5×10^9 km) away from the sun. Let us consider another example. The solar system belongs to the Milky Way galaxy. The Milky Way is a barred spiral galaxy that has a diameter more than 100,000 light-years, i.e., it takes more than 100,000 years for light to pass through the Milky Way galaxy.

On the other hand, the universe also comprises much smaller objects such as neutron stars. The structure of a neutron star is drawn in Fig. 9.11. The radius of neutron stars is about $10 \sim 12$ km, but its weight is almost the same as that of the sun. Imagine the size difference of the sun compared to a neutron star is 10,000 times! This means the density of neutron stars can reach and exceed the nuclear saturation density. This is a very curious and fascinating feature to make a bridge between nuclear physics and astrophysics. In other words, the properties of a nucleus contain a lot of information about neutron stars, i.e., their structure, size and weight through the nuclear EoS.

9.5.1 Tolman–Oppenheimer–Volkoff Equation

Let us consider a neutron star as a static, perfectly spherically symmetric fluid, i.e., an isotropic material without viscosity. The Tolman-Oppenheimer-Volkoff (TOV) equation constrains the structure of such a spherically symmetric body of isotropic material in a static gravitational equilibrium. The TOV equation is derived from the Einstein general relativity equations and reads

$$\frac{dP(r)}{dr} = -\frac{G}{r^2}\left[\rho(r) + \frac{P(r)}{c^2}\right]\left[M(r) + 4\pi r^3 \frac{P(r)}{c^2}\right]\left[1 - \frac{2GM(r)}{c^2 r}\right]^{-1}, \quad (9.26)$$

where G and r are the gravitational constant and the radial coordinate, while $\rho(r)$ and $P(r)$ are the density and pressure, respectively, at the radius r. The terms $\frac{P(r)}{c^2}$ and $4\pi r^3 \frac{P(r)}{c^2}$ in the brackets on the r.h.s. in Eq. (9.26) are relativistic corrections to the density and the total mass, respectively, and the last term $\frac{2GM(r)}{c^2 r}$ comes from the Schwarzschild metric of general relativity. In the non-relativistic limit ($1/c^2 \to 0$), the TOV equation becomes the Newtonian hydrostatic equation, used to find the equilibrium structure of a spherically symmetric body of isotropic material. The total mass $M(r)$ inside the radius r satisfies an equation determined by a geometrical constrain of a sphere with the thickness dr,

$$\frac{dM(r)}{dr} = 4\pi\rho(r)r^2. \quad (9.27)$$

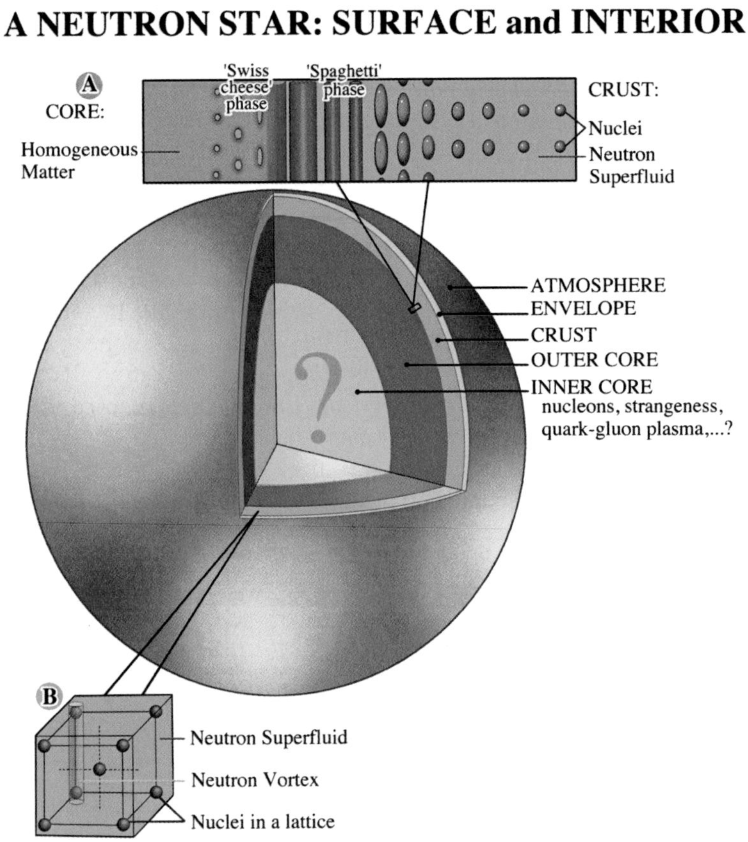

A NEUTRON STAR: SURFACE and INTERIOR

Fig. 9.11 Structure of a neutron star. The radius of the neutron star is about $10 \sim 12$ km, while the weight is almost the same as that of the sun. Models for neutron stars explore exotic structure of the crust and the core. See Sects. 9.5.1 and 9.5.4 for details

When supplemented with an equation of state, which relates density to pressure, the TOV equation completely determines the structure of a spherically symmetric neutron star.

Problem

9.1 Derive the Newtonian hydrostatic equation for a spherically symmetric body of isotropic material from TOV Eq. (9.26) in the non-relativistic limit and obtain a relation between the pressure and the height difference from the reference point.

The TOV equation is solved for a static spherical non-rotating star with different J values in Fig. 9.12. The predicted mass and radius relation is plotted for different symmetry energy value $E_{sym}(\equiv J)$. The estimated radius for a neutron star with the mass $1.4\,M_{\odot}$ varies depending on the value of the symmetry energy. The uncertainty

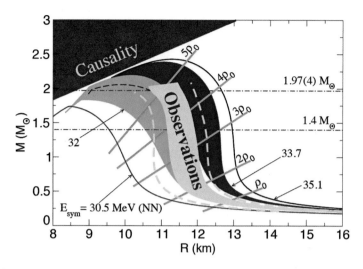

Fig. 9.12 Mass and radius plot of a neutron star calculated by solving the TOV equation with different EoSs of neutron matter, obtained by Green's function Quantum Monte Carlo (GFMC) calculations with realistic two-body AV8'+three-body UIX potentials (J and L values are shown in Fig. 9.10). Each curve is obtained by using the EoS with different symmetry energy value $J = E_{sym}$. The numbers indicate the value of J for the various EoSs. The yellow band corresponds to observations of neutron stars. The blue and green shaded area show the uncertainty due to the L value for a given $J = 32$ and 33.7 MeV, respectively. Brown tilted lines indicate the central densities $\rho_{central} = \rho_0, 2\rho_0, 3\rho_0, 4\rho_0, 5\rho_0$ realized in these neutron stars. The red region is forbidden by the causality relation. The figure is reprinted with permission from [9]. ©2021 by Springer

in the symmetry energy ± 2MeV leads to an uncertainty of about 3 km for the radius. In order to reproduce the recent observed massive neutron stars with twice the solar mass ($2\,M_\odot$), the symmetry energy should be $J > 32$ MeV and $L > 40$ MeV. The central density of neutron star becomes larger than 3 times of the saturation density for the stars with $M \geq 1.5\,M_\odot$. In this high density region, three-body and four-body forces might have contributions to the EoS as well as exotic constituents such as hyperons. It is also expected that the core of a neutron star can be composed of a quark-gluon plasma.

9.5.2 Ligo–Virgo Neutron Star Merger Observation and NICER Experiment

The experimental determination of neutron star masses and radii has been pursued with great effort. Empirical radius estimations have been made using models of the X-ray emission from quiescent neutron stars, from neutron stars during thermonuclear X-ray bursts, and from accretion-powered millisecond pulsars, with inferred radii typically ranging from $\sim$10 to $\sim$14 km, consistent with most theoretical predictions.

However, these estimates are susceptible to significant systematic errors as shown by the yellow region in Fig. 9.12.

Recently, the experimental study of neutron stars was advanced by two new measurements of mass and radius relation. One is the observation of tidal deformation of a neutron star merger of the GW170817 event, and another is the waveform analysis of the millisecond pulsar PSR J0030+0451 observed by using the Neutron Star Interior Composition Explorer (NICER). While the terrestrial experiments summarized in Sect. 9.4 are able to test and constrain the cold EOS at densities below and near the saturation density of nuclei, astrophysical measurements of neutron star offer information about the EOS at the density several times larger than the nuclear saturation density. The most prominent effect on matter during the observed binary inspiral comes from the tidal deformation which is induced by each star's gravitational field on its companion. In general in the astronomy, a deformation of an object induced by the gravitational field of another object is called "tidal deformation". As an example on the Earth, tides are caused by the Sun and the Moon and produce a deformation of the surface of the oceans with consequent daily fluctuation in ocean level. A neutron star in a binary system will be also squeezed by the gravity of its nearby companion. The amount of this squeezing, tidal deformability, depends on the size and the equation of state of the neutron star and increases as the two stars spiral closer and closer together. The tidal deformability induces a change in the gravitational potential, which in turn modifies the gravitational-wave signal. Therefore, the gravitational waves from NS merger will provide information about the properties of the neutron star and its matter through the tidal deformability.

The tidal deformability parameter Λ is defined by a ratio,

$$\Lambda = \frac{2}{3} \frac{k_2}{C^5}, \tag{9.28}$$

where k_2 is the induced quadrupole tidal Love number, and $C = GM/Rc^2$ is the inducing gravitational field expressed by R and M being the radius and the mass of the NS, respectively. From the GW170817 event, the tidal deformability of a NS with the mass of $1.4\,M_\odot$ is estimated to be $\Lambda = 190^{+390}_{-120}$ at 90% confidential level, which prefers soft EoS over stiff EoS. The estimation of NS mass and size are also reported based on the first binary NS merger gravitational wave event, GW170817. The gravitational waveform can be translated by means of EOS model assumptions, into constraints on mass and radius. It is found that their radii are nearly equal and have the common value 11.9 ± 1.4 km (for the 90% credible interval) with the total mass of the binary NSs $2.82^{+0.47}_{-0.09}\,M_\odot$ as is shown in Fig. 9.13.

In 2019, the mass and radius of neutron star were obtained in a good accuracy from the energy-dependent thermal X-ray waveform from the isolated 205.53 Hz millisecond pulsar PSR J0030+0451 observed by the NICER apparatus. NICER performed the relativistic ray-tracing of thermal emission from hot regions of the pulsar's surface. The waveform is analyzed by a Bayesian inference approach and the inferred mass M and equatorial radius R_{eq} are obtained, respectively, $1.34^{+0.16}_{-0.15}\,M_\odot$ and $12.71^{+1.19}_{-1.14}$ km. This is the most precise NS radius measurement so far.

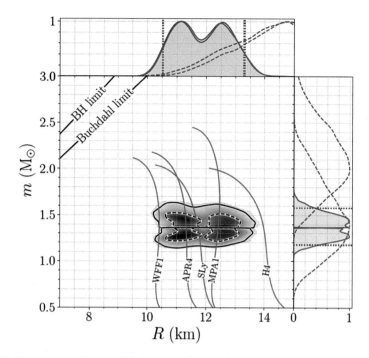

Fig. 9.13 Mass m and radius R of NSs determined by the observation of tidal deformation of binary neutron states GW170817. A parametrized EOS imposing a lower limit on the maximum mass of $1.97\,M_\odot$. The top blue (bottom orange) posterior corresponds to the heavier (lighter) NS. Example mass-radius curves for selected EOSs are overplotted in gray. The lines in the top left denote the Schwarzschild Black Hole, R (km) $= 2\,m$, and Buchdahl compactness R (km) $= 9m/4$ limits, where m is the mass of neutron star in units of the solar mass $M_\odot$. In the one-dimensional plots, solid lines are used for the posteriors, while dashed lines are used for the corresponding parameter priors. Dotted vertical lines are used for the bounds of the 90% credible intervals. Figure reprinted with permission from [8]. ©2021 by the American Physical Society

9.5.3 Hyperon Puzzle

In free space, hyperons are unstable and decay into nucleons through the weak interaction. On the contrary, in the degenerate dense matter forming the inner core of a neutron star, Pauli blocking prevents hyperons from decaying by limiting the phase space available to nucleons. When the nucleon chemical potential is large enough, the creation of hyperons from nucleons is energetically favorable. This new degree of freedom (appearance of hyperons among nucleons) leads to a reduction of the pressure by the baryons and, as a consequence, to a softening of the EoS beyond a certain high density $\rho > 2\rho_0$. With this soft EoS, heavy neutron star with mass of $2\,M_\odot$ is hardly predicted, while the maximum observed mass exceeds $2\,M_\odot$. This is called "the Hyperon puzzle".

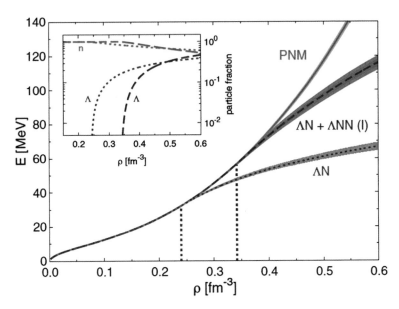

Fig. 9.14 Equations of state of pure neutron matter (PNM) and hyperon nuclear matter (HNM) calculated by Green's function Monte Carlo (GFMC) model. The green solid curve refers to the PNM EoS calculated with the two-body AV8'+three-body UIX potentials. The red dotted curve represents the EoS of hyperon matter with hyperons interacting via the two-body ΛN force alone. The blue dashed curve is obtained including a three-body hyperonnuclear potential. Shaded regions represent the uncertainties on the results coming from Monte Carlo technique. The vertical dotted lines indicate the threshold densities ρ_Λ^{th}. Neutron and lambda fractions corresponding to the two HNM EoSs are shown in the inset. Figure reprinted with permission from [10]. ©2021 by the American Physical Society

The EoS of hyperon nuclear matter (HNM) was investigated by employing the Green's function Monte Carlo (GFMC) model, which has been successfully applied to study properties of pure neutron matter (PNM) in the previous section. The phenomenological two-body ΛN interaction, which includes central and spin-spin components, has been fitted on the available hyperon-nucleon scattering data.

The repulsive three-body $\Lambda N N$ force is introduced to simultaneously reproduce the hyperon separation energy of $^5_\Lambda$He and $^{17}_\Lambda$O. The hyperon-nucleon potential is attractive, and reproduces not only data of light hypernuclei but also the structure of medium mass Λ hypernuclei.

In Fig. 9.14, the EoS for PNM (green solid curve) and HNM using the two-body ΛN interaction alone (red dotted curve) and two- and three-body hyperonnucleon $\Lambda N N$ forces (blue dashed curve) are displayed. As expected, the presence of hyperons makes the EoS softer. In particular, a low threshold density of hyperon $\rho_\Lambda^{th} = 0.24$ fm^{-3} is noticed if hyperons only interact via the two-body ΛN potential. Within the two-body interaction, hypernuclei turn out to be strongly overbound and the inclusion of the repulsive three-body force helps to obtain more reasonable binding energies. This three-body force stiffens the EoS and pushes the threshold density to $\rho_\Lambda^{th} = 0.34$ fm^{-3}. In the inset of Fig. 9.14, the neutron and lambda fractions are shown for the two EoSs for HNM. With the ΛN interaction only, the EoS is too soft. However, the ΛNN force is repulsive and pushes the chemical potential of Λ particles to higher densities. The net effect recovers largely the original EoS with NN interactions only so that the Λ strangeness has less impact on the EoS with the inclusion of the ΛNN force.

Currently, there is no general agreement among the predicted results for the EoS and the maximum mass of neutron stars including hyperons. One needs more data to constrain the two- and three-body hyperon-nucleon interactions. On top of that, hyperon-hyperon forces might play a role in the high density neutron matter in which other baryons such as Ξ, Σ, π condensation, K condensation and quarks may also appear in the inner core of neutron stars. The hyperon puzzle is an open and interesting problem which should be addressed in the near future.

9.5.4 Crust of Neutron Star

In the previous section, the size and the weight of neutron stars were discussed in the context of realistic nuclear EoS. We study now the structure of neutron stars from the surface to the interior. Neutron stars accompany often a strong magnetic field outside surrounded by atoms and molecules. The strong magnetic field is considered as a result from a magnetohydrodynamic dynamo process in the turbulent, extremely dense conducting fluid that exists before the neutron star settles into its equilibrium configuration. These fields then persist due to no resistance currents in a proton-superconductor phase of matter that exists at an intermediate depth within the neutron star.[4] The crust of neutron stars extends down to about 1 km below the surface, with densities ranging from a $\rho \sim$ a few g/cm^3 on the ground up to 10^{14} g/cm^3 at the bottom where the density is close to the nuclear matter saturation density $\rho_0 = 0.16$ nucleons/fm^3, equivalently, 2.7×10^{14} g/cm^3. The crust thickness is largely determined by the slope of the symmetry energy L, the incompressibility K_∞, and the neutron star compactness M/R, i.e., the transition from the crust to the core is strongly influenced by these quantities, especially, the L value. In the outer crust, below the density $\rho \sim 4 \times 10^{11}$ g/cm^3, which is still much lower than the nuclear matter density, all nuclei are separated and immersed in a relativistic degenerate electron gas, which gives the dominant contribution to the pressure of the outer crust. In this domain, the Coulomb interaction dominates and makes a lattice structure of nuclei, mainly with Fe, in an analogous way to the crystal structure of condensed matter. As pressure increases with depth, the pressure fuses more and more electrons and protons into neutrons and makes the nuclei more neutron-rich, so called "exotic nuclei". Eventually, in the inner crust, the nuclei cannot accept any more neutrons,

---

[4] Neutron stars, which have even much stronger magnetic fields than normal neutron star, are called "Magnetars". Their magnetic fields are of the order of $\sim 10^9$ to 10^{11} T, which are a hundred million times stronger than any man-made magnet on the earth.

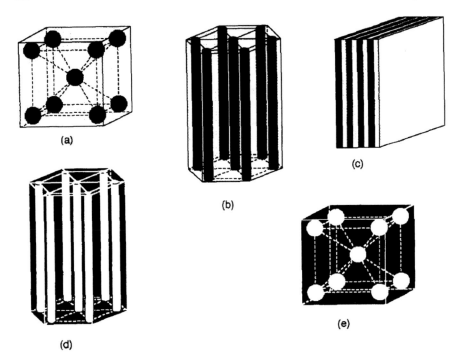

Fig. 9.15 Schematic pictures of nuclear "pasta". The black regions show the liquid phase of nuclear matter, where protons and neutrons coexist. Nuclear shapes are **a** sphere, **b** cylinder, **c** slab, **d** cylindrical hole, and **e** spherical hole. The lattice type to minimize the Coulomb energy for each nuclear shape is; **a** and **e** the spherical nuclei as well as the spherical hole nuclei form the body-centered cubic lattice. **b** and **d** The cylindrical nuclei and the cylindrical hole nuclei form the two-dimensional hexagonal lattice, i.e., any cross section perpendicular to their sides shows a pattern of regular hexagon. **c** The slab nuclei form the one-dimensional lattice, i.e., they are all parallel and equally spaced. Reprinted with permission from [11]. ©2021 by Elsevier

and free neutrons form a superfluid that permeates the lattice. At a critical density $\rho \sim 4 \times 10^{11}$ g/cm^3, where the nuclear force can not bind all the nucleons, the neutrons drip from nuclei and move freely in the crust. Above this critical density, the nuclear lattice structure is surrounded by a fluid of dripped neutrons. These neutrons are not localized and are floating in the crust region. Inside of this dense medium, the EoS is dominated by the pressure from nuclear interactions. A cartoon illustrating the crust layer is shown in Fig. 9.11.

Under the condition of terrestrial matter, atomic nuclei are spherical or modestly deformed (the deformation parameter $\beta < 0.5$). However, this common feature does not necessarily hold for matter in supernovae and neutron stars. There, the matter density is very high and it can reach the same value as inside the atomic nuclei themselves, i.e., the normal saturation nuclear density. In such a high-density environment, nuclei will adopt various shapes.

More specifically, the inner crust between densities $\rho = 4 \times 10^{11} \sim 10^{14}$ g/cm^3 has a solid structure of heavy neutron-rich nuclei surrounded by neutron fluid which is expected to give rise to the superfluid phase. In the region of densities higher than $\rho = 10^{14}$ g/cm$^3 \sim 0.4\rho_0$, there is the mantle where the matter with exotic shapes may exist, i.e., the competition between the nuclear surface and the Coulomb energies drives the formation of exotic geometry so-called nuclear "pasta" phase, which proceeds through a sequence of different shapes: sphere $\rightarrow$ cylindrical (spaghetti) $\rightarrow$ slab (lasagna) $\rightarrow$ cylindrical bubble $\rightarrow$ spherical bubble (see Fig. 9.15). The lattice type to minimize the Coulomb energy for each nuclear shape is; (a) and (e) are the spherical nuclei as well as the spherical hole nuclei forming the body-centered cubic lattice. (b) and (d) are the cylindrical nuclei and the cylindrical hole nuclei forming the two-dimensional hexagonal lattice, i.e., any cross section perpendicular to their sides shows a pattern of regular hexagon. (c) is the slab nuclei forming the one-dimensional lattice, i.e., they are all parallel and equally spaced. Nuclear shape changes in the sequence from (a) to (e) with increasing density.

9.6 Nucleosynthesis

How matter is created in the universe is one of the most fundamental and fascinating questions whatever asked. Our universe was created by the Big Bang, 13.7 billion years ago. During the few minutes after the Big Bang, light elements with masses $A = 1 \sim 7$ were created under extremely hot and dense circumstances. Then, much later, stellar nucleosynthesis started to create heavier elements associated with the star evolutions, supernova explosions and neutron star mergers. Nuclear physics information such as masses and beta decay rates gives essential ingredients to understand the nucleosynthesis in the universe. A dramatic turn of the nucleosynthesis study has recently taken with the observations of gravitational waves and photons from the neutron star merger "GW170817" and the "kilonovae" event afterwards. These events provide decisive information on one of the astronomical sites of the rapid process (r process) nucleosynthesis.

9.6.1 Solar System Abundances

The solar system abundances give a direct evidence of "nucleosynthesis" in the universe. The solar system abundances are measured by two types of data; one is the spectral analysis of the sun and the second is the direct measurement of chemical elements on the Earth, the Moon, meteorites and interstellar matters. The abundances outside of the solar system are also available by the analysis of spectra of other stars and gaseous nebulae and external galaxies. It is based on the fact that each element produces a set of characteristic spectral lines from the de-excitation of atomic levels, which can be used to identify each element. The relative intensity of spectral lines

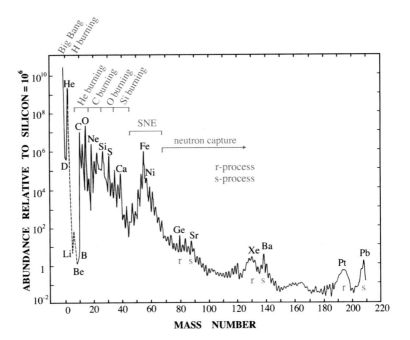

Fig. 9.16 The abundance distribution of nuclear species, as a function of mass number A. The abundances are given relative to the Si abundance which is set to 10^6. Possible process for each mass region is indicated in the figure. SNE stands for the supernovae explosions. Peaks due to the r- and s-process are also indicated. Each process is discussed in the text

gives the information of abundance of each element. Abundance of the chemical elements is also obtained in the analysis of materials in Earth's upper continental crust. The observed abundances are shown in Fig. 9.16.

The abundances of other stars and galaxies show qualitatively similar abundances to the solar system, but in some cases significant differences are observed. These differences are related to the special stage of evolution history of these systems.

One of the characteristic features of solar abundances is the great preponderance of ^{1}H and ^{4}He: the mass abundance is 70.6% and 27.5% for ^{1}H and ^{4}He, respectively. In general, the abundances decrease rapidly for larger mass nuclei. The empirical abundances for $A > 50$ show two striking features. The first one is a pronounced peak in the region of Fe and the second is a gradual decrease toward the heavier elements on which a number of small of peaks are superimposed. The peak near Fe can be explained as the influence of the most stable nucleus ^{56}Fe and the neighboring nuclei. These nuclei would acquire the maximum abundance under the conditions of nuclear processes in thermal equilibrium. It is envisaged that the steps leading to equilibrium involve first the collection of primordial matter from the early universe and the nuclear fusion of hydrogens to ^{4}He at the temperature of around 10^7 K ($\sim$1 keV).[5] This stage is called *primordial nucleosynthesis* or *Big Bang nucleosynthesis*, and it takes place during the first few minutes after the Big Bang. The next stage of nucleosynthesis is the *stellar nucleosynthesis* which occurred a few hundred million years after the Big Bang. After the exhaustion of hydrogen in the core, the star contracts and the temperature is raised. This makes it possible for α particles to fuse and to form ^{12}C and ^{16}O at the temperature about 10^8 K (~ 10 keV). With a further increase in temperature, these nuclei react again and ultimately produce the most stable nuclei in the mass region $A \sim 60(Z \sim 26)$. Beyond $A \sim 60$, reactions with charged particles are not efficient to produce heavier elements because of the high Coulomb barrier; since the Coulomb barrier is high, α particles cannot penetrate into nuclei efficiently. For heavier nuclei $A > 60$, plausible reactions are neutron capture reactions, so-called slow (s-) process and rapid (r-) process networks with consecutive β-decay processes.

9.6.2 Big Bang Nucleosynthesis

Big Bang nucleosynthesis (BBN) refers to the production of nuclei other than those of the lightest isotope of hydrogen (^1H) during the early phases of the universe, well before the existence of stars. BBN is believed to have taken place in a short time interval.

There are several important characteristics of BBN:

- The initial conditions (neutron-proton ratio) were set in a few seconds after the Big Bang.
- The universe was very close to homogeneous at this time, and strongly radiation-dominated.
- The fusion of nuclei occurred between roughly 10 s to 20 min after the Big Bang; this corresponds to the temperature range in which the universe was cool enough for deuterium to survive, but hot and dense enough for fusion reactions to occur at a significant rate. Prior to this time, the temperature of the universe was so high that light nuclei formed by nucleons broken up by photodisintegration as soon as they were formed.

The key parameters of BBN are the proton-neutron (p-n) ratio and the baryon-photon ratio. The p-n ratio is determined by the following thermal equilibrium reactions induced by the weak interaction:

$$n + e^+ \rightleftharpoons p + \bar{\nu}_e,$$
$$n + \nu_e \rightleftharpoons p + e^-. \tag{9.29}$$

At a temperature T, the n-p ratio is given by the Boltzmann distributions,

---

[5] A thermal energy $kT = 1$ eV corresponds to a temperature of 1 eV $= 1.16 \times 10^4$ K in units of Kelvin, in which k is the Boltzmann constant.

$$\frac{n_n(T)}{n_p(T)} = \left(\frac{m_n}{m_p}\right)^{3/2} exp\left(-\frac{(m_n - m_p)c^2}{kT}\right), \tag{9.30}$$

where $(m_n - m_p)c^2 = 1.29\,\text{MeV}$. For extremely high temperatures $T > 10^{11}\,\text{K}$ ($kT > 10\,\text{MeV}$) just after the Big Bang, the n-p ratio is in a thermal equilibrium phase and closer to 1 as is expected from Eq. (9.30). When the temperature is going down, the n-p ratio decreases gradually. When $T < T_{crit} \simeq 10^{10}\,\text{K}$, weak interactions become slower and the ratio n/p freezes out to be a constant value. The weak interaction freeze-out is reached at $t \sim 2$ s after the Big Bang, at which the neutron abundance by mass becomes $n_n/n_p \sim 0.2$ according to Eq. (9.30). The T_{crit} is called the freeze-out temperature. Since free neutrons are unstable and have the mean life time $\tau = 886.7 \pm 1.9$ s, some neutrons decay within few minutes before fusing with any nucleus. Eventually, the n-p ratio after BBN is considered to be about 1:7.

Another key parameter to calculate the effects of BBN is the baryon-photon number ratio, which is a small number of order 6×10^{-10}. This parameter controls the rates of nuclear fusion and also of the dissociation of the created nuclei. It is possible to calculate abundances of light elements after BBN ends, as a function of this parameter. Experimentally, the baryon-photon number ratio was determined by the observation of cosmic microwave background (CMB)[6] in the Universe. A starting reaction to create ^{4}He is

$$p + n \to d + \gamma. \tag{9.31}$$

Since the photodisintegration cross section of d is large, the destruction of d is very likely because of the intense photon background until the temperature has dropped down below the binding energy of d, which is 2.22 MeV, corresponding to 2.6×10^{10} K. In fact, d fusion remains stable against photodisintegration when the temperature falls to $T = 9 \times 10^8$ K or $kT \sim 80$ keV, which happens $t \sim 270$ s after the Big Bang.

Once the deuteron is created, the next step is the fusion reaction to create $A = 3$ system, ^{3}He and $t\,(^3\text{H})$;

$$p + d \to {}^3\text{He} + \gamma,$$
$$d + d \to {}^3\text{He} + n,$$
$$d + d \to t + p,$$
$$^3\text{He} + n \to t + p. \tag{9.32}$$

The following reactions are the fusion reactions leading to ^{4}He;

---

[6] The cosmic microwave background (CMB) is the electromagnetic radiation, remnant from the early stage of the universe, found in the 1940s by American astronomers Arno Penzias and Robert Wilson. The CMB was observed as faint cosmic background radiation filling all space. It is an important source of data on the early universe, and proved the scenario of Big Bang cosmology of the early universe.

$$^3\text{He} + d \rightarrow {}^4\text{He} + p,$$
$$t + d \rightarrow {}^4\text{He} + n. \tag{9.33}$$

BBN results in mass abundances of about 75% of ^1H, about 25% of ^4He, about 0.01% of d and ^3He, tiny amounts (on the order of 10^{-10}) of Li, and negligible heavier elements. The abundance of ^4He depends largely on the p-n ratio. The mass abundance n_α of ^4He (α) is defined as[7]

$$Y \equiv \frac{m_\alpha n_\alpha}{m_p n_p + m_n n_n} = \frac{4 n_\alpha}{n_p + n_n}, \tag{9.34}$$

where ^4He is a nucleus with 2 neutrons and 2 protons and $m_p \sim m_n$. Essentially, all neutrons are converted into ^4He to the first-order estimation so that $n_\alpha = n_n/2$. Inserting this ratio and $n_n/n_p = 1/7$, the α abundance by mass is estimated as

$$Y \sim 0.25. \tag{9.35}$$

This value is consistent with the observed mass abundance of ^4He. The remaining mass abundance is hydrogen H.

Since there are no stable nuclei with mass $A = 5$, an addition of a proton to ^4He does not play a role in the network. A fusion of ^4He with d is also not efficient because the d abundance is very low due to the efficient follow-up reactions. The effective reaction to create nuclei beyond mass $A = 5$ is fusing ^4He with t and ^3He to build up elements with $A = 7$,

$$t + {}^4\text{He} \rightarrow {}^7\text{Li} + \gamma, \tag{9.36}$$
$$^3\text{He} + {}^4\text{He} \rightarrow {}^7\text{Be} + \gamma, \tag{9.37}$$

followed by

$$^7\text{Be} + e^- \rightarrow {}^7\text{Li} + \nu_e. \tag{9.38}$$

Some amount of ^7Li is destroyed by a reaction

$$^7\text{Li} + p \rightarrow {}^8\text{Be} \rightarrow {}^4\text{He} + {}^4\text{He}, \tag{9.39}$$

since ^8Be is unstable and decays to two ^4He. Further, a fusion of ^8Be with ^4He can rarely occur because the low density of the reaction partners essentially excludes that a ^8Be nucleus meets a ^4He nucleus during its lifetime. Thus, the absence of stable elements with $A = 8$ prohibits any primordial element fusion beyond ^7Li.

The detailed abundances of the light elements as produced by the primordial fusion must be calculated solving rate equations based on the respective fusion cross

---

[7] The fraction of mass abundance is written by a symbol Y, so that 25% ^4He means that ^4He atoms account for 25% of the mass, but less than 7% of the nuclei in number would be ^4He nuclei.

sections. To this end, the precise values of the cross sections and their energy dependence, and the precise lifetime of the free neutrons are important. The primordial nucleosynthesis happens during the radiation era of the Big Bang, and the radiation density is precisely determined by the expansion rate in the Big Bang model. In fact, the only parameter defining the primordial abundances is the ratio η between the number densities of baryons and photons. The ^{4}He abundance depends only very weakly on the η because the largest fraction of free neutrons is swept up into ^{4}He without strong sensitivity to the detailed conditions. On the other hand, the abundances of d, ^{3}He and ^{7}Li depend largely on the baryon density η. With increasing η, they can more easily be burned to ^{4}He, and so their abundances drop as η increases; at low η, a large proton density causes ^{7}Li to be destroyed by the reaction (9.39) so that ^{7}Li abundance decreases for increasing η from 10^{-11} to $\sim (2-3) \times 10^{-10}$. On the other hand, the precursor nucleus ^{7}Be is more easily produced if the baryon density increases further to $\eta \sim 10^{-9}$ and the abundance of ^{7}Li increases also by the reaction (9.38). These circumstances create a characteristic valley of the predicted ^{7}Li abundance near $\eta \sim (2-3) \times 10^{-10}$.

The right panel of Fig. 9.17 shows a comparison between the primordial abundances inferred from observation and the predictions based upon the BBN model calculation as a function of the baryon to photon ratio η. Shaded bands show the uncertainty in the BBN abundances while the horizontal lines show the uncertainties in the observation-inferred primordial abundances. Vertical lines show the ratio $\eta = (6.10 \pm 0.04) \times 10^{-10}$ deduced from the analysis of the cosmic microwave background (CMB) in the universe. One of the amazing triumphs of BBN is the overall agreement between the predictions of BBN and the observed primordial abundances over nine orders of magnitude in abundances. In particular the observed D/H abundance agrees almost perfectly with the BBN prediction within the CMB inferred value of η. Similar agreement exists for Y_p and the ^{3}He/H ratio, although the observational uncertainty is much larger. There remains, however, one serious open problem. The calculated and observed ^{7}Li/H ratio differ by about a factor of 3. This is known as the *lithium problem*. Besides this issue, the fact, that the observed abundances in the universe are generally consistent with these abundance numbers predicted by the BBN model, is considered a strong evidence for the Big Bang scenario.

9.6.3   Stellar Nucleosynthesis

Stellar nucleosynthesis is the scenario explaining the creation (nucleosynthesis) of chemical elements by nuclear fusion reactions between nuclei in the stars. The synthesis has started after stars are formed in the universe, a few hundreds million years after the Big Bang, in which hydrogen, helium and lithium were originally created. The stellar nucleosynthesis explains why the observed abundances of elements in the universe grow over time and why some elements and their isotopes are much more abundant than others. The theory was initially proposed by Fred Hoyle in 1946.

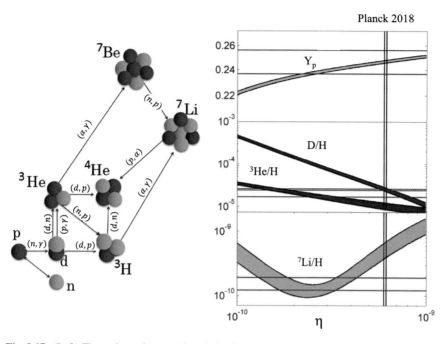

Fig. 9.17 (Left) The main nuclear reaction chains for Big Bang nucleosynthesis. At the universal temperature $T = 10^9$ K, the trigger fusion reaction starts to create deuterium from proton and neutron. Then sequential reactions continue to create from ^{4}He to ^{7}Li. (Right) relative abundances of ^{4}He (Y_p), deuterium (D), ^{3}He and ^{7}Li to hydrogen (H) as a function of the baryon density (baryon-photon ratio η). For the observed value of the ratio ^{3}He/H, only the upper limit is shown. Shaded bands correspond to the 2σ (95% confidential level) uncertainties. Horizontal lines show the range of the uncertainties in the observed primordial abundances. Vertical lines indicate the value of η deduced in the recent analysis of the Cosmic Microwave Background (CMB) in the universe. The abundances of all elements except ^{7}Li are consistent with the observed baryon-photon rate η determined by CMB data (see also [12]). Courtesy G. J. Mathews, Notre Dame

Further, advances were made, especially to nucleosynthesis by neutron capture of the elements heavier than iron, by Margaret Burbidge, Geoffrey Burbidge, William Alfred Fowler and Fred Hoyle in their famous paper in 1957 denoted B2FH, which became one of the most heavily cited papers in astrophysics history.

In a star, the gravity pressure tends to decrease the star's volume, while the pressure of the gas inside tends to increase the volume. In the sun, the pressure at the center is about 2×10^{10} bar (2×10^{15} Pa) and the temperature $T = 1.6 \times 10^7$ K $= 16$ MK ($kT = 1.4$ keV). Under these circumstances, atoms are completely ionized, resulting in a mixture of free electrons and bare nuclei. This is the origin of gas pressure in the star. The internal pressure is created by the nuclear reaction that provides the energy for the stellar radiation. As long as this reaction continues, the gravitational and gas pressure balance and the star is in equilibrium. However, when the star exhausts most of the fuel, i.e., when most of the hydrogen is used and the p-p cycle stops, the star will start to contract by gravity and the central temperature and pressure will increase. At some higher temperature, new reactions start to occur and new elements are created. There are alternate stages of nuclear burnings and creations of heavier elements in stars or supernovae.

In stars of mass larger than $0.1 \times M_\odot$ (the solar mass is denoted as a symbol $M_\odot = 1.989 \times 10^{30}$ kg), the process called "Hydrogen burning (p-p cycle)" is started at the temperature $T \sim 10^7$ K. The net effect is given by

$$4H \rightarrow{}^4 He + 2e^+ + 2\nu_e, \tag{9.40}$$

which is not a single process, but a combination of reactions,

$$H + H \rightarrow d + e^+ + \nu_e + Q,$$
$$d + H \rightarrow {}^3He + \gamma,$$
$${}^3He + {}^3He \rightarrow {}^4He + 2H. \tag{9.41}$$

The p-p chain is an exothermic reaction and the main origin of the energy of stars. This reaction has last 5 billion years in the sun and will continue for the next 5 billion years. In heavier stars, for example, with a mass of $15\,M_\odot$, the hydrogen burning process is much shorter, about 10 million years.

Problem

9.2 The energy for the solar luminosity is mostly provided by the exothermic p-p cycle,

$$4H \rightarrow {}^4He + 2e^+ + 2\nu_e + Q,$$

where the Q value is $Q = 26.72$ MeV.
(a) The solar luminosity is $L_\odot = 3.844 \times 10^{33}$ erg s^{-1} (The energy unit "erg" is often adopted in the astrophysics). How many neutrinos are liberated in the sun each second?
(b) How much is the solar neutrino flux on the earth? The distance (1AU) between the sun and the earth is $1 \text{ AU} = 1.496 \times 10^{13}$ cm.

After the Hydrogen burning, the star expands and becomes a Red Giant star whose composition is dominated by helium gas. In the center of such stars, a helium core is created and its temperature increases by gravitational contraction. At the temperature $T \sim 10^8$ K, the "helium burning" process starts. The starting reaction is

$$^4He + {}^4He \rightarrow {}^8 Be, \tag{9.42}$$

where the ^{8}Be is unstable against the breakup into two ^{4}He. However, the life time $T_{1/2} = 3 \times 10^{-16}$ s is much longer than the reaction time of ^{4}He+^{4}He ($t \sim 10^{-22}$ s with a reaction energy E $\sim$ 10 MeV). This time scale is enough to create a small concentration of ^{8}Be, which enables a further reaction through a resonance state ^{12}C*,

$$^8\text{Be} +{}^4\text{He} \rightarrow {}^{12}\text{C}^* \rightarrow {}^{12}\text{C} + \gamma. \tag{9.43}$$

The existence of this resonance at around $^8\text{Be}+{}^4\text{He}$ threshold energy, $E_x \sim 7.3\,\text{MeV}$, was predicted by Fred Hoyle in 1954 and called "Hoyle" state in ^{12}C. Without this resonance, there would be neither carbon nor more heavier elements in the universe! Experimentally, the Hoyle state was observed a few years later utilizing the beta decay of ^{12}B through $^{12}\text{C}^*$ (at $E_x = 7.653\,\text{MeV}$) to $^4\text{He}+{}^8\text{Be}$.

After ^{12}C is created, ^{16}O is formed dominantly by the fusion reaction

$$^{12}\text{C} +{}^4\text{He} \rightarrow {}^{16}\text{O} + \gamma. \tag{9.44}$$

In massive stars $M \sim 4\,\text{M}_\odot$, also sometimes called *exploding stars*, the *carbon burning* process will start at $T > 6 \times 10^8$ K,

$$^{12}\text{C} +{}^{12}\text{C} \rightarrow {}^{24}\text{Mg}^* \rightarrow {}^{20}\text{Ne} + \alpha, \tag{9.45}$$

$$^{12}\text{C} +{}^{12}\text{C} \rightarrow {}^{24}\text{Mg}^* \rightarrow {}^{23}\text{Na} + p. \tag{9.46}$$

These carbon burning reactions are considered to account for the abundances of nuclei with mass $20 \leq A \leq 32$. Similarly, "oxygen burning"

$$^{16}\text{O} +{}^{16}\text{O} \rightarrow {}^{32}\text{S}^* \rightarrow {}^{28}\text{Si} + \alpha, \tag{9.47}$$

$$^{16}\text{O} +{}^{16}\text{O} \rightarrow {}^{32}\text{S}^* \rightarrow {}^{31}\text{P} + p, \tag{9.48}$$

occurs at a temperature of about 2×10^9 K and accounts for the abundance of some of nuclei with $32 \leq A \leq 42$. Furthermore, "silicon burning" takes place at the temperature $T \geq 3 \times 10^9$ K. Silicon burning is not a direct reaction of $^{28}\text{Si}+{}^{28}\text{Si}$, but goes through (α, γ) reactions:

$$^{28}\text{Si}(\alpha, \gamma)^{32}\text{S}(\alpha, \gamma)^{36}\text{Ar}(\alpha, \gamma)^{40}\text{Ca}(\alpha, \gamma)^{44}\text{Ti}(\alpha, \gamma)^{48}\text{Cr}(\alpha, \gamma)^{52}\text{Fe}(\alpha, \gamma)^{56}\text{Ni}. \tag{9.49}$$

The silicon burning explains the formation of many nuclei up to ^{56}Ni. Stars evolve because of changes in their composition (the abundance of their constituent elements) over their life spans, first by burning hydrogen (main sequence star), then helium (red giant star), and progressively burning higher elements. With all these processes, the core of the star takes an onion-like structure following the hierarchy of nuclear burning thresholds, as illustrated in Fig. 9.18.

Problem

9.3 After C burning, O burning $^{16}\text{O} + {}^{16}\text{O}$ is expected to occur before Ne burning since ^{16}O is the lightest remaining nucleus. However, Ne burning $^{20}\text{Ne}+\alpha$ occurs at lower temperature that O burning. Why is it so?

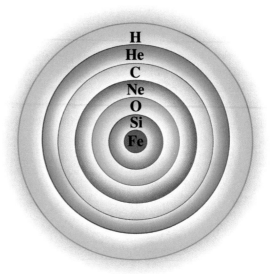

Fig. 9.18 Nuclear burning in heavier stars. The core of stars takes on an onion-like structure of nuclear burnings

Stellar evolution

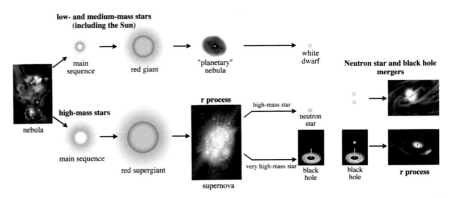

Fig. 9.19 Chart of stellar evolution. Depending on the mass of the star, its lifetime can range from a few million years for the most massive to trillions of years for the least massive. All stars are born from collapsing clouds of gas and dust, so called nebulae or molecular clouds. Over the course of millions of years, these protostars settle down into a state of equilibrium, becoming "main-sequence stars". Nuclear fusion gives the energy for a star, starting the fusion of hydrogen which changes the core contents to helium atoms and heavier elements. This process causes the star to gradually grow in size until it reaches the red giant. Once a star like the sun has exhausted its nuclear fuel, its core collapses into a dense white dwarf and the outer layers are expelled as a planetary nebula. Stars with masses of ten or more times the solar mass can explode in a supernova as their inert iron cores collapse into an extremely dense neutron star or black hole

An illustration of star evolution is shown in Fig. 9.19. In the evolution of stars, a low-mass star slowly ejects its atmosphere via stellar wind, forming a planetary nebula later in its life, while a higher mass star ejects mass via a sudden catastrophic event called "supernova". The term supernova nucleosynthesis is used to describe the creation of elements during the evolution and explosion of a pre-supernova massive star with mass of $(12–35) M_\odot$. Those massive stars are the productive source of new isotopes from carbon $(Z = 6)$ to nickel $(Z = 28)$ and also some heavier elements. Namely, the burning fuel of nuclei is driven by gravitational pressure and its associated heating, resulting in the subsequent burning of carbon, oxygen and silicon. Most of the nucleosynthesis in the mass range $A = 28 - 56$ (from silicon to nickel) takes place in the upper layers of the star and becomes a trigger for the collapsing of layers onto the core. The core collapsing creates a compressional shock wave rebounding outward and the shock front briefly raises temperatures by roughly 50%, thereby causing furious burning for about a second. This final burning in massive stars, called explosive nucleosynthesis or supernova nucleosynthesis, is the final epoch of stellar nucleosynthesis.

9.7 Neutron Capture Nucleosynthesis

Reactions with charged particles are not efficient to produce elements beyond ^{56}Fe because of the high Coulomb barrier between nuclei which prevents fusion to occur. More effective reactions are the neutron capture reaction and the consecutive β-decay process;

$$(N, Z) + n \rightarrow (N + 1, Z) + \gamma, \tag{9.50}$$

$$(N + 1, Z) \rightarrow (N, Z + 1) + e^- + \nu_e. \tag{9.51}$$

The β-decay time of nuclei close to the "valley of stability" in the nuclear chart is mostly of the order of hours and long enough to capture thermal neutrons with very low energy. The low energy neutron has an advantage for the capture rate since the thermal neutron capture cross section is proportional to the inverse of neutron velocity v,

$$\sigma(v) \sim \frac{1}{v}. \tag{9.52}$$

This process is called "s-process" where "s" stands for "slow". This process takes place along the valley of stability in the nuclear chart.

There is another neutron-capture process under the extreme high temperature and high density which takes place in a region of unstable neutron-rich nuclei close to the neutron drip line at which the neutron separation energy becomes zero. In this region, the β-decay is fast so that only energetic neutrons are captured. This process is called r-process where "r" stands for "rapid" (see Fig. 9.20). These two processes

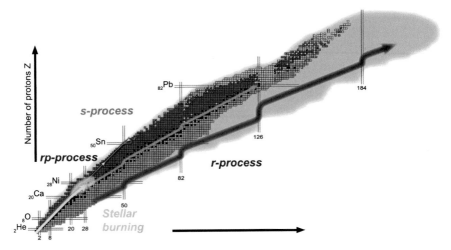

Fig. 9.20 Nuclear chart with various processes of nucleosynthesis. Black squares show stable nuclei, while yellow and green zones correspond to unstable nuclei already known and still unknown, respectively. On one hand, the s-process takes place along the stability line, while the r-process holds along unknown unstable neutron-rich nuclei. The rp-process on the other hand proceeds along the region of the proton drip line. Magic numbers are indicated by horizontal and vertical lines in the chart

are considered to be responsible for the majority of synthesis of nuclei with $A > 56$. There are also minor processes: one is induced by proton capture reactions (p, γ) and (γ, p) and produces neutron-deficient nuclei near the proton drip line. This is called "rp"-process, as indicated in Fig. 9.20. Another one, called "p"-process, also feeds some stable proton-rich nuclei which are not reached by neither s-process nor r-process. Intensive neutrino flux from the supernova explosion can also feed some exotic nuclei along both the neutron and proton drip lines of the nuclear chart. The former is called the weak r-process and the latter is named "νp"-process.

9.7.1 Neutron Capture Cross Section

In Sect. 8.5.1 of Chap. 8, neutron-induced resonant scattering was discussed. To evaluate the neutron capture cross section, we consider here a square well optical potential. For simplicity, the s-wave ($l = 0$) neutron is considered. The square well optical potential is defined as (see Sect. 8.2.6 in Chap. 8 for the optical potential)

$$V(r) = \begin{cases} V_0 + i W_0, & r \leq a \\ 0, & r > a, \end{cases} \qquad (9.53)$$

Fig. 9.21 Slow neutron capture in a strong attractive square well potential. The optical potential parameters are set to be $V_0(<0)$ and $W_0(<0)$

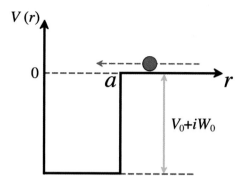

where $V_0(<0)$ is an attractive central potential, and $W_0(<0)$ is an imaginary potential as illustrated in Fig. 9.21. The Schrödinger equation in 3D space is given by

$$-\left[\frac{\hbar^2}{2\mu}\frac{d^2}{dr^2} + V(r)\right]u(r) = Eu(r), \tag{9.54}$$

where the wave function for the s-wave is written as $\Psi(r) = u(r)/r$ and μ is the reduced mass. The radial dependence of (9.54) is exactly the same as that of the 1D tunneling problem (see (1.54) in Chap. 1). At $r = 0$, the wave function $\Psi(r)$ is finite so that $u(r) = r\Psi(r) = 0$ at $r = 0$. In the regions $r < a$, the wave function is expressed by a *sine* function, and the wave function has an incoming and an outgoing waves for $r > a$, respectively given by

$$u_1(r) = A\sin Kr \quad (r < a), \tag{9.55}$$
$$u_2(r) = e^{-ikr} - Se^{ikr} \quad (r > a), \tag{9.56}$$

where K is a complex number $K = \sqrt{2\mu(E - V_0 - iW_0)/\hbar^2}$, $k = \sqrt{2\mu E/\hbar^2}$, and S is the scattering amplitude. The continuity condition of the wave function and its derivative at $r = a$ can be set by the logarithmic derivative

$$L = r\frac{d\log u(r)}{dr}\bigg|_{r=a} = r\frac{du(r)}{dr}\bigg/u(r)\bigg|_{r=a}, \tag{9.57}$$

which gives a boundary condition for the wave function

$$Ka\cot Ka = -ika\frac{e^{-2ika} + S}{e^{-2ika} - S}. \tag{9.58}$$

From Eq. (9.58), the scattering amplitude S is extracted as

$$S = e^{-2ika}\frac{Ka\cot Ka + ika}{Ka\cot Ka - ika}. \tag{9.59}$$

See Chap. 8 for more details about the S matrix. Now we define the quantity

$$L(Ka) = Ka \cot Ka \equiv L_R + iL_I. \tag{9.60}$$

The S-matrix is then expressed as

$$S = e^{-2ika}\left(1 + \frac{2ika}{L_R + i(L_I - ka)}\right), \tag{9.61}$$

and the absorption cross section is obtained as

$$\sigma = \frac{\pi}{k^2}(1 - |S|^2) = \frac{\pi}{k^2}\frac{4ka|L_I|}{L_R^2 + (L_I - ka)^2}. \tag{9.62}$$

To study the resonance feature of the wave function, we expand the function L_R around an energy E_R

$$L_R(Ka) = L^{(1)} + L^{(2)}(E - E_R) + \cdots. \tag{9.63}$$

The functions $L^{(1)}$ and $L^{(2)}$ are given by

$$L^{(1)} = L_R(E_R) = a\sqrt{2\mu(E_R - V_0)/\hbar^2} \cot a\sqrt{2\mu(E_R - V_0)/\hbar^2}, \tag{9.64}$$

$$L^{(2)} = \frac{dL_R}{dE}\bigg|_{E=E_R} = -\frac{\mu a^2}{\hbar^2}. \tag{9.65}$$

The imaginary term L_I can be also obtained in the case of $|W_0| < |V_0|$ as

$$L(E - V_0 - iW_0) \approx L_R(E - V_0) + iL_I, \tag{9.66}$$

where

$$L_I = -W_0 \frac{dL_R}{dE}. \tag{9.67}$$

The resonance energy is obtained by a condition $L^{(1)} = 0$ in the S-martix (9.61), i.e., $L^{(1)} = 0 \rightarrow \cot a\sqrt{2\mu(E_R - V_0)/\hbar^2} = 0$ is satisfied at

$$a\sqrt{2\mu(E_R - V_0)/\hbar^2} = \left(n + \frac{1}{2}\right)\pi, \qquad n = 0, 1, 2, \ldots, \tag{9.68}$$

which gives

$$E_R = \frac{\hbar^2\pi^2}{2\mu a^2}\left(n + \frac{1}{2}\right)^2 + V_0, \qquad n = 0, 1, 2, \ldots. \tag{9.69}$$

The shape of the resonance (9.62) is expressed as a Breit–Wigner function of single-particle resonance substituting L_R in Eq. (9.62),

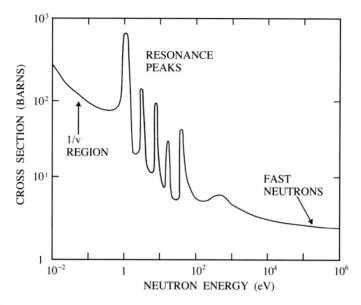

Fig. 9.22 Typical neutron capture cross section as a function of neutron energy. The region $E_n < 1$ eV is called $1/v$ region, where the cross section is inversely proportional to the velocity of neutron. Many resonances appear above the very low energy region. In the energy $E_n > 1$ keV is called fast neutron region

$$\sigma = \frac{\pi}{k^2} \frac{\Gamma_\alpha \Gamma_r}{(E - E_R)^2 + \Gamma_t^2/4},\qquad(9.70)$$

where $\Gamma_r = -2W_0$ is the width from the imaginal part of the optical potential, $\Gamma_\alpha = -2ka/dL_R/dE = 2\hbar^2 k/\mu a$ is the width of particle emission, and $\Gamma_t = \Gamma_r + \Gamma_\alpha$ is the total single-particle resonance width.

For very low-energy neutrons, $k^2/(2\mu) << E_R$, the velocity dependence of the cross section is obtained from Eq. (9.62) as

$$\sigma \sim \frac{4\pi a L_l}{k(L_R^2 + (L_l - ka)^2)} \sim \frac{1}{v},\qquad(9.71)$$

since k is the free momentum and proportional to the velocity v. This result is also shown in Eq. (8.233) in Chap. 8 for compound reaction cross sections.

A typical neutron capture cross section is illustrated in Fig. 9.22. In the thermal-neutron energy regime ($E_n < 1$ eV), the cross section decreases steadily with increasing neutron energy or increases inversely with the neutron velocity as $\sim 1/v$, as expressed in Eq. (9.71). In this region, the cross section is very high. After the "$1/v$" regime, "resonance peaks" occur at certain energies. These resonances take place in the so-called *resonance energy regime* and are a result of neutrons whose energies closely match their discrete quantum energy levels. One example is shown in Eq.

(9.69) in the case of a square well optical potential. Typical heavy nuclei such as uranium isotopes have many closely-spaced resonances starting in the low energy (eV) range (see Fig. 8.29). For higher neutron energies, the absorption cross section steadily decreases as the energy of the neutron increases. This region is called the "fast neutron region" and the absorption cross sections are usually less than 10 barns.

Problem

9.4 Derive $L^{(2)}$ in Eq. (9.65) from L_R in Eq. (9.60).

9.7.2 s-Process, r-Process and Supernova Explosion

Iron group elements (Fe, Ni, Co, Mn) (see Fig. 9.16) originate mostly from a nuclear-statistical equilibrium (NSE) process in thermonuclear supernova explosions. In the relative abundance of nuclei shown in Fig. 9.16, the peak near Fe reflects essentially the NSE process of most stable nuclei (^{56}Fe and neighboring nuclei) of all nuclear species. These nuclei would therefore acquire a maximum abundance under the conditions of nuclear processes in the thermal equilibrium. The formation of nuclei heavier than Fe involves the successive capture of neutrons followed by β-decay processes. The isotopes produced in the neutron capture chain depend on the rate of the capture process as compared with the β-decay rates, which are typically in orders of hours to days near the stability line and which decrease to a fraction of a second near the neutron drip line. Under the condition of low neutron flux, β stability is established at every step before the next neutron capture can occur. This process is named s-process, in which the abundance depends inversely on the neutron capture cross section. Since these capture cross sections are extremely small for nuclei with neutron closed shell, a characteristic feature of the s-process appears as peaks in the abundance curves at neutron numbers $N = 50$ ($^{88}_{38}$Sr$_{50}$), 82($^{138}_{56}$Ba$_{82}$), and 126($^{208}_{82}$Pb$_{126}$) in Fig. 9.16. The systematic measurements of neutron capture cross sections have been made to provide a detailed correlation with the abundance data of nuclei produced in the s-process. This correlation in turn provides information on the astrophysical conditions (such as temperature and neutron density) in which these elements were created.

The s-process abundance $N_A(t)$ is determined by a network equation;

$$\frac{dN_A(t)}{dt} = -\langle\sigma v\rangle_A n_n(t) N_A(t) + \langle\sigma v\rangle_{A-1} n_n(t) N_{A-1}(t), \qquad (9.72)$$

where n_n is the neutron density and $\langle\sigma v\rangle_A$ is the thermal averaged neutron-nucleus reaction rate,

$$\langle\sigma v\rangle_A \equiv \int_0^\infty \sigma v \phi(v) dv, \qquad (9.73)$$

with $\phi(v)$ being a Maxwell–Boltzmann distribution. In Eq. (9.72), the first term on the r.h.s is the neutron capture rate of a nucleus with A which changes $A + n \rightarrow (A + 1)$, while the second term represents the reaction $(A - 1) + n \rightarrow A$. For low energy neutrons, the cross section is proportional to v^{-1} so that $\langle\sigma v\rangle_A$ can be considered as a constant;

$$\langle\sigma v\rangle_A \sim C \int_0^\infty \phi(v)dv \equiv v_T \langle\sigma\rangle_A, \tag{9.74}$$

where v_T is the thermal velocity defined by $v_T = \sqrt{2kT/\mu}$ with $\mu = m_n m_A/(m_n + m_A)$. The network equation then becomes

$$\frac{dN_A(t)}{dt} = v_T n_n \left(-\langle\sigma\rangle_A N_A(t) + \langle\sigma\rangle_{A-1} N_{A-1}(t)\right). \tag{9.75}$$

This equation can be solved with the initial conditions

$$N_A(t = 0) = N_{56}(t = 0) \quad \text{for} \quad A = 56,$$
$$= 0 \qquad\qquad \text{for} \quad A > 56. \tag{9.76}$$

where $N_{56}(t = 0)$ is the abundance of Ni isotope with $A = 56$ at the initial stage $t = 0$. About 150 coupled equations of $A = 56 \sim 209$ should be solved to obtain s-process abundances with realistic ingredients for neutron capture cross sections which are well established by experiments.

If the neutron captures proceed more rapidly than the corresponding β decays, this process leads to isotopes on the neutron-rich side of the stability line. This process can produce many neutron-rich nuclei which can not be reached by s-process. This process is called "r-process". The appreciable abundance observed for these neutron-rich nuclei indicates the importance of the r-process in the evolution of material found in the universe.

For a sufficiently large neutron flux, neutron captures continue until approaching the neutron drip line at a given Z. The relative abundances depend on the β-decay lifetimes of these extremely neutron-rich nuclei, since these lifetimes determine the waiting time for successive neutron captures. In comparison with the s-process, the assumption of r-process is that the beta decay time scale is long compared to the neutron capture time scale. In this case, the capture chain continues at fixed Z from the path of beta stability to the neutron drip line where the neutron binding energy becomes so small (almost zero) that further neutron captures are rejected. Then, for each charge Z, there will be a waiting point at which a beta decay must occur in order to continue the chain to larger A. The abundance at each Z is governed then by the equation

$$\frac{dN_Z(t)}{dt} = \lambda_{Z-1}^\beta N_{Z-1}(t) - \lambda_Z^\beta N_Z(t), \tag{9.77}$$

where λ_Z^β is the beta-decay rate at the waiting point for a nucleus with Z protons. These equations are similar to those of the s-process (9.75), except that abundances are proportional to the slowness of beta-decay rate rather than the neutron-capture rate. Since the slowest beta decay of the isotopes may appear in nuclei with closed shells, the highest abundances will be associated also with magic numbers as can be seen in Fig. 9.20.

In Fig. 9.20, a tentative capture path is indicated for the r-process. The shell structure at $N = 82$, implying an exceptionally weak binding of the 83rd neutron, manifests itself by the vertical break in the capture path. This is called "waiting point" of the r-process nucleosynthesis. As a consequence, the path approaches to the stability line near $A \sim 130$. The relative low β-decay energies at the waiting point imply a relative long waiting time and large abundance of nuclei with this mass region. In this way, the observed abundance peaks at $A \sim 80, 130$ and 194 in Fig. 9.16 can be understood as the correlations with the neutron shell closures at $N = 50, 82$ and 126, respectively. At the end of the r-process, it is impossible to make a bigger nucleus by the neutron capture reaction and the big nuclei with $N \sim 170$ result unstable for spontaneous fission as well as neutron induced fission. Thus, the r-process is terminated by fission whose fragments will re-seed the r-process.

As explained in this section, the observed abundance for the elements above Fe can be understood by two different synthesizing processes, s- and r-processes. Another mechanism is required for the production of some very neutron-deficient (proton-rich) nuclei, shown by yellow color in Fig. 9.23, through the so-called p-process or γ-process, i.e., (p, γ) or (γ, p) reactions. There are 32 stable proton-rich nuclei in nature which are shielded against neutron capture. While the p-process possibly occurs in Type II supernova explosions,[8] the accretion onto neutron star surfaces resulting from Type I X-ray bursts has been shown to produce some of these proton-rich nuclei as well. Furthermore, observed timescales of these bursts bear a good relationship to expected p-process timescales.

The astronomical environment of these nucleosynthesis is an important problem under active investigations. The s-process environment can be identified in the Red Giant stage of star evolution. In these highly evolved stars, hydrogen has been exhausted in the core and main source of energy is α burning. The neutrons for the s-process can be provided by exothermic (α, n) reactions, such as $^{13}C(\alpha, n)^{16}O$ and $^{21}Ne(\alpha, n)^{24}Mg$, and also by the endothermic reaction $^{22}Ne(\alpha, n)^{25}Mg$, as well as in the heavy ion reactions $^{12}C(^{12}C,n)^{23}Mg$ and $^{16}O(^{16}O,n)^{31}S$. The short duration and violent conditions of the nuclear statistical equilibrium site responsible for the Fe peak have been implemented in explosions of white dwarfs as well as supernovae.

There exist uncertainties concerning the possible sites where the r-process takes place. Three major astronomical sites of the r-process nucleosynthesis are proposed:

---

[8] Supernovae are classified according to the different chemical elements that appear in their spectra. The first element for division is the presence of hydrogen. If a supernova's spectrum contains no line of hydrogen spectrum, it is classified as Type I; Type II has hydrogen in its spectrum. Type I is further classified by the existence of silicon and helium spectrum lines. Type Ia represents an ionized silicon line. Type Ib show a helium (He I) line, while Type Ic has neither Si nor helium line.

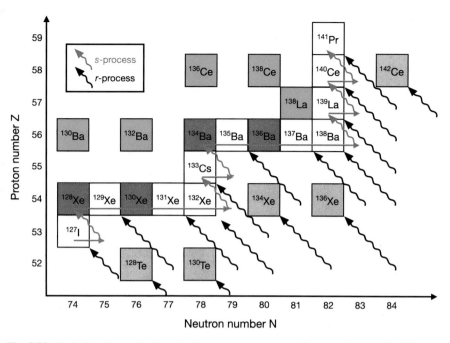

Fig. 9.23 Typical nucleosynthesis paths by neutron capture reactions associated with β decays. On the one hand, the s process proceeds along the stability line and never goes beyond it. On the other hand, the r process goes along near the neutron drip line and produces β unstable nuclei which decay until they reach the stability line. Some stable nuclei are never reached by neither the s nor the r processes, but synthesized by the p process. Nuclei marked by purple and yellow colors are synthesized only by s or r process, while those marked by white color are produced by both processes. Nuclei marked in green are reached only by the p-process

- Neutrino-driven wind (NDW) of core-collapse supernovae (CCSNe). This process is responsible to create nuclei with a lighter mass of $A < 120$.
- Magnetohydrodynamic jet (MHDJ) from CCSNe. This process can explain elements $A > 110$, and considered as a promising process besides NSM (neutron-star merger).
- Neutron star mergers (NSMs). Nuclei with mass $A > 150$ are created by this process, but the occurrence is considered less than CCSNe.

Current core collapse supernova models do not account for the solar system abundance pattern, in particular, nuclei with $A > 110$. New astronomical data about the r-process was discovered in 2017 when the LIGO and Virgo gravitational-wave observatories discovered a merger of two neutron stars "GW170817". Studying the optical counterpart of the merger, spectroscopic information of r-process materials are found thrown out from the merging neutron stars. The bulk of this material seems to consist of lower-mass-range heavy nuclei ($A < 140$) as well as higher mass-number r-process nuclei ($A > 140$) rich in actinides (such as uranium, thorium, and californium). These spectroscopic features support the NSMs as a prominent astronomical site of the r-process. The evolution of nuclear flow in a NSM r-process is

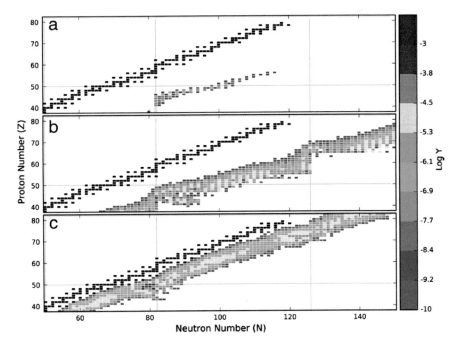

Fig. 9.24 The evolution of nuclear flow in a neutron-star-merger r-process. The simulation is calculated including fission recycling. Snapshots taken at (**a**) the first hundred milliseconds (**b**) neutron exhaustion to reach the neutron drip line (neutron-to-seed ratio equal to unity) and (**c**) the final stage of the r-process undergoing fission recycling. The black squares show stable nuclei and the abundance of r-process elements is classified by color shown along the right axis. Reprinted with permission from [13]. ©2021 by Elsevier

shown in Fig. 9.24. The fission recycling process at the final stage of the r-process is also included.

The nucleosynthesis analysis was performed by a NSM model based on general relativity including the effect of neutrino transport, i.e., the absorptions of neutrinos on free nucleons are considered. The masses of NSs are considered equal and taken as M $= 1.3\,$M$_\odot$. The merger ejecta exhibit a wide range of nuclei with different electron fractions

$$Y_e = \frac{n_{e^-}}{n_B} = \frac{N_p}{N_n + N_p} \sim (0.1 - 0.4), \tag{9.78}$$

where n_{e^-} and n_B are the number of densities of electrons and baryons, respectively, while N_n and N_p are the total neutron and proton numbers, respectively. The fractions Y_e and $1 - Y_e$ determine the mass fractions of protons and neutrons, respectively, at the initial stage. The nuclear reaction network calculations are performed with 6300 species between the β-stability and neutron-drip lines, i.e., all the way from single neutrons and protons up to the $Z = 110$ isotopes. Experimental rates of neutron capture and beta decay are adopted, when available. Otherwise, the theoret-

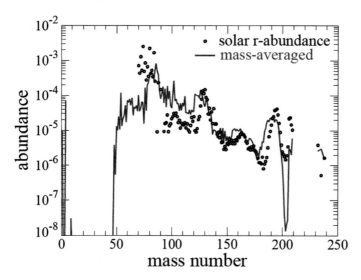

Fig. 9.25 The nuclear abundances obtained by a NS-NS merger model. The results are compared with the solar r-process abundance distribution over the full mass range of r-process $A \sim 90$–240. The 36 representative r-process trajectories with different electron fractions Y_e are averaged with the weight obtained by the NS-NS merger simulations for the final results. This figure is reprinted with permission from [14]. ©2021 by AAS

ical estimates of capture rates and β-decays are adopted based on the microscopic nuclear mass model with Skyrme interaction. Theoretical fission properties adopted are also estimated on the basis of the same mass model. Figure 9.25 displays the averaged nuclear abundances for different trajectories with wide distributions of Y_e as the outcome of NSM simulations. There is a variety of nucleosynthetic peaks: iron-peak and $A \sim 90$ abundances made in nuclear quasi-equilibrium for $Y_e \geq 0.4$. The second peak ($A \sim 130$) and the rare-earth element peak ($A \sim 160$) abundances are dominated by direct productions from trajectories of $Y_e \sim 0.2$. The results reproduce also the abundance ratio between the second ($A \sim 130$) and the third ($A \sim 195$) peaks as well. The nuclear flow for the lowest $Y_e(=0.09)$ trajectory reaches $A \sim 280$, the fissile point by neutron-induced fission, only at the freeze-out of r-processing. Spontaneous fission plays an important role for forming the A ~ 130 abundance peak, but only for $Y_e < 0.15$.

As the model is representative of NSMs, the merger ejecta exhibit a wide range of electron fraction that led to the nucleosynthetic abundance distribution being in excellent agreement with the solar r-process pattern. The result gives an important implication; the dynamical component of the NSM ejecta can be considered as one of the important origins of the cosmic r-process nuclei. Recent GW170817 measurements are consistent with this conclusion.

Neutrino-driven winds from proto-neutron stars, which follow core-collapse supernova explosions, or winds after a neutron star merger may enhance r-process abundances. The neutrino-driven winds can create a variety of neutron-rich and

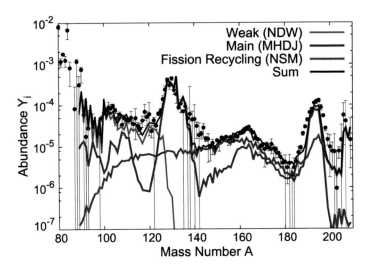

Fig. 9.26 Abundance patterns for NSM (red curve), the *r*-process abundance from the MHDJ model (blue curve) and from NDW (green curve). The black curve shows the sum of all contributions. These curves are compare with the observed *r*-process abundance in the solar system (black filled circles with thin error bars). Figure reprinted with permission from [15]. ©2021 by AAS

proton-rich nuclei, leading to the production of the lighter heavy elements up to silver. In Fig. 9.26, the abundance pattern for NSM, MHDJ model ad NDW of CCSNe are shown. In this figure, the relative abundance of each site is 79% from NDW, 16% from MHDJ and 5% from NSM. The summed results show a fine agreement with the observed *r*-process abundance.

There is a controversial argument on the NDW model. Recent sophisticated simulations pointed that the neutrino flux of supernova explosions is not strong enough and a successful *r*-process up to the heaviest elements have become unlikely. At the present understanding of neutrino-nucleus interaction, it is still uncertain how much importance the neutrino-driven wind has for the *r*-process problem.

From a nuclear physics perspective, there will be more measurements on neutron-rich nuclei which are involved in the site of the *r*-process. New-generation radioactive beam facilities, such as RIBF at RIKEN, Japan, FAIR at GSI, Germany, and FRIB at Michigan State University, USA will identify the properties of nuclei (e.g., nuclear masses and beta-decays) along the nucleosynthesis path. These measurements help to improve the uncertainties for a quantitative understanding of *r*-process abundances, since the theoretical simulations are sensitive to these properties.

Summary

The incompressibility of nuclear matter is defined as the second derivative of the binding energy per particle with respect to the density at the saturation point;

$$K_\infty = 9\rho^2 \frac{\partial^2 (E/\rho)}{\partial \rho^2}|_{\rho=\rho_0},$$

in Eq. (9.15) and determined to be

$$K_\infty = 225 \pm 20 \text{ MeV},$$

by the analysis of ISGMR experiments.

The symmetry energy E_{sym} in Eq. (9.3) is expanded around the saturation density ρ_0 as

$$E_{sym}(\rho) \equiv S(\rho) = J + L\frac{(\rho - \rho_0)}{3\rho_0} + \frac{1}{2}K_{sym}\frac{(\rho - \rho_0)^2}{9\rho_0^2}.$$

The values J and L are constrained to be

$$J = 32.3 \pm 0.5 \text{ MeV},$$
$$L = 53.5 \pm 15 \text{ MeV},$$

by the mass model FRDM. These values are consistent with the extracted ones from the dipole polarizability, HIC experiments and also from the analysis of IAS.

The precise measurement of a neutron star was performed by NICER apparatus, which extracts a radius of 12.7 ± 1 km for a mass of $m = 1.34 \pm 0.15 \, M_\odot$. The observed maximum neutron star mass is reported to be twice the solar mass by astronomical observations and gives a critical constraint for the nuclear matter EoS.

Hyperons may exist in the core of neutron stars at high density. The attractive ΛN interaction makes the EoS softer and predicts a lower neutron star mass than the observed maximum $2\,M_\odot$. This is called the "Hyperon puzzle". To cure this puzzle, a strong repulsive three-body ΛNN interaction is introduced.

Nuclei with exotic shapes, so called pasta phases, may occur in the inner crust of neutron stars with the density higher than $\rho = 10^{14}$ g/cm$^3 \sim 0.4\rho_0$. The competition between the nuclear surface and the Coulomb energies is responsible to create these exotic shapes in the mantle of the neutron star. The nuclear "pasta" phase evolves a sequence of different shapes: sphere $\rightarrow$ cylindrical (spaghetti) $\rightarrow$ slab (lasagna) $\rightarrow$ cylindrical bubble $\rightarrow$ spherical bubble with increasing density (see Fig. 9.15).

Big Bang nucleosynthesis (BBN, also called primordial nucleosynthesis) starts at a few seconds after the Big Bang. The BBN is considered to be responsible for the formation of most of the helium isotope ^{4}He, along with small amounts of the hydrogen isotope deuterium (d or D), the helium isotope ^{3}He, and a very small amount of the lithium isotope ^{7}Li. In addition to these stable nuclei, two unstable or radioactive isotopes were also produced: the heavy hydrogen isotope, tritium ^{3}H (t), and the beryllium isotope ^{7}Be.

All of the elements heavier than lithium have been created constantly in evolving and exploding stars, since stars exist in the universe, i.e., a few hundred million years after the Big Bang. This process is called stellar nucleosynthesis. The most important reactions in stellar nucleosynthesis start from light elements, i.e., hydrogen fusion and deuteron fusion. Sequential processes after deuteron fusion are helium burning, carbon burning, oxygen burning, neon burning and silicon burning processes. These fusion reactions produce elements up to Fe isotopes.

Production of elements heavier than iron is implemented mainly by two neutron capture processes; the s-process along the stability valley and the r-process close to the neutron-drip line. Remaining neutron-deficient nuclei which are not reached by neutron capture processes are produced by the proton capture reaction, the p-process. Recent observations of gravitational waves GW170817 from the NS-NS merger and the kilonovae event afterward give the strong indication of astronomical sight of a r-process to create heavy elements $A > 150$, while NDW of CCSNe and MHDJ from CCSNe are also important to understand the full r-process abundance.

Solutions of Problems

9.1 In the non-relativistic limit $(1/c^2 \to 0)$, the TOV Eq. (9.26) becomes

$$\frac{dP(r)}{dr} = -\frac{G}{r^2}\rho(r)M(r). \tag{9.79}$$

The hydrostatic pressure is expressed in the integral form,

$$P(r) - P(r') = -\int_{r'}^{r} \frac{G}{r^2}\rho(r)M(r)dr. \tag{9.80}$$

In the circumstance of the earth, the gravitational force is given by the gravitational constant $g = GM/R^2$, where M and R are the mass and the radius of the Earth, respectively. Consider a cylinder filled with a fluid with a constant density ρ. For a small area with acting force of the weight of the fluid above it, the hydrostatic pressure can be calculated from the reference point z_0 to the point z as

$$P(z) - P(z_0) = gR^2\rho\left[\frac{1}{R-h} - \frac{1}{R}\right] \simeq g\rho h, \tag{9.81}$$

where h is the height difference between z_0 and z.

9.2 (a) The p-p cycle (9.42) provides the energy of 26.73 MeV and 2 neutrinos. The number of neutrinos N_ν produced by the p-p cycle is estimated by using the identity $1\,\mathrm{eV} = 1.60\times10^{-12}$ erg as

$$N_\nu = 2 \times \frac{3.844 \times 10^{33}}{26.73 \times 10^6 \times 1.60 \times 10^{-12}} = 1.80 \times 10^{38}. \tag{9.82}$$

(b) The neutrino flux on the earth is estimated as

$$\nu \ flux = \frac{1.80 \times 10^{38}}{4\pi \times (1.496 \times 10^{13})^2} = 6.40 \times 10^{10} \, cm^{-2}s^{-1}. \tag{9.83}$$

The empirical neutrino flux from the sun is $6.6 \times 10^{10} \, cm^{-2}s^{-1}$.
9.3 ^{16}O is a doubly magic nucleus and extremely stable. The threshold energy of the reaction,

$$^{16}O(\gamma, \alpha)^{12}C,$$

is 7.2 MeV, while that for

$$^{20}Ne(\gamma, \alpha)^{16}O, \tag{9.84}$$

is 4.7 MeV. The main Ne burning processes $^{20}Ne + \gamma \rightarrow {}^{16}O + \alpha$ and $^{20}Ne + \alpha \rightarrow {}^{16}O + \gamma$ require a temperature of around 1.2×10^9 K and a density of about 4×10^9 kg/m^3. The fusion reaction of O burning

$$^{16}O + {}^{16}O \rightarrow {}^{32}S^* \tag{9.85}$$

ignites at a higher temperature of about 2×10^9 K and a higher density of about 4×10^{12} kg/m^3
9.4 The derivative of $L_R = Ka \cot Ka$ in Eq. (9.60) with respect to the energy E can be evaluated with a help of the variable $X \equiv Ka = \sqrt{\frac{2\mu}{\hbar^2}(E-V)a}$ as,

$$\frac{dL_R}{dE} = \frac{dL_R}{dX}\frac{dX}{dE} = \frac{\cos X \sin X - X}{\sin^2 X} a \frac{\mu}{\hbar^2} \frac{1}{K}. \tag{9.86}$$

At the resonance energy $E = E_R$, $\cos X = 0$ and $\sin X = 1$ since $X = aK$ is given by Eq. (9.68). Then, $L^{(2)}$ is obtained as

$$L^{(2)} = \frac{dL_R}{dE}\bigg|_{E=E_R} = -aKa\frac{\mu}{\hbar^2}\frac{1}{K} = -\frac{a^2\mu}{\hbar^2}. \tag{9.87}$$

Books for Further Readings

"Cauldrons in the cosmos" by C. E. Rolfs and W. S. Rodney (University of Chicago press, 2005) gives an excellent overview to nuclear astrophysics, including a description of nucleosynthesis processes and related nuclear-physics experiments.

"The Physics and Astrophysics of neutron stars" edited by L. Rezzolla, P. Pizzochero, D. I. Jones, N. Rea, and I. Vidana (Springer, 2018) summarizes the recent progress in the physics and astrophysics of neutron stars. It identifies and develops effective strategies to explore, both theoretically and observationnaly, the many remaining open questions in this field.

"Compact star physics" by J. Schaffner-Bielich (Cambridge, 2020) provides important insights on the basic concepts of compact stars, discusses white dwarfs, neutron stars, quark stars and hybrid stars. The book discusses in depth the theory underpinning understanding of general relativity and the thermodynamics of dense matter.

References

1. M. Uchida et al., Phys. Rev. C **69**, 051301 (2004)
2. D.H. Youngblood et al., Phys. Rev. Lett. **82**, 691 (1999) and Y. -W. Lui, private communication
3. X. Roca-Maza et al., Phys. Rev. Lett. **106**, 252501 (2011)
4. A. Tamii et al., Phys. Rev. Lett. **107**, 062502 (2013)
5. X. Roca-Maza et al., Phys. Rev. C **88**, 024316 (2011)
6. P. Möller, A.J. Sierk, T. Ichikawa, H. Sagawa, Nuclear ground-state masses and deformations: FRDM (2012). Atomic Data Nucl. Data Tables 109–110, 1–204
7. D. Adhikari et al., PREX Collaboration, Phys. Rev. Lett. **126**, 172502 (2021)
8. B.P. Abbott et al., The LIGO Scientific Collaboration and The Virgo Collaboration, Phys. Rev. Lett. **121**, 161101 (2018)
9. S. Gandolfi, J. Carlson, S. Reddy, A.W. Steiner, R.B. Wiringa, Eur. Phys. J. A **50**, 10 (2014)
10. D. Lonardoni, A. Lovato, S. Gandolfi, F. Pederiva, Phys. Rev. Lett. **114**, 092301 (2015)
11. K. Oyamatsu, Nucl. Phys. A **561**, 431 (1993)
12. G.J. Mathews, M. Kusakabe, T. Kajino, Int. J. Mod. Phys. E **26**, 5 (2017)
13. M.R. Mumpower et al., Prog. Part. Nucl. Phys. **86**, 86 (2016)
14. S. Wanajo, Y. Sekiguchi, N. Nishimura, K. Kiuchi, K. Kyutoku, M. Shibata, Astr. J. Lett. **789**, L39 (2014)
15. S. Shibagaki, T. Kajino, G.J. Mathews, S. Chiba, S. Nishimura, G. Lorusso, Astr. J. **816**, 79 (2016)

Chapter 10
Nuclear Physics and Standard Model of Elementary Particles

Abstract In this chapter, we present how nuclear physics contributes to establish the standard model of elementary particles and test its validity. We also discuss recent progresses of lattice quantum chromodynamics (LQCD) in calculating nucleon-nucleon interactions and few-body bound nuclei.

Keywords Standard model · Lattice quantum chromodynamics · Neutrino mass · Super-allowed Fermi beta decay · Cabbibo–Kobayashi–Maskawa matrix · Double-β decay

10.1 Standard Model

10.1.1 Overview

The standard model of elementary particles is a theory of the electroweak and strong interactions as well as all the subatomic particles, baryons and mesons. The current version of the model was established in the middle of the 70s by the experimental confirmation of quarks. The validity of the standard model was further strengthened by the discoveries of the top quark in 1995, the τ neutrino in 2000, and the Higgs boson in 2012. From a theoretical point of view, the standard model is a paradigm of quantum field theory embedding spontaneous symmetry breaking (SSB), anomalies, and non-perturbative behavior. The model was started by the discovery of a way to unify the electromagnetic and weak interactions by Sheldon Glashow in 1961. In 1967, Steven Weinberg, and Abdus Salam introduced the Higgs mechanism to give rise to masses of hadrons in the model together with the W and Z bosons for the electroweak interaction, and quarks and gluons for the strong interaction. After the discovery of the neutral weak currents by Z-boson exchange in 1973, the electroweak theory became widely accepted and Glashow, Weinberg and Salam got the Nobel prize in Physics in 1979. The W boson was discovered in 1981. The W and Z bosons

© Springer Nature Singapore Pte Ltd. 2021
A. Obertelli and H. Sagawa, *Modern Nuclear Physics*, UNITEXT for Physics,
https://doi.org/10.1007/978-981-16-2289-2_10

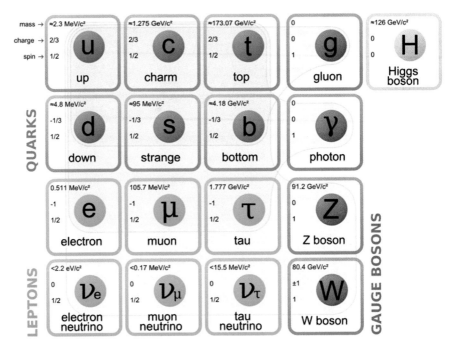

Fig. 10.1 Elementary particles of the standard model. The approximate mass, electric charge, and spin of each particle are given inside the box. The purple and green colors denote quarks and leptons, respectively, while orange and yellow colors correspond to gauge bosons and the Higgs boson, respectively. There are three generations of quarks: the first generation is up ⇔ down, while the second and third generations are charm ⇔ strange and top ⇔ bottom quarks. There are also three generations of leptons e, μ, and τ in the figure and corresponding three neutrinos. The strong interaction is mediated by gluons, while the electromagnetic and weak (electroweak) interactions are mediated by photons, and Z and W bosons, respectively. The mass of particles is created by the Higgs boson

were measured having very heavy masses of 80.4 GeV and 91.2 GeV, respectively, which are almost 100 times heavier than that of the proton, and consistent with the predictions of the standard model.

Figure 10.1 shows different classes of elementary particles such as fermions, gauge bosons, and the Higgs bosons. The standard model contains 12 fermions of spin 1/2, referred also 12 different "flavor" particles: six quarks (up, down, charm, strange, top, and bottom quarks) and six leptons (electron e, muon μ, tau τ, and their corresponding neutrinos, ν_e, ν_μ, ν_τ). An antiparticle is associated to each of these 12 particles. Each quark carries "color charge", electric charge, and isospin. There are three types of color charges: green, red, and blue which, obviously, have nothing to do with the colors of visible lights. The color charge is an extra degree of freedom of quarks to explain how quarks coexist inside some hadrons in otherwise identical quantum states without violating the Pauli exclusion principle. This theory is called "quantum chromodynamics (QCD)". The hadrons (baryons and mesons) should be color-neutral,[1] i.e., baryons are made by three different colors of quarks (qqq), and mesons consist of a quark and an anti-quark with the same color ($q\bar{q}$) (See Table 10.1). There are 9 combinations of color and anti-color charges of three colors. It is the case of the color combinations of gluons. There are 9 types of gluons which are labeled by a combination of color and anti-color charges such as red-anti-green ($r\bar{g}$) and blue-anti-red ($b\bar{r}$). The gluon state is described by a linear combination of these pairs of color-anti-color states. However, there is no color singlet state,

$$(r\bar{r} + b\bar{b} + g\bar{g})/\sqrt{3}, \qquad (10.1)$$

for gluons and only eight types of combinations are allowed for gluons. This is related to the confinement of mediating gluons within a short range.[2]

Just as particles with electric charges interact by exchanging photons, the color-charged particles interact by exchanging gluons. The strong binding of color neutral protons and neutrons in nuclei is due to strong interactions mediated by gluons between their color charged constituents. In effective theories of strong interactions, this mechanism of quarks and gluons is replaced by the exchanges of mesons between baryons in a hadronic picture. As discussed in Chap. 2, one example is ChEFT which is based on the QCD, but adopts the meson exchange picture for the strong interaction.

10.1.2 Confinement and Experimental Evidence of Quarks

Quarks are confined in mesons and baryons and cannot be isolated from color neutral particles. The reasons for quarks confinement are somewhat complicated. The current theory is that the confinement is a result of exchange of gluons between color-charged constituents. In QCD, as a quark-antiquark pair separates, the gluon field is visualized in Fig. 10.2 to form a narrow tube (or string) of color field between them. This is an analogy of electric or magnetic line fields to visualize the electromagnetic fields. Because of this behavior of the gluon field, a strong force between the quark pair acts constantly regardless of their distance. This is quite different from the electric field in which the field becomes weaker at longer distance. When two quarks become separated, at some point, it is more energetically favorable to create a new quark-antiquark pair, than to allow the tube to extend further.

As a result, when quarks are produced in the laboratory, physicists see "jets" of many color-neutral clustered particles (mesons and baryons), instead of observing the

---

[1] Mathematically it is a color-single state, which is analogous to the spin singlet state $S = 0$ in the coupling of spins.

[2] Confinement requires all naturally occurring particles to be a color singlet, such as a proton and a neutron and mesons. If a color-singlet state exists for gluons, it exists as a free particle and it could be exchanged between two color singlets, for example between a proton and a neutron, giving rise to a long-range strong interactions. However, in nature, the strong interaction is very short range and proves no color-singlet gluon state.

Fig. 10.2 Quark
confinement in mesons. The
external energy transferred to
a meson cannot separate
quarks, but creates a new
quark-antiquark pair, another
meson

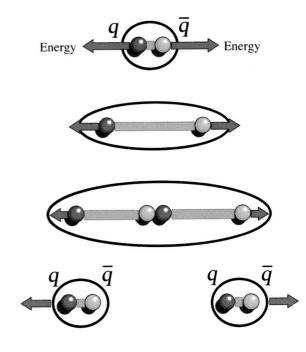

individual quarks. This process is called "hadronization of quarks". This mechanism
is successfully used to explain the high energy experiment in terms of the quark
confinement, but can not be proved directly from the standard model. Such an event
measured by the ATLAS experiment at the Large Hadron Collider (LHC), CERN, is
illustrated in Fig. 10.3. This is one of the weak points of the standard model.

Despite the non-observability of individual quarks because of the hadronization
process, the existence of quarks and the number of flavors can be experimentally
proven by the hadron production cross section from e^+e^- annihilations as a function
of the center of mass energy $\sqrt{s}$. A detailed demonstration would require the evaluation of process amplitudes from Feynman diagrams, beyond the scope of the present
book. Instead, we present a semi-quantitative derivation of the expected hadron production cross section from e^+e^- collisions. Starting with the production of a muon
pair, the cross section scales with the charge of the electron (initial state) and the
muon (final state) as

$$\sigma_{e^+e^-\to\mu^+\mu^-} = q_e^2 q_\mu^2 \sigma_0, \qquad (10.2)$$

where q_e and q_μ are the electric charges of electron and muon, respectively, σ_0
contains kinematical factors and the fine structure constant $\alpha = e^2/\hbar c$. σ_0 depends
on the center of mass energy of the collision. In the particular case of the muon
production from e^+e^- collisions, all involved particles have a charge $q^2 = 1$ so the
cross section reduces to σ_0 with the above notations. Considering now the production
of a quark-antiquark $q\bar{q}$ from the same mechanism, the production cross section will
be

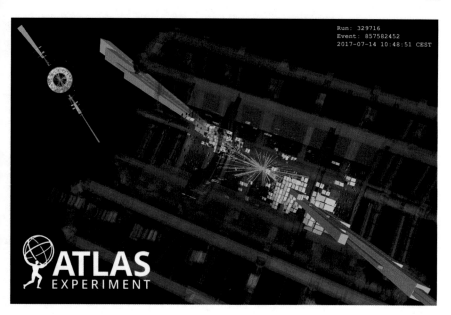

Run: 329716
Event: 857582452
2017-07-14 10:48:51 CEST

Fig. 10.3 Display of a di-jet event (Run = 329716, Event = 8575822452) with an estimated di-jet mass of $m_{jj} = 9.3$ TeV, produced in pp collisions at $\sqrt{s} = 13$ TeV data in 2017. The two high-perpendicular momentum ($p_\perp$) jets both have $p_\perp = 2.9$ TeV. The two yellow cones represent the reconstructed jets. ATLAS Experiment ©2021 CERN

$$\sigma_{e^+e^- \to q\bar{q}} = \Big(\sum_f q_f^2\Big)\sigma_0, \tag{10.3}$$

where the index f denotes the quark flavor and q_f is the electric charge of the quark flavour f.

From Fig. 10.1, it is noticeable that the six quark flavors can be divided in four mass regions: the u, d and s quark have small masses. The following quark, in terms of mass, is the charm with a mass of 1.275 GeV/c^2 leading to an expected production threshold for a $c\bar{c}$ pair at 2.55 GeV/c^2. The bottom quark has a mass of ~ 4.18 Gev/c^2 leading to a production threshold around 9.5 Gev/c^2. Finally, the *top* quark is much heavier with a mass of ~ 173 Gev/c^2. From e^+e^- collisions, it is therefore expected that the production of hadrons increases when a new quark channel is open, i.e., at the production threshold.

We define the ratio R of the production cross section of quark-antiquark pairs from e^+e^- collisions to the production cross section of $\mu^+\mu^-$ lepton pairs. From Eqs. (10.2) and (10.3), this threshold is given by

$$R = \frac{\sigma_{e^+e^- \to q\bar{q}}}{\sigma_{e^+e^- \to \mu^+\mu^-}} = \sum_f q_f^2. \tag{10.4}$$

Table 10.1 Baryons and mesons in the quark model. There are about 120 types of baryons and 140 types of mesons. There are corresponding anti-particles, such as an anti-proton and an anti-neutron, for all the particles

Symbol	Name	Quark content	Electric charge	Mass (MeV/c^2)	Spin	Strangeness
p	Proton	uud	+1	938.3	1/2	0
n	Neutron	udd	0	939.6	1/2	0
Δ^{++}	Delta	uuu	+2	1222	3/2	0
Δ^+	Delta	uud	+1	1222	3/2	0
Δ^0	Delta	udd	0	1222	3/2	0
Δ^-	Delta	ddd	−1	1222	3/2	0
Λ	Lambda	uds	0	1116	1/2	−1
Ξ^0	Cascade	uss	0	1315	1/2	−2
Ξ^-	Cascade	dss	−1	1321	1/2	−2
Ω	Omega	sss	−1	1672	3/2	−3
π^+	pion	$u\bar{d}$	+1	139.6	0	0
K^-	kaon	$s\bar{u}$	−1	493.7	0	−1
ρ^+	rho	$u\bar{d}$	+1	776	+1	0
ϕ	phi	$s\bar{s}$	0	1020	+1	0
B^0	B-zero	$d\bar{b}$	0	5279	0	0
η_c	eta-c	$c\bar{c}$	0	2980	0	0

Below a center of mass energy of 2.5 GeV, the charm production threshold, only three flavors of quarks can be produced, and Eq. (10.4) gives

$$R = q_u^2 + q_d^2 + q_s^2 = \left(\frac{2}{3}\right)^2 + \left(\frac{1}{3}\right)^2 + \left(\frac{1}{3}\right)^2 = \frac{6}{9} = \frac{2}{3}. \qquad (10.5)$$

Above the charm quark-antiquark pair production threshold and below the bottom production threshold, one finds

$$R = q_u^2 + q_d^2 + q_s^2 + q_c^2 = \frac{10}{9}. \qquad (10.6)$$

At the bottom quark-antiquark pair production threshold, R is expected to rise to

$$R = q_u^2 + q_d^2 + q_s^2 + q_c^2 + q_b^2 = \frac{11}{9}, \qquad (10.7)$$

while, based on the same considerations, a value of $5/3$ is expected above the top quark-antiquark pair production threshold.

The experimental values for R as a function of the center of mass energy $\sqrt{s}$ (Mandelstam variable) from 300 MeV to 200 GeV is given in Fig. 10.4. Despite the

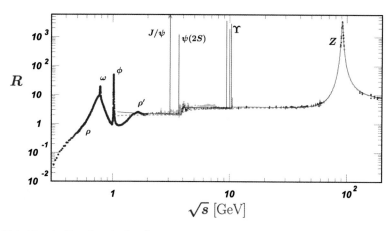

Fig. 10.4 Threshold regions in the e^+e^- hadron production. The onset due to the opening of the charm and bottom quark-antiquark production channels are visible at the center of mass energy $\sqrt{s} = 3.5 - 4$ Gev and ~ 10.6 GeV, respectively. The data (blue and green markers) are compared to two sets of calculations: (green line) assuming a naive quark model and (red markers and line) a three-loop perturbative QCD prediction. The red dots are the calculations and the red line is a fit to guide the eye. Figure reprinted from [1]

existence of resonances, the plateaus corresponding to the different quark flavors are clearly visible. This experimental observation is a strong proof in favor of the existence of quarks, although not observable. The experimental values of R and the threshold energies are different from the above simple model since the hadronization involves the creation of massive hadrons and that the production mechanism can not be reduced to a simple virtual photon exchange. Experimentally, one finds the first plateau at $R \simeq 2.2$ for $\sqrt{s}$ from 2 to 3.5–4 Gev for the three lightest quarks. The second plateau at $R \simeq 3.6$ is found for $\sqrt{s} = 4 \sim 10.6$ GeV corresponding to the addition of the charm quark. The third plateau at $R \simeq 3.8$ is found for $\sqrt{s} > 10.6$ GeV, where the bottom quark production channel is open. No data exist yet to evidence the expected fourth plateau at the top quark-antiquark production threshold.

10.1.3 Gauge Bosons

In the standard model, "gauge bosons" mediate the electromagnetic, weak (or unified to be electroweak), and strong forces between particles. All gauge bosons have spin 1 and thus do not follow the Pauli exclusion principle. The different types of gauge bosons are (see also Fig. 10.1):

1. Photons for the electromagnetic force. Photons are massless.
2. W^+, W^-, and Z bosons for the weak interaction. $W^\pm$ bosons carry positive and negative electric charges, and couple to the electromagnetic interaction. Namely,

W bosons mediate the electromagnetic interactions and the weak interaction simultaneously.

3. The eight gluons for the strong interaction between the color charged particles (quarks). Gluons are massless. The eightfold multiplicity of gluons is labeled by a combination of color and anti-color charges.

The Higgs boson plays an unique role in the standard model. It creates mass for elementary particles other than photons and gluons. In particular, the Higgs boson explains why the photon has no mass, but the W and Z bosons have very heavy masses. The Higgs boson also creates the masses of leptons and quarks. Because the Higgs boson has a very heavy mass and a very short decay time, it has been extremely difficult to prove its existence although many efforts have been made. Eventually, in July 2012, the finding of the Higgs boson was reported by experiments of the Large Hadron Collider (LHC) at CERN (European Organization for Nuclear Research) with a mass of 125 GeV (about 130 times of the proton mass).

Problem

10.1 Up and down quarks have isospin $I = 1/2$, and isospin z-components I_3 of $1/2$ and $-1/2$, respectively. All other quarks have $I = 0$. The isospin z-component of hadron is determined by the number of up and down quarks in the constituents,

$$I_3 = \frac{1}{2}(n_u - n_d). \tag{10.8}$$

Determine the isospin of proton, neutron, delta, and cascade particles.[3]

10.2   Lattice Quantum Chromodynamics

10.2.1   Framework

Quantum Chromodynamics (QCD) is the quantum field theory (QFT) for the strong interaction between quarks and gluons. The fields are composed from quarks and gluons. The quark fields have six flavors and three colors,

$$\psi_{c\mu}^{(f)}(x) \equiv \psi_{c\mu}^{(f)}(\mathbf{r}, t), \quad f \in \{u, d, s, c, t, b\}, \quad c \in \{b, g, r\}, \tag{10.9}$$

where f and c stand for the flavor and the color, respectively.[4] The quarks are spin $1/2$ particles and thus the fields have Dirac index $\mu = 1, 2, 3, 4$ corresponding to the

---

[3] In this chapter, the isospin and its z-component are denoted as I and I_3, which are commonly adopted in the elemental particle physics community, instead of T and T_z in the previous chapters. While $I = T$, the z-component I_3 is $+1/2$ for protons and $-1/2$ for neutrons, which are opposite signs to T_z.

[4] Needless to say, the symbol c in the flavor stands for "*charme*" as shown in Fig. 10.1.

freedoms of spin up/down and particle/anti-particle. Because the quarks carry three colors, QCD is called a $SU(3)$ gauge field. Contrarily, photons have no color and the gauge field is simply described by $U(1)$ for the electromagnetic field. $SU(3)$ and $U(1)$ groups have 8 and 1 generators, respectively, corresponding to the number of freedom in each field. The force field is a massless gluon field which has the $SU(3)$ index ($a = 1, 2, \ldots, 8$) ($SU(3)$ gauge index) and Lorentz or space-time index ($\mu = 0, 1, 2, 3$).

$$A_\mu(x) \equiv A_\mu(\mathbf{r}, t) = \frac{A_\mu^a \lambda^a}{2}, \tag{10.10}$$

where λ^a is the 3×3 Gell–Mann matrix for $SU(3)$ gauge field and satisfies the commutation relation

$$[\lambda^a, \lambda^b] = i\frac{f^{abc}\lambda^c}{2}, \tag{10.11}$$

where f^{abc} is the totally anti-symmetric $SU(3)$ metric analogous to the Levi–Civita symbol ε^{abc} for the $SU(2)$ group (see Sect. 1.2.3 in Chap. 1 for a discussion on Lie algebras of the SU(2) and SU(3) groups).

QFT is an equivalent theory to quantum mechanics based on the idea of Feynman's path integral. In the following, the basic idea of the path integral is illustrated by considering a single degree of freedom. The time evolution of a system from the position $q_a(t_a) \to q_b(t_b)$ in the Heisenberg picture is expressed by an action integral $S(q)$ as

$$\langle q_b | e^{-iH(t_b - t_a)/\hbar} | q_a \rangle = \int D(q(t)) e^{iS(q)/\hbar}, \tag{10.12}$$

where $D(q)$ represents integrals of all possible paths from q_a to q_b and $S(q)$ is defined by

$$S(q) = \int_{t_a}^{t_b} \mathcal{L}(q, \dot{q}) dt, \tag{10.13}$$

for the Lagrangian $\mathcal{L} = m\dot{q}^2/2 - V(q)$.

We now derive the formula (10.12) for an infinitesimal time interval Δt. The l.h.s. of Eq. (10.12) is expressed for the time interval Δt as

$$\langle q_{i+1} | e^{-iH\Delta t/\hbar} | q_i \rangle = \int \frac{dp_j}{2\pi} \langle q_{i+1} | p_j \rangle \langle p_j | e^{-iH(p,q)\Delta t/\hbar} | q_i \rangle, \tag{10.14}$$

where p and q are conjugate operators satisfying the anti-commutation relation $[q, p] = i\hbar$ and an identity

$$1 = \int \frac{dp}{2\pi} |p\rangle\langle p|, \tag{10.15}$$

Fig. 10.5 Various quantum paths that contribute to the path integral for a particle moving from the point q_a at time t_a to the point q_b at time t_b. The points q_i and q_{i+1} are characterized by an infinitesimal space interval together with an infinitesimal time interval Δt

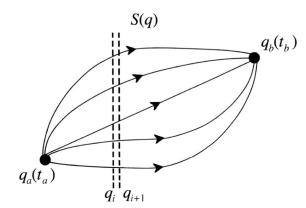

for the completeness of eigenstates for the operator p. Equation (10.14) is rewritten in the limit $\Delta t \to 0$ to be (see Fig. 10.5)

$$
\begin{aligned}
\langle q_{i+1}|e^{-iH\Delta t/\hbar}|q_i\rangle &= \int \frac{dp_j}{2\pi} e^{i(p_j q_{i+1} - p_j q_i)/\hbar - iH(p_j, q_i)\Delta t/\hbar} \\
&= \int \frac{dp_j}{2\pi} e^{i(p_j \dot{q}_i - H(p_j, q_i))\Delta t/\hbar},
\end{aligned} \tag{10.16}
$$

where the $\langle q|p\rangle = e^{ipq/\hbar}$ is used.

Problem

10.2 Derive the equation $\langle q|p\rangle = e^{ipq/\hbar}$ for the eigenvectors $|p\rangle, |q\rangle$ for the operators p, q, respectively.

Equation (10.16) is further transformed for the Hamiltonian $H = \frac{p^2}{2m} + V(q)$ as

$$
\begin{aligned}
\langle q_{i+1}|e^{-iH\Delta t/\hbar}|q_i\rangle &= \int \frac{dp_j}{2\pi} e^{i(-(p_j - m\dot{q}_i)^2/2m + m\dot{q}_i^2/2 - V(q_i))\Delta t/\hbar} \\
&= I_c e^{i(m\dot{q}_i^2/2 - V(q_i))\Delta t/\hbar} = I_c e^{iL(q_i, \dot{q}_i)\Delta t/\hbar}, \tag{10.17}
\end{aligned}
$$

where the integral of dp_j can be performed analytically as an extended Gauss integral and gives a constant I_c. Then, the time evolution of the wave function can be expressed as

$$
\begin{aligned}
U(q_a, q_b, T) &= \langle q_b|e^{-iHT/\hbar}|q_a\rangle = \int dq_1 dq_2 \cdots dq_n e^{i\sum_{j=1}^{n} L(q_j, \dot{q}_j)\Delta t} \\
&= \int Dq(t) e^{i\int_{t_a}^{t_b} L(q, \dot{q})dt} \equiv \int Dq(t) e^{iS(q)}, \tag{10.18}
\end{aligned}
$$

where $T = t_b - t_a$ and $Dq(t)$ expresses all the paths between q_a and q_b. In Eq. (10.18), the action $S(q)$ is defined for a "classical" field $(q, \dot{q})$.

We will consider a generalized field in QFT. For a scalar field φ, the action integral is given by

$$S(\varphi) = \int d^4x \left(-\partial_\mu \varphi \partial^\mu \varphi - m^2 \varphi^2 - \frac{\lambda}{4!} \varphi^4 \right), \tag{10.19}$$

where the first term is the kinetic energy, the second term is the mass, and the third term is the field with the coupling constant λ. In QFT, the correlation between n particles is represented by the n-point correlator (correlation function)

$$\langle \Omega | \varphi(x_1)\varphi(x_2) \cdots \varphi(x_n) | \Omega \rangle = \frac{\int D(\varphi(x)) \varphi(x_1)\varphi(x_2) \cdots \varphi(x_n) e^{iS(\varphi)}}{\int D(\varphi(x)) e^{iS(\varphi)}}, \tag{10.20}$$

where $D(\varphi(x))$ represents all the possible configurations in the scalar field and $|\Omega\rangle$ is the vacuum. It is practical to introduce a generating function

$$Z[\eta] = \int D(\varphi(x)) e^{iS(\varphi)+i \int d^4x \eta(x)\varphi(x)}, \tag{10.21}$$

where $\eta(x)$ is called a source term. Then, the two-point correlator can be expressed as

$$\langle \Omega | \varphi(x_1)\varphi(x_2) | \Omega \rangle = \frac{1}{Z[\eta = 0]} \frac{-i\delta}{\delta\eta(x_1)} \frac{-i\delta}{\delta\eta(x_2)} Z(\eta) \Big|_{\eta=0}, \tag{10.22}$$

where the functional derivative with $\eta(x_1)$ reads

$$\frac{\delta}{\delta\eta(x_1)} \eta(x) = \delta^4(x_1 - x). \tag{10.23}$$

The QCD action depends on the quark fields $\psi^{(f)}$ as well as the gluon field A_μ^a. The quark part reads

$$S_{\text{quark}}[\psi, \overline{\psi}, A] = \int d^4x \sum_f \overline{\psi}^{(f)} [i\gamma^\mu D_\mu - m^{(f)}] \psi^{(f)}$$

$$= \int d^4x \sum_f \overline{\psi}_{c\alpha}^{(f)} [i(\gamma^\mu)_{\alpha\beta} (\delta^{cd} \partial_\mu - igA_\mu^a (\lambda^a)^{cd}/2) - \delta^{cd} \delta_{\alpha\beta} m_f] \psi_{d\beta}^{(f)}, \tag{10.24}$$

where $\overline{\psi} = \psi^\dagger \gamma^0$ is the Dirac adjoint wave function, $D_\mu = \partial_\mu - igA_\mu^a \lambda^a/2$ is the covariant derivative with the QCD coupling constant g between quark and gluon, and γ_μ is the 4×4 Dirac matrix. The Dirac adjoint wave function is commonly adopted in the covariant formalism of QCD. (see Sect. 1.8.2 for details of covariant formalism). The repeated indices indicate the sum of all components

$(\alpha, \beta = 1, \ldots, 4, \mu = 0, \ldots, 3, c, d, e = 1, \ldots, 3, a = 1, \ldots, 8)$. Notice the similarity between the QCD action and the Dirac equation $(i\gamma^\mu \partial_\mu - m)\psi = 0$ for the relativistic QM.

The gluon part of the QCD action reads

$$S_{\text{gluon}} = -\frac{1}{4} \int d^4x \, F^{\mu\nu,a} F^a_{\mu\nu}, \tag{10.25}$$

where the gluon field $F^a_{\mu\nu}$ has a self-interaction term

$$F^a_{\mu\nu} = \partial_\mu A^a_\nu - \partial_\nu A^a_\mu - g f^{abc} A^b_\mu A^c_\nu. \tag{10.26}$$

The gluon action (10.25) is also written as

$$S_{\text{gluon}} = -\frac{1}{2} \int d^4x \, \text{Tr}[F^{\mu\nu} F_{\mu\nu}], \tag{10.27}$$

where

$$F_{\mu\nu} = F^a_{\mu\nu} \lambda^a / 2 = \partial_\mu A_\nu - \partial_\nu A_\mu - ig[A_\mu, A_\nu]. \tag{10.28}$$

We should notice that the gluon QCD action (10.25) has a similar structure to the electromagnetic fields of quantum electrodynamics (QED),

$$S_{\text{QED}} = -\frac{1}{4} \int d^4x \, F^{\mu\nu} F_{\mu\nu}, \quad F_{\mu\nu} = \partial_\mu A_\nu - \partial_\nu A_\mu, \tag{10.29}$$

where A_μ is the electromagnetic field. The QED action is invariant under the Lorentz gauge transformation $U(1)$ since there is only the colorless photon field in QED, while the QCD action must be invariant under the $SU(3)$ gauge transformation due to the three colors of quarks (g, b, r). Moreover, the gluon action has the self-interaction terms of gluons, which has no counterpart in QED.

Although the QCD action looks complicated with several indices, there are only a few free parameters (quark masses m_u, m_d, and the field coupling g) which determine all the properties of mesons and baryons and also dynamics of quantum many-body problems.

10.2.2 Expectation Value of Observables

The expectation value of an observable $O[\psi, \overline{\psi}, A]$ in QCD is given by the path integral

$$\langle \Omega | O[\psi, \overline{\psi}, A] | \Omega \rangle = \frac{1}{Z} \int D\psi \, D\overline{\psi} \int DA \, O[\psi, \overline{\psi}, A] e^{i S_{\text{QCD}}[\psi, \overline{\psi}, A]} \tag{10.30}$$

with $S_{\text{QCD}} = S_{\text{quark}} + S_{\text{gluon}}$ and

$$Z = \int D\psi\, D\overline{\psi} \int DA\, e^{i S_{\text{QCD}}[\psi, \overline{\psi}, A]}, \tag{10.31}$$

where $\int D\psi\, D\overline{\psi}$ are the integrations of all possible quark field configurations $\psi^{(f)}(x)$ and $\int DA = \int \prod_x \prod_{\mu=0}^{3} dA_\mu(x)$ are the integrations of all possible gluon configurations $A_\mu(x)$. To determine a hadron mass, one needs to calculate a two point correlation function for an appropriate hadron creation operator. For example, to determine the pion mass, one needs to evaluate the correlation function

$$\langle \Omega | O_\pi(t_2) \overline{O}_\pi(t_1) | \Omega \rangle = \frac{1}{Z} \int D\psi\, D\overline{\psi} \int DA\, O_\pi(t_2) \overline{O}_\pi(t_1) e^{i S_{\text{QCD}}[\psi, \overline{\psi}, A]}, \tag{10.32}$$

where the pion operator has a pseudo-scalar property and is expressed by a product of the quark and anti-quark wave functions (left panel of Fig. 10.6) with the pseudo-scalar operator γ_5

$$O_\pi(t) = \int d\mathbf{r}\, \overline{\psi}^{(d)}(x) \gamma_5 \psi^{(u)}(x), \tag{10.33}$$

$$\overline{O}_\pi(t) = \int d\mathbf{r}\, \overline{\psi}^{(u)}(x) \gamma_5 \psi^{(d)}(x). \tag{10.34}$$

For Baryons, one needs a correlation function for the operator with three quarks (right panel of Fig. 10.6). The proton operator is given by

$$O_p(t) = \int d\mathbf{r}\, \varepsilon_{cde} (C\gamma_5)_{\beta\gamma} \psi_{\alpha c}^{(u)}(x) (\psi_{\beta d}^{(u)}(x) \psi_{\gamma e}^{(d)}(x) - \psi_{\gamma e}^{(d)}(x) \psi_{\beta d}^{(u)}(x)), \tag{10.35}$$

where $C = \gamma_0 \gamma_2$ is the charge conjugation operator. The neutron operator $O_n(t)$ is expressed by interchanging of $\psi_{\alpha c}^{(u)}(x)$ with $\psi_{\alpha c}^{(d)}(x)$ in Eq. (10.35).

10.2.3 Algorithms

The path integral has, in principle, infinite dimensions, but suites for numerical evaluation on high performance computing (HPC). To solve the path integral numerically, one has to convert it to a feasible multi-dimensional integral on the lattice points. To solve QCD on the finite lattice points is known as Lattice QCD (LQCD), which allows first principle, gauge invariant and non-perturbative calculations of strongly interacting quarks and gluons. There are several basic ideas to implement LQCD on HPC:

1. The space-time is discretized by mesh points N with constant mesh size a. A periodic condition is required for the space-time with the size $L = aN_L$ and the time $T = aN_T$. The resulting path integral reads

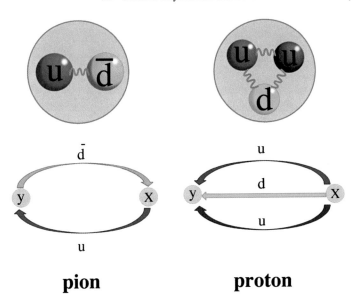

pion   **proton**

Fig. 10.6 Quark compositions and propagators for pion and proton

$$\int D\psi D\overline{\psi} \int DA \rightarrow \prod_{n=1}^{N-1} d\psi(an)d\overline{\psi}(an)dU(an), \tag{10.36}$$

where U is called a link variable, i.e., the lattice equivalent of the gluon field A_μ assigned to each link between sites x and $x + \hat{\mu}$ of the space-time lattice,

$$U_\mu(x) = \exp\left(igaA_\mu(x + \hat{\mu}/2)\right), \tag{10.37}$$

where $\hat{\mu}$ is the vector pointing the direction μ with a length a. The link variable is introduced to preserve the gauge invariance for the LQCD calculations. The basic idea of LQCD is that the quark field is defined on the lattice point as an anti-commuting Grassmann variables (see Sect. 10.6.2 for more details on the Grassman algebra), but the gluon field is defined on the link between the lattice points.

2. The oscillating weight $e^{iS(x)}$ is changed to the exponential weight by the Wick rotation $t \rightarrow -it$. The variable t does not have the meaning of time, but it is just a mathematical parameter. The action integral $S(x)$ is changed by the Wick rotation from the Minkowski space-time to the Euclidean space-time

$$\int Dx e^{iS(x)}, \quad \text{with} \quad S(x) = \int dt \left(\frac{m}{2}\dot{x}^2 - V(x)\right) \quad \text{(Minkowski)}$$

$$\rightarrow \int Dx e^{-S(x)}, \quad \text{with} \quad S(x) = \int dt \left(\frac{m}{2}\dot{x}^2 + V(x)\right) \quad \text{(Euclidean)}.$$

The real positive weight $e^{-S(x)}$ exponentially suppresses any paths $x(t)$ with respect to the minimum of the action and allows for efficient stochastic methods to evaluate these path integrals numerically. The dimension of the path integral can be estimated as follows

- $O((L/a)^4)$ for 4D-lattice,
- $d_f \times d_c$ (3)$\times$ Lorentz indices (4) for the quark field,
- $SU(3)$ gauge $\times$ Lorentz indices (4) for the gluon field.

Taking 10 lattice points for each dimension of the lattice size L, the number of variables would be

$$10^4 \times (d_f \times 3 \times 4 + 8 \times 4) = 10^4 \times (12 d_f + 32). \tag{10.38}$$

Taking u, d, and s quarks for the flavor, the dimension is 68×10^4.

The correlation function for a hadron creation operator $O_H(t)$ is computed as

$$C_H(\Delta t) = \langle \Omega | O_H(t_2) \overline{O}_H(t_1) | \Omega \rangle \tag{10.39}$$

$$= \frac{1}{Z} \int D\psi \, D\overline{\psi} \int DA \, O_H(t_2) \overline{O}_H(t_1) e^{-S_{QCD}(\psi, \overline{\psi}, A)}, \tag{10.40}$$

where $\Delta t = t_2 - t_1$. The time evolution in Eq. (10.39) implies

$$C_H(\Delta t) = \sum_{n=0}^{\infty} \langle \Omega | e^{H \Delta t} O_H(t_1) e^{-H \Delta t} | n \rangle \langle n | \overline{O}_H(t_1) | \Omega \rangle$$

$$= \sum_{n=0}^{\infty} |\langle n | O_H(t_1) | \Omega \rangle|^2 e^{-(E_n - E_0)\Delta t} \xrightarrow[\Delta t \to \infty]{} |a_0|^2 e^{-m_0 \Delta t}, \tag{10.41}$$

where n stands for all the states of the hadron spectra and $a_0 = \langle n = 0 | O_H(t_1) | \Omega \rangle$. In the limit of $\Delta t \to \infty$, the ground state of the hadron spectrum dominates the correlation function and gives the mass of the hadron ground state m_0. The correlation function involves a huge dimensional integration and it is practically impossible to solve it in any standard numerical algorithm. The only plausible method is the Monte Carlo simulation, which is a statistical way to evaluate integrals. Many advanced techniques suitable for supercomputers have been invented to solve the correlation functions in the last three decades and unravel the nature of strong interactions from the non-perturbative calculations under the chiral invariance. There is still an open problem to overcome in the Monte Carlo simulations, the so-called "sign problem". This means that at the moment the method can be applied only for attractive interactions, but repulsive interactions are excluded from the simulation study because of a non-convergence nature of the action integral.

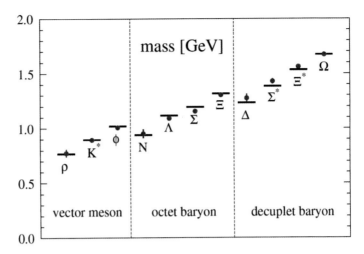

Fig. 10.7 The $N_f = 2 + 1$ light hadron spectrum extrapolated to the physical point using m_π, m_K and m_Ω as inputs. Horizontal bars denote the experimental values. Reprinted with permission from [2]. ©2021 by the American Physical Society

10.2.4 Masses

Light hadron masses are calculated with the (2+1) flavor LQCD simulations in [2], where the u, d, and s quark masses are taken as $m_u = m_d \neq m_s$. Numerical simulations are carried out at the lattice spacing of $a = 0.0907$ fm in $32^3 \times 64$ lattice. The inputs are observed masses of mesons $m_\pi = 135.0$ MeV, $m_K = 497.6$ MeV and a baryon $m_\Omega = 1672.5$ MeV. The quark masses, the pseudoscalar pion coupling constant f_π and the kaon coupling constant f_K are determined at the physical point as

$$m_u = m_d = 2.527 \text{ MeV},$$
$$m_s = 72.72 \text{ MeV},$$
$$f_\pi = 134.0 \text{ MeV}, \quad [f_\pi(exp) = 130.7 \text{ MeV}],$$
$$f_K = 159.4 \text{ MeV}, \quad [f_K(exp) = 159.8 \text{ MeV}].$$

with a large cutoff energy of $a^{-1} \simeq 2$ GeV for the simulation. The physical quark masses and the pseudoscalar meson decay constants show a good consistency with the experimental values.

The light hadron spectrum extrapolated to the physical point is shown in Fig. 10.7. The figure shows an agreement with the experimental values at most 3% level of errors. The discrepancy between the LQCD results and the experimental values are not small for the ρ meson, the nucleon and the Δ baryon, albeit there remain possible cutoff effects $O((a\Lambda_{QCD})^2)$.

Fig. 10.8 Lattice QCD calculations for the central part of the NN potential in the $^1S_0$ ($^3S_1$) channel with quark masses corresponding to an effective ratio of the pion and ρ masses of $m_\pi/m_\rho = 0.595$. The inset shows a zoom. The solid lines correspond to the one-pion exchange potential (OPEP) given in Eq. (2.10). Reprinted with permission from [3]. ©2021 by the American Physical Society

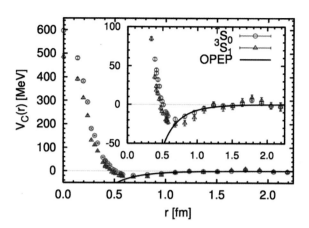

10.2.5 NN Potentials

Let us next study the nucleon-nucleon (N-N) potential in LQCD. In QM, the two-body scattering is studied by solving the Bethe–Salpeter equation as was shown in Chap. 2. From the phase shift analysis of two-body scattering, the N-N potential is derived. We will follow this procedure in QM also in LQCD calculations. Firstly, we consider a N-N system. To this end, we introduce a 4-point correlation function

$$\psi(\mathbf{r}) \equiv \langle \Omega | O_N(\mathbf{r}) O_N(\mathbf{r} = 0) \overline{O}_N(\mathbf{k}) \overline{O}_N(-\mathbf{k}) | \Omega \rangle, \tag{10.42}$$

where $\mathbf{k}$ is the initial momentum of nucleons in the center of mass system, while $\psi(\mathbf{r})$ is the two-body scattering wave function with the relative coordinate between two nucleons. The operator O_N stands for nucleon and it is expressed by three quarks as Eq. (10.35). The wave function ψ is equivalent to the Bethe–Salpeter wave function and satisfies

$$(\nabla^2 + k^2)\psi(\mathbf{r}) = 0, \tag{10.43}$$

in the free space outside the interaction range at $r \to \infty$. The phase shift is determined from the asymptotic behavior of the wave function as

$$\psi(\mathbf{r}) \sim \frac{\sin(kr - l\pi/2 + \delta(k))}{kr}, \tag{10.44}$$

in a standard 2-body scattering theory. As will be discussed hereafter, one of the strategies proposed by the HAL QCD (Hadrons to Atomic Nuclei from Lattice QCD) method for the two-body interaction has an advantage to obtain physical results regardless of the impurity of excited states of baryons in LQCD calculations. In the HAL QCD method, the wave function in the interaction range (typically $r < 2$ fm) is taken as

$$(\nabla^2 + k^2)\psi(\mathbf{r}) = m_N \int d\mathbf{r}' U(\mathbf{r}, \mathbf{r}')\psi(\mathbf{r}'), \qquad (10.45)$$

where $U(\mathbf{r}, \mathbf{r}')$ is a non-local N-N potential and m_N is the nucleon mass. In the lowest-order approximation, the non-local potential is replaced by a local potential as $U(\mathbf{r}, \mathbf{r}') = V_{NN}(\mathbf{r})\delta(\mathbf{r} - \mathbf{r}')$. The potential is then obtained through an equation

$$V_{NN}(r) = E + \frac{\nabla^2 \psi(r)}{m_N \psi(r)}, \qquad (10.46)$$

at a given energy E. Noriyoshi Ishii and collaborators obtained the central potentials $V_C(r)$ for spin-singlet $S = 0$ and spin-triplet $S = 1$ channels with the angular momentum $L = 0$, i.e., $^{2S+1}L_J = {}^1S_0, {}^3S_1$ channels, respectively. A tensor coupling between the $^3S_1$ and $^3D_1$ channels is renormalized in an effective potential V_C. The simulation carried out on a 32^4 lattice in the quenched approximation (see Appendix 10.6.4 for details).[5] The mesh size is taken to be $a = 0.137$ fm, which leads to the lattice size L $= 4.4$ fm. The LQCD results reproduce well the radial dependence of one-pion exchange potential in the region $r > 0.7$ fm, while the short-range repulsion dominates in the region $r < 0.5$ fm as seen in Fig. 10.8.

10.3   Cabibbo–Kobayashi–Maskawa Matrix and Superallowed β Decay

In the standard model, the Cabibbo–Kobayashi–Maskawa matrix (CKM matrix, quark mixing matrix, sometimes also called KM matrix) is a unitary matrix[6] which contains information on the strength of flavor-changing weak decays. Technically, it specifies the mismatch of quantum states of quarks when they propagate freely and when they take part to the weak interactions. The 2×2 matrix was introduced by Nicola Cabibbo to understand the universality of weak decays within the two generation quark model ($u \leftrightarrow d$ and $c \leftrightarrow s$). Observing that Charge-Parity (CP) violation could not be explained in a two-generation quark model, Makoto Kobayashi and Toshihide Maskawa generalized the Cabbibo matrix into the 3×3 CKM matrix to explain CP violation in the weak decays. The CKM matrix of three generations of quarks reads

$$\begin{bmatrix} u' \\ c' \\ t' \end{bmatrix} = \begin{bmatrix} V_{ud} & V_{us} & V_{ub} \\ V_{cd} & V_{cs} & V_{cb} \\ V_{td} & V_{ts} & V_{tb} \end{bmatrix} \begin{bmatrix} d \\ s \\ b \end{bmatrix}. \qquad (10.47)$$

---

[5] This approximation neglects the quark-antiquark creations from the vacuum for the physical process.

[6] An unitary matrix is defined by a condition $UU^\dagger = U^\dagger U = I$, where $U^\dagger$ is the Hermitian conjugate of a matrix U and I is the identity matrix.

On the left is the weak interaction doublet partners of up-type quarks, u, c, and t quarks, and on the right is the CKM matrix along with a vector of mass eigenstates of down-type quarks, d, s, and b quarks. The pairs (u and d), (c and s), and (t and b) are the weak interaction doublets, respectively. The CKM matrix describes the probability of a transition from one quark i to another quark j. These transitions are proportional to $|V_{ij}|^2$. For example, $|V_{ud}|^2$ gives a decay probability between up and down quarks in $\beta^{\pm}$ decays (proton to neutron or neutron to proton decays).

Four independent parameters are required to fully define the CKM matrix. A "standard" parameterization of the CKM matrix uses three Euler angles ($\theta_{12}, \theta_{23}, \theta_{13}$) and one CP-violating phase δ_{13}. The Cabibbo angles θ_{ij} determine the couplings between quark generations i and j, and the coupling vanishes if $\theta_{ij} = 0$. The CKM matrix is thus expressed as

$$\begin{bmatrix} 1 & 0 & 0 \\ 0 & c_{23} & s_{23} \\ 0 & -s_{23} & c_{23} \end{bmatrix} \begin{bmatrix} c_{13} & 0 & s_{13}e^{-i\delta_{13}} \\ 0 & 1 & 0 \\ -s_{13}e^{i\delta_{13}} & 0 & c_{13} \end{bmatrix} \begin{bmatrix} c_{12} & s_{12} & 0 \\ -s_{12} & c_{12} & 0 \\ 0 & 0 & 1 \end{bmatrix}$$

$$= \begin{bmatrix} c_{12}c_{13} & s_{12}c_{13} & s_{13}e^{-i\delta_{13}} \\ -s_{12}c_{23} - c_{12}s_{23}s_{13}e^{i\delta_{13}} & c_{12}c_{23} - s_{12}s_{23}s_{13}e^{i\delta_{13}} & s_{23}c_{13} \\ s_{12}s_{23} - c_{12}c_{23}s_{13}e^{i\delta_{13}} & -c_{12}s_{23} - s_{12}c_{23}s_{13}e^{i\delta_{13}} & c_{23}c_{13} \end{bmatrix}, \quad (10.48)$$

where $s_{ij} \equiv \sin(\theta_{ij})$ and $c_{ij} \equiv \cos(\theta_{ij})$. Essentially, the 2×2 unitary matrix introduced by Cabibbo was determined by a single parameter θ_{12} and does not leave any room for the CP violation phase δ_{13}, while the CKM matrix has 4 parameters including the CP violation phase which makes the matrix elements complex.

There are a few processes to test the unitarity of this matrix in particle and nuclear physics. In nuclear physics, the free neutron decay, the pion beta decay, and the superallowed Fermi beta decay[7] are to be considered. Among them, the superallowed Fermi beta decay between the members of isobaric analog states (IAS) with ($J^{\pi} = 0^+ \rightarrow J^{\pi} = 0^+$) provides the best opportunity to test the unitarity of CKM matrix in the electroweak standard model. The superallowed transition has been the subject of intense studies for several decades both experimentally and theoretically. The determination of the matrix elements is also related to our current understanding of the CP violation mechanism within or beyond the standard model. The related element of the CKM matrix with the superallowed Fermi Beta decay is V_{ud}. According to the unitarity condition, the sum of the square of the first row in the CKM matrix obeys the condition

$$|V_{ud}|^2 + |V_{us}|^2 + |V_{ub}|^2 = 1. \quad (10.49)$$

The weak interactions are mediated by vector mesons Z and $W^{\pm}$ (see Fig. 10.1), analogous to the electromagnetic interactions mediated by photons. The decay amplitude of beta decay is thus given by the vector current V_{μ} as

---

[7] The β decay of ($J^{\pi} = 0^+ \rightarrow J^{\pi} = 0^+$) is named the "superallowed beta decay" or "superallowed Fermi decay". The beta decay of ($J^{\pi} = 0^+ \rightarrow J^{\pi} = 1^+$) is called "allowed Gamow–Teller decay".

$$T = \frac{g}{\sqrt{2}} V^{\mu} \frac{1}{M_W^2 - q^2} \frac{g}{\sqrt{2}} V_{\mu}^{\dagger} \approx \frac{g^2}{2M_W^2} V^{\mu} V_{\mu}^{\dagger}, \tag{10.50}$$

where M_W and q are the mass and the momentum of the weak boson and g is the coupling constant of the weak decay. Here, we assume $M_W^2 \gg q^2$, which is the case of beta decay. The vector coupling constant G_V is defined as $G_V/\sqrt{2} = g^2/(8M_W^2)$ and it is a replacement of the fine structure constant $\alpha^2 = (e^2/\hbar c)^2$ of the electromagnetic interaction. The electromagnetic and weak vector currents are written as

$$J_{\mu} = \frac{1}{2}\overline{\Psi}\gamma_{\mu}(1 - \tau_z)\Psi, \tag{10.51}$$

$$V_{\mu} = \frac{1}{2}\overline{\Psi}\gamma_{\mu}\tau_{\pm}\Psi, \tag{10.52}$$

where Ψ and $\overline{\Psi}$ are spinor wave functions of nucleons $\overline{\Psi} = \Psi^{\dagger}\gamma^0 = (\overline{u}_n, \overline{u}_p)$. The conserved Vector Current (CVC) hypothesis assumes that the isovector part of J_{μ} and V_{μ} are just the different components of the same current which satisfy the conservation law together with the contributions of a pion cloud around the neutron. This is the same as the conservation of the electromagnetic current (10.51), i.e., the conservation of charges.

The superallowed beta decay $(J^{\pi} = 0^+, T = 1 \rightarrow J^{\pi} = 0^+, T = 1)$ between the mother and the daughter nuclei is induced by the vector current and the experimental ft value is directly related to the vector coupling constant G_V,

$$ft = \frac{K}{G_V^2 |M_F|^2}, \tag{10.53}$$

where f is the statistical rate function depending on Z and the decay transition energy Q_{EC}, t is the partial half-life for the transition, and K is the known constant

$$K/(\hbar c)^6 = 2\pi^3\hbar \ln 2/(m_e c^2)^5 = (8120.2776 \pm 0.0009) \times 10^{-10} \text{ s/GeV}^4. \tag{10.54}$$

The Fermi transition matrix element $|M_F|$ has several nuclear structure corrections so that these corrections should be subtracted from the experimental ft values. First of all, we have to consider the isospin breaking effects. For the mother state with $T = 1$, $T_{z0} = 0$ or -1, the Fermi matrix is expressed as

$$|M_F|^2 = |\langle T = 1, T_z = T_{z0} + 1|T_+|T = 1, T_{z0}\rangle|^2 = 2(1 - \delta_C), \tag{10.55}$$

where δ_C is the isospin breaking correction due to the Coulomb, isospin symmetry and isospin independence breaking interactions.

Considering the radiative correction terms, we define a "corrected Ft" value as

$$Ft = ft(1 + \delta_R')(1 + \delta_{NS} - \delta_C) = \frac{K}{2G_V^2(1 + \Delta_R^V)}, \tag{10.56}$$

where Δ_R^V is the transition-independent part of the radiative correction. The terms δ_R' and δ_{NS} comprise the transition-dependent part of the radiative correction, the former being a function only of the electron's energy and the Z of the daughter nucleus, while the latter depends on its evaluation on the details of nuclear structure as does δ_C. In consequence, in the middle of Eq. (10.56), the first correction term is independent from the nuclear structure and the second term is related to the nuclear structure. Typical scales of the correction terms are around 1.5% for δ_R' and 0.5–1.7% for $\delta_C - \delta_{NS}$. Each measured ft value establishes an independent value for G_V to the renormalization in the nuclear medium under the CVC hypothesis, which predicts that all the Ft values should be identical regardless of the specific nuclei involved in the β decay.

Twenty-one super allowed β decays were observed until 2020. The observed ft values and the fully corrected Ft values are plotted as a function of the proton number Z of the daughter nuclei in Fig. 10.9. While the values of ft are scattered, the Ft values show only a small fluctuation around the mean value. This feature is expected from the CVC hypothesis and is an essential prerequisite for further probes of the standard model. The average $\overline{Ft}$ in Eq. (10.56) from this study becomes

$$\overline{Ft} = 3072.24 \pm 0.57_{stat} \pm 0.36_{\delta_R'} \pm 1.73_{\delta_{NS}} \text{ s},$$
$$= 3072.24 \pm 1.85 \text{ s}, \tag{10.57}$$

where the error stems from the statistical uncertainty of experimental data and the theoretical ones associated to evaluate the radiative corrections δ_R' and the nuclear structure correction δ_{NS}.

In the electroweak theory, the relationship between the Fermi and the vector coupling constant is given by $G_V = G_F V_{ud}$. The Fermi coupling constant G_F is obtained from the pure leptonic muon beta decay and the matrix element V_{ud} is now determined by a formula

$$V_{ud}^2 = \frac{K}{2G_F^2(1 + \Delta_R^V)\overline{Ft}} = \frac{2915.64}{\overline{Ft}}, \tag{10.58}$$

where G_F is taken as

$$G_F/(\hbar c)^3 = 1.1663787 \times 10^{-5} (\text{GeV})^{-2}, \tag{10.59}$$

from the Review of Particle Physics 2020 [1]. Substituting the result for $\overline{Ft}$ in Eq. (10.57), the V_{ud} value is determined to be

$$|V_{ud}| = 0.97373 \pm 0.00031. \tag{10.60}$$

Another CMK matrix element in the top row V_{us} is determined from the leptonic and semi-leptonic decays of kaons and the extracted value is

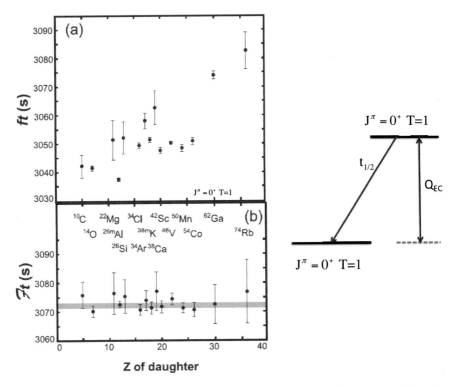

Fig. 10.9 In the right panel, the superallowed Fermi decay is shown from the mother state $J^\pi = 0^+$, $T = 1$ to the daughter state $J^\pi = 0^+$, T=1. The life time $t_{1/2}$ and the decay transition energy Q_{EC} measurements are essential ingredients for the determination of the ft value. In the left top panel **a**, the uncorrected experimental ft values for the 15 precisely known superallowed transitions are plotted as a function of the proton number Z of the daughter nucleus. In the bottom panel **b**, the corrected Ft values are plotted; they differ from the ft values by the inclusion of the correction terms δ'_R, δ_{NS} and δ_C. The corresponding mother nucleus is written for each Ft value in the lower panel. The horizontal grey bar in the bottom panel gives one standard deviation around the average value $\overline{Ft} = 3072.24$. Reprinted with permission from [4]. ©2021 by the American Physical Society

$$|V_{us}| = 0.2245 \pm 0.0008. \tag{10.61}$$

The third element of the top row V_{ub} is very small

$$|V_{ub}| = (3.83 \pm 0.24) \times 10^{-3}, \tag{10.62}$$

compared with the other two elements. Combining these values, the sum of the squares of elements in the first raw of CMK matrix is

$$|V_u|^2 \equiv |V_{ud}|^2 + |V_{us}|^2 + |V_{ub}|^2 = 0.9985 \pm 0.0005, \tag{10.63}$$

compiled by the Particle Data Group 2020 [1]. The other two rows of CKM matrix are also determined experimentally. These results suggest that there is no sign of violation of the unitarity in the CKM matrix. The three generation quark model is quite successful and suggests no need for any modification of the standard model. The currently best known values for the standard parameters of the CMK matrix are:

$$\sin \theta_{12} = 0.22650 \pm 0.00048,$$
$$\sin \theta_{13} = 0.00361^{+0.00011}_{-0.00009},$$
$$\sin \theta_{23} = 0.04053^{+0.00083}_{-0.00061}, \tag{10.64}$$

and the parity violation phase is

$$\delta_{13} = 1.196^{+0.045}_{-0.0431} \text{ rad.} \tag{10.65}$$

10.4 Neutrinos

10.4.1 Flavors and Masses

Neutrinos are produced in nuclear β decay driven by the weak interaction. In the standard model, there are three flavors of mass-less neutrinos: ν_e, ν_μ and ν_τ, associated to the three leptons (electron, muon and tau), together with their associated antiparticles.

In reality, neutrinos have a small mass, although not yet measured experimentally, which is one of the first pieces of evidence of physics beyond the standard model. The discovery of the neutrino mass comes from a long-standing puzzle from the 1960s solved only 30 years later: the so-called *solar neutrino anomaly*. Electronic neutrinos are produced in large amount during capture reactions and decays in the core of the sun. Raymond Davis and collaborators measured an upper limit of the solar neutrino flux of 3 SNU (Solar Neutrino Units defined as 10^{-36} captures per target atom per second), while the estimated flux is close to 8 SNU [5]. An updated value of the observed electronic neutrino flux is 2.65 SNU [6], while recent predictions give an expected value of 7.65. Only one third of the electronic neutrinos produced in the sun arrive on earth. Other flavors of neutrinos were later discovered, and the anomaly has been understood as the following: neutrinos have masses. The mass eigenstates are not aligned with the flavor eigenstates and neutrinos can therefore oscillate from one flavor to another. The electronic neutrinos produced in the sun "disappear" on their way to the earth by changing flavor. Since the experiment of Davis could only detect electronic neutrinos because of the selective reaction, $\nu_e +^{37} \text{Cl} \rightarrow^{37} \text{Ar} + e^-$, and as proposed by Bruno Pontecorvo in 1946, it was blind to the other flavors. Davis together with Masatoshi Koshiba obtained the Nobel prize for physics in 2002 for their contributions to the detection of cosmic neutrinos.

Neutrino oscillation is a quantum mechanical phenomenon, in which a neutrino created with a specific lepton flavor is changed to a mixture of different flavors at a later time. The probability of measuring a particular flavor for neutrino varies periodically as it propagates through the Minkovski space-time. Neutrino oscillations have been of great theoretical and experimental interest, since the observation of the phenomenon implies that the neutrino has a non-zero mass, which was not included as part of the original Standard Model. Takaaki Kajita at the Super-Kamiokande Observatory and Arthur McDonald at the Sudbury Neutrino Observatory won the 2015 Nobel prize for physics by the first measurement and discovery of neutrino oscillations, and thus the proof that neutrinos have masses.

The flavor neutrinos can be expressed as a superposition of the three neutrino mass eigenstates, usually quoted as ν_1, ν_2 and ν_3. The superposition is driven by the unitary leptonic mixing Pontecorvo–Maki–Nakagawa–Sakata (PMNS) matrix (the unitary transformation between the neutrino states in the weak decay process and the mass basis)

$$
\begin{bmatrix} \nu_e \\ \nu_\mu \\ \nu_\tau \end{bmatrix} = \begin{bmatrix} U_{e1} & U_{e2} & U_{e3} \\ U_{\mu1} & U_{\mu2} & U_{\mu3} \\ U_{\tau1} & U_{\tau2} & U_{\tau3} \end{bmatrix} \begin{bmatrix} \nu_1 \\ \nu_2 \\ \nu_3 \end{bmatrix}, \tag{10.66}
$$

where the vector on the left represents a neutrino state in the weak decay process, and on the right the PMNS matrix is multiplied by a vector representing the neutrino state in the mass basis. A neutrino of a given flavor $\alpha = (e, \mu,$ or $\tau)$ is thus a "mixed" state of neutrinos with different masses. If neutrino masses are measured, it would give an empirical information of mass m_i with probability $|U_{\alpha i}|^2$ (see Eq. (10.70)). The PMNS matrix is specified by exactly the same parameters as the CKM matrix (10. 48), 3 mixing angles $(\theta_{12}, \theta_{23}, \theta_{13})$ and the CP-violation phase δ_{13} with a different physics context.

It is interesting to notice that the PMNS matrix is a neutrino version of the CKM matrix which contains information on the mismatch of quantum states of neutrinos when they propagate in free space and when they take part to the weak interactions. This matrix was introduced in 1962 by Ziro Maki, Masami Nakagawa, and Shoichi Sakata to explain the neutrino oscillations predicted by Bruno Pontecorvo. This matrix was introduced more than ten years earlier than the CKM matrix for quarks proposed in 1973. All Japanese physicists involved in PMNS and CKM matrices belonged to the Sakata school of Nagoya university. The Standard Model of particle physics contains three generations or "flavors" of neutrinos, ν_e, ν_μ, and ν_τ as is shown in Fig. 10.1.

The mixing matrix elements can be determined experimentally from different types of experiments: solar neutrino and atmospheric neutrino measurements, measurements at reactors and at accelerator neutrino beams. The masses of the three types (m_1, m_2, m_3) are different. Experiments observing the oscillations of neutrinos produced in the sun have determined the squared difference of the masses m_2 and m_1, $\Delta m_{21}^2 = m_2^2 - m_1^2$. The squared difference between the masses $\Delta m_{31}^2 = m_3^2 - m_1^2$ has been measured using the oscillations of neutrinos produced in the Earth's atmosphere. From a global fit of several results from neutrino oscillation experiments [7], they are determined as

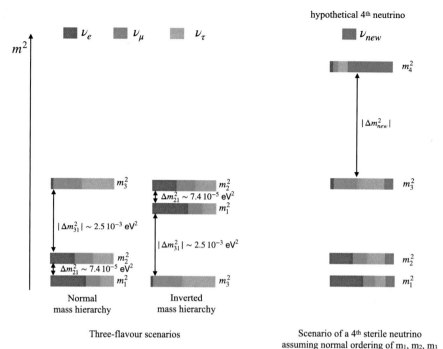

Fig. 10.10 (Left) Flavor composition of neutrino mass eigenstates in a three-flavor scenario. The *normal* and *inverted* mass hierarchy scenarios are represented. The absolute mass of neutrinos is not known. The current state of the art gives only an upper limit of $m \leq O(1)$ eV. (Right) Representation of a four-flavour scenario implied by the existence of a *sterile* neutrino which does not interact with matter via the weak interaction, and is therefore not detectable. The vertical mass axis is not at scale

$$\Delta m_{21}^2 = 7.39 \pm 0.21 \; 10^{-5} \; \text{eV}^2,$$
$$|\Delta m_{31}^2| = 2.525 \pm 0.033 \; 10^{-3} \; \text{eV}^2. \tag{10.67}$$

Sun oscillation experiments determined that Δm_{21}^2 is positive. However, since oscillation experiments can only probe the squared difference of the masses, the absolute values of m_1, m_2, and m_3 as well as the question of whether or not m_2 is heavier than m_3 remains unknown. The latter question is called the *neutrino mass hierarchy problem*. If m_2 is lighter than m_3, the hierarchy is said to be *normal*, but if it is heavier the hierarchy is called *inverted*. These mass differences provide also the information of absolute values of the mixing matrix U_{ei}. The possible mass-hierarchy scenarios are illustrated on the left of Fig. 10.10.

The absolute masses of the neutrinos cannot be probed by oscillations. Today, only an upper limit for the electron neutrino mass of 1.1 eV is known [8]. It was determined by the KATRIN experiment which measures with high precision the electron energy spectrum from the beta decay of tritium $^3\text{H} \rightarrow \, ^3\text{He}^+ + e^- + \bar{\nu}_e$. Tritium has the advantage of low-energy endpoint (18.57 keV) and a favorable half life of 12.32 years.

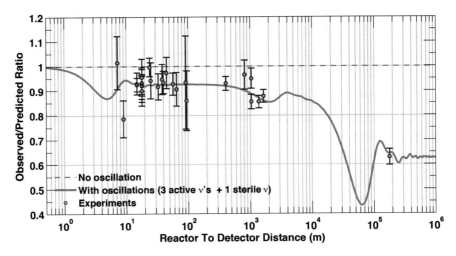

Fig. 10.11 Baseline anti-neutrino anomaly. The experimental results are compared to the prediction (the solid curve) without oscillation, taking into account the new anti-neutrino spectra, the corrections of the neutron mean lifetime, and the off-equilibrium effects. Published experimental errors and anti-neutrino spectra errors are added in quadrature. The mean-averaged ratio including possible correlations is 0.927 ± 0.023. As an illustration, the dashed line shows a three active neutrino mixing solution fitting the data, with $\sin^2(2\theta_{13}) = 0.15$. The solid line displays a solution including a new neutrino mass state, such as $|\Delta m^2_{\text{new}}| >> 2\,\text{eV}^2$, $\sin^2(2\theta_{\text{new}}) = 0.12$ and $\sin^2(2\theta_{13}) = 0.085$, where θ_{new} is the mixing angle of the 4th generation neutrino mass

10.4.2 Search for a 4th Neutrino

Recently observed anomalies in neutrino oscillation data could originate in a hypothetical fourth neutrino separated from the three standard neutrinos by a squared mass difference of a few eV^2. Indeed, the reanalysis of reactor anti-neutrino experimental data using updated calculation of $\bar{\nu}_e$ flux shows a deficiency of more than 3% compared to expectations. This deficiency is systematically found across independent sets of data, leading to a nearly 3σ indication of neutrino disappearance at baselines of less than 100 m, as illustrated in Fig. 10.11. This observation is commonly called the *reactor anti-neutrino anomaly* and may be interpreted as the existence of yet-undiscovered mixing between the electron anti-neutrino and a new, 4th neutrino flavor that is *sterile*, i.e., does not interact through the Standard Model weak interactions but only with gravitation. Sterile neutrinos are therefore undetectable through direct measurements. To be compatible with the observed deficit, this sterile neutrino would need to have a mass splitting $|\Delta m^2_{\text{new}}| \geq 0.1\,\text{eV}^2$. Such a scenario is represented on the r.h.s. of Fig. 10.10.

Other experiments aiming at the measurement of neutrino oscillations, such as the gallium solar neutrino experiments, LSND and MinoBooNE, provide consistent results with this hypothesis, while other similar experiments give negative results. The five-year data of Planck on the cosmological background lead to the conclusion

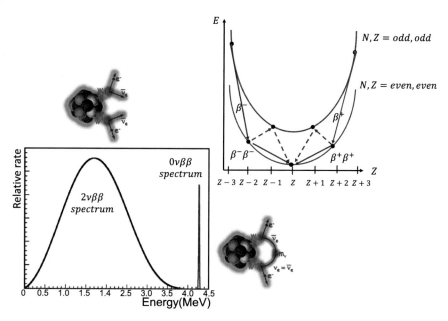

Fig. 10.12 Double beta decay in nuclei. In the right panel, mass parabola for isobaric nuclei with even atomic mass number A. It is split into two parabolas due to the nuclear pairing energy. The β decays that change the nuclear charge Z by a value of 1 can only occur if the energy of the daughter nuclei is smaller than the energy of the parent nuclei. If this condition is not fulfilled, two consecutive β decays (double beta decays) become possible (dashed lines). In the left panel, typical energy spectra of two electrons are shown for two neutrino double beta decay (black) and zero-neutrino double beta decay (red). The spectra of zero-neutrino double beta decay is very much hindered and shows a sharp single peak while the spectrum of two-neutrino double beta decays is large and much broader, Courtesy S. Umehara, RCNP, Osaka University

that the number of neutrino families is $2.99^{+0.34}_{-0.33}$ [9], which is not consistent with a 4th neutrino hypothesis at 95% level confidence. It provides a strong argument against the existence of a fourth neutrino which mixes with neutrinos of the standard model.

If confirmed, the presence of a 4th family of sterile neutrinos would impact the extraction of the neutrino mixing angle θ_{13} as well as measurements of the CP-violating phase in the leptonic sector.

10.4.3 Double Beta Decay

In nuclear physics, the double beta decay has been often discussed in relation with the neutrino mass. The double beta decay is a radioactive decay in which two protons are simultaneously transformed into two-neutrons inside an atomic nucleus as illustrated in Fig. 10.12. As a result of this decay process, the nucleus emits two detectable beta particles, which are electrons or positrons. There are two types of double beta decay: two-neutrino double beta decay and neutrino-less double beta decay. In the two-neutrino double beta decay, which has been observed in 11 isotopes until 2020, two neutrons (protons) in the nucleus are converted to protons (neutrons) and two electrons (positrons), and two electron anti-neutrinos (neutrinos) are emitted in the following process

$$A(Z, N) \rightarrow A(Z \pm 2, N \mp 2) + 2e^{\mp} + 2\overline{\nu}_e(2\nu_e). \tag{10.68}$$

On the other hand, in the neutrino-less double beta decay, only two electrons would be emitted. This neutrino-less process has never been observed until 2020, but only the lower limit of the lifetime was suggested.

For the double beta decay in general, the final daughter nucleus must have a larger binding energy than the original mother nucleus. For some nuclei, such as $^{76}_{32}$Ge, the nucleus with one atomic number higher, $^{76}_{33}$As, has a smaller binding energy, preventing single beta decay. However, the nucleus with the atomic number two higher, $^{76}_{34}$Ce, has a larger binding energy, so the double beta decay is allowed.

It has been established by the neutrino oscillation experiments that the neutrino has non-zero mass. If the neutrino is a Majorana particle, meaning that the anti-neutrino and the neutrino are the same particle, which has non-zero mass, neutrino-less double beta decay is possible. On the other hand, if the neutrino and anti-neutrino are different particles, so-called Dirac particles, they never annihilate at the same time so that the neutrino-less double beta decay can not occur. In the simple theoretical conjecture for the Majorana particle, the two neutrinos may annihilate with each other, or equivalently, a nucleon absorbs the neutrino emitted by another nucleon since neutrino and anti-neutrino can merge and disappear. The neutrinos in this process behave as virtual particles. Since only two electrons exist in the final state of the neutrino-less double beta decay, the total kinetic energy of electrons would be the binding energy difference of the initial and final nuclei with a very small correction due to the recoil of the daughter nucleus. In a very good approximation, the electrons are emitted back-to-back and the decay rate for this process is given by

$$\Gamma = \frac{1}{T^{0\nu}_{1/2}} = |\langle m_{\beta\beta}\rangle|^2 |M^{0\nu}|^2 G^{0\nu}(E_0, Z), \tag{10.69}$$

where $G^{0\nu}(E_0, Z)$ and $M^{0\nu}$ are the known kinematic phase-space factor (E_0 is the energy release) and the nuclear matrix element, respectively. The value $M^{0\nu}$ depends on the nuclear structure of the particular isotopes (A, Z), $(A, Z \pm 1)$, and $(A, Z \pm 2)$ under study. The effective Majorana mass of the electron neutrino $m_{\beta\beta}$ is defined by

$$m_{\beta\beta} = \sum_{i=1}^{3} m_i U_{ei}^2, \tag{10.70}$$

where m_i are the neutrino masses of the ith mass eigenstate, and the U_{ei} are elements of the lepton-mixing PMNS matrix.

One immediately notices from Eq. (10.69) that the decay rate is directly related to the Majorana neutrino mass, i.e., as the neutrino mass trends toward zero, the decay rate will also be zero. Thus, observing neutrino-less double beta decay, in addition to confirming the Majorana neutrino nature, would give information on the absolute neutrino mass scale at the level of a few tens of meV. Double beta decays are low-energy second-order weak processes with $Q_{\beta\beta} \sim 2$–3 MeV. The decay rates of the $2\nu\beta\beta$ decay are of the order of 10^{-20} per year and the rates of the $0\nu\beta\beta$ decay are even predicted to be many orders of magnitudes smaller than the $2\nu\beta\beta$ decay depending on the $Q_{\beta\beta}$ value and the effective Majorana neutrino mass $m_{\beta\beta}$. The $0\nu\beta\beta$ decay half lives are predicted of the order of $T_{1/2} \sim 10^{27-29}$ years. For the experimental studies of such rare decays, large detectors containing ton-scale double beta decay isotope quantities are needed to obtain $0\nu\beta\beta$ signals. Background signals are huge in the energy region $E_{\beta\beta} \leq 3$ MeV. Therefore, it is crucial to build ultra-low background detectors to find the rare and small $0\nu\beta\beta$ signals. Experimental studies of $0\nu\beta\beta$ decay have been carried out on several double-β decay nuclei. Early experiments did claim neutrino-less decay, but modern searches have set limits refuting those results. Some of them are counter measurements such as Heidelberg-Moskow, IGEX project, while spectroscopic measurements are performed in ELEGANT V and NEMO III projects. The zero-neutrino double-beta decay of ^{128}Te was studied by a geochemical method, and the half life and the neutrino mass limits are reported as 1.9×10^{24} year and 1.1–1.5 eV, respectively. Calorimetric detectors have been used for isotopes such as ^{48}Ca, ^{116}Cd, and ^{130}Te. Currently, the most stringent limit on the effective Majorana mass comes from the measurement in KamLAND-Zen experiment.

The nuclear matrix element (NME) associated with the Majorana neutrino exchange $M^{0\nu}$ consists of the Fermi, Gamow–Teller and tensor parts. The interpretation of existing results as a measurement of the neutrino mass $\langle m_{\beta\beta} \rangle$ depends crucially on the knowledge of the nuclear matrix elements that govern the decay rate. The calculation of the $0\nu\beta\beta$ decay nuclear matrix elements is a difficult problem because the ground and many excited states (if closure approximation is not adopted) of open-shell nuclei with complicated nuclear structure have to be considered. In the last few years the reliability of the calculations has greatly improved. Five different many-body methods have been applied for the calculation of the $0\nu\beta\beta$-decay NME

1. The Large Scale Shell Model (LSSM),
2. Quasiparticle Random Phase Approximation (QRPA),
3. The Projected Hartree–Fock–Bogoliubov Method (PHFB),
4. The Energy Density Functional Method (EDF),
5. Interacting Boson Model (IBM).

The differences among the listed methods of NME calculations for the $0\nu\beta\beta$ decay are due to the following reasons
(i) The mean field is used in different ways. As a result, single particle occupancies of individual orbits of various methods differ significantly from each other.
(ii) The residual interactions are of various origin and renormalized in different ways.
(iii) Various sizes of the model space are taken into account.

(iv) Different many-body approximations are used to calculate the nuclear Hamiltonian.

Each of the applied methods has some advantages and drawbacks, whose effects in the evaluations of the NME should be explored. While the advantage of the LSSM calculations is their full treatment of the nuclear correlations, the drawback of the LSSM is the limited number of orbits in the valence space and as a consequence the violation of the Gamow–Teller Ikeda sum rule and underestimation of the NMEs. On the contrary, the QRPA, the EDF, and the IBM underestimate the many-body correlations in different ways and tend to overestimate the NMEs, while they conserve the Ikeda sum rule.

As discussed above, direct neutrino mass measurements in nuclear β decay experiments and cosmology yield $m_i \leq O(1)$ eV. The neutrino-less double beta decay has been actively searched for but it has not been experimentally discovered up to now. The available data allow to assess an upper limit for the effective Majorana neutrino mass $\langle m_{\beta\beta} \rangle$. The best current limits come from neutrino-less double beta decay experiments of ^{136}Xe and ^{76}Ge (90% confidential level):

$$\langle m_{\beta\beta} \rangle < 0.14\text{--}0.38 \text{ eV} \quad \text{for } ^{136}\text{Xe},$$
$$\langle m_{\beta\beta} \rangle < 0.2\text{--}0.4 \text{ eV} \quad \text{for } ^{76}\text{Ge}, \tag{10.71}$$

where the ranges are due to the uncertainties in the values of the nuclear matrix elements, i.e., the calculated half-lives for neutrino-less double β decays in nuclei are rather spread out and the spread of half-lives for given isotope reaches up to the factor of 4–5. The present study is based on the so-called "light" neutrino mass hypothesis. There are also other models for neutrinos, for examples, based on the "heavy" neutrino mass hypothesis and the supersymmetry (SUSY) particle model for neutrinos. These models could be experimentally justified or denied if neutrino-less double β decays are systematically observed in several isotopes.

10.5   Non-zero Electric Dipole Moment and Octupole Deformation

In Sect. 7.9 of Chap. 7, the octupole deformation was discussed as a shape deformation in nuclei. It is known that atoms with octupole-deformed nuclei provide a good playground in the search for permanent atomic electric dipole moments (EDMs). The observation of a non-zero EDM would indicate time-reversal (T) or equivalently charge-parity (CP) violation due to physics beyond the standard model. The electric dipole moment must vanish if the system is invariant under the parity transformation (P) for which $\mathbf{r} \rightarrow -\mathbf{r}$. Under the assumption of the CPT theorem, if T is violated then CP must be also violated, and the observation of a non-zero EDM would indicate CP violation, which will be an evidence of the physics beyond the standard model. A measurement of such CP violation is strongly motivated to account for the observed cosmological dominance of baryons over anti-baryons. To create the baryon dominance in the universe, one of Sakharov conditions,[8] C and CP violation should be confirmed by reliable experimental data since Sakharov condition requires several order of magnitudes larger C and CP violation rates than those observed by the neutral kaon decay and B meson decays.

In fact, experimental limits on EDMs provide important constraints on many proposed extensions to the standard model. The EDM of odd-mass nucleus will provide the information of *nuclear Schiff moment* as will be discussed below. For a neutral atom in its ground state, the nuclear Schiff moment S can be measured by the atomic EDM as a whole. The Schiff moment is the lowest-order observable nuclear moment, which is given by the operator corresponding to the electric-dipole distribution weighted by radius squared

$$\mathbf{S} = \frac{1}{10} \left(\int e\rho(\mathbf{r})r^2\mathbf{r}d\mathbf{r} - \frac{5}{3}\mathbf{d}\frac{1}{Z} \int \rho(\mathbf{r})r^2d\mathbf{r} \right), \tag{10.72}$$

where $\rho(\mathbf{r})$ is the charge density, $\int \rho(\mathbf{r})d\mathbf{r} = Z$, and $\mathbf{d}$ is the EDM of the nucleus defined by $\mathbf{d} = e \int \rho(\mathbf{r})\mathbf{r}d\mathbf{r}$.

Note that the radial dependence of the Schiff moment is the same as the IS dipole compression mode. In the absence of T- and P-violating interactions the electric dipole moment of an atom is equal to zero. If the T- or P-violating interactions exist between atomic electrons and also the T-, or P-odd part of the nuclear potential, the atomic states of the opposite parities are mixed and thus generate an atomic EDM

$$D_z = -e\langle\widetilde{\psi}|r_z|\widetilde{\psi}\rangle = -2e\sum_m \frac{\langle i|r_z|m\rangle\langle m|\phi^S|i\rangle}{E_i - E_m}, \tag{10.73}$$

where $\widetilde{\psi}$ is the perturbed wave function of the electron in the atom, $|i\rangle$ is the unperturbed electron ground state and $|m\rangle$ is the opposite parity excited state mixed by the perturbation ϕ^S, which is proportional to the Schiff moment

$$\phi^S = e4\pi\mathbf{S} \cdot \nabla(\delta(\mathbf{R})), \tag{10.74}$$

where $\mathbf{R}$ indicates the coordinate of the electrons. Equation (10.73) indicates that the atomic EDM is induced entirely by the perturbation of the TP violation effect due to the nuclear Schiff moment. That is, the T- and P-violating interactions at the nuclear level gives the significant effect of the atomic EDM.

If a deformed nucleus in the intrinsic (body-fixed) frame is reflection asymmetric such as octupole deformed, it can have collective T- and P-odd moments. As a

---

[8] In 1967, Andrei Sakharov proposed a set of necessary conditions that a baryon interaction must satisfy to realize matter dominance in the universe over antimatter. These conditions were inspired by the recent discoveries of the cosmic background radiation and CP violation in the neutral kaon decays. The three necessary "Sakharov conditions" are: (1) Baryon number violation. (2) Violations of C and CP symmetries. (3) Interactions out of thermal equilibrium.

consequence of this reflection asymmetry, parity-doublet rotational bands appear in the laboratory system. Without T-, P-violation effect, T-, P-odd moments (dipole, octupole or spin-quadrupole moment) vanish exactly in the laboratory frame. Consider a nearly degenerate rotational parity doublet in the case of an axial symmetric nucleus. The wave functions of the members of the doublet are the eigenstates of parity and are written as

$$|\Psi^{\pm}\rangle = \frac{1}{\sqrt{2}}(|IMK\rangle \pm |IM-K\rangle), \tag{10.75}$$

where I, M are the nuclear spin and its z-component of the laboratory frame, while $K = \mathbf{I} \cdot \mathbf{n}$ is the projection of $\mathbf{I}$ on the intrinsic z-axis denoted by a unit vector $\mathbf{n}$ (see Fig. 7.5). The intrinsic dipole and Schiff moments are defined along the intrinsic z-axis as

$$\mathbf{d}_{\text{intr}} = d_{\text{intr}}\mathbf{n}, \tag{10.76}$$

$$\mathbf{S}_{\text{intr}} = S_{\text{intr}}\mathbf{n}. \tag{10.77}$$

For the good parity states (10.75), the expectation values of two moments vanish since

$$\langle \Psi^{\pm}|\mathbf{n}|\Psi^{\pm}\rangle = 0. \tag{10.78}$$

This is the outcome of $|\Psi^{\pm}\rangle$ in which K and $-K$ have equal probabilities. In other words, this is a consequence of time-reversal invariance and parity conservation since the operator $\mathbf{I} \cdot \mathbf{n}$ is T-, and P-odd.

A P-, T-violation interaction V^{PT} will mix the parity doublet state in the first-order perturbation theory as

$$|\widetilde{\Psi}\rangle = |\Psi^{+}\rangle + \alpha|\Psi^{-}\rangle = \frac{1}{\sqrt{2}}[(1+\alpha)|IMK\rangle + (1-\alpha)|IM-K\rangle], \tag{10.79}$$

where α reads

$$\alpha = \frac{\langle \Psi^{-}|V^{PT}|\Psi^{+}\rangle}{E^{+} - E^{-}}. \tag{10.80}$$

Here, $E^{+} - E^{-}$ is the energy difference between the members of parity doublets with the same I. For the mixed state, the K value is finite

$$\langle \widetilde{\Psi}|\mathbf{I} \cdot \mathbf{n}|\widetilde{\Psi}\rangle = \langle \widetilde{\Psi}|K|\widetilde{\Psi}\rangle = 2\alpha K, \tag{10.81}$$

and, therefore, the dipole and Schiff moments become also a non-zero value for the state with $M = K = L$

$$\langle \widetilde{\Psi} | d_z | \widetilde{\Psi} \rangle = 2\alpha \frac{I}{(I+1)} d_{intr}, \tag{10.82}$$

$$\langle \widetilde{\Psi} | S_z | \widetilde{\Psi} \rangle = 2\alpha \frac{I}{I+1} S_{intr}, \tag{10.83}$$

where $I/(I+1)$ is a geometrical factor from the projection of the intrinsic frame to the laboratory frame.

Octupole-deformed nuclei with odd nucleon number A will have enhanced nuclear Schiff moments owing to the presence of the large octupole collectivity and the occurrence of nearly degenerate parity doublets that naturally arise if the deformation is static.

For a long time, non-octupole-enhanced systems such as ^{199}Hg provided the most stringent limit for atomic EDM measurements. The best limits on CP-violating interactions originating within the nucleus are derived from the limit on the atomic EDM of ^{199}Hg: $d(^{199}$Hg$) < 3.1 \times 10^{-29} e$ cm. Recently, the radioactive radium ^{225}Ra atom has been studied to search for a permanent electric dipole moment. Because of its strong nuclear octupole deformation and large atomic mass, ^{225}Ra is particularly sensitive to interactions in the nuclear medium that violate both time-reversal and parity symmetries. ^{225}Ra has parity doublets with $I = 1/2$ separated by about 50 keV and will have large enhancement of its Schiff moment (see Eq. (10.80)), while the Schiff moment is zero for non-octupole deformed nucleus ^{199}Hg. A cold-atom technique has been developed to study the spin precession of ^{225}Ra atoms held in an optical dipole trap, and demonstrated the first measurement of its atomic electric dipole moment, reaching an upper limit of $d(^{225}$Ra$) < 5.0 \times 10^{-22} e$ cm (95% confidence) [10]. The upper limit of the atomic dipole moment of $d(^{225}$Ra$)$ is shown to be several orders of magnitude larger than that of a non-octupole deformed nucleus $d(^{199}$Hg$)$, so that experimental studies of the nucleus ^{225}Ra provide better opportunities to clarify the magnitude of CP violating interaction.

10.6 Appendix

10.6.1 Dirac γ Matrices

Dirac matrices satisfy the anti-commutation relation

$$\{\gamma^\mu, \gamma^\nu\} \equiv \gamma^\mu \gamma^\nu + \gamma^\nu \gamma^\mu = 2g^{\mu\nu}, \tag{10.84}$$

where $g^{\mu\nu}$ is the diagonal matrix with the Minkowski metric $g^{\mu\nu} = g_{\mu\nu} = (+1, -1, -1, -1)\delta_{\mu,\nu}$ for $\mu = 0, 1, 2, 3$. The gamma matrix can be expressed by a 4×4 matrix analogous to the α, β matrices for the Dirac equation in Sect. 1.8.1 of Chap. 1;

$$\gamma^0 = \beta = \begin{pmatrix} +I & 0 \\ 0 & -I \end{pmatrix}, \quad \gamma^j = \gamma^0 \alpha_j = \begin{pmatrix} 0 & \sigma_j \\ -\sigma_j & 0 \end{pmatrix} \quad for \ j = 1, 2, 3. \quad (10.85)$$

The chiral matrix γ_5 is defined as

$$\gamma_5 = i\gamma^0\gamma^1\gamma^2\gamma^3 = -i\gamma_0\gamma_1\gamma_2\gamma_3 = \begin{pmatrix} 0 & I \\ I & 0 \end{pmatrix}, \quad (10.86)$$

where $\gamma_\mu = g_{\mu\nu}\gamma^\nu$ and I is 2×2 unit matrix. Then, we have $\gamma_0 = \gamma^0$, $\gamma_j = -\gamma^j$ and $\mathrm{Tr}\{\gamma_\mu\} = 0$. For the Minkowski space-time metric $dx = (dt, d\mathbf{r})$ and $\partial_\mu = (\partial/\partial t, \partial/\partial \mathbf{r})$), the Dirac equation is written as

$$i\gamma^\mu \partial_\mu \psi(x) - m\psi(x) = 0. \quad (10.87)$$

We can rewrite Eq. (10.87) to the α, β representation of the Dirac equation in Chap. 1 multiplying γ^0 from the right:

$$\gamma^0 i\gamma^\mu \partial_\mu \psi(x) - m\gamma^0 \psi = \gamma^0 (i\gamma^0 \frac{\partial}{\partial t} \psi + i\boldsymbol{\gamma} \cdot \boldsymbol{\nabla}) - m\gamma^0 \psi$$

$$= i\frac{\partial}{\partial t}\psi + i\boldsymbol{\alpha} \cdot \boldsymbol{\nabla}\psi - m\beta\psi = 0, \quad (10.88)$$

which is identical to Eq. (1.94).

10.6.2 Grassmann Algebra

The Grassmann algebra is invented for anti-commutable numbers associated with systems of fermions. Grassmann numbers (G-numbers) ξ, η satisfy the anti-commutati relation

$$\{\xi, \eta\} = 0, \quad \xi^2 = 0. \quad (10.89)$$

Any function of ξ, η can be expressed as a linear combination of two numbers

$$f(\xi, \eta) = a + b\xi + c\eta + d\xi\eta, \quad (10.90)$$

where a, b, c, d are C-numbers.[9] Differentiations are defined as two-fold, i.e., from the left and from the right

$$\frac{d^L f}{d\xi} = b + d\eta, \quad \frac{d^R f}{d\xi} = b - d\eta. \quad (10.91)$$

---

[9] The C-numbers refer both real and complex numbers.

The integrals are defined as

$$\int d\xi = 0, \quad \int d\xi \xi = 1. \tag{10.92}$$

For two variables ξ, η, the ordering of G-numbers changes the sign of the integral

$$\int d\xi d\eta \xi \eta = 1 = -\int d\eta d\xi \xi \eta. \tag{10.93}$$

It is straightforward to extend the algebra for complex G-numbers ξ, ξ^*

$$\int d\xi d\xi^* \xi \xi^* = 1 = -\int d\xi d\xi^* \xi^* \xi, \quad \int d\xi = \int d\xi^* = 0. \tag{10.94}$$

By definition, the Taylor expansion of the exponent turns out to be

$$e^{-a\xi\xi^*} = 1 - a\xi\xi^*, \tag{10.95}$$

because of $\xi^2 = (\xi^*)^2 = 0$. Thus, the integral of the Gauss function becomes

$$\int d\xi d\xi^* e^{-a\xi\xi^*} = -a. \tag{10.96}$$

Notice that a Gauss integral for a real number x is

$$\int_{-\infty}^{\infty} dx e^{-ax^2} = \sqrt{\frac{\pi}{a}}, \tag{10.97}$$

and for complex C-numbers z, z^*, the generalized Gauss integral gives

$$\int dz dz^* e^{-azz^*} = \frac{2\pi i}{a}. \tag{10.98}$$

Let us consider the Gauss integral of four-fermion fields, $\{\overline{\psi}_i, \psi_i, i = 1, 2\}$

$$\int \prod_{i=1}^{2} d\psi_i d\overline{\psi}_i e^{-\sum_{i,j} \overline{\psi}_i D_{ij} \psi_j}$$

$$= \int \prod_{i=1}^{2} d\psi_i d\overline{\psi}_i [\overline{\psi}_1 D_{11} \psi_1 \overline{\psi}_2 D_{22} \psi_2 + \overline{\psi}_1 D_{12} \psi_2 \overline{\psi}_2 D_{21} \psi_1]$$

$$= D_{11} D_{22} - D_{12} D_{21} = \det D. \tag{10.99}$$

The integral (10.99) can be generalized for the $2n$ multi-variable case as

$$\int \prod_{i=1}^{n} d\psi_i d\overline{\psi}_i e^{-\sum_{i,j} \overline{\psi}_i D_{ij} \psi_j} = \det D. \tag{10.100}$$

10.6.3 Gauge Theory and Wilson Action

It is useful to study the gauge theory with a simple example. The simplest example is the $U(1)$ gauge. Let us consider a Lagrangian for a complex scalar field

$$\mathcal{L} = -\partial_\mu \psi^* \partial^\mu \psi - V(\psi^* \psi), \tag{10.101}$$

where V is a potential. In the case of the local gauge transformation

$$\widetilde{\psi}(x) = e^{i\alpha(x)} \psi(x), \tag{10.102}$$

the derivatives in $\mathcal{L}$ yield new terms

$$\partial_\mu \widetilde{\psi}^* \partial^\mu \widetilde{\psi} = \left[(\partial_\mu + i(\partial_\mu \alpha)) \psi \right]^* \left[(\partial^\mu + i(\partial^\mu \alpha)) \psi \right]. \tag{10.103}$$

For the principle of local gauge invariance, the theory should be independent of the arbitrary phase $\alpha(x)$. This principle is repaired by adding a compensating field to A_μ such that

$$\widetilde{A}_\mu(x) = A_\mu(x) + \frac{1}{g} \partial_\mu \alpha(x). \tag{10.104}$$

At the same time, one has to introduce the covariant derivative

$$D_\mu \equiv \left(\partial_\mu - ig A_\mu(x) \right), \tag{10.105}$$

in the Lagrangian. One can see that the transformation A_μ exactly cancels with the undesired derivative of $\alpha(x)$ and gives an invariant Lagrangian since

$$D_\mu \widetilde{\psi}(x) = e^{i\alpha} \left(\partial_\mu + i(\partial_\mu(\alpha)) \right) \psi(x) - ig e^{i\alpha} \left(A_\mu(x) + \frac{1}{g} \partial_\mu(\alpha) \right) \psi(x)$$

$$= e^{i\alpha} \left(\partial_\mu - ig A_\mu(x) \right) \psi(x), \tag{10.106}$$

and

$$\mathcal{L} = -D_\mu \widetilde{\psi}^* D^\mu \widetilde{\psi} - V(\widetilde{\psi}^* \widetilde{\psi})$$

$$= -D_\mu \psi^* D^\mu \psi - V(\psi^* \psi). \tag{10.107}$$

It is known that the electromagnetic field

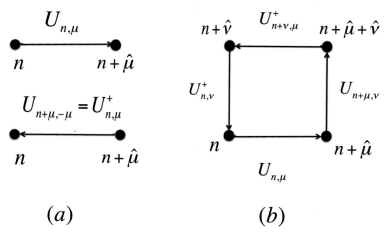

Fig. 10.13 a The link variables $U_{n,\mu}$ and $U_{n,\mu}^{\dagger}$ which connect the neighboring sites n and $n + \mu$, and $n + \mu$ and n, respectively. **b** An elementary plaquette of link variable for gluons

$$F_{\mu\nu} = \partial_\mu A_\nu - \partial_\nu A_\mu, \tag{10.108}$$

is gauge invariant

$$\partial_\mu \tilde{A}_\nu - \partial_\nu \tilde{A}_\mu = \partial_\mu \left(A_\nu + \frac{1}{g} \partial_\nu \alpha(x) \right) - \partial_\nu \left(A_\mu + \frac{1}{g} \partial_\mu \alpha(x) \right)$$
$$= \partial_\mu A_\nu - \partial_\nu A_\mu. \tag{10.109}$$

For the lattice theory, the appropriate operator to compensate the gauge field at two different points is the link variable. Let us consider a four dimensional lattice with a lattice spacing a and the four dimensional volume L^4. Each lattice site is specified by the index n_μ corresponding the Euclidian coordinate $x_\mu = an_\mu$. The link variable is an $SU(3)$ matrix connecting the neighboring sites n and $n + \hat{\mu}$

$$U_{n,\mu} = e^{iag\frac{1}{2}\lambda^c A_\mu^c(n)} \equiv e^{iagA_\mu(n)}, \tag{10.110}$$

where λ^c are Gell–Mann metrics for $SU(3)$ and c is the color indices. The repeated indices c run over the $3^2 - 1$ generators λ^c. The vector $\hat{\mu}$ implies a vector pointing the direction μ with a length a (see Fig. 10.13a). The link variable pointing to the opposite direction is given by $U_{n+\mu,-\mu} = [U_{n,\mu}]^{\dagger}$.

The fundamental building blocks of the lattice action are the products of discrete link variables taken anti-clockwise around a plaquette of the lattice. A typical example is sketched in Fig. 10.13b

$$U_{\mu\nu}(n) = \prod_{\text{plaquette}} U_{n,\nu}^{\dagger} U_{n+\nu,\mu}^{\dagger} U_{n+\mu,\nu} U_{n,\mu}. \tag{10.111}$$

Under the local gauge transformation with a arbitrary $SU(3)$ matrix $V(n)$, we have

$$U_{n,\mu} \to U_{n,\mu}^V = V(n)U_{n,\mu}V^\dagger(n+\mu),$$
$$U_{\mu\nu}(n) \to U_{\mu\nu}^V(n) = V(n)U_{\mu\nu}(n)V^\dagger(n), \tag{10.112}$$

where we use an identity $V(n)V^\dagger(n) = I$. In the naive continuum limit, we have

$$U_{n,\mu} = 1 + igaA_\mu + O(a^2), \tag{10.113}$$

and, applying the Baker–Campbell–Hausdorf identity $e^{\hat{x}}e^{\hat{y}} = e^{\hat{x}+\hat{y}+\frac{1}{2}[\hat{x},\hat{y}]+\cdots}$ for no-commutable variables $\hat{x}$ and $\hat{y}$, the plaquette action is rewritten to be

$$
\begin{aligned}
U_{\mu\nu} &= e^{igaA_\mu(n)}e^{igaA_\nu(n+\hat{\mu})}e^{-igaA_\mu(n+\hat{\nu})}e^{-igaA_\nu(n)} \\
&\approx e^{igaA_\mu}e^{iga(A_\nu+a\partial_\mu A_\nu)}e^{-iga(A_\mu+a\partial_\nu A_\mu)}e^{-igaA_\nu} \\
&\approx e^{iga(A_\mu+A_\nu)+iga^2\partial_\mu A_\nu+(iga)^2[A_\mu,A_\nu]/2}e^{-iga(A_\mu+A_\nu)-iga^2\partial_\nu A_\mu+(-iga)^2[A_\mu,A_\nu]/2} \\
&\approx e^{iga^2(\partial_\mu A_\nu-\partial_\nu A_\mu)+(iga)^2[A_\mu,A_\nu]} = e^{iga^2 F_{\mu\nu}} \approx 1 + iga^2 F_{\mu\nu} - \frac{1}{2}g^2 a^4 F_{\mu\nu}^2,
\end{aligned}
\tag{10.114}
$$

where $F_{\mu\nu} = \partial_\mu A_\nu(x) - \partial_\nu A_\mu(x)) + ig[A_\mu, A_\nu]$. Notice that the $SU(3)$ gauge creates the self-interaction terms in the field. Since $\text{tr}(F_{\mu\nu}) = 0$, the trace of $U_{\mu\nu}$ reads

$$\text{tr}(U_{\mu\nu}) = \text{tr}\left(1 - \frac{1}{2}g^2 a^4 F_{\mu\nu}^2 + O(a^5)\right). \tag{10.115}$$

The action for the gauge field is defined as

$$
\begin{aligned}
S_G &= \beta_g \sum_n \sum_{plaquette}\left[1 - \frac{1}{N_c}\text{Re tr}U_{\mu\nu}(n)\right] \\
&= \beta_g \sum_n \sum_{\mu<\nu}\text{tr}\left[1 - \frac{1}{2N_c}\text{tr}(U_{\mu\nu}(n) + U_{\mu\nu}^\dagger(n))\right] \\
&= \frac{\beta_g}{N_c}\sum_n \sum_{\mu<\nu}\text{tr}(1 - U_{\mu\nu}(n)) = \frac{\beta_g}{2N_c}\sum_n \sum_{\mu\neq\nu}\text{tr}(1 - U_{\mu\nu}(n)),
\end{aligned}
\tag{10.116}
$$

where β_g is the inverse coupling constant of the gluon field $\beta_g = 2N_c/g^2$ with $N_c = 3$ being the number of colors in QCD. The trace is taken for the 3×3 matrix for the color degree of freedom. In the continuum limit $a \to 0$, the gluon action reads

$$S_G \approx \frac{1}{g^2}\frac{1}{2}g^2\sum_n \sum_{\mu\neq\nu}a^4 4F_{\mu\nu}^2(n) \xrightarrow{a\to 0} \frac{1}{2}\int d^4x F_{\mu\nu}^2(x). \tag{10.117}$$

10.6.4 *Correlation Function*

Let us formulate the correlation function for QCD action on the lattice. We introduce a generating function (10.21) for the quark field

$$Z[\psi, \overline{\psi}] = \int dU d\psi d\overline{\psi} e^{-(S_{QCD} + \int dx \overline{\psi} \eta + \int dx \overline{\eta} \psi)}. \tag{10.118}$$

The correlation function for the operator O in Eq. (10.30) is then given for the action $S_{QCD} = S_{quark} + S_{gluon}$ in the Euclidean metric as

$$\langle \Omega | O[\psi, \overline{\psi}, U] | \Omega \rangle = \frac{1}{Z} \int D\psi D\overline{\psi} \int DU \, O[\psi, \overline{\psi}, U] e^{-S_{QCD} + \int dx \overline{\psi} \eta + \int dx \overline{\eta} \psi}$$

$$= \frac{1}{Z} \int DU e^{-S_{gluon}} O[\frac{\delta}{\delta \eta} \frac{\delta}{\delta \overline{\eta}}, U] \int D\psi D\overline{\psi} e^{-(S_{quark} + \int dx \overline{\psi} \eta + \int dx \overline{\eta} \psi)} \Bigg|_{\eta = \overline{\eta} = 0}.$$
$$\tag{10.119}$$

The integrals of quark field involves anti-commutable numbers, Grassmann number, so that numerical calculations are impossible. However, one can perform the analytical integration for Grassmann numbers. The exponent of the generating function is integrable by rewriting it in a form

$$S_{quark} + \int dx \overline{\psi} \eta + \int dx \overline{\eta} \psi = \int dx dy \overline{\psi}(x) D(x, y) \psi(y) + \int dx \overline{\psi} \eta + \int dx \overline{\eta} \psi$$

$$= \int dx dy \left(\overline{\psi}(x) + \int dz D^{-1}(z, x) \overline{\eta}(z) \right) D(x, y) \left(\psi(y) + \int dz' D^{-1}(y, z') \eta(z') \right)$$

$$- \int dx dy \overline{\eta}(x) D^{-1}(x, y) \eta(y), \tag{10.120}$$

where $D(x, y) = D(x)\delta^4(x - y)$ and $D(x) = \gamma_\mu D_\mu + m$. The inverse of $D(x, y)$ is defined by an identity

$$\int dz D^{-1}(x, z) D(z, y) = \delta^4(x - y). \tag{10.121}$$

For the rescaled quark field $\psi(y)' = \psi(y) + \int dz' D^{-1}(y, z') \eta(z')$, the action integral of the quark field can be calculated as

$$\int D\psi D\overline{\psi} e^{-\int dx dy \overline{\psi}'(x) D(x, y) \psi'(y) - \int dx dy \overline{\eta}(x) D^{-1}(x, y) \eta(y)}$$

$$= \det D e^{-\int dx dy \overline{\eta}(x) D^{-1}(x, y) \eta(y)}, \tag{10.122}$$

where the Grassmann algebra is used to integrate in terms of $D\psi D\overline{\psi}$.

The two-point correlation function for the pion operator (10.34) is expressed by using the generating function as

$$\langle \Omega | O_\pi(x) \overline{O}_\pi(y) | \Omega \rangle = \langle \Omega | \overline{\psi}^{(u)}(x) \gamma_5 \psi^{(d)}(x) \overline{\psi}^{(d)}(y) \gamma_5 \psi^{(u)}(y) | \Omega \rangle$$

$$= -\frac{1}{Z} \int DU \det D(U) e^{-S_{\text{gluon}}} \text{Tr} \left[D_d^{-1}(U) \gamma_5 D_u^{-1}(U) \gamma_5 \right]. \tag{10.123}$$

To make numerical evaluations easier, the determinant $\det D(U)$ is taken often only the diagonal matrix elements $\det D(U) = 1$. This approximation is called "quenched approximation", which corresponds physically to neglect the creation and annihilation of quark-antiquark pair from the vacuum.

For the Wilson fermion with $m = m_u = m_d$, the action S on the lattice can be expressed as

$$S_{\text{quark}}^{\text{naive}} = \int dx^4 \overline{\psi}(x) \left[\gamma_\mu D_\mu + m \right] \psi(x) \tag{10.124}$$

$$= a^4 \sum_{\mathbf{n}} \left(\frac{1}{2a} \sum_{\hat{\mu}} \left[\overline{\psi}(\mathbf{n}) \gamma_\mu U_{n,\mu} \psi(\mathbf{n} + \hat{\mu}) - \overline{\psi}(\mathbf{n}) \gamma_\mu U_{n-\hat{\mu},\mu}^\dagger \psi(\mathbf{n} - \hat{\mu}) \right] + \overline{\psi}(\mathbf{n}) m \psi(\mathbf{n}) \right), \tag{10.125}$$

where $\mathbf{n}$ represents all the mesh points in 4D space-time coordinates and $\hat{\mu}$ is the vector in μ-direction with the length a.

Problem

10.3 Prove the lattice expression of quark action (10.125) is equivalent to (10.124) in the lowest order expansion of U_μ in terms of a.

It is known that the naive action has the so-called "doubling problem" of eigenenergies. To avoid these unphysical solutions, the naive action is combined with a second derivative term $S' = -ar \int dx \psi(x) \Delta^2 \psi(x)$ which gives zero contribution in the continuum limit $a \to 0$. This is called Wilson action. The value r is a parameter and usually taken to be $r = 1$. Then the modified action is expressed avoiding the doubling problem as,

$$S_W = a^4 \sum_{\mathbf{n}} \psi(\mathbf{n}) M \psi(\mathbf{n})$$

$$- a^4 \frac{1}{2a} \sum_{\mathbf{n},\hat{\mu}} \left[\psi(\mathbf{n})(r - \gamma_\mu) U_{n,\mu} \psi(\mathbf{n} + \hat{\mu}) - \psi(\mathbf{n})(r + \gamma_\mu) U_{n-\hat{\mu},\mu}^\dagger \psi(\mathbf{n} - \hat{\mu}) \right], \tag{10.126}$$

where $M = m + \frac{4r}{a}$ is considered as a mass term in the Wilson fermion action. For a rescaled quark field

$$\Psi = (Ma^4)^{1/2} \psi, \tag{10.127}$$

the action has the form

$$S_W = \sum_n \overline{\Psi}(\mathbf{n})\Psi(\mathbf{n})$$

$$- \kappa \sum_\mu \left[\overline{\Psi}(\mathbf{n})(r - \gamma_\mu)U_{n,\mu}\Psi(\mathbf{n} + \hat{\mu}) - \overline{\Psi}(\mathbf{n})(r + \gamma_\mu)U^\dagger_{n-\hat{\mu},\mu}\Psi(\mathbf{n} - \hat{\mu}) \right],$$

$$(10.128)$$

where κ is defined as

$$\kappa \equiv \frac{1}{2Ma} = \frac{1}{2ma + 8r},$$

$$(10.129)$$

called as the *hopping parameter*.

Summary

The standard model of elementary particles consists of three generations of quarks and leptons, their antiparticles, and four gauge bosons and one Higgs boson. The model is successful to describe phenomena involving baryons and mesons, including the origin of mass.

Precise $N_f = 2 + 1$ flavor LQCD simulations have been performed successfully to obtain baryon masses. The N-N interactions are also calculated by LQCD simulations successfully not only for the short range repulsive part, but also for the medium and long range attractive parts.

The observation of the neutrino oscillation establishes finite masses for neutrinos.

The superallowed beta decay provides an opportunity to determine the CKM matrix element V_{ud} experimentally.

The neutrino-less double beta decay could be possible if the neutrino has a finite mass and it is a Majorana particle (neutrino and anti-neutrino are the same particle). However, no experimental evidence has been observed, but only the lower limit of the lifetime was indicated.

The nuclear octupole deformation will enhance the atomic electric dipole moment (EDM), whose measurements can indicate time-reversal (T), or equivalently charge-parity (CP) violation due to the physics beyond the standard model.

Solutions of Problems

10.1 The proton has 2 up quarks and 1 down quark and the neutron has 1 up quark and 2 down quarks so that the isospin z-component will be

$$I_3 = \frac{1}{2}(2-1) = \frac{1}{2} \quad \text{for proton,}$$

$$I_3 = \frac{1}{2}(1-2) = -\frac{1}{2} \quad \text{for neutron.} \tag{10.130}$$

The isospin of these fermions will be $I = 1/2$. The quark constituents of Δ^{++} is (uuu) so that $I_3 = 3/2$. The constituents of other Delta particles are given in Table 10.1 and give the values of $I_3 = 1/2, -1/2, -3/2$ for $\Delta^+, \Delta^-, \Delta^{--}$, respectively. These 4 Delta particles have the isospin $I = 3/2$ with different z components. In Table 10.1, Ξ^0 has one up quark, while Ξ^- has one down quark so that the value of $I_3 = 1/2, -1/2$, respectively. The isospin then is assigned to be $I = 1/2$.

10.2 For the operator $p = \frac{\hbar}{i}\frac{\partial}{\partial q}$, the expectation value is given as

$$\langle q|p|p\rangle = p\langle q|p\rangle = \frac{\hbar}{i}\frac{\partial}{\partial q}\langle q|p\rangle, \tag{10.131}$$

which is satisfied by the identity

$$\langle q|p\rangle = e^{ipq/\hbar}. \tag{10.132}$$

10.3 The lowest order expansion of $U_{n,\mu} = e^{iagA_\mu(n+\frac{1}{2}\hat{\mu})}$ is given by $U_{n,\mu}(n) \simeq 1 + iagA_\mu(n+\frac{1}{2}\hat{\mu})$. Inserting this expansion into Eq. (10.125), one obtains

$$\frac{1}{2a}\sum_{\hat{\mu}}\left[\overline{\psi}(n)\gamma_\mu U_{n,\mu}\psi(n+\hat{\mu}) - \overline{\psi}(n)\gamma_\mu U^\dagger_{n-\hat{\mu},\mu}\psi(n-\hat{\mu})\right]$$

$$\simeq \frac{1}{2a}\sum_{\hat{\mu}}[\overline{\psi}(n)\gamma_\mu(1+iagA_\mu(n+\frac{1}{2}\hat{\mu}))(1+a\partial_\mu)\psi(n)$$

$$-\overline{\psi}(n)\gamma_\mu(1-iagA_\mu(n-\frac{1}{2}\hat{\mu}))(1-a\partial_\mu)\psi(n)]$$

$$\simeq \overline{\psi}(n)\gamma_\mu(igA_\mu(n)+\partial_\mu)\psi(n) = \overline{\psi}(n)\gamma_\mu D_\mu\psi(n),$$

where $A_\mu(n) \sim (A_\mu(n+\frac{1}{2}\hat{\mu})) + A_\mu(n-\frac{1}{2}\hat{\mu})))/2$, and $D_\mu = \partial_\mu + igA_\mu(n)$.

Books for Further Readings

"Quantum Chromodynamics on the Lattice—a first look." by C. Gattringer and C. B. Lang (Lecture Notes in Physics, Springer, 2009) presents quantum chromodynamics starting from the basic concepts and calculations in a clear and pedagogical style accessible to those who are new to the field.

"Quantum Fields on a Lattice" by I. Montvay and G. Munster (Cambridge Monographs on Mathematical Physics, 2010) provides a comprehensive and coherent account of the theory of quantum fields on a lattice, an essential technique for the study of the strong and electroweak nuclear interactions.

References

1. P.A. Zyla et al. (Particle Data Group), Rev. Particle Phys. 2020, Prog. Theor. Exp. Phys 083C01 (2020)
2. S. Aoki et al., PACS-CS Collaboration, Phys. Rev. D **79**, 034503 (2009)
3. N. Ishii, S. Aoki, T. Hatsuda, Phys. Rev. Lett. **99**, 022001 (2007)
4. J.C. Hardy, I.S. Towner, Phys. Rev. C **102**, 045501 (2020)
5. R. Jr. Davis, D.S. Harmer, K.C. Hoffman, Phys. Rev. Lett. **20**, 1205 (1968)
6. B. Cleveland et al., Astrophys. J. **496**, 505 (1998)
7. I. Esteban, M.C. Gonzales-Garcia, A. Hernandez-Cabezudo, M. Maltoni, T. Schwetz, J. High Energ. Phys. **2019**, 106 (2019)
8. M. Aker et al., KATRIN collaboration, Phys. Rev. Lett. **123**, 221802 (2019)
9. N. Aghanim et al., Planck collaboration, A&A **641**, A1 (2020)
10. R.H. Parker et al., Phys. Rev. Lett. **115**, 233002 (2015)

Printed in the United States
by Baker & Taylor Publisher Services